Nanowires

This comprehensive resource covers the fundamentals of synthesis, characterizations, recent progress, and applications of nanowires for many emerging applications. Early chapters address their unique properties and morphology that enable their electronic, optical, and mechanical properties to be tuned. Later chapters address future perspectives and future challenges in areas where nanowires could provide possible solutions. All chapters are written by global experts, making this a suitable textbook for students and an up-to-date handbook for researchers and industry professionals working in physics, chemistry, materials, energy, biomedical, and nanotechnology.

- Covers materials, chemistry, and technologies for nanowires.
- Covers the state-of-the-art progress and challenges in nanowires.
- Provides fundamentals of the electrochemical behavior of various electrochemical devices and sensors.
- Offers insights on tuning the properties of nanowires for many emerging applications.
- Provides a new direction and understanding to scientists, researchers, and students.

Nanowires

Applications, Chemistry, Materials, and Technologies

Edited by
Ram K. Gupta

CRC Press
Taylor & Francis Group
Boca Raton London New York

CRC Press is an imprint of the
Taylor & Francis Group, an **informa** business

Designed cover image: © Shutterstock

First edition published 2023
by CRC Press
6000 Broken Sound Parkway NW, Suite 300, Boca Raton, FL 33487-2742

and by CRC Press
4 Park Square, Milton Park, Abingdon, Oxon, OX14 4RN

CRC Press is an imprint of Taylor & Francis Group, LLC

© 2023 selection and editorial matter, Ram K. Gupta

Library of Congress Cataloging-in-Publication Data
Names: Gupta, Ram K, editor.
Title: Nanowires : applications, chemistry, materials and technologies / Ram K. Gupta.
Other titles: Nanowires (CRC Press : 2023)
Description: First edition. | Boca Raton : CRC Press, 2023. |
Includes bibliographical references and index. |
Identifiers: LCCN 2022042651 (print) | LCCN 2022042652 (ebook) |
ISBN 9781032283852 (hardback) | ISBN 9781032283906 (paperback) |
ISBN 9781003296621 (ebook)
Classification: LCC TK7874.85 .N3588 2023 (print) |
LCC TK7874.85 (ebook) | DDC 621.3815–dc23/eng/20221205
LC record available at https://lccn.loc.gov/2022042651
LC ebook record available at https://lccn.loc.gov/2022042652

ISBN: 9781032283852 (hbk)
ISBN: 9781032283906 (pbk)
ISBN: 9781003296621 (ebk)

DOI: 10.1201/9781003296621

Typeset in Times
by Newgen Publishing UK

Contents

About the Editor

Dr. Ram Gupta is an Associate Professor at Pittsburg State University. Before joining Pittsburg State University, he worked as an Assistant Research Professor at Missouri State University, Springfield, MO, then as a Senior Research Scientist at North Carolina A&T State University, Greensboro, NC. Dr. Gupta's research focuses on green energy production and storage using conducting polymers and composites, electrocatalysts, fuel cells, supercapacitors, batteries, nanomaterials, optoelectronics and photovoltaics devices, organic-inorganic hetero-junctions for sensors, nanomagnetism, bio-based polymers, bio-compatible nanofibers for tissue regeneration, scaffold and antibacterial applications, bio-degradable metallic implants. Dr. Gupta has published over 250 peer-reviewed articles, made over 350 national/international/regional presentations, chaired many sessions at national/international meetings, wrote several book chapters (55+), worked as editor for many books (20+) for American Chemical Society, CRC, and Elsevier publishers, and received over two million dollars for research and educational activities from external agencies. He has served as associate editor, guest editor, and editorial board member for various journals.

Contributors

Elizabeth Adzo Addae
Department of Materials Science and
Engineering, School of Engineering
Sciences, College of Basic and Applied
Sciences, University of Ghana, Ghana

Benjamin Agyei-Tuffour
Department of Materials Science and
Engineering, University of Ghana, Ghana
Center of Excellence in Energy, African
Research Universities Alliance (ARUA),
University of Stellenbosch, Stellenbosch,
South Africa

Stefania Akromah
Department of Aerospace Engineering,
University of Bristol, UK

Reuben Amedalor
Department of Materials Science and
Engineering, University of Ghana, Ghana
Institute of Photonics, University of Eastern
Finland, Finland

Chohdi Amri
Department of Physics, College of Science,
United Arab Emirates University, United
Arab Emirates; Photovoltaic Laboratory,
Energy Research and Technology Center,
Tunisia

G. Anand
Department of Mechanical Engineering,
Achariya College of Engineering
Technology, Pondicherry, India

Vy Anh Tran
Department of Chemical and Biological
Engineering, Gachon University, Korea

V. Anusooya
Department of Electronics and Communication
Engineering, Amrita College of Engineering
and Technology, India

Joseph Asare
Department of Physics, School of Physical
and Mathematical Sciences, College of
Basic and Applied Sciences, University of
Ghana, Ghana,

Magdalene A. Asare
Department of Chemistry, Pittsburg State
University, Pittsburg, USA
National Institute for Materials Advancement,
Pittsburg State University, Pittsburg, USA

Fevzihan Basarir
Department of Chemistry and Materials
Science, Aalto University, Finland

Shuja Bashir Malik
MINOS-EMaS, Universitat Rovira i Virgili,
Tarragona, Spain

Shiva Bhardwaj
Department of Physics, Pittsburg State
University, Pittsburg, USA
National Institute for Materials Advancement,
Pittsburg State University, Pittsburg, USA

Rebecca Boamah
Department of Materials Science and
Engineering, University of Ghana, Legon-
Accra, Ghana

Benachir Bouchikhi
Biosensors and Nanotechnology Group, Faculty
of Sciences, Moulay Ismaïl University of
Meknes, Morocco

N. Chidhambaram
Department of Physics, Rajah Serfoji
Government College (Autonomous), India

Hyunjin Cho
Institute of Advanced Composite Materials,
Korea Institute of Science and Technology,
Republic of Korea

Tapas Das
Department of Chemical Engineering &
  Technology, Indian Institute of Technology,
  Banaras Hindu University, Varanasi, India

Felipe de Souza
National Institute for Materials Advancement,
  Pittsburg State University, Pittsburg, USA

David Dodoo-Arhin
Department of Materials Science and
  Engineering, University of Ghana, Ghana

Yukou Du
College of Chemistry, Chemical Engineering
  and Materials Science, Soochow University,
  PR China

Nezha El Bari
Biosensors and Nanotechnology Group,
  Department of Biology, Faculty of
  Sciences, Moulay Ismaïl University,
  Morocco

Fatima Ezahra Annanouch
MINOS-EMaS, Universitat Rovira i Virgili,
  Tarragona, Spain

Qiang Gao
School of Chemistry and Chemical
  Engineering, Yangzhou University,
  Yangzhou, China

Neill J. Goosen
Center of Excellence in Energy, African
  Research Universities Alliance (ARUA),
  University of Stellenbosch, Stellenbosch,
  South Africa

Frank Güell
MINOS-EMaS, Universitat Rovira i Virgili,
  Tarragona, Spain

Ram K. Gupta
Department of Chemistry, Pittsburg State
  University, Pittsburg, USA
National Institute for Materials
  Advancement, Pittsburg State University,
  Pittsburg, USA

Abderrahmane Hamdi
Institute of Electronics, Microelectronics and
  Nanotechnology (IEMN UMR CNRS 8520),
  Polytechnic University of Hauts-de-France
  (UPHF), France

Ibrahim Issah
Faculty of Engineering and Natural Science,
  Photonics, Tampere University, Finland

S. Jasmine Jecintha Kay
Department of Physics, Rajah Serfoji
  Government College (Autonomous), India

Milton B. F. Junior
Department of Physics and Chemistry,
  Universidade Estadual Paulista (UNESP),
  São Paulo, Brazil

Michael R. Koblischka
Saarland University, Experimental Physics,
  Saarbrücken, Germany

Saurabh Kumar
Department of Electronics and Communication
  Engineering, NIT, Himachal Pradesh, India

Vikas Kumar Pandey
Department of Chemical Engineering &
  Technology, Indian Institute of Technology,
  Banaras Hindu University, Varanasi, India

Jie Li
College of Chemistry, Chemical Engineering
  and Materials Science, Soochow University,
  PR China

Eduard Llobet
MINOS-EMaS, Universitat Rovira i Virgili,
  Tarragona, Spain

Zahra Madani
Department of Chemistry and Materials
  Science, Aalto University, Finland

Satish P. Mardikar
Department of Chemistry, SRS College,
  Sant Gadge Baba Amravati University,
  Amravati, India

Kwadwo Mensah Darkwa
Department of Materials Engineering, College
of Engineering, Kwame Nkrumah University
of Science and Technology, Kumasi, Ghana

Soukaina Motia
Biosensors and Nanotechnology Group,
Department of Biology, Faculty of Sciences,
Moulay Ismaïl University, Morocco

Adel Najar
Department of Physics, College of Science,
United Arab Emirates University, United
Arab Emirates

Daniel Nframah Ampong
Department of Materials Engineering, College
of Engineering, Kwame Nkrumah University
of Science and Technology, Kumasi, Ghana

N. Ponpandian
Department of Nanoscience and Technology,
Bharathiar University, India

Ujjwal K. Prajapati
Department of Physics, Maulana Azad National
Institute of Technology, India

A. Rajapriya
Department of Nanoscience and Technology,
Bharathiar University, Tamil Nadu, India

Jyoti Rani
Department of Physics, Maulana Azad National
Institute of Technology, India

A. Rebekah
Department of Nanoscience and Technology,
Bharathiar University, Tamil Nadu, India

T. Sangavi
Department of Nanoscience and Technology,
Bharathiar University, India

N. Santhosh
Department of Mechanical Engineering,
MVJ College of Engineering, Whitefield,
Bangalore, India

Hamidreza Daghigh Shirazi
Department of Chemistry and Materials
Science, Aalto University, Finland

Ashokkumar Sibiya
Nanobiosciences and Nanopharmacology
Division, Biomaterials and Biotechnology
in Animal Health Lab, Department of
Animal Health and Management, Alagappa
University, India

S. Sivaselvam
Department of Nanoscience and Technology,
Bharathiar University, India

Weronika Smok
Silesian University of Technology, Department
of Engineering Materials and Biomaterials,
Gliwice, Poland

Mohit Solanki
Department of Physics, Maulana Azad National
Institute of Technology, India

Endersh Soni
Department of Physics, Maulana Azad
National Institute of Technology.
Bhopal, India

Rishabh Srivastava
Department of Physics, Pittsburg State
University, Pittsburg, USA
National Institute for Materials
Advancement, Pittsburg State
University, Pittsburg, USA

Ming Tan
College of Science, Henan Agricultural
University, Zhengzhou, China

Tomasz Tański
Silesian University of Technology, Department
of Engineering Materials and Biomaterials,
Gliwice, Poland

Arun Thirumurugan
Sede Vallenar, Universidad de Atacama,
Vallenar, Chile

Thu-Thao Thi Vo
Department of Food Science and
   Biotechnology, Gachon University,
   Korea

Santosh J. Uke
Department of Physics, JDPS College,
   Sant Gadge Baba Amravati University,
   Amravati, India

Jaana Vapaavuori
Department of Chemistry and
   Materials Science, Aalto University,
   Finland

Baskaralingam Vaseeharan
Nanobiosciences and Nanopharmacology
   Division, Biomaterials and Biotechnology
   in Animal Health Lab, Department of
   Animal Health and Management, Alagappa
   University, India

Anjela Koblischka-Veneva
Saarland University, Experimental Physics,
   Saarbrücken, Germany

Bhawna Verma
Department of Chemical Engineering &
   Technology, Indian Institute of
   Technology, Banaras Hindu University,
   Varanasi, India

Sanjeev Verma
Department of Chemical Engineering &
   Technology, Indian Institute of
   Technology, Banaras Hindu University,
   Varanasi, India

Shivani Verma
Department of Chemistry, CBSH, G. B. Pant
   University of Agriculture and Technology,
   Uttarakhand, India

S. Vishvanathperumal
Department of Mechanical Engineering,
   S.A. Engineering College, India

Giang N. L. Vo
Faculty of Pharmacy, University of Medicine
   and Pharmacy at Ho Chi Minh City, Vietnam

Cheng Wang
College of Chemistry, Chemical Engineering
   and Materials Science, Soochow University,
   PR China

Jianjun Wei
The Department of Nanoscience, Joint School
   of Nanoscience and Nanoengineering,
   University of North Carolina at
   Greensboro, USA

Ziyu Yin
The Department of Nanoscience, Joint School
   of Nanoscience and Nanoengineering,
   University of North Carolina at
   Greensboro, USA

Marta Zaborowska
Silesian University of Technology, Department
   of Engineering Materials and Biomaterials,
   Gliwice, Poland

Rafael Zadorosny
Department of Physics and Chemistry,
   Universidade Estadual Paulista (UNESP),
   São Paulo, Brazil

Omar Zaim
Biosensors and Nanotechnology Group,
   Department of Biology, Faculty of Sciences,
   Moulay Ismaïl University, Morocco

Yucheng Zhou
Department of Mechanical and Aerospace
   Engineering, University of Virginia,
   Charlottesville, USA

Jiadeng Zhu
Chemical Sciences Division, Oak Ridge
   National Laboratory, USA
Smart Devices and Printed Electronics Foundry,
   Brewer Science Inc., Springfield, USA

1 Introduction to Nanowires

Felipe de Souza
National Institute for Materials Advancement, Pittsburg State University, Pittsburg, KS, USA

Ram K. Gupta
National Institute for Materials Advancement and Department of Chemistry, Pittsburg State University, Pittsburg, KS, USA

CONTENTS

1.1 INTRODUCTION

The advent of nanotechnology enabled the development of a plethora of nanomaterials that find applications in many emerging areas, such as energy, sensor, and biomedical fields. Nanotechnology consists of materials that present at least one of its dimensions that is equal to or smaller than 100 nm. Nanomaterials can present enhanced properties, such as higher surface area, conductivity, optical properties, magnetism, reactivity, etc. compared to their bulky counterparts. Recently, nanomaterials have attracted significant attention for energy generation and storage devices, such as fuel cells, solar cells, electrolyzers, batteries, and supercapacitors. Nanomaterials are suitable for such applications due to their (a) high surface area, which allows the permeation of electrolyte within the nanomaterial's structure, leading to an increase in energy storage capability; (b) high conductivity that facilitates the electron transfer step during charging and discharging processes; and (c) high reactivity that prompts fast reversibility for redox mechanism defined by pseudocapacitance.

For sensors, nanoengineering techniques can be employed to design accurate nanostructures with defined pore sizes and active sites, which enable the selective adsorption of the desired analytes. In this sense, nanomaterials can be used as sensors for the identification of gases, hazardous chemicals, and biological compounds, among others. Through that, nanomaterials can also be incorporated into the biomedical field since their enhanced conductibility along with a defined structure can be used to identify certain biomolecules. Nanostructures can also be decorated with biological probes that enable the identification of specific biological materials, such as DNA fragments, disease markers, and antigens, among many others. Notably, nanomaterials play a critical role in the further advancement of technology in several fields, which justifies the interest of the scientific community in this area.

Several types of nanostructures can be synthesized, which can be three-dimensional (3D), two-dimensional (2D), one-dimensional (1D), or zero-dimensional (0D). The 3D structure presents its

DOI: 10.1201/9781003296621-1

dimension on the microscale, which includes microparticles, micronetworks, and porous structures that may include MXenes, metal-organic frameworks (MOFs), and covalent organic frameworks (COFs). The 2D structure includes materials in which at least one dimension is within the nanoscale, which includes nanolayers, nanosheets, and nanofilms, such as graphene, layered transition metal oxides or sulfides, and boron nitride. Through the same logic 1D nanomaterials present two of its dimensions within the nanoscale, which involve carbon, polymeric, or metallic-based nanomaterials in the form of nanowires, nanorods, nanofibers, or nanotubes. Lastly, 0D, also known as quantum dots, are nanomaterials in which all their dimensions are within the nanoscale. Some examples of 0D nanomaterials can be based on carbon, and transition metal sulfides, among others. The representation of the decrease in dimension of the structures is illustrated in Figure 1.1.

Under this perspective, this chapter is focused on nanowires in terms of synthesis and applications in energy, sensor, and biomedical fields. Nanowire-based structures provide a pathway for the transport of electrons as well as ions to shorten the diffusion length, which can enhance the rate capability in energy storage devices. Nanowires also present a high surface area, which increases the adsorption of electrolytes and leads to an enhancement of energy storage through the electric double-layer capacitance (EDLC) mechanism. Another important property of nanowires within the field of energy storage is related to their relatively high chemical stability in terms of volume expansion and mechanical strain, which is directly related to the long-term stability of electronic devices. Nanowires can

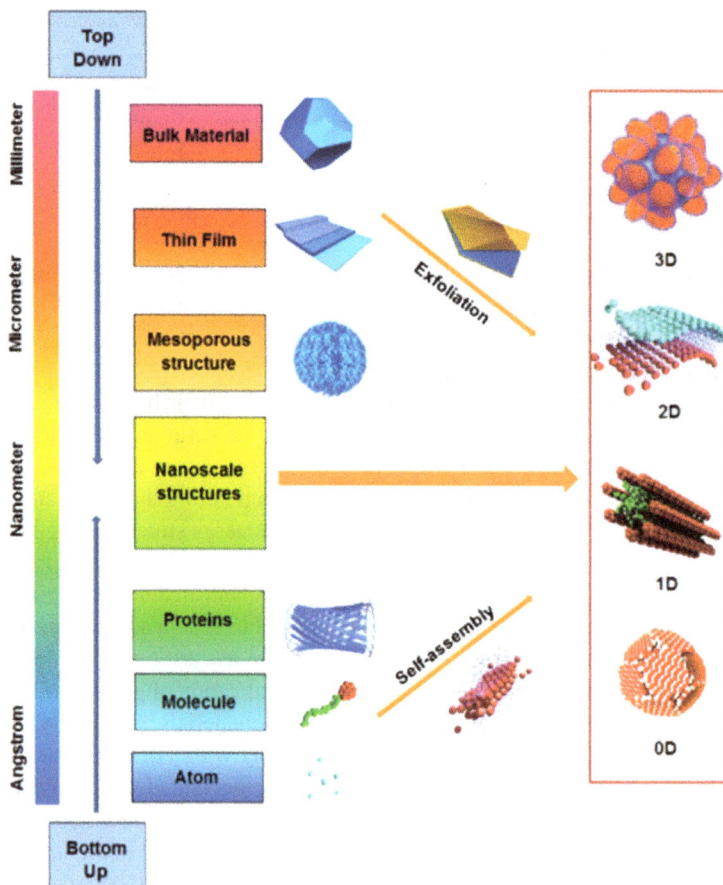

FIGURE 1.1 Representation of materials' dimensions from the macro to the nanoscale. Adapted with permission [1]. Copyright (2016), Springer Nature.

be grown in several substrates, such as carbon or metallic surfaces through binder- and additive-free methods along with yielding uniform structure arrays. Finally, the high surface area of nanowires allows them to be functionalized with different types of materials such as nanoparticles or biological probes, which can direct their applications toward the desired field. Thus, it is notable that nanowire-based materials can acquire valuable properties, which are desired in several areas of interest.

1.2 SYNTHESIS OF NANOWIRES

The synthesis of nanowires can be defined by the growth of atoms in the lengthwise direction to form a 1D structure, which can occur through space confinement or spontaneous growth. The latter can be performed on some metallic elements such as Mo, Ti, Sb, and V that can form 1D nanostructures through liquid-phase reactions [2–4]. Space-confinement growth, on the other hand, is performed with the aid of extra tools such as molds and templates, which function as a gradient for the 1D growth direction for materials that do not form this type of structure spontaneously [5,6]. Under this scope, the 1D nanomaterials can be synthesized through wet or dry chemistry strategies. The former includes hydro or solvothermal, electrodeposition, electrospinning, coprecipitation, and sol-gel. The latter includes high-temperature solid state, chemical vapor deposition (CVD), and physical vapor deposition (PVD). Based on that, some of these synthetical routes are discussed.

One of the most employed wet chemistry strategies is the hydro or solvothermal approach, which can be defined as a spontaneous growth method in which the reaction takes place by dissolving the reagents and placing them in a pressurized autoclave at a high temperature where several parameters can be varied such as temperature, time, pH, and solvent, etc., which can offer a change in morphology and properties. The advantages of this approach are related to its facile procedure, relatively low cost, and high efficiency. Through that, several types of materials such as TiO_2, Sb_2O_3, $NaV_3O_4 \cdot 1.63H_2O$, $H_2V_3O_4$ have been obtained in the form of nanowires [7–9]. Luo and co-workers performed the synthesis of ultralong Sb_2Se_3 nanowires through the hydrothermal method [10]. It was observed that the anisotropic growth of the Sb_2Se_3 was highly influenced by the reaction time as the longest of 36 hours led to the formation of nanowires with 90 nm of diameter along with a length within the micrometer scale. The schematics for the synthesis approach are depicted in Figure 1.2a. Along with that, the effect of 6, 12, 24, and 36 hours in the growth of the Sb_2Se_3 can be seen in Figure 1.2b-e.

The solvothermal method is based on the same principle as the hydrothermal with the difference of employing an organic solvent to improve the reagent solubility, which can lead to more organized nanostructures. Liao and colleagues performed the solvothermal method to obtain Se nanowires [11]. For that, SeO_2 was dissolved in a water/ethanol mixture with cethyltrimethylammonium bromide (CTAB) as a surfactant, glucose as a green reducing agent, and ammonia. The process was performed for 12 hours at 150 °C. Through that, highly crystalline Se nanowires were obtained. Alongside that, it was observed that the growth of nanowires was directly related to the reaction time as the mechanism was based on a solid-solution-solid mechanism.

The sol-gel technique usually consists of hydrolysis and a condensation polymerization process in the presence of a metal-based compound, such as metal halides, which have a strong tendency to hydrolyze. Then, the process is followed by a drying and sintering step to obtain the desired material. One of the main advantages of this technique is that it leads to materials with high surface area. An example of this approach was performed by Wang *et al.* [12] who synthesized nanowires composed of WO_3 doped with Nb. For that, WCl_6 and $NbCl_5$ were strongly mixed in a solution containing a non-ionic copolymeric surfactant and ethanol. The process formed a sol that was deposited over a fluorine-doped tin oxide (FTO) glass. Then, a high-temperature treatment was performed to remove the polymeric surfactant as well as promote the oxidation of WO_3. Through that, an Nb-doped WO_3 with mesoporous structure was obtained, which presented viable applications as a cathode for supercapacitors as the high surface area could promote a better permeation of electrolytes within the material's structure.

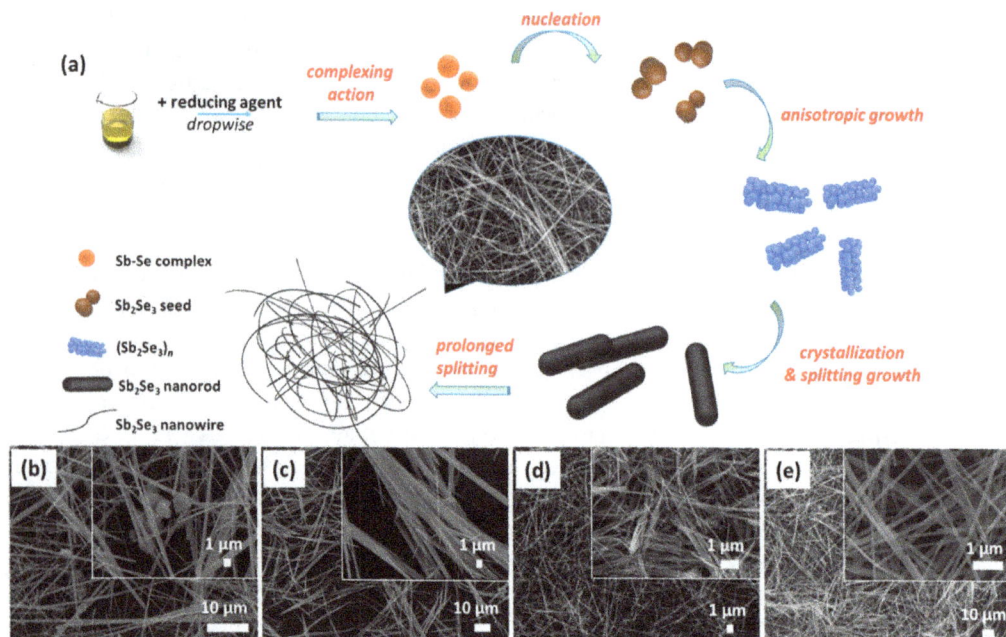

FIGURE 1.2 (a) Scheme for the hydrothermal synthesis of Sb_2Se_3 ultralong nanowires. Scanning electron microscopy (SEM) images for the Sb_2Se_3 obtained at different times of (b) 6, (c) 12, (d) 24, and (e) 36 hours, respectively. Adapted with permission [10]. Copyright (2016), American Chemical Society.

The template approach is another method that was performed to obtain nanowire-based structures. It is a convenient method as it can be used to embed the desired nanomaterial over a surface. Zhang *et al.* [13] synthesized a composite based on an N-doped hollow carbon nanotube (NHCNT) functionalized with electrodeposited Fe_3O_4. In this case, the NHCNT served as a proper substrate to incorporate the Fe_3O_4 to optimize its surface area as well as diminish its volume expansion during the lithiation and delithiation process related to the charge and discharge. The synthesis of the composite can be divided into three steps. First, polypyrrole (PPy) was synthesized through traditional wet synthesis using $FeCl_3$ as initiator and methyl orange as a soft template. Second, the PPy nanotubes were coated with FeOOH, which was formed by an *in-situ* process from the reaction between $Fe(NO_3)_3 \cdot 9H_2O$ with Na_2CO_3. Third, a calcination process was performed to convert the PPy/FeOOH precursor into the NHCNT/Fe_2O_3 composite. The synthetical scheme for the NHCNT/Fe_2O_3 is presented in Figure 1.3.

The dry chemistry approaches are another set of techniques that allow the synthesis of several types of nanostructures, such as nanowires without the need for solvents. In that sense, CVD is a widely employed technique in this regard, as it is relatively facile, cheaper, and applicable for several materials for the synthesis of nanowires such as Fe_3O_4, ZnO, Ge, GaN, silicon carbide, conducting polymers, among others [14–16]. Yet, the CVD often requires high temperature and can lead to the release of toxic fumes, which are matters that imply higher safety requirements. An example of the use of CVD for the fabrication of a supercapacitor electrode was performed by Zhang *et al.* [17]. In their work carbon nanotubes (CNTs) were grown through CVD at a carbon cloth substrate. For that, $Ni(NO_3)_2$, which functioned as catalysts, was coated over the carbon cloth. Then, the reactive gases such as C_2H_2 and H_2 along with Ar were flown through the system to promote the growth of the CNT at 700 °C for 30 min. Then, Fe_2O_3 was coated over the CNT through direct current magnetron sputtering by using Fe, Ar, and O_2 at room temperature, 1 Pa, and 16 W of power. Through that, the CNT/Fe_2O_3 composite was obtained as the process is described in Figure 1.4.

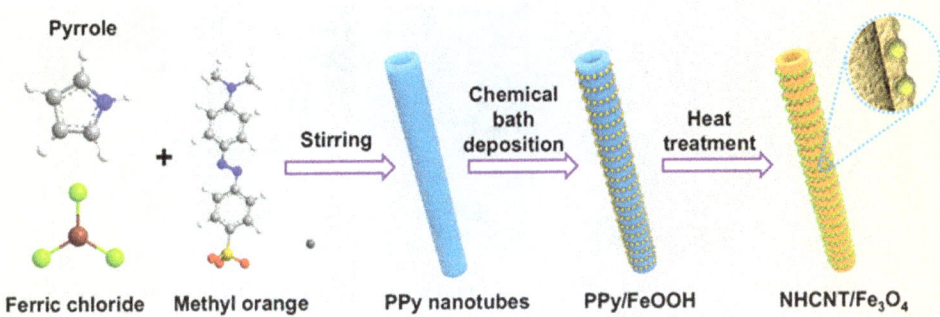

FIGURE 1.3 Schematic for the synthesis of a nanowire-like composite based on NHCNT/Fe$_3$O$_4$ obtained through a template approach. Adapted with permission [13]. Copyright (2020), Springer Nature.

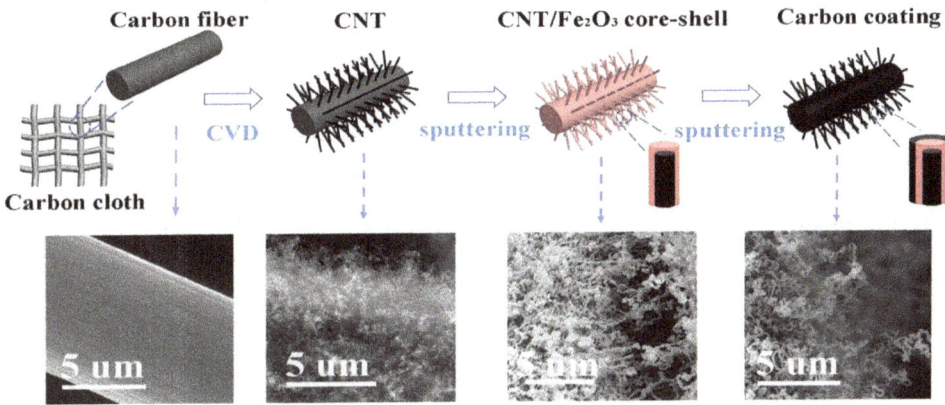

FIGURE 1.4 Scheme for the composite fabrication based on CNT synthesized through CVD and Fe$_2$O$_3$ obtained through magneton sputtering and the variation in morphology demonstrated in the SEM images below. Adapted with permission [17]. Copyright (2017), Elsevier.

1.3   APPLICATIONS OF NANOWIRES

1.3.1   Nanowires for Energy Applications

Nanowires are materials that can cover a vast field of applications within the energy storage field as they can be used in metal-ion, metal-sulfur, metal-air, supercapacitors, solar cells, and fuel cells. Based on that, it is important to understand the role of nanowires in these technologies. Under this perspective, one of the main challenges that surround Li-ion battery (LIB) technology is related to the volume expansion and Li dendrimer formation, which can lead to permanent capacity loss as well as short-circuiting the cell. Nanowire-based materials are an interesting candidate to address this issue as they possess a high surface area to incorporate Li$^+$ during the charging/discharging process without causing much disruption in their morphology. Xie *et al.* [18] proposed a study of α-MoO$_3$ nanobelts concerning the effect of the lithiation/delithiation process. It was observed that during the first voltammetric cycle there was a conversion of MoO$_3$ into metallic Mo in the form of nanograins that were inside the Li$_2$O matrix formed during lithiation. This process led to an expansion of the nanobelts. In addition, it was found that two other reactions were taking place during the delithiation process. First, the nanograins of Mo were converted into crystalline Li$_{1.66}$Mo$_{0.66}$O$_2$ along with the consumption of Li$_2$O. Because of that, there was a shrinkage of the nanobelts along with a conversion of the crystalline Li$_{1.66}$Mo$_{0.66}$O$_2$ into amorphous Li$_2$MoO$_3$. Since this process was

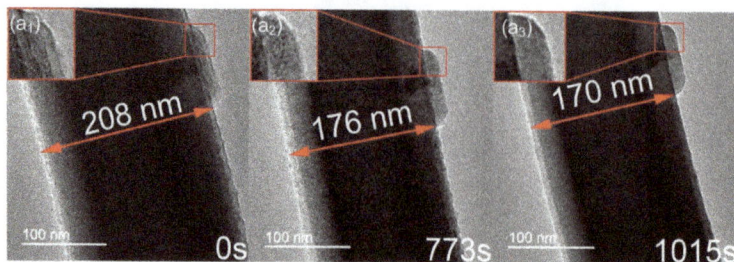

FIGURE 1.5 Nanostructure variation of the MoO_3 through the first intercalation with Li. (a_1) Lithiation process. (a_2) Intermediate delithiation process. (a_3) Fully delithiation process. Adapted with permission [18]. Copyright (2016), American Chemical Society.

irreversible it could explain the initial capacity loss that was observed in the first cycle. On the other hand, there was also a reversible process among crystalline Mo and Li_2O that formed amorphous Li_2MoO_3. Under this scope during the first stage of delithiation, there was a shrinkage of the anode's width from 208 to 176 nm due to the dissolution of the Li_2O layer as Li^+ was extracted as seen in Figure 1.5. Yet, the Mo nanograins were still observed as seen in Figure 1.5a_2. Further delithiation occurred as it could be observed through the nanograins that disappeared, which led to a smoother morphology as seen in Figure 1.5a_3.

The reactions for the lithiation/delithiation mechanism could be summarized through Equations 1.1 and 1.2, followed by the global reaction in Equation 1.3. Under this perspective, this work demonstrated a mechanistic study that described the variation in volume expansion of LIBs, which, for this case, was attributed to the conversion of MoO_3 with a specific capacitance (C_{sp}) of 1117 mAh/g to Li_2MoO_3 that presented a C_{sp} of 734 mAh/g.

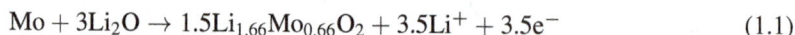

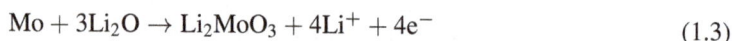

$$Mo + 3Li_2O \rightarrow 1.5Li_{1.66}Mo_{0.66}O_2 + 3.5Li^+ + 3.5e^- \tag{1.1}$$

$$1.5Li_{1.66}Mo_{0.66}O_2 \rightarrow Li_2MoO_3 + 0.5Li^+ + 0.5e^- \tag{1.2}$$

$$Mo + 3Li_2O \rightarrow Li_2MoO_3 + 4Li^+ + 4e^- \tag{1.3}$$

Another technology that has been highly regarded and studied within the scientific community is Li-S batteries, which can surpass the performance of status-quo LIBs due to the high theoretical C_{sp} of S, which is around 1673 mAh/g, high availability of starting materials, and relatively low cost. Yet, several unsolved challenges are on the way to introducing this technology into the market. Some of these issues are related to the inherently high resistance of S with a value of 5×10^{-30} S/cm at room temperature, the drastic variation of volume that S-based electrodes can suffer, and the shuttle effect due to the formation of polysulfides that can lead to a permanent decrease in capacitance due to the loss of active material. Under this perspective, one of the most employed strategies relied on using carbon-based materials due to their high conductivity, electrochemical stability, and low cost. In this sense, carbon-based materials can serve as a viable matrix to control the S expansion, improve its conductivity to facilitate electron transfer steps, and therefore provide longer-term stability.

Zheng *et al.* [6] reported the fabrication of a composite based on an array of a core-shell structure of sulfur and hollow carbon nanofibers. The carbon nanofibers were obtained through the thermal carbonization process of polystyrene over an anodic aluminum oxide (AAO). Then, melted sulfur was added to the AAO carbon nanofiber array to make the sulfur coat only the nanofiber's interior. After that, the AAO template was etched to obtain the desired array of carbon nanofiber internally coated with sulfur. The scheme for this process is presented in Figure 1.6. Based on this approach, it is notable that designing an optimal structure is inherently challenging as the internal

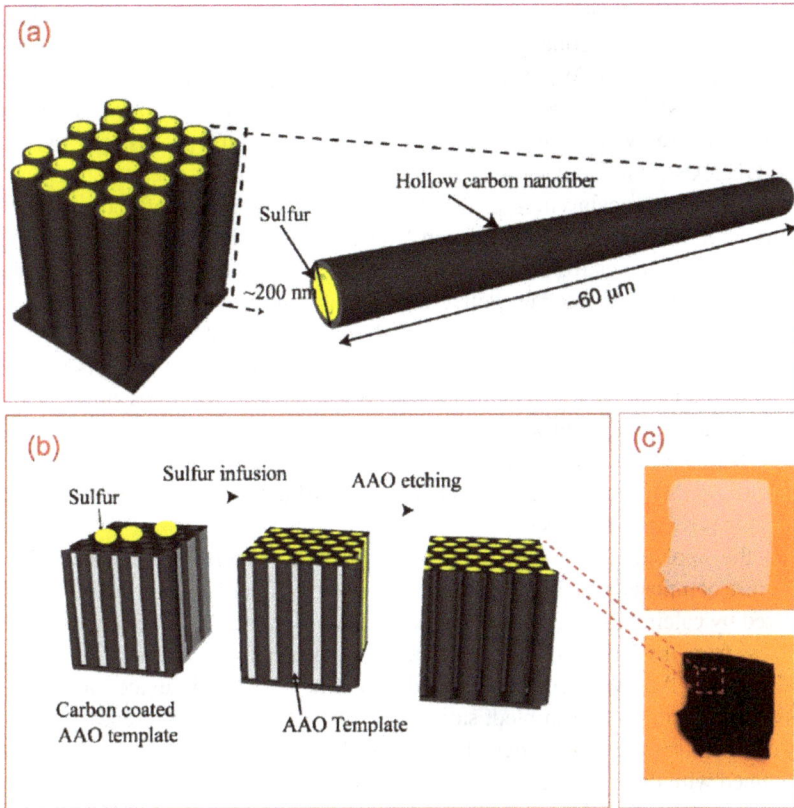

FIGURE 1.6 Scheme for the fabrication of hollow carbon nanofibers internally coated with sulfur. (a) Parameters related to the carbon nanofiber structure's array internally coated with sulfur. (b) The fabrication process for the composite. (c) Photocopies for the AAO template before (upper) and after (lower) the carbon nanofiber growth and sulfur insertion process. Adapted with permission [6]. Copyright (2011), American Chemical Society.

coating of sulfur can trap the polysulfide's shuttle effect, which is likely to improve the long-term stability. In addition, this type of nanofiber structure provides intimate contact between the carbon and sulfur, which shorten the transport path for the electrons as well as Li$^+$. However, it simultaneously decreases the active area of sulfur that can be exposed to the electrolyte, which is the region where the lithiation/delithiation process occurs. On top of that, the inner structure of the carbon nanofiber needs to present enough room to accommodate the formation of Li$_2$S$_2$ and Li$_2$S. Based on these propositions, the authors obtained a satisfactory performance for the developed cathode as a reversible C$_{sp}$ of 730 mAh/g after 150 charge/discharge cycles at C/5. It is also worth mentioning that the presence of additives in the electrolyte, such as LiNO$_3$, can passivate the surface of Li and therefore diminish the shuttle effect. Through that, the initial discharge capacity for the composite after the addition of LiNO$_3$ was in the range of 1560 mAh/g, which was close to sulfur's theoretical capacity. Even though it did not improve the stability it functioned as a tool to compensate for the composite's reduced active area.

The metal-air batteries (MABs) are another branch within the energy generation field that is gaining considerable attention due to their promising environmental credentials. Oxygen evolution reaction (OER) and oxygen reduction reaction (ORR) are the two main processes that can provide high-efficiency MABs. One widely used transition metal oxide for this technology is NiCo$_2$O$_4$-based nanowire [19,20]. Yet, it presents relatively low tunability to further improve its catalytic performance. To address that, Yin and colleagues performed an annealing process of $NiCo_2O_4$ nanowires along with carbon fiber paper at 380 °C in an NH_3 atmosphere [21]. Through that, a nanowire-based structure composed of NiO/CoN was obtained. The presence of NH_3 functioned as a doping agent to insert N into the structure, which is important to improve the catalytic process due to higher adsorption of O_2 and OH⁻. Also, the incorporation of N led to a disruption of the structure by introducing more defects that can act as catalytic sites. On top of that, the decrease of coordination number in Co atoms along with the nanointerface between NiO/CoN were also responsible for improving the catalytic activity and vacancies for oxygen, respectively. Through that, an overpotential of 300 mV at 10 mA/cm² for OER was obtained as well as appreciable stability for over 48 hours of continuous operation. Also, a satisfactory stability towards ORR was achieved along with an energy and power densities of 945 Wh/kg and 79.6 mW/cm², respectively. Based on the concepts presented in this work it was reinforced that, nanowires are highly versatile nanostructures that can considerably improve the surface area along with withstanding aggressive reaction process that can lead to further optimization of its properties. For this case, there was a synergy among the nanowire structure along with the high porosity, N-doped active sites, and lower coordination number for Co, which enhanced its reactivity towards oxygenated species.

Aiming for the application of supercapacitors, Chen and colleagues fabricated a core-shell nanowire structure composed of $NiCo_2O_4$@Co-Fe synthetized over Ni foam through a hydrothermal method followed by calcination [22]. The obtained structure enabled a better exposure of active site for redox process from the $NiCo_2O_4$ along with better conductivity from Co-Fe. In that sense, a high C_{sp} of 1557.5 F/g at 1 A/g was achieved. In addition, when a hybrid device based on $NiCo_2O_4$@Co-Fe // activated carbon was assembled, satisfactory values of power and energy densities of 950 W/kg and 28.94 Wh/kg were also achieved, respectively. On top of that, the nanowire structure of $NiCo_2O_4$ combined with the nanoflakes of Co-Fe layered double hydroxide played an important role in the energy storage process. The scanning electron microscope (SEM) images for the composite's morphology are presented in Figure 1.7.

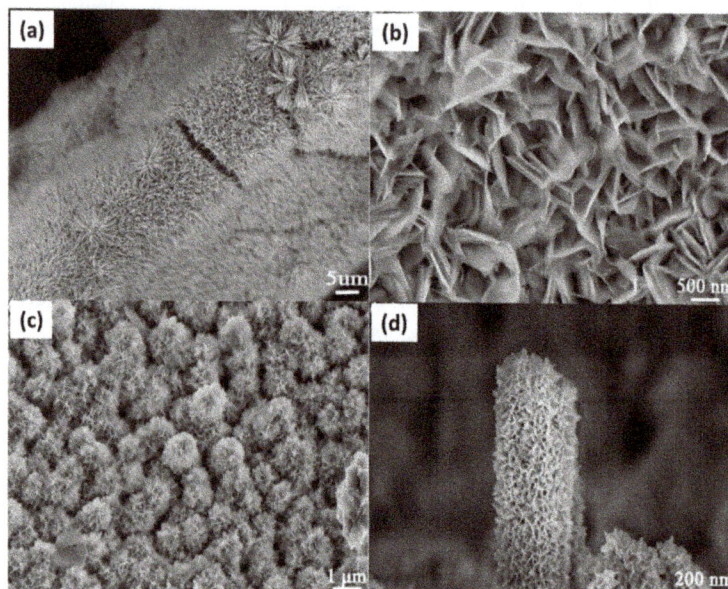

FIGURE 1.7 SEM images over Ni foam for the (a) nanowire arrays of $NiCo_2O_4$, (b) nanoflakes of Co-Fe double layered hydroxide, and (c) nanowire arrays for the $NiCo_2O_4$@Co-Fe. Reproduced with permission from [22]. Copyright (2018), Elsevier.

1.3.2 NANOWIRES FOR ENVIRONMENTAL AND SENSING APPLICATIONS

The progress of society is directly linked to the incorporation of technologies that are eco-friendly as well as efficient, which have demonstrated to be a challenging combination. Developing new technologies to address the current environmental issues are highly required. In this sense, nanowire-based materials can provide one of the answers to this complex topic as they can be hierarchically organized along with presenting a high surface area that enables their use to filtrate or separate toxic components. Zheng *et al.* [23] fabricated a nanowire-based composite that consisted of polyester fibrous structures embedded with TiO_2 nanoparticles and percolated Ag nanowires. Through that, a dual-functioning air filter was obtained due to its high air permeability, capability to remove small particulates through electrostatic charges at its surface, and to decompose formaldehyde through the photocatalysis process. The composite fabrication was performed by using a non-woven polyester fiber. First, a layer-by-layer (LbL) was performed at the polyester fiber by an alternate dip solution process of poly(diallyldimethylammonium) (PDDA) polycation followed by poly(sodium 4-styrenesulfonate) (PSS) polyanion. In this sense, the polyester and PDDA possess negatively and positively charged groups, respectively, which promotes the self-assembly process. Then, PSS, which presents negative charges was naturally deposited over the fibrous polyester-PDDA surface to form PDDA/PSS by-layers that adhered to small particulates through electrostatic forces. Following that, the TiO_2 nanoparticles were embedded into the polyester fiber through electrophoresis deposition (EPD). Lastly, the Ag nanowires were introduced through a modified polyol process in which ethylene glycol was used as a solvent and reducing agent to convert $AgNO_3$ into Ag at a higher temperature. Along with that, polyvinylpyrrolidone (PVP) was also used as a capping agent to control the Ag nanowire dimensions, which presented diameters of around 200 nm along with a length of 70–80 µm. The schematic for this composite's fabrication is described in Figure 1.8. Throughout this approach, highly satisfactory performance was achieved in terms of particulate removal efficiency, which was around 99.5%, while enabling 50.8 g/m² of dust hold capacity. In addition, the decorated TiO_2 nanoparticles and Ag nanowires demonstrated an efficient approach to both adhering and degrading organic volatile chemicals (OVC), such as formaldehyde when under UV light. Hence, it could be noted that using a low-cost matrix such as polyester non-woven fibers allowed the proper deposition of different types of nanomaterials that can greatly improve the quality of air by removing dust as well as degrading volatile toxic chemicals.

Even though water is an abundant resource, making it potable can be a multi-step and highly demanding process. In addition to that, waste and contaminated waters can contain toxic chemicals that may require a different type of treatment to allow their consumption. One of the concerns related to the contamination of water is the presence of heavy metals, which can lead to complicated environmental concerns, such as bioaccumulation, high levels of toxicity, and being hard to remove from the body's system once ingested. As a means to address such issues, Alizadeh *et al.* [24] proposed the use of SnO_2 nanowires as an adsorbent for the removal of Pb^{2+} and Cu^{2+} in wastewater. It was observed that even with the presence of interferent ions the removal efficiency of SnO_2 nanowires maintained a relatively efficient process, which was around 96.8% (Pb^{2+}) and 85.28% (Cu^{2+}) in 7 min. Along with that, the SnO_2 nanowires were deemed to be reusable, which is a good indication of scalability. The SnO_2 was obtained through the reaction between tin chloride and ammonia at 95 °C for 15 min as the ratio between these reagents influenced the nanowire's aspect ratio. Then, high-speed centrifugation was performed to decant the nanowires followed by an annealing process at 300 °C for 5 min.

Aside from heavy metals, the presence of toxic organic compounds is another issue that can impose a strain on the environment, which may require aggressive chemical treatment to fully decompose such substances. Within this line, one of the known approaches for the decomposition of organic matter is the Fenton process, which is a relatively easy and low-cost methodology. The working principle of this technology is based on the decomposition of H_2O_2, which leads

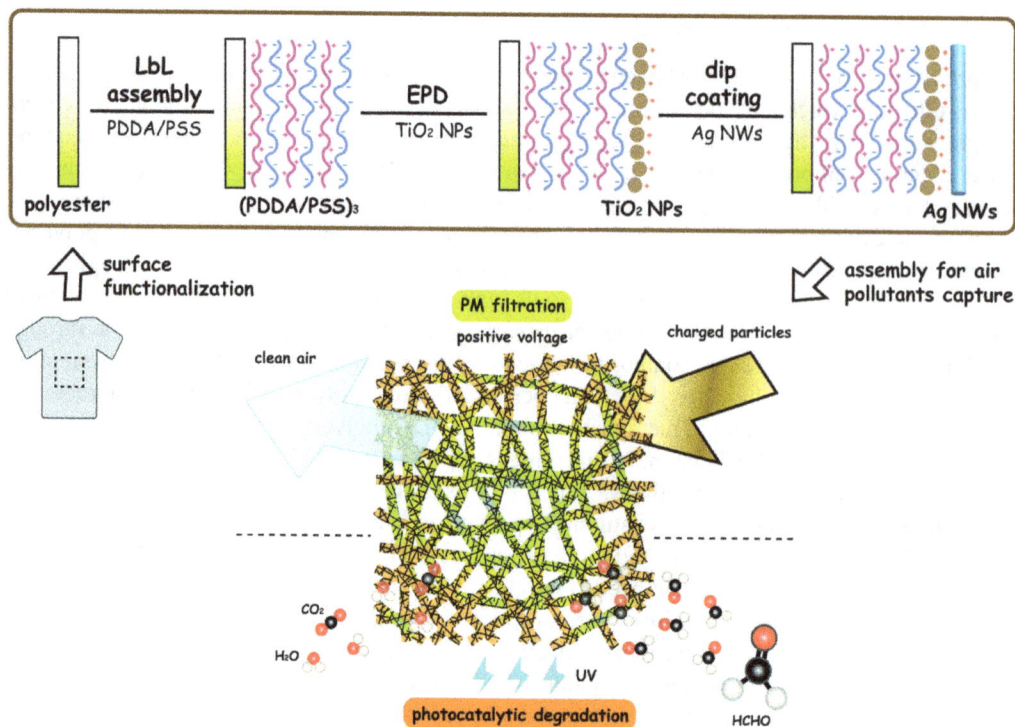

FIGURE 1.8 Scheme for the fabrication process of functionalized polyester air filters at which the fiber networks were obtained through LbL assemble process to embed polyelectrolytes, followed by TiO_2 nanoparticle's deposition through EPD and dip-coating for percolation of Ag nanowires. Adapted with permission [23]. Copyright (2020), Royal Society of Chemistry.

to the formation of OH$\cdot$, which reacts violently with organic matter [25,26]. In addition, iron-based materials can be used as catalysts to form radical hydroxyl. Based on this concept, Su and colleagues synthesized $Fe_2O_3 \cdot 3.9MoO_3$ nanowires through a natural decantation process of the reaction between $FeCl_3$ and Na_2MoO_4 carried out in 7 days [27]. The nanowire was studied in terms of the degradation of congo red dye. The performance of the $Fe_2O_3 \cdot 3.9MoO_3$ nanowires was compared to a 3D $Fe_2O_3 \cdot 3MoO_3$ bowknot-like catalyst obtained through the hydrothermal method. The nanowires presented a degradation rate of 92.6% along with 231.4 mg/g of degradation capacity. Such results surpassed the performance of the 3D $Fe_2O_3 \cdot 3MoO_3$ bowknot structure, which could be attributed to the higher surface area of the nanowires that facilitated the contact between the dye and active sites at the nanomaterial. On top of that, the high performance could be attributed to the synergy effect between the Fe^{3+} as well as MoO_4^{3-}, which could catalyze the decomposition of H_2O_2 as well as recycle the reactive pairs due to the reaction between Fe^{3+} with MoO_4^{2-}. That approach demonstrated the improvement in property based on the material's morphology for the case where the naturally obtained nanowires outperformed the hydrothermally synthesized 3D structure, showing the advantages of this type of 1D nanostructure. The scheme for the synthetical process is presented in Figure 1.9.

1.3.3 NANOWIRES FOR BIOMEDICAL APPLICATIONS

The accurate measurement of biological species is an important process to study the diseases, provide a precise diagnosis, and understand the mechanism of a cell's adaptability as well as the effect of different environments on its behavior, for instance. In this sense, nanowires can serve as viable

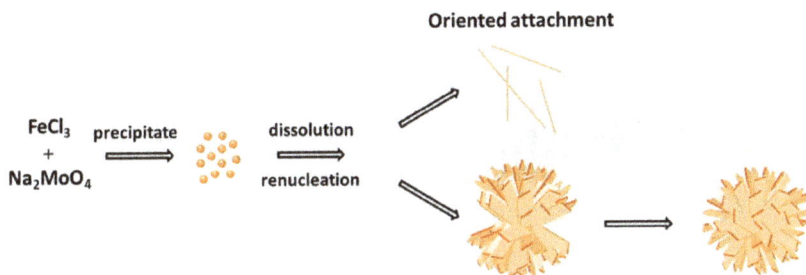

FIGURE 1.9 Scheme for the synthesis of $Fe_2O_3 \cdot 3.9MoO_3$ nanowires through natural precipitation and 3D $Fe_2O_3 \cdot 3MoO_3$ bowknot-like nanostructure obtained through the hydrothermal method. Adapted with permission [27]. Copyright (2018), Royal Society of Chemistry.

nanomaterials to enable bio-detection, which can be attributed to their wide tunability in terms of electrical, optical, and morphological properties [28–30]. Also, it has been observed that vertically aligned nanowire arrays tend to present a better response when compared to free-standing nanowires, which is partially attributed to their higher spatial resolution along with a higher surface area. Through that, a higher number of active sites for detection can be used along with the arrangement of a multichannel detection matrix as the nanowires can be organized in an aligned pattern. Under this perspective, several materials can be obtained as nanowires and employed in biomedical applications, such as transition metal oxides, i.e., TiO_2, ZnO_2, Al_2O_3, and IrO_2 as well as CNTs [31,32].

Under this scope, the proper identification of DNAs and RNAs is an important analytical process for the tracking of diseases and understanding cellular behavior. In this sense, the polymerase chain reaction (PCR) is a widely used method for their identification, however, it is a time-consuming analysis. To counter that drawback nanowires have been employed as potential materials for the detection of fragments of genetic material, which can take place through several different mechanisms like fluorescence, photonic resonance, photoluminescence, electrical diode effect, and cyclic voltammetry [33,34]. Under this perspective, Kim and colleagues fabricated a biosensor based on a diode [35]. It consisted of a Si wafer at which a metal-assisted chemical etching (MaCE) with an Au film was performed for the growth of Si nanowires with hydroxyl groups attached to the side of its walls. Then, 3-aminopropyl triethoxysilane (3-APTES) was used to bind the Si with -OH to form a Si-O bond. Following that, the $-NH_2$ groups were used to functionalize glutaraldehyde at the side of the Si NWs. Finally, graphene was coated over the functionalized Si nanowires to promote a conductive pathway for the signal. The composite was then sandwiched between two electrodes to reinforce electrical contact. The schematic of this procedure is described in Figure 1.10. Such functionalization over the Si nanowire's surface promoted an increase in current flow in the biosensor, which could be attributed to the increase in doping effect over the Si nanowires. Through that, a quantitative analysis of DNA could be performed as the signal varied according to the DNA's concentration. In this sense, when the concentration went from 100 pM to 500 nM there was an increase in current from 19 to 120%. On top of that, the biosensor demonstrated proper robusticity, which allowed a constant working performance for a longer time.

One of the main requirements for materials to be used in biomedical applications is biocompatibility. The incorporation of materials that are not recognizable by the body's system may lead to inflammatory responses, bacterial infection, or rejection. To pave the way for addressing this challenge, Li and colleagues reported the synthesis of $Cu_{2-x}S$ nanowires (CSNW) that were reinforced with poly(citrate-siloxane) (PCS) named PCS-CSNW [36]. The hot-injection method was employed to cap the CSNW with oleylamine. They were then dispersed in a mixture containing DMSO and PCS. Following that, both CSNW and PCS were crosslinked using hexamethylene diisocyanate (HDI). Since both oleylamine-capped CSNW and PCS present a hydrophobic nature

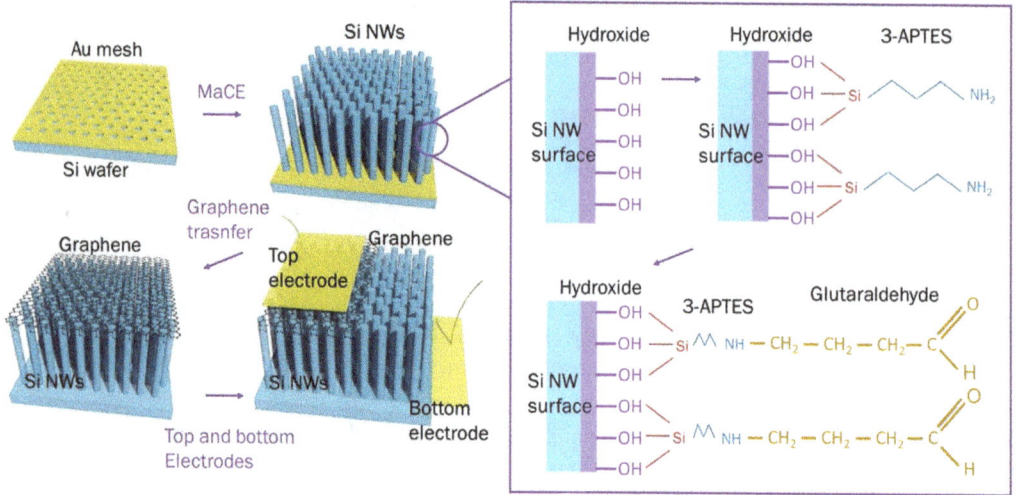

FIGURE 1.10 Scheme for the fabrication of functionalized Si nanowire array for DNA sensing. Along with that, graphene was coated on the array's top along with a sandwiching with a top and bottom electrode to improve the electric contact. Adapted from reference [35]. Copyright (2016), authors, Springer Nature. This is an open access article distributed under the terms of the Creative Commons CC BY license.

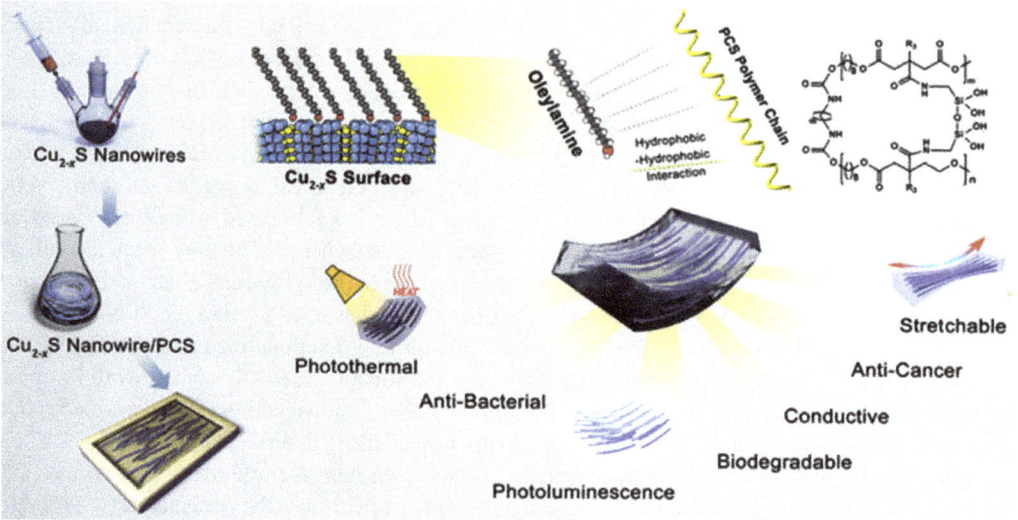

FIGURE 1.11 Scheme for the synthetical approach for the fabrication of PSC-CSNW suitable for biomedical applications due to stretchability, anti-cancer, anti-bacterial, conductivity, and biodegradable properties. Adapted with permission [36]. Copyright (2019), Elsevier.

the PCS-CSNW nanocomposite presented a uniform structure. In addition, the nanowire composite presented several desired properties for biomedical applications, such as effective antibacterial activity, relatively elasticity for proper accommodation within the body, and photothermal properties that can enable therapy for cancer treatment. The scheme for the synthetical procedure as well as biomedical-related properties is displayed in Figure 1.11.

Other materials that have been widely used in the biomedical field are Ag-based nanomaterials, which present antibacterial properties. In this regard, some studies were made to analyze the effect

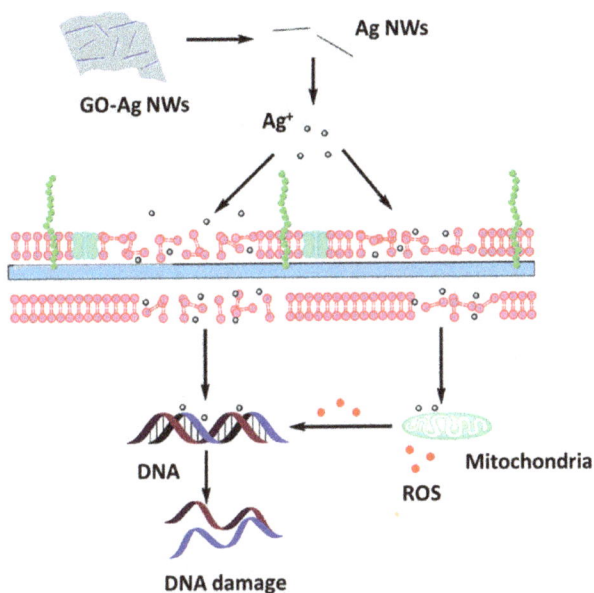

FIGURE 1.12 Proposed mechanism for the action of GO-Ag nanowire composite for the antibacterial process. Reproduced with permission [38]. Copyright (2015), Royal Society of Chemistry.

of the size and shape of Ag-based nanoparticles on their antibacterial activities. It was observed that the main factor for the improvement of such a property was related to the release rate of Ag^+. Yet, the Ag nanomaterial's shape can influence that effect to some degree as observed in the study of Hong *et al.* [37], which analyzed the difference among Ag nanowires (length around 2–4 μm, diameter around 60 nm), nanospheres (diameter of 60 nm) and nanocubes (dimensions around 55 nm). It was observed that nanocubes presented a better interaction with bacteria compared to both nanospheres and nanowires, which was related to the reactivity of crystal facets. In this sense, the Ag nanocubes presented (100) facets that presented a more effective antibacterial activity when compared to nanospheres and nanowires, which presented (111) exposed facets. Despite that, Cui *et al.* [38] demonstrated the enhancement in the antibacterial activity when Ag nanowires were composited with GO sheets (GO-Ag NWs) as the composite's performance surpassed those of pristine Ag nanowires and GO sheets separately. Based on the results, it was proposed that the GO-Ag NW interacted with the bacterial cell's membrane based on electrostatic forces. After that, the Ag^+ could diffuse to the cytoplasm, leading to the formation of reactive oxygen species, which are responsible for degrading the bacteria's cell inner components and preventing the DNA replication process. The proposed mechanism for that process can be seen in Figure 1.12.

1.3.4 OTHER EMERGING APPLICATIONS OF NANOWIRES

The technological push for the development of novel materials and devices along with the high demands from society has led to the need for flexible electronic devices, which can undergo bending and twisting without much loss of their performance. Nanowires can be used as suitable components for flexible devices as they can be grown over a flexible substrate in an aligned pattern. Through that, the nanowires can withstand a certain degree of bending without losing their properties. One example of this design in an electronic device has been studied by Dai *et al.* [39] who fabricated a flexible light-emitting diode (LED) device based on an InGaN/GaN nanowires arranged in a core-shell structure, respectively. The InGaN/GaN core-shell nanowires were grown through metal-organic CVD (MOCVD). The nanowires were incorporated into polydimethylsiloxane (PDMS)

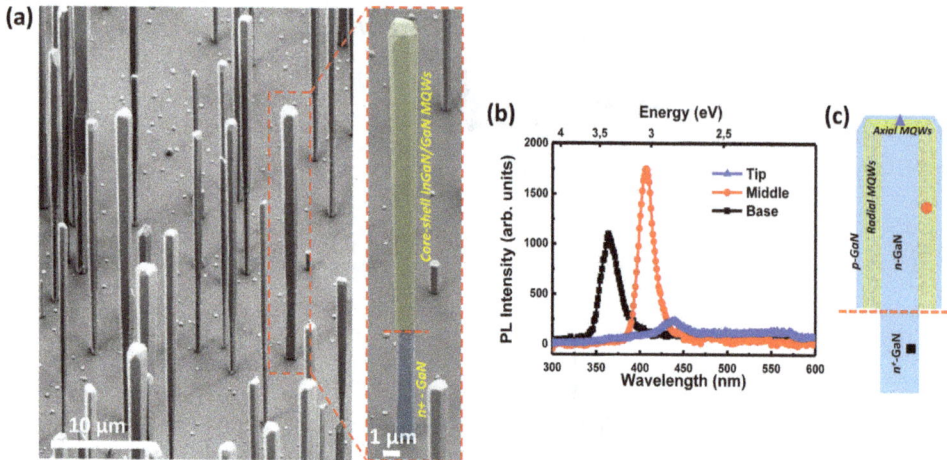

FIGURE 1.13 (a) SEM along with a zoomed image of the InGaN/GaN nanowires with artificial colors displaying the n⁺ GaN as a base and the InGaN/GaN core-shell structure. (b) Plot for the micro luminescence profile for a single nanowire at various excitation regions. (c) Nanowire's structure demonstrates the excitation regions related to the photoluminescence process. Adapted with permission from [39]. Copyright (2015), American Chemical Society.

and lifted from the substrate through mechanical force. Then, transparent and conducting Ag nanowire electrodes were employed to provide electrical contact for the device. The base section of the nanowire was grown on c-sapphire at 1040 °C by using ammonia and trimethylgallium as precursors. Along with that, a flux of silane was employed to promote the n⁺-doping process and form the nanowire structure. The top part of the nanowire was covered with a p-doped GaN shell. The morphology and schematics for this process are presented in Figure 1.13. The LED-based flexible device demonstrated stable electroluminescence even after several bending cycles of around 3 mm of curvature. Hence, based on the morphology and synthetical approach it was observed that the presence of the polymeric matrix along with Ag nanowires promoted the device's flexibility and proper electrical contact, respectively. On top of that, varying the In composition on the InGaN/GaN and combining two layers of this nanomaterial allowed the fabrication of LEDs with wavelengths in the green and blue regions.

Nanowires are promising materials that can function as important components for the latest and near-future electronic devices. In this sense, it is important to propose facile and fast synthetical procedures along with high efficiency to enable their use in large-scale applications. Under this perspective, Bi_2O_3 is a p-type semiconductor that presents appreciable electrical and optical properties, which can make it suitable for applications such as photovoltaic cells, fuel cells, supercapacitors, optical coatings, photocatalysts, and sensors [40,41]. Nanowires can present desired properties concerning their application as sensors due to their high surface-to-volume ratio along with dimensions within the Debye length. Because of these factors, nanowires can be highly sensitive to materials that are adsorbed to their surface. Gou *et al.* [42] proposed a room temperature synthesis of Bi_2O_3 nanowires through the reaction between $Bi(NO_3)_3$ and NaOH. Yet, the presence of oleic acid was a core component to direct the growth of nanowires. The formation of nanowires was important as it allowed enhancement of the sensitivity of Bi_2O_3 towards the adsorption of NO_2 in a nearly exclusive matter, hence making it a selective sensor for this gas. It is worth noting that the Bi_2O_3 was highly selective towards NO_2 due to its specific diameter. It was observed that the increase in temperature during synthesis caused an increase in the diameter of Bi_2O_3. Hence, maintaining the same reaction parameters was an important aspect to maintain the sensing properties of the Bi_2O_3 nanowires. Based on these observations the authors analyzed the nanowire's sensitivity to several

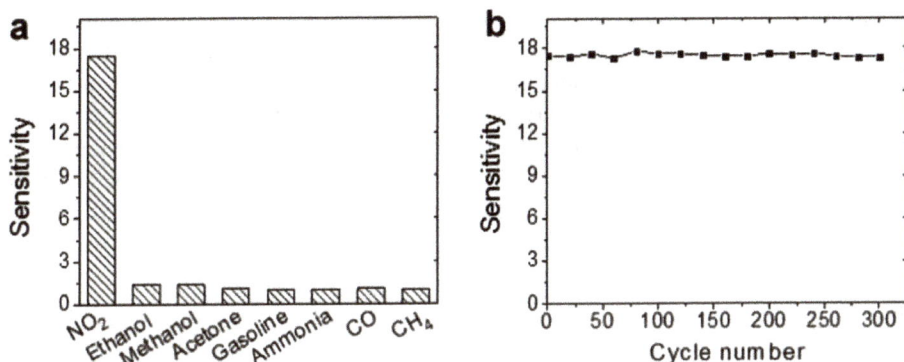

FIGURE 1.14 (a) Analysis of Bi_2O_3 nanowire selectivity towards several gases at a concentration of 100 ppm. (b) Nanowire sensor's stability at 100 ppm of NO_2. Reproduced with permission from [42]. Copyright (2009), IOP Science.

other gases, which demonstrated higher selectivity as well as stability for NO_2 (Figures 1.14a and b). It was notable that relatively simple procedures can be adopted to fabricate nanowires with high-end applications such as gas sensors, which can serve in several areas for monitoring and control.

1.4 CONCLUSION AND PERSPECTIVES

Throughout the discussion in this chapter, it has been observed that nanowires are becoming highly influential nanomaterials within the scientific community and conquering more space in several sectors, which further attracts the research interest in finding novel materials that can be obtained in this type of nanostructure. The synthesis of 1D nanomaterials can be performed through a plethora of methods, which can include, PVD, CVD, hydrothermal, solvothermal, soft or hard template, and so on. Through that, some methods form the nanowires spontaneously or by inducing their formation through the influence of the reaction's system, such as in the case of the latter two approaches, i.e., soft and hard templates. In that sense, the number of methods as well as materials that can be obtained as 1D nanostructures enables a wide range of possibilities leading to materials with different properties. Some of the appreciable properties of nanowires are related to their high surface area as well as their morphology, which can provide a conducting pathway for the transport of electrons and/or ionic species. These aspects are highly desired for applications as supercapacitors as it functions on the principle of the EDLC process. Also, nanowires can be embedded with nanoparticles to enhance some of their properties. In this sense, transition metal-based nanomaterials can be either synthesized as nanowires or incorporated into the surface of nanowire materials with better conductivity to add specific electrochemical properties, such as redox processes based on pseudocapacitance. Through that, a synergism can be created by taking advantage of the nanowires' conductivity along with the incorporated nanoparticles by providing a decrease in the electron transfer processes and further enhancing the energy storage properties. In another case, nanowires can be employed as biosensors as their surface can be functionalized with specific probes that can allow selective identification of biological species such as DNA, RNA, or disease makers. Through that, viruses, pathogenic agents, or antigens can be analyzed qualitatively as well as quantitatively, in some cases, enabling a more accurate diagnosis. Within this line, nanowires that possess appreciable conductivity can deliver an electric signal when there is a binding between the probe with the analyte, providing the identification of the desired specie. Also, nanowires can serve as suitable materials for the fabrication of highly selective sensors of different substances such as gases. Such an application is possible due to the high degree of tunability during its synthetical processes as it allows control over its length and diameter, which can function as an accurate site for adsorption of a specific type of analyte.

The development of flexible devices is another sector that takes advantage of the inherent properties of nanowires as they can be grown in several types of flexible substrates while withstanding appreciable bending and twisting without deterioration of their properties. Based on this discussion it has been noted that nanowires are materials with great potential for high-end and value-added applications. Yet, some recurring challenges such as nanowire's tendency to agglomerate, obtaining aligned nanostructures through relatively cheaper methods, and accurate control of nanowire's morphology along with enabling large-scale processing are some of the main points that the current generation of scientists is required to address.

REFERENCES

[1] Z. Sun, T. Liao, L. Kou, Strategies for designing metal oxide nanostructures, Sci. China Mater. 60 (2017) 1–24.

[2] G. Zhou, L. Xu, G. Hu, L. Mai, Y. Cui, Nanowires for electrochemical energy storage, Chem. Rev. 119 (2019) 11042–11109.

[3] L.Q. Mai, B. Hu, W. Chen, Y.Y. Qi, C.S. Lao, R.S. Yang, Y. Dai, Z.L. Wang, Lithiated MoO_3 nanobelts with greatly improved performance for lithium batteries, Adv. Mater. 19 (2007) 3712–3716.

[4] J.-Y. Liao, D. Higgins, G. Lui, V. Chabot, X. Xiao, Z. Chen, Multifunctional TiO_2–C/MnO_2 core–double-shell nanowire arrays as high-performance 3D electrodes for lithium ion batteries, Nano Lett. 13 (2013) 5467–5473.

[5] C. Niu, J. Meng, X. Wang, C. Han, M. Yan, K. Zhao, X. Xu, W. Ren, Y. Zhao, L. Xu, Q. Zhang, D. Zhao, L. Mai, General synthesis of complex nanotubes by gradient electrospinning and controlled pyrolysis, Nat. Commun. 6 (2015) 7402.

[6] G. Zheng, Y. Yang, J.J. Cha, S.S. Hong, Y. Cui, hollow carbon nanofiber-encapsulated sulfur cathodes for high specific capacity rechargeable lithium batteries, Nano Lett. 11 (2011) 4462–4467.

[7] P. Hu, T. Zhu, X. Wang, X. Wei, M. Yan, J. Li, W. Luo, W. Yang, W. Zhang, L. Zhou, Z. Zhou, L. Mai, Highly durable Na2V6O16·1.63H2O nanowire cathode for aqueous zinc-ion battery, Nano Lett. 18 (2018) 1758–1763.

[8] Y. Meng, D. Wang, Y. Zhao, R. Lian, Y. Wei, X. Bian, Y. Gao, F. Du, B. Liu, G. Chen, Ultrathin TiO2-B nanowires as an anode material for Mg-ion batteries based on a surface Mg storage mechanism, Nanoscale. 9 (2017) 12934–12940.

[9] K. Li, H. Liu, G. Wang, Sb_2O_3 nanowires as anode material for sodium-ion battery, Arab. J. Sci. Eng. 39 (2014) 6589–6593.

[10] W. Luo, A. Calas, C. Tang, F. Li, L. Zhou, L. Mai, Ultralong Sb_2Se_3 nanowire-based free-standing membrane anode for lithium/sodium ion batteries, ACS Appl. Mater. Interfaces. 8 (2016) 35219–35226.

[11] F. Liao, X. Han, Y. Zhang, H. Chen, C. Xu, CTAB-assisted solvothermal synthesis of ultralong t-selenium nanowires and bundles using glucose as green reducing agent, Mater. Lett. 214 (2018) 41–44.

[12] W.Q. Wang, Z.J. Yao, X.L. Wang, X.H. Xia, C.D. Gu, J.P. Tu, Niobium doped tungsten oxide mesoporous film with enhanced electrochromic and electrochemical energy storage properties, J. Colloid Interface Sci. 535 (2019) 300–307.

[13] G. Zhang, X. Li, H. Liu, D. Wei, Engineering capacitive contribution in dual carbon-confined Fe_3O_4 nanoparticle enabling superior Li+ storage capability, J. Mater. Sci. 56 (2021) 5100–5112.

[14] D. Wang, H. Dai, Low-temperature synthesis of single-crystal germanium nanowires by chemical vapor deposition, Angew. Chemie Int. Ed. 41 (2002) 4783–4786.

[15] P.-C. Chang, Z. Fan, D. Wang, W.-Y. Tseng, W.-A. Chiou, J. Hong, J.G. Lu, ZnO nanowires synthesized by vapor trapping CVD method, Chem. Mater. 16 (2004) 5133–5137.

[16] Q. Su, S. Wang, Y. Xiao, L. Yao, G. Du, H. Ye, Y. Fang, Lithiation behavior of individual carbon-coated Fe_3O_4 nanowire observed by in situ TEM, J. Phys. Chem. C. 121 (2017) 3295–3303.

[17] Z. Zhang, H. Wang, Y. Zhang, X. Mu, B. Huang, J. Du, J. Zhou, X. Pan, E. Xie, Carbon nanotube/hematite core/shell nanowires on carbon cloth for supercapacitor anode with ultrahigh specific capacitance and superb cycling stability, Chem. Eng. J. 325 (2017) 221–228.

[18] W. Xia, Q. Zhang, F. Xu, L. Sun, New insights into electrochemical lithiation/delithiation mechanism of α-MoO$_3$ nanobelt by in situ transmission electron microscopy, ACS Appl. Mater. Interfaces. 8 (2016) 9170–9177.

[19] C. Xia, Q. Jiang, C. Zhao, M.N. Hedhili, H.N. Alshareef, Selenide-based electrocatalysts and scaffolds for water oxidation applications, Adv. Mater. 28 (2016) 77–85.

[20] X. Gao, H. Zhang, Q. Li, X. Yu, Z. Hong, X. Zhang, C. Liang, Z. Lin, Hierarchical NiCo$_2$O$_4$ hollow microcuboids as bifunctional electrocatalysts for overall water-splitting, Angew. Chemie Int. Ed. 55 (2016) 6290–6294.

[21] J. Yin, Y. Li, F. Lv, Q. Fan, Y.-Q. Zhao, Q. Zhang, W. Wang, F. Cheng, P. Xi, S. Guo, NiO/CoN porous nanowires as efficient bifunctional catalysts for Zn–air batteries, ACS Nano. 11 (2017) 2275–2283.

[22] W. Chen, J. Wang, K.Y. Ma, M. Li, S.H. Guo, F. Liu, J.P. Cheng, Hierarchical NiCo2O4@Co-Fe LDH core-shell nanowire arrays for high-performance supercapacitor, Appl. Surf. Sci. 451 (2018) 280–288.

[23] R.-Y. Zhang, G.-W. Hsieh, Electrostatic polyester air filter composed of conductive nanowires and photocatalytic nanoparticles for particulate matter removal and formaldehyde decomposition, Environ. Sci. Nano. 7 (2020) 3746–3758.

[24] R. Alizadeh, R.K. Kazemi, M.R. Rezaei, Ultrafast removal of heavy metals by tin oxide nanowires as new adsorbents in solid-phase extraction technique, Int. J. Environ. Sci. Technol. 15 (2018) 1641–1648.

[25] S.-S. Lin, M.D. Gurol, catalytic decomposition of hydrogen peroxide on iron oxide: kinetics, mechanism, and implications, Environ. Sci. Technol. 32 (1998) 1417–1423.

[26] S.H. Tian, Y.T. Tu, D.S. Chen, X. Chen, Y. Xiong, Degradation of Acid Orange II at neutral pH using Fe2(MoO4)3 as a heterogeneous Fenton-like catalyst, Chem. Eng. J. 169 (2011) 31–37.

[27] Y. Su, X. Zhao, Y. Bi, C. Li, Y. Feng, X. Han, High-concentration organic dye removal using Fe2O3·3.9MoO3 nanowires as Fenton-like catalysts, Environ. Sci. Nano. 5 (2018) 2069–2076.

[28] X. Li, J. Mo, J. Fang, D. Xu, C. Yang, M. Zhang, H. Li, X. Xie, N. Hu, F. Liu, Vertical nanowire array-based biosensors: device design strategies and biomedical applications, J. Mater. Chem. B. 8 (2020) 7609–7632.

[29] E. Borberg, M. Zverzhinetsky, A. Krivitsky, A. Kosloff, O. Heifler, G. Degabli, H.P. Soroka, R.S. Fainaro, L. Burstein, S. Reuveni, H. Diamant, V. Krivitsky, F. Patolsky, Light-controlled selective collection-and-release of biomolecules by an on-chip nanostructured device, Nano Lett. 19 (2019) 5868–5878.

[30] C. Y., C. H., Q. R., H. M., M. E., H. M., Z. A., L.R. S., W.J. C., M.N. A., Universal intracellular biomolecule delivery with precise dosage control, Sci. Adv. 4 (2022) eaat8131.

[31] W. Kim, J.K. Ng, M.E. Kunitake, B.R. Conklin, P. Yang, Interfacing silicon nanowires with mammalian cells, J. Am. Chem. Soc. 129 (2007) 7228–7229.

[32] A.M. Xu, A. Aalipour, S. Leal-Ortiz, A.H. Mekhdjian, X. Xie, A.R. Dunn, C.C. Garner, N.A. Melosh, Quantification of nanowire penetration into living cells, Nat. Commun. 5 (2014) 3613.

[33] Y. Zhang, T. Jiang, L. Tang, Near-infrared photoluminescence biosensing platform with gold nanorods-over-gallium arsenide nanohorn array, Biosens. Bioelectron. 97 (2017) 278–284.

[34] P. Serre, C. Ternon, V. Stambouli, P. Periwal, T. Baron, Fabrication of silicon nanowire networks for biological sensing, Sensors Actuators B Chem. 182 (2013) 390–395.

[35] J. Kim, S.-Y. Park, S. Kim, D.H. Lee, J.H. Kim, J.M. Kim, H. Kang, J.-S. Han, J.W. Park, H. Lee, S.-H. Choi, Precise and selective sensing of DNA-DNA hybridization by graphene/Si-nanowires diode-type biosensors, Sci. Rep. 6 (2016) 31984.

[36] Y. Li, N. Li, J. Ge, Y. Xue, W. Niu, M. Chen, Y. Du, P.X. Ma, B. Lei, Biodegradable thermal imaging-tracked ultralong nanowire-reinforced conductive nanocomposites elastomers with intrinsic efficient antibacterial and anticancer activity for enhanced biomedical application potential, Biomaterials. 201 (2019) 68–76.

[37] X. Hong, J. Wen, X. Xiong, Y. Hu, Shape effect on the antibacterial activity of silver nanoparticles synthesized via a microwave-assisted method, Environ. Sci. Pollut. Res. 23 (2016) 4489–4497.

[38] J. Cui, Y. Liu, Preparation of graphene oxide with silver nanowires to enhance antibacterial properties and cell compatibility, RSC Adv. 5 (2015) 85748–85755.

[39]   X. Dai, A. Messanvi, H. Zhang, C. Durand, J. Eymery, C. Bougerol, F.H. Julien, M. Tchernycheva, Flexible light-emitting diodes based on vertical nitride nanowires, Nano Lett. 15 (2015) 6958–6964.

[40]   L. Zhou, W. Wang, H. Xu, S. Sun, M. Shang, Bi$_2$O$_3$ hierarchical nanostructures: controllable synthesis, growth mechanism, and their application in photocatalysis, Chem. – A Eur. J. 15 (2009) 1776–1782.

[41]   A. Cabot, A. Marsal, J. Arbiol, J.R. Morante, Bi$_2$O$_3$ as a selective sensing material for NO detection, Sensors Actuators B Chem. 99 (2004) 74–89.

[42]   X. Gou, R. Li, G. Wang, Z. Chen, D. Wexler, Room-temperature solution synthesis of Bi$_2$O$_3$ nanowires for gas sensing application, Nanotechnology. 20 (2009) 495501.

2 Semiconductor Nanowires

Weronika Smok, Marta Zaborowska, and Tomasz Tański
Silesian University of Technology, Department of Engineering Materials
and Biomaterials, Konarskiego, Gliwice, Poland

CONTENTS

2.1 INTRODUCTION

Semiconductor nanowires (SNs) are a group of quasi-one-dimensional (1D) nanostructures that are characterized by diameters ranging from a few to about 300 nm, thus exhibiting specific and shape-dependent interactions with the electromagnetic field, especially in the ultraviolet (UV) region. The second key driver of the popularity of SNs is their free-standing nature, which defines many application possibilities in future electronic and photonic devices, among others, and which can be consciously designed during the fabrication process. The high sensitivity of nanowires to various environmental stimuli and interactions affects their optical and electrical properties, which are successfully used to build bio- and chemisensors. Another key advantage of SNs is the ability to fabricate highly crystalline nanowires, even for strongly mismatched substrate networks. It is also worth mentioning that the characteristic properties are mainly due to their morphology, which are not exhibited by two-dimensional nanomaterials constrained by their geometry and nanoparticles, which exhibit a high agglomeration capacity.

Over the past several years, with the ever-increasing development of nanotechnology, the work of researchers in the area of SNs has also intensified. The market for 1D nanostructures is estimated to reach USD 6.7 billion by 2026 with compound annual growth rates (CAGR) of 25.1% [1]. The current knowledge of SNs allows us to accurately classify this type of material based on structure, chemical composition, and specific properties of nanostructures. The first level of subdivision is based on the number of elements the nanowire crystal structure consists of. Two groups are distinguished: elementary, consisting of single elements, such as silicon (Si) or germanium (Ge), and heterostructures, i.e. nanowires with alloy structure. The second group of SNs includes binary nanostructures composed of chemical elements from a combination of groups of the periodic table elements III-V, II-VI, IV-IV, and IV-VI, ternary and quaternary, e.g. InGaAs and InGaZnO and metal oxides, i.e. ZnO, SnO_2, or TiO_2 (**Figure 2.1**).

DOI: 10.1201/9781003296621-2

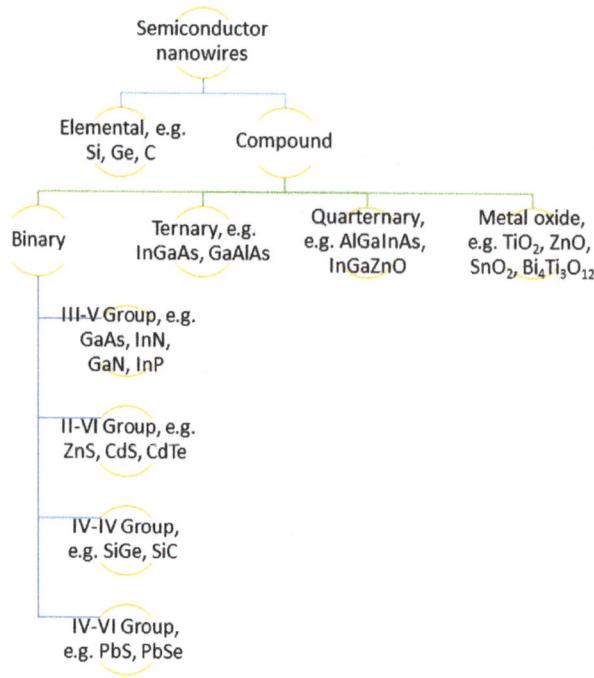

FIGURE 2.1   Classification of SNs.

Among elemental SNs, silicon nanowires were obtained the earliest, as early as 1964, whose application potential was to revolutionize the electronics industry in applications, such as solar cells, field effect transistors (FETs), and biosensors. With years of research on the properties of Si nanowires, it became apparent that proper crystalline quality consistent with crystallographic orientation was an extremely important element of their electrical properties. However, based on direct theoretical calculations and indirect experimental analyses by measuring I-V characteristics, it has been shown that Si nanowires exhibit much lower or similar charge carrier mobility to bulk Si [2]. In contrast to silicon nanowires, germanium nanostructures exhibit 2.5- to 3.5-fold higher charge carrier mobility, which makes them suitable for the fabrication of high-performance transistors with nanometric gate dimensions [3]. Another feature that differentiates Ge and Si nanowires is the difference between direct and indirect band gap, which is 140 meV and 2.4 eV, for germanium and silicon, respectively. This property allows for a direct band gap in the germanium crystal by applying external strain or doping with other elements, which is ideal for optoelectronic applications. Both groups of nanowires, due to their similarities in the crystal structure, are most often fabricated by VLS [4] and VLS-type mechanisms [5], in which researchers are still making further attempts to bypass the involvement of a catalyst, whose presence negatively affects the crystal purity of the final nanowires.

The second group of SNs is compound nanowires, which due to the combination of key semiconductors show much better physical properties, compared to elemental. An extremely important property of binary nanowires of group III-V, which include GaSb, InAs, or InSb, among others, is their small energy gap width of 0.7 eV providing broadband photodetection from the visible to the infrared range. An important group is the family of group III-V ternary nanowires, whose key feature is the ability to adjust the energy gap width to obtain desired electrical properties, e.g. InGaAs, INAsP, and GaAsSb.

Group II-VI SNs are mainly binary nanostructures, which include ZnS, ZnSe, ZnTe, CdSe, CdTe, CdS, and ZnO, however, quaternary nanowires such as ZnCdSSe are also distinguished here.

The most important properties of this group of nanowires are their large exciton binding energy, photoluminescence resulting from their direct band gap nature, and absorption of electromagnetic radiation covering a broad spectrum from ultraviolet to near-infrared. Their main applications are optoelectronic devices. By forming quaternary alloy nanowires, nanostructures with much better properties can be obtained compared to the properties of binary nanowires. Such nanostructures can find applications for the construction of full-spectrum solar cells, displays, multispectral detectors, and spectrometer-on-a-chip to super broadly tunable nanolasers [6].

The combination of group IV elements allows SiGe or SiC nanowires to be obtained. SiGe materials exhibit exceptional thermoelectric properties and have been used to convert heat to electricity in spacecraft designed for National Aeronautics and Space Administration (NASA) missions in space since 1976. However, conventional SiGe alloys show reduced values of κ, due to alloy-induced scattering of high-frequency phonons, which can be improved with the reduction of their dimensions to nanometer size. Using a combination of CVD and VLS methods, SiGe nanowires were also obtained and incorporated into microplatforms as thermoelectric material. The energy harvesting tests carried out showed a level of 7.1 $\mu W/cm^2$ without any additional heat exchanger, which is a very good result [7]. Nevertheless, IV-VI semiconductors are an important group due to their properties such as large exciton Bohr radius (PbS – 18 nm, PbSe – 47 nm, and PbTe – 150 nm), and are widely studied for application in the construction of nanocrystalline solar cells or FETs. The last but not least important family of SNs are metal oxide (MO) nanostructures. Here we include a very rich group of materials, e.g. ZnO, SnO_2, In_2O_3, TiO_2, ZrO_2, Fe_2O_3, CuO, or Al_2O_3.

An important advantage of MO-based SNs is the multitude of fabrication methods, using both *bottom-up* and *top-down* approaches. Due to their unique properties, applications of MO nanowires can be found in FETs, transparent electronics, lasers and waveguides, piezoelectric nanogenerators, solar cells, photocatalysis, or chemical sensors. Nanostructures of SnO_2 demonstrate great potential for applications in wearable or bio-implantable devices, based on the results of the performance analysis with high field-effect mobilities of 10^2 $cm^2/V.s$ and small hysteresis [8]. The use of ZnO or In_2O_3 nanowires in the role of active channels in nanowire transistors on flexible plastic substrate allows optical transparency of 82% to be achieved. Moreover, they exhibit transfer and current versus voltage comparable to transistors based on non-transparent nanostructures. Nanowires obtained in this way can be used as pixel-switching and driving transistors in AMOLED displays [9].

ZnO SNs are ideal candidates for the construction of innovative optical waveguides. The work [10] describes the synthesis of 1D ZnO nanostructures on Si substrate using thermal evaporation from Zn powder. The photoluminescence results showed that upon excitation with 325 nm PL light can be coupled into a single ZnO nanowire and the existence of a single electromagnetic mode in the nanostructures. Titanium dioxide nanostructures are most commonly used for nanostructured solar cells and photocatalysis processes. Porous nanowires of Nb_2O_5, obtained by solvothermal method with reduced energy gap width in comparison with niobium oxide bulk were investigated for application in photodegradation processes of an aqueous solution of rhodamine B under UV light. The analytical results indicated that the porous Nb_2O_5 nanowire achieved a hydrogen evolution rate of 243.8 $\mu mol/g.h$ [11]. In_2O_3, which is a semiconductor with an energy gap width of 3.55–3.75 eV, in its nanostructured form is an ideal sensing material. Using the hydrothermal-annealing route, indium oxide nanowires with diameters in the range of 30 to 50 nm were prepared and its sensing properties were investigated in the detection of NO_2. The results clearly confirmed the ability to detect 1 ppm of gaseous nitrogen dioxide at 250°C with a response value of 2,57 [12].

2.2   GROWTH AND SYNTHESIS METHODS OF SNS

SNs are an extremely important component of modern electronic, optoelectronic, and thermal devices, whose sophistication is constantly increasing, and thus the demands on the quality and characteristics of nanowires are increasing. It is desirable to fabricate high-density, ordered nanowire architectures in which periodic structure, anisotropic parameters, and size may be tuned [13]. So far, many methods have been developed for the fabrication of SNs and nanowire arrays, which, depending on the synthesis mechanism, can be assigned to bottom-up or top-down approaches.

2.2.1 BOTTOM-UP APPROACHES

Bottom-up semiconductor nanowire synthesis methods are based on a single molecule growth strategy that uses chemical reactions to control growth. One of the main advantages of bottom-up nanowire growth over top-down is the possibility of in situ doping during nanostructure fabrication. The doping procedure can be performed by incorporating dopant precursors into the main nanowire growth procedure. With this approach, SNs do not require destructive ion implantation techniques to generate additional charge carriers that may negatively affect the atomic ordering in the semiconductor crystal region [14].

One of the more widely used bottom-up methods for the growth of nanowires with semiconducting properties currently includes the VLS mechanism or vapor-liquid-solid, which is based on the presence of a precursor nanowire material in the gas phase bombarding liquid phase crystal seeds and unidirectional growth of the nanostructure (**Figure 2.2**). The selection of a suitable substrate material allows the control of the nanowire diameter and its crystalline quality. An important issue affecting the performance of VLS processes is also the precursor material, which should have the least effect on the purity of the final nanowires. For this purpose, metal hydrides are usually used, which have the advantage of a high purity by-product in the form of hydrogen gas. In recent years, there has also been tremendous interest, especially in the case of SNs such as InP, GaN, and SnO_2 in self-seeded VLS-type growth methods in which seed layers are formed on the surface of metallic substrate materials to grow nanowires [15].

Another method for the growth of SNs is metal-organic vapor phase epitaxy (MOVPE), which mainly allows the control of the crystallographic orientation of nanostructures by selecting a suitable substrate [16]. In the molecular beam epitaxy (MBE) process, the growth of SNs under ultrahigh vacuum conditions on a heated crystalline substrate takes place through the interaction of adsorbed molecules delivered by atomic beams. The key advantage of nanostructures obtained in this way is

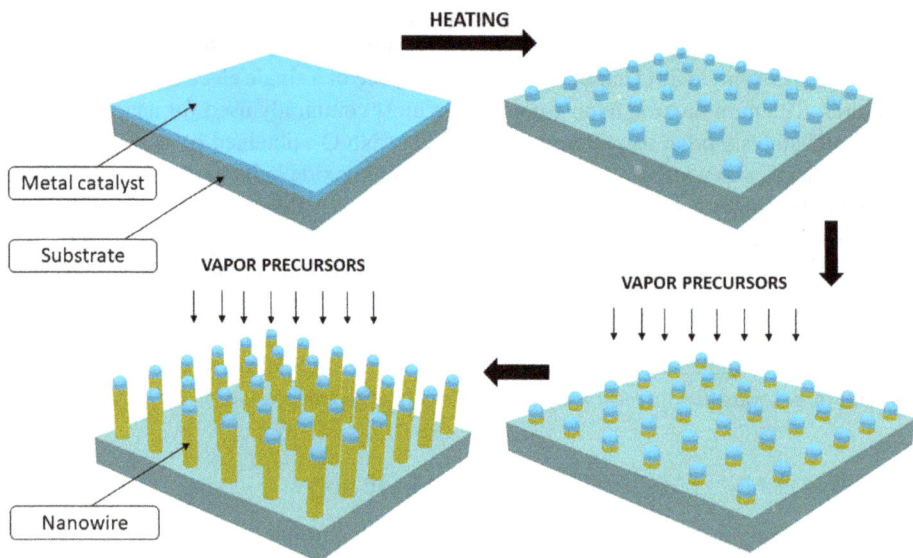

FIGURE 2.2 Vapor-liquid-solid mechanism of SN growth.

the high crystallinity and preferentially form in-plane nanowires [17]. In addition to strictly controlled processes for the growth of crystalline SNs, self-assembly methods are also distinguished, which involve the spontaneous aggregation of molecules into stable, non-covalently organized structures presenting three-dimensional order [18].

Another family of methods for the synthesis of SNs is template-assisted methods, the key element of which is a matrix that can take any form with nanostructural characteristics, whose size, morphology, and charge distribution have a great influence on the final material. In the first step of template-assisted methods, a template is prepared, most often from AAO (anode aluminum oxide). A thin layer of a conductive material such as gold is applied to the prepared template to serve as a cathode. The template is then immersed in the solution, and the applied current or potential causes ions to migrate from the solution toward the mouth of the pore, where electrochemical reactions occur on the gold layer and nanowires grow along the pore [19]. In the final step, the template can be removed by dissolution or calcination. One of the template-assisted methods is the soft template method, the key element of which is a template in the form of flexible nanostructures connected by intermolecular interactions. Soft templates usually consist of surfactants, flexible organic molecules, or block copolymers. Another type of modification of template-assisted methods can be the one-step template-assisted electrodeposition technique, which allows the fabrication of crystalline ZnCuTe nanowire on indium oxide substrate [20].

Nevertheless, an important group of SNs fabrication techniques is solvo- and hydrothermal methods. The distinguishing feature of this type of crystalline nanowire growth approach is the ability to subject the initial materials to specific conditions to obtain nanostructures, and control over their properties and morphology can be achieved by changing process factors such as solvents, surfactants, organometallic compounds, reaction temperature, or pH value [21]. Using an economic and low-cost solvothermal method it is possible to obtain MOs nanowires with controlled electrical properties, morphology, and density [22].

Another approach for the fabrication of crystalline SNs is electrospinning, the essence of which is an electrostatic field. In this process, a spinning solution consisting of suitable organometallic precursors, polymer, and solvents is placed in the nozzle of a device to which a high voltage is applied, and the generated electrostatic field causes stretching and final formation of fibrous structures (**Figure 2.3**) [23]. The use of a polymeric material to obtain the spinning solution allows the control of the morphology and structure of the micro- and nanofibers produced. In the post-process treatment, the polymer is degraded, usually by high-temperature thermal treatment, which also aims to react to form MOs from the precursors. Using the electrospinning method, semiconducting 1D nanostructures based on MOs are obtained, including but not limited to ZnO [24], TiO_2 [25], and SnO_2 [26,27].

2.2.2 Top-Down Methods

The set of bottom-up methods has many advantages such as high quality of nanowires and precise diameter control, however, still, the limitations of this method include alignment of nanowires or

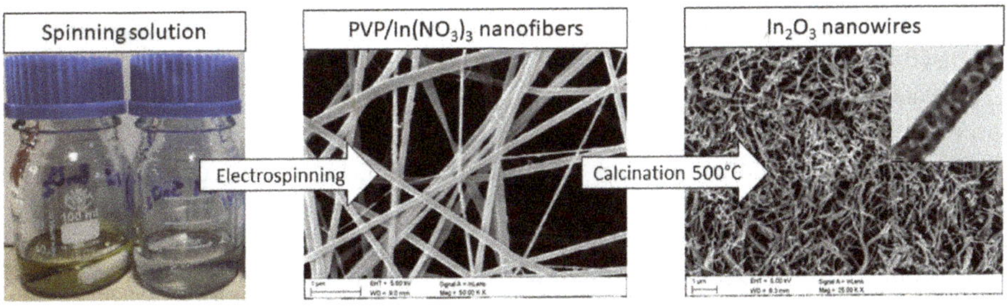

FIGURE 2.3 Methodology approach for electrospun In_2O_3 nanowires.

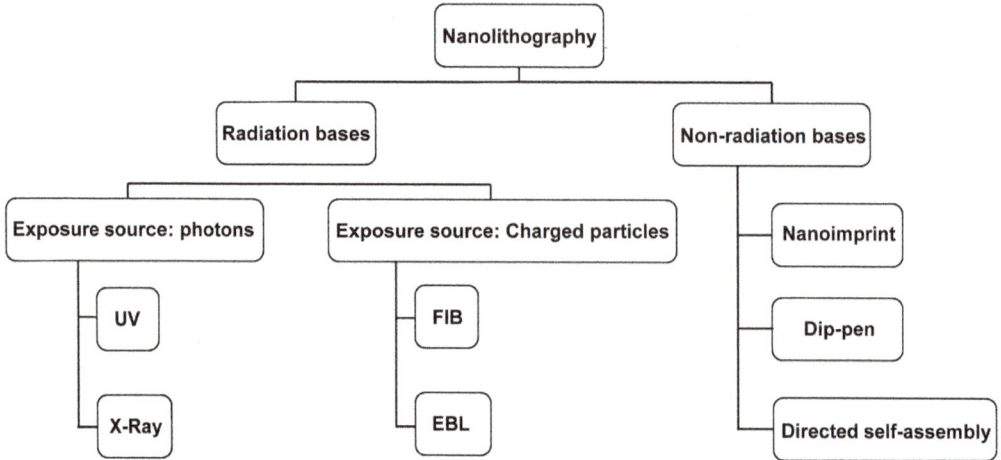

FIGURE 2.4 Types of nanolithography used for the fabrication of SNs.

contamination during wire growth. Hence, the top-down strategy has been dominating industrial-scale electronic device fabrication for decades. The top-down strategy for fabrication of SNs and underlying devices with complex geometries involves the etching or sculpting of a larger semiconductor element causing loss of material and leaving nanostructures with the desired morphology. Top-down methods for nanowire fabrication include lithography and its variations (**Figure 2.4**) and dry and wet etching.

Currently, the most widely used method for large-scale fabrication of SNs in a top-down approach is photolithography [13]. This method is based on the reproduction of patterns using a mask on a thin layer of photosensitive emulsion covering a semiconductor substrate, resulting in the desired pattern on the substrate. The smallest critical dimension of the obtained nanostructures depends linearly on the wavelength of the irradiation used, hence the aim is to reduce the wavelength by moving away from visible light to ultraviolet and X-rays. The disadvantages of photolithography, such as high cost, time consumption, and limited resolution have been overcome by using extreme ultraviolet (EUV) lithography, which uses 13.5 nm light, combining the simplicity of laser interference lithography with high resolution [28].

However, the need to fabricate sub-100 nm SNs requires improved lithography techniques such as electron beam lithography (EBL) and focused ion beam lithography (FIB). The EBL technique involves scanning, with a highly focused electron beam, a sample coated with an electron-sensitive film (or resist), which is usually applied by spin-coating. This is followed by developing, which is the selective removal of electron beam exposed or unexposed regions of the resist. After developing, the resist layer on top of the sample can be used as a mask or template for transferring the pattern. Transfer can be accomplished by etching material underneath the voids in the resist or depositing a layer of material over the entire sample and removing the resist. FIB instead of electrons, uses a highly focused beam of high mass ions, usually Ga^+ to scan and remove material from a semiconductor substrate in a vacuum chamber leaving a finished pattern on the substrate. This method has been successfully used to fabricate single Si, ZnO, CdS, InAs, and SiO_2 nanowires. However, EBL and FIB have significant limitations such as low throughput, high cost, complexity, and are also much slower than current optical lithography tools, so they are currently used at a laboratory scale.

Fabrication of SNs of Si, ZnO, SiC, InP at a considerably lower cost and with a much higher throughput compared to EBL is also possible using nanoimprint lithography (NIL) [29]. NIL consists in covering the substrate with a layer of molded material and pressing an appropriately shaped mold to it, which determines the form of nanostructures left after its removal. In the case

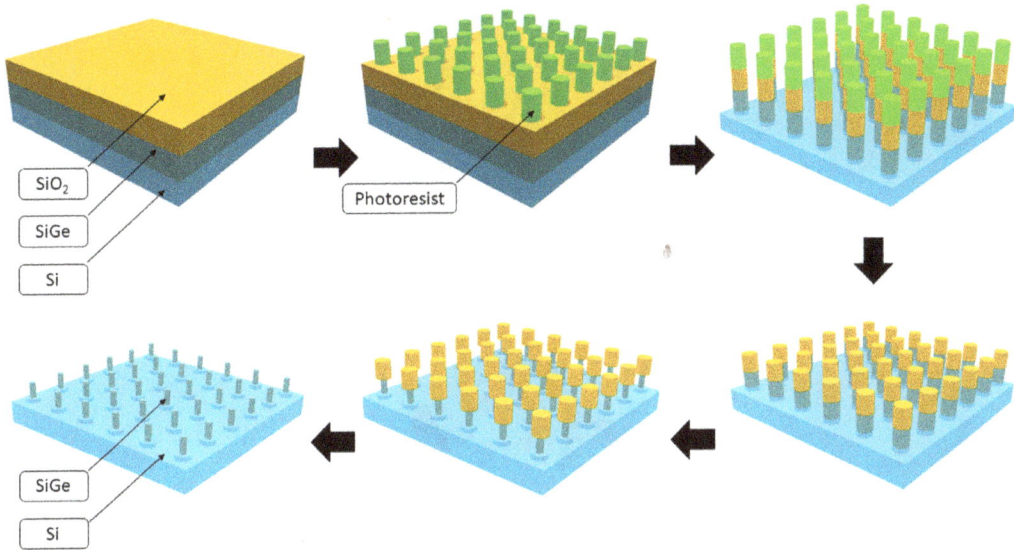

FIGURE 2.5   Atomic layer etching of vertical SiGe nanowires.

of semiconductor materials, the UV-assisted nanoimprint lithography (UV-NIL) method is used, in which after pressing the mold to the UV-curable precursor liquid, UV irradiation occurs, and a solid nanomaterial is obtained [30]. Dry atomic layer etching (ALE) is also an interesting method to fabricate SNs. ALE is an advanced technique that removes material layer by layer using sequential self-limiting chemical reactions. For example, atomic layer etching of vertical SiGe nanowires performed by J. Li et al. [31] was mainly two self-limiting processes: cyclic self-limited oxidation and selective etching of SiGe oxides, which were repeated until the nanowires of the required diameter were obtained (**Figure 2.5**). The advantages of this method include the ability to fabricate devices with complex shapes while maintaining high aspect ratios and minimization of line edge roughness.

Wet etching technique-metal-assisted chemical etching (MACE) is used for the fabrication of Si, GaAs, GaN, and SiC nanowires, among others, which is a cheap and versatile technique that allows precise control of the morphology of the fabricated nanowires [32–35]. MACE involves etching a semiconductor substrate by coating it with a noble metal in the form of a thin film with uncoated areas left behind and immersing it in a solution of HF, H_2O_2. The catalyst lowers the activation energy required for the chemical etching reaction, which leads to an increased etching rate at the metal-semiconductor interface, resulting in leaving unetched nanowires and noble metal.

Despite its disadvantages such as limited resolution, lack of defect number control, and wasteful nature, the techniques belonging to the top-down approach still dominate among industrial methods of nanowire fabrication for semiconductor devices. Due to the increasing demands of the semiconductor industry, it may be necessary to combine the advantages of both strategies to achieve the latest goals of the industry. Hybrid methods of semiconductor nanowire fabrication include dielectrophoresis, directed self-assembly of block copolymer lithographic patterning, and VLS nanowire growth on the top-down patterned metal catalyst seed particles [36].

2.3   APPLICATION OF SEMICONDUCTOR NANOWIRES

SNs, due to their unique properties and wide range of synthesis possibilities described in the previous subsections, can revolutionize such industries as electronics, computing, medicine, energy, or safety and environmental protection (**Figure 2.6**).

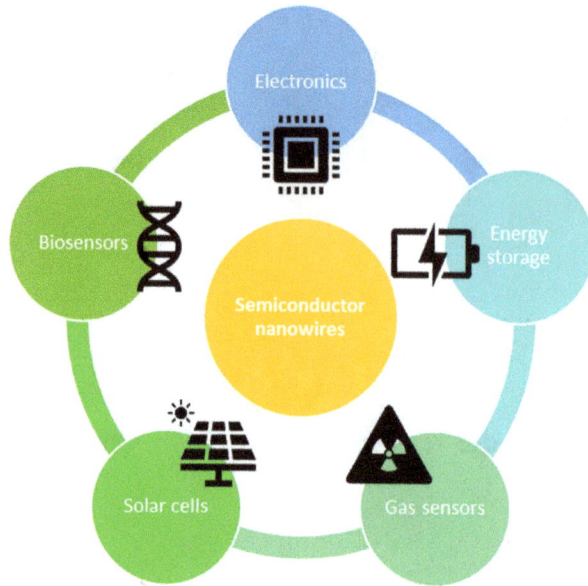

FIGURE 2.6   The most important applications of 1D semiconductor nanostructures.

2.3.1   GAS SENSORS

The development of gas sensors plays a significant role in industrial plant safety, environmental monitoring, smart home construction, food quality control, and medical diagnostics. For this reason, research on reliable portable wearable sensors with small form factors has been continuously conducted. In this regard, researchers' attention has focused on sensors based on SNs. Nanowire-based sensors although they detect gases based on the same mechanism (change in resistance of the sensing layer upon the adsorption/desorption of gas molecules) as conventional sensors based on thin films exhibit much better sensing properties, such as fast response, high sensitivity, low power consumption, and high selectivity. The extremely advantageous properties of SNs are due to their large specific surface area allowing just a few gas molecules to change their electrical properties, which consequently leads to a significant reduction in the minimum gas concentration threshold needed for sensor response. The selection of a suitable semiconductor material due to its chemical composition, structure, and electrical properties plays a key role in the development of a modern gas sensor [37]. Research carried out in this field indicates that among SNs, the ones based on MOs stand out because they show stability during the increased operating temperature of the sensor, they are characterized by unique optical properties, and they also have defects in the crystal structure and on the surface, which positively influence the sensor properties. In the context of manufacturing new generation sensors with high sensitivity, selectivity, and stability, the most extensively studied are SnO_2, NiO, ZnO, CuO, Fe_2O_3 for the fabrication of which methods such as electrospinning, CVD, hydro- and solvothermal are used. The generally used methods for fabrication of sensors based on MOs nanowire are multistep because they require the synthesis of wires and collecting and depositing them on a substrate with electrodes. Hence, the latest direction of sensor development is to grow the nanowire directly on the chip, which can increase the scale of sensor fabrication and reduce the fabrication cost generated by post-synthesis techniques [38].

Pure MOs nanowires cannot meet the requirements of high-performance gas sensors, so researchers are dynamically working to intensify the sensing properties of MOs nanowire-based sensors by fabricating nanowires with n-n, p-p, or p-n heterojunction. Fabricating a heterojunction allows the semiconductor to change its energy structure and increase the number of electron-depleted layers, which allows for increased sensor performance, sensitivity, and selectivity. Alternative strategies used by researchers include, but are not limited to, doping or surface modification of nanowires with rare earth metals, noble metals, and carbon materials [39].

Despite the advanced research conducted in the field of semiconductor nanowire application in modern gas sensors, there are still areas that require development. Particular attention should be paid to the development of sensors with the lowest possible detection threshold while maintaining the lowest possible energy consumption by lowering the operating temperature of the sensor [38].

2.3.2 PHOTOCATALYSIS

Increasing urbanization and industrialization are constantly reducing access to clean water by introducing organic pollutants, fertilizers, dyes, and heavy metals. These pollutants adversely affect the entire ecosystem as well as the health of humans, who are at increased risk for infections, cancer, or Alzheimer's. Photocatalysis is one of the most promising, economical, and sustainable avenues to address aquatic pollution. Recent years have seen a surge of interest in the synthesis of SNs and their potential application as photocatalysts. Nanowires, compared to nanoparticles or bulk materials, show several advantages due to their intrinsic structural features. They enable fast and efficient charge transport, promote separation of electron-hole pairs by directing charge transport in two opposite directions limiting electron-hole recombination, have a large specific surface area, and have a high length-to-diameter ratio allowing for low optical reflection and enhanced light absorption and scattering.

TiO_2 is most often mentioned as photocatalysts in the form of SNs due to its low cost, low toxicity, high refraction index, wide band gap (3.2 eV), and high chemical and photostability. However, other semiconductors based on ZnO, SiO_2, Si, sulfides (CdS, ZnS, PbS), and bismuth compound (BiOI) have also been carefully studied. These materials are excellent in degrading rhodamine B, methylene blue, and heavy metals. Various techniques are currently used to produce SNs for photocatalytic applications, which can be classified by the nature of the product obtained, the nanowires can be in the form of powder/suspension or can be deposited on a semiconductor substrate. Due to the concern of abandoning the photocatalysts after performing their function in the environment, it seems more advantageous to restrain them on the substrate, however, this significantly reduces the photocatalytically active surface area. To fabricate nanowires with high photocatalytic activity, researchers most often turn to methods such as CVD, MACE, chemical bath deposition (CBD), self-assembly, hydrothermal growth, and electrospinning [40,41].

Unfortunately, single-phase nanowires are not able to efficiently utilize visible light, often undergo photodegradation, and have low quantum efficiency, which significantly limits their use. To exploit the full photocatalytic potential of SNs, researchers take the following approaches: precise control of morphology (shape, dimension, surface area, pore size), doping, defect generation, and formation of heterojunction and hybrid materials. The control of morphology is achieved by choosing an appropriate method of nanowire fabrication and selecting process parameters, but this is sometimes insufficiently effective [42,43].

Doping with a transition metal, rare earth, and non-metal ions is an effective way to produce new energy levels located within the semiconductor's band gap and to shift the minimum and maximum of the conduction band. The doping allows a decrease of the energy gap of the semiconductor and shifts the absorption edge towards visible light, which increases the efficiency of the photocatalysis process. It should be kept in mind that, too high a concentration of midgap states obtained by doping can induce recombination of electron-hole pairs deteriorating the photocatalytic activity of nanowires [42,43].

Alternatively, photoactivity can be improved by inducing surface defects, such as oxygen vacancies (OVs) in the material, which promote electron-hole pair separation. Generating OV on the surface of nanowires similarly to doping allows the formation of new energy levels below the conduction band, which reduces electron-hole pair recombination. OVs can also provide additional electrons for the formation of reactive oxygen species thus contributing to the improved photocatalytic activity of the nanowire. To create OVs on the surface of SNs, hydrogen treatment, high-energy bombardment, heat treatment in a reducing atmosphere, or treatment with reducing aqueous solutions are used. The formation of the desired number of vacancies seems to be a problematic issue at present; too high a concentration of vacancies, as in the case of doping, decreases the photocatalysis ability of nanowire [42].

An effective method to enhance the photocatalytic activity of SNs is the formation of heterojunctions between semiconductors with appropriately matched bands and the development of hybrid materials conjugated with metals or carbon materials. These treatments are aimed at improving the separation of electron-hole pairs, increasing the adsorption surface area of impurities, photosensitizing, and absorbing more of the solar spectrum by reducing the energy gap width of the semiconductor. These types of materials are obtained by decorating nanowires with nanoparticles, applying thin films to nanowires, or fabricating heterojunctions directly in the nanowire structure [42].

Despite the great potential of SNs and the rapid progress that is being made in using them as photocatalysts to solve the problem of environmental pollution, there are still many challenges that can be solved by further research to understand their electronic structure and to precisely design nanowires with the desired properties.

2.3.3 ENERGY

The destructive impact of fossil fuels on the environment, and the production of greenhouse gases, CO_2 and methane are intensifying current climate problems, forcing scientists to look for sustainable energy solutions. Hence, devices for efficient energy conversion, harvesting, and storage are of great interest. A wide range of nanomaterials are continuously being explored to provide high-performance photovoltaic cells, nanogenerators, batteries, and supercapacitors, however, among them, SNs deserve special attention.

Among the available solar cells, dye-sensitized solar cells (DSSCs) are currently of the greatest interest to researchers due to their low fabrication cost, low toxicity, environmental friendliness, and transparency. DSSC cells consist of a photoanode deposited on an FTO glass surface, a dye, an electrolyte, and a counter electrode. TiO_2, ZnO, SnO_2, GaAs, and CdS nanowires are most often indicated as materials for the heart of next-generation photovoltaic cells – the photoanode, whose task is to collect and transfer excited electrons and adsorb dye. SNs meet high requirements for photoanodes, such as high transparency to prevent photon scattering, large dye adsorption area, high electron mobility, and non-reactivity with electrolytes. They can be applied to the FTO substrate in paste form by screen printing or grown directly on the substrate by hydrothermal, solvothermal, or self-assembly. Treatments such as doping with metals and non-metals and precise control of dimensions, geometry, crystallinity, and porosity through appropriate control of fabrication and heat-treatment parameters are applied to photoanodes to increase the efficiency of solar energy conversion to useful electricity. Despite these efforts, it is still necessary to improve the efficiency and outdoor/indoor stability of DSSCs so that they can be introduced on a large scale for commercial sale [44].

So far, polymer fibers and lead zirconate titanate (PZT) have been the most popular materials in the conversion of mechanical energy to electrical energy using nanogenerators, however, due to their drawbacks such as brittleness, low deformation, lead toxicity, modern piezoelectric materials are constantly being sought. Due to higher electrical outputs, large specific surface area, good mechanical properties such as high hardness and strength, and good thermal resistance, SNs exhibit the ability to efficiently generate potential differences due to mechanical deformation, which includes

ZnO, SnO$_2$, GaN, ZnS, and SbSI, are gaining popularity. A contemporary proposal is the fabrication of composite structures on a polymer matrix with the addition of piezoelectric nanowires, which provides flexibility, active stress transfer, and greater durability to the overall piezoelectric architecture [45].

A disadvantage of many renewable energy sources (RES) is their inconsistency in the amount of energy generated, which means that energy generated under favorable conditions should have a chance to be used at any time, for example, during reduced RES efficiency. Hence, an important area besides energy conversion where the introduction of semiconductor nanowire can prove to be extremely beneficial is energy storage through devices such as batteries, rechargeable batteries, and supercapacitors. The incorporation of nanowires in the construction of electrodes of lithium-ion (LIB), potassium-ion (KIB) supercapacitors and batteries have significant benefits in terms of increasing their capacity and performance, reducing charging time, and increasing lifetime. This is due to increased electron and ion transport, reduced electrode degradation mechanism, and large specific surface area that promotes greater contact with the electrolyte. As of today, the beneficial electrochemical properties have been confirmed for, among others, nanowires of MnO$_2$, Fe$_2$O$_3$, MoS$_2$, SiC, SnO$_2$, V$_2$O$_5$, ZnMn$_2$O$_4$ in the application to the construction of supercapacitors. Further development of energy storage technologies is key to realizing the full potential of intermittent alternative energy, electric vehicles, or creating portable, personalized energy sources. Hence, researchers often turn to such ways of improving the electrochemical properties of SNs as doping them with carbon, graphene, and graphene oxide. Device engineering based on SNs offers many ways to exploit their favorable electrical, optical, and mechanical properties for a wide range of innovative energy applications.

2.3.4   OTHER APPLICATIONS

The progressive miniaturization of electronics and photonics would not be possible without SNs. They are an integral part of devices such as integrated circuits, FETs, lasers, storage media, diodes, and biosensors. However, further progress in electronics based on SNs will require a move away from conventional top-down methods to a variety of chemical methods that allow precise control of the geometry, morphology, and chemical composition of the nanowire. A possible path for the development of semiconductor devices is also related to the substitution of silicon by compound nanowires of GaN, CdSe, CdS, CdTe, or InP, for this purpose, it will be necessary to develop not only appropriate methods for their fabrication but also to fully understand their physicochemical properties through simulations and experimental studies.

2.4   FUTURE PERSPECTIVE

SNs have become key building blocks in nanotechnology research and the analysis of novel concepts in fields traditionally dominated by non-nanotechnological approaches. This undoubted success is due to their versatility, both in terms of growth capabilities and their functionality, which opens up new possibilities for existing and entirely new applications.

With the growing popularity of SNs, researchers are proposing further modifications to the growing methods that should ensure the highest quality of the fabricated nanostructures.

Further research on designed growth methods and syntheses of SNs for large-scale devices will undoubtedly have a huge impact on nanotechnology as a whole and individual areas. Key applications including photocatalysis, sensing, transistors, photodetectors, nanogenerators, and spintronics are likely to revolutionize functionality using SNs. However, the future of these technologies largely depends on whether researchers succeed in minimizing the cost of nanowire growth while maintaining high quality and above all performance and stability of nanowires and nanodevices.

REFERENCES

1. Global Markets and Technologies for Nanofibers, March 2022 https://www.researchandmarkets.com/reports/4733527/global-markets-and-technologies-for-nanofibers?gclid=Cj0KCQjwxIOXBhCrARIsAL1QFCYaBGZnkmknHL7_XWrIok6tvOZhyy0A9gLY3ediQalfwCFHTUohpK4aAj_cEALw_wcB
2. J. Ramanujam; D. Shiri; A. Verma. Silicon nanowire growth and properties: A review. Mater. Express 2011, 1, 105–126.
3. A. Garcia-Gil; S. Biswas; A.D. Holmes. A review of self-seeded germanium nanowires: Synthesis, growth mechanisms and potential applications. Nanomaterials 2021, 11, 2002.
4. D. Wang. Synthesis and properties of germanium nanowires. Pure Appl. Chem. 2007, 79, 55–65.
5. R.G. Hobbs; S. Barth; N. Petkov; M. Zirngast; C. Marschner; M.A. Morris; J.D. Holmes. Seedless growth of sub-10 nm germanium nanowires. J. Am. Chem. Soc. 2010, 132, 13742–13749.
6. A. Pan; R. Liu; M. Sun; C.Z. Ning. Spatial composition grading of quaternary ZnCdSSe alloy nanowires with tunable light emission between 350 and 710 nm on a single substrate. ACS Nano 2010, 4, 671–680.
7. I. Domnez Noyan; G. Gadea; M. Salleras; M. Pacios; C. Calaza; A. Stranz; M. Dolcet; A. Morata; A. Tarancon; L. Fonseca. SiGe nanowire arrays based thermoelectric microgenerator. Nano Energy 2019, 57, 492–499.
8. G. Shin; C.H. Yoon; M.Y. Bae; Y.C. Kim; S.K. Hong; J.A. Rogers; J.S. Ha. Stretchable field-effect-transistor array of suspended SnO_2 nanowires. Small 2011, 7, 1181–1185.
9. S. Ju; A. Facchetti; Y. Xuan; J. Liu; F. Ishikawa; P. Ye; C. Zhou; T.J. Marks; D.B. Janes. Fabrication of fully transparent nanowire transistors for transparent and flexible electronics. Nat. Nanotechnol. 2007, 2, 378–384.
10. D.Y. Jiang; J.X. Zhao; M. Zhao; Q.C. Liang; S. Gao; J.M. Qin; Y.J. Zhao; A. Li. Optical wave-guide based on ZnO nanowires prepared by a thermal evaporation process. J. Alloys Compd. 2012, 532, 31–33.
11. Y. Zhang; H. Zhao; X. Zhao; J. Lin; N. Li; Z. Huo; Z. Yan; M. Zhang; S. Hu. Narrow-bandgap Nb_2O_5 nanowires with enclosed pores as high-performance photocatalyst. Sci. China Mater. 2019, 62, 203–210.
12. P. Xu; Z. Cheng; Q. Pan; J. Xu; Q. Xiang; W. Yu; Y. Chu. High aspect ratio In2O3 nanowires: Synthesis, mechanism and NO_2 gas-sensing properties. Sensors Actuators, B Chem. 2008, 130, 802–808.
13. S. Ramadan; L. Bowen; S. Popescu; C. Fu; K.K. Kwa; A. O'Neill. Fully controllable silicon nanowire fabricated using optical lithography and orientation dependent oxidation. Appl. Surf. Sci. 2020, 523, 146516.
14. R. Duffy; M. Shayesteh; B. McCarthy; A. Blake; M. White; J. Scully; R. Yu; A.M. Kelleher; M. Schmidt; N. Petkov; L. Pelaz; L.A. Marques. The curious case of thin-body Ge crystallization. Appl. Phys. Lett. 2011, 99, 2–5.
15. R.L. Woo; L. Gao; N. Goel; M.K. Hudait; K.L. Wang; S. Kodambaka; R.F. Hicks, Kinetic control of self-catalyzed indium phosphide nanowires, nanocones, and nanopillars. Nano Lett. 2009, 9, 2207–2211.
16. K. Tomioka; J. Motohisa; S. Hara; T. Fukui, T. Control of InAs nanowire growth directions on Si. Nano Lett. 2008, 8, 3475–3480.
17. R. Bansen; J. Schmidtbauer; R. Gurke; T. Teubner; R. Heimburger; T. Boeck. Ge in-plane nanowires grown by MBE: Influence of surface treatment. CrystEngComm 2013, 15, 3478–3483.
18. Y.P. Rakovich; F. Jäckel; J.F. Donegan; A.L. Rogach. Semiconductor nanowires self-assembled from colloidal CdTe nanocrystal building blocks: Optical properties and application perspectives. J. Mater. Chem. 2012, 22, 20831–20839.
19. Sisman. Template-assisted electrochemical synthesis of semiconductor nanowires. Nanowires - Implementations Appl. IntechOpen. 2011.
20. S. Kumar; A. Vohra,; S.K. Chakarvarti. Synthesis and morphological studies of ZnCuTe ternary nanowires via template-assisted electrodeposition technique. J. Mater. Sci. Mater. Electron. 2012, 23, 1485–1491.
21. G. Zou; H. Li; Y. Zhang; K. Xiong; Y. Qian. Solvothermal/hydrothermal route to semiconductor nanowires. Nanotechnology 2006, 17, S313.

22. S. Boubenia; A.S. Dahiya; G. Poulin-Vittrant; F. Morini; K. Nadaud; D. Alquier. A facile hydrothermal approach for the density tunable growth of ZnO nanowires and their electrical characterizations. Sci. Rep. 2017, 7, 1–10.

23. W. Matysiak; T. Tański; W. Smok. Electrospinning as a versatile method of composite thin films fabrication for selected applications. Solid State Phenom. 2019, 293, 35–49.

24. W. Matysiak; T. Tański; M. Zaborowska. Manufacturing process, characterization and optical investigation of amorphous 1D zinc oxide nanostructures. Appl. Surf. Sci. 2018, 442, 382–389.

25. T. Tański; W. Matysiak. Synthesis of the novel type of bimodal ceramic nanowires from polymer and composite fibrous mats. Nanomaterials 2018, 8, 179.

26. W. Matysiak; T. Tański; W. Smok. Study of optical and dielectric constants of hybrid SnO2 electrospun nanostructures. Appl. Phys. A Mater. Sci. Process. 2020, 126, 115.

27. W. Matysiak; T. Tański; W. Smok; Polishchuk, O. Synthesis of hybrid amorphous/crystalline SnO$_2$ 1D nanostructures: investigation of morphology, structure and optical properties. Sci. Rep. 2020, 10, 14802.

28. J. Huang; D. Fan; Y. Ekinci; C. Padeste. Fabrication of ultrahigh resolution metal nanowires and nanodots through EUV interference lithography. Microelectron. Eng. 2015, 141, 32–36.

29. C. Gao; Z.C. Xu; S.R. Deng; J. Wan; Y. Chen; R. Liu; E. Huq; X.P. Qu. Silicon nanowires by combined nanoimprint and angle deposition for gas sensing applications. Microelectron. Eng. 2011, 88, 2100–2104.

30. D. Resnick. Nanoimprint lithography. Nanolithography Art Fabr. Nanoelectron. Nanophotonic Devices Syst. 2014, 315–347.

31. J. Xiang; C. Li, X. Yin; Y. Li, X. Wang; H. Yang; X. Ma; J. Han; J. Zhang; T. Hu; T. Yang; J. Li; H. Yin; H. Zhu; W. Wang; H.H. Radamson. A novel dry selective isotropic atomic layer etching of SiGe for manufacturing vertical nanowire array with diameter less than 20 nm. Mater. 2020, 13, 771.

32. Q. Wang; G. Yuan; S. Zhao; W. Liu; Z. Liu.; J. Wang; J. Li. Metal-assisted photochemical etching of GaN nanowires: The role of metal distribution. Electrochem. commun. 2019, 103, 66–71.

33. M. Dejarld; J.C. Shin; W. Chern; D. Chanda; K. Balasundaram; J.A. Rogers; X. Li. Formation of high aspect ratio GaAs nanostructures with metal-assisted chemical etching. Nano Lett. 2011, 11, 5259–5263.

34. Y. Liao; S.H. Shin; M. Kim, M. Ultraviolet antireflective porous nanoscale periodic hole array of 4H-SiC by photon-enhanced metal-assisted chemical etching. Appl. Surf. Sci. 2022, 581, 152387.

35. X. Leng; C. Wang; Z. Yuan. Progress in metal-assisted chemical etching of silicon nanostructures. Procedia CIRP 2020, 89, 26–32.

36. P.C. McIntyre; A. Fontcuberta i Morral. Semiconductor nanowires: To grow or not to grow? Mater. Today Nano 2019, 100058.

37. V. Galstyan; A. Moumen; G.W.C. Kumarage; E. Comini. Progress towards chemical gas sensors: Nanowires and 2D semiconductors. Sensors Actuators B Chem. 2022, 357, 131466.

38. M.M. Arafat; B. Dinan; S.A. Akbar; A.S.M.A. Haseeb. Gas sensors based on one dimensional nanostructured metal-oxides: A Review. Sensors (Basel). 2012, 12, 7207.

39. W. Smok; T. Tański. A short review on various engineering applications of electrospun one-dimensional metal oxides. Mater. 2021, 14, 5139.

40. Z. Li; S. Wang; J. Wu; W. Zhou. Recent progress in defective TiO$_2$ photocatalysts for energy and environmental applications. Renew. Sustain. Energy Rev. 2022, 156, 111980.

41. J. Qiu; M. Li.;M. Ding; J. Yao. Cellulose tailored semiconductors for advanced photocatalysis. Renew. Sustain. Energy Rev. 2022, 154, 111820.

42. M. Samadi; M. Zirak; A. Naseri; M. Kheirabadi; M. Ebrahimi; A.Z. Moshfegh. Design and tailoring of one-dimensional ZnO nanomaterials for photocatalytic degradation of organic dyes: A review. Res. Chem. Intermed. 2019, 45, 2197–2254.

43. B. Weng; S. Liu; Z.R. Tang; Y.J. Xu. One-dimensional nanostructure based materials for versatile photocatalytic applications. RSC Adv. 2014, 4, 12685–12700.

44. H. Mohammadian-Sarcheshmeh; R. Arazi; M. Mazloum-Ardakani. Application of bifunctional photoanode materials in DSSCs: A review. Renew. Sustain. Energy Rev. 2020, 134, 110249.

45. B. Zaarour; L. Zhu; C. Huang; X.Y. Jin; H. Alghafari; J. Fang; T. Lin. A review on piezoelectric fibers and nanowires for energy harvesting. J. Ind. Text. 2019, 51, 297–340.

3 Superconducting Nanowires

Rafael Zadorosny and Milton B. F. Junior
Department of Physics and Chemistry, Universidade Estadual Paulista (UNESP), São Paulo, Brazil

Anjela Koblischka-Veneva and Michael R. Koblischka
Saarland University, Experimental Physics, Saarbrücken, Germany

CONTENTS

3.1 INTRODUCTION

Superconductivity is a macroscopic quantum phenomenon as in a homogeneous superconductor all the superconducting electrons (Cooper pairs) have to be considered as a single quantum mechanical entity. Superconductivity itself is characterized by two critical lengths, which are material-dependent parameters, i.e., the London penetration depth, $\lambda_L(T)$, and the coherence length, $\xi(T)$, so superconducting nanowires may have at least one dimension below one of these characteristic lengths. Such nanowires are considered mesoscopic one-dimensional (1D) objects if their diameter is smaller than ξ. Thus, achieving mesoscopic scales may create new and interesting quantum effects [1,2]. Nanostructuring superconducting materials may alter the superconducting properties, quantum fluctuations may dominate [3,4] and show up effects that are not known from the respective bulk materials, e.g., size-dependent breakdowns of superconductivity or enhanced transition temperatures, T_c [5,6].

The present-day superconducting materials can be classified as conventional or low-T_c superconductors (LTSc), where the transition temperature, T_c, is below 30 K [7,8]. All elemental and metallic superconductors belong to this class; the only exception being MgB_2, discovered in 2001 with a T_c of ~ 38 K. In contrast, the high-T_c superconductors (HTSc) have T_c above 40 K; the most prominent case is materials with T_c above 77 K (liquid nitrogen temperature), which enables a simple and cheap cooling process. However, the border of 30 K must be seen as an artificial one as several members of HTSc families may have T_c lower than 30 K but retain other HTSc properties like high critical fields [7,8].

Here, we also must note that $\lambda_L(T)$ and $\xi(T)$ are temperature-dependent – nearly constant at $T \rightarrow 0$, and diverging for $T \rightarrow T_c$. For this reason, the fabrication of superconducting nanowires is possible for

DOI: 10.1201/9781003296621-3

all types of materials operating close to their transition temperatures. Several different approaches to fabricating them are described in the literature, including patterning techniques, templating, and spinning processes. The latter techniques, namely solution blow-spinning and electrospinning, enabled a fully new class of superconducting materials, called fibrous non-woven fabrics or short "nanofiber mats" with entirely new properties and possible applications.

Current applications of superconducting nanowires operate mostly at $T = 4.2$ K; the temperature of liquid helium. Modern cryo-cooling systems also enable this temperature to be reached effectively, so the need for cryogenic liquids is reduced. However, for various new and still uncommon applications of superconductors, there is a strong demand to work at higher temperatures, e.g., 20 K (as envisaged for MgB_2) or even 77 K (the temperature of liquid nitrogen). Thus, as the development of new materials like nanowires and nanofiber fabrics of the ceramic HTSc materials has seen enormous progress in recent years, we focus on this contribution mainly on the fabrication techniques and the applications of these HTSc nanowires.

3.2 MATERIAL PRODUCTION TECHNIQUES ON SMALL SCALES

Nowadays, there is a large scientific and technological demand for materials at small scales due to their unusual properties inherent to a large surface/volume ratio. Thus, the development of new methods for synthesizing these materials has been widely studied to understand the technical variables that allow their morphology and physical properties to be controlled.

The techniques used for the production of materials on small scales can be distinguished into two classes, "top-down" and "bottom-up" (**Figure 3.1a**). The "top-down" class refers to nanomaterials obtained from the successive slicing or cutting of bulk specimens, but also to thin films prepared on substrates and requiring structuring. Some other examples of top-down processes are friction and milling techniques. The techniques considered "bottom-up" refer to materials obtained from joining smaller particles, such as atom to atom, which will form molecules that will later come together to

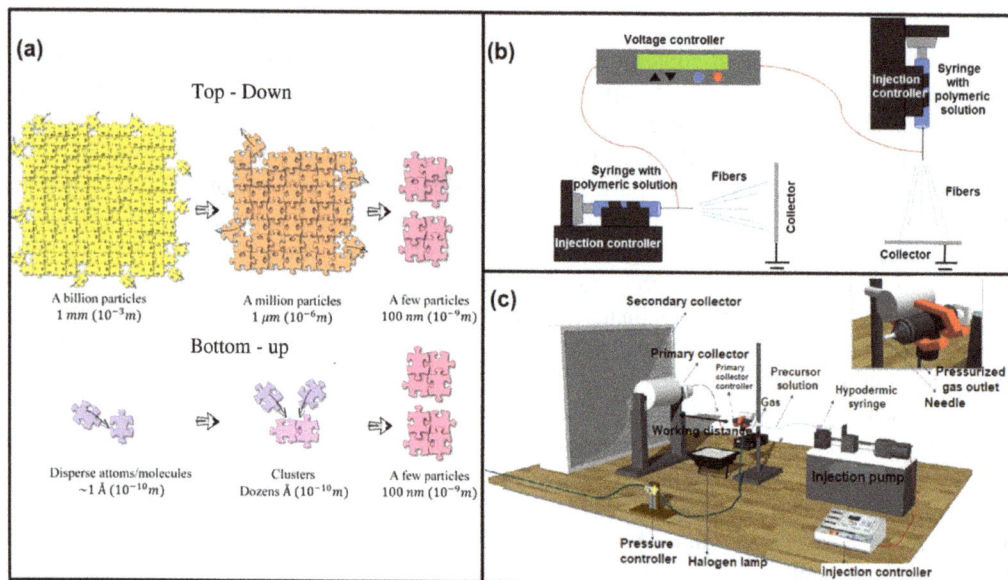

FIGURE 3.1 (a) Types of nanoparticle synthesis. (b) Example of electrospinning apparatus. (c) Configuration of the solution blow spinning (SPS) experimental apparatus. The inset presents an enlargement of the concentric needle nozzle, where the precursor solution undergoes shear by the pressurized gas.

form particles of nanometric sizes. Excellent examples of this type of synthesis are the colloidal dispersion and chemical routes.

From a technological point of view, bottom-up techniques are more advantageous when compared to top-down methods. In general, the production cost is lower, and the possibility of reaching the nanoscale is more straightforward since the tools available for physical processing at the nanoscale are limited and expensive [9]. Superconducting nanowires for applications are commonly prepared by structuring thin film materials using lithography/etching or ion milling. All such nanowires are located on substrates, which may influence their superconducting properties via stress/strain induced by crystallographic misfit to the substrate. This places limits on the possible choice of materials regarding the necessity to integrate the superconducting films into semiconductor devices.

Thus, in this contribution, the focus is placed on the discussion of two bottom-up techniques recently employed to produce superconducting fibers on reduced scales. One of them has been consolidated for almost a century, the electrospinning (ES) technique (depicted in **Figure 3.1b**, showing the possible vertical and horizontal configurations), and the other method emerged in 2009, the solution blown spinning (SBS) technique (**Figure 3.1c**). However, other techniques, such as melt blowing, wet spinning, template synthesis, self-assembly, direct drawing, and phase separation, are also widely used to produce materials at small scales. The template synthesis of superconducting materials was recently reviewed in [10]. In common, all cited techniques are of low productivity, making the process unfeasible from a commercial point of view [11].

Specifically, superconducting materials are commonly produced as wires, tapes, bulks, and thin films, depending on the desired type of application. When applying those methods, a textured growth is expected to reach higher critical current densities, J_c. However, texturing a ceramic superconductor material demands specific apparatuses and several processing steps. Thus, it is costly and time-consuming. The scalability of the sample sizes is also a problem for bulk specimens employed in, e.g., levitation systems, since the amount of magnetic field trapped in the sample depends on its size. Additionally, the oxygenation time increases by several weeks as the sample size increases. Another issue is that the oxygenation process can produce (micro-)cracking, thereby reducing the mechanical stability of bulk ceramic samples. For film production, the evaporation chambers limit their reachable size, and to obtain long-length tapes, their width has to be limited.

A possible alternative to contour the aforementioned situation is the production of porous ceramic superconductors. The pores facilitate the oxygen flow through the sample, and its diffusion length is considerably reduced. From the application point of view, the cryogenic liquid or gas (e.g., liquid nitrogen) can also flow by the pores leaving a more effective cooling process due to the small size of the superconducting regions. Additionally, substitutions can be used, and porous structures can be tailored to reduce the costs of applying superconductors and porous materials that present less weight and lower material content, which can be essential for applications. Besides different reported approaches to producing such porous samples, the present work focuses mainly on the production of ceramic, HTSc superconductors by ES and SBS techniques, as such samples will be important for the further development of this field.

3.3 NANOFIBER SPINNING TECHNIQUES

3.3.1 ELECTROSPINNING

The first reports about the ES technique occurred in the late 19th century and gained prominence in the early 20th century. The publication of a series of patents by John Francis Cooley (1900, 1902, and 1903) described the method used for producing fibrous materials for different purposes, from electrostatic principles. Over the century, several other researchers made significant contributions toward improving the technique and understanding the parameters that interfere in the production process.

Several other patents were registered to produce different material classes, so nowadays, many universities and research centers explore the technique for the most diverse types of materials [13].

ES-produced materials have applications in many different areas, such as nanoelectronics, nanofiltration, nanosensors, polymer matrix nanocomposites and blends, human tissue engineering, agriculture, water purification, and also in the processing of ceramic nanostructures. However, most of the studies focus on polymeric fibers. The technique consists of applying a significant voltage (of the order of tens of kilovolts) between the injection nozzle of the polymeric precursor solution, with a high dielectric constant value, and the collector. Upon reaching a critical voltage, the electrical repulsion force acting on solution particles (in the shape of a spherical drop when leaving the injection nozzle) overcomes the solution's surface tension, stretching it (Taylor cone). Thus, the forming fibers can have diameters that vary from several micrometers to 2 nm [12].

The wires' or fibers' morphology depends on solution parameters (viscosity, polymer concentration, polymer molecular mass, conductivity, and surface tension), processing parameters (voltage, solution flow rate, and working distance – distance between the solution injection nozzle to the collector) and environmental ones (humidity and temperature). Both the polymer concentration and molecular weight affect the viscosity of the solution. Increasing concentration tends to increase the viscosity. Solutions with low viscosities do not produce fibers and/or present many beads. As the viscosity increases, it is possible to verify fibers' formation with larger diameters and lengths. Still, the fibers have an optimal concentration to form continuously and uniformly [14].

The molecular weight of the polymer also affects the viscosity, as it reflects the number of polymer chain entanglements in the solution. Generally, polymers with lower molecular weights tend to generate more grains and fibers with smaller diameters. Polymers with higher molecular weights produce more uniform fibers with larger diameters. Still, the molecular weight of the polymer is not always essential for electrospinning because the molecular interactions of the reactants in the solution can play the role of bonds offered by polymer entanglements [15]. Solutions with higher conductivities guarantee the formation of fibers with smaller diameters and are more uniform when compared to fibers produced with solutions of lower conductivity. The type of polymer, solvents, and the number of ions available determine the solution's conductivity. Alternatively, the salts in the precursor solution increase the charge density imposed by such a process, increasing the stretching force, which causes a decrease in the fibers' diameter and elongation [16].

Both the solvents used and the concentration interfere with the surface tension of the solution. Solutions with high surface tensions make stretching the wires challenging, requiring the application of a higher voltage so that the Coulomb force can overcome the surface tension [17]. The voltage used in the process can reach 60 kV, and higher values enhance the grain's formation probability. The balance between voltage and fiber morphology depends on the solution parameters (e.g., concentration) and external ones (e.g., the collector distance) [18].

The solution injection rate and the distance between the injection nozzle and collector (working distance) interfere with the morphology of the fibers. Lower rates allow the evaporation of solvents, preventing the fibers from sticking together and presenting smaller diameters. Likewise, the working distance must be such as to allow the evaporation of the solvents [12]. Environmental factors also change the fibers' characteristics. The viscosity of the solution depends directly on the temperature. Higher temperatures cause a decrease in viscosity and, consequently, a reduction of the diameter. In addition, the evaporation of solvents is accelerated at higher temperatures and low humidity, leading to clogging of the injection nozzle and interfering with the electrospinning process. On the other hand, moisture can form pores along the fibers, and the increase in humidity destabilizes the process, producing non-uniform structures [12].

Applying ES for producing ceramic nanofibers, both the precursor solution and heat treatments influence the sample's morphology. As an example, an ES precursor solution is obtained from bismuth, strontium, calcium, and copper acetates in the molar ratio Bi:Sr:Ca:Cu = 2:2:1:2 dissolved in propionic acid. An excess of calcium and copper acetates and the addition of tin acetate can also be used to suppress the formation of impurities during the heat-treatment processes, as described in [19-21]. In the case presented here, the solution polymerization was done by adding polyvinyl-pyrrolidone (PVP) (MW 1,300,000 g/mol). The as-collected polymer sample was heat treated at the final step of 800 °C for 1 h in air with additional oxygenation at 500 °C/10 h [19]. Thus, a non-woven BSCCO ceramic sample was obtained consisting of nanowires with an average diameter of 317.6 nm (**Figure 3.2a–d**). Characterizing the sample regarding the $R(T)$ curves, it was observed that such responses are due to intragrain material, weak links, and interconnections between the wires. Thermally activated phase-slips (TAPS) were also verified [20]. **Figure 3.2a–e** compiles the described information. In **Figure 3.2f**, the $M(H)$ hysteresis loops are shown for several temperatures, and in panel (g), $M(T)$ curves after zero-field cooling (ZFC) and field cooling (FC) are presented. T_c is recorded at 85 K, and basically, the magnetic flux is trapped by grain defects and, mainly, by the wire's interconnections. Besides, the modeling of the nanowires network is complex, in [21] an extended critical state model was used to fit the experimental data. A relationship between the average nanowire diameter and J_c was shown.

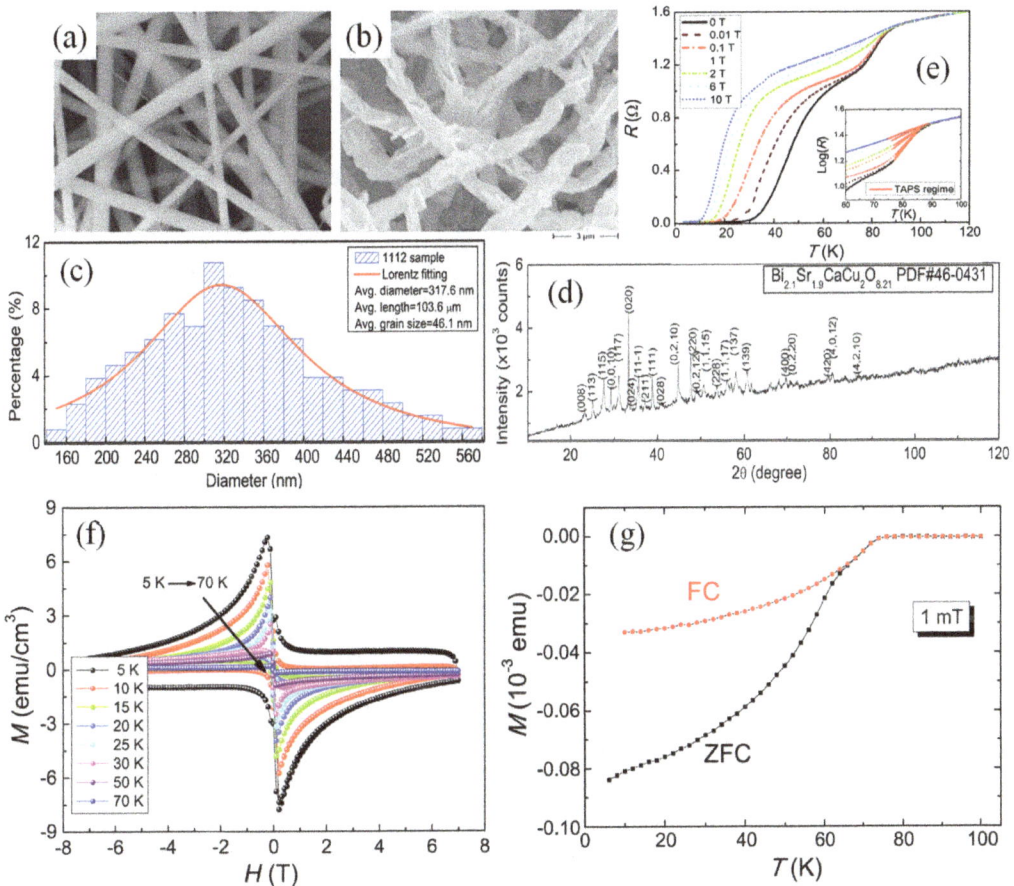

FIGURE 3.2 (a) As collected non-woven BSCCO nanofibers, (b) the ceramic composite after the heat treatments. (c) The wire diameter distribution with an average value of about 318 nm. (d) XRD diffractogram showing the formation of BSCCO-2212 phase, (e) $R(T)$ response, and the temperature range, where thermally activated phase-slips (TAPS) take place. Adapted with permission. Copyright (2016), IEEE. (f) Granular characteristic $M(H)$ loops of other BSCCO non-woven nanofibers, and (g) $M(T)$ response following the ZFC and FC measurements procedures in an external applied magnetic field of 1 mT. Adapted with permission. Copyright (2016), IEEE.

3.3.2 SOLUTION BLOW SPINNING

In 2009, a group of researchers reported the production of nanofibers for the first time through a new technique based on the fundamentals of electrospinning and melted blowing. The method, called solution blow spinning (SBS), has arisen as an excellent alternative for improving the productivity rate of micro- and nanometric fibers [22]. Since then, several researchers have used the technique for producing fibers from different material classes on small scales. It highlights the simplicity of the process compared to ES, as it does not require the application of voltages, or solutions with high dielectric constants. Furthermore, the production rate can be more than 30 times higher when compared to ES [23].

For fibering by SBS technique, principles of fluid dynamics are applied in an apparatus constituted by an injection pump [where a syringe connected to an inner needle (of a set of concentric needles) is installed], a pressure regulator connected to an outer needle, and a collector. The pump controls the solution's flow to the central needle, while the eccentric needle expels a continuous stream of pressurized gas. This compressed air flux causes a pressure drop when leaving the needle and shearing at the gas/solution interface, thus promoting the stretching of the fibers, which are deposited in a collector. The SBS apparatus can be adapted due to its simplicity and versatility. Another detail that can be implemented is using a halogen lamp somewhere within the working distance. The light increases the local temperature maximizing the solvents' evaporation. It is worth mentioning that the collector is the definitive sample deposition, thus, a specific organ can be dressed with a sample for medical purposes [12,24].

As with ES, the fibers' morphology produced by SBS depends on the control of solution parameters (viscosity, polymer concentration, polymer molecular mass, and surface tension), processing parameters (gas pressure, working distance, and flow rate solution), and working environment parameters (humidity and temperature). The viscosity of the solution depends on the polymer concentration, its molar mass, and the solutes' concentration and used solvents. The different properties of polymers lead to different interactions between viscous forces, capillary inert forces, and polymer relaxation time, making it difficult to formulate a theory that discusses the influence of so many factors on solution viscosity [24]. Thus, researchers have analyzed the effects of experimental variables on process success [27–29]. It is a consensus that continuous fibers are obtained when the polymer concentration is sufficient to completely entangle the long polymeric chains with the solvent molecules. There is a critical viscosity value for successful fiber production. The solution viscosity as a function of the polymer concentration can be divided into three regimes, i.e., diluted, semi-diluted, and concentrated. Solution viscosities from concentrated regimes to higher ones produce fibers of greater diameters but smoother, uniform, and with fewer grains [29].

The SBS technique has been widely used in biological and health areas, tissue engineering, textiles, and smart fabrics, among others, where there is no need for heat treatments. However, it has also been applied for producing fiber mats of $YBa_2Cu_3O_7$ (YBCO) ceramic superconductors. As a complex ceramic, the precursor solution of YBCO has been obtained by a one-pot acetate route, where the polymer used, polyvinylpyrrolidone (PVP), acts as a reducer of metallic cations and also as a stabilizer, isolating them from each other [31]. **Figures 3.3a–e** present such YBCO nanofiber mats prepared by the SBS technique. Besides the exciting phenomena related to superconducting ceramic nanofibers, such materials are fragile and brittle. A possible way to deal with such an issue is the addition of silver. Thermal analyses showed a decrease of about 30 °C in the crystallization and partial melting temperatures of YBCO nanofibers produced by SBS. Also, the sample's shrinkage occurs when heat treated at the same temperatures as the ones used for material without silver addition [26]. **Figures 3.3e–g** present the thermal analysis and optical and SEM images of YBCO nanofibers with and without silver. As described above, the gas pressure must be adjusted to produce a shear force sufficient to stretch the fibers. There is a threshold for pressures because fiber production is harmed at very low intensities, forming many beads and even clogging the needle. As the pressure increases, the fibers tend to become thinner. However, very high pressures can cause

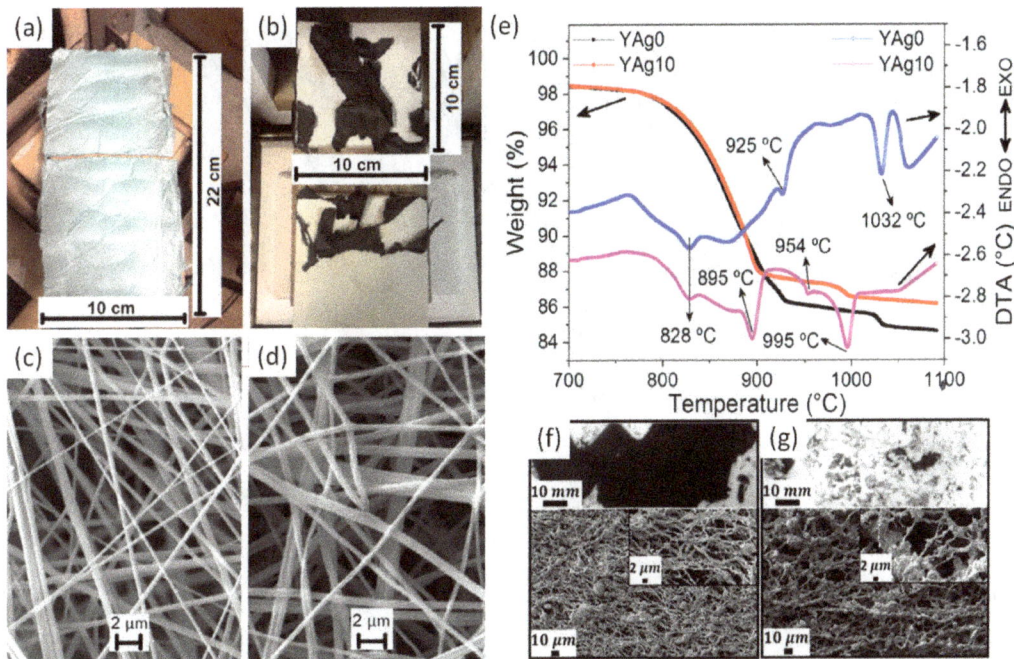

FIGURE 3.3 YBCO nanofiber mats produced by the SBS technique. (a) As-collected mat and (b) after burning out the polymer. SEM images (c) before and (d) after the heat treatment. (e) The thermogravimetric analysis of YBCO nanofibers with and without Ag addition. (f), (g) Optical images before and after the final heat treatment, respectively. Panels (h) and (i) are SEM images of a sample with and without Ag after the last heat treatment, respectively. Figures 3.3(e)-(g) adapted with permission. Copyright (2020), Elsevier.

a decrease in temperature due to the abrupt drop in pressure of the compressed gas, increasing the solution viscosity and, consequently, joining the fibers before reaching the collector [12]. Besides, increasing turbulence causes loss of collected material due to its dispersion in the environment.

Using electron microscopy techniques like SEM, TEM, and electron backscatter diffraction (EBSD), the microstructure of the nanowires can be characterized in detail [30]. The results obtained by EBSD demonstrate that the Bi-2212 grains within a nanowire are not randomly oriented, but their *c*-axis is oriented parallel to the nanowire direction (see the microstructure analysis given in **Figure 3.4**). This points to the fact that the dimensions of the as-spun polymer nanowire set the growth limits for the ceramic material.

The precursor solution flow rate must also be carefully adjusted. Low injection rates produce fibers of smaller diameters when compared to fibers produced at higher injection rates [29]. However, very high injection rates hinder the processability of the fibers, and this is an empirical adjustment that must be made for each production process with different solutions. The working distance must be such that it allows the evaporation of the solvents and the consequent collection of the fibers without their fusion. Therefore, there is a need to use solutions with high volatility, e.g., by adding about 60% of methyl alcohol. Although the SBS technique is more straightforward than other fiber production techniques on reduced scales, the knowledge of experimental adjusting parameters is necessary. What is discussed here are general behaviors concerning these parameters. A precursor solution synthesized with the same elements, concentrations, and control parameters can present different morphologies depending on the room temperature and humidity.

All the as-synthesized nanowires by SBS and ES are usually randomly distributed. When reacting to the ceramic material, there will be numerous interconnects between the individual nanowires, which is the base for the specific electrical properties. However, for most optical and sensor

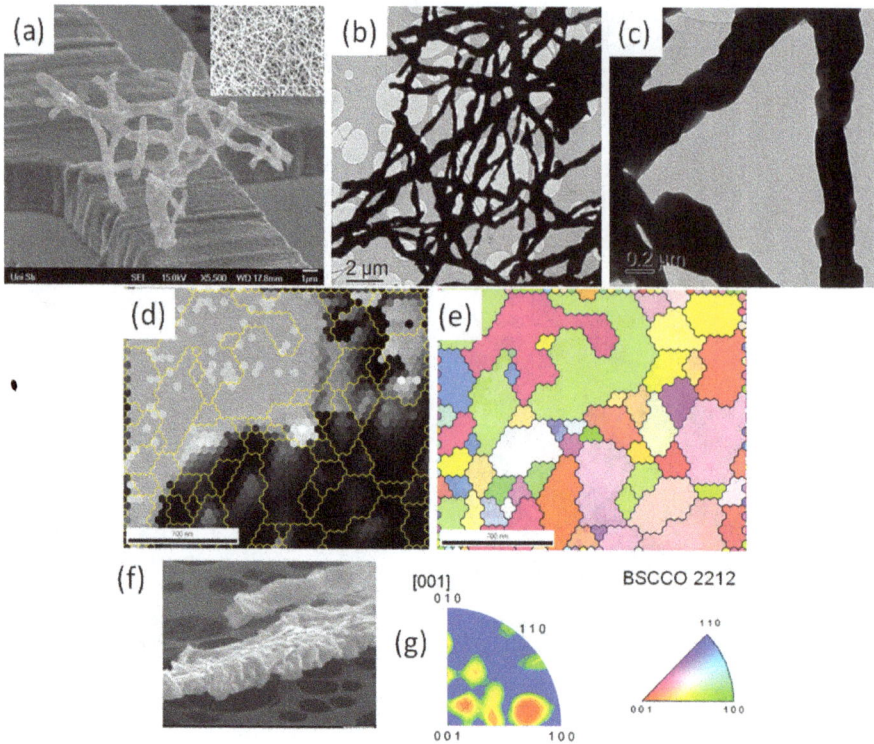

FIGURE 3.4 Microstructure of ES-fabricated Bi-2212 nanowires. (a) SEM image of a selected piece from the nanofiber fabric (the inset gives the original arrangement of the nanofiber fabric). (b), (c) TEM images at different magnifications. (d) Image quality mapping with the grin boundaries indicated in yellow, (e) inverse pole figure orientation map, (f) SEM image of the selected nanowire section, and (g) the inverse pole figure in [001]-direction and the stereographic triangle for Bi-2212. Figure adapted with permission. Copyright The Authors, some rights reserved; exclusive licensee [IOP Publishing]. Distributed under a Creative Commons Attribution-NonCommercial-ShareAlike 3.0 license.

applications, this is not an ideal configuration. Thus, to align the "messy" as-spun nanowires, many strategies have been proposed in the literature, for instance, the application of electric/magnetic fields, contact/roll printing techniques, electrostatic interaction, Langmuir-Blodgett (LB), bubble-blown techniques, etc. These techniques can be summarized by the keyword "grow and place" [32]. The principles of the three most common approaches (field-assisted, the use of a rotating collector drum, and near-field are illustrated in **Figure 3.5**. This alignment of the nanowires is especially important for their applications as photodetectors as discussed in Sec. 4.2 below. The placed as-spun nanowires require undergoing the ceramic reaction, so a relatively low heating temperature will lead to a larger number of aligned ceramic nanowires. Thus, the growth temperature of the ceramic material must be carefully controlled.

It has been shown that an ideal heat treatment to produce submicron fibers from this material is to work with slow heating rates, between 3 and 1 °C/min, with sintering at 925 °C/1 h (which ensures better connectivity between the grains). Under these conditions, ceramic grains grow following the direction of the wire in a successive Ostwald ripening-like process [34]. In contrast to ES-produced BSCCO-2212 nanofibers, YBCO nanofibers present a paramagnetic Meissner effect (PME) in the μT-field range. This is probably because the J_c of BSCCO-2212 nanowires is much lower at higher temperatures than the YBCO ones [35]. On the other hand, the PME of YBCO nanofibers is very similar to that presented by a YBCO artificial granular sample produced by e-beam lithography.

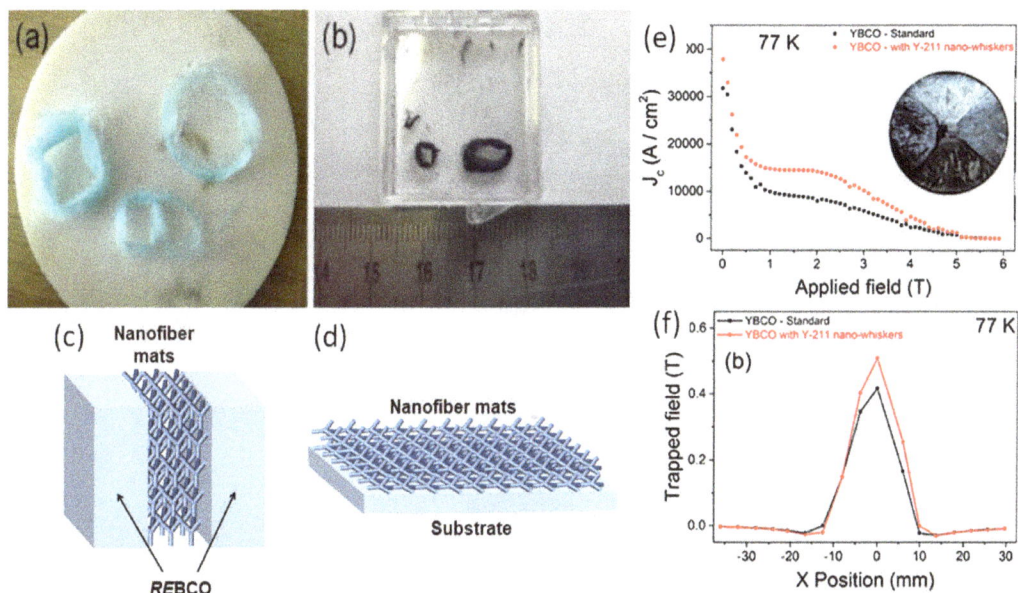

FIGURE 3.5 Schematic diagram for the alignment of the nanowires in the spinning process for both ES and SBS. The examples given are field-assisted growth, the use of a rotating collector drum, and near field. Adapted with permission. Copyright (2016), Springer Nature. Upper right image: the aligned nanowires are placed across a groove in the Si/SiO$_2$ substrate.

3.4 APPLICATIONS OF SUPERCONDUCTING NANOWIRES AND NANOFIBER MATS

3.4.1 NANOFIBER MATS

The production of high critical temperature ceramic superconducting materials (HTCs), such as YBa$_2$Cu$_3$O$_y$ (YBCO), GdBa$_2$Cu$_3$O$_y$ (GdBCO), and Bi$_2$Sr$_2$CaCu$_2$O$_y$ (Bi-2212) in the form of wires or fibers at nano- or submicrometer scales has been successfully reported through ES and SBS techniques [36,37]. For this purpose, acetate solutions of the material compounds with volatile solvents and polymers are prepared to obtain fibrous mats, which after heat treatment, the polymer is burned, inducing the ceramic reaction. The second heat treatment is carried out with a continuous oxygen flow to form the ceramic superconductor.

The resulting samples produced by ES and SBS are highly porous due to the numerous interconnections between the individual nanowires. Thus, the nanofiber mats form a new class of superconducting materials with unprecedented superconducting/magnetic properties. In addition, there is the possibility of modeling relatively complicated shapes in the green stage (as-spun fibers), where the nanofibers are still fully flexible. In this way, sample geometries like the donut-shaped coils for superconducting shielding (**Figures 3.6a, b**) can be created. These additional features are interesting from the application point of view. Besides carrying superconducting currents in high-intensity fields (10 T), they also have low weight and high porosity. This can facilitate the flow of cryogenic fluid required to cool the material, which is important for several types of applications.

Superconducting fiber mats can also be used as an adhesive between superconducting parts (bulk) to produce larger specimens (**Figure 3.6c**) for use in motors and electrical machines. Due to its structure and dimension, the nanofiber mat has a lower melting point than the bulk material. Thus, it starts to melt before the bulk, providing a good connection between the elements when it

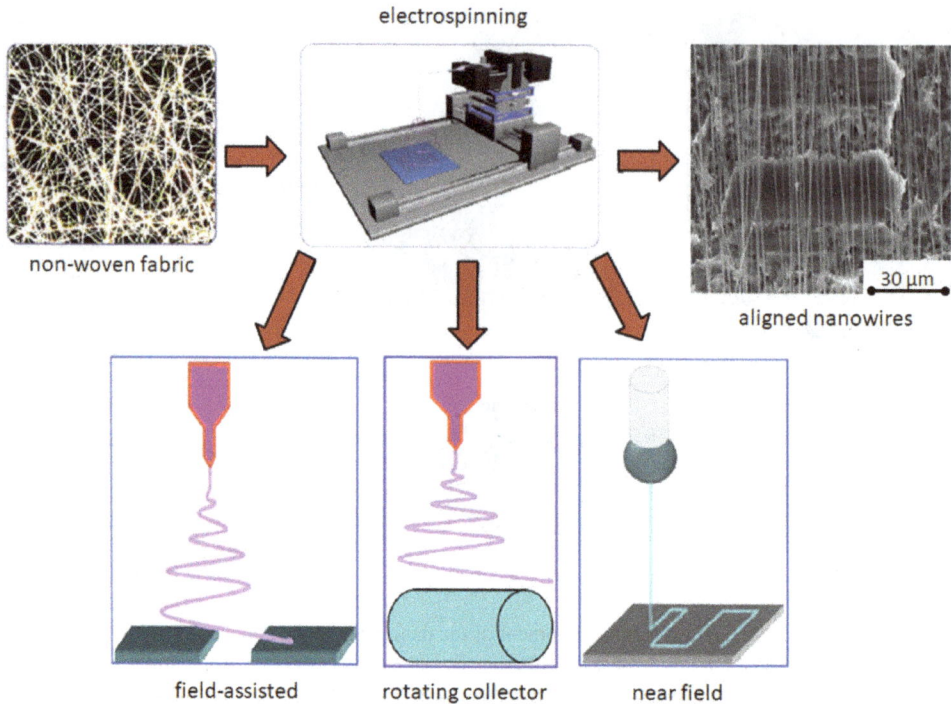

FIGURE 3.6 Donut-shaped nanofiber coils for magnetic shielding. (a) Green stage and (b) after heat treatment. (c) Superconducting mat used to bind REBCO superconductors and (d) deposited on a substrate for levitation rails. Adapted with permission. Copyright (2020), Elsevier. (e) $J_c(H)$ curves of single-grain samples (see the inset) with and without Y-211 nanowires. (f) Trapped field as a function of the same sample's position. Adapted with permission. Copyright (2019), Elsevier.

cools down again. By using a substrate as support (**Figure 3.3d**), the superconducting fiber mats can equip transport system rails like the ones that require movement and manipulation of objects without contact using levitation, such as a mechanical arm, ideal for sectors that need high levels of cleanliness [36]. Copper as a substrate is also an excellent alternative serving as a thermal contact to cool the superconducting material.

Finally, large, bulk superconducting materials, like YBCO single-grains, can have their critical current density, J_c, and the trapped field ability increased via the addition of Y_2BaCuO_5 (Y-211) nanofibers. Such a powder with nanofibers is mixed with commercial YBCO-123 powder. This mixture is compacted into a pellet and heat treated following a top-seed melting-growth process [38]. The inset of **Figure 3.6e** shows the single-grain sample produced with Y-211 nanofibers. In panels (e) and (f), the J_c and the recorded trapped magnetic field are compared between a reference sample and one with nanofibers, where the enhancement is visible.

Other possibilities for superconductors at micro- and nanoscales are gaining prominence in photonics. They are an excellent alternative to metamaterials such as gold and silver since their properties are easily adjustable by magnetic fields, temperature, light, and applied currents. Also, superconductors behave like plasmonic materials since their electromagnetic response occurs through Cooper pairs, analogous to electrons in plasmonic metals at optical frequency. As metamaterials, several effects have been investigated using superconductors. Among applications, YBCO rings can be used to create and control resonances that are adjusted by changing the device temperature or designing asymmetric rings [39].

3.4.2 NANOWIRES

The two most prominent applications for superconducting nanowires are interconnected [40], superconducting field detectors (SQUIDs), superconducting temperature detectors (nanoSQUIDs) [41], and single-photon detectors (SNSPDs). The fabrication techniques and properties of these detectors were reviewed in [42,43]. Thus, we point here to the principle of operation and the main demands of the superconducting materials to emphasize the necessary future developments in this field.

In the operation of an SNSPD [44], the superconductor is kept well below the superconducting transition temperature, and an applied current is biased just below the critical current, J_c [**Figure 3.7a(i)**]. The incoming infrared photons provide enough energy to break up hundreds of Cooper pairs within the nanowire, thus forming a "hot spot" (ii), which is not large enough to span the entire width of the nanowire. So, the currents must flow around this hot spot (iii), and the resulting current density in this path is now above J_c, causing a resistive barrier across the nanowire. This forces the resistivity to increase and generates a measurable output voltage pulse (iv). The final steps are (v) the Joule heating (via the DC bias), which aids in the growth of the resistive region along the axis of the nanowire until the current flow is blocked and the bias current is shunted by the external circuit. In (vi) the re-cooling process is depicted, which enables the resistive region to subside and the nanowire becomes fully superconducting again. The bias current through the nanowire returns to

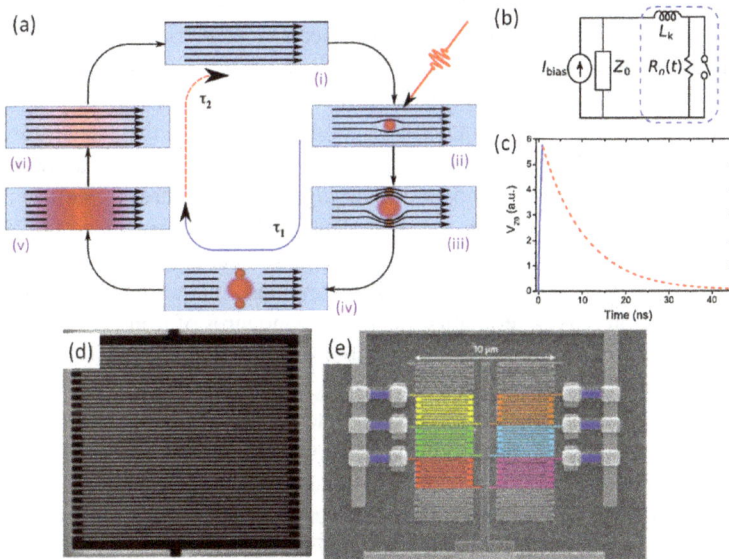

FIGURE 3.7 (a) Detection cycle of a SNSPD via the creation of a hot spot. (b) Electrical equivalent circuit of a SNSPD. L_k is the kinetic inductance of the superconducting nanowire and R_n is the hotspot resistance of the SNSPD. The SNSPD is current biased at I_{bias}. Opening and closing the switch simulate the absorption of a photon. An output pulse is measured across the load resistor Z_0. (c) Simulation of the output voltage pulse of the SNSPD. (d) SEM image of a NbTiN detector. The detector is 10×10 μm² and the wire has dimensions 100 nm wide, 6 nm thick. Adapted with permission. Copyright (2008), AIP Publishing. Multiple nanowire elements are biased in parallel via independent resistors resulting in a photon-number-resolving SNSPD (PNR-SNSPD) (e) Ultra-thin nanowires (30 nm wide) are connected in parallel to improve the sensitivity (registering efficiency) of the SNSPD – this device is known as a superconducting nanowire avalanche photodetector (SNAP). Figures (a)–(c), (e) adapted with permission. Copyright the authors, some rights reserved; exclusive licensee [IOP Publishing]. Distributed under a Creative Commons Attribution-NonCommercial-ShareAlike 3.0 licence.

its original value (i), allowing a new detection. The hotspot concept was originally developed by Skocpol *et al.* to study self-heating effects in superconducting micro-bridges [46]. The detection limit energy is given by the equation

$$hc/\Lambda \sim N_0 \Delta^2 wd\sqrt{\pi D \tau_{th}} \left(1 - I_{bias}/I_c\right) \tag{3.1}$$

where h denotes Planck's constant, c the speed of light, Λ the cutoff wavelength, N_0 the density of states at the Fermi level, Δ the superconducting energy gap, w and d describe the width and thickness of the nanowire, D the electronic diffusivity, τ_{th} the electronic thermalization time, I_{bias} the applied bias current, and I_c the critical current of the nanowire. This equation relates basic superconducting parameters with the detection limit, so, therefore, a material having small N_0, small Δ, small D, and short τ_{th} is preferable for longer wavelength detection.

Single-photon detection at visible wavelength was reported for Nb-based SNSPDs ($T_c = 9.25$ K) fabricated using sputtering and e-beam lithography. However, the Nb-SNSPD is easily latched because of the long electron-phonon relaxation time, τ_{e-ph}. Also for NbC, NbSi, and NbRe single-photon detection was reported in the literature. Another development was the use of amorphous superconductors (WSi, MoSi, MoGe), which enable an easy, straightforward fabrication of nanowires by lithographic processes. The most employed superconducting material is NbN, having a relatively high T_c of 17 K, and its derivative, NbTiN with T_c ~16 K. The use of these LTSc materials for SNSPDs was recently reviewed in [47].

MgB_2 is expected to become an SNSPD that can operate at a higher temperature and higher count rate than NbN-SNSPDs [47]. MgB_2-SNSPD detects at 11 K for a photon in the visible range and 13 K for single biomolecular ions. Recently, MgB_2-SNSPD was found to work more than 10 times faster than NbN-SNSPD. Although these results are very promising for the future of MgB_2-SNSPDs, the major problem is the low internal detection efficiency in the near-infrared region [47]. The HTSc materials exhibit a much higher T_c as well as very high upper critical fields, which makes such nanowires interesting for SNSPDs operating at higher temperatures as well as in high applied magnetic fields, which is desirable in the area of nuclear and high-energy physics [47]. Furthermore, there are several more areas of applications opening up like quantum optics and quantum communication, high energy particle detection, and neutron and dark matter detection [48,49]. Thus, one can find in the literature already several attempts to use nanowires of iron-based HTSc (abbreviated IBS) [48] and the cuprate HTSc materials [50,51], but still based on the use of top-down fabricated nanowires from thin film materials.

To summarize this section, one can state that the Nb(Ti)N-SNSPDs and amorphous-based SNSPDs are currently showing the highest performance single-photon detectors, and thus, significantly contribute to the progress achieved in the development of SNSPD detectors. To further expand the use of SNSPDs in other fields, the increase in the operating temperature seems to be the most important factor, but also the use of SNSPD detectors in high applied magnetic fields is a real challenge. Thus, it is desirable to employ HTSc nanowire materials for SNSPDs without degrading the performance achieved by the LTSc materials in the literature. Here, we must note that currently the HTSc nanofiber mats are not yet tested for their performance as SNSPDs, which is a task for future work.

3.5   SUMMARY AND CONCLUSIONS

In the present contribution, we have discussed the various fabrication techniques and important applications for superconducting nanowires, for both high- and low-T_c superconductors. Whereas many successes are seen in superconducting electronics and detectors based on the "classic" thin film technology, the recent development of the spinning techniques (ES and SBS) enabling the fabrication of ceramic HTSc nanowires has led to a fully new class of superconducting materials, the nanofiber fabrics. Further, we have discussed the applications of superconducting nanowires, the most prominent of which are the interconnects and the single-photon detectors. To the present day, the applications are mainly using metallic LTSc, but for future applications, the use of MgB_2 and the HTSc materials is desirable allowing new directions.

3.6 ACKNOWLEDGMENTS

The work in Brazil was possible thanks to the scholarship granted from the Brazilian Federal Agency for Support and Evaluation of Graduate Education (CAPES), in the scope of the Program CAPES-PrInt, process number 88887.194785/2018-00. RZ and MBFJ acknowledge São Paulo Research Foundation (FAPESP), grants 2016/12390-6, and 2021/08781-8, CAPES - Finance Code 001, and National Council of Scientific and Technological Development (CNPq, grant 310428/2021-1). The work in Germany funded by Volkswagen Stiftung ("Experiment!") and grants Ko2323-8 (DFG) and Ko2323-10/ANR-17-CE05-0030 (DFG-ANR).

REFERENCES

[1] Bezryadin, Superconductivity in Nanowires; Wiley-VCH: Weinheim, Germany, 2013.

[2] F. Altomare, A.M. Chang, One-Diemsional Superconductivity in Nanowires; Wiley-VCH: Weinheim, Germany, 2013.

[3] M.Tinkham, C.N. Lau, Quantum limit to phase coherence in thin superconducting wires. **Appl. Phys. Lett**. 2002, 80, 2946–2948.

[4] D. Haviland, Quantum phase slips. **Nat. Phys**. 2010, 6, 565–566.

[5] Y. Guo, Y.F. Zhang, X.J. Bao, T.Z. Han, Z. Tang, L.X. Zhang, W.G. Zhu, E. Wang, Q. Niu, Z.Q. Qiu, J.-F. Jia, Z.-X. Zhao, Q.-K. Xue, Superconductivity modulated by quantum size effects. **Science** 2004, 306, 1915–1917.

[6] M. Zgirski, K.P. Riikonen, V. Touboltsev, K. Arutyunov, Size dependent breakdown of superconductivity in ultranarrow nanowires. **Nano Lett**. 2005, 50, 1029–1033.

[7] A.V. Narlikar, Superconductors; Oxford University Press: Oxford, UK, 2014.

[8] K.H. Bennemann, J.B. Ketterson (Eds.), Superconductivity; Springer: Berlin/Heidelberg, Germany, 2008.

[9] G. Pandey, D. Rawtani, Y. K. Agrawal, Aspects of Nanoelectronics in Materials Development. In: Nanoelectronics and Materials Development, A. Kar (ed.). IntechOpen, London, 2016.

[10] M.R. Koblischka, A. Koblischka-Veneva, Fabrication of superconducting nanowires using the template method. **Nanomaterials** 2021, 11, 1970.

[11] G.C. Dadol, A. Kilic, L.D. Tijing, Kramer Joseph A. Lim, L.K. Cabatingan, N.P.B. Tan, E. Stojanovska, Y. Polat, Solution blow spinning (SBS) and SBS-spun nanofibers: Materials, methods, and applications. **Materials Today Commun.**, 25, 2020, 101656.

[12] N. Bhardwaj, S.C. Kundu, Electrospinning: A fascinating fiber fabrication technique. **Biotechnol. Adv.**, 28, 2010, 325–347.

[13] M.S. Islam, B.C. Ang, A. Andriyana, A. Muhammad Afifi. A review on fabrication of nanofibers via electrospinning and their applications. **SN Appl. Sci**. 1, 1248 (2019).

[14] S. Sukigara, M. Gandhi, J. Ayutsede, M. Micklus, F. Ko Regeneration of *Bombyx mori* silk by electrospinning—part 1: processing parameters and geometric properties. **Polymer**, 44, 2003, 5721–5727.

[15] Burger, B.S. Hsiao, B. Chu, Nanofibrous materials and their applications. **Annu. Rev. Mater. Res.**, 36, 2006, 333–368.

[16] X. Zong, K. Kim, D. Fang, S. Ran, B.S. Hsiao, B. Chu. Structure and process relationship of electrospun bioadsorbable nanofiber membrane. **Polymer**, 439, 2002, 4403–4412.

[17] A.K. Haghi, M. Akbari. Trends in electrospinning of natural nanofibers. **Phys. Stat. Sol. A**, 204, 2007, 1830–1834.

[18] O.S. Yördem, M. Papila, Y.Z. Menceloğlu, Effects of electrospinning parameters on polyacrylonitrile nanofiber diameter: An investigation by response surface methodology. **Materials & Design**, 29, 2008, 34–44.

[19] X.L. Zeng, M.R Koblischka, T. Karwoth, T. Hauet, U. Hartmann. Preparation of granular Bi-2212 nanowires by electrospinning. **Supercond. Sci. Technol.** 30, 2017, 035014.

[20] M.R. Koblischka, X.L. Zeng, T. Karwoth, T. Hauet and U. Hartmann, Transport and magnetic measurements on $Bi_2Sr_2CaCu_2O_8$ nanowire networks prepared via electrospinning, **IEEE Trans. Appl. Supercond.**, 26, 1800605.

[21] X.L. Zeng, T. Karwoth, M.R. Koblischka, U. Hartmann, D. Gokhfeld, C. Chang, T. Hauet. Analysis of magnetization loops of electrospun nonwoven superconducting fabrics. **Phys. Rev. Mater.** 1, 2017, 044802.

[22] E.S. Medeiros, G.M. Glenn, A.P. Klamczysnki, W.J. Orts, L.H.C. Mattoso, Solution blow spinning: A new method to produce micro- and nanofibers from polymer solutions. **J. Appl. Polymer Sci.** 113, 2009, 2322–2330.

[23] M.W. Lee, S.S. Yoon, A.L. Yarin, Solution-blown core-shell self-healing nano- and microfibers. **ACS Appl. Mater. Interfaces**, 8, 2016, 4955–4962.

[24] J.Song, Z. Li, H. Wu, Blowspinning: A new choice for nanofibers. **ACS Applied Materials & Interfaces**, 12, 2020, 33447–33464.

[25] A.M. Caffer, D. Chaves, A.L. Pessoa, C.L. Carvalho, W.A. Ortiz, R. Zadorosny, M. Motta, Optimum heat treatment to enhance the weak-link response of Y123 nanowires prepared by solution blow spinning. **Supercond. Sci. Technol.**, 34, 2021, 025009.

[26] A.L. Pessoa, M.J. Raine, D.P. Hampshire, D.K. Namburi, J.H. Durrell, R. Zadorosny, Successful production of solution blow spun YBCO+Ag complex ceramics. **Ceram. Int.**, 46, 2020, 24097–24101.

[27] Y. Polat, E.S. Pampal, E. Stojanovska, R. Simsek, A. Hassanin, A. Kilic, A. Demir, S. Yilmaz, solution blowing of thermoplastic polyurethane nanofibers: A facile method to produce flexible porous materials. **J. Appl. Polym. Sci.** 133, 2016, 43025.

[28] E. Hofmann, K. Krüger, C. Haynl, T. Scheibel, M. Trebbin, S. Förster, Microfluidic nozzle device for ultrafine fiber solution blow spinning with precise diameter control. **Lab on a Chip**, 18, 2018, 2225–2234.

[29] F. Liu, R.J. Avena-Bustillos, R. Woods, B.S. Chiou, T.G. Williams, D.F. Wood, C. Bilbao-Sainz, W. Yokoyama, G.M. Glenn, T.H. Mchugh, F. Zhong, Preparation of Zein fibers using solution blow spinning method. **J. Food Sci**. 2016, 81, N3015–N3025.

[30] A. Koblischka-Veneva, M.R. Koblischka, X.L. Zeng, J. Schmauch, U. Hartmann, TEM and electron backscatter diffraction analysis (EBSD) on superconducting nanowires. **J. Phys.: Conf. Ser.** 1054 (2018) 012005.

[31] M. Rotta, M. Motta, A.L. Pessoa, C.L. Carvalho, C.V. Deimling, P.N. Lisboa-Filho, w. A. Ortiz, R. Zadorosny, One-pot-like facile synthesis of $YBa_2Cu_3O_{7-\delta}$ superconducting ceramic: Using PVP to obtain a precursor solution in two steps. **Mater. Chem. Phys.**, 243, 122607, 2020.

[32] Z. Zheng, L. Gan, T. Zhai, Electrospun nanowire arrays for electronics and optoelectronics. **Sci. China Mater**. 2016, 59(3): 200–216.

[33] X.L. Zeng, Properties of electrospun superconducting and magnetoresistive nanowires. PhD thesis, Saarland University, Saarbrücken, Germany, 2019 (unpublished).

[34] Z. Wang, X. Liu, M. Lv, P. Chai, Y. Liu, X. Zhou, J. Meng, Preparation of one-dimensional $CoFe_2O_4$ nanostructures and their magnetic properties **J. Phys. Chem. C** 112, 2008, 15171–15175.

[35] A.L. Pessoa, A. Koblischka-Veneva, C.L. Carvalho, R. Zadorosny, M. R. Koblischka, Microstructure and paramagnetic Meissner effect of $YBa_2Cu_3O_y$ nanowire networks. **J. Nanopart. Res.** 22, 2020, 360.

[36] M.R. Koblischka, A. Koblischka-Veneva, The possible applications of superconducting nanowire networks, **Materials Today**: Proc., 54, 2022, 125–130.

[37] D.M. Gokhfeld, A. Koblischka-Veneva, M.R. Koblischka, Highly porous superconductors: synthesis, research, and prospects. **Physics of Metals and Metallography**, 2020, 121, 936–948.

[38] M. Rotta, D.K. Namburi, Y. Shi, A.L. Pessoa, C.L. Carvalho, J.H. Durrell, D.A. Cardwell, R. Zadorosny, Synthesis of Y_2BaCuO_5 nano-whiskers by a solution blow spinning technique and their successful introduction into single-grain, YBCO bulk superconductors. **Ceram. Int.**, 45, 2019, 3948–3953.

[39] R. Singh, N. Zheludev, Materials superconductor photonics. **Nature Photonics** 8, 2014, 206.

[40] D. Awschalom, K.K. Berggren, H. Bernien, S. Bhave, L.D. Carr, P. Davids, S.E. Economou, D. Englund, A. Faraon, M. Fejer, G. Saikat, M.V. Gustafsson, E. Hu, L.Jiang, J. Kim, B. Korzh, P. Kumar, P.G. Kwiat, M. Lončar, M.D. Lukin, D.A.B. Miller, C. Monroe, S. W. Nam, P. Narang,

J.S. Orcutt, M.G. Raymer, A.H. Safavi-Naeini, M. Spiropulu, K. Srinivasan, S. Sun, J. Vučković, E. Waks, R. Walsworth, A.M. Weiner, Z. Zhang, Development of quantum interconnects for next-generation information technologies. **PRX Quantum** 2021, 2, 017002.

[41] R.L. Fagaly, Superconducting quantum interference device instruments and applications. **Rev. Sci. Instrum.** 77, 101101, 2006.

[42] C.M. Natarajan, M.G. Tanner, R.H. Hadfield, Superconducting nanowire single-photon detectors: Physics and applications. **Supercond. Sci. Technol**. 25 (2012) 063001.

[43] R.H. Hadfield, Single-photon detectors for optical quantum information applications. **Nature Photonics**, 2009, 3, 696.

[44] A.D. Semenov, G.N. Gol'tsman, A.A. Korneev, Quantum detection by current carrying superconducting film. **Physica C** 2001, 351, 349–56.

[45] S.N. Dorenbos, E.M. Reiger, U. Perinetti, V. Zwiller, T. Zijlstra, T.M. Klapwijk, Low noise superconducting single photon detectors on silicon. **Appl. Phys. Lett.** 93, 131101 (2008).

[46] W.J. Skocpol, M.R. Beasley, M. Tinkham, Self-heating hotspots in superconducting thin-film microbridges. **J. Appl. Phys**. 45, 1974, 4054–66.

[47] H. Shibata, Review of superconducting nanostrip photon detectors using various superconductors. **EICE Trans. Electron.**, 2021, E104-C, 429–434.

[48] S. Pagano, N. Martucciello, E. Enrico, E. Monticone, K. Iida, C. Barone. Microfabrication of NdFeAs(O,F) thin films and evaluation of the transport properties, **Supercond. Sci. Technol**. 33, 074001, 2020.

[49] T. Polakovic, W. Armstrong, G. Karapetrov, Z.-E. Meziani, V. Novosad, Unconventional applications of superconducting nanowire single photon detectors. **Nanomaterials** 2020, 10, 1198.

[50] F. Paolucci, F. Giazotto, GHz superconducting single-photon detectors for dark matter search. **Instruments** 2021, 5, 14.

[51] D.F. Santavicca, Prospects for faster, higher-temperature superconducting nanowire single-photon detectors, **Supercond. Sci. Technol**. 31, 040502, 2018.

4 Characterization Techniques of Nanowires

Endersh Soni, Ujjwal K. Prajapati, Mohit Solanki, and Jyoti Rani
Department of Physics, Maulana Azad National Institute of Technology
Bhopal, India

CONTENTS

4.1 INTRODUCTION

Major progress in the field of nanoscience and nanotechnology has been done in the last two decades. Nanotechnology is the science of manipulating a material, structure, or particle at the nanoscale. Nanoparticles like quantum dots, nanowires, carbon nanotubes (CNTs), and graphene are being used in multiple fields of science and engineering. Among these, the one-dimensional (1D) nanostructure is one of the most promising among the other structures due to its physical properties and its application in semiconductor devices. Nanowires are 1D nanostructures with a diameter range of ~10^{-9} meters and a length of a few nanometers to some millimeters. It is very similar to the normal wires other than its diameter in the nanorange. A vapor-liquid-solid process developed by Dr. R. S. Wagner at Bell Laboratories in 1964 is the foundation of today's nanowire research. At that time, Wagner suggested that gold can be used as a catalyst in the vapor-liquid-solid process to grow silicon microrods. Today, gold is used as a catalyst to synthesize silicon, copper, titanium, and aluminum nanowires [1,2].

In the 1990s, carbon nanotubes were the hot topic of research, and many researchers dedicated their time to synthesizing and characterizing nanowires of metal oxides, III−V compounds, and elemental semiconductors. The synthesis method was very much similar to Wagner's vapor-liquid-solid process. Today, nanowires look promising due to their wide application in the field of science and engineering [3]. The properties of materials at the nanoscale are found to be altered more than the properties of materials at the microscale or as bulk. The optical, chemical, catalytic, and electronic properties of nanomaterials depend upon the shape, size, and aspect ratio (length/diameter ratio). Nanomaterials of different shapes and sizes can be utilized for different applications according to their properties. Hence, determining the properties of nanomaterials is of paramount importance. There are many characterization techniques like scanning electron microscopy (SEM), transmission electron microscopy (TEM), X-ray diffraction (XRD), energy dispersive x-ray spectroscopy (EDX), UV-visible spectroscopy, and atomic force microscopy (AFM), etc. This helps us to determine the properties of nanomaterials. This chapter explains the different characterization techniques in detail with examples and suggests the field of application based on the determined properties [4,5].

Every nanoparticle requires a characterization technique at a particular stage of the growth process, including that for particle size and surface area analysis microscopy and XRD, and for surface composition and topography FTIR and spectroscopy, etc.

4.2 CHARACTERIZATION TECHNIQUES FOR NANOWIRES

4.2.1 Scanning Electron Microscopy

Scanning electron microscopy (SEM) is electron microscopy that uses focused electron beams with higher energy instead of light to produce an image of a solid object's surface. In the early 1950s, SEM had developed new areas of study in the medical, chemical, and physical science communities. SEM uses the characterized technique to determine the topography of the material surface and the chemical composition size of the constituent particle. SEM is a traditional microscopy technique and has many advantages, As a result of its wide depth of field, more of the specimen is in focus at once. In addition, SEM uses electromagnets instead of lenses, so close specimens can be magnified at higher levels due to its much higher resolution. These advantages, as well as clear images of the object's surface, make scanning electron microscopy one of the most useful instruments nowadays [6].

Using scanning electron microscopy, kinetic energy is applied to produce signals on electron interactions. This kinetic energy of the electron beam is dissipated as a variety of signals like secondary electrons (SE), backscattered electrons (BSE), diffracted backscattered electrons (EBSD), photons (X-rays), auger electrons, visible light, and heat on the surface of the sample shown in **Figure 4.1**. Secondary electrons and backscattered electrons are mainly used to produce the SEM image of a sample. Specimen topography and morphology are detected using the secondary electron, which one emitted on the surface of the specimen while backscattered electrons are the most valuable for illustrating contrasts in the composition of the elements of the specimen and rapid phase discrimination, and backscattered electrons generate the deep part of the specimen [7]. Through the inelastic collisions of the incident electrons and the discrete shell electrons of atoms in the specimen, X-rays are generated. X-rays are emitted when the electron beam removes an inner shell electron from the sample, causing a higher-energy electron to fill the shell and release energy. The SEM consists of the following instruments: a combination electron gun, which generates the electron beam with energy ranging from 0.2 to 40 keV, an anode coil used for focusing the electron beam and also accelerating the beam, magnetic lens that converges and defines the resolution of the electron beam, scanning coils are used to raster the beam onto the sample, backscattered electron detector, secondary electron detector, sample stage, and a TV scanner here a final image of the SEM display (**Figure 4.2**) [8].

SEM was used to study the morphology of ZnO nanowire arrays grown via the ultra-fast microwave method by Mahpeykar et al. [9]. Nanowires are generally spherical structures, large in size, and vertically aligned. Different microwave power ranges are used to grow ZnO nanowire arrays without any solution refreshment. By increasing the microwave power level, we can increase the aspect ratio and growth rate enhancement of ZnO nanowires shown in **Figure 4.3.1**. When the power level of the nanowire array increases, the growth rate of the nanowires should decrease since the power level is proportional to its growth rate [9].

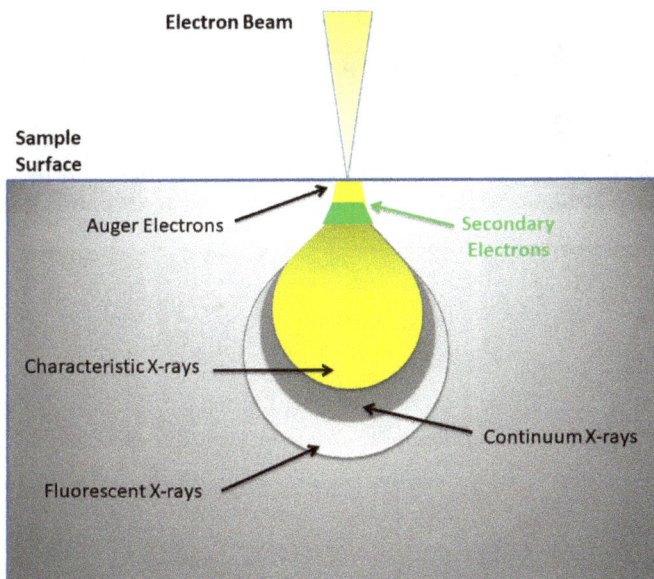

FIGURE 4.1   Schematic interaction of electron beams in SEM.

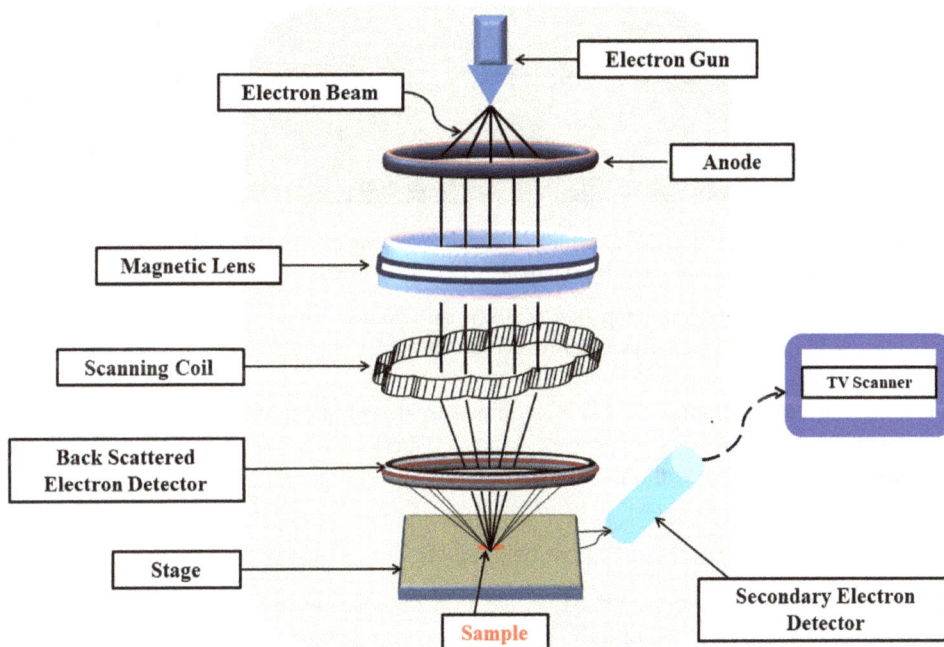

FIGURE 4.2   Set-up of an SEM instrument.

Xia et al. reported an enhanced film conductance of silver (Ag) nanowire based on flexible transparent and conductive networks. **Figure 4.3.2** is a surface morphological image of the Ag nanowire synthesized by the one-step solvothermal reactions and here the scale bar in the SEM image is set to 100 μm. In this process, the solution was filtered 4–5 times to remove the unwanted nanoparticles. After that, this solution was diffused in ethanol to make the wires propagate. If these Ag nanowires

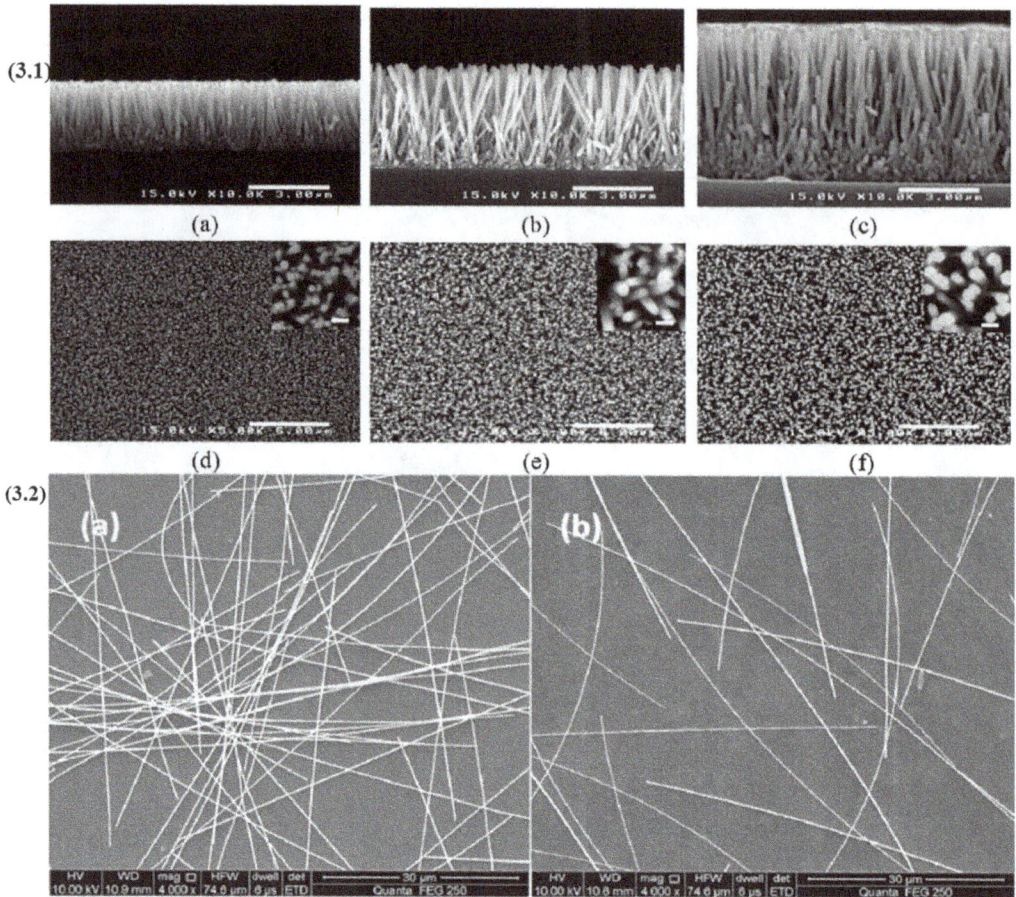

FIGURE 4.3.1 SEM images of ZnO nanowires at different microwave power and without any growth solution refreshment. Adapted with permission [9]. Copyright 2012, IOP Science.

FIGURE 4.3.2 SEM images show a top view of silver nanowires, which are coated on (a) pristine PET and (b) PET substrate by TiO$_x$ hydrosol. Adapted with permission [10]. Copyright 2015, IOP Publishing.

use a coating, such as titanium oxide (TiO$_x$) hydrosol-treated on the uniform Ag nanowire networks, then the wetting ability will be improved, which is shown as a morphological SEM image in **Figure 4.3.2** (SEM top view with the scale bar set to 30 μm)[10].

4.2.2 Transmission Electron Microscopy (TEM)

Transmission electron microscopy (TEM) is electron microscopy that uses focused and electron beams with higher energy instead of light, this electron beam passed through the specimen and gets the image of the specimen's internal behavior. In the TEM specimen, thickness is much less compared to other microscopies around 100 nm. The first transmission electron microscope was discovered in 1931 by Ernst Ruska and Max Knolls. TEM was used to study specimen structure, crystallization, morphology, and stress. TEM has a very powerful magnification as compared to light microscopy [11]. TEM can be used for biology, nanotechnology education, and industries, and can also obtain huge amounts of information on compounds and their structures. These are all advantages as well as the ability to produce high-quality images with a high-clarity, permanent image, therefore, very efficient image transmission electron microscopy is a useful technique to characterize specimens.

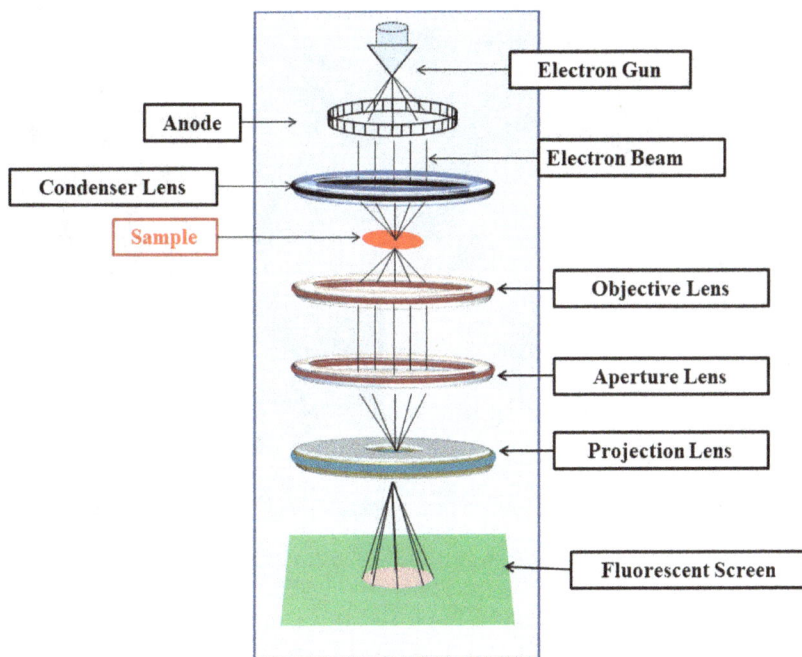

FIGURE 4.4 Set-up of a TEM instrument.

TEM operates by passing electrons through the specimen and forming an image on a fluorescent screen using either the transmitted beam or the diffracted beam (**Figure 4.4**). By illuminating the specimen with electrons that have shorter wavelengths, the resolution power and wavelength of electron transmission increase. Electrons have a wavelength of 0.005 nm, which is shorter than light. As compared to light microscopy, TEM provides a better image [12].

There are some instruments included in TEM. There is an electron gun equipped with a tungsten filament, which is heated and releases electron beams. There is also an electromagnetic coil, or anode, which accelerates electron beams to higher speeds with high voltage. Using a high-aperture condenser lens, all high-angle electrons are eliminated and the electron beams are focused into a thin, small beam for transmission [13]. After that, an objective lens is focused on the transmitted electron beams into an image. About 1–5 mm is the focal length of the objective.

Projector lenses come in two types: the intermediate lens, which magnifies the image to a great extent, and the projector lens, which magnifies the image to a greater extent than the intermediate lens. TEM needs to have high voltage power supplies that are stable and provide high resolution to produce images of a high standard. Using a fluorescent screen and a digital camera that records the images permanently after viewing, images can be captured and permanently stored. Images produced on a fluorescent screen are monochromatic and images are black or grayish and white.

Wang et al. have used TEM for TiO_2 nanowire characterization for morphology and crystal structure. TEM images have been shown in **Figure 4.5.1**. Using TEM, Wang has confirmed that TiO_2 nanowire is a single crystal structure with a diameter of 7–8 nm. **Figure 4.5.1**(a) is the TEM image of the selected area electron diffraction (SAED) of TiO_2 nanowires. **Figures 4.5.1**(b) and (c) represent the high-resolution TEM image of the same nanowire. Based on the interplane distance of a single crystal nanowire, 32.5 nm is the measured distance between adjacent lattice fringes as shown in **Figure 4.5.1**(d). The growth plane (001) and the rutile plane (110) are found to be at 90° [14].

Cao et al. have also used TEM for the characterization of silver nanowires for morphological study. The TEM image of the silver nanowire has been shown in **Figure 4.5.2**. The top and side view

(5.1)   (5.2)

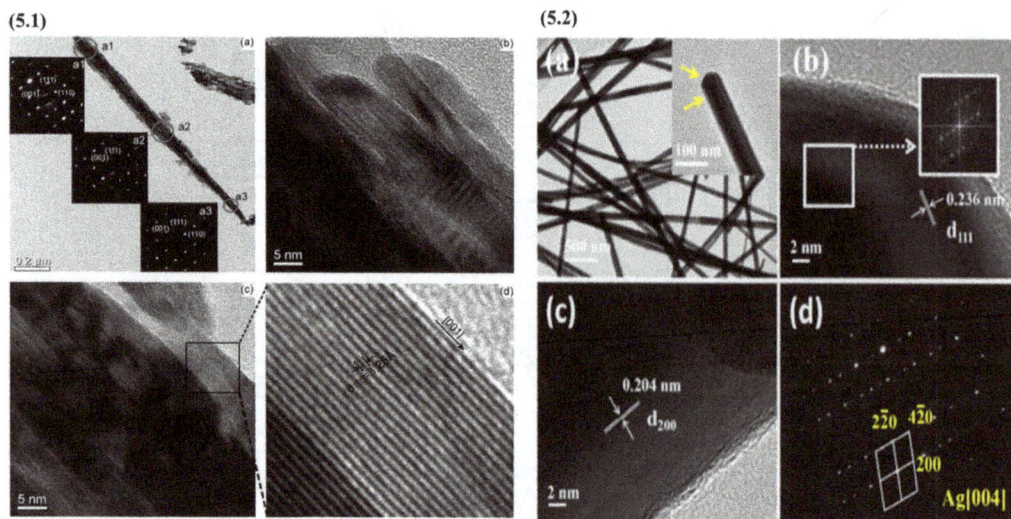

FIGURE 4.5.1 TEM image of the TiO$_2$ nanowires. (a) TEM image and SAEDs of nanowires, (b) and (c) TEM of TiO$_2$ nanowires at a1 and a2 positions, respectively, and (d) TEM image with a lattice fringe of 3.25Å. Adapted with permissions [14]. Copyright 2012, Royal Society of Chemistry.

FIGURE 4.5.2 (a) TEM image of a single Ag (silver) nanowires, (b),(c) TEM image of the tip and side regions of a single Ag nanowire, (d) SAED pattern of the tip region. Adapted with permission [15]. Copyright 2020, MDPI.

has been investigated as represented in **Figure 4.5.2**(b), (c). **Figure 4.5.2**(a) shows a single crystal of silver nanowire, sharp end head with a pentagon shape with five (111) planes can be seen. The lattice spacings were 0.236 nm and 0.204 nm, respectively, which was in good agreement with the *d* value of the (111) plane and (100) plane of FCC Ag [15].

4.2.3 X-RAY DIFFRACTION

X-ray diffraction means diffraction of light i.e. bending of light around the corner of a sample (**Figure 4.6**). This characteristic technique is more efficient, pure, and powerful. This technique based on the specimen size is used for different analyses. For large size specimens, such as inorganic compounds and macromolecules, then XRD is used to determine the structure of the atom in a specimen. If the specimen size is too small, then XRD is used for specimen crystallinity, phase purity, and also composition. In the XRD sample, size should be equal to the wavelength of light used, the size of the sample is around a few angstroms, which is approximately the wavelength of an X-ray. Crystal structures, atomic spacing, the orientation of a single crystal, and atomic spacing can also be studied with XRD, which is known as constructive interference [16].

In the XRD measurements, X-rays are incident on the sample, which are in the form of powder, and then measuring the intensities and scattering angles of X-rays leave the material. In XRD elastic scattering occurs, which means the incident X-ray scatters the crystal atoms, primarily through interaction with the atoms' electrons and these electrons are known as the scatter. Then an array of spherical waves is produced because crystals are a regular array of atoms and through this scattering occurs in a regular array of scatters. XRD's basic principle is Bragg's law:

$$2d \sin \theta = n\lambda \tag{4.1}$$

In equation (4.1) *d* is the spacing between diffracting planes, θ is the incident angle, n is an integer, and λ is the beam wavelength.

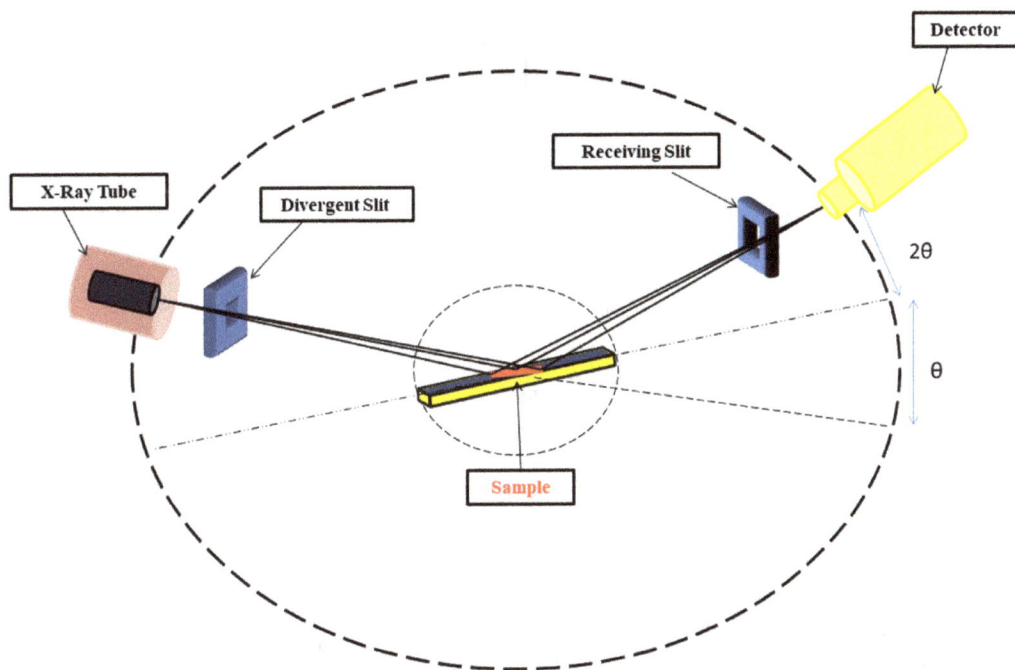

FIGURE 4.6   Set-up of an XRD instrument.

The XRD instrument's main element is the X-ray cathode tube, which generates the X-ray by applying heat to a filament, and after that applies a voltage to direct the electrons towards the sample. The sample holder holds the sample, and rotates with continuous speed. When the conditions of Bragg's law are met for the sample/specimen being analyzed, interference occurs constructively from both detector and source, causing X-ray one peak intensity [17]. The detector processes this signal and records it, and also converts it into a count rate for output to a computer.

Li et al. reported the gold doped ZnO nanowires (Au/ZnO) and studied these photocatalytic properties. Using XRD we collected the structural analysis of the specimen. **Figure 4.7.1** shows the XRD patterns of the Au/ZnO and ZnO nanowire samples [18]. According to **Figure 4.7.1**, ZnO nanowires have mainly five peaks at $2\theta = 31.72°, 34.42°, 36.25°, 47.54°$, and $62.86°$, which determine the crystal plane (1 0 0), (0 0 2), (1 0 1), (1 0 2), and (1 0 3) of the ZnO crystal hexagonal structure. In the XRD pattern, peaks are clear meaning that the ZnO nanowires have a fine crystalline structure. This is the same as in Au/ZnO nanowires, which have one peak around 38.61 that is assigned to the (1 1 1) crystal plane of Au [18]. A structural study has been performed using XRD by Qin et al. on silver nanowires. As shown in **Figure 4.7.2**, XRD analysis of bulk nanowires showed some diffraction peaks at $2\theta = 38.08°, 44.42°, 64.40°$, and $77.84°$ that could be assigned crystalline planes (1 1 1), (2 0 0), (2 2 0), and (3 1 1) of a face-centered cubic structure of silver crystals [19].

4.2.4   ENERGY DISPERSIVE X-RAY SPECTROSCOPY

An energy-dispersive X-ray spectrometer (EDS, EDX) is an analytical tool used for the analysis of elements, chemicals, or images of samples. It is also sometimes called energy dispersive X-ray microanalysis (EDXMA). SEM and TEM are combined with the EDX instrument during electron microscopy, where the microscope's imaging capabilities identify the specimens of interest.

The beam of X-rays used in EDX simulates the characteristic X-rays from a specimen when a high energy beam of charged particles is incident on it, such as photons, electrons, or X-rays. An

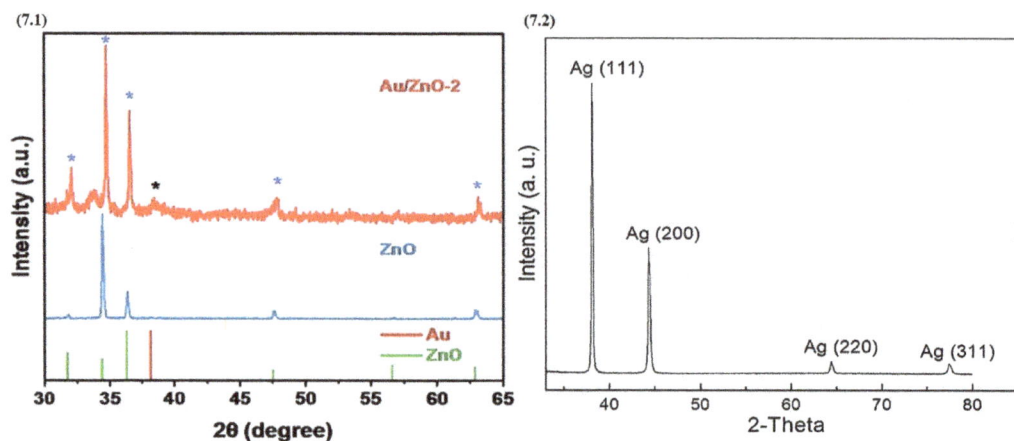

FIGURE 4.7.1   XRD of the Au and ZnO nanowires. Adapted with permission [18]. Copyright 2022, Elsevier.

FIGURE 4.7.2   XRD patterns of the silver (Ag) nanowires. Adapted with permission [19]. Copyright 2011, Elsevier.

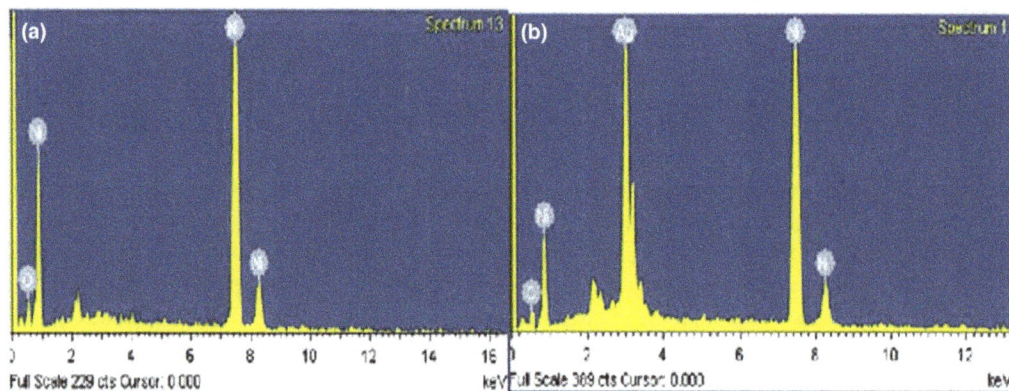

FIGURE 4.8   EDX spectrum of (a) Ni nanowires, (b) core-shell-type Ni/Ag nanostructures. Adapted with permission [22]. Copyright 2012, Royal Society of Chemistry.

electron in an inner shell may be stimulated by the incident beam, ejecting it from the shell and creating electron-hole pairs [20]. X-rays are emitted when an electron from the specimen's outer shell fills this hole. An energy dispersive spectrometer, which determines the elemental composition of a specimen, can measure the number and energy of X-rays that a specimen emits [21].

Senapati et al. reported element analysis of the nickel (Ni) and Ag nanowires using the EDX technique shown below in **Figure 4.8.** The EDX spectrum of Ni nanowires in **Figure 4.8**(a) shows that nickel peaks occur at a very low-intensity peak of oxygen (~ 3%) because of the air oxidation being high during the washing of the final product. **Figure 4.8**(b) is the EDX spectrum of the Ni/Ag core-shell nanostructures and shows that Ni and Ag have a very weak peak of oxygen [22].

4.2.5   Ultraviolet-Visible Spectroscopy

In UV-Vis spectroscopy, the wavelengths of light transmitted and absorbed by a specimen are measured separately against a blank sample, causing electrons to move from their ground state to higher energy states (**Figure 4.9**). UV-Vis spectroscopy, also called absorbance spectroscopy, covers

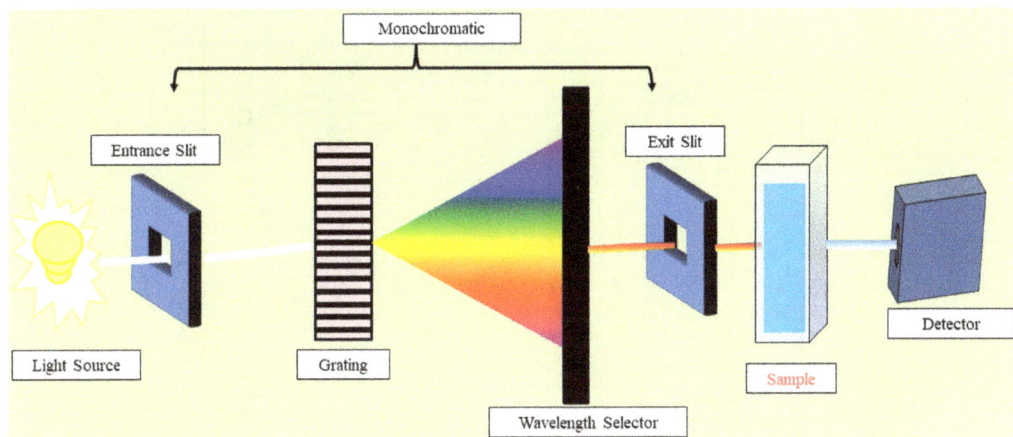

FIGURE 4.9   Set-up of a UV-Vis spectroscopy.

a wavelength range of around 800 to 200 nm when relating to 1.5 eV to 6.2 eV in the electromagnetic spectrum. [23] UV-Vis spectroscopy is based on the Beer-Lambert law:

$$A = \varepsilon bc \tag{4.2}$$

In equation (4.2), A is absorbance (unit less), ε is the molar absorptivity of the compound or molecule in solution (M^{-1} cm^{-1}), b is the path length of the sample holder or cuvette (normally 1 cm), and c is the concentration of the solution (M).

A UV-Vis spectroscopy instrument consists mainly of a four-part, light source that produces white light, after that a monochromatic unit that consists of an entrance slit (narrow the light beam to wide) and exit slit (select the desired monochromatic wavelength beam), the grating unit, which split the white light into a light prism (VIBGYOR), and the wavelength selector, which selects the specific wavelength light, after the sample holder unit, and finally the detector to detect the monochromatic wavelength light beam, which is passed through the sample [24].

Wang at el. reported the UV-Vis light spectrum of the gold (Au) nanowire, which is connected to the zinc oxide (ZnO) nanowire arrays shown in **Figure 4.10.** According to the UV-Vis spectrum, it can be seen that the pure ZnO nanowire has the step absorption profile at the shorter wavelength (400 nm), which indicates the ZnO nanowire arrays have transparent properties in the visible region. Another side Au nanowire shows that varying the wavelength from 400 nm to 750 nm decreases transmittance properties. **Figure 4.10**(a) Au/ZnO nanowire arrays with only Au nanoparticles and the (b), (c), (d) Au/ZnO nanowire array with different deposition, (e) shows the HAuCl$_4$ used in the reaction [25].

4.2.6   ATOMIC FORCE MICROSCOPY

The atomic force microscopy (AFM) characteristics' technique produced the three-dimension (3D) image of the specimen surface at a sub-nanometer scale using a microscale probe (**Figure 4.11**). The AFM basic principle measures the repulsive and attractive forces with the help of the probe on the surface of the specimen. Here the probe is fixed on the cantilever, and interacting forces are balanced by stain on the cantilever and surface interaction of the probe. The flexible cantilever will bend due to the force acting on it according to Hooke's law. The specimen surface is scanned using either the probe or the sample moved by piezoelectric elements [26]. This allows for the measuring of the interacting forces' entire specimen surface. The force applied by the probe has the potential to change the surface by simply moving loosely bound surface or etching features [27].

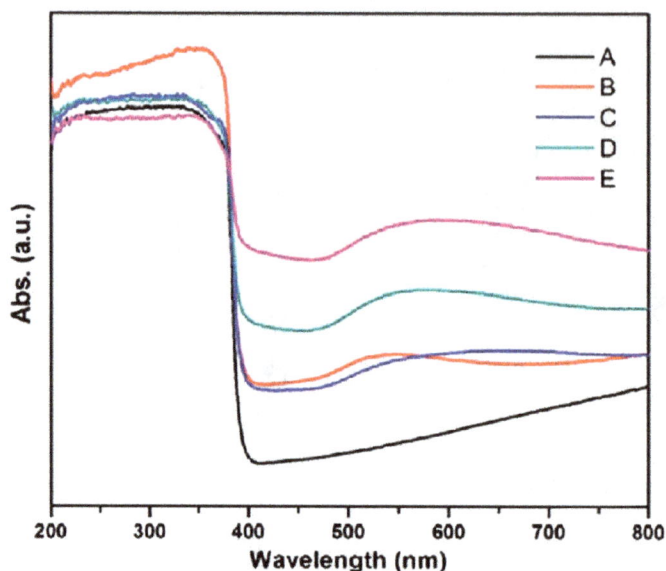

FIGURE 4.10   A UV-visible spectrum of (a) Au/ZnO nanowire array (b), (c), and (d) Au/ZnO nanowire array and (e) HAuCl$_4$ was used in the reaction. Reproduced with permission [25]. Copyright 2014, Royal Society of Chemistry.

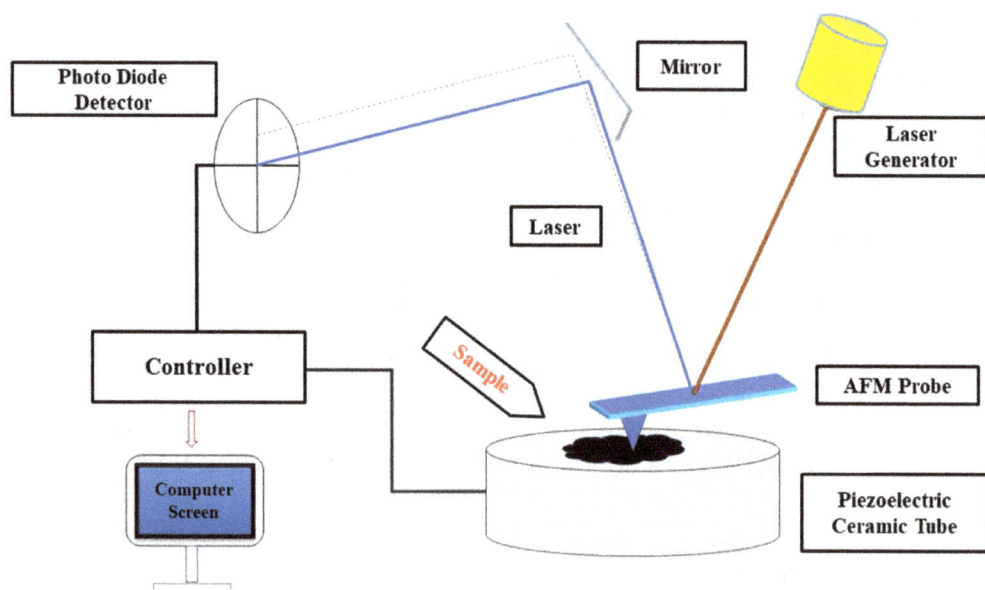

FIGURE 4.11   Set-up of an AFM instrument.

AFM has a different mode of operation, contact mode (the probe has a proper contact with the surface of the specimen and constant potential for vertical piezoelectric), frication mode (the probe is in contact with the sample resistance acting on the probe horizontal motion, and strain proportional to the resistance, friction), adhesion mode/non-contact mode (here attractive forces measure the probe pulled from the specimen), and tapping mode (cantilever works as a harmonic oscillator with a specific frequency proportional to the spring constant).

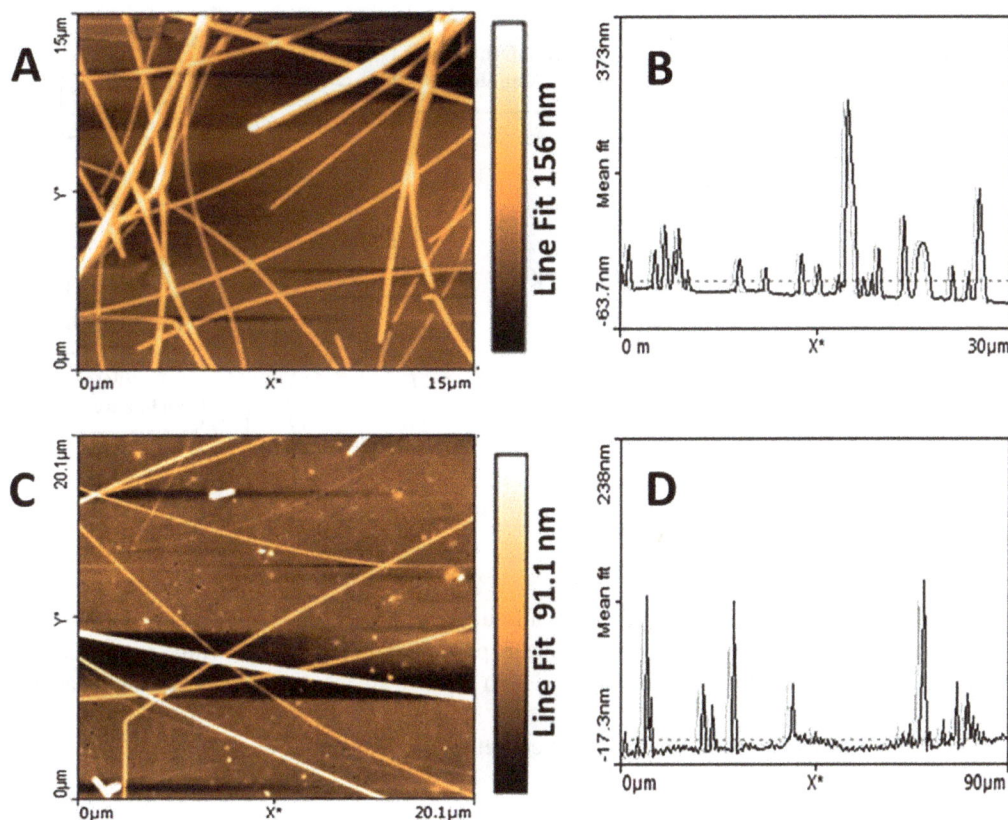

FIGURE 4.12 (a), (c) AFM image of silver nanowires and (b), (d) AFM height profile of Ag nanowires. Adapted with permission [29]. Copyright 2016, Royal Society of Chemistry.

AFM instruments consist of equipment set-up, like a laser generator, position-sensitive photo-detector, controller, AFM probe or cantilever, and piezoelectric ceramic tube; the last one is a computer screen where we get the AFM 3D image [28].

Bari et al. did a surface topography and roughness study of silver nanowire using AFM as shown in **Figure 4.12**. **Figures 4.12**(a) and (b) represent the AFM image and height profile of silver nanowires with a diameter greater than 100 nm, respectively, whereas **Figures 4.12**(c) and (d) represent the AFM image and height profile of silver nanowires with a lower surface roughness than the nanowire [29].

4.2.7 NUCLEAR MAGNETIC RESONANCE (NMR)

NMR is also an important analytical method for determining the structural and quantitative properties of nanoscale materials (**Figure 4.13**). A critical analytical tool for organic chemists is nuclear magnetic resonance spectroscopy (NMR). In this technique, using a strong magnetic field, it is possible to observe the spin-up and spin-down states of nuclei that possess non-zero spin, which causes a small energy difference between the two states. Diamagnetic or antiferromagnetic nanowires are typically studied using NMR to study the interaction between the ligand and the surface [30]. Capping ligands are also investigated by NMR in the determination of particle shape. Chemical conversion of nanowire precursors can be screened with NMR under different reaction conditions and for diverse metal identities, with a high spatial and chemical resolution; this helps to understand reaction mechanisms for nanowire synthesis [31].

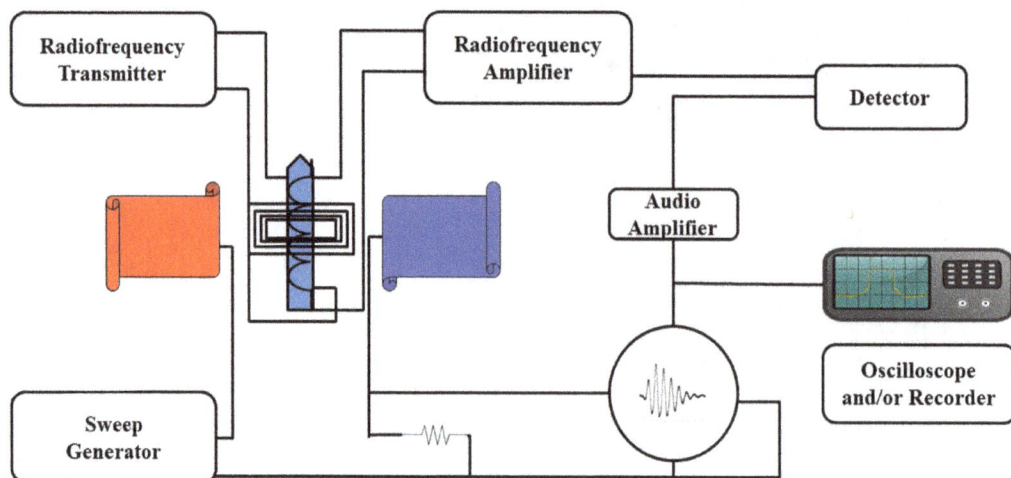

FIGURE 4.13   Set-up of an NMR instrument.

In NMR, the fundamental principle is that all nuclear spins are disordered at the ground state, as well as that they have the same energy. Nuclear spins are either against or with a strong external magnetic field when it is applied. Nuclei aligned with the field are always over those pointing against it (population excess).

Basic NMR spectroscopy has some instrument setup.

1. Sample holder – it is a glass tube.
2. Magnetic coils – when current flows through the magnetic coil, a magnetic field is generated.
3. Permanent magnet – a homogeneous magnetic field is provided by it.
4. Sweep generator – this makes the magnetic field stronger by modifying its strength.
5. Radiofrequency transmitter – radio waves are produced in a powerful but short pulse.
6. Radiofrequency – detect the received RF.
7. RF detector – determine the RF, which is unabsorbed.
8. Recorder – receive the RF from the detector and record the NMR.
9. Readout system – a computer.

A magnetic field should be applied to the sample. Radio waves are used to generate NMR signals from the nuclei sample. Sensitive radio receivers detect these NMR signals. Intermolecular magnetic fields change the resonance frequency of atoms in molecules. Functional groups and electronic structures of molecules are described in this document. Monomolecular organic compounds can be identified definitively by NMR spectroscopy. In NMR, a molecule's reaction state, structure, chemical composition, and environment are all described.

Chuan et al. reported the self-catalyzed sensitization of CuO nanowires via a solvent-free click reaction. **Figure 4.14** shows the 13_C ss-NMR spectrum of the CuO nanowires, ZnO nanorods, and prop-2-ynoic acid in the D_2O. This comparison clearly shows that the experiment confirms the prediction of a downfield shift when comparing the monodentate carboxylate ligand on the CuO surface with the bidentate ligand on the ZnO surface [32].

4.2.8   DYNAMIC LIGHT SCATTERING (DLS)

DLS is a modern technique used to measure the size or diameter down to 1 nm range. It is also known as photon correlation spectroscopy **or** quasi-elastic light scattering (**Figure 4.15**). DLS works on

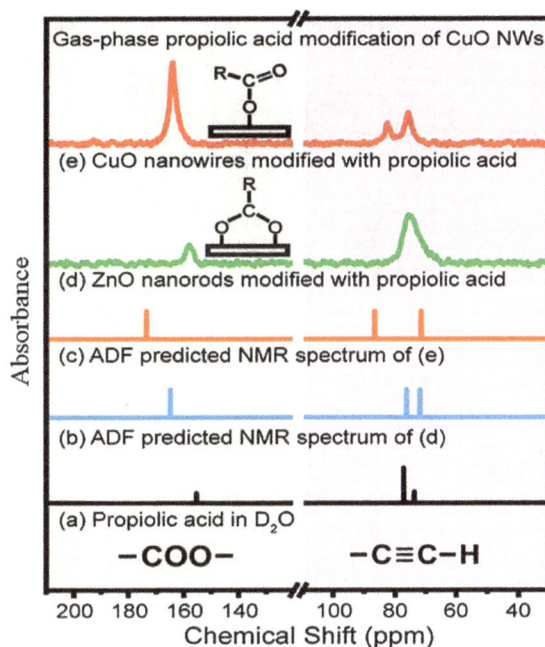

FIGURE 4.14 Liquid-phase NMR spectrum of prop-2-ynoic acid in D_2O provided as a reference (a) and computationally predicted isotropic 13_C as the chemical shift of a bidentate surface intermediate (b) corresponding to the ZnO modification with prop-2-ynoic acid (d) and a monodentate intermediate (c) corresponding to the CuO modified with prop-2-ynoic acid (based on the Cu_2O model) (e). Adapted with permission [32]. Copyright 2020, ACS publications.

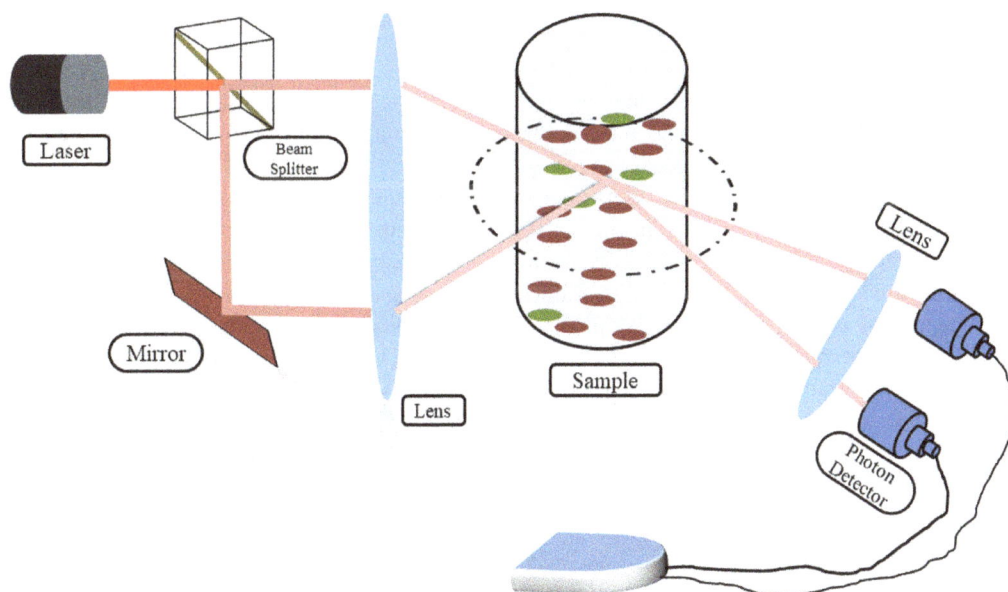

FIGURE 4.15 Set-up of DLS instrument.

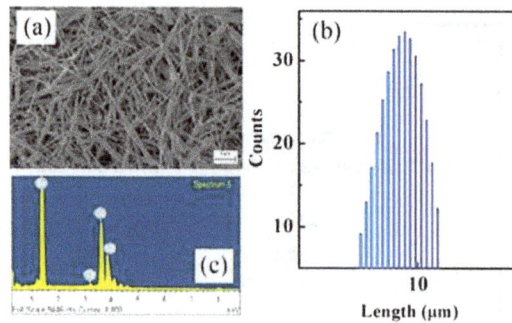

FIGURE 4.16 Image (a), (b), and (c) SEM, DLS Spectra, and EDS spectra of the Sb_2Se_3 nanowires. Adapted with permission [33]. Copyright 2017, Elsevier.

the principle of the Brownian motion of a dispersed particle. The principle of Brownian motion of dispersed particles states that when some particle is dispersed in some solvent, particles will have random motion in all directions. This random motion of the particle results in a collision with the solvent particle. There will be some energy transfer in this collision. This transferred energy may be of small or large amount but it may impact small particles greatly. It results in a higher velocity of small particles than that of the larger ones. If all other parameters influencing the particle movement are known, particle size/diameter can be estimated based on the particle velocity.

The Stokes-Einstein equation relates the velocity of a particle directly to the size or diameter of the particle. Velocity is represented by translational diffusion coefficient D. Equation also includes parameters like viscosity and temperature because these parameters directly impact the velocity of the particle. Equation (4.3) is given below

$$D = \frac{K_B T}{6\pi\mu R_H} \tag{4.3}$$

where K_B is the Boltzmann constant, T represents temperature in K, μ is the viscosity of the solvent, and R_H is the hydrodynamics radius of the particle.

DLS consists of a single frequency laser beam and a detector. The instrumental setup is very simple. The sample is dispersed in some dispersive liquid and put in a transparent vessel as shown in the figure. The laser beam is made to fall on the sample through some focusing lens. The incident laser beam gets scattered in all directions.

Using a detector, these scattered laser lights are detected at some predefined angle for a time. Using the detector translational diffusion coefficient D is estimated and further using the Stokes-Einstein equation, hydrodynamic radius is calculated. The detection angle is a very important parameter. In modern DLS instruments, instruments automatically decide the measuring/detection angle depending upon the sample being processed. Most typical detection angles are 15°, 90°, and 175°. This angle can be adjusted manually if the nature of the sample is well-known.

Rajesh Kumar Yadav et al. used the DLS technique in their study on Sb_2Se_3 nanowires to confirm the average length of the grown nanowires shown in **Figure 4.16**. The DLS method has confirmed that the average length of the Sb_2Se_3 nanowires is about 10μm SEM and DLS images confirm that grown nanowires are of good aspect ratio with a diameter of about 50–100 nm and a length of about 10μm [33].

4.3 SUMMARY

To make those ideas of fabricating different types of nanowires, the search for a material that possesses all of the required characteristics is a great problem. During the research, many groups of researchers are looking and observing the different materials for different properties according to their application for the fabrication of nanowires. For the examination of properties and characteristics, there are many characterization tools and techniques. These techniques are used to examine the materials to make sure whether the material possesses the required properties or not. This process is called characterization. Hence characterization can be defined as "the step-by-step process by which the material's structure and properties can be probed and measured". It is a fundamental process in the field of materials science without which no scientific understanding of engineering material could be ascertained.

REFERENCES

1. R Wagner, W Ellis, "Vapor-Liquid-Solid Mechanism of Single Crystal Growth (New Method Growth Catalysis from Impurity Whisker Epitaxial + Large Crystals Si E)", Appl. Phys. Lett., 4, 1964, 89–90.
2. M Yazawa, M Koguchi, A Muto, K Hiruma, "Semiconductor Nanowhiskers", Adv. Mater., 5, 1993, 577–580.
3. T Trentler, K Hickman, S Goel, A Viano, P Gibbons, W E Buhro, "Solution-Liquid-Solid Growth of Crystalline III-V Semiconductors - an Analogy to Vapor-Liquid-Solid Growth", Science, 270, 1995, 1791–1794.
4. D Titus, J J Samuel, S Roopen, (2019), "Green Synthesis, Characterization and Applications of Nanoparticles and Nanoparticle Characterization techniques", Micro and Nano Technologies, 2019, 303–319.
5. E Soni, U Prajapati, A Katariya, J Rani, R Gupta, T Nguyen, "Smart and Flexible Energy Devices and Characterization Techniques of Flexible Energy Devices", CRC Press Taylor and Francis group, 2022, 59–80.
6. A Argast, C Tennis, "A Web Resource for the Study of Alkali Feldspars and Perthitic Textures Using Light Microscopy, Scanning Electron Microscopy and Energy Dispersive X-ray Spectroscopy", Journal of Geoscience Education, 52(3), 2004, 213–217.
7. R Beane, "Using the Scanning Electron Microscope for Discovery Based Learning in Undergraduate Courses", Journal of Geoscience Education, 52(3), 2004, 250–253.
8. D Moecher, "Characterization and Identification of Mineral Unknowns: A Mineralogy Term Project", Journal of Geoscience Education, 52(1), 5–9, 2018.
9. S Mahpeykar, J Koohsorkhi, H Ghafoori-fard, "Ultra-Fast Microwave-Assisted Hydrothermal Synthesis of Long Vertically Aligned ZnO Nanowires for Dye-Sensitized Solar Cell Application", Nanotechnology 23, 165602(7pp), 2012.
10. X Xia, B Yang, X Zhang, C Zhou, "Enhanced Film Conductance of Silver Nanowire-Based Flexible Transparent & Conductive Networks by Bending", Materials Research Express, 2(7), 2015.
11. T Kirschling, P Golas, J Unrine, K Matyjaszewski, K Gregory, G Lowry, and R Tilton, "Microbial Bioavailability of Covalently Bound Polymer Coatings on Model Engineered Nanomaterials", Environ. Sci. Technol., 45, 2011, 5253–5259.
12. Y Cheng, L Yin, S Lin, M Wiesner, E Bernhardt, J Liu, "Toxicity Reduction of Polymer-Stabilized Silver Nanoparticles by Sunlight" J. Phys. Chem. C, 115, 2011, 4425–4432.
13. L Jun, W Tianpin, A Khalil, "State-of-the-Art Characterization Techniques for Advanced Lithium-Ion Batteries", Nature Energy 2, 2017, 17011.
14. X Wang, Y Liu, X Zhou, B Li, H Wang, W Zhao, H Haung, C Liang, X Yu, Z Liu, H Shen, "Synthesis of Long TiO$_2$ Nanowire Arrays with High Surface Areas via Synergistic Assembly Route for Highly Efficient Dye-Sensitized Solar Cells", Journal of Materials Chemistry, 22, 2012, 17531–17538.
15. L Cao, Q Huang, J Cui, H Lin, W Li, Z Lin, P Zhang, "Rapid and Facile Synthesis of High-Performance Silver Nanowires by a Halide-Mediated, Modified Polyol Method for Transparent Conductive Films", Nanomaterials, 10, 2020, 1139.
16. J Brady, S Boardman, "Introducing Mineralogy Students to X-ray Diffraction Through Optical Diffraction Experiments Using Lasers", Jour. Geol. Education, (43)5, 1995, 471–476.
17. S Dann, E Abel, "Reactions and Characterization of Solid, Characterization of Solids", 2000, Royal Society of Chemistry.
18. H Li, J Ding, S Cai, W Zhang, X Zhang, T Wu, C Wang, M Foss, R Yang, (2022), "Plasmon-Enhanced Photocatalytic Properties of Au/ZnO Nanowires", Applied Surface Science, 538, 2022, 152539.

19. X Qin, H Wang, Z Miao, X Wang, Y Fang, Q Chen, X Shao, (2011), "Synthesis of Silver Nanowires and their Applications in the Electrochemical Detection of Halide", Talanta, 84, 2011, 673–678.
20. J Goldstein, P Echlin, C Lyman, D Newbury, D Joy, E Lifshin, L Sawyer, J Michael, "Scanning Electron Microscopy and X-Ray Microanalysis" Third Edition, Germany, Springer, 2003,US.
21. J Orasugh, S Ghosh, D Chattopadhyay, B Han, S Sharma, T Nguyen, L Longbiao, K Bhat, "Fiber-Reinforced Nanocomposites: Fundamentals and Applications, Nanofiber-Reinforced Biocomposites", Micro and Nanotechnologies, Chapter 10, 2020, 199–233.
22. S Senapati, S Srivastava, S Singh, H Mishra, "Magnetic Ni/Ag Core–Shell Nanostructure from Prickly Ni Nanowire Precursor and its Catalytic and Antibacterial Activity" Journal of Materials Chemistry, 22(14), 2012, 6899.
23. C Ojeda, F Rojas, "Recent Applications in Derivative Ultraviolet/Visible Absorption Spectrophotometry: 2009–2011: A Review", Microchemical Journal, 106, 2013, 1–16.
24. D Harries, "Quantitative Chemical Analysis", Seventh Edition, (2007), W.H. Freeman and Company, New York.
25. T Wang, B Jin, Z Jiao, G Lu, J Ye, Y Bi, "Photo-directed Growth of Au Nanowires on ZnO Arrays for Enhancing Photoelectrochemical Performances", Journal of Materials Chemistry A, 2, 2014, 15553.
26. R Jagtap, A Ambre, "Overview Literature on Atomic Force Microscopy (AFM): Basics and its Important Applications for Polymer Characterization", Indian J. Eng. Mater. Sci, 13, 2006, 368–384.
27. N Jalili, K Laxminarayana, "A Review of Atomic Force Microscopy Imaging Systems: Application to Molecular Metrology and Biological Sciences", Machatronics, **14**, 2004, 907–945.
28. S Achary, "Ph.D. thesis, Characterization of Individual NPs and Applications of NPs in Mass Spectrometry", 2010, Texas A&M University.
29. B Bari, J Lee, T Jang, P Won, S Ko, K Alamgir, M Arshad, L Guo, (2016), "Simple Hydrothermal Synthesis of Very-Long and Thin Silver Nanowires and their Application in High Quality Transparent Electrodes", Journal of Materials Chemistry A, 29, 2016.
30. L Lu, "Ph.D. thesis, Water-dispersible Magnetic NPS for Biomedical applications: Synthesis and Characterisation", 2011, University of Liverpool.
31. L Marbella, J Millstone, "NMR Techniques for Noble Metal Nanoparticles", Chem. Mater., **27**, 2015, 2721–2739.
32. C He, X Cai, S Wei, A Jantti, A Teplyakov, "Self-Catalyzed Sensitization of CuO Nanowiers via a Solvent-free Click Reaction" Langmuir, 36(48), 2020, 14539–14545.
33. R Yadav, R Sharma, G Omar, J Aneesh, K Adarsh, "Ultrafast Broadband Saturable Absorption in Sb_2Se_3 Nanowires", Procedia Engineering, 216, 2017, 168–174.

5 Nanowires for Metal-Ion Batteries

Santosh J. Uke
Department of Physics, JDPS College, Sant Gadge Baba Amravati
University, Amravati, India

Satish P. Mardikar
Department of Chemistry, SRS College, Sant Gadge Baba Amravati
University, Amravati, India

CONTENTS

5.1 INTRODUCTION

With the ever-increasing demand for electrical energy, the development of metal-ion batteries with high volumetric capacity, high gravimetric capacity, high energy density, high power density, long cycling life, safe and low cost gained enormous attention. Metal-ion batteries have a key role in renewable energy systems. Over the years, lithium-ion batteries (LIBs) have been the common source of electrical energy for portable electronics like mobile phones and laptops. Also, it fulfills the energy needs of electric and hybrid vehicles. Interestingly, recently due to the limited resources of lithium metal in the Earth's crust and its high cost, the different metal-ion batteries viz. the sodium-ion battery, magnesium-ion battery, aluminum-ion battery, etc., have gained enormous attention as a secondary energy source to fulfill the need as a power back system and alternative power source in the global energy market. The Na-, Mg-, and Al-ion batteries are the most promising and low-cost alternative solutions for global energy needs [1,2]

Na is the fourth most abundant element on the Earth, second to that Al and Mg metal are also abundant in nature. Recently, due to the similarity in energy storage mechanism and low cost, batteries based on Na-, Mg-, and Al- metals are highly studied. The metal-ion (Na, Mg, and Al) batteries show a higher specific and volumetric capacity than LIBs. Moreover, in addition to the advantages associated with metal-ion batteries, there are different issues related to these batteries [3,4]. The different components of batteries such as substrate, electrode, electrolyte, separator, etc. have their special importance in the demonstration of high volumetric capacity, high gravimetric capacity, high energy density, high power density, and long cycle life at the external load. Recently, to enhance the performance of metal-ion batteries in the form of energy density, power density, and cycle life, many research communities have adopted many advanced strategies. This includes synthesis of new electrode and electrolyte materials, modification in electrode fabrication techniques, advancement in complete cell fabrication, use of solid electrolytes, etc.

The energy stored in metal-ion batteries solely depends on the active material used for the fabrication of electrodes in it. During the charging of the metal-ion batteries, the electrolyte ions are stored in the pores of the active material used for the fabrication of electrodes in it [5–8]. Also, the specific capacity and energy density of the metal-ion batteries depend on the shape, size, porosity, and structure of the electrode material used in it. Many recent reports demonstrated that advanced synthesis techniques can easily control the shape, size, porosity, and structure of the electrode material. In consequence, the output parameters such as high energy density, high power density, and long cycle life of the battery can be controlled via tailoring the electrode material. Therefore, for enhancement of output parameters of batteries, and fabrication of high-performance electrodes for metal-ion batteries, for many years the researchers majorly focused on the development of advanced materials for anode and cathode for metal-ion batteries [8–10].

There are various active materials, such as layered and tunnel-type transition metal oxides, ternary metal oxides, transition metal hydroxides, transition metal fluorides, oxyanionic compounds, polymers, Prussian blue analogs, etc., on the other hand, graphitic and non-graphitic carbon materials are one of the most successful candidates as the anode for metal-ion batteries. In addition, tin-based material, silicon-based material, transition metal oxides, transition metal sulfides, and organic materials like Schiff bases, etc., have been used as the anode material for metal-ion batteries [3,6,11].

5.2 NANOCOMPOSITES FOR METAL-ION BATTERIES

In most of the MIBs, graphite has been used as an anode. The graphite is widely sourced and abundant in nature, and low cost. In addition, graphite shows excellent electrochemical performance with stable behavior in solid as well as liquid electrolytes. Also, the graphite anode in MIBs demonstrated equal theoretical and practical volumetric capacitance. However, MIBs containing graphite anode are unstable, which shows side reactions during the charging and discharging process. Also, an accommodation of metal ions is very difficult and is still a challenge for metal- (Na-, Al-, and Mg-) ion batteries containing graphite material as an anode. In the case of commercial graphite, the intercalation of metal ions into it is very difficult. As a consequence, metal-ion batteries with commercial graphite result in a very low volumetric capacity, energy density, and cycle life. Therefore, it is a very prime requirement to find a suitable anode material for metal- (Na-, Al-, and Mg-) ion batteries [10–12].

To date, there are many anode materials for MIBs. This includes carbon-based materials, metal oxide, metal sulfides, the nanocomposite of carbon-based materials with different metal oxides and metal sulfides, etc. The carbon-based nanocomposites are an efficient anode material for MIBs. The carbon metal oxides and carbon metal sulfide nanocomposites have been excessively used and demonstrated to be excellent anode materials for MIBs. The MIBs with carbon nanocomposites are reported to be high volumetric capacity, high energy density, and excellent cyclic stability [5,13]. Also, the synthesis of carbon-based nanocomposites is an easy and simple process. The existence of nanosized carbon in nanocomposites retains the high electrical conductivity, and high mechanical strength and has high structural stability in anode material. To date, the nanocomposites of nanostructured metal oxides and/or metal sulfides with different carbon nanostructured material viz. carbon nanotubes, single-walled carbon nanotubes, multiwalled carbon nanotubes, etc. have been heavily explored, and excessively used as the anode material for MIBs [5,9,11,14]. The latest development in the field of nanotechnology makes an easy synthesis of a variety of nanostructures, viz, nanotubes, nanowires, nanoneedles, nanobelts, etc. These attractive nanostructured materials have unique geometry with a high aspect ratio, due to that, these materials exhibit excellent physical and electrochemical properties [15–18].

5.3 TYPES OF MIBS AND THEIR WORKING PRINCIPLE

5.3.1 THE ENERGY STORAGE MECHANISM OF LIBs

Using $LiCoO_2$ as a cathode material and graphite (C) as the anode material, equations 5.1–5.3 demonstrate the energy storage mechanism of typical LIBs. During the charging of LIBs the lithium metal ions (Li^+) de-intercalate at the cathode material $LiCoO_2$. After de-intercalation Li^+ diffused into the electrolyte present between the cathode and anode in the LIB cell. Further, Li^+ passes through the nanopores and separates toward the graphite anode. Simultaneously, to maintain the charge neutrality through an external circuit the electrons move in the opposite direction. While during the discharging the Li^+ ion moves from an anode (C) to cathode $LiCoO_2$ [3,9]. The schematics of the charge-discharge mechanism of LIBs are demonstrated in **Figure 5.1**.

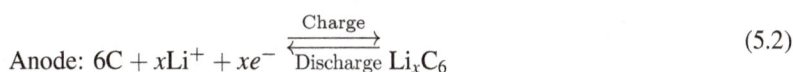

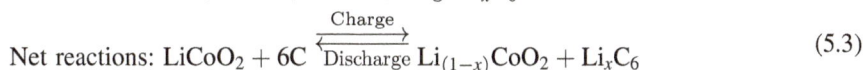

$$\text{Cathode:} \ LiCoO_2 \underset{\text{Discharge}}{\overset{\text{Charge}}{\rightleftharpoons}} Li_{(1-x)}CoO_2 + xLi^+ + xe^- \qquad (5.1)$$

$$\text{Anode:} \ 6C + xLi^+ + xe^- \underset{\text{Discharge}}{\overset{\text{Charge}}{\rightleftharpoons}} Li_xC_6 \qquad (5.2)$$

$$\text{Net reactions:} \ LiCoO_2 + 6C \underset{\text{Discharge}}{\overset{\text{Charge}}{\rightleftharpoons}} Li_{(1-x)}CoO_2 + Li_xC_6 \qquad (5.3)$$

5.3.2 CHARGE STORAGE MECHANISM IN NIBs

In NIB_S the Na^+ can be inserted into the host of active materials at electrodes via three types of charge storage mechanisms viz. intercalation, alloying, and conversion. The intercalation can also be termed insertion. In the intercalation process, the Na^+ can be inserted into the host active material or the anode materials. The structural changes of the anode material have been retained in the intercalation process. The alloying mechanism can also be termed a solid-state reaction mechanism. In this mechanism, the Na^+ metal can be inserted into the host electrode material via reaction 5.4. Where A is the active material or metal electrode. In the alloying mechanism, the no phase transformation occurred [19]. The conversion mechanism is a relatively new mechanism that is still under development. The conversion mechanism results in a very high volumetric capacity of NIBs. In the

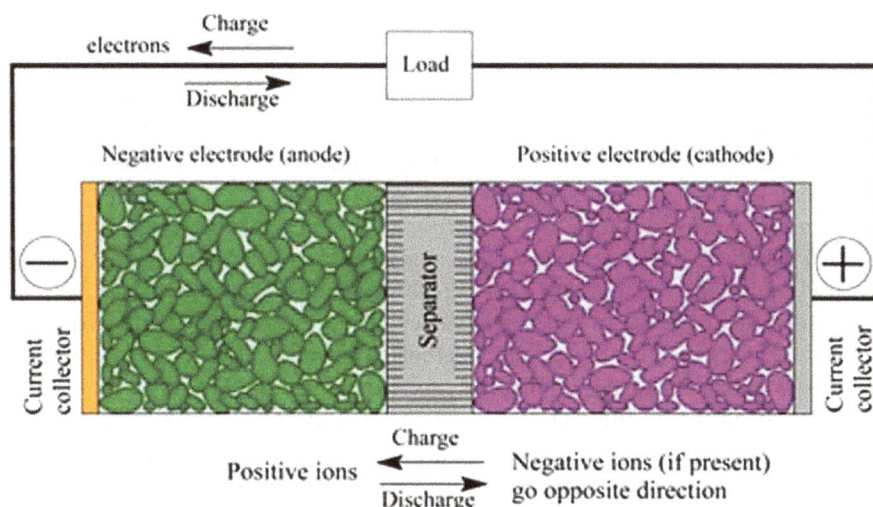

FIGURE 5.1 Schematics of the charge-discharge mechanism of LIBs. Adapted with permission. Copyright © 2020 by the authors. Licensee MDPI, Basel, Switzerland. Distributed under the terms and conditions of the Creative Commons Attribution (CC BY) license (http://creativecommons.org/licenses/by/4.0/).

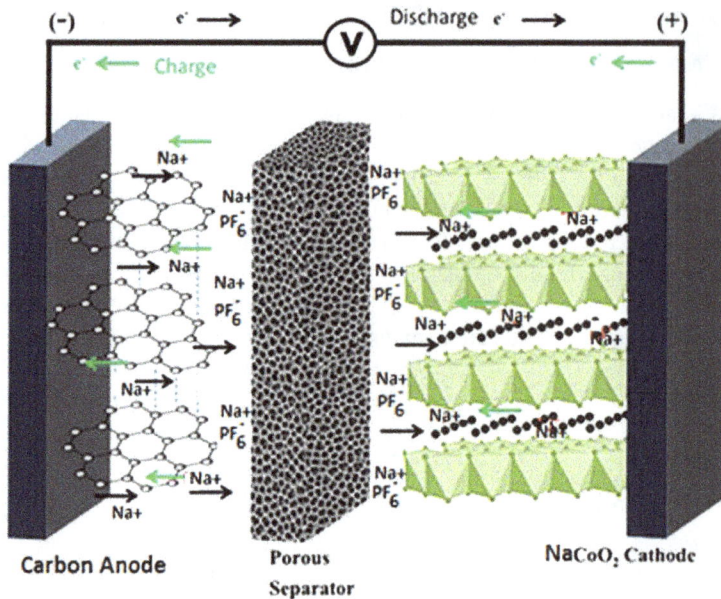

FIGURE 5.2 Schematics of charge–discharge mechanism of NIBs. Adapted with permission. Copyright (2020) American Chemical Society.

conversion mechanism, compared with the intercalation and alloying mechanism, the large volume expansion and heavy voltage hysteresis are persisted. These are the two major shortcomings of the conversion mechanism [20,21]. The schematics of the charge–discharge mechanism of NIBs are shown in **Figure 5.2**.

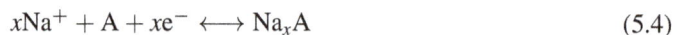

$$x\mathrm{Na}^+ + A + xe^- \longleftrightarrow \mathrm{Na}_x A \qquad (5.4)$$

5.4 NWS IN MIBS

5.4.1 NANOWIRES IN LITHIUM-ION BATTERIES (LIBS)

The volume expansion, pulverization, and shifting effect are the major serious issues of MIBs. To avoid such problems, recently different strategies have been introduced in the literature. The application of different carbon composites with ternary metal sulfide and metal oxide is one of the important remedies. The nanostructure materials have a good aspect ratio, high mechanical strength, high surface area, large electrolyte electrode interface area, etc., such outstanding characteristics of nanostructured material reduce the charging-discharging time, provide the excellent electron transmission path decrease the electrode-electrolyte pathway to enhance the electrochemical activity of metal ion in MIBs [2,23,24]. To enhance the charge capacity, cycle life, and enhanced electrochemical characteristics of MIBs the different carbon structures such as nanostructure, nanorods, nanotubes, nanowire, nanobelt, and nanofiber are highly appreciated [8,17,18].

However, the synthesis of different carbon structures required complex reactions, multi-step processes, and high temperatures. Also, during synthesis, it is a very difficult task to control the carbon structure and maintain structural uniformity. Also, the low electrical conductivity and huge volume expansion of metal sulfides hinder its practical applications. Recently, the highly conductive and moderate ion size metal insertion in metal sulfide and metal oxide is one of the important remedies to these problems in MIBs. The application of metal-inserted metal sulfide nanowire electrodes is one of the important remedies to reduce cycle loss and improve the electrochemical

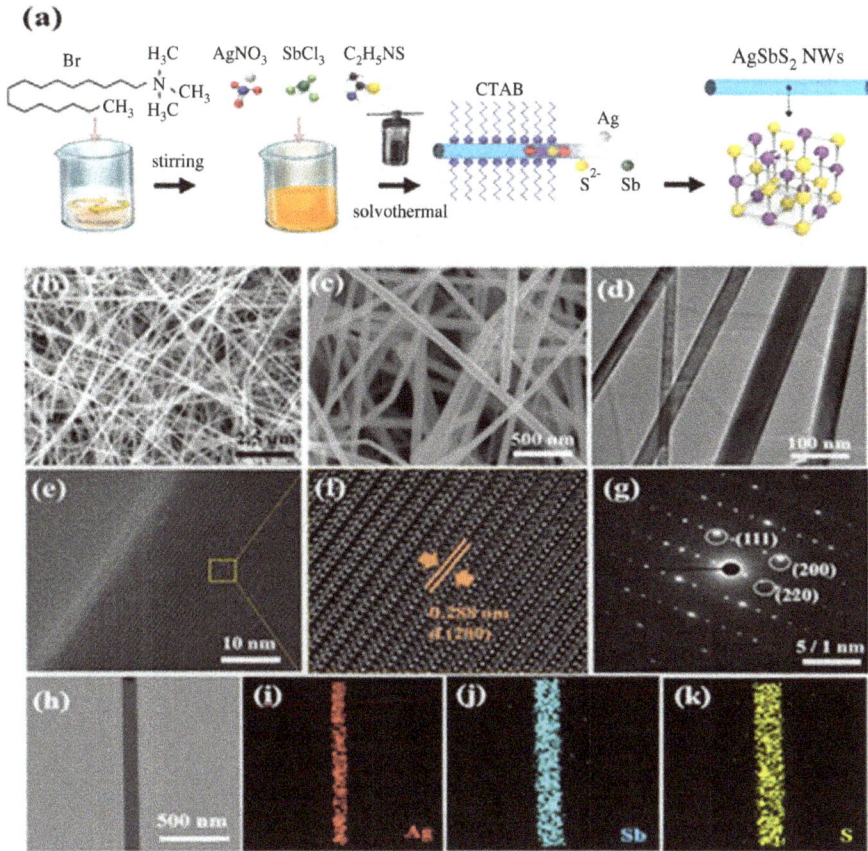

FIGURE 5.3 (a) Schematic demonstration of synthesis AgSbS$_2$ nanowires via hydrothermal method. (b, c) SEM images scanning electron microscope images (SEM), (d) Transmission electron microscope images, (e, f) High-resolution transmission electron microscope images, images, and (g) selected area diffraction pattern of AgSbS$_2$ nanowires. (h-k) energy dispersive X-ray spectroscopy mapping images of Ag, S, and Sb in AgSbS$_2$ nanowires. Adapted with permission. Copyright (2022) Elsevier.

behavior of metal sulfide and metal oxide-based MIBs [4,25,26]. To demonstrate this fact and for the enhancement of volumetric capacitance and cycle life of LIBs, Ho *et al.* [26] reported the insertion of silver-doped SbS$_2$ nanowires via the hydrothermal method as the anode material for LIBs. **Figure 5.3a** demonstrates the schematic of the synthesis AgSbS$_2$ nanowires via the hydrothermal synthesis route. A hydrothermal method is one of the simple and low-cost synthesis methods for uniform and hierarchical nanostructured materials [12,27,28]. Further, **Figures 5.3b–c** show the scanning electron microscope (SEM) images, demonstrating the morphology and average diameter (10–40 nm) of AgSbS$_2$ nanowires. **Figure 5.3d** shows transmission electron microscopy (TEM) images that depict the straight morphology of AgSbS$_2$ nanowires. The high-resolution transmission electron microscope (HRTEM) images (**Figures 5.3e–f**), and the selected area diffraction pattern (SAED) (**Figure 5.3g**) of AgSbS$_2$ nanowires demonstrated the single crystal of AgSbS$_2$ nanowires. Moreover, the uniform distribution of Ag, Sb, and S elements are conformed via energy dispersive X-ray spectroscopy mapping images and demonstrated in **Figures 5.3h–k**. In this report, the presence of Ag metal in SbS$_2$, enhances the electrical conductivity of the host material, reduced the volume change during cycling, constrains the shuttling effect of sulfur, and enhanced the lithium-ion absorption to improve the electrochemical performance of metal sulfide in LIBs. Using the LiNi$_5$Co$_3$Mn$_2$ cathode, AgSbS$_2$ anode and electrolyte 1 M LiPF$_6$, the full cell of LIBs demonstrated a high discharge capacity of 904.5 mAh/g with an initial Coulombic efficiency of 62.3%. More interestingly, due to reaching a Coulombic efficiency of 99% per cycles and capacity retention of 90.6% over 7000 charge discharge cycle at 2A/g makes the $AgSbS_2$ anode material a super stable material for LIBs. Similarly, Lan *et al.* synthesized the silver-doped one-dimensional (1D) attapulgite (hydrated magnesium-aluminum rich silicate mineral) via in-situ reduction of silver nitrate onto the attapulgite as anode material LIBs The discharge specific capacity for silver-doped attapulgite anode-based LIBs is reported to be 133.0 mAh/g at a current density of 0.1 A/g after 50 cycles [29].

Guo *et al.* [30] reported the synthesis of ternary Fe_7S_8/SiO_x/nitrogen-doped carbon matrix by hydrothermal method and subsequent sulfur. The matrix has been utilized as an anode for LIBs. The presence of the reported matrix SiO_x/nitrogen-doped carbon in Fe_7S_8 enhances the electrical conductivity and provides improved electrical performance, cycle life, and reverse capacity of LIBs fabricated using Fe_7S_8/SiO_x/nitrogen-doped carbon as the anode with a charge capacity of 1060.2 mAh/g at 200 cycles along with an excellent cyclic performance of 415.8 mAh/g at the 1000th cycle at 5 A/g. Li *et al.* [31] reported the antimony sulfide (Sb_2S_3) nanowires on reduced graphene oxide composite synthesized by the self-assembly method. The as-reported three-dimensional (3D) architecture demonstrated an excellent electrochemical behavior as an anode in LIBs and effectively reduced volume change and improved the conductivity of Sb_2S_3. The morphology and structure of reduced graphene oxide (rGO), antimony sulfide (Sb_2S_3) nanowires and antimony sulfide (Sb_2S_3) nanowires reduced graphene oxide (rGO) composite are shown in **Figure 5.4a–f**. Further, the

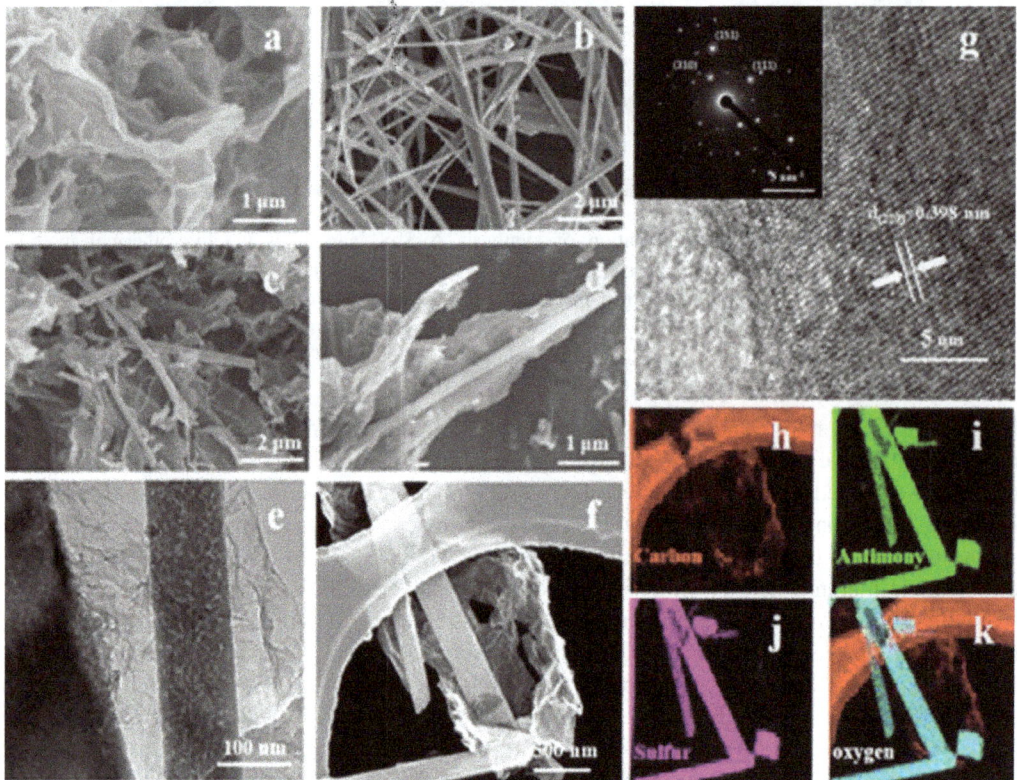

FIGURE 5.4 (a) Scanning electron microscopy (SEM) images of reduced graphene oxide (rGO), (b) SEM images of antimony sulfide (Sb_2S_3) nanowires, (c) low magnification, (d) high magnification SEM images of antimony sulfide (Sb_2S_3) nanowires, (e, f), high-resolution transmission electron microscopy (HRTEM) images, (g) corresponding SAED pattern (inset) of antimony sulfide (Sb_2S_3) nanowires reduced graphene oxide (rGO) composite, and (h), (i), (j), (k) elemental mapping of C, Sb, S, and O, respectively. Adapted with permission. Copyright (2021) Elsevier.

elemental mapping of C, Sb, S, and O is confirmed via EDS elemental mapping (**Figure 5.4h–k**). The 3D architecture as the reported composite shows the reversible capacity of 505 mAh/g at 3 A/g and long cycle stability 697 mAh/g after 100 cycles at 100 mA/g. The electrochemical performance of the antimony sulfide (Sb_2S_3) nanowire reduced graphene oxide (rGO) composite is demonstrated in **Figure 5.5**. The antimony sulfide (Sb_2S_3) nanowire reduced graphene oxide (rGO) shows superior electrochemical performance over reduced graphene oxide (rGO) and antimony sulfide (Sb_2S_3)

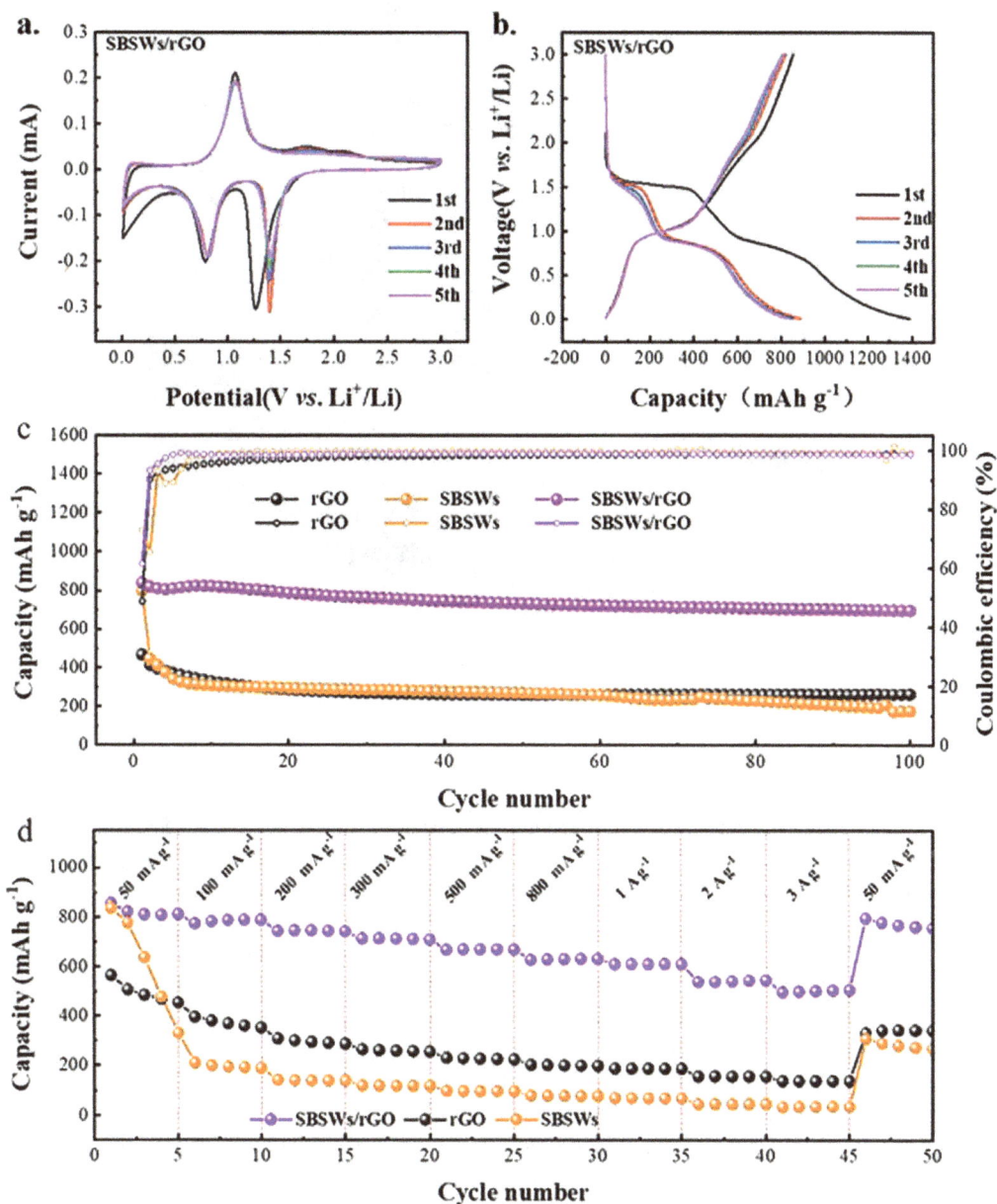

FIGURE 5.5 Cyclic voltammetry curves (0.01–3.00 V at 0.1 mV/s) (a), voltage profiles of antimony sulfide (Sb_2S_3) nanowire reduced graphene oxide (rGO) composite at 50 mA/g (b), cycle performances at 100 mA/g (c) and rate capability (d) for nanowires reduced graphene oxide (rGO) antimony sulfide (Sb_2S_3) nanowires and antimony sulfide (Sb_2S_3) nanowire reduced graphene oxide (rGO) composite electrodes. Adapted with permission. Copyright (2021) Elsevier.

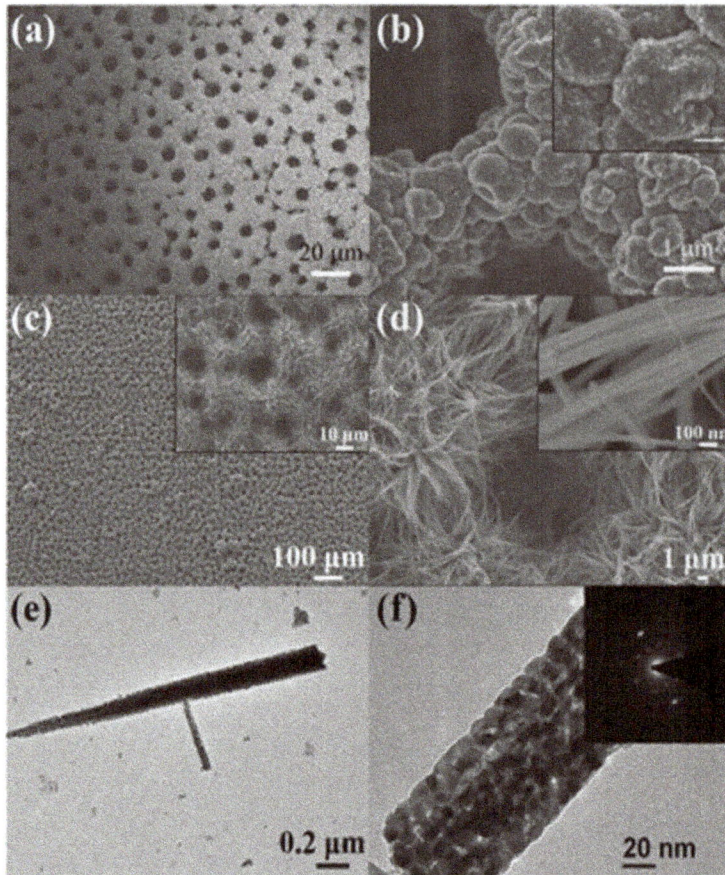

FIGURE 5.6 (a, b) Scanning electron microscopy images with low and high magnification of the Ni-Co$_3$O$_4$ dense film; (c, d) SEM images with low and high magnification of the Ni-Co$_3$O$_4$ nanowires; (e, f) transmission electron microscopy images with low and high magnification of the Ni-Co$_3$O$_4$ nanowires (selected area diffraction pattern (SAD) shown in inset). Adapted with permission. Copyright (2016) Elsevier.

nanowires. Moreover, the presence of reduced graphene oxide in antimony sulfide (Sb$_2$S$_3$) nanowires reduces the Li$^+$ diffusion path, can provide the space for shortening the Li$^+$ diffusion path, reduce the volume change, improve the electrical conductivity of the composite, and enhance the electron transfer between electrode-electrolyte interface and improve the rate performance.

Xiong *etal.* [32] prepared the 3D Ni-Co$_3$O$_4$ with nanowire porous branches and interconnected pore material by electrodeposition followed by the hydrothermal method. **Figure 5.6** shows the porous branches and interconnected pores of Ni-Co$_3$O$_4$ dense film Ni-Co$_3$O$_4$ nanowires. Compared with the dense Co$_3$O$_4$ film, the reported nanowire shows excellent electrochemical characterization. The Ni-Co$_3$O$_4$ with nanowire electrode demonstrates higher discharge capacities and high cycling stability of 714 mAh/g at 0.5 A/g after 100 cycles.

Usually, the binder used for the fabrication of electrodes introduces its resistivity into the electrode, which further influences the electrochemical activities of the anode in LIBs. Therefore, the fabrication of binder-free electrodes gains enormous attention. Many advanced strategies are used for the fabrication of thin films of active material electrodes. In this context, Wu *et al.* [33] synthesized the porous nanowire of porous NiO nanowires and coated it with Zr-based metal-organic gel (Zr-MOG), and used it as an electrode in LIBs. The LIBs with this electrode showed an excellent electrochemical performance with a specific capacity of 1816.3 mAh/g at a current density of 100 mA/g,

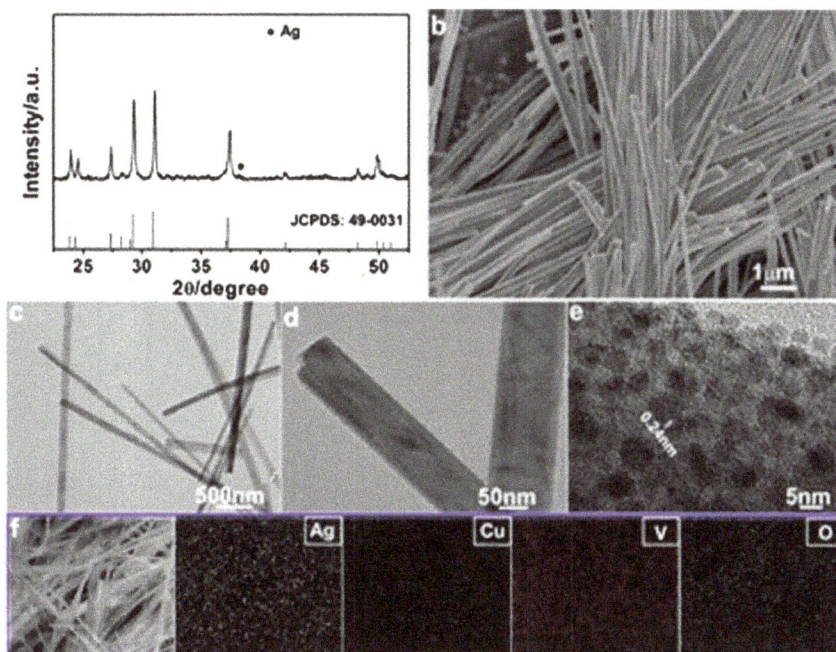

FIGURE 5.7 $Ag_2Cu(VO_3)_4/Ag$ nanowires (a) XRD pattern, (b) FESEM, (c–e) TEM and (f) SEM-EDX mapping images. Adapted with permission. Copyright (2021), Elsevier.

and capacitance retention of 1318.7 mAh/g over 150 charge-discharge cycles. Similarly, Zhang *et al.* synthesized quaternary transition-metal vanadium oxide $Ag_2Cu(VO_3)_4/Ag$ nanowires (**Figure 5.7**) by a simple hydrothermal method used as the cathode material for LIBs. Electrochemical measurement shows that the $Ag_2Cu(VO_3)_4/Ag$ nanowires are a new cathode material with good electrochemical performance and the discharge capacity can be stabilized at 132.4 mAh/g after 50 cycles at 200 mA/g. Moreover, the recent advancement in the nanostructured material used as an electrode for LIBs is illustrated in **Table 5.1**.

5.4.2 NANOWIRES IN SODIUM-ION BATTERIES (SIBs)

Sodium ion batteries (SIBs) are another promising energy storage system in comparison to LIBs. SIBs have low cost and equivalent electrochemical principle to LIBs. However, major issues like a large radius of Na^+ (1.02 Å) than Li^+ vs. Li^+ (0.76 Å), lower electrochemical activity than LIBs, and serious safety issues (fire or explosion due to the high flammability of organic solvents), etc., need more research and development in electrodes and electrolytes used in NIBs. Therefore, to make the availability of user-friendly NIBs to fulfill global energy needs, the development of NIBs is searching for suitable anode, cathode, electrode, binder, and separator materials to improve their electrochemical performance [48,49].

To tackle the issues and remove the existing problems of the NIBs, recently many recent reports suggested solutions and reported advanced strategies. The use of nanostructure single/binary/ternary metal sulfide, metal oxide, and their composites with different carbon materials as an electrode is one of the important remedies to the problem associated with NIBs. For example, Cao *et al.* [50] reported the single crystalline $Na_4Mn_9O_{18}$ nanowires synthesized by template-assisted sol-gel method. The reported electrode material shows high crystallinity, pure phase, and homogeneous size, using the $Na_4Mn_9O_{18}$ nanowires as the cathode, the as-fabricated NIBs show excellent

TABLE 5.1

Recent Advancement in the Nanostructured Material used as an Electrode for LIBs

Sr. No.	Materials	Method of synthesis	Operating voltage	Electrolyte	Specific capacitance	Retention of capacity	Reference
1.	Li_2MnO_3	Molten-Salt method.	2.0–4.8 V.	Dimethyl carbonate (DMC)	194.4 mAh/g	88.2% after 20 cycles at 0.1C	[34]
2.	Si/Cu	The pulsed electrical discharge method and chemical etching.	0.01–2 V	$LiPF_6$ 1 mol/L	1456 mAh/g	86.5% after 500 cycles.	[35]
3.	$CeVO_4$–V_2O_5	Cation-exchange and heat-treatment route	0.01–3 V	1 M $LiPF_6$	487 mAh/g		[36]
4.	Cu/Cu_2O@ Ppy	One-step hydrothermal method	0.01–3.0 V.	$LiPF_6$ in EC + DMC1 mol/L	787 mAh/g		[37]
5.	SiC	Electro-deoxidation	0–3.0 V	1 M $LiPF_6$	1000 mAh/g	99% over hundreds of cycles.	[38]
6.	GeSe	Rapid thermal processing method	0.01–3.0 V	1 M $LiPF_6$	~815.49mAh/g		[39]
7.	MnO/Sb@NC	Hydrothermal-mixing-Calcination	0.01–3.0 V	1.0 M $LiPF_6$	664 mAh/g	89.15% after 1100 cycles	[40]
8.	$ZnFe_2O_4$@ polypyrrolE	Electrospinning technique with gas-phase polymerization	0.01–3.0 V	1 M $LiPF_6$	~881 mAh/g		[41]
9.	GaZnON	Chemical vapor deposition	0.01–3.0 V	1 M $LiPF_6$	878.2 mAh/g		[42]
10.	AgNWs@Si@GO	Thermal reduction, Hummers' method	0–3.0 V	1 M $LiPF_6$	830 mAh/g	94% after 70 cycles	[43]
11.	$MnMoO_4$/C	Top-down tailoring strategy	0.01–3.0 V.	1M $LiPF_6$	994 mAh/g	90% after 100 cycles	[44]
12.	CeB_6	Low-temperature solution combustion method	0.01–3.0 V	1M $LiPF_6$	200 mAh/g	~99% . After the 80 or 100 cycles	[45]
13.	Fe_3O_4	A facile deposition immersion calcination process	0.05–3.0 V.	1M $LiPF_6$	100 mAh/g	93.6% after100 cycles	[46]
14.	TiO_2	Simple hydrothermal reaction	2.6–0.8 V a	1 mol L^{-1} $LiPF_6$ and ethylene carbonate (EC)/dimethyl carbonate	280 mAh/g	~98% after 40 cycles	[47]

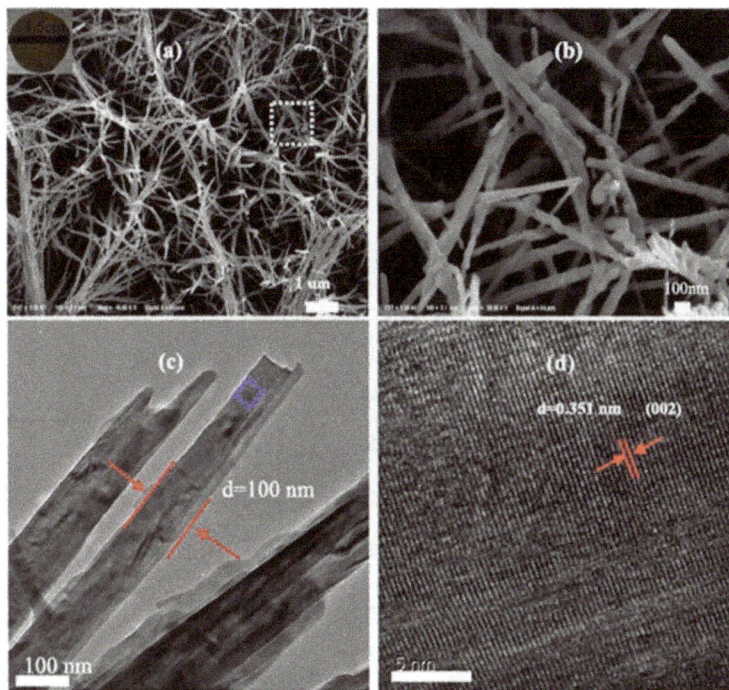

FIGURE 5.8 (a) SEM images of the samples on Ti foil annealed at 250 °C, inset in (a) is a digital photo of the entire sample; (b) high-resolution SEM image from the marked in area (a); (c) TEM image of several nanowires; (d) HR-TEM image of a $Na_5V_{12}O_{32}$ nanowire. Adapted with permission. Copyright (2019) Elsevier.

electrochemical performance, high capacity of 109 mAh/g at 0.5C, and capacity retention of 77% of this value even after 1000 charge/discharge at 0.5 C. Similarly, Ta et $al.$ reported the synthesis of $Na_4Mn_9O_{18}$ nanowires via simple and low-cost hydrothermal method [51]. Using $Na_4Mn_9O_{18}$ nanowires as a cathode in NIBs demonstrated excellent electrochemical performance, delivering a high capacity of 90 mAh/gand Coulombic efficiencies of greater than 91% at a rate of 0.2C during 30 charge/discharge. Interestingly, Cao et $al.$ [52] reported the first time use of the sodium vanadate ($Na_5V_{12}O_{32}$) array nanowire (**Figure 5.8**) on titanium foil synthesis via the hydrothermal synthesis route. The electrochemical performance of sodium vanadate nanowires is illustrated in **Figure 5.9**. The sodium vanadate nanowires demonstrated excellent electrochemical performance and resulted in 166 and 161 mAh/g charge/discharge capacities at a current density of 30 mA/g, respectively, and retained a capacity of 130 mAh/g after 40 cycles. Moreover, the recent advancement in the nanostructured material used as an electrode for NIBs is illustrated in **Table 5.2**.

5.5 CONCLUSIONS

In conclusion, the recent advancements in designing and fabrication of NW electrodes for MIBs have been summarized. The electrode materials for MIBs countenance various factors viz. material properties and volume, etc. Providentially, the NWs have given rise to MIBs with improved performance. The unique structural feature of the NWs offers their high performance. Compared to conventional electrodes, NW electrodes tender various advantages as electrodes for MIBs, which includes the following: (i) compliant volume changes during the metal-ion insertion/extraction; (ii) simple ion diffusion mechanism; (iii) effective charge transfer; (iv) tough structural permanence.

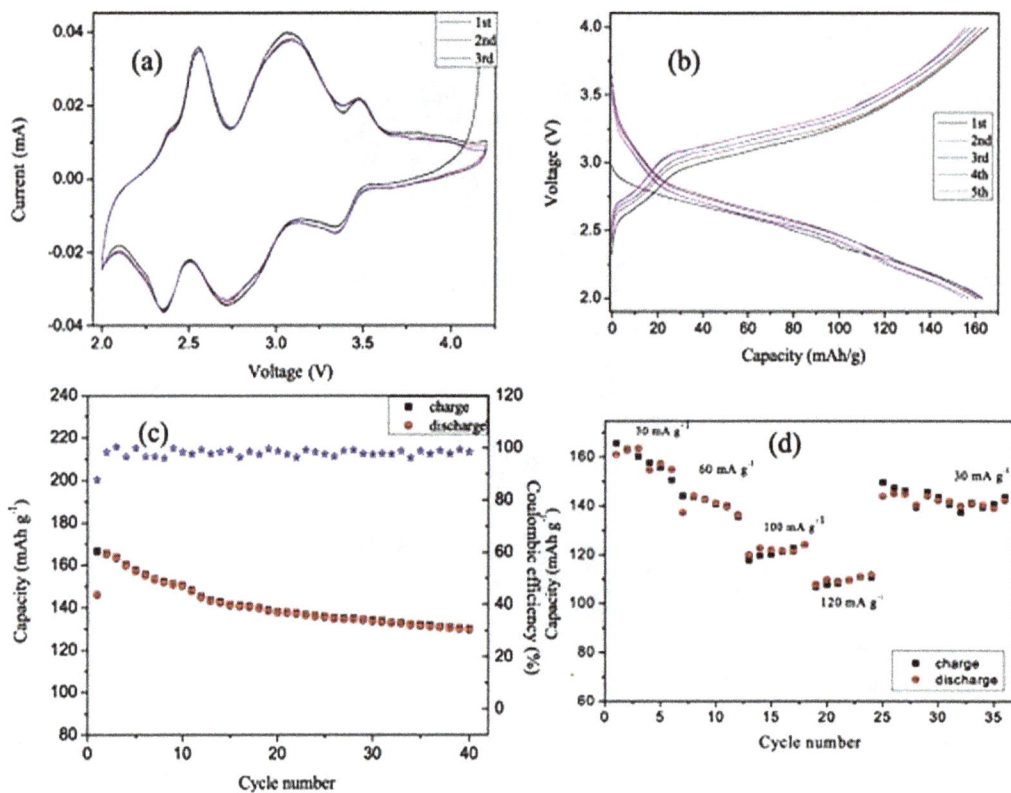

FIGURE 5.9 (a) Cyclic voltammograms of Na$_5$V$_{12}$O$_{32}$ nanowire arrays at a scan rate of 0.5 mV/s between 2.0 and 4.0 V; (b) the first five galvanostatic charge/discharge cycles at a current density of 30 mA/g; (c) capacity retention of the galvanostatic test at a current density of 30 mA/g; (d) charge-discharge capacities at various current densities of 30–120 mAh/g. Adapted with permission. Copyright (2019) Elsevier.

TABLE 5.2

Recent Advancements in the Nanostructured Material used as an Electrode for NIBs

Sr. No.	Material	Method of synthesis	Electrolyte	Operating voltage	Capacitance	Coulombic efficiencies	Reference
1.	TiO_{2-x}/Sb	Chemical de-alloying method	0.1 to 2.0 V	1 M $NaClO_4$	591.9 mAh/g	96.4% after 200 cycles	[53]
2.	MnO_2@rGO	Mno_2@rgo	0.01 to 3.0 V	1 M $NaClO_4$	250 mAh/g	81.7% over 400 cycles	[53]
3.	hydrogen titanate	Alkali hydrothermal Process, thermal reduction	0.0 to 2.5 V	1.0 M $NaClO_4$, PC/EC	238 mAh/g	80.6% after 4200 cycles	[54]
4.	$Na_3V_2(PO_4)_3/C.$	Agar-gel combined freeze-drying method	1.0 to 2.5 V	0.8 M $NaPF_6$/EMC + PC + FEC	113.4 mAh/g	88.2% after 1000 cycles	[55]
5.	VS_4@L-Ti_3C_2Tx	Hydrothermal	0.01–4.0 V	1 M $LiPF_6$	599 mAh/g	67% after 6700 cycles	[56]
6.	Fe_3O_4@C/rGO	In-situ plantation	1.0 M $LiPF_6$	0.01–3.0 V	429 mAh/g		[57]
7.	WS_2/hollow carbon composite	Hydrothermal	1 M $NaPF_6$	0.01–3 V.	575 mAh/g	76.6%after 80 cycles	[58]
8.	MWCNT@ polyimide	Thermal imidization	1.0 M Na_2SO_4	-1.0–0.0 V	209.3 mAh/g	77.8%. after 100 cycles	[59]
9.	MnO_2@rG	Hydrothermal method	1 M KPF_6	0.01–3 V	389 F/g	81.7% over 500 cycles	[60]
10.	VO_2 (A)/graphene nanostructure	Hydrothermal	1.0 M $NaClO_4$	1–4 V	—	115 mAh/g over 100 cycles at a current density of 100 mA/g.	[61]

REFERENCES

[1]   N. Singh, K. Banerjee, M. Gupta, Y.K. Bainsla, V.U. Pandit, P. Singh, S.J. Uke, A. Kumar, S.P. Mardikar, Y. Kumar, Concentration dependent electrochemical performance of aqueous choline chloride electrolyte, Mater. Today Proc. 53 (2022) 161–167.

[2]   X. Cao, X. Ren, L. Zou, M.H. Engelhard, W. Huang, H. Wang, B.E. Matthews, H. Lee, C. Niu, B.W. Arey, Monolithic solid–electrolyte interphases formed in fluorinated orthoformate-based electrolytes minimize Li depletion and pulverization, Nat. Energy. 4 (2019) 796–805.

[3]   J. Verma, D. Kumar, Metal-ion batteries for electric vehicles: current state of the technology, issues and future perspectives, Nanoscale Adv. 3 (2021) 3384–3394.

[4]   Z. Pang, B. Ding, J. Wang, Y. Wang, L. Xu, L. Zhou, X. Jiang, X. Yan, J.P. Hill, L. Yu, Metal-ion inserted vanadium oxide nanoribbons as high-performance cathodes for aqueous zinc-ion batteries, Chem. Eng. J. 446 (2022) 136861.

[5]   C. Shi, K.A. Owusu, X. Xu, T. Zhu, G. Zhang, W. Yang, L. Mai, 1D carbon-based nanocomposites for electrochemical energy storage, Small. 15 (2019) 1902348.

[6]   M. Mallik, M. Saha, Carbon-based nanocomposites: Processing, electronic properties and applications, in: Carbon Nanomater. Electron. Devices Appl., Springer, 2021: pp. 97–122.

[7]   D. Majumdar, M. Mandal, S.K. Bhattacharya, V_2O_5 and its carbon-based nanocomposites for supercapacitor applications, ChemElectroChem. 6 (2019) 1623–1648.

[8]   R. Febrian, N.L.W. Septiani, M. Iqbal, B. Yuliarto, Recent advances of carbon-based nanocomposites as the anode materials for lithium-ion batteries: Synthesis and performance, J. Electrochem. Soc. 168 (2021) 110520.

[9]   Y. Chen, Y. Kang, Y. Zhao, L. Wang, J. Liu, Y. Li, Z. Liang, X. He, X. Li, N. Tavajohi, A review of lithium-ion battery safety concerns: The issues, strategies, and testing standards, J. Energy Chem. 59 (2021) 83–99.

[10] N. Singh, K. Banerjee, Y.K. Bainsla, M.K. Singh, M. Gupta, A. Kumar, P. Singh, S.J. Uke, S.P. Mardikar, V.U. Pandit, Preparation of electrochemically stable choline chloride-sugar based sustainable electrolytes and study of effect of water on their electrochemical behaviour, Mater. Today Proc. 53 (2022) 179–184.

[11] S.J. Uke, S.P. Mardikar, A. Kumar, Y. Kumar, M. Gupta, Y. Kumar, A review of π-conjugated polymer-based nanocomposites for metal-ion batteries and supercapacitors, R. Soc. Open Sci. 8 (2021) 210567.

[12] S.J. Uke, D.S. Thakre, V.P. Akhare, G.N. Chaudhari, S.P. Meshram, PEG-600 assisted hydrothermal synthesis of $NiCo_2O_4$ as an electrode material for supercapacitor application, Proc. Int. Meet. Energy Storage Devices Ind.-Acad. Conclave, 2018.

[13] R. Sanjinés, M.D. Abad, C. Vâju, R. Smajda, M. Mionić, A. Magrez, Electrical properties and applications of carbon based nanocomposite materials: An overview, Surf. Coat. Technol. 206 (2011) 727–733.

[14] Y. Zhao, L.P. Wang, M.T. Sougrati, Z. Feng, Y. Leconte, A. Fisher, M. Srinivasan, Z. Xu, A review on design strategies for carbon based metal oxides and sulfides nanocomposites for high performance Li and Na ion battery anodes, Adv. Energy Mater. 7 (2017) 1601424.

[15] M. Li, J. Lu, K. Amine, Nanotechnology for sulfur cathodes, ACS Nano. 15 (2021) 8087–8094.

[16] P. Chen, C. Zhang, B. Jie, H. Zhang, K. Zhang, Y. Song, Large-scale preparation of cobalt niobate/ reduced graphene oxide composite materials for high-performance lithium-ion battery anodes, J. Alloys Compd. 908 (2022) 164542.

[17] K. Yu, X. Pan, G. Zhang, X. Liao, X. Zhou, M. Yan, L. Xu, L. Mai, Nanowires in energy storage devices: Structures, synthesis, and applications, Adv. Energy Mater. 8 (2018) 1802369.

[18] G. Zhou, L. Xu, G. Hu, L. Mai, Y. Cui, Nanowires for electrochemical energy storage, Chem. Rev. 119 (2019) 11042–11109.

[19] M. Al-Gabalawy, N.S. Hosny, S.A. Hussien. Lithium-ion battery modeling including degradation based on single-particle approximations, Batteries 2020, 6 (3), 37.

[20] Z. Wang, X. Yan, F. Wang, T. Xiong, M.-S. Balogun, H. Zhou, J. Deng, Reduced graphene oxide thin layer induced lattice distortion in high crystalline MnO_2 nanowires for high-performance sodium- and potassium-ion batteries and capacitors, Carbon. 174 (2021) 556–566.

[21] L. Gao, D. Lu, Y. Yang, R. Guan, D. Zhang, C. Sun, S. Liu, X. Bian, Amorphous TiO_{2-x} modified Sb nanowires as a high-performance sodium-ion battery anode, J. Non-Cryst. Solids. 581 (2022) 121396.

[22] K.M. Abraham, How comparable are sodium-ion batteries to lithium-ion counterparts? ACS Energy Lett. 5 (2020) 3544–3547.

[23] M. Ma, S. Zhang, L. Wang, Y. Yao, R. Shao, L. Shen, L. Yu, J. Dai, Y. Jiang, X. Cheng, Harnessing the volume expansion of MoS_3 anode by structure engineering to achieve high performance beyond lithium-based rechargeable batteries, Adv. Mater. 33 (2021) 2106232.

[24] W. An, B. Gao, S. Mei, B. Xiang, J. Fu, L. Wang, Q. Zhang, P.K. Chu, K. Huo, Scalable synthesis of ant-nest-like bulk porous silicon for high-performance lithium-ion battery anodes, Nat. Commun. 10 (2019) 1–11.

[25] H.-Y. Wang, H.-Y. Chen, Y.-Y. Hsu, U. Stimming, H.M. Chen, B. Liu, Modulation of crystal surface and lattice by doping: achieving ultrafast metal-ion insertion in anatase TiO_2, ACS Appl. Mater. Interfaces. 8 (2016) 29186–29193.

[26] S.-F. Ho, Y.-C. Yang, H.-Y. Tuan, Silver boosts ultra-long cycle life for metal sulfide lithium-ion battery anodes: Taking $AgSbS_2$ nanowires as an example, J. Colloid Interface Sci. 621 (2022) 416–430.

[27] S.J. Uke, G.N. Chaudhari, A.B. Bodade, S.P. Mardikar, Morphology dependant electrochemical performance of hydrothermally synthesized NiCo2O4 nanomorphs, Mater. Sci. Energy Technol. 3 (2020) 289–298.

[28] S.J. Uke, G.N. Chaudhari, Y. Kumar, S.P. Mardikar, Tri-ethanolamine-ethoxylate assisted hydrothermal synthesis of nanostructured $MnCo_2O_4$ with superior electrochemical performance for high energy density supercapacitor application, Mater. Today Proc. 43 (2021) 2792–2799.

[29] Y. Lan, D. Chen, Fabrication of silver doped attapulgite aerogels as anode material for lithium ion batteries, J. Mater. Sci. Mater. Electron. 29 (2018) 19873–19879.

[30] X. Guo, S. Wang, B. Yang, Y. Xu, Y. Liu, H. Pang, Porous pyrrhotite Fe7S8 nanowire/SiO$_2$/nitrogen-doped carbon matrix for high-performance Li-ion-battery anodes, J. Colloid Interface Sci. 561 (2020) 801–807.

[31] Z. Li, H. Du, J. Lu, L. Wu, L. He, H. Liu, Self-assembly of antimony sulfide nanowires on three-dimensional reduced GO with superior electrochemical lithium storage performances, Chem. Phys. Lett. 771 (2021) 138529.

[32] Q.Q. Xiong, H.Y. Qin, H.Z. Chi, Z.G. Ji, Synthesis of porous nickel networks supported metal oxide nanowire arrays as binder-free anode for lithium-ion batteries, J. Alloys Compd. 685 (2016) 15–21.

[33] D. Wu, L. Zhang, J. Zhang, Z. Zhang, F. Liang, L. Jiang, B. Tang, Y. Rui, F. Liu, Novel self-supporting multilevel-3D porous NiO nanowires with metal-organic gel coating via "like dissolves like" to trigger high-performance binder-free lithium-ion batteries, Microporous Mesoporous Mater. 328 (2021) 111483.

[34] Y. Sun, M. Guo, S. Shu, D. Ding, C. Wang, Y. Zhang, J. Yan, Preparation of Li2MnO$_3$ nanowires with structural defects as high rate and high capacity cathodes for lithium-ion batteries, Appl. Surf. Sci. 585 (2022) 152605.

[35] J. Hong, K. Cheng, G. Xu, M. Stapelberg, Y. Kuai, P. Sun, S. Qu, Z. Zhang, Q. Geng, Z. Wu, M. Zhu, P.V. Braun, Novel silicon/copper nanowires as high-performance anodes for lithium ion batteries, J. Alloys Compd. 875 (2021) 159927.

[36] X. Xu, S. Chang, T. Zeng, Y. Luo, D. Fang, M. Xie, J. Yi, Synthesis of $CeVO_4$-V_2O_5 nanowires by cation-exchange method for high-performance lithium-ion battery electrode, J. Alloys Compd. 887 (2021) 161237.

[37] Y. Wang, L. Cao, J. Li, L. Kou, J. Huang, Y. Feng, S. Chen, Cu/Cu2O@Ppy nanowires as a long-life and high-capacity anode for lithium ion battery, Chem. Eng. J. 391 (2020) 123597.

[38] D. Sri Maha Vishnu, J. Sure, H.-K. Kim, R.V. Kumar, C. Schwandt, Solid state electrochemically synthesised β-SiC nanowires as the anode material in lithium ion batteries, Energy Storage Mater. 26 (2020) 234–241.

[39] K. Wang, M. Liu, D. Huang, L. Li, K. Feng, L. Zhao, J. Li, F. Jiang, Rapid thermal deposited GeSe nanowires as a promising anode material for lithium-ion and sodium-ion batteries, J. Colloid Interface Sci. 571 (2020) 387–397.

[40] B. Wang, Y. Xia, Z. Deng, Y. Zhang, H. Wu, Three-dimensional cross-linked MnO/Sb hybrid nanowires co-embedded nitrogen-doped carbon tubes as high-performance anode materials for lithium-ion batteries, J. Alloys Compd. 835 (2020) 155239.

[41] L. Hou, R. Bao, D. kionga Denis, X. Sun, J. Zhang, F. uz Zaman, C. Yuan, Synthesis of ultralong $ZnFe_2O_4$@polypyrrole nanowires with enhanced electrochemical Li-storage behaviors for lithium-ion batteries, Electrochimica Acta. 306 (2019) 198–208.

[42] Y. Han, C. Sun, K. Gao, S. Ding, Z. Miao, J. Zhao, Z. Yang, P. Wu, J. Huang, Z. Li, A. Meng, L. Zhang, P. Chen, Heterovalent oxynitride GaZnON nanowire as novel flexible anode for lithium-ion storage, Electrochimica Acta. 408 (2022) 139931.

[43] J. Wei, C. Qin, X. Pang, H. Zhang, X. Li, One-dimensional core-shell composite of AgNWs@Si@GO for high-specific capacity and high-safety anode materials of lithium-ion batteries, Ceram. Int. 47 (2021) 4937–4943.

[44] L. Zhang, J. Bai, F. Gao, S. Li, Y. Liu, S. Guo, Tailoring of hierarchical $MnMoO_4$/C nanocauliflowers for high-performance lithium/sodium ion half/full batteries, J. Alloys Compd. 906 (2022) 164394.

[45] Z. Wang, W. Han, Q. Kuang, Q. Fan, Y. Zhao, Low-temperature synthesis of CeB6 nanowires and nanoparticles as feasible lithium-ion anode materials, Adv. Powder Technol. 31 (2020) 595–603.

[46] K. Cheng, F. Yang, K. Ye, Y. Zhang, X. Jiang, J. Yin, G. Wang, D. Cao, Highly porous Fe_3O_4–Fe nanowires grown on C/TiC nanofiber arrays as the high performance anode of lithium-ion batteries, J. Power Sources. 258 (2014) 260–265.

[47] Y. Wang, M. Wu, W.F. Zhang, Preparation and electrochemical characterization of TiO_2 nanowires as an electrode material for lithium-ion batteries, Electrochimica Acta. 53 (2008) 7863–7868.

[48] X. Chen, Y. Zhang, The main problems and solutions in practical application of anode materials for sodium ion batteries and the latest research progress, Int. J. Energy Res. 45 (2021) 9753–9779.

[49] M. Chen, Q. Liu, S.-W. Wang, E. Wang, X. Guo, S.-L. Chou, High-abundance and low-cost metal-based cathode materials for sodium-ion batteries: Problems, progress, and key technologies, Adv. Energy Mater. 9 (2019) 1803609.

[50] Y. Cao, L. Xiao, W. Wang, D. Choi, Z. Nie, J. Yu, L.V. Saraf, Z. Yang, J. Liu, Reversible sodium ion insertion in single crystalline manganese oxide nanowires with long cycle life, Adv. Mater. 23 (2011) 3155–3160.

[51] A.T. Ta, T.T.O. Nguyen, H.C. Le, D.T. Le, T.C. Dang, M.T. Man, S.H. Nguyen, D.L. Pham, Hydrothermal synthesis of $Na_4Mn_9O_{18}$ nanowires for sodium ion batteries, Ceram. Int. 45 (2019) 17023–17028.

[52] Y. Cao, J. Wang, X. Chen, B. Shi, T. Chen, D. Fang, Z. Luo, Nanostructured sodium vanadate arrays as an advanced cathode material in high-performance sodium-ion batteries, Mater. Lett. 237 (2019) 122–125.

[53] Z. Wang, X. Yan, F. Wang, T. Xiong, M.-S. Balogun, H. Zhou, J. Deng, Reduced graphene oxide thin layer induced lattice distortion in high crystalline MnO_2 nanowires for high-performance sodium- and potassium-ion batteries and capacitors, Carbon. 174 (2021) 556–566.

[54] M.-Y. Sun, F.-D. Yu, Y. Xia, L. Deng, Y.-S. Jiang, L.-F. Que, L. Zhao, Z.-B. Wang, Trigger Na+-solvent co-intercalation to achieve high-performance sodium-ion batteries at subzero temperature, Chem. Eng. J. 430 (2022) 132750.

[55] G. Cui, H. Wang, F. Yu, H. Che, X. Liao, L. Li, W. Yang, Z. Ma, Scalable synthesis of $Na_3V_2(PO_4)_3$/C with high safety and ultrahigh-rate performance for sodium-ion batteries, Chin. J. Chem. Eng. (2021).

[56] H. Wang, P. Wang, W. Gan, L. Ci, D. Li, Q. Yuan, VS_4 nanoarrays pillared $Ti_3C_2T_x$ with enlarged interlayer spacing as anode for advanced lithium/sodium ion battery and hybrid capacitor, J. Power Sources. 534 (2022) 231412.

[57] J.-L. Xu, X. Zhang, Y.-X. Miao, M.-X. Wen, W.-J. Yan, P. Lu, Z.-R. Wang, Q. Sun, In-situ plantation of Fe_3O_4@C nanoparticles on reduced graphene oxide nanosheet as high-performance anode for lithium/sodium-ion batteries, Appl. Surf. Sci. 546 (2021) 149163.

[58] W. Liu, M. Wei, L. Ji, Y. Zhang, Y. Song, J. Liao, L. Zhang, Hollow carbon sphere based WS_2 anode for high performance lithium and sodium ion batteries, Chem. Phys. Lett. 741 (2020) 137061.

[59] B. Cho, H. Lim, H.-N. Lee, Y.M. Park, H. Kim, H.-J. Kim, High-capacity and cycling-stable polypyrrole-coated MWCNT@polyimide core-shell nanowire anode for aqueous rechargeable sodium-ion battery, Surf. Coat. Technol. 407 (2021) 126797.

[60] Z. Wang, X. Yan, F. Wang, T. Xiong, M.-S. Balogun, H. Zhou, J. Deng, Reduced graphene oxide thin layer induced lattice distortion in high crystalline MnO_2 nanowires for high-performance sodium- and potassium-ion batteries and capacitors, Carbon. 174 (2021) 556–566.

[61] X. Hu, Z. Zhao, L. Wang, J. Li, C. Wang, Y. Zhao, H. Jin, VO_2 (A)/graphene nanostructure: Stand up to Na ion intercalation/deintercalation for enhanced electrochemical performance as a Na-ion battery cathode, Electrochimica Acta. 293 (2019) 97–104.

[62] S. Zhanga, L. Ci, W. Mu, Hydrothermal synthesis of $Ag_2Cu(VO_3)_4$/Ag nanowires: A new cathode material for lithium-ion batteries, Scripta Materialia, 210, 2022, 114431

6 Nanowires for Metal-Air Batteries

Shiva Bhardwaj and Rishabh Srivastava
Department of Physics and National Institute for Materials Advancement, Pittsburg State University, Pittsburg, KS, USA

Felipe de Souza
National Institute for Materials Advancement, Pittsburg State University, Pittsburg, KS, USA

Ram K. Gupta
Department of Chemistry and National Institute for Materials Advancement, Pittsburg State University, Pittsburg, KS, USA

CONTENTS

6.1 INTRODUCTION

Energy is the core part of modern civilization, and fossil fuel is not enough to meet the requirement of energy for an advanced society. Continuous usage of fossil fuels mainly oil (gasoline) for various transportation devices and industrial application results in global warming due to the emission of carbon dioxide (CO_2) as the by-product. The growing demand for transportation and industries requires more energy, which tends researchers to look forward to the electrification of vehicles and increasing the usage of electricity in industrial applications. To run these transportation devices, a continuous supply of electricity is required, which can only be provided using energy storage devices (ESDs). There are various types of ESDs like batteries, supercapacitors, and fuel cells. Among them, fuel cells are the new as well as green ESDs that scientists are looking for and are currently in the developing phase, which may replace the use of fossil fuels for many applications. Supercapacitors are being used continuously in devices where power density is the significant factor and lacks the energy density on which scientists are still working. Besides all these devices, batteries are the oldest ESDs and are still in use due to their implacable amount of energy density. Various types of batteries like lead-acid, nickel-cadmium (Ni-Cd), nickel-metal hydride (Ni-MH), metal-ion (MIB), and metal-air battery (MAB) have been developed [1].

Lead-acid batteries have major applications in the automotive industry where a 12 V battery is used. It contains metallic lead, lead dioxide, lead sulfate, and sulfuric acid, which are categorized as hazardous materials according to international health and environmental studies [2]. However, other ESDs such as Ni-Cd batteries have taken a significant part in industries due to their higher

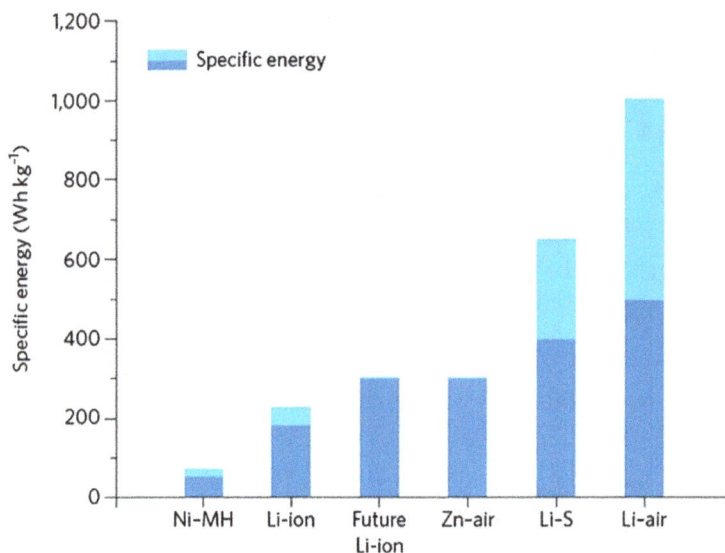

FIGURE 6.1 Specific energies for the different types of rechargeable batteries. Adapted with permission. Copyright (2016), Springer Nature.

energy density and longer life [3]. However, Ni-Cd batteries have several disadvantages such as the use of a strong alkaline electrolyte, which causes skin problems and toxicity of cadmium. Therefore, researchers transfer to Ni-MH batteries, which outperform the above two batteries in terms of performance with their specific energy of over 100 Wh/kg. Ni-MH batteries have several advantages like flexibility, high temperature operating range, and cost-effectiveness but capacity deterioration due to the alloy pulverization is a major drawback [4]. The next-generation battery is a MIB, which has specific energy more than double of Ni-MH batteries and their low-cost processing gradually replaces Ni-MH batteries. Researchers find the future of ESDs in MABs as they have the highest theoretical specific energy (Figure 6.1). In addition to high energy density, they have various advantages, such as high faradaic efficiency, fast charging, and an improved life cycle. MABs can meet the necessity of zapped transportation and fulfill the high energy and power density needs.

A MAB typically comprises an air cathode, a metal anode, and a separator loaded up with an electrolyte. The MAB works on metal oxidation where metal anode M gets oxidized to M^+ and oxygen reduction reaction (ORR). This M^+ comes via electrolyte and reacts with O_2 at the cathode to yield the product as metal oxides. MAB has advantages as well as some challenges affecting their practicality. The electrolyte reacts with the metal anode, which forms a passivation layer referred to as solid electrolyte interface (SEI), causing the loss in performance of the battery. The dendrite formation on the anode leads to the internal short-circuiting, and uncontrollable dissolution occurs. The electrolyte with desired properties is required, including low volatility, high stability, high oxygen solubility, non-toxicity, and a wide operational window. The cathode itself has various challenges like instability, large overpotential, low rate capability, and depends on the structure type. However, these limitations are to be resolved before the practicality of MAB. The historical development of MABs is given in Figure 6.2.

6.2   METAL-AIR BATTERIES

MABs have four major parts: an air electrode referred to as a cathode, a metal electrode (anode), an electrolyte, and the separator. The cathode is the major component of MAB and is used for air diffusion into the electrodes. Generally, the cathode materials must be porous and electrochemically active. A variety of materials has been used to improve the performance of cathodes like carbon-based materials (ketjen black, acetylene black, carbon nanowires), graphene, and nitrogen-doped carbon.

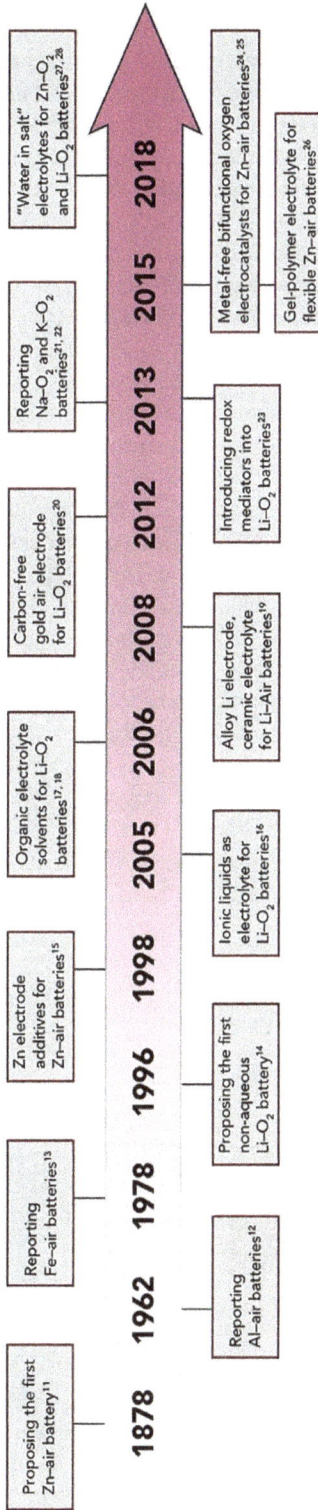

FIGURE 6.2 Progressive report for the MAB from 1878 to 2018. Adapted with permission. Copyright (2019), Elsevier.

These materials provide more reaction sites due to the large surface area and fortify gas diffusion and charge transfer, which results in high electrical conductivity [7]. On the other hand, the negative electrode (anode) uses a wide range of metallic materials like aluminum, zinc, lithium, magnesium, and potassium. However, applying additives (carbon-based materials) and surface coatings (nanoporous $CaCO_3$) on anodes can enhance electrolyte flux performance and uniformity, intensifying the potential gradient. The combination of cathode and anode requires a medium for the transportation of ions provided by the electrolyte. Depending on the type of MAB, there are various types of electrolytes used, such as aqueous electrolytes, like water, potassium hydroxide, non-aqueous electrolytes, like lithium aluminum chloride ($LiAlCl_4$), tetraethyl ammonium tetrafluoroborate, ionic liquid electrolytes, like lithium-tetra glyme-bis-(trifluoromethanesulfonyl)amide) [Li(G4)(TFSA)], solid-state electrolytes (Ge film-coated lithium aluminum germanium phosphate (LAGP)), gel-polymer electrolytes (sodium polyacrylate hydrogel), solid polymer electrolytes, like polyacrylonitrile (PAN) and polyethylene oxide (PEO), and ceramic electrolytes, like lithium superionic conductor (LISICON, perovskite, garnet, and sulfides) [8]. To prevent short-circuiting inside the batteries, a separator is used, which plays a vital role in a battery during electrochemical reactions. The separator acts as a semi-permeable membrane between the sections of the two electrodes and behaves as a partial barricade. As a result, a selective transfer of ions take place from both sides, maintaining the electrochemical configuration. Furthermore, the separator solves many other issues like stability and ionic conductivity. Generally, polymers are utilized as the separator inside the battery. These polymers are polyolefin-based materials such as polyethylene, and polypropylene, having a semi-crystalline morphology leading to their porous structure and allowing them to improve the ionic conductivity [9].

Various types of MABs like lithium-air (LAB), aluminum-air (AAB), zinc-air (ZAB), iron-air (FAB), sodium-air (NAB), and magnesium-air (MGAB) batteries have been developed [10,11]. The performance of MABs depends on the type of metal anode used, like LAB, NAB and KAB anodes are unstable in contact with water and cannot be used in an aqueous system. Diffusion of moisture and atmospheric carbon dioxide in LAB distorts its capability, cyclability, and efficiency in an aqueous electrolyte, leading to a drop in specific energy by ~ 30% through the discharge products [12]. Figure 6.3 compares various kinds of MAB.

To better understand MAB, it is necessary to comprehend its working principle. The working principle of MAB resembles an oxygen reduction reaction and oxygen evolution reaction (OER).

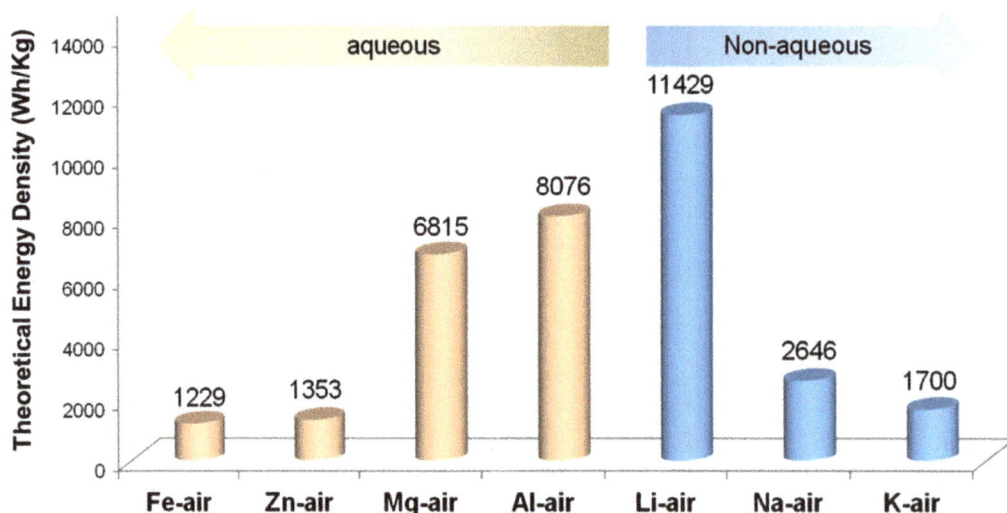

FIGURE 6.3   Comparison between the various theoretical energy density of metal-air batteries. Adapted with permission. Copyright (2017), American Chemical Society.

Anode: M ↔ M⁺ + e⁻
Cathode: xM+ O₂ +xe⁻ ↔ MₓO₂
Overall: xM + O₂ ↔ MₓO₂

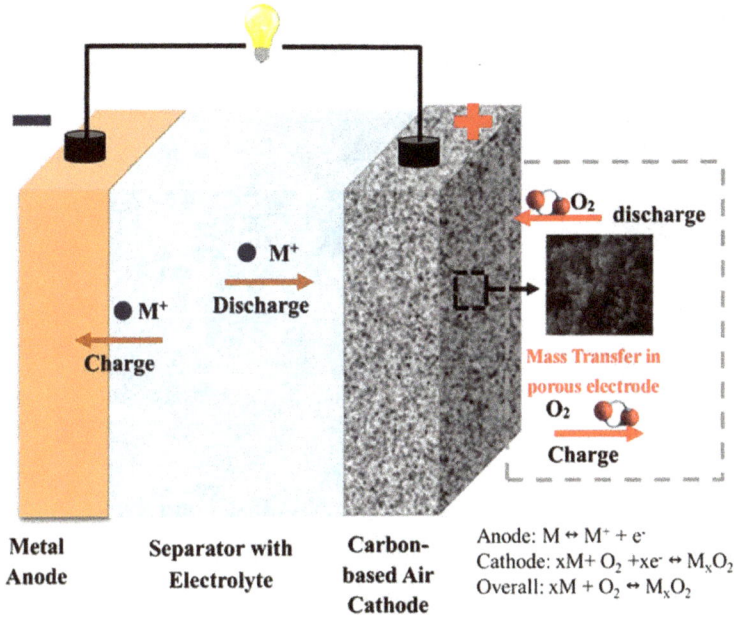

FIGURE 6.4 Schematic illustration of the structure and working mechanism of MAB. Adapted with permission. Copyright (2022), American Chemical Society.

The discharge phenomenon refers to the diffusion of atmospheric oxygen into the system, which occurs via a porous material as a gas diffusion layer in the cathode region. However, the amount of O_2 depends on the type of electrolyte, where O_2 transforms into O_2^- during its working in the non-aqueous electrolyte. At the same time, it turns into hydroxide (OH^-) in aqueous electrolytes. On the other hand, metal ions' aggregation occurs at anode electrodes. Therefore, metallic ions migrate from the negative electrode to the positive electrode, and subsequently, electrons are produced, and the metallic ions vanish into the electrolyte [14]. Electrons follow the external electrical network and generate electricity. Therefore, ORR administers the discharge operation. OER conducts the charging phenomenon and is the reversal of ORR [15]. Thus, electrical energy is stored as chemical energy. Figure 6.4 describes the working mechanism of the MAB. The general reaction formula is given below in equations (6.1–6.3):

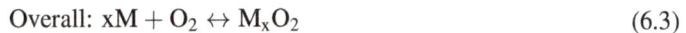

$$\text{At anode: } M \leftrightarrow M^+ + e^- \tag{6.1}$$

$$\text{At cathode: } xM + O_2 + xe^- \leftrightarrow 4OH^- \tag{6.2}$$

$$\text{Overall: } xM + O_2 \leftrightarrow M_xO_2 \tag{6.3}$$

For better understanding, LAB working is discussed in detail. LAB obeys ORR and OER, during which lithium-oxide (Li_2O) (minor occurrence in electrolytes) or lithium dioxide (Li_2O_2) (main product) is formed as the discharge product of lithium and oxygen electrochemically, which are non-soluble and causes degradation of cell in the aprotic electrolyte. Figure 6.5a shows the discharging process in LAB [17]. The chemical reactions are as follows:

During discharging

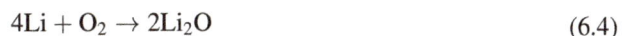

$$2Li + O_2 \rightarrow Li_2O_2 \tag{6.3}$$

$$4Li + O_2 \rightarrow 2Li_2O \tag{6.4}$$

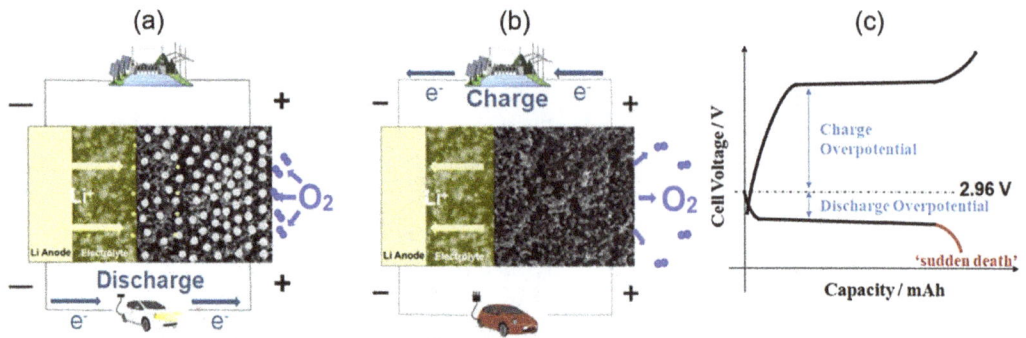

FIGURE 6.5 (a) Schematic depiction of discharge process of LAB, (b) charging process of LAB, (c) potential of O_2/Li_2O_2 v/s Li^+ between 2.96 V couple is marked as shown. Adapted with permission. Copyright (2020), American Chemical Society.

The reversible voltage can be derived using Gibbs' free energy, which is given below in equation (6.5).

$$E^o = -\Delta G/nF \qquad (6.5)$$

where n is the number e^- transferred in one molar reaction. For example, the E^o for equations (6.3) and (6.4) is 2.96 V and 2.91 V, respectively, with an outer circuit voltage (OCV) ~ 3.7 V [18]. The half-cell reaction at both the electrodes is given below:

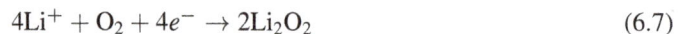

$$\text{At anode: } Li \rightarrow Li^+ + e^- \qquad (6.5)$$

$$\text{At cathode: } 2Li^+ + O_2 + 2e^- \rightarrow Li_2O_2 \qquad (6.6)$$

$$4Li^+ + O_2 + 4e^- \rightarrow 2Li_2O_2 \qquad (6.7)$$

The OER process retrogresses the discharged products, or Li-ion retraces its path to the Li anode, and oxygen is transferred back to the atmosphere [7]. The below given chemical reaction represents the charging process:

During charging

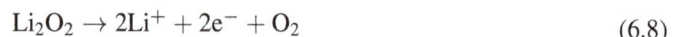

$$Li_2O_2 \rightarrow 2Li^+ + 2e^- + O_2 \qquad (6.8)$$

Figures 6.5(b–c) show the charging and potential window range of MAB. Many approaches have also been made to aqueous LAB, including Li-ion solid membrane near anode that prevents lithium from corrosion, aqueous electrolyte, and carbon-based cathode. However, the discharged products are soluble in electrolytes and produce low energy density, which explains their non-commercialization [19]. The actual reaction taking place in aqueous electrolyte media is given below.

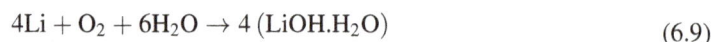

$$4Li + O_2 + 6H_2O \rightarrow 4\,(LiOH.H_2O) \qquad (6.9)$$

6.3 MATERIALS FOR METAL-AIR BATTERIES

MABs are recognized for their high energy density but the improvement in surface phenomena and electrolyte chemistry could further enhance their performance. In recent times, innumerable research has been taking place for the fabrication and synthesis of oxygen electrocatalyst on the air anode, metal electrode, and electrolytes of MABs. Carbon and transition-metal-based materials have been designed to improve the internal activity of elusive OER/ORR reactions in aqueous

MABs [6]. Transition metal-based nanomaterials help overcome the difficulty in fabricating bifunctional electrocatalyst on the anode to the least overpotential of the discharge and charge process. Graphene among the carbon-based materials such as carbon nanotubes (CNTs), carbon nanofibers (CNFs), and carbon microfibers, is selected for the air electrode due to its large surface area, elevated conductivity, thermal and chemical stability, and abundance of the three-dimensional (3D) triple-phase electrochemical zone [21]. It is observed that doping of non-metallic elements, such as nitrogen and oxygen, can influence electrocatalytic activity due to defects.

Moreover, it is observed that the morphology of active electrode materials could affect the performance of MABs. Nanomaterials have a wide range of morphology like nanofibers, nanorods, and nanowires (NWs). All these morphologies belong to the one-dimensional (1D) class of materials. Among these materials, NW shows the advanced characteristics for MABs. Furthermore, these NWs show distinctive electronic, thermal, mechanical, optical, and magnetic properties [22]. It also enables the direct electron exchange to the electrodes and less radial diffusion length, providing maximum electrode contact area with the electrolyte. Therefore, decreasing mechanical and physical disintegration enhances the volume expansion and minimizes the charge-discharge time with an overpotential electrode. Moreover, NWs enable the geometrical merits for in-situ probing, high optical structures, and electronic revolution can be studied by optimizing NWs in the electrodes of the MAB operations.

Furthermore, the metal-organic framework (MOF)-based NWs bring a revolutionary uplift to the MAB due to more electroactive sites and high surface area. Therefore, a group of researchers, along with Zou *et al.* [21] produces high power density MABs in an aqueous electrolyte where the cobalt nanoparticles embedded into N-doped carbon nanotubes grown on carbon NWs (N-Co@CNT-NW-700). The electrocatalytic activity results were noticeable, and ORR performance was noted at 0.87 V *vs.* reversible hydrogen electrode (RHE), lower than the 7 mV of Pt/C electrocatalyst. In addition, they exhibit a high acme in power density of 200 mV/cm² for a MAB. The scanning electron microscopy (SEM) and transmission electron microscopy (TEM) images are shown in Figure 6.6, along with the polarization curve and corresponding power density plot. The porous structure influences the mass transfer of oxygen molecules from the atmosphere. In comparison, there are different types of porous anatomy: macropores, micropores, and mesopores, which help increase the reactivity. Likewise, it is also required for a discharge capability and can be subjugated to discordance at current densities. The surface area, pore size, and pore volume can ameliorate the electrocatalytic activity and oxygen transfer [23]. Macropores and micropores increase the pore volume and turn into a large surface area that provides maximum reaction sites to the oxygen for OER/ORR. Pore volume and surface area are the two factors that adversely affect the performance of MAB. However, both are affixed with the rationalized configuration of pore size as it is responsible for transmitting electrons, ions, gas molecules, and other species.

In ideal electrodes, the specific capacity is regulated by the surface passivation and least mass loading of the discharge products when the electrocatalyst is grown or coated on the conductive

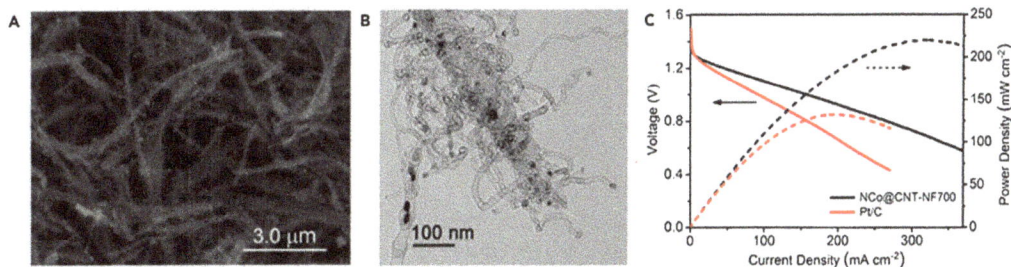

FIGURE 6.6   SEM (A), TEM (B) images, and polarization curve/power density plot (C) of primary ZAB. Adapted with permission. Copyright (2018), American Chemical Society.

proliferated substrates. For instance, Lyu *et al.* [24] fabricated a 3D-printed porous structure air cathode for a LAB. This hierarchical micropores' morphology can exhibit enough sites for accumulating Li_2O_2 with better integration as electrocatalysts and its disintegration into Li_2O_2. The meso and microporous morphology help in O_2 diffusion. The carbonized 3D-printed cobalt-based MOF (Co-MOF), along with Co nanoparticles, which are embedded into nitrogen nanoporous carbon flakes act as an active material for cobalt doped (3DP-NC-Co) NWs [24]. The presence of Co as a nanocatalyst enhances the rate of OER/ORR. Hence, 3DP-NC-NW-Co demonstrated the dominant performance for overpotential, and specific energy than the NC-Co packed on carbon paper, and NC-Co grew on carbon paper. Figures 6.7(a–d) show the complete passivation along with the design of the cathode, and Figures 6.7(e–g) offer their electrochemical performance. Still, cyclic stability and real-time hindrance such as distortion in the shape of metal electrodes takes place due to the differences in the physical and chemical behavior of the metal anodes used.

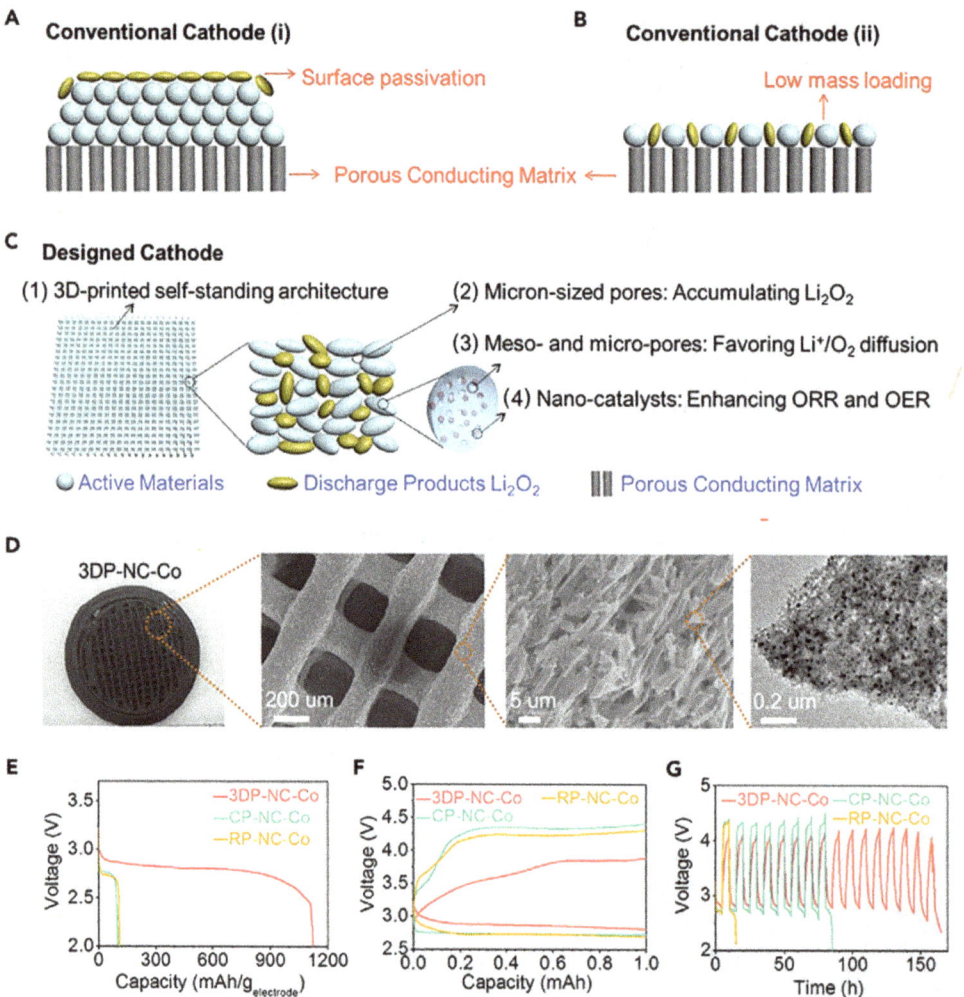

FIGURE 6.7 Schematic illustration of Li_2O_2 growth on air electrodes with their conventional structure (A and B). Description and advantages of 3D-printed self-standing and hierarchically porous catalyst framework as the air electrode (C). The optical, SEM, and TEM images of 3DP-NC-Co (D). Figure (E) shows the first discharge cycle at 0.05mA/cm² along with (F) shows the charge-discharge cycles till 1 mAh/g and (G) represents the cycle ability at 0.1 mA/cm²(E–G). Adapted with permission. Copyright (2018), John Wiley and Sons.

The formation of dendrites in LAB results in the depletion of specific capacitance, coulombic efficiency, and cyclic stability. Impediments of Li dendrites can be monitored by using a surface coating, amalgamating the electrode composition and 3D Li material, which can help regulate the current distribution and ion exchange by affecting the nucleation sites and deposition attribute of Li [25]. Thus, work has been done by Guo *et al.* [26] in the same direction and accomplished similar strategies to overcome Li dendrites to stabilize Li-metal electrodes by applying lithiophilic MOF synthesized by N-doped carbon-based cubes with Co/Co$_4$N NWs proportionally. Henceforth, it controls the nucleation and deposition of Li and therefore serves 70 cycles at a current density of 200 mA/g and a specific capacity of 500 mAh/g better than the pure Li electrode, which elucidates only 34 cycles in the same testing criteria.

6.4   NANOWIRES FOR METAL-AIR BATTERIES

MAB brought a new generation into an electrochemical society due to the use of air-cathode. It makes MAB reliable and more prominent compared to other classes of batteries. Researchers across the globe are continuously working on the improvement of MABs. The introduction of suitable 1D material plays a significant role in bringing optimum goals of MABs into existence, such as high energy density, power density, safety, and extended durability [22]. Among 1D materials, NWs are the structured material that can readily exhibit the required properties for the energy demand. These 1D materials possess a high crystalline structure, well-managed dimensional configuration, electronic radial transport, and functional characteristics leading towards designing a nanoscale morphological system for MABs. The merits of the NWs are already discussed in the previous section, as morphology plays a vital role in the functioning of MABs. Therefore, various types of NWs based on their structure, such as normal, core or shell coated, hollow design, porous, arrays, networks, and bundles are developed [27].

Moreover, the high surface area of NWs provides excellent stability. Due to composite structures like core or shell and hierarchal or heterostructural NWs generate swift and non-stop electron or ion insertion and good reversibility during electrochemical measurements. On the other hand, NWs are beneficial for enhancing the material strain and reaction potential of MABs. Therefore, nanosize also inculcates mechanical flexibility and exceptional Young's modulus [28]. These properties are considered immense for fabricating micro-flexible electronic parts and significant potent momentum to MABs.

Hybrid or combined NWs have a variety of uses due to their advanced structure and different morphology, which further provides no degradation of the material. Therefore, a plethora of morphology leads to a vast range of performance, so various research based on the synthesis of the different types of NWs have been done, including calcination, hydrothermal reaction, electrolytic deposition, electrospinning, chemical vapor deposition (CVD), template method, pyrolysis, vapor-liquid deposition, and many more. However, scientists are still working on in-situ inspections and looking towards a better way to observe the mechanism instead of ex-situ. Since NWs portray the effective charge-discharge due to dynamic changes brought during the electrochemical testing are an excellent advantage for electrode material [29]. Nowadays, transition metal and metal oxide-based NWs are synthesized as they are feasible, abundant in nature, and available at a low cost. They also show good performance in enhancing the traits of MAB electrodes and the electrocatalyst such as CuO-, ZnO-, MnO-, and CoO-based NWs. Bi-functional electrocatalysts are in supreme demand for both charge-discharge cycles in the same system. Therefore, scientists are currently focusing on fabricating materials that enumerate better performance in MABs. Figure 6.8 shows the schematic view for developing various materials for the MAB.

ZAB has the theoretical specific energy of 1084 Wh/Kg, with cell voltage 1.667 and 1.35 V. Therefore, researchers are working to alleviate the performance of ZABs and develop the distinctly acting bifunctional catalyst for OER and ORR reaction to improve the blunt mechanism reaction

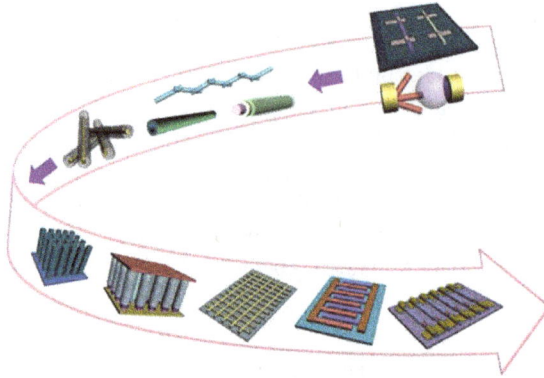

FIGURE 6.8 Schematic structure representing the development of the MAB using a variety of materials. Adapted with permission. Copyright (2014), American Chemical Society.

of ZAB [23]. A group of scientists created porous nickel oxide-doped cobalt nitrate nanocomposite (NiO/CoN), whose interface appears to be a NW array through the morphological studies and has both capabilities to make an oxygen vacancy on the surfaces of NiO/CoN and coupled interfaces are reasonable for electrocatalytic durability and stability [23]. The results for the air cathode were anomalous since it records a high-power density of about 79.6 mW/cm^2, at a higher open circuit potential of 1.335 V, good cycling stability, and an unexpected energy density of 945 Wh/Kg closer to its theoretical value. However, other parameters might degrade its performance during the higher number of cycles. Therefore, it allows researchers to still work on a cost-effective alternative of noble metals to enhance the OER and ORR mechanism in MABs. Hence, Chen *et al.* [30] synthesized a Co-N/C framework on bamboos like MnO NWs leading toward a promising high potential and charge-discharge cycle for solid-state ZAB in which NWs assist feasible diffusion of electrolytes to the active sites. The thermal pyrolysis synthesis method obtained a highly porous MnO@Co-N/C composite whose morphology confirms its NW structure. Therefore, MnO and porous Co-N/C demonstrate better electrocatalytic activity due to a combined effect (synergism) and the transference of e$^-$ via MnO/C.

However, MOF-derived Co-N/C around MnO exhibits better electronic conduction and mass transfer. Therefore, it results in comprehensive ORR performance. The noted CV in O$_2$ and N$_2$ saturated medium at 0.1M KOH is 0.85 V, and no reduction peak providing the ORR catalyst with a specific current density of 5.25 mA/cm^2, lowest Tafel slope 78 mV/dec indicates fast e$^-$ transfer kinetics as compared to pristine MnO (243 mV/dec), Co-N/C (103 mV/dec) and platinum-carbon based (Pt/C-95 mV/dec). Therefore, MnO@Co-N/C-NWs show better performance when compared to pristine RuO$_2$ (1.68 V), Co-N/C (1.82 V), and Pt/C (1.89 V). The results obtained after CV, LSV, and CA curves were evident for the bifunctional nature of MnO@Co-N/C and show approximately the same current density as commercial RuO$_2$ catalyst. Moreover, the Tafel slope obtained for MnO@Co-N/C was lower among the other prepared electrodes manifesting the durability and stability of fabricated material. They also noticed a commendable polarization curve and power density curve for MnO@Co-N/C is 300 mA/cm^2 at 0.4 V and 130.3 mW/cm^2, respectively. Thereby, casting its superior performance after using MnO NW and they further performed the cyclic durability and rechargeability test for the practical existence of MAB. Henceforth, a noticeable degradation was observed in Pt/C, RuO$_2$ batteries after 980 cycles for 327 h, whereas MnO@Co-N/C delineated stable results for 1900 cycles (327 h) at the same current density of 5 mA/cm^2. Perhaps, the obtained open-circuit voltage (OCV) for the synthesized material was 1.429 V, which is equivalent to the theoretical value for high-performance liquid and solid-state ZAB.

Furthermore, the ZAB was replaced by a NAB as it gained focus because of the sixth most bountiful element on Earth. Moreover, its energy density is 1.5 times higher than the ZAB and three to five times less than LAB [31], [27]. Lithium and sodium are also chemically placed close in a periodic table but share a different bonding between each other. Na in combination with O_2 forms sodium superoxide (NaO_2), which is the noblest form of its existence as it does not corrode the electrode during discharge. In addition, it helps in the inversion of products obtained during discharge. The theoretical figures reported by a scientist for energy density are 3164 Wh/kg and C_{sp} of 1166 mAh/g for NABs. Therefore, a variety of research is going on to achieve the apogee in energy density and specific capacitance with low overpotential and better stability. Peled *et al.* [32] were the first to evaluate NAB at higher than room temperature (105 °C) in 2012, the revolutionary year for the NAB. Researchers operated NAB at room temperature, which enumerated 85% coulombic efficiency. However, the research was provoked after the evolution of nanoscience, and scientists brought NWs into the picture, thereby fabricating gold-doped manganese oxide (Au/MnO_2) NW air cathode (Au/MnO_2) for NAB to execute in-situ inspection [34]. Coating with Au helps the NW 18 times more volume extension on the upper layer of the MnO_2 NW; hence due to nucleation NaO_2 formed and ORR takes place, which hardly executes in the bare MnO_2. NaO_2 is abruptly converted into Na_2O_2 and O_2, which disturbs the rectitude of the system, and accumulation of discharge products takes place on the surface of Au nanoparticles, but due to the presence of Au/MnO_2 NWs, which has a higher density than the discharge products. Hence, volume diminution of discharge products occurred. Therefore, it was suggested as a new approach to examining better electrocatalysts in the real world by the in-situ method. The above two batteries consist of Zn, Na as their metal anode but these are hard to recycle. Therefore, researchers find aluminum-air batteries (AABs) as a suitable replacement due to their high energy density, low price, safety, and being environmental friendly. AAB has a high theoretical specific energy density, which is three times higher than NAB and six times greater than ZAB. Aluminum is an abundant resource. Along with this, it can be recycled at a low cost. Co-based materials have some advantages as well as disadvantages and are preferred as they have good electrocatalytic activity, are inexpensive, and provide more excellent stability.

Co-based materials have restrictions like their catalytic performance is affected by electronic state and the surface area. Therefore, doping and fabricating Co-based material on some nanocrystalline structures enhances its performance. Liu *et al.* [35] synthesized the 3D graphene on nickel foam, which was prepared by chemical vapor deposition (CVD) technique and further targeted by the reduced mesoporous (RM) Co-oxide (Co_3O_4) NWs via thermal heat treatment along with a hydrothermal method, which is further followed by the annealing for its reduction. The prepared RM-Co_3O_4/3D NW is then transferred to electrochemical measurements for the battery functionality and the catalytic ORR activity. The higher the CV peaks, the better the ORR catalytic activity. The best sample RM-Co_3O_4/3D peaks at 0.76 V (*vs.* RHE) suggest better catalytic activity and the lowest possible Tafel slope (71.45 mV/dec), indicating the faster kinetic reactions. These results are attributed to the mesoporous NW structure with 1D and 3D morphology, providing large active sites and more charge-transfer routes. The synergistic effect is also the reason for the better activity of NW-based composite. Furthermore, the retention in current density after 12000 s at a constant voltage was 97.7%, comparatively higher than platinum-based electrodes (Pt/C). The group uses RM-Co_3O_4/3D as a binder-free cathode for AAB via fabricating the Al-air coin cells, exhibiting the OCV ~ 1.53 V with C_{sp} 422.74 mAh/g, which is traditionally much higher than Pt/C-based cathodes. O_2 creates vacancies at defected sites primarily manifested for the reduction mechanism and depicting the 3D characteristics of transition metals.

Gu *et al.* [36] fabricated defect-based manganese oxide (α-MnO_2) NWs using a facile hydrothermal synthesis route followed by washing and drying overnight. Doping creates defects and increases the electrical conductivity for which α-MnO_2 NW is loaded with argon plasma in a quartz tube to obtain A-MnO_2. The morphology study states that the as-synthesized A-MnO_2 NWs are thinner due to the etching effect of argon plasma. The as-prepared α-MnO_2 and A-MnO_2 are then subjected to electrochemical characterization through ORR and real-time fabrication of Al-air coin cells. The α-MnO$_2$ has the Tafel slope of 139 mV/dec, which gets improved after etching the MnO$_2$ to A-MnO$_2$ (105 mV/dec), indicating the faster kinetics. The superior performance leads the researcher to fabricate the Al-air coin cell to test the performance whose OCV is 1.90 V higher than that of pristine MnO$_2$ (1.74 V), leading to its power density of 159 mW/cm^2 at 157 mA/cm^2 higher than that of pristine MnO$_2$. The work has its advantage as the byproduct obtained is Al(OH)$_3$ which immediately gets dissolved in the aqueous NaOH to generate NaAlO$_2$, which gets dissolved in the electrolyte, solving the issue of byproducts. The practical application of AAB is clear when it shows the two-series connection can light up 65 LEDs. Apart from this, scientists are still developing these batteries to match with LAB as the theoretical capacity of LAB is 11249 Wh/kg. Despite the high theoretical capacity, LAB still faces high overpotential and sluggish kinetics reactions, leading to bad stability and poor cycling performance [27].

Researchers worldwide are developing new nanomaterials, among which NW enhances LAB's performance. Lu *et al.* [37] fabricated the conducting polyaniline (PANI) NWs, which are dispersible in water. The PANI-NWs were prepared using the synthetic approach. The monomer aniline is subjected to nitric acid in a dilute solution followed by etching through ammonium hydroxide (NH$_4$OH) to form the polymeric structure and dispersible in H$_2$O. The morphology of obtained wire shows its thin structure with a high surface area of ~ 70 m^2/g, which is four times higher than pristine PANI (15 m^2/g). The PANI-NWs are then subjected to electrochemical measurements for CV and discharge capacity. The PANI-NWs at the current density of 0.05 mA/cm^2 shows the discharge capacity of 3280 mAh/g at the first cycle between the voltage window range of 1.75 to 4.2 V. The LAB during the testing shows the degradation due to the morphological and chemical changes of nanowires in battery cathode. These H$_2$O dispersed conducting PANI-NW deployed in LAB are directly used as the cathode without an additional catalyst. Apart from this, perovskites also show better durability and enhanced performance due to their hierarchical mesoporous structure. The group of researchers at the State Key Laboratory of China [38] fabricates the perovskite lanthanum strontium cobalt oxide (La$_{0.5}$Sr$_{0.5}$CoO$_{2.91}$) (LSCO) NWs with tri-continuous tunnels for O$_2$ transport and ion diffusion. The LSCO is synthesized through a multi-step micro-emulsion self-assembly method, then annealing. The fabricated coin cell for practical application is tested, which maintains the specific capacity of 1000 mAh/g at 0.1 A/g even after 25 cycles indicating the higher electrolyte stability. The catalytic process and the rate control process are governed through the chemisorption of O$_2$ (i.e., adsorption and desorption), resulting in sluggish kinetics, which can be easily elucidated through the σ-π bonding allowing the O$_{ads}$ enough force through coordination that enables the higher efficiency for electron transfer. The low-degree electrolyte easily forms the Li$_2$O$_2$ through in-situ reactions and decomposes them for good cycling performance.

Furthermore, advancement has been made, and the amalgamation of graphene oxide (GO) and α-MnO$_2$ NWs (10 wt.%, 30 wt.%, 50 wt.%, and 70 wt.%) was fabricated by the vacuum filtration method [39]. Whereas α-MnO$_2$ NWs were synthesized by microwave hydrothermal method. The maximum specific discharge capacity of 2900 mAh/g was obtained by adding 70 wt.% of α-MnO$_2$ NWs in the flexible α-MnO$_2$$^+$ GO composite for LAB. After the electrochemical cyclic life of cathode testing, it has been theorized that ORR was disintegrating rather than OER, which shows an acme when it increases the wt.% of synthesized NWs. Moreover, concentration also affects the catalytic performance then probed for CV test, which resembled overlapping of charge-discharge cycles since CV increases with the concentration of wt.% of NWs in a composite, decreasing the polarization curve. However, the current peak intensity is significantly noticed for 70 wt.%, and high-capacity retention was recorded as 405 mAh/g and evidence of successful synthesized reinforced material for LABs. However, a lot of efforts are still needed to overwhelm the impediments of MAB, and continuous progress is going on with advanced research to enhance the electrochemical activity. The battery life cycle is also a significant target to evaluate for better running of a MAB, taking the stability and durability of the material into account. Therefore, Xing *et al.* [40] fabricated a LAB with the long-run aim of introducing a harmonized quasi-solid electrolyte. Highly active Pt$_3$Co

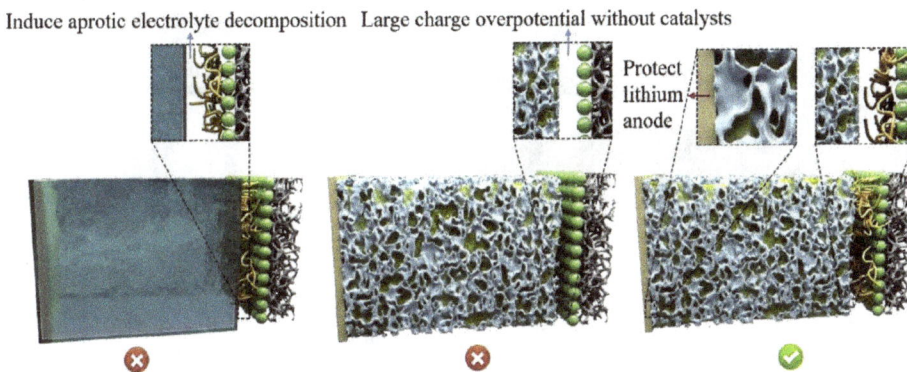

Induce aprotic electrolyte decomposition Large charge overpotential without catalysts

Protect lithium anode

FIGURE 6.9 Schematic diagram of roles of SiO_2-IL quasi-solid electrolyte and Pt_3Co NWs catalyst in LAB. Adapted with permission. Copyright (2019), Elsevier.

NWs act as a catalyst because of the optimistic synergistic effect achieved by synthesized NWs by a conventional method and quasi-solid electrolyte SiO_2-ionic liquid (IL) was prepared by in-situ sol-gel process. Synthesized NWs decrease the charge voltage under 3.2 V due to the high ionic conductivity of the electrolyte. As a result, LAB deliberately shows reversible charge and discharge operations above 300 cycles (> 3000 h) at a current density of 100 mA/g with a capability of 500 mAh/g at 60 °C because of the presence of both the Pt_3Co NWs and SiO_2-IL. Figure 6.9 depicts the role of SiO_2-IL in the formation of NWs. The life span of a battery has been enhanced. Therefore, it helps to acquire the persistent cyclic stability for a LAB. However, under air conditions, the reported reversible total charge and discharge specific capacity was approximately 6000 mAh/g (1.8 mAh/cm^2) and maintained a good performance at the sixth cycle of 1.5 mAh/cm^2. The results showed that the modified LAB provides a long-lasting battery because of the synergistic electrolyte and highly active catalyst.

The next generation is looking for a battery that offers high specific capacity and low electrochemical driven potential as LAB is five–ten times more efficient than other MAB. On the other hand, the safety is still a significant concern due to the growth of dendrites on the surface of lithium in LAB, which adversely affects the long-run capacity and charge-discharge process. Lu *et al.* [41] pragmatically designed a Cu-NWs 3D nanostructure network over the Li-metal to enhance the property such as current collector and coulombic stability. The developed CuNWs were synthesized by solvent evaporation method at ambient conditions and activated by H_2/Ar (5%/95) pass at 450 °C in the vessel. The group had found that after using the adequate solid-electrolyte interface (SEI) such as hydrogen fluoride (HF), lithium fluoride (LiF), copper acetate ($Cu(CH_3COO)_2$), vinylene carbonate (VC)-lithium nitrate ($LiNO_3$), lithium polysulfide and infusion of Cs^+ and Rb^+ ions in the electrolyte have produced the protective covering on the anode surface but unable to prevent it from the undesirable growth of dendrites. It is responsible for the inhomogeneity of charge distribution, which was overcome by introducing Cu-NWs over the Li electrode. It provides a nanoporous structure for Li growth to infuse. Figure 6.10 shows both the planar Cu and Cu-NW network fabricated on Li-ion flux.

Moreover, it restricted the ion flux density and averted the dendrites. Therefore, responsible for the better current collector. The stripping of Li was observed by considering the different plating capacities of Cu-NWs from 2.5 mAh/cm^2 to 7.5 mAh/cm^2. The observation was also carried out for 10 mAh/cm^2, but dendrite growth was observed under SEM, so 7.5 mAh/cm^2 is considered high and better for growth inhibition. Furthermore, it was reported that due to an excellent synergistic effect and rational structure, coulombic efficiency raised from 99.2% after 50 cycles to 98.6% during 200 cycles with a stable potential hysteresis of approximately 0.04 V and best for practical application for solid-state LAB.

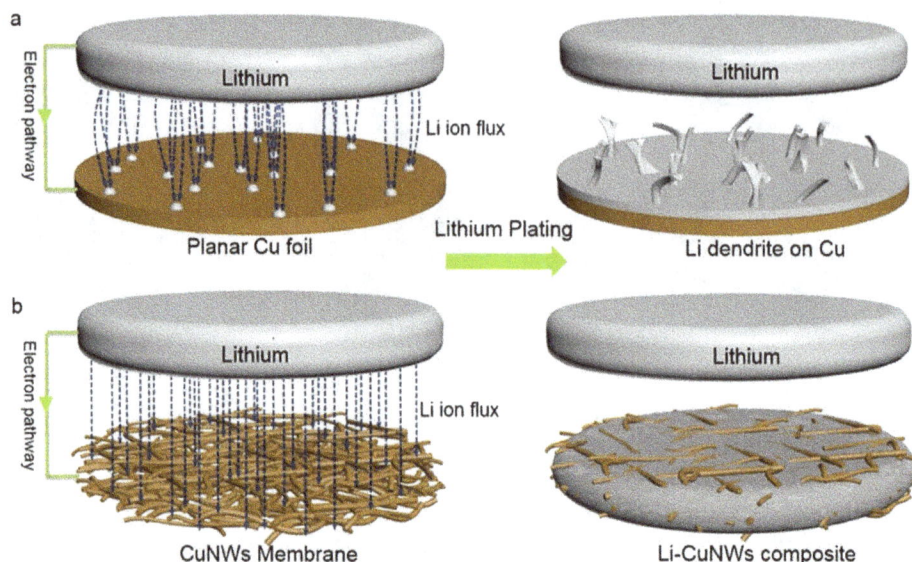

FIGURE 6.10 Schematic illustration of Li-ion flux distribution and Li-metal pelting along with (a) shows the planar Cu foil and (b) shows the Cu-NW network fabricated over the coin cell. Adapted with permission. Copyright (2016), American Chemical Society.

Similarly, a study was done by Liu *et al.* [42] to exhibit high conductivity, porosity, and lipophilicity by developing a 3D triple-layer integrated TiC/C core or shell-shaped NW arrays as a skeleton on the surface of the metal electrode by CVD method on the Ti_6Al_4V substrate. The unique NWs' matrix presents the interspatial sites for the Li anode. It also generates a pathway for electron or ion transmission at low current densities, leading to a barricade for dendrites' growth. Herein, the LAB's lifetime with a very effective coulombic efficiency of 98.5% for 100 cycles at 1.0 mA/cm². The stable TiC/C framework helps to reduce current and Li+ ion flux distribution, which affects the usage of Li metal and results in a high C_{sp} of 3110 mAh/g at a current density of 0.5 mA/cm² and a cut-off potential of 2.0V. The interfacial impedance of synthesized material is three–five times lower than the resistance offered by electrolyte, SEI layer, and due to charge transfer. Therefore, the overwhelming performance was shown by LAB with the least hysteresis of 85 mV after 200 cycles at 3 mA/cm². Hence, the research showed the best possible outcome and better material with strong mechanical support and high intervention of charge species with lower hysteresis and increased cyclic stability and Coulomb efficiency.

6.4 CONCLUSION

Through the discussion in this book chapter, it was noticeable that the implementation of MABs had plenty of challenges that would hinder their applicability as an efficient alternative to MIBs. The main challenges related to this technology include the decrease in performance due to insoluble byproducts. The back and forth motion of charge carriers during charge and discharge is the main reason for the formation of dendrites in the metal anode during the working of MAB, leading to a significant fall in its ability to store charges. This dendrite formation also shows in the volume expansion and is the leading cause of the battery's short-circuiting. Apart from all of that, significant research is going on to overcome these drawbacks yielding satisfactory results that move toward the feasible application of MAB. For instance, the development of several types of metal-anode and cathode with NW morphology can be quickly helpful in absorbing a greater amount of oxygen from the air. Furthermore, the carbon-based NWs provide enough room for the high charge storage capacity due to their comparatively higher surface area. Lastly, effective properties to prevent the formation of a dendritic layer have been successfully employed using the coated layer of NWs on the substrates, which does not allow them to grow much and reach the other electrode. These NWs provide stability and decompose the degradation of the MABs. Likely, MABs will soon compose the battery market due to their low-cost, high energy density, being environmentally friendly, and having a versatile range of anodes that can be used along with the abundant materials.

REFERENCES

[1]   G. Girishkumar, B. McCloskey, A.C. Luntz, S. Swanson, W. Wilcke, Lithium−air battery: promise and challenges, J. Phys. Chem. Lett. 1 (2010) 2193–2203.

[2]   U. Koehler, General overview of non-lithium battery systems and their safety issues, Elsevier B.V., 2018.

[3]   M. Beaudin, H. Zareipour, A. Schellenberg, W. Rosehart, Energy storage for mitigating the variability of renewable electricity sources, Elsevier Inc., 2015.

[4]   Z. Melhem, Electricity transmission, distribution and storage systems, Woodhead Publishing Limited, 2013.

[5]   D. Aurbach, B.D. McCloskey, L.F. Nazar, P.G. Bruce, Advances in understanding mechanisms underpinning lithium–air batteries, Nat. Energy. 1 (2016) 16128.

[6]   H.-F. Wang, Q. Xu, Materials design for rechargeable metal-air batteries, Matter. 1 (2019) 565–595.

[7]   Y. Li, J. Wang, X. Li, D. Geng, R. Li, X. Sun, Superior energy capacity of graphene nanosheets for a nonaqueous lithium-oxygen battery, Chem.Commun. 47 (2011) 9438–9440.

[8]   T. Zhang, H. Zhou, From Li-O-2 to Li-air batteries: carbon nanotubes/ionic liquid gels with a tricontinuous passage of electrons, ions, and oxygen, Angew. Chem. Int. Ed. Engl. 51 (2012) 11062–11067.

[9]   Y. Wu, Y. Zhang, Y. Ma, J.D. Howe, H. Yang, P. Chen, S. Aluri, N. Liu, Ion-sieving carbon nanoshells for deeply rechargeable Zn-based aqueous batteries, Adv. Energy Mater. 8 (2018).

[10]   X. Zhang, X.-G. Wang, Z. Xie, Z. Zhou, Recent progress in rechargeable alkali metal–air batteries, Green Energy Environ. 1 (2016) 4–17.

[11]   B.T. Hang, T. Watanabe, M. Egashira, I. Watanabe, S. Okada, J. Yamaki, The effect of additives on the electrochemical properties of Fe/C composite for Fe/air battery anode, J. Power Sources. 155 (2006) 461–469.

[12]   D. Geng, N. Ding, T.S.A. Hor, S.W. Chien, Z. Liu, D. Wuu, X. Sun, Y. Zong, From lithium-oxygen to lithium-air batteries: challenges and opportunities, Adv. Energy Mater. 6 (2016) 1502164.

[13]   Y. Li, J. Lu, Metal–air batteries: will they be the future electrochemical energy storage device of choice?, ACS Energy Lett. 2 (2017) 1370–1377.

[14]   Q. Liu, Z. Chang, Z. Li, X. Zhang, Flexible metal-air batteries: progress, challenges, and perspectives, Small Methods. 2 (2017) 1700231.

[15]   N. Parveen, Z. Khan, S.A. Ansari, S. Park, S.T. Senthilkumar, Y. Kim, H. Ko, M.H. Cho, Feasibility of using hollow double walled Mn2O3 nanocubes for hybrid Na-air battery, Chem. Eng. J. 360 (2019) 415–422.

[16]   F. Wang, X. Li, Y. Xie, Q. Lai, J. Tan, Effects of Porous Structure on Oxygen Mass Transfer in Air Cathodes of Nonaqueous Metal–Air Batteries: A Mini-review, ACS Appl. Energy Mater. 5 (2022) 5473–5483.

[17]   M. Klanchar, B.D. Wintrode, J.A. Phillips, Lithium−water reaction chemistry at elevated temperature, Energy & Fuels. 11 (1997) 931–935.

[18]   T. Zhang, N. Imanishi, S. Hasegawa, A. Hirano, J. Xie, Y. Takeda, O. Yamamoto, N. Sammes, Li/polymer electrolyte/water stable lithium-conducting glass ceramics composite for lithium–air secondary batteries with an aqueous electrolyte, J. Electrochem. Soc. 155 (2008) A965.

[19]   S. Sunahiro, M. Matsui, Y. Takeda, O. Yamamoto, N. Imanishi, Rechargeable aqueous lithium–air batteries with an auxiliary electrode for the oxygen evolution, J. Power Sources. 262 (2014) 338–343.

[20]   T. Liu, J.P. Vivek, E.W. Zhao, J. Lei, N. Garcia-Araez, C.P. Grey, current challenges and routes forward for nonaqueous lithium–air batteries, Chem. Rev. 120 (2020) 6558–6625.

[21] L. Zou, C.-C. Hou, Z. Liu, H. Pang, Q. Xu, Superlong single-crystal metal–organic framework nanotubes, J. Am. Chem. Soc. 140 (2018) 15393–15401.

[22] L. Huang, Q. Wei, R. Sun, L. Mai, Nanowire electrodes for advanced lithium batteries, Front. Energy Res. 2 (2014).

[23] J. Yin, Y. Li, F. Lv, Q. Fan, Y.-Q. Zhao, Q. Zhang, W. Wang, F. Cheng, P. Xi, S. Guo, NiO/CoN porous nanowires as efficient bifunctional catalysts for Zn–air batteries, ACS Nano. 11 (2017) 2275–2283.

[24] Z. Lyu, G.J.H. Lim, R. Guo, Z. Kou, T. Wang, C. Guan, J. Ding, W. Chen, J. Wang, 3D-printed mof-derived hierarchically porous frameworks for practical high-energy density Li–O$_2$ batteries, Adv. Funct. Mater. 29 (2019) 1–8.

[25] X.-B. Cheng, T.-Z. Hou, R. Zhang, H.-J. Peng, C.-Z. Zhao, J.-Q. Huang, Q. Zhang, Dendrite-free lithium deposition induced by uniformly distributed lithium ions for efficient lithium metal batteries, Adv. Mater. 28 (2016) 2888–2895.

[26] Z. Guo, F. Wang, Z. Li, Y. Yang, A.G. Tamirat, H. Qi, J. Han, W. Li, L. Wang, S. Feng, Lithiophilic Co/Co$_4$N nanoparticles embedded in hollow N-doped carbon nanocubes stabilizing lithium metal anodes for Li-air batteries, J. Mater. Chem. A. 6 (2018) 22096–22105.

[27] L. Mai, X. Tian, X. Xu, L. Chang, L. Xu, Nanowire electrodes for electrochemical energy storage devices, Chem. Rev. 114 (2014) 11828–11862.

[28] F. Li, J. Du, H. Yang, W. Shi, P. Cheng, Nitrogen-doped-carbon-coated SnO$_2$ nanoparticles derived from a SnO$_2$@MOF composite as a lithium ion battery anode material, RSC Adv. 7 (2017) 20062–20067.

[29] L.-Q. Mai, F. Yang, Y.-L. Zhao, X. Xu, L. Xu, Y.-Z. Luo, Hierarchical MnMoO$_4$/CoMoO$_4$ heterostructured nanowires with enhanced supercapacitor performance, Nat. Commun. 2 (2011) 381.

[30] Y.-N. Chen, Y. Guo, H. Cui, Z. Xie, X. Zhang, J. Wei, Z. Zhou, Bifunctional electrocatalysts of MOF-derived Co–N/C on bamboo-like MnO nanowires for high-performance liquid- and solid-state Zn–air batteries, J. Mater. Chem. A. 6 (2018) 9716–9722.

[31] P. Hartmann, C.L. Bender, M. Vračar, A.K. Dürr, A. Garsuch, J. Janek, P. Adelhelm, A rechargeable room-temperature sodium superoxide (NaO$_2$) battery, Nat. Mater. 12 (2013) 228–232.

[32] E. Peled, D. Golodnitsky, H. Mazor, M. Goor, S. Avshalomov, Parameter analysis of a practical lithium- and sodium-air electric vehicle battery, J. Power Sources. 196 (2011) 6835–6840.

[33] N. Chawla, M. Safa, Sodium batteries: A review on sodium-sulfur and sodium-air batteries, Electron. 8 (2019).

[34] Q. Liu, L. Geng, T. Yang, Y. Tang, P. Jia, Y. Li, H. Li, T. Shen, L. Zhang, J. Huang, In-situ imaging electrocatalysis in a Na-O$_2$ battery with Au-coated MnO$_2$ nanowires air cathode, Energy Storage Mater. 19 (2019) 48–55.

[35] Y. Liu, F. Zhan, N. Zhao, Q. Pan, Z. Li, Y. Du, Y. Yang, Reduced mesoporous Co$_3$O$_4$ nanowires grown on 3D graphene as efficient catalysts for oxygen reduction and binder-free electrodes in aluminum–air batteries, J. Mater. Sci. 56 (2021) 3861–3873.

[36] Y. Gu, G. Yan, Y. Lian, P. Qi, Q. Mu, C. Zhang, Z. Deng, Y. Peng, MnIII-enriched α-MnO2 nanowires as efficient bifunctional oxygen catalysts for rechargeable Zn-air batteries, Energy Storage Mater. 23 (2019) 252–260.

[37] Q. Lu, Q. Zhao, H. Zhang, J. Li, X. Wang, F. Wang, Water dispersed conducting polyaniline nanofibers for high-capacity rechargeable lithium-oxygen battery, ACS Macro Lett. 2 (2013) 92–95.

[38] C. Shi, J. Feng, L. Huang, X. Liu, L. Mai, Electrocatalyst for rechargeable Li-air battery, 8 (2013) 8924–8930.

[39] S. Ozcan, M. Tokur, T. Cetinkaya, A. Guler, M. Uysal, M.O. Guler, H. Akbulut, Free standing flexible graphene oxide + α-MnO$_2$ composite cathodes for Li–air batteries, Solid State Ionics. 286 (2016) 34–39.

[40] Y. Xing, N. Chen, M. Luo, Y. Sun, Y. Yang, J. Qian, L. Li, S. Guo, R. Chen, F. Wu, Long-life lithium-O$_2$ battery achieved by integrating quasi-solid electrolyte and highly active Pt$_3$Co nanowires catalyst, Energy Storage Mater. 24 (2020) 707–713.

[41] L.-L. Lu, J. Ge, J.-N. Yang, S.-M. Chen, H.-B. Yao, F. Zhou, S.-H. Yu, Free-standing copper nanowire network current collector for improving lithium anode performance, Nano Lett. 16 (2016) 4431–4437.

[42] S. Liu, X. Xia, Y. Zhong, S. Deng, Z. Yao, L. Zhang, X.-B. Cheng, X. Wang, Q. Zhang, J. Tu, 3D TiC/C core/shell nanowire skeleton for dendrite-free and long-life lithium metal anode, Adv. Energy Mater. 8 (2018) 1702322.

7 Nanowires for Metal-Sulfur Batteries

Yucheng Zhou[†]
Department of Mechanical and Aerospace Engineering,
University of Virginia, Charlottesville, VA, USA

Hyunjin Cho[†]
Institute of Advanced Composite Materials, Korea Institute of Science
and Technology, Wanju-gun, Jeollabuk-do, Republic of Korea

Qiang Gao
School of Chemistry and Chemical Engineering, Yangzhou University,
Yangzhou, China

Jiadeng Zhu
Chemical Sciences Division, Oak Ridge National Laboratory,
Oak Ridge, TN, USA
Smart Devices and Printed Electronics Foundry, Brewer Science Inc.,
Springfield, MO, USA

[†]The first two authors contributed equally to this work.

CONTENTS

DOI: 10.1201/9781003296621-7

7.1 INTRODUCTION

Nanowires have received tremendous attention during the past few decades due to their unique features, including extremely small size, high surface area to volume ratio, *etc.* [1–3]. Benefiting from those characteristics, nanowires have been widely explored and applied in various areas, such as energy storage and conversion, filtration, biosensing, insulation, *etc.* [4, 5].

It is important to note that lithium-ion batteries (LIBs) have been considered as the most advanced and appealing electrochemical energy storage system, which have been widely utilized in electronics, mobile devices, grid energy, *etc.*, during the last four decades [6–9]. However, they are suffering from providing enough energy density when being used for systems that require high energy like electric vehicles (EVs) [10–12]. Therefore, a general goal for battery development is to increase the energy density to meet the increasing demand for world energy consumption. Nevertheless, a traditional Li-ion cell, which typically has a graphite anode and an oxide cathode (*e.g.*, $LiCoO_2$) or a phosphate cathode (*e.g.*, $LiFePO_4$), can store up to ~200 Wh/kg [13–15]. As it is well known that the graphite anode has a capacity of 372 mAh/g and the capacity of these oxide cathodes is generally lower than 180 mAh/g, thus, alternative electrodes including both cathode and anode materials with high capacities should be further explored and developed [16, 17].

As a representative of metal-sulfur batteries, the lithium-sulfur (Li-S) battery is a promising candidate on this point since both sulfur and lithium have high theoretical capacities, which are 1675 and 3800 mAh/g, respectively [18–23]. Besides, sulfur is abundant and environmentally friendly. Moreover, lithium metal has a low negative potential, which enables a remarkable enhancement of the cell's energy density [24]. However, the practical application of Li-S batteries has been retarded by the following main challenges: (a) poor conductivity of sulfur and its intermediates, (b) volume expansion of sulfur electrodes, (c) shuttle effects of polysulfides, and (d) safety concerns of organic electrolytes and the lithium metal electrode [25–28]. So far, tremendous efforts have been taken to address these issues by creating and developing novel sulfur host materials, solid-state electrolytes, multifunctional separators, and interlayers by using nanowires owing to their considerable surface-to-volume ratio, flexibility in surface functionalities, and robust mechanical properties [29, 30].

In this chapter, the advancements and progress of nanowires for obtaining high-performance Li-S batteries will be reviewed and discussed to provide readers with a better understanding of applying nanowires in this field. It starts with an introduction of Li-S batteries' working mechanism followed by their challenges. Then, the advantages of utilizing nanowires in Li-S batteries are explored in detail from four aspects: (a) cathode design; (b) separators and interlayers; (c) electrolytes; and (d) anode protection. Some perspectives related to the remaining challenges, opportunities, and directions are presented at the end to accelerate the development of nanowires for practical Li-S cells.

7.2 OVERVIEW OF METAL-SULFUR BATTERIES

Metal-sulfur batteries represent one of the most attractive energy storage systems that utilize metal (*e.g.*, Li and Na) as anodes and sulfur-embedded composites as cathodes attributed to their high energy density and low cost [31]. Nonetheless, the use of active metallic anode and the solubility of sulfur species in the electrolyte render these batteries difficult to achieve practical realization. Li-S battery is representative of the metal-sulfur battery species and will be used to introduce the working mechanism and potential challenges.

7.2.1 WORKING MECHANISM OF LI-S BATTERIES

As the flourishing research interest in Li-S batteries keeps increasing, their internal mechanism has been widely discussed. The most recognized one is a stepwise reaction from crystalline S_8 to solid Li_2S to achieve discharging, and from Li_2S back to S_8 to attain charging [32]. During the process, electrons are released or collected as shown in $S_8 + 16Li^+ + 16e^- \leftrightarrow 8Li_2S$. When discharging, the stepwise process is summarized as the conversion from (a) S_8 to Li_2S_8 and Li_2S_6 (solid to liquid), (b) Li_2S_8 and Li_2S_6 to Li_2S_4 (liquid to liquid), (c) Li_2S_4 to Li_2S_2 (liquid to solid), and (d) Li_2S_2 to Li_2S (solid to solid), which is represented by the double-stage discharge curve (**Figure 7.1a**) [33]. The opposite happens when the battery is charging, where Li_2S is eventually converted back to S_8 to be stored and ready to perform discharging again.

7.2.2 CHALLENGES

Li-S batteries are known to have outstanding theoretical energy capacity and energy density while having relatively low costs. However, the commercial viability is hindered by four major challenges: (a) the extremely low conductivity of sulfur and its intermediates; (b) the large volume expansion of sulfur during battery operation; (c) the undesirable shuttle effect of polysulfides; and (d) the possibility of safety issues and short lifespan due to lithium dendrite growth. These challenges will be discussed in detail below.

7.2.2.1 Insulating Nature of Sulfur and its Intermediates

To ensure batteries' performance, especially when a high current rate is applied and allowing electrons to travel through, a highly conductive electrode is needed [19]. Nevertheless, sulfur is an insulator as its conductivity is only 5×10^{-30} S/cm at room temperature [34]. This means it is almost impossible for sulfur to be utilized by itself and thus, a highly conductive framework is necessary. The conductivity is usually the first factor to be considered when researching Li-S batteries' cathodes. Carbon-based materials is one of the most commonly used conductive materials to combine with sulfur, such as carbon black, graphite, and carbon nanotubes (CNTs) [35]. Recently, other types of materials are also considered as excellent support matrices, including conductive polymers and metal-organic frameworks [36].

7.2.2.2 Volume Expansion

Despite the insulation, because of the conversion between S_8 and Li_2S and their density difference, sulfur exhibits a significant volume expansion of around 78% during battery operation (**Figure 7.1b**) [37]. This will potentially lead to the collapse of cathodes' structure. Without proper management, the active materials will be randomly redistributed, causing multiple issues such as large polarization and poor cycling performance. The redistribution will also result in the detachment of sulfur species from the conductive support and a loss of active materials. To overcome the challenge, cathode material structure design has been introduced and studied. A framework that is flexible and has excellent mechanical properties along with a large surface area is preferred as sulfur support to cope with the volume expansion.

7.2.2.3 Shuttle Effect

Another inevitable issue in Li-S batteries is the polysulfide shuttle effect (**Figure 7.1c**) [38]. Polysulfides are a series of Li_2S_x ($4 \leq x \leq 8$) that are soluble in most organic electrolytes used in Li-S batteries. They are generated during battery operation and even when a battery is not in use since Li ions disperse in the electrolyte and react with sulfur automatically. These polysulfides are so small that they can easily diffuse through common separators to reach the Li metal and corrode it, leading to the waste of active materials and eventually, a fast decay of capacity and poor lifespan [18]. Studies have been focusing on this issue since the beginning and several general ideas to suppress it include designing cathode materials, adding interlayers, decorating separators, and optimizing electrolytes [32].

7.2.2.4 Lithium Dendrite Growth

Apart from the challenges mentioned above relating to sulfur, safety issues have been hindering the application of Li-S batteries as well. Dendritic Li growth is one of the most mentioned challenges

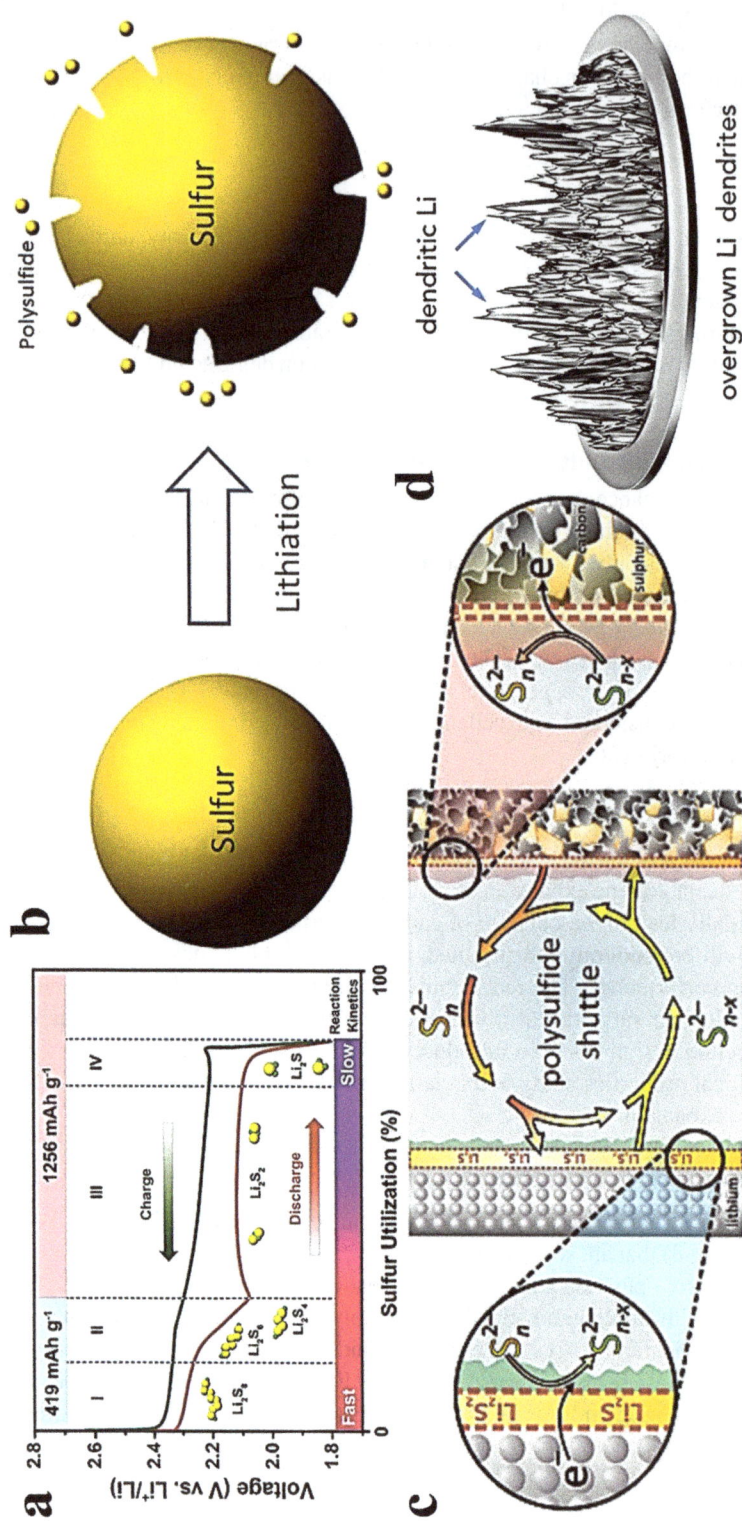

FIGURE 7.1 (a) Typical charge/discharge profiles and their mechanism. Adapted with permission [33]. Copyright 2021, Wiley Online Library. (b) Sulfur volume expansion during battery operation. Adapted with permission [37]. Copyright 2016, UWM Digital Commons. (c) Polysulfide shuttle effect causing self-discharging. Adapted with permission [38]. Copyright 2014, Elsevier. (d) Dendritic lithium growth on the anode. Adapted with permission [39]. Copyright 2018, Elsevier.

because of the high activity of Li metals and flammable liquid electrolytes (**Figure 7.1d**) [39]. Although the reason for dendritic Li growth stays debatable, many studies have shown that the formulation of electrolytes plays a crucial role in this case, especially for the form of solid-electrolyte interfaces (SEI). When the dendritic Li grows long enough, it will probably pierce the separator in the battery, causing a localized short circuit, which largely threatens the lifespan of the batteries and even leads to fire hazards. A potential route to address the issue is improving the composition of electrolytes and optimizing cathode surface structure to enable an even formation of SEI layers within the batteries. Another way to minimize the damage caused by dendritic Li is to insert an interlayer between the metal and the separator [40].

7.3 ADVANTAGES OF NANOWIRES IN METAL-SULFUR BATTERIES

Nanowires have been considered as one of the most promising materials for achieving high-performance Li-S batteries because of their above mentioned features, which are favorable for being used as cathode host materials, exceptional polysulfide absorbents/adsorbents, multifunctional interlayers, and solid-state electrolytes to enhance cell performance [32]. For instance, multitudinous carbon nanowires, fabricated by electrospinning followed by thermal treatment, could be performed as the host for free-standing sulfur cathodes with outstanding electrical conductivity, unique three-dimensional (3D) porous structure, and high surface area, which not only enhanced the sulfur utilization but obtained good cycling stability [36]. Besides, compared to conventional polyolefin separators, functionalized membranes by utilizing nanowires enable cells to have better performance due to their outstanding barrier effect on polysulfide migration [41]. Moreover, an effective nanowire-based interlayer can be inserted between a cathode and a separator to enhance the utilization of those polysulfide intermediates [42]. What is more, nanowire-based solid-state electrolytes are another research hotspot to replace liquid organic electrolytes, which can both restrain the polysulfide shuttle effect and avoid safety concerns [9, 43].

It has been noticed that nanowires with decent size distribution and large specific surface area to volume ratio have been comprehensively explored by incorporating sulfur electrodes, modifying the separators, serving as interlayers and electrolytes, *etc.*, to achieve high-performance Li-S cells.

7.4 APPLICATIONS OF NANOWIRES IN METAL-SULFUR BATTERIES

Nanowires with many advantages mentioned above have been extensively utilized in metal-sulfur batteries. These nanowires' morphology and functionality heavily depend on their growing methods (*i.e.*, wet or dry synthesis processes), by which they can be synthesized in different forms according to the specific purpose, including coaxial structure, nanowire arrays, hierarchical structure, *etc.* [44-47]. More details regarding their applications in Li-S batteries will be introduced and discussed from four aspects: (a) cathodes; (b) separators/interlayers; (c) electrolytes; and (d) anode protection.

7.4.1 CATHODES

Sulfur electrodes suffer from their extremely low electrical conductivity and volume expansion during the discharge process. Carbon-based materials including CNTs, carbon nanofibers (CNFs), and metal oxides have attracted much interest in this regard due to their high specific surface area and superior adsorption/absorption capability of polysulfides.

Zhao *et al.* designed a new tube-in-tube structured carbon nanomaterial (TTCN) on pristine multi-walled carbon nanotubes (MWCNTs) using an encapsulation method [48]. Briefly, a solid SiO_2 layer and a porous SiO_2 layer were coated on the acid-treated MWCNTs followed by trapping an organosilicon compound (octadecyltrimethoxysilane, $C_{18}TMS$) within the porous SiO_2 shells. After chemical treatment, high-temperature calcination, and an etching process of the SiO_2 layer, the TTCN with MWCNTs confined into hollow porous CNTs was prepared. Finally, an S-TTCN composite with a sulfur content of 71 wt.% could be fabricated by a simple melt infiltration method. The cell with such prepared electrodes could achieve a high reversible capacity, good cycling performance, as well as excellent rate capability.

Nanowires made by hybridizing various metals or ceramic materials on the CNTs' surface have also been reported. These carbon/inorganic scaffold structures have high electrical conductivity and specific surface area, which possess efficient electron transport, mass diffusion, and structural stabilization [49]. As can be seen from **Figure 7.2a**, Li *et al.* designed and developed a free-standing pie-like paper electrode *via* electrospinning, which was further applied as a cathode [50]. Polyacrylonitrile (PAN) and polystyrene (PS) were used as the precursors in this study. After the carbonization process, CNFs were prepared in the hollow area where PS nanofibers were decomposed. Then, sulfur was introduced into these spaces. The resultant structure was composed of a 3D nanofiber membrane, which could have excellent electrical conductivity, and be able to maximize the loading and efficiency of active sulfur material, showing excellent cycling stability.

Pu *et al.* developed a unique ceramic nanowire composed of Co_3S_4@S nanotube and applied it as a cathode for Li-S batteries [51]. The hydrothermal treatment of aqueous urea and $CoCl_2$ solutions was utilized for fabricating Co_3S_4 nanotubes, and then sulfur was introduced into the Co_3S_4 nanotubes *via* a melt-diffusion method (**Figure 7.2b**). This nanowire structure had superior absorption and catalytic effect on sulfur species thanks to its high specific surface area with the help of hosting sulfur species. In addition, the cell with it could have a high capacity and a slow capacity decay rate

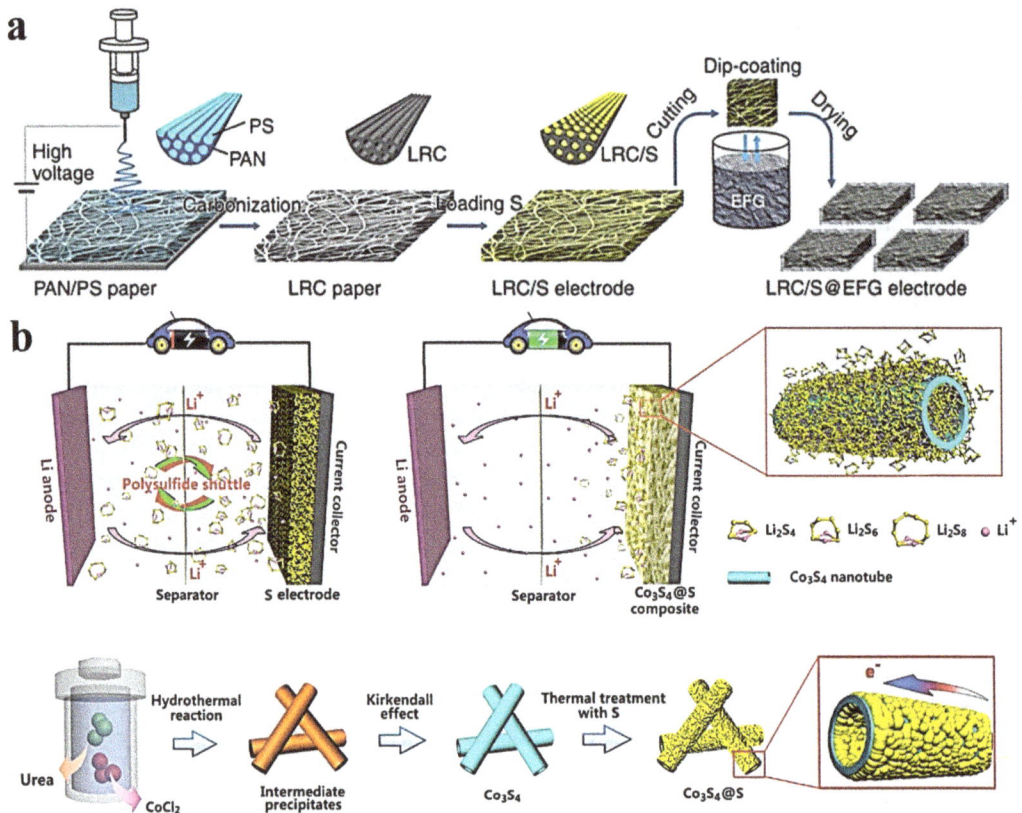

FIGURE 7.2 (a) The procedure of the lotus root-like multichannel carbon (LRC)/S@ethylenediamine-functionalized reduced graphene oxide (EFG) electrode preparation. Adapted with permission [50]. Copyright 2015, Nature Publisher. (b) The preparation of the sulfur-coated Co_3S_4 and its working mechanism. Adapted with permission [51]. Copyright 2017, Elsevier.

of 0.041% over 1000 cycles. Li *et al.* synthesized another ceramic nanowire structure, freestanding carbon encapsulated conductive mesoporous vanadium nitride nanowires, which served as a high-performance sulfur host material since this structure had good mechanical flexibility and short electron/ion transfer paths along with a high areal mass loading [52]. Therefore, tremendous efforts have been devoted to applying nanowires to Li-S batteries because of their outstanding properties. As various types of metal-sulfur batteries are continuously being developed, nanowires are expected to be more promising in the future.

7.4.2 SEPARATORS AND INTERLAYERS

Nanowires have also been utilized as separators and interlayers in metal-sulfur batteries to minimize the diffusion of polysulfides. Particularly, the nanowire structure composed of carbon and ceramic nanomaterials associated with excellent polysulfide adsorption/absorption capability has attracted extraordinary attention.

Zhu *et al.* introduced a PAN/graphene oxide (GO) composite nanofiber membrane *via* electrospinning, which was further applied as a multifunctional separator in Li-S batteries [28]. **Figure 7.3a** shows a schematic image of a Li-S cell containing the above mentioned highly porous membrane, which could successfully suppress the diffusion of polysulfides due to the interaction between the negatively charged functional groups on GO and polysulfides, enabling significant enhancement in terms of the utilization of the active material. Inspired by the previous work, Zhu *et al.* also developed a bi-functional double-layer polyvinylidene fluoride (PVDF) based composite nanofiber separator [27]. PVDF, a polymer with excellent chemical and thermal stability, was combined with electrically conductive reduced GO (rGO) to finally synthesize a double-layer separator. The film composed of these nanofibers could trap polysulfides in the cathode region (**Figure 7.3b**), therefore, the cycling stability and rate capability of Li-S batteries could be remarkably improved.

In addition, ceramic nanowires have been investigated and demonstrated as effective interlayers. Kim *et al.* manufactured N-doped carbon-embedded one-dimensional (1D) TiN nanowires (NC/TiN

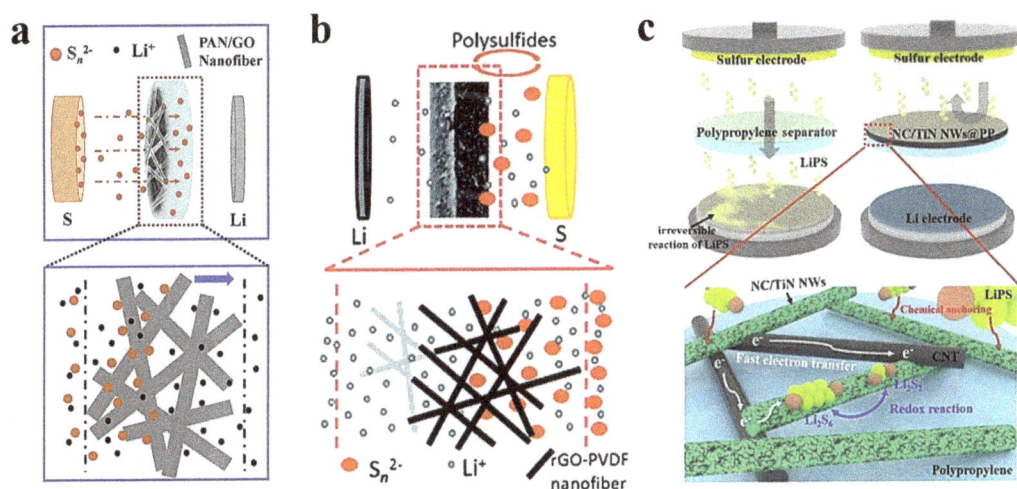

FIGURE 7.3   (a) Schematic illustration of a Li-S cell with the PAN/GO composite separator. Adapted with permission [28]. Copyright 2016, Elsevier. (b) Schematic illustration of a Li-S cell with the PVDF/rGO-PVDF composite separator. Adapted with permission [27]. Copyright 2017, Royal Society of Chemistry. (c) Schematic images showing the Li-S cells with traditional PP and NC/TiN nanowires coated PP separators. Adapted with permission [53]. Copyright 2021, Elsevier.

NWs) *via* electrospinning followed by a vacuum filtration approach, and they fabricated a novel ceramic nanowire interlayer by depositing them on a polypropylene (PP) separator (**Figure 7.3c**) [53]. This material exhibited good electrochemical performance and high reversible discharge capacity because of its unique structural features. Huang *et al.* obtained a novel functional interlayer based on the sodium-containing titanium oxide nanowires/nanosheets (STO-W/S) hybrid, which was deposited on the PP separator [54]. The prepared novel ceramic nanowires were chemically stable and had a strong adsorption capacity for lithium polysulfides (LPSs). Thus, it effectively blocked LPSs and improved cyclic stability, the capacity, and the rate capability of Li-S batteries.

7.4.3 ELECTROLYTES

Nanowires have been widely utilized to improve the electrical conductivity of electrodes, suppress their volume expansion, and address the shuttle effect of polysulfides. However, the growth of lithium dendrites has not been solved, which is mainly because of the uneven nucleation and inferior reversibility during the lithium plating/stripping process [16]. In addition to that, the traditional flammable organic liquid electrolytes might cause some other issues such as leakage and explosion. In this regard, nanowire-based solid-state electrolytes have been further explored for Li-S cells, which cannot only avoid using the flammable electrolytes but suppress the lithium dendrite growth.

For instance, Wan *et al.* introduced $Li_7La_3Zr_2O_{12}$ (LLZO) nanowires to polyethylene oxide (PEO) to enhance the ionic conductivity and mechanical strength of the solid-state composite electrolyte [55]. Briefly, the LLZO nanowire precursor was prepared by electrospinning, which was peroxided at 280 °C for 2 h in air and then further heated up to 700 °C for another 2 h to fabricate LLZO nanowires. LLZO nanowires were later mixed with PEO and lithium bis(trifluoromethylsulphonyl) imide (LiTFSI) with the assistance of acetonitrile. The existence of LLZO nanowires could not only provide continuous ionic conductive paths but also improve the effective conductive interface between LLZO and PEO due to their high surface area to volume ratio. The prepared hybrid solid-state electrolyte reached an ionic conductivity of 2.39×10^{-4} S/cm even at room temperature with a strength of 1 MPa. Attributed to the outstanding properties of the integrated solid electrolytes, the symmetric Li cells with PEO/LiTFSI/LLZO nanowires (PLLN) exhibited much better stability compared with PEO/LiTFSI (PL) and PEO/LiTFSI/LLZO microparticles (PLLM), which could be cycled stably even up to 1000 h at 60 °C without short circuit. Additionally, a large extended electrochemical stability window of up to 6.0 V could be achieved, suggesting that this unique composite solid electrolyte could meet the higher voltage requirements of most lithium metal batteries.

PEO has been extensively studied as a polymer matrix for solid electrolytes mainly because of its good thermal/mechanical properties and excellent processability. However, the relatively low ionic conductivity mainly due to the formation of crystallites in PEO systems has hindered its practical applications, thus, other polymers (*i.e.*, PVDF, PAN, *etc.*) have also been investigated [56]. Yao *et al.* synthesized palygorskite $((Mg, Al)_2Si_4O_{10}(OH))$ nanowires cooperated with PVDF to generate composite solid electrolytes for solid-state Li batteries (**Figure 7.4a, b**) [57]. It should be noticed that the yield strength of the prepared solid-state electrolyte membrane could be up to 4.7 MPa (**Figure 7.4c**) with a Li-ion transfer number of 0.54. The effect of drying temperatures on ionic conductivity has been explored. **Figure 7.4d** shows that an ionic conductivity of 1.2×10^{-4} S/cm of the electrolyte with an addition of 5 wt.% palygorskite $((Mg, Al)_2Si_4O_{10}(OH))$ nanowires could be obtained under a temperature of 60 °C, which was significantly higher than that of 80 °C (2.0×10^{-6} S/cm) and 100 °C (1.1×10^{-6} S/cm). Nevertheless, the current density used in this work should be further increased since it could not meet the requirement of practical applications. What is more, the prepared solid-state electrolytes by using nanowires have not been performed in Li-S cells, which should be comprehensively evaluated.

In summary, advanced 1D nanowire-based solid-state electrolytes have been successfully synthesized, which can be assembled into two-dimensional (2D) and/or 3D architectures

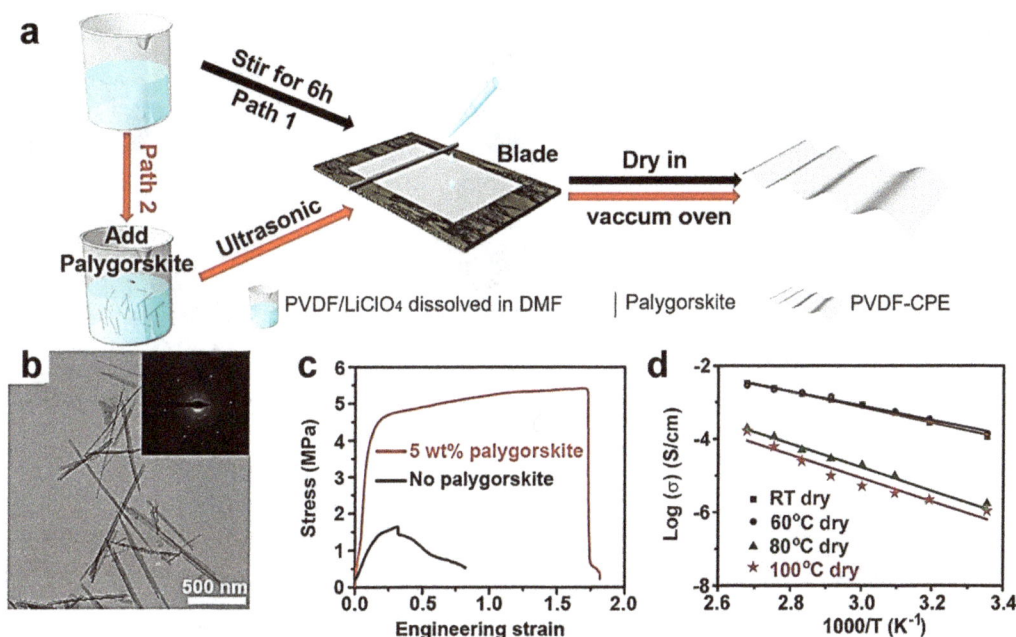

FIGURE 7.4 (a) Schematic illustration of the PVDF/palygorskite composite electrolyte preparation. (b) A transmission electron microscopy (TEM) image of palygorskite nanowires. (c) Mechanical properties of the electrolyte membranes with or without palygorskite. (d) Ionic conductivities of the composite electrolyte at different temperatures. Adapted with permission [57]. Copyright 2018, American Chemical Society.

with multiple embedded phases. However, their applications in Li-S cells should be thoroughly investigated to demonstrate their applicability and capability.

7.4.4 ANODE PROTECTION

Li has been considered as a promising anode candidate for next-generation rechargeable batteries attributed to its high theoretical capacity and low negative potential. While the uncontrollable Li dendrite growth has tremendously impeded its practical applications [9]. Additionally, in a Li-S cell, those polysulfide intermediates could react with Li to form an inactive layer, shortening its longevity along with low efficiency. Therefore, nanowires have been performed to not only minimize the diffusion of polysulfides but suppress the Li dendrite growth, which can both protect the Li anode.

A low-cost roll-press approach has been utilized to prepare the Li coated copper nitride (Cu_3N) nanowires [58]. Cu foil was first immersed in a solution of ammonium persulfate and sodium hydroxide for 10 min to form $Cu(OH)_2$ nanowires, which were heat-treated at 350 °C for 1.5 h under a mixed gas of Ar and NH_3 to fabricate Cu_3N nanowires. Cu_3N nanowires with a thickness of 3 μm covered the pristine Li *via* the roll-press method. A symmetric Li cell with such a unique configuration achieved an excellent cycling performance along with low overpotentials even at 5 mA/cm², suggesting its superior capability in protecting Li anodes.

A tailored 3D Li_2O coated by Cu nanowires has also been established to protect the Li anode [59]. **Figure 7.5a** exhibits the procedure of a 3D skeleton of Li_2O@Cu composite nanowires array on Cu form (Li_2O@CuNA/CF). By combining the advantages of a porous 3D skeleton and outstanding Li compatibility, it has been investigated that the cell with such a prepared electrode could accelerate the formation of a stable passivation film, suppressing the lithium dendrite growth to achieve high Coulombic efficiencies (98.5% after 300 cycles) (**Figure 7.5b**). Meanwhile, it not only

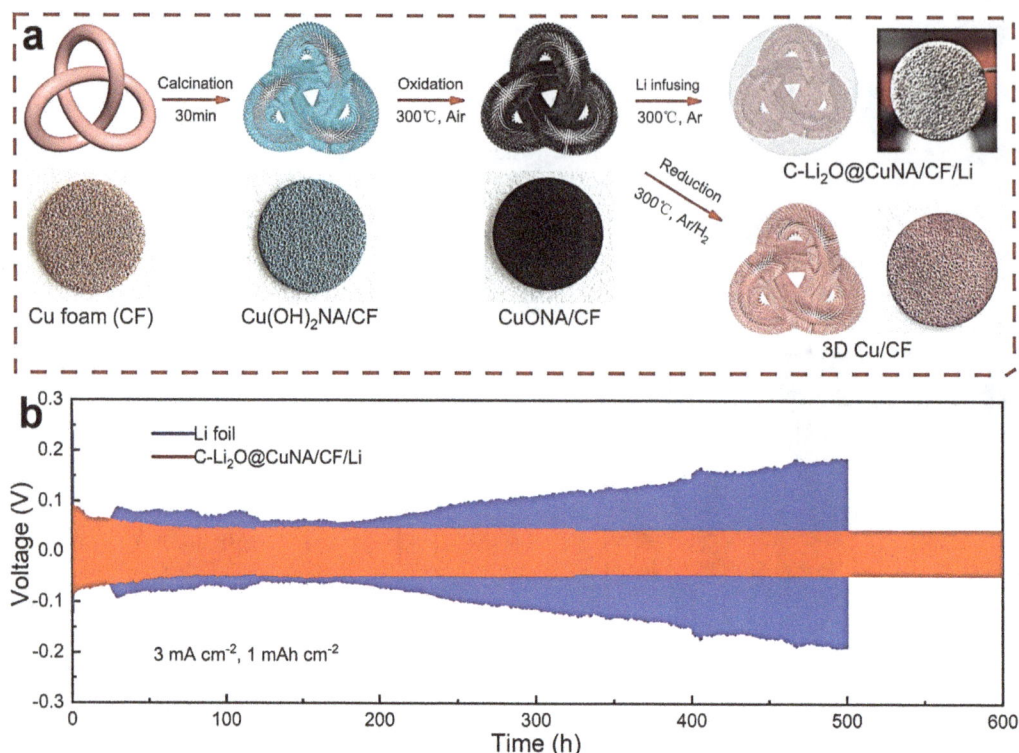

FIGURE 7.5 (a) Schematic image showing the procedure of 3D Cu/CF composite. (b) Cycling performance of the symmetric Li cells with the Li foil and C-Li$_2$O@CuNA/CF/Li electrodes. Adapted with permission [59]. Copyright 2020, Elsevier.

tremendously lowered the interfacial resistance between electrolytes and electrodes but remarkably shortened the ion/electron diffusion distance. As a result, the surface of Li after cycling became relatively rougher compared with that of fresh Li, and there were no Li dendrites observed, which could be ascribed to its excellent lithiophilic property, providing enough Li nucleation sites for regulating the distribution of Li ions and homogenizing the repeated Li plating/stripping. Even though the 3D structure can facilitate the ion and electron supply, it sacrifices the energy density of the cell because of introducing more inactive mass. Therefore, optimizing the mixing ratio of active/inactive materials is critical, which enables the cell to achieve the highest active material loading.

Overall, we have discussed how to obtain high-performance Li-S cells by nanowires *via* various approaches with remarkable progress in this field. However, more intensive studies should be explored to further enhance the energy density and rate capability of Li-S cells with the successful suppression of Li dendrite growth.

7.5 CONCLUSIONS AND PERSPECTIVES

Metal-sulfur batteries, especially Li-S batteries, are recognized as the next-generation energy storage devices due to their high theoretical capacity and energy density. Sulfur is also highly accessible, economical, and environmentally friendly. However, several challenges are hindering the commercial viability including the insulation and volume expansion of sulfur, the polysulfide shuttle effect, the Li dendrite growth, *etc*. Nanowires are a promising candidate to be used in metal-sulfur batteries to address current challenges because of their high surface-to-volume ratio and versatility. Applications of nanowires in metal-sulfur batteries have been discussed in detail. They can be processed into host materials for cathodes, adsorbents towards polysulfides, multi-functional interlayers, and even solid-state electrolytes. Although many efforts have been devoted to nanowires within metal-sulfur batteries, more fundamental research is demanded in this field to further expand possibilities. Meanwhile, low-cost synthesis methods are highly demanded to enable better cost-effectiveness and efficiency. Nanowire-based composites should also be explored to have both signature features of the nanowires and the couple materials, and serve multi-function in metal-sulfur batteries. With such expanding research, nanowires with high potential can be performed to solve the challenges and pave the way toward commercially viable metal-sulfur batteries.

ACKNOWLEDGMENTS

This work was partially supported by the Brain Pool program funded by the Ministry of Science and ICT through the National Research Foundation of Korea (2020H1D3A1A04080324).

REFERENCES

1. C. M. Lieber, Z. L. Wang. Functional nanowires. *MRS Bull.* 32, **2007**, 99–108.
2. W. Lu, C. M. Lieber. Semiconductor nanowires. *J. Phys. D* 39, **2006**, R387.
3. N. Wang, Y. Cai, R. Q. Zhang. Growth of nanowires. *Mater. Sci. Eng. R Rep.* 60, **2008**, 1–51.
4. Y. Huang, X. Duan, C. M. Lieber. Nanowires for integrated multicolor nanophotonics. *Small*, 1, **2005**, 142–147.
5. X. Duan, C. M. Lieber. General synthesis of compound semiconductor nanowires. *Adv. Mater.* 12, **2000**, 298–302.
6. J. Zhu, Y. Lu, C. Chen, Y. Ge, S. Jasper, J. D. Leary, D. Li, M. Jiang, X. Zhang. Porous one-dimensional carbon/iron oxide composite for rechargeable lithium-ion batteries with high and stable capacity. *J. Alloys Compd.* 672, **2016**, 79–85.
7. S. T. Mahmud, R. Mia, S. Mahmud, S. Sha, R. Zhang, Z. Deng, M. Yanilmaz, L. Luo, J. Zhu. Recent developments of tin (II) sulfide-based materials for achieving high-performance lithium-ion batteries: a critical review. *Nanomaterials* 12, **2022**, 1246.
8. J. Zhu, C. Yan, X. Zhang, C. Yang, M. Jiang, X. Zhang. A sustainable platform of lignin: from bioresources to materials and their applications in rechargeable batteries and supercapacitors. *Prog. Energy Combust. Sci.* 76, **2020**, 100788.
9. C. Yan, P. Zhu, H. Jia, Z. Du, J. Zhu, R. Orenstein, H. Cheng, N. Wu, M. Dirican, X. Zhang. Garnet-rich composite solid electrolytes for dendrite-free, high-rate, solid-state lithium-metal batteries. *Energy Storage Mater.* 26, **2020**, 448–456.
10. H. Hamed, S. Yari, J. D'Haen, F. U. Renner, N. Reddy, A. Hardy, M. Safari. Demystifying charge transport limitations in the porous electrodes of lithium-ion batteries. *Adv. Energy Mater.* 10, **2020**, 2002492.
11. E. C. Evarts. Lithium batteries: To the limits of lithium. *Nature* 526, **2015**, S93-S95.
12. N. Yabuuchi, T. Ohzuku. Novel lithium insertion material of $LiCo_{1/3}Ni_{1/3}Mn_{1/3}O_2$ for advanced lithium-ion batteries. *J. Power Sources* 119, **2003**, 171–174.
13. J. Zhu, C. Chen, Y. Lu, Y. Ge, H. Jiang, K. Fu, X. Zhang. Nitrogen-doped carbon nanofibers derived from polyacrylonitrile for use as anode material in sodium-ion batteries. *Carbon* 94, **2015**, 189–195.
14. L.-X., Yuan, Z.-H. Wang, W.-X. Zhang, X.-L. Hu, J.-T. Chen, Y.-H. Huang, J. B. Goodenough. Development and challenges of $LiFePO_4$ cathode material for lithium-ion batteries. *Energy Environ. Sci.* 4, **2011**, 269–284.
15. Y.-H. Huang, J. B. Goodenough. High-rate $LiFePO_4$ lithium rechargeable battery promoted by electrochemically active polymers. *Chem. Mater.* 20, **2008**, 7237–7241.
16. J. Zhu, H. Cheng, P. Zhu, Y. Li, Q. Gao, X. Zhang. Electrospun nanofibers enabled advanced lithium-sulfur batteries. *Acc. Mater. Res.* 3, **2022**, 149–160.

17. T. Li, X. Bai, U. Gulzar, Y.-J. Bai, C. Capiglia, W. Deng, X. Zhou, Z. Liu, Z. Feng, R. P. Zaccaria. A comprehensive understanding of lithium–sulfur battery technology. *Adv. Funct. Mater.* 29, **2019**, 1901730.

18. J. Zhu, M. Yanilmaz, K. Fu, C. Chen, Y. Lu, Y. Ge, D. Kim, X. Zhang. Understanding glass fiber membrane used as a novel separator for lithium-sulfur batteries. *J. Membrane Sci.* 504, **2016**, 89–96.

19. J. Zhu. Advanced separator selection and design for high-performance lithium-sulfur batteries. Ph.D thesis, North Carolina State University, **2016**.

20. R. K. Selvan, P. Zhu, C. Yan, J. Zhu, M. Dirican, A. Shanmugavani, Y. S. Lee, X. Zhang. Biomass-derived porous carbon modified glass fiber separator as polysulfide reservoir for Li-S batteries. *J. Colloid Interface Sci.* 513, **2018**, 231–239.

21. Y. Li, J. Zhu, P. Zhu, C. Yan, H. Jia, Y. Kiyak, J. Zang, J. He, M. Dirican, X. Zhang. ZrO_2 confined in porous nitrogen-doped carbon nanofiber used as a novel separator in lithium-sulfur batteries with a long lifespan. *Chem. Eng. J.* 349, **2018**, 376–387.

22. J. Zhu, E. Yildirim, K. Aly, J. Shen, C. Chen, Y. Lu, M. Jiang, D. Kim, A. E. Tonelli, M. A. Pasquinelli, P. D. Bradford, X. Zhang. Hierarchical multi-component nanofiber separators for lithium polysulfide capture in lithium-sulfur batteries: an experimental and molecular modeling study. *J. Mater. Chem.A* 4, **2016**, 13572–13581.

23. P. Zhu, C. Yan, J. Zhu, J. Zang, Y. Li, H. Jia, X. Dong, Z. Du, C. Zhang, N. Wu, M. Dirican, X. Zhang. Flexible electrolyte-cathode bilayer framework with stabilized interface for room-temperature all-solid-state lithium-sulfur batteries. *Energy Storage Mater.* 17, **2019**, 220–225.

24. Y. Li, J. Zhu, P. Zhu, C. Yan, H. Jia, Y. Kiyak, J. Zang, J. He, M. Dirican, X. Zhang. Glass fiber separator coated by porous carbon nanofiber derived from immiscible PAN/PMMA for high-performance lithium-sulfur batteries derived from polymer blends. *J. Membrane Sci.* 552, **2018**, 31–42.

25. P. Zhu, J. Zang, J. Zhu, Y. Lu, C. Chen, M. Jiang, C. Yan, M. Dirican, R. K. Selvan, D. Kim, X. Zhang. Effect of rGO reduction degree on the performance of polysulfide rejection in lithium-sulfur batteries. *Carbon* 126, **2018**, 594–600.

26. J. Zhu, Y. Ge, D. Kim, Y. Lu, C. Chen, M. Jiang, X. Zhang. A novel separator coated by carbon for achieving exceptional high-performance lithium-sulfur batteries. *Nano Energy* 20, **2016**, 176–184.

27. P. Zhu, J. Zhu, J. Zang, C. Chen, Y. Lu, M. Jiang, C. Yan, K. R. Selvan, X. Zhang. A novel bi-functional double-layer rGO-PVDF/PVDF composite nanofiber membrane separator with enhanced thermal stability and effective polysulfide inhibition for high-performance lithium-sulfur batteries. *J. Mater. Chem. A* 5, **2017**, 15096–15104.

28. J. Zhu, C. Chen, Y. Lu, J. Zang, M. Jiang, D. Kim, X. Zhang. Highly porous polyacrylonitrile/graphene oxide membrane separator exhibiting excellent anti-self-discharge feature for high-performance lithium-sulfur batteries. *Carbon* 101, **2016**, 272–280.

29. Z. Li, Q. He, X. Xu, Y. Zhao, X. Liu, C. Zhou, D. Ai, L. Xia, L. Mai. A 3D nitrogen-doped graphene/ tin nanowires composite as a strong polysulfide anchor for lithium–sulfur batteries with enhanced rate performance and high areal capacity. *Adv. Mater.* 30, **2018**, 1804089.

30. L. Luo, S.-H. Chung, H. Y. Asl, A. Manthiram. Long-life lithium–sulfur batteries with a bifunctional cathode substrate configured with boron carbide nanowires. *Adv. Mater.* 30, **2018**, 1804149.

31. K. Fu, Y. Li, M. Dirican, C. Chen, Y. Lu, J. Zhu, Y. Li, L. Cao, P. D. Bradford, X. Zhang. Sulfur gradient-distributed CNF composite: a self-inhibiting cathode for binder-free lithium-sulfur batteries. *Chem. Commun.* 50, **2014**, 10277–10280.

32. J. Zhu, P. Zhu, C. Yan, X. Dong, X. Zhang. Recent progress in polymer materials for advanced lithium-sulfur batteries. *Prog. Polym. Sci.* 90, **2019**, 118–163.

33. L. Zhou, D. L. Danilov, R.-A. Eichel, P. HL Notten. Host materials anchoring polysulfides in Li–S batteries reviewed. *Adv. Energy Mater.* 11, **2021**, 2001304.

34. H. Xu, Q. Jiang, B. Zhang, C. Chen, Z. Lin. Integrating conductivity, immobility, and catalytic ability into high-N carbon/graphene sheets as an effective sulfur host. *Adv. Mater.* 32, **2020**, 1906357.

35. M. Li, R. Carter, A. Douglas, L. Oakes, C. L. Pint. Sulfur vapor-infiltrated 3D carbon nanotube foam for binder-free high areal capacity lithium–sulfur battery composite cathodes. *ACS Nano* 11, **2017**, 4877–4884.

36. P. Zhu, J. Zhu, C. Yan, M. Dirican, J. Zang, H. Jia, Y. Li, Y. Kiyak, H. Tan, X. Zhang. In-situ polymer-ization of nanostructured conductive polymer on 3D sulfur/carbon nanofiber composite matrix as a cathode for high-performance lithium-sulfur batteries. *Adv. Mater. Interfaces* 5, **2018**, 1701598.

37. X. Chen. Rational design of cathode materials for high performance lithium-sulfur batteries. University of Wisconsin Milwaukee, **2016**.
38. M. R. Busche, P. Adelhelm, H. Sommer, H. Schneider, K. Leitner, J. Janek. Systematical electrochemical study on the parasitic shuttle-effect in lithium-sulfur-cells at different temperatures and different rates. *J. Power Sources* 259, **2014**, 289–299.
39. S. Bai, Y. Sun, J. Yi, Y. He, Y. Qiao, H. Zhou. High-power Li-metal anode enabled by metal-organic framework modified electrolyte. *Joule* 2, **2018**, 2117–2132.
40. Y. Zhao, X. Yang, Q. Sun, X. Gao, X. Lin, C. Wang, F. Zhao, Y. Sun, K. R. Adair, R. Li, M. Cai. Dendrite-free and minimum volume change Li metal anode achieved by three-dimensional artificial interlayers. *Energy Storage Mater.* 15, **2018**, 415–421.
41. K. Chen, G. Zhang, L. Xiao, P. Li, W. Li, Q. Xu, J. Xu. Polyaniline encapsulated amorphous V_2O_5 nanowire-modified multi-functional separators for lithium–sulfur batteries. *Small Methods* 5, **2021**, p.2001056.
42. K. Gao, D. Xia, H. Ji, Y. Chen, L. Zhang, Y. Wang, S. Yi, Y. Zhou, Z. Zhang, D. Chen, Y. Gao. Linear-PEI-derived hierarchical porous carbon nanonet flakes decorated with MoS_2 as efficient polysulfides stabilization interlayers for lithium–sulfur battery. *Energy Fuels* 35, **2021**, 10303–10314.
43. P. Zhu, C. Yan, M. Dirican, J. Zhu, J. Zang, R. K. Selvan, C.-C. Chung, H. Jia, Y. Li, Y. Kiyak, N. Wu, X. Zhang. $Li_{0.33}La_{0.55}TiO_3$ ceramic nanofiber-enhanced polyethylene oxide-based composite polymer electrolyte for all-solid-state lithium batteries. *J. Mater. Chem. A* 6, **2018**, 4279–4285.
44. K. Yu, X. Pan, G. Zhang, X. Liao, X. Zhou, M. Yan, L. Xu, L. Mai. Nanowires in energy storage devices: Structures, synthesis, and applications. *Adv. Energy Mater.* 8, **2018**, 1802369.
45. X. Liu, Y. Li, X. Xu, L. Zhou, L. Mai. Rechargeable metal (Li, Na, Mg, Al)-sulfur batteries: Materials and advances. *J. Energy Chem.* 61, **2021**, 104–134.
46. T. Liu, H. Hu, X. Ding, H. Yuan, C. Jin, J. Nai, Y. Liu, Y. Wang, Y. Wan, X. Tao. 12 years roadmap of the sulfur cathode for lithium sulfur batteries (2009–2020). *Energy Storage Mater.* 30, **2020**, 346–366.
47. M. Cheng, R. Yan, Z. Yang, X. Tao, T. Ma, S. Cao, F. Ran, S. Li, W. Yang, C. Cheng. Polysulfide catalytic materials for fast-kinetic metal–sulfur batteries: Principles and active centers. *Adv. Sci.* 9, **2022**, 2102217.
48. Y. Zhao, W. Wu, J. Li, Z. Xu, L. Guan. Encapsulating MWNTs into hollow porous carbon nanotubes: a tube-in-tube carbon nanostructure for high-performance lithium-sulfur batteries. *Adv. Mater.* 26, **2014**, 5113–5118.
49. Q. Fan, W. Liu, Z. Weng, Y. Sun, H. Wang. Ternary hybrid material for high-performance lithium–sulfur battery. *J. Am. Chem. Soc.* 137, **2015**, 12946–12953.
50. Z. Li, J. T. Zhang, Y. M. Chen, J. Li, X. W. D. Lou. Pie-like electrode design for high-energy density lithium–sulfur batteries. *Nature Commun.* 6, **2015**, 1–8.
51. J. Pu, Z. Shen, J. Zheng, W. Wu, C. Zhu, Q. Zhou, H. Zhang, F. Pan. Multifunctional Co_3S_4@ sulfur nanotubes for enhanced lithium-sulfur battery performance. *Nano Energy* 37, **2017**, 7–14.
52. X. Li, K. Ding, B. Gao, Q. Li, Y. Li, J. Fu, X. Zhang, P. K. Chu, K. Huo. Freestanding carbon encapsulated mesoporous vanadium nitride nanowires enable highly stable sulfur cathodes for lithium-sulfur batteries. *Nano Energy* 40, **2017**, 655–662.
53. Y. Kim, Y. Noh, J. Bae, H. Ahn, M. Kim, W. B. Kim. N-doped carbon-embedded TiN nanowires as a multifunctional separator for Li–S batteries with enhanced rate capability and cycle stability. *J. Energy Chem.* 57, **2021**, 10–18.
54. Z. D. Huang, M. T. Yang, J. Q. Qi, P. Zhang, L. Lei, Q. C. Du, L. Bai, H. Fu, X. S. Yang, R. Q. Liu, T. Masese. Mitigating the polysulfides "shuttling" with TiO_2 nanowires/nanosheets hybrid modified separators for robust lithium-sulfur batteries. *Chem. Eng. J.* 387, **2020**, 124080.
55. Z. Wan, D. Lei, W. Yang, C. Liu, K. Shi, X. Hao, L. Shen, W. Lv, B. Li, Q. H. Yang, F. Kang. Low resistance–integrated all-solid-state battery achieved by $Li_7La_3Zr_2O_{12}$ nanowire upgrading polyethylene oxide (PEO) composite electrolyte and PEO cathode binder. *Adv. Funct. Mater.* 29, **2019**, 1805301.
56. R. Prasanth, N. Shubha, H. H. Hng, M. Srinivasan. Effect of poly(ethylene oxide) on ionic conductivity and electrochemical properties of poly(vinylidenefluoride) based polymer gel electrolytes prepared by electrospinning for lithium ion batteries. *J. Power Sources* 245, **2014**, 283–291.

57. P. Yao, B. Zhu, H. Zhai, X. Liao, Y. Zhu, W. Xu, Q. Cheng, C. Jayyosi, Z. Li, J. Zhu, K. M. Myers. PVDF/palygorskite nanowire composite electrolyte for 4 V rechargeable lithium batteries with high energy density. *Nano Lett.* 18, **2018**, 6113–6120.

58. D. Lee, S. Sun, J. Kwon, H. Park, M. Jang, E. Park, B. Son, Y. Jung, T. Song, U. Paik. Copper nitride nanowires printed Li with stable cycling for Li metal batteries in carbonate electrolytes. *Adv. Mater.* 32, **2020**, 1905573.

59. L. Tan, X. Li, M. Cheng, T. Liu, Z. Wang, H. Guo, G. Yan, L. Li, Y. Liu, J. Wang. In-situ tailored 3D Li$_2$O@Cu nanowires array enabling stable lithium metal anode with ultra-high coulombic efficiency. *J. Power Sources* 463, **2020**, 228178.

8 Nanowires for Electrochemical Energy Storage Applications

Kwadwo Mensah Darkwa and Daniel Nframah Ampong
Department of Materials Engineering, College of Engineering, Kwame Nkrumah University of Science and Technology, Kumasi, Ghana

Rebecca Boamah
Department of Materials Science and Engineering, University of Ghana, Legon-Accra, Ghana

Stefania Akromah
Department of Aerospace Engineering, University of Bristol, UK

Reuben Amedalor
Department of Materials Science and Engineering, University of Ghana, Legon-Accra, Ghana
Institute of Photonics, University of Eastern Finland, Joensuu, Finland

Benjamin Agyei-Tuffour
Department of Materials Science and Engineering, University of Ghana, Legon-Accra, Ghana
Center of Excellence in Energy, African Research Universities Alliance (ARUA), University of Stellenbosch, Stellenbosch, South Africa

David Dodoo-Arhin
Department of Materials Science and Engineering, University of Ghana, Legon-Accra, Ghana

Neill J. Goosen
Center of Excellence in Energy, African Research Universities Alliance (ARUA), University of Stellenbosch, Stellenbosch, South Africa

Ram K. Gupta
Department of Chemistry and National Institute for Materials Advancement, Pittsburg State University, Pittsburg, KS, USA

CONTENTS

DOI: 10.1201/9781003296621-8

113

8.1   INTRODUCTION

Following the growing concerns regarding the health and wellbeing of our environment and the preservation of our natural resources, energy storage has become one of the most important topics of the 21st century. The global population is exponentially increasing and technological systems are becoming increasingly sophisticated, resulting in the pressing need for state-of-the-art sustainable energy storage devices. While reducing the consumption rate of fossil fuels (oil, coal, and natural gas) and preserving our limited natural resources, energy storage also supports the mitigation of the environmental pollution arising from the combustion of non-renewable energy sources. Additionally, it supports renewable energy systems to make them more stable and continuous, and improves the quality of the power from these electrical systems by homogenizing loads, regulating frequencies, and damping energy oscillations.

There are various types of energy storage systems, and they each have distinct storage principles and characteristics (e.g., energy density and power density, short-term or long-term storage), benefits and disadvantages, as well as distinct applications. Generally, they are classified as (a) mechanical storage systems, which include compressed air, flywheel, and pumped hydro systems; (b) electrical and electrochemical energy storage systems; (c) chemical storage systems, and (d) thermal storage systems. Electrical and electrochemical energy storage (EES) comprises batteries and supercapacitors (SCs) and has gained much attention in recent years for their ability to satisfy the demands of modern society in terms of cost-efficiency, eco-friendliness, flexibility, and being lightweight. The performance of EESs is closely related to the properties of their basic electrode materials. Despite batteries being very popular in industrial and household applications, supercapacitors have attracted the attention of both the scientific and industrial communities because of their extraordinary characteristics such as large energy storage capacity, high power density, and longer lifespan. Nevertheless, there has been a huge investment in both technologies, resulting in the wide variety of EESs currently available.

8.2   TYPES OF ELECTROCHEMICAL ENERGY DEVICES AND THEIR WORKING PRINCIPLE

8.2.1   BATTERIES

Batteries can be classed as primary batteries, secondary batteries, and battery systems for grid-scale energy supply, such as flow batteries and sodium-sulfur batteries. Primary batteries are typically single-use, non-rechargeable, and non-recyclable; however, they are the simplest and most convenient, with minimal maintenance requirements. Depending on their electrolyte system, these batteries can further be classified as aqueous and non-aqueous. Examples of primary batteries include zinc-carbon batteries, alkaline batteries, and lithium primary cells. Secondary batteries, on the other hand, are rechargeable and can undergo multiple charge/discharge cycles; they are used in car ignition and portable electronic devices, electric and hybrid vehicles, and modern electric households. Examples of secondary batteries include lead-acid, lithium-ion, nickel-metal hydride, and nickel-cadmium batteries [1].

Li-ion batteries (LIBs) have gained huge popularity among batteries for their high energy storage capacity. Since their commercialization in 1971, a lot of research has been conducted to improve their properties. Like all other electrochemical storage devices, LIBs consist of an anode, a cathode, and an electrolyte in a system that operates by intercalation and de-intercalation processes. During charging, ions are adsorbed on the anode from the cathode; the reverse occurs during discharge. The transport of Li^+ ions across the electrolyte forces electrons to flow through the external circuit. The cathode is typically made of graphite, although, recently, carbon nanotubes (CNTs) have been widely researched and proposed as an alternative due to the large surface area for intercalation and excellent electrical conductivity of CNT electrodes. Furthermore, aligned CNTs have proven to display a large specific capacity even at high discharge rates [2]. The anode is usually a Li-based material, typically lithium iron phosphate or lithium manganese oxide, while the electrolyte may be either an aqueous or a solid-state lithium-based electrolyte, e.g., lithium hexafluorophosphate [3,4]. Despite the remarkable performance and popularity of LIBs, the gradual depletion of lithium reserves and the resulting increased cost have driven current research to identify more sustainable alternatives.

8.2.2 SUPERCAPACITORS

Supercapacitors are electrochemical capacitors with the advantages of higher power densities and higher energy densities compared to batteries and conventional capacitors, respectively. SCs have higher power densities (up to 10 kW/kg), faster charge/discharge rates, and longer cycle stability compared to batteries due to their characteristic storage mechanisms [5]. They are classified as electrostatic double-layer capacitors (EDLCs), pseudocapacitors, and hybrid capacitors. The first conceptualization of supercapacitors is dated back to the 18th century with the invention of the Leyden Jar, a glass jar, and two metal foils (connected via a metallic wire) acting as the dielectric and electrodes, respectively. Following that, despite the first electrolytic capacitor being invented in the 1920s, the first EDLC based on activated charcoal was only patented in 1957 [6]. Since then, the progression of SC technology has grown exponentially in parallel with technology. These devices are generally used in applications that require that a huge amount of energy be stored or released within a very short time, for example: in hybrid vehicles, electric vehicles, fuel cell vehicles, and uninterruptible power supplies. Additionally, due to their tunable dimensions, weight, and flexibility, they can be used to supply power to portable devices like mobile, phones, notebook computers, and digital cameras [6].

EDLCs are based on carbon materials, usually activated carbons (ACs). These conductive porous electrodes store charges electrostatically; thus, no charge transfer reaction occurs between the electrodes and the electrolyte. In EDLCs, energy is stored by ion adsorption and desorption at the electrode/electrolyte interface: when the electrodes are charged, solvated ions of opposing charges are accumulated at the interface, forming an electric double layer. Several models have been developed to describe this interaction, the most popular ones being (a) the Helmholtz model, (b) the Gouy-Chapman model, and (c) the Stern model (**Figure 8.1**) [5]. The Helmholtz theory is the most basic and commonly used: according to this model, the charge on the electrode is neutralized by ions of opposite charge at a distance d, the thickness of the double-layer. According to the Gouy-Chapman model, these electrolyte ions are not rigidly bound within this layer d but are distributed within a distance known as the "diffuse layer". This model is often used for its simplicity; however, it assumes that the ions are point charges and, hence, are not limited by their size (which does not represent reality). Therefore, it is usually complemented by the Stern model, which takes the ion size into account [5]. EDLCs have higher capacitance because they have (a) a shorter double-layer distance d, (b) a larger electrode/electrolyte interface, (c) excellent electrical conductivity, and (d) good wettability of the carbon electrode [7].

Pseudocapacitors, also known as faradaic SCs, are based on redox reactions, thus, involving electron charge transfer at the surface and in the bulk of the electroactive material. These reactions

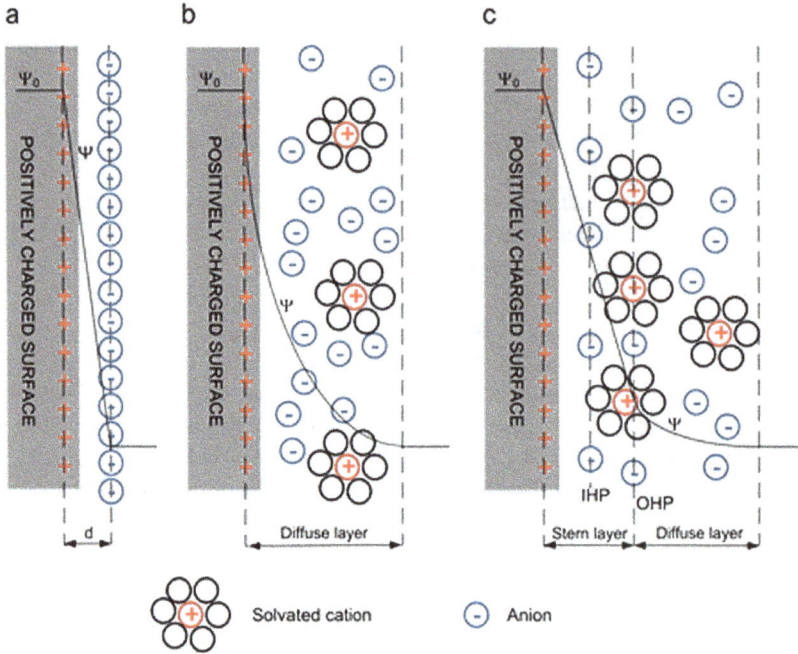

FIGURE 8.1   EDLC models of (a) the Helmholtz model, (b) the Gouy-Chapman model, and (c) the Stern model. Ψ = potential, Ψ_0 = electrode potential, IHP = inner Helmholtz plane, OHP = outer Helmholtz plane. Adapted with permission [5]. Copyright (2016) Elsevier.

follow Faraday's law and include reversible adsorption, exchange of ions through electric double layers, and reversible redox reactions. Owing to their storage mechanism, pseudocapacitors have a higher theoretical specific capacitance than EDLCs. However, this also generates high internal resistance; hence, the comparatively lower performance. Moreover, they often suffer from higher rates of degradation, lower conductivity, and limited reversibility [8]. Pseudocapacitor electrodes are typically based on metal oxides (e.g., RuO_2, IrO_2, and Co_3O_4), transition metal sulfates, and conductive polymers (polypyrrole, polyaniline, and polythiophene) [8]. These materials may be intrinsically pseudocapacitive and, thus, display this behavior regardless of the particle size and morphology, or extrinsically pseudocapacitive exhibiting this behavior only under specific conditions [5].

Hybrid SCs (HSCs) are made of electrode materials based on carbonaceous and pseudocapacitive materials, thus, synergistically combining the characteristics of both SC types (i.e., EDLCs and pseudocapacitors) to overcome their characteristic limitations [8]. HSCs have improved cell voltage, energy, and power densities; they can be divided into (a) asymmetric HSCs, based on either two electrodes of the same materials but different thicknesses, two different EDLC or pseudocapacitive materials, or an EDLC-type and pseudocapacitive-type electrodes within the same system; (b) battery/ supercapacitor hybrids, with batteries and supercapacitors together in series; and (c) self-charging supercapacitors, defined by their self-charging mechanisms, i.e., self-charging triboelectric, piezo-electric, thermal, photo capacitors, hydro capacitors, and chemical HSCs [5,9].

8.3   MATERIALS AND THEIR ARCHITECTURAL ASPECTS FOR ELECTROCHEMICAL ENERGY STORAGE

Materials for electrochemical energy storage can be classified based on morphology as zero-dimensional (0D), one-dimensional (1D), two-dimensional (2D), and three-dimensional (3D) [10]. The use of monodispersed 0D nanomaterials is often limited by their characteristic self-assembly into denser aggregates as a result of the high surface energy of 0D nanomaterials induced by the high surface-area-to-volume ratio of the particles [11]. This characteristic increased the contact resistance of the particles, thus, reducing the electrical conductivity of the material. 0D nanomaterials are often decorated on the surface of conductive 1D and 2D materials, which often enhances the storage and conductivity properties. The limitation of 0D nanomaterials can also be limited by modifying the morphology of the particles: for example, conductive particles with onion morphology (e.g., carbon onions) have been reported to display high conductive properties. Nevertheless, many disadvantages are reported in the literature for 0D nanomaterials: (a) even conductive onion particles often form Schottky junctions [12]; (b) dichalcogenides with 0D morphology are highly reactive and require surface treatments [13]; (c) semiconductors and metallic 0D nanoparticles tend to form a passivation surface layer, which hinders charge transport [14].

On the other hand, 1D nanoparticles such as carbon nanotubes and nanowires can be arranged into highly conductive continuous networks with improvements in mechanical properties as well: for example, the elastic modulus of aligned CNTs can be as high as 0.6 TPa [15]. Additionally, 1D nanomaterials produce flexible structures and porous electrodes, making them suitable for wearable electrodes and zero-expansion electrodes, respectively, with high mechanical, thermal, and conductive stability [16]. The integration of 0D and 1D nanomaterials shows some advantages in 1D-based electrode materials. 1D electrodes are limited by their low packing density, which reduces their volumetric performance; the incorporation of 0D particles results in improved packing density as these particles fill the pores between the 1D nanoparticles [16]. Despite the overall benefits of 1D materials, they are still produced on a small scale because of the high production costs.

Like 1D materials, 2D materials are characterized by their intrinsic structural porosity, which promotes fast electrolyte diffusion through the porous channels and high in-plane charge conductivity [17,18]. 2D electrodes have higher packing density and volumetric performance and do not require additives or binders; however, this high packing density may also reduce electrolyte diffusion and the material's out-of-plane conductive properties, limiting 2D nanomaterials to thin electrodes [19–21]. 2D nanomaterials are often hybridized with 1D and 0D materials to control the packing density and maximize electrolyte diffusion and the number of reaction sites [22]. Sometimes, the improvement can be achieved by interlayer spacing and chemistry modification [22]. The large-scale production of 2D via bottom-up routes is very expensive, while top-down methods such as the liquid-phase exfoliation (LPE) method, the wet chemical synthesis, and the selective etching and dealloying methods offer a relatively cheaper option [10]. 3D structures include metal-organic frameworks and aerogels. 3D nanostructures are often produced by combining 0D, 1D, and 2D nanomaterials and are most suitable for thick electrodes with up to 200 μm thickness and large areal and volumetric properties [10].

8.4 METHODS TO SYNTHESIZE NANOWIRES

Among 1D nanomaterials, nanowires have gained a lot of attention in energy storage because of the following: (a) their morphology facilitates electrode/electrolyte interaction due to increased specific surface area, providing more active sites for electrochemical reactions; (b) they provide continuous and shorter pathways for charge transfer, thus, improving the electrode conductivity; (c) they can be assembled into hierarchical and complex architectures, and 3D networks for maximum charge storage capacity; (d) they display fast strain relaxation during electrochemically driven volume expansion/contraction [23]. These advantages are related to their high crystallinity and morphology, which facilitate the absorption of high-energy photons and separation of charge carriers, providing 1D electronic pathways for efficient charge transport. Nanowires are characterized by diameters ranging from 1 to 100 nanometers and high aspect ratios. Additionally, their high strain relaxation capacity and resilience to volumetric expansion improve their mechanical stability and cycle life [24].

Based on their morphologies, nanowires are classified as simple nanowires, core-shell coated nanowires, hierarchical nanowires, porous nanowires, and hollow nanowires, as shown in Figure 8.2(a–j). These can further be assembled into nanowire arrays, nanowire networks, and

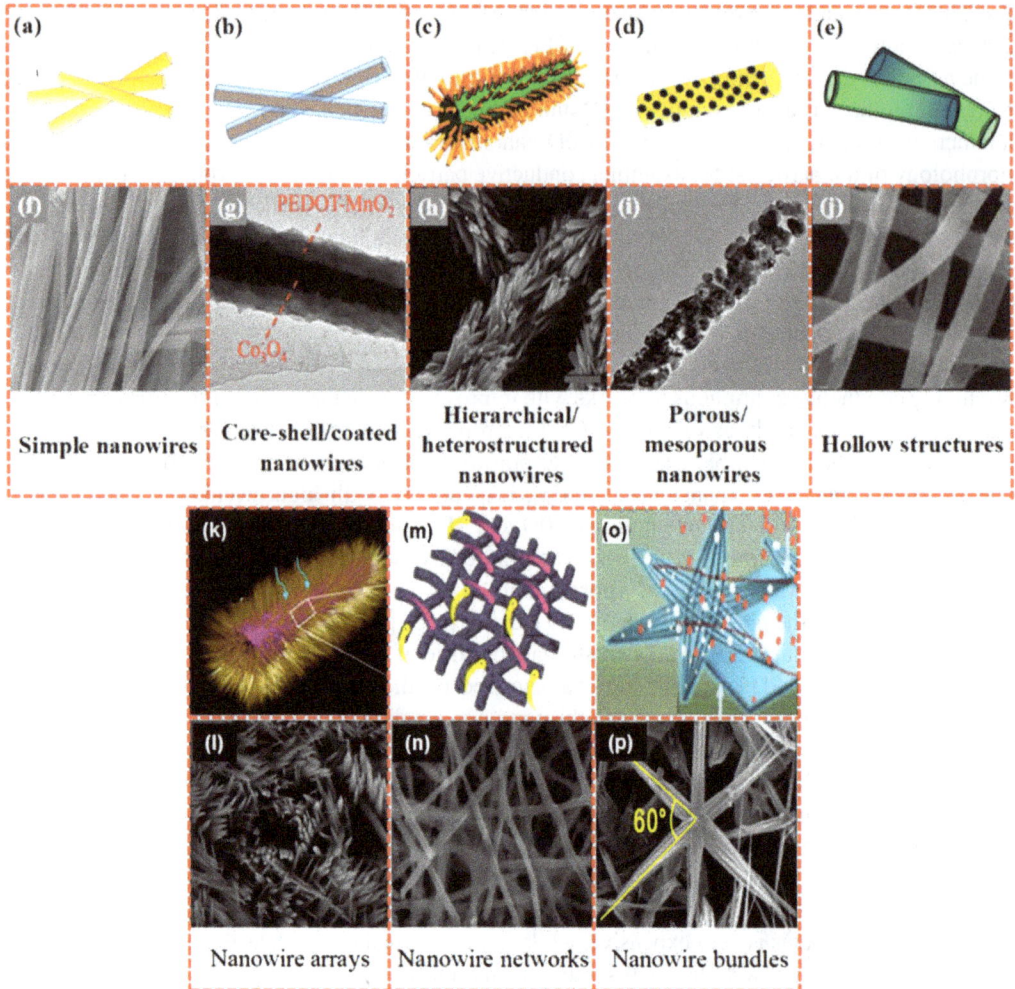

FIGURE 8.2 Schematic diagrams and respective SEM/TEM micrographs of the various nanowire morphologies. Adapted with permission [24]. Copyright (2018) Wiley.

nanowire bundles as seen in Figure 8.2(k–p). Nanowires and their nanostructures are usually synthesized via two-step processes including hydro-thermal processes, electrolytic deposition, calcination, electrospinning, microemulsion, templating, chemical vapor deposition (CVD), and vapor-liquid-solid (VLS) growth processing, among others [25].

Electrophoretic deposition (EPD) has been widely explored mainly because it does not require any sacrificial templates or high temperatures, allowing the desired parameters (morphology and composition) to be easily tuned, compared to the templating methods. EPD is a relatively simple process that involves the deposition of particles dispersed in a colloidal solution onto a compatible substrate resulting in well-defined nanostructures (Figure 8.3c) with the desired composition. Hydrothermal synthesis is another widely used method that offers improved crystallinity, narrow size distribution, and selective growth orientation enhanced by the specific processing environment (Figure 8.3a). Solution-phase methods such as oil-in-water microemulsion are simple and scalable routes that have been demonstrated to efficiently control the growth kinetics of the nanoparticles (Figure 8.3b), despite the unclear nature of the growth mechanisms [23–25].

Simple nanowires are 1D solid nanostructures. They have been typically synthesized by CVD metal substrates (e.g., silicon, germanium, and zinc) under vacuum, two-step VLS-CVD method, template

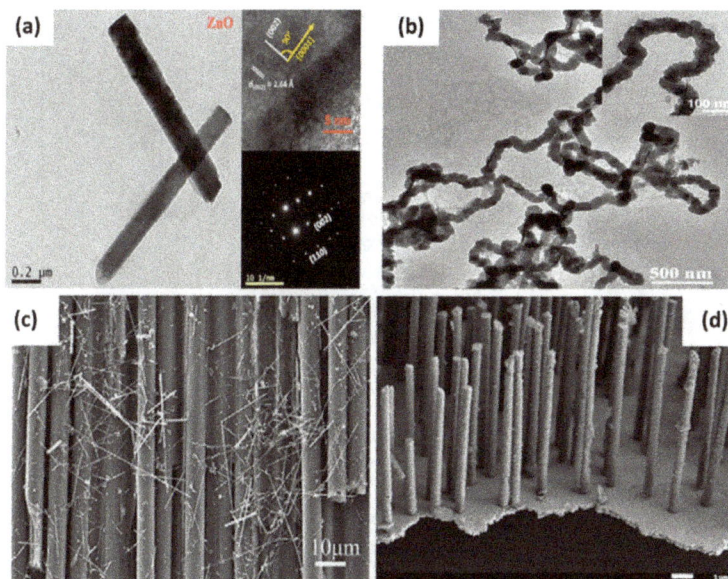

FIGURE 8.3 TEM micrographs of (a) ZnO nanorods synthesized by hydrothermal method and (b) self-assembled silica nanowires produced by microemulsion. Vertically aligned (c) SiC nanowires synthesized by EPD and (d) Co nanowires synthesized by electrochemical deposition. Adapted with permission [26]. Copyright (2020) Elsevier.

methods, and electrospinning. Despite the general benefits of simple nanowires, their use is limited by their poor stability in electrochemical systems and rapid charge/discharge cycles, resulting in detrimental structural expansions. Core-shelled and coated nanowires have been used as alternatives to mitigate this challenge: the outer shell controls the rate of release and intercalation of ions, enhancing the stability of electrochemical devices based on this charge/discharge mechanism [24].

Simple and core-shell nanowires are very unstable and tend to agglomerate. As a result, electrodes based on these materials have poor electrode/electrolyte contact areas and cycle stability. Hierarchical, porous, and hollow nanowires, on the other hand, have larger surface areas for charge transfer and surface reactions, higher ionic and electron conductivity, and improved strain adaptation capacity; hence, electrodes based on these materials have improved reaction kinetics and electrochemical stability.

2D nanowire arrays can be synthesized via "bottom-up" processes involving the assembly of structures from atoms and molecules, or via "top-down" processes whereby bulk materials are chemically etched to obtain nanowire structures. Nanowire arrays appear as a unique layered mesoporous structure that exhibits good surface properties and enhanced electrode/electrolyte interactions. It has been reported that electrodes based on these structures have higher volumetric capacitance and energy density than conventional materials, even in hybrid systems. Similarly, nanowire-based networks have a stable layered framework structure, promote rapid ion diffusion through their channels, and provide 3D pathways for electron transport. They are typically synthesized by electrospinning, but other routes can be used such as the hydrothermal, spin-coating, and annealing multi-step method proposed by Liu et al. [27].

8.5 NANOWIRES FOR ELECTROCHEMICAL ENERGY STORAGE

Electrochemical energy storage devices, such as batteries and supercapacitors have good storage performance and are less affordable such that, they are being used in portable electronics, EVs, and large-scale storage grids. Nanostructured materials such as nanorods, nanowires, nanoflowers, nanoribbons, nanotubes, etc., with their respective physical and chemical features, have significantly

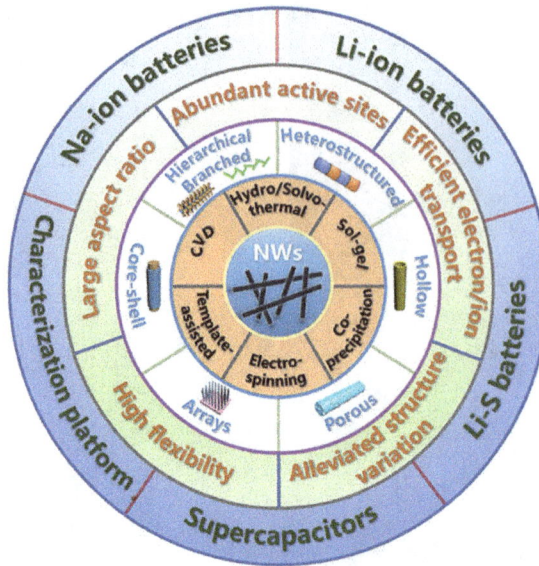

FIGURE 8.4 Nanowires' energy storage systems, synthesis methods, nanostructure types, and characteristics. Adapted with permission [23]. Copyright (2019) American Chemical Society.

contributed to the progress of energy storage systems and have sparked interest in recent decades [23]. Due to their large surface-to-volume ratio and simple synthesis approach, nanowires are promising for energy-related applications [28]. Electrodes made of nanowires are influenced by chemical composition, morphology, phase presence and stability, and surface defect patterns. For electrodes with increased capacity and cycle stability, monitoring electrode behavior during electrochemical cycling is required. Recent applications of nanowires' electrode materials include LIBs, Li-S batteries, Na-ion batteries, and supercapacitors [29].

Nanowires as common energy storage materials have many advantages in the operation of energy storage systems. The nanowire configuration expands the electrode-electrolyte contact area and improves the electrode active regions due to the electrolyte available increasing material application and capacity. Nanowires create an easy pathway that helps in the transport of electrons as well as improving the electrode kinetics and conductivity through the reduction of resistance in its charge transfer. Nanowires are an excellent model system that provides a unique platform for in situ electrochemical investigations at the microscopic scale with high precision, revealing the natural medium without the impact of external environmental factors (Figure 8.4) [23]. Despite the numerous benefits of nanowires, low coulombic effectiveness and poor cycling life are some of the challenges of nanowires. To achieve a low-cost production of nanowires in greater quantities, much effort is needed in the optimization of the methods for effective implementation.

8.5.1 Nanowires for Batteries

Generally, batteries store a large amount of energy compared to other energy storage systems. In batteries, nanowires have been used as electrodes to store a large amount of energy due to their large surface area. Nanowires have been utilized in many battery systems as electrode materials.

8.5.1.1 Lithium-Ion Battery

Lithium-ion batteries have a long cycle life, higher energy density with low self-discharge, and cost rate. Since 1991, the LIBs introduced by Sony have influenced market sales. LIBs comprise

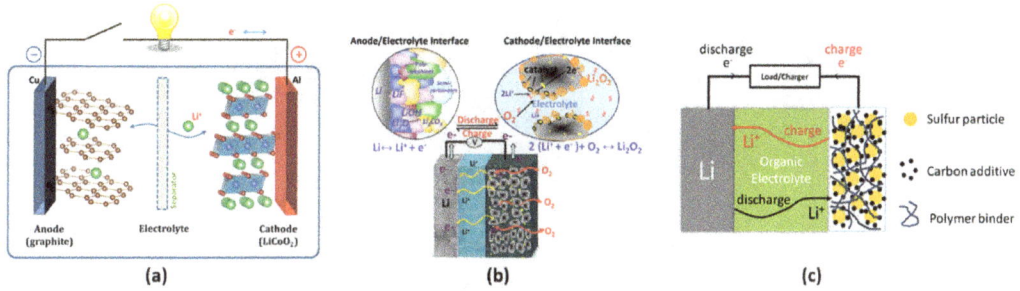

FIGURE 8.5 Schematic representation of the structure and working of (a) Li-ion battery (Adapted with permission [31]. Copyright (2013) American Chemical Society). (b) Li-air battery (Adapted with permission [32]. Copyright (2010) American Chemical Society). (c) Li-S battery (Adapted with permission [33]. Copyright (2014) American Chemical Society).

a cathode (such as $LiCoO_2$), a separator, and a graphitic anode immersed in an electrolyte (non-aqueous). The charging process of LIBs involves the detachment of Li ions from the $LiCoO_2$ cathode and being transported through the polymeric separator and electrolyte to intercalate into the graphite anode as store energy. This process causes electrons to be extracted by the external load. The discharging process is a reverse of the above. The chemical reactions that occur during intercalation at both the anode and the cathode of a Li-ion battery are shown below:

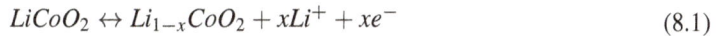

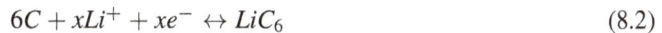

$$LiCoO_2 \leftrightarrow Li_{1-x}CoO_2 + xLi^+ + xe^- \tag{8.1}$$

$$6C + xLi^+ + xe^- \leftrightarrow LiC_6 \tag{8.2}$$

In addition, overcharging at high voltage and supersaturating the cathode lead to the formation of irreversible by-products that lead to cell destruction. Each chemical reaction is shown below:

$$LiCoO_2 \rightarrow CoO_2 + e^- + Li^+ \tag{8.3}$$

$$LiCoO_2 + e^- + Li^+ \rightarrow Li_2O + CoO \tag{8.4}$$

Other lithium-based batteries, which work on the same principle as lithium-ion batteries, include lithium-air and lithium-sulfur batteries and are represented in Figure 8.5 [30].

8.5.1.2 Nanowires as an Anode in LIBs

Lithium-ion batteries with their high energy density rate are being developed as batteries that can be recharged in portable electronic devices, power grids, EVs, etc. Developing such advanced batteries requires electrodes of high performance. Several electrode materials such as germanium, silicon, metal oxides, etc., have been reported to be good electrode materials to replace graphite anodes for lithium-ion batteries. Below are some nanowire anodes for LIBs:

Silicon (Si)-based anode: Silicon is a promising material for next-generation lithium battery anodes since it has a low discharge potential and a theoretical capacity 10 times higher than commercial graphite. Silicon is Earth's second-most abundant element [34]. Nanostructured Si electrodes stand out as a special candidate for anode materials for use in LIBs. During charging, silicon expands by 400% as it intercalates lithium, which leads to material deterioration. A few examples of nanostructures are yolk-shell structures, hollow Si spheres, 1D Si NWs, and Si nanotubes. A typical representation of Si NWs as anodic material is shown in Figure 8.6 [35]. Silicon NWs have an insignificant diameter that can contain large volume variations, rapid electronic and ionic transport pathways with constructed void spaces by adjacent NWs which allows

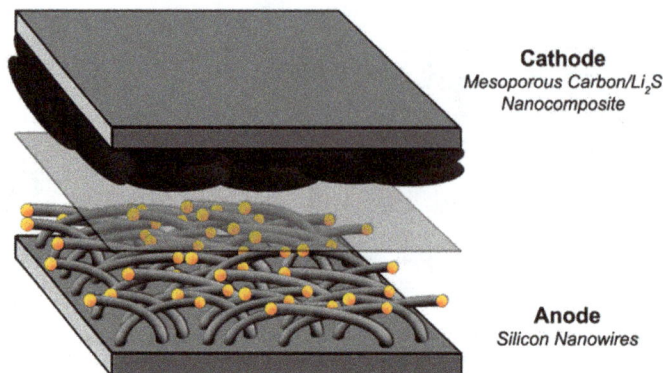

FIGURE 8.6 Schematic diagram of battery structure; the cathode contains lithium sulfide (Li_2S) encapsulated within ordered mesoporous carbon, and the anode consists of Si NWs. Adapted with permission [35]. Copyright (2010) American Chemical Society.

for fast penetration and storage of electrolytes and can offer high material utilization due to its high surface-to-volume ratio.

Nanowires coated with carbon improve the stability in the formation of solid electrolytes with much more stable interfaces. The formed solid electrolyte interface has the potential to undergo many cyclical dissolution and reformation in battery cycles. This implies that having sustainable solid electrolyte interphase is much preferred to avoid loss of capacity when the battery is in use. Silicon nanowires coated with carbon can retain ~89% of their battery capacity after a 200-cycle experiment, which is equal to that of graphitic anodes presently [34].

Germanium (Ge)-based anode: Germanium nanowire anodes have the potential to augment the cycle durability and energy density of LIB. Its high theoretical strength or capacity causes expansion in the course of charging and disintegrates over time after further cycles. Ge nanowires used as lithium anodes are about 400 times more effective during intercalating compared to Si anodes. Research conducted between Si and Ge showed that the Ge anodes retained a battery capacity of 900 mAh/g after 1100 cycles [34]. A publication by Chan et al., with the aid of chemical vapor deposition (CVD) via a vapor liquid solid (VLS) mechanism, first demonstrated the direct growth of Ge nanowires onto metal current collector substrates, which were used as high-capacity lithium battery anodes [23]. The electrode was designed to enhance effective electrical contact between the nanowires and the current collector. This helps to contain the large strain arising from charging and discharging processes and facilitates charge transport and electrolyte access, and hence, resulting in a stable cycling life with reversible capacities of about 1141 mAh/g at the C/20 rate and a high rate of 2C with a capacity of about 600 mAh/g and coulombic efficiency exceeding 99% [36].

However, there is still the need for integrating carbon materials with Ge nanowires for optimum performance. Carbon-based material such as graphene with its accompanied good physical, chemical, and electrical characteristics is of interest to be used as coating layers, supporters, and additives in Ge-based anodes for LIBs. As an example, the growth of a single and multi-layer graphene directly Ge nanowires has been demonstrated via a metal-catalyst-free CVD method [37]. This method yields a high-quality graphene layer, comparable to the metal catalytic growth of graphene. The graphene-germanium nanowires (Gr-Ge NWs) show a high specific capacity of about 1059 mAh/g and a good life cycle and capacity retention of about 90% for over 200 cycles. This result was reported to be an improvement to the bare Ge NWs [37,38].

Transition metal oxides (TMO)-based anode: Research into many TMO revealed their great potential as substitute anode materials to the conventional cell materials for batteries and other energy storage systems. Some of these notable metal oxides include Cr_2O_3, Fe_2O, PbO_2, Co_3O_4, and MnO_2. These metal oxides possess high theoretical energy capacity, and are non-toxic, environmentally friendly, and naturally abundant. Since the introduction of nanostructures for battery electrodes, many researchers have conducted numerous experiments on the possibility of TMO-based nanowires as electrode materials [39]. These materials, however, are confronted with bad electrical properties, quickness at agglomerating, charge transfer kinetics, brittleness resulting in contraction and expansion problems during Li extraction/insertion, resulting in a low rate of performance and cycling stability. These challenges limit their use as LIB anode materials [40].

To solve these problems, the design of different nanowire-based structures such as hollow-porous nanowires, core-shell nanowires, coated nanowires, branched and hierarchical nanowires, and arrays of nanowires have been proposed and adopted to enhance the electrical conductivity, shorten the charge transport path, de-stress/strain the material, and also enhance the electrochemical performance of TMO nanowires. The hollow or porous metal oxide NW structures are constructed to contain high volume change and de-stress/strain the material, thereby preventing fast capacity fading [23].

Lead oxide (PbO_2)-based anodes: Although PbO_2 is an abundant and cheap raw material for cell production, it suffers from reduced specific energy. The problem of volumetric expansion during its working cycle limits electrolyte flow. These issues hinder the ability of the cell to operate successfully during energy-intensive cycles. In 2014, researchers successfully developed PbO_2 NW via a rather simple and easy template electrodeposition method. Its performance as an anode in LIB was evaluated. The cell could operate at a fairly constant capacity of about 190 mAh/g for over 1000 cycles due to the increased surface area of the PbO_2 nanowires [29,41,42]. The result indicates or points to the use of PbO_2 nanowires as a good substitute for the conventional lead-acid anode.

Manganese oxide (MnO_2)-based anodes: MnO_2 is another potential metal oxide that has been noted to have the potential for energy storage because it is hazard-free, has high energy capacity, and is economically viable. Yet, it faces volumetric expansion during charging/discharging operation when lithium ion is inserted into its matrix. Li-enriched MnO_2 nanowires as the alternative anode materials for LIBs have been proposed as the solution. With this, the LIB could reach an energy capacity of 1279 mAh/g at 500 mA for over 500 cycles, which is higher than that of pure MnO_2 nanowire anode cells [42].

8.5.1.3   Nanowires as a Cathode in LIBs

Lithium cobalt oxide ($LiCoO_2$)-based cathodes: Currently, $LiCoO_2$ is a commercial cathode material for LIBs, with a relatively high theoretical strength of 274 mAh/g without lithium ions. Of this capacity, only about half is reversible through intercalation reaction. Extracting more than 70% of Li will destabilize the crystal structure and cause serious dissolution of cobalt. This causes rapid capacity decay [29]. An alternative to improving the electrochemical performance of $LiCoO_2$ is to employ $LiCoO_2$ nanowires. The $LiCoO_2$ nanowire will offer a high surface-to-volume ratio, improved stress/strain relief, and rapid Li-ion transfer pathways. Despite the huge patronage of $LiCoO_2$ in the battery market, it is relatively expensive, hazardous, and in limited supply. As such, the quest for an alternative solution is at the heart of current research.

Spinel lithium manganese oxide ($LiMn_2O_4$)-based cathodes: Spinel $LiMn_2O_4$ is another promising material as a cathode. It is cheaper, not harmful, readily available, and has a relatively good energy storage capacity of 148 mAh/g. Manganese oxide shows phase transformation from cubic to tetragonal as a result of deep discharge. In addition, Li-ion has slow diffusion kinetics. These issues result in capacity deterioration and poor rating potential of $LiMn_2O_4$ electrodes [28]. In an attempt to solve the problem, Kim et al. employed a combined hydrothermal solid-state synthesis to form single-crystalline $LiMn_2O_4$ nanorods. These rods offer a relatively high energy storage capacity of ~100 mAh/g at a current density of ~148 mA/g. This value is about twice the capacity of the commercially available $LiMn_2O_4$ powders [34]. Moreover, over 85% starting capacity is reserved after 100 cycles. This is ascribed to the small change in lattice constant when Li ions transform into the cubic phase of the $LiMn_2O_4$ nanorod and the simultaneous coexistence of two similar cubic phases around 3.5–4.3 V [41]. Usually, highly crystalline $LiMn_2O_4$ is best for obtaining optimum performance, which is mostly achieved with high-temperature sintering [42]. In a publication by Hosono et al., they reported the synthesis of a high-quality single-crystal $LiMn_2O_4$ nanowire via a self-sacrificing template approach from $Na_{0.44}MnO_2$. This is an effort at preventing problems of agglomeration, big grain sizes, and accompanied poor battery efficiency during the calcination process. The as-synthesized $LiMn_2O_4$ nanowires' electrodes exhibit superb rate capacities of about 108 and 88 mAh/g at 5 and 20 A/g, respectively [23].

Vanadium oxides (V_2O_5)-based cathodes: Vanadium oxide is another potential alternative for LIB cathode. Reasons are its high specific capacity, availability, relatively cheaper, and ease of synthesis [43,44]. On the other hand, this material is faced with poor electrical conductivity, unstableness, easy agglomeration during the operation cycle, and slow reaction kinetics. The effect of these challenges is limited long-term cycling stability and low-rate capability [45]. The challenge has been overcome by Lee et al., who reported a V_2O_5 nanowire-graphene composite paper with superb rate capabilities. The capacity of the said designed material is 255.5 mAh/gat 10 A/g with 15 wt% of V_2O_5. This implies a charging and discharging time of 1.5 minutes. Moreover, there is a retained capacity of 94.4 mAh/g after cycling about 100000 cycles at a large current density of 10 A/g [44,45]. These good results are ascribed to the high aspect ratio of V_2O_5 nanowires with decreased lithium-ion diffusion lengths and the improved contacts between V_2O_5 nanowires and graphene resulting in enhanced electron transfer and limiting structural changes. The V_3O_7 nanowire templated graphene scrolls is about 30 μm in length, having interior cavities that permit free volume expansion of V_3O_7, offer ion transport paths, and limit self-agglomeration of V_3O_7 during its operation cycle [44].

8.5.2 Nanowires for Supercapacitors

The ever-rising utilization of non-renewable energy sources and the stresses associated with the global warming phenomenon have compelled the scientific community to explore other reliable energy storage materials. SCs are reliable energy storage materials because of their inherent merits in terms of performance. SCs conserve energy by using electric charges produced by Faraday pseudocapacitors and double-layer-capacitor-based electrochemical reactions. SCs store energy in two ways; either ion uptake or redox reaction. The electrode material used for the supercapacitor storage device should have high specific volume, long cycle stability, fast charge/discharge, and high rate function. The adoption of SCs can be enhanced when the properties of the material for the electrodes as well as their designs are optimized. Based on the properties of the electrode materials, increased electron/ion conductivity is essential to reduce capacitance loss, especially when the sampling rates, as well as current densities, are increased. Therefore, conductive material with structurally nanoporous properties is required for supercapacitor applications.

Based on the design of the electrode, there must be integration with the collector (conductive metal) that will achieve high stability at the collector-electrode interface without any binder [46,47]. This guarantees a long service life and high electron/ion conductivity. In addition, increasing the surface area of the electrode increases the amount of charge that accumulates on the surface of the electrode, thus increasing the specific volume. Recently, much effort has been focused on developing nanostructured electrode materials to meet the above needs by bridging those excellent characteristics of the materials and the design of the electrodes [48]. For instance, a collection of active one-dimensional nanowires integrated directly into a current collector has several advantages. Their large surface area with small diameters provides easy access to electrolyte ions and a short diffusion path, which can improve capacity [44]. Furthermore, by directly integrating nanowires (electrode active materials) into metal substrates (current collectors) without the use of bonding materials, it is possible to improve mechanical robustness while reducing capacitive loss at the

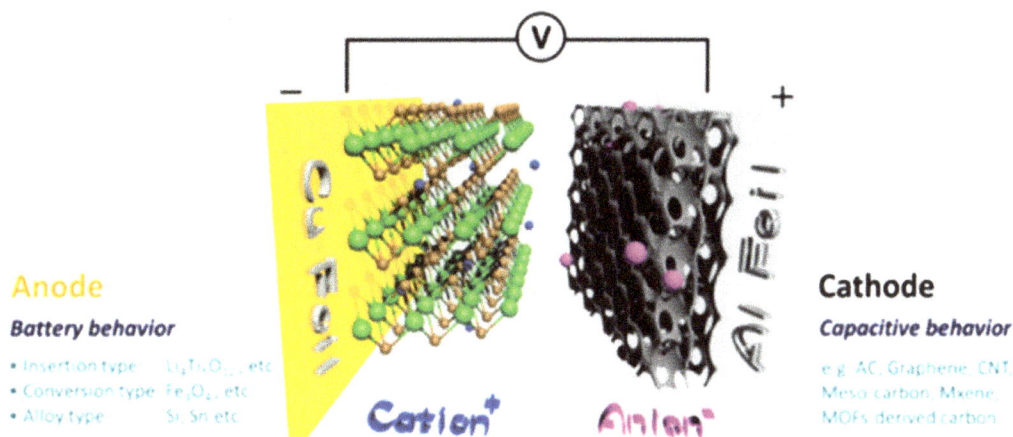

FIGURE 8.7 A cell structure and electrode materials of a non-aqueous hybrid capacitor. Adapted with permission [50]. Copyright (2017) John Wiley and Sons.

interface. In addition, 2D transition metal dichalcogenides (TMDs) such as molybdenum disulfide (MoS_2) and tungsten disulfide (WS_2) have currently become promising capacitive materials due to their structural advantages [44,49,50].

1D NW electrode materials have been extensively studied in SCs. The inherent 1D morphology of the NWs provides a fast pathway for efficient electron transportation, which is very beneficial to the improvement of the rate performance and power density of SCs [51]. The larger length-diameter ratio of NWs brings a higher specific surface area, increases the contact area of the electrode interface/electrolyte, and then adsorbs a large number of electrolyte ions on the surface of the electrode, allowing the electrode reaction kinetics to proceed rapidly [51,52]. In addition, carbon materials with a high specific surface area based on the EDLC storage mechanism are faced with agglomeration at the nanometer level, and the 1D NW structure can solve this problem well and improve the specific surface area extensively. Pseudocapacitor NW materials on the other hand can reduce the volume expansion during cycling and then increase the stability of the electrode [41,53]. A schematic characteristic cell structure of a non-aqueous hybrid capacitor and its anodic and cathode electrode materials is shown in Figure 8.7 [23].

8.5.2.1 Carbon-Based NWs for EDLCs

Carbon-based materials have become the main commercial electrode material due to their excellent cycling stability, availability, and cheap cost. Also, the large specific surface area and porous structures are important parameters needed for electrochemical performance optimization. The 1D porous carbon NWs have a large specific surface area and high porosity, which reduces the electron transfer resistance and also increases the adsorption area of the electrolyte ions. Some typical carbon-based nanowires include CNTs and CNFs.

CNTs are one of the primary researched carbon-based SC electrode materials with a hollow structure able to absorb a large amount of electrolyte ions and has a highly conductive graphitized wall that allows the effective transport of electrons and shows the excellent specific capacity and rate performance in supercapacitor applications [55]. Frackowiak et al. produced multiwall CNTs with a specific surface area of 410 m²/g and a specific capacitance reached 137 F/g using the CVD method. An et al. increased the specific capacitance to 180 F/g, and the energy density reached 6.5 Wh/kg by using single-walled CNTs as electrode materials [23,56]. In recent times, research reports on CNT as supercapacitor electrode materials have been minimal due to their high cost and use as excellent conductive additives [57]. Another form of 1D carbon material is carbon nanofibers

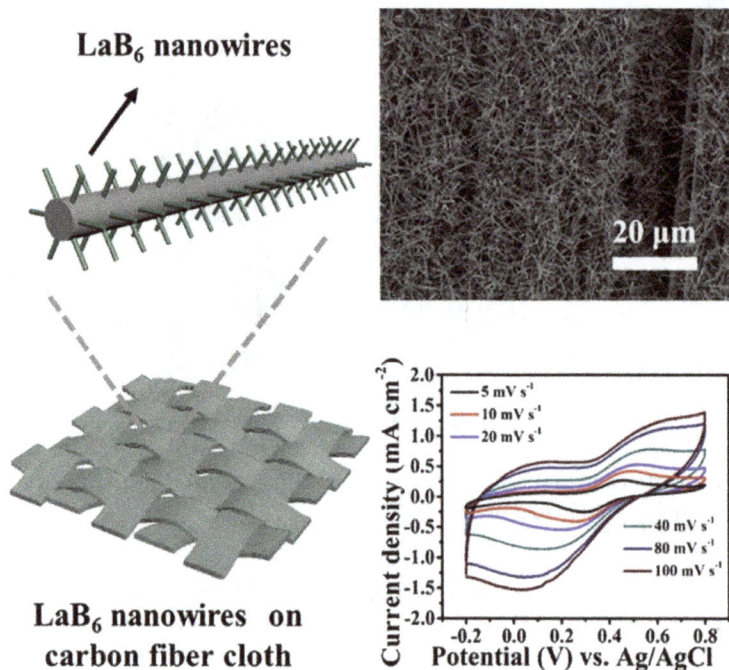

FIGURE 8.8   Schematic illustration of LaB$_6$ nanowires grown directly on CFC for supercapacitors. Adapted with permission [49]. Copyright (2018) Elsevier.

(CNFs), which have excellent flexibility properties and easy synthesis techniques. Moreover, the cross-linked structures, structural stability, and large specific surface area of CNFs make them an excellent candidate as energy storage electrode materials. One common method widely adapted to produce nanofibers is electrospinning. In an experiment, Chen et al. developed an efficient method of synthesizing N-doped CNFs by using ZIF-8 as a precursor. The CNFs with high specific surface area (418 m^2/g) show large specific capacitance (307 F/g at 1 A/g) and enhanced energy density (10.96 Wh/kg) [58]. Xue et al. also prepared LaB$_6$ nanowires grown directly on the surface of carbon fiber cloth (CFC) for supercapacitor application (Figure 8.8). The LaB$_6$ NWs electrode exhibited an important areal capacitance of 17.34 mF/cm^2 in 1.0 M Na$_2$SO$_4$ solution at a current density rate of 0.1 mA/cm^2 [53].

8.5.2.3   NWs for Pseudocapacitors

Another form of SCs is pseudocapacitors, which store electron charges in a reversible faradaic process near the surface of the active material. Some pseudocapacitive electron materials include conducting polymers, metal oxides, and metal sulfides, which have properties such as higher specific capacitance and energy density compared to carbon materials [7]. However, they are mainly hindered by poor rate capability and inferior cycling stability. Owing to the high aspect ratio of NWs, pseudocapacitive electrodes with NW structures have large electrode surface areas and efficient conducting pathways to facilitate fast redox reactions, resulting in better rate capability and improved cycling stability [59].

Metal oxides and sulfide-based nanowires for pseudocapacitors: Metal oxides and sulfides have been extensively studied and proved to be excellent energy storage electrode materials. Ruthenium (RuO$_2$) is known as an ideal pseudocapacitive material due to its high specific capacitance, conductivity, and excellent cycling performance. Yet, its high cost has resulted in the study of other low-cost metal oxides (such as MnO$_2$ and Co$_3$O$_4$) for energy storage [60,61]. But, these low-cost metal oxides are faced with energy conductivity challenges. Hence, producing metal oxides with nanowire structures can improve the conducting pathways for the faradaic reactions for excellent electrochemical performance [62]. For instance, Yang and co-workers reported MnO_2 nanowires grown on carbon fiber clothes, which exhibited a high specific capacitance of 150 mF/ cm^2 at 1 mA/cm^2 and an excellent capability rate of 73.6% at 10 mA/cm^2 [63]. Rakhi and co-workers also synthesized Co_3O_4 brush-like nanowires, which showed a high specific capacitance of 1525 F/ g at 1 A/g carried out in a three-electrode configuration [60].

Additionally, fabricating pseudocapacitive or carbon composites (metal doping), hybrid core-shell (coaxial), etc. can improve the conductivity and surface area of the metal oxide nanowire electrodes. A research experiment conducted by Zhou and co-workers showed a successfully designed CoO@ PPy core-shell structure that exhibited an ultrahigh specific capacitance of 2223 F/g at 1 mA/cm^2 and excellent capacitance retention [23]. Xia et al. also achieved a double increase in capacitance and rate capability by doping Ag nanoparticles into MnO_2 nanowires [52]. Recent works show that, combining metal oxides (such as $NiCo_2O_4$, $ZnCo_2O_4$, $MnCo_2O_4$ etc.) as an SC electrode has excellent electrochemical performance compared to the single used metal oxides. For instance, in an experiment by Shen et al., a mesoporous $NiCo_2O_4$ nanowire array deposited on a carbon cloth substrate exhibited a high specific capacitance of 1283 F/g at a current density of 1 A/g, a rate capability of 79% at 20 A/g, and cycling stability of up to 5000 cycles. Also, a $NiCo_2O_4$@PPy nanowire array was produced, which showed a high capacitance of 2244 F/g due to the synergistic effect between the two pseudocapacitive electrodes [23]. Transition metal sulfide nanowire electrodes such as CoS_2, $NiCo_2S_4$, and $ZnCo_2S_4$ on the other hand have been studied to have excellent conductivity and capacitance properties compared to metal oxides. Li et al. adopted a hydrothermal and ion-exchange method to synthesize Ni-Co sulfide nanowire array on a nickel foam substrate, which showed high areal capacitance 6 F/cm^2 and gravimetric capacitance of 2415 F/g[47].

Conductive polymer-based nanowires for pseudocapacitors: Polymers with alternating single and double bonds commonly referred to as conjugated (conducting) polymers such as polypyrrole (PPy), polyacrylonitrile (PANI), and polyethylenedioxyethane (PEDOT), which when used in pseudocapacitor fabrication demonstrates high specific capacitance and excellent conductivity characteristics. However, they possess low ion transport potential and this leads to undesirable stabilities during cycling [39]. To ensure effective utilization of the polymer nanowires to reduce bulk resistance, 1D conducting polymer nanowire arrays with no binder is constructed on current collector substrates [42]. For example, Wang et al. in a facile one-step galvanostatic deposition method synthesized 50 nm thick PANI nanowires on conductive substrates, which showed a specific capacitance of 950 F/g at a current density and retention rate of 40 A/g and 82%, respectively [23]. Combining highly conductive carbon nanomaterials with conducting polymer NW materials will increase the cycling stability and capacitances of the conducting polymer significantly. Using a 5 nm carbonaceous shell, Liu et al. increased the cycling performance of PANI (from 20 to 95%) and PPy (from 25 to 85%) NWs. In comparison to individual graphene sheets and PANI NWs, Xu et al. successfully synthesized 1D PANI NWs on a 2D graphene nanosheet hierarchical nanocomposite with higher electrochemical performance. Yu et al. also used an in situ synthesis approach to generate well-distributed PPy NWs on graphene nanosheets, with a high capacitance of 434.7 F/g at 1 A/g and 83% capacitance retention at 20 A/g [23].

8.5.2.4 NWs for Hybrid Capacitors (HCs)

One way of increasing the energy density of supercapacitors is to build hybrid capacitors. The cathode of the HCs is made of capacitor-type electrode material, and the anode is made of a battery-type electrode material with insertion-type materials. Based on the electrolyte, HCs can be divided into two types: aqueous and non-aqueous. Aqueous HCs attract a lot of attention because of their increased ionic concentration and reduced resistance. Harilal et al., for example, discovered that CuO-Co_3O_4 nanocomposite nanowires when immersed in an aqueous alkaline electrolyte have a high capability rate but lower energy density due to their lower working voltage [64]. Non-aqueous systems, on the other hand, give higher energy density, which has recently aroused the interest of many research works, such as Li-ion and Na-ion capacitors. For battery electrodes, stable and rapid intercalation/de-intercalation of cations is required [65]. Nanostructured materials, particularly NWs, can promote their high-rate properties as active electrodes. To fully utilize the merits of materials in energy storage applications, NWs are of major interest [66]. Many studies have confirmed that nanowire type of nanostructured materials with such attributes is best for HCs. Xu et al. manufactured LICs with WO_{3-x}/N-doped carbon NWs showing an ultrahigh-energy thickness of 195.58 Wh/kg with 90.7% maintenance over 6000 cycles [54].

8.5.3 Nanowire-Based Flexible Electrochemical Energy Devices

Flexible electronics have attracted intensive attention due to the rapidly growing interest in the 21st century for portable/wearable devices, roll-up displays, artificial skins, and implantable medical devices. As a result, to power these flexible devices, there are increasing need for developing flexible and high energy/power density energy-storage systems that are lightweight, safe, mechanically stable, and able to integrate with other energy-harvesting systems. To date, flexible energy-storage systems, such as LIBs, and SCs have shown tremendous progress. Emerging systems, such as sodium-ion batteries (SIBs), lithium-sulfur (Li_2S) batteries, metal-air, and alkaline batteries, have also sparked increasing research interests in recent years [65]. To achieve high performance in flexible energy-storage devices, the design of flexible electrodes with outstanding conductivity, good mechanical properties under various deformations, and those that are lightweight are crucial. However, the most commonly used electrode materials containing particles are usually rigid and hard to cater to the requirements of flexible electrodes [23]. Thus, the conventional electrode is usually prepared by coating a slurry that contains active materials, conductive additives, and polymer binders on metallic current collectors. The slurry casting method usually makes the electrode rigid, thick, bulky, and heavy. In this case, the electrode materials are easily delaminated from the current collector under prolonged cycling and deformation tests, which, in turn, leads to the degradation of the cell. Moreover, the insulating and electrochemically inactive binders decrease the electrical conductivity, with the heavy weight of the binders, conductive additives, and current collectors also reduce the overall volumetric and gravimetric energy densities of the entire electrode [67].

NWs with a high aspect ratio, superior electrical properties, fast charge transfer, and outstanding mechanical flexibility offer great opportunities for building flexible electrodes in the area of energy storage [23]. Also, flexible freestanding electrodes with an integrated structure are usually made from a variety of active materials built onto or into flexible conductive substrates with no binder, conductive additive, or even metal current collector; thus, the drawbacks of conventional fabrication processes could be effectively overcome. As a result, the freestanding electrodes are usually lightweight, flexible, and bendable, and thus, represent favorable candidates for high-performance flexible electronics and devices [68]. Due to their intrinsically flexible structure, NWs can be easily constructed into interconnected 3D freestanding membranes with better mechanical stability against bending/twisting/stretching than that of the corresponding bulk materials or sphere-like NP-based membranes; hence being particularly suitable for the fabrication of flexible freestanding electrodes [65]. Considering the unique characteristics of NWs in energy storage and their advantages in the fabrication of flexible freestanding electrodes, it is worthwhile to summarize the fabrication methods for flexible freestanding NW-based electrodes and versatile NW-based materials that have been used for electrode construction [69]. Flexible electrodes can be classified either as metal-based, carbon-based, polymer-based, or micro-patterned as shown in Figure 8.9 [69].

Researchers over some time now have sought to grow active energy storage material directly onto a metal substrate to create an electrode that is free of binders and conductive additives. Various methods such as in situ conversion, hydrothermal growth, and electrodeposition are being

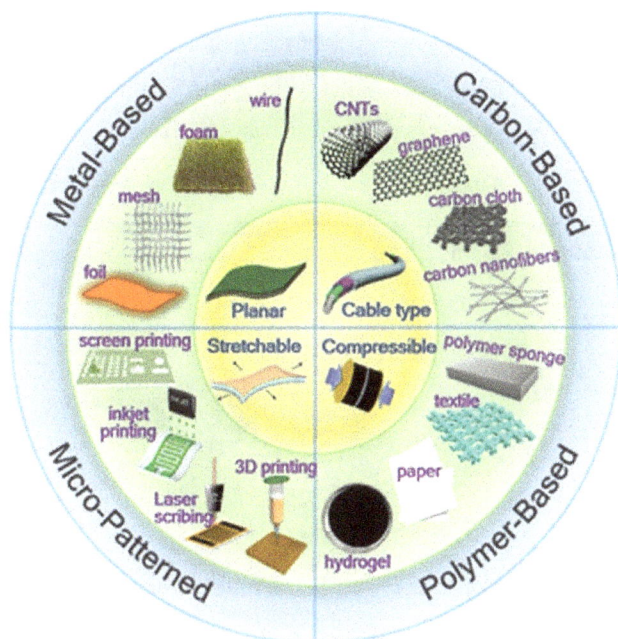

FIGURE 8.9 Schematic diagram of flexible electrode materials/substrates categorized metal-based, carbon-based, polymer-based, and micro-patterned flexible electrodes. Adapted with permission [66]. Copyright (2021) Wiley.

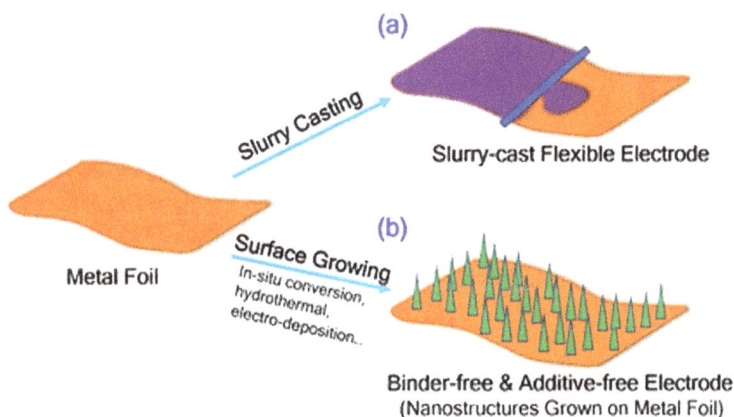

FIGURE 8.10 Schematic illustration of the fabrication of flexible electrodes by coating active materials on metal foils through methods of (a) slurry casting and (b) surface growing. Adapted with permission [66]. Copyright (2021) Wiley.

investigated to construct nanostructures on metal substrates (Figure 8.10). Nanostructures fixed at "zero distance" on the metal substrate show close interactions between the active material and the current collector, which not only prevents the active material from peeling off but also benefits fast charge transport and promotes energy and power density [69]. However, the high density and rigidity of metal leaves limit their use. The actual disadvantages of the use of flexible electrodes are that they are not malleable enough to withstand severe mechanical deformations like repeated bending, folding, and twisting. During deformation, the electrode and solid electrolyte have

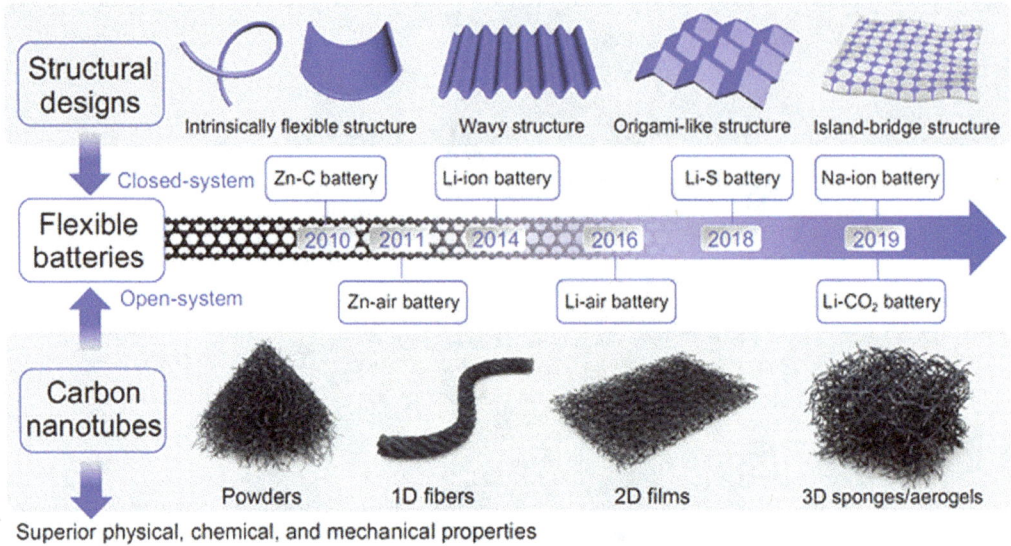

FIGURE 8.11 Schematic diagram of assembly CNTs into different structures and architectures for flexible battery applications. Adapted from reference [71]. This is an open access article distributed under the terms of the Creative Commons CC BY license 4.0 (CC BY).

insufficient interfacial contact, resulting in increased contact resistance and reduced durability [67]. Shuang et al. in their research fabricated a flexible and porous CuO nanorod arrays (CNAs) electrode engraved in commercial copper (Cu) foils in alkaline solution using an in situ approach and heat treated at 200 °C. The foil was used directly as an anode with no polymer binder or any further coating methods. The porous CAN binder-free electrode exhibited an excellent electrochemical performance with a good cycling stability, high capacitance above over 640 mAh/g at a current density of 200 mA/g and high rate capability at room temperature [70].

By adopting the template-assisted and self-assembly syntheses' techniques, graphene, as well as CNTs, can be fabricated in the monolithic form due to the large aspect ratio of their 1D (CNTs) and 2D (graphene) fibrous structures [53]. For instance, Figure 8.11 summarizes the different types of flexible or stretchable CNT fibers prepared through the wet-spinning method [67]. These unsupported structures are perfect substrates for fabricating adaptable cathodes that can be used for the storage of different forms of energy.

8.6 CONCLUSION AND FUTURE REMARKS

Nanomaterials (nanospheres, nanowires, nanorods, nanosheets, etc.) have recently been of great interest for electrochemical energy storage applications due to their unique features and characteristics. Nanowires as 1D nanomaterials have proven to be more effective as energy storage electrode materials due to their long length-diameter ratio with large surface area for ionic adsorption and desorption, to address the problems and limitations associated with conventional electrochemical energy storage devices. They have high conductivity with cross-linked network structures, which leads to high-rate capability as their charge transport rate increases. The methods of syntheses also contribute to the development of these nanowires greatly as electrode materials for energy storage in batteries and supercapacitors. Although these nanowire structures have great acknowledgment and exhibited excellent application properties, they are still faced with some challenges. Among these challenges is that their fabrication methods are very expensive relating to the desired structures (core-shell, hierarchical, branched, etc.), hence, an inexpensive process of fabricating nanowires with desirable structures and properties is needed. Nanowire electrodes cannot independently be used in the energy storage system, which makes it a great challenge, therefore, nanowire-based 3D arrays or porous current collectors can be used to provide large active surface areas for material storage and fast ionic adsorption and desorption. The large surface area of nanowires at the electrode-electrolyte interface provides an avenue for dangerous occurrences of undesirable reactions. Despite these challenges, nanowire-based electrodes remain reliable and advanced electrode materials for electrochemical energy storage devices.

REFERENCES

[1]   A.R. Dehghani-Sanij, E. Tharumalingam, M.B. Dusseault, R. Fraser, Study of energy storage systems and environmental challenges of batteries, Renew. Sustain. Energy Rev. 104 (2019) 192–208.

[2]   R.N. Gayen, V.S. Avvaru, V. Etacheri, Carbon-based integrated devices for efficient photo-energy conversion and storage, Carbon Based Nanomater. Adv. Therm. Electrochem. Energy Storage Convers., Elsevier Inc., 2019: pp. 357–374.

[3]   J.B. Goodenough, Y. Kim, Challenges for rechargeable Li batteries, Chem. Mater. 22 (2010) 587–603.

[4]   C. Zhang, Y.L. Wei, P.F. Cao, M.C. Lin, Energy storage system: Current studies on batteries and power condition system, Renew. Sustain. Energy Rev. 82 (2018) 3091–3106.

[5]   A. González, E. Goikolea, J. Andoni, R. Mysyk, Review on supercapacitors: Technologies and materials, 58 (2016) 1189–1206.

[6]   Poonam, K. Sharma, A. Arora, S.K. Tripathi, Review of supercapacitors: Materials and devices, J. Energy Storage. 21 (2019) 801–825.

[7]   B.K. Kim, S. Sy, A. Yu, J. Zhang, Electrochemical supercapacitors for energy storage and conversion, Handb. Clean Energy Syst., 2015: pp. 1–25.

[8]   L. Chang, Y.H. Hu, U. States, Supercapacitors, Compr. Energy Syst., 2018: pp. 663–695.

[9]   D. Gao, Z. Luo, C. Liu, S. Fan, A survey of hybrid energy devices based on supercapacitors, Green Energy Environ. (2022).

[10] E. Pomerantseva, F. Bonaccorso, X. Feng, Y. Cui, Y. Gogotsi, Energy storage: The future enabled by nanomaterials, Science (80). 366 (2019) eaan8285.

[11] E. V Shevchenko, D. V Talapin, N.A. Kotov, S.O. Brien, C.B. Murray, Structural diversity in binary nanoparticle superlattices, Nat. Lett. 439 (2006) 55–59.

[12] Y. Song, X. Li, C. Mackin, X. Zhang, W. Fang, T. Palacios, H. Zhu, J. Kong, Role of interfacial oxide in high-efficiency graphene-silicon Schottky barrier solar cells, Nano Lett. 15 (2015) 2104–2110.

[13] J. Gao, B. Li, J. Tan, P. Chow, T. Lu, N. Koratkar, Aging of transition metal dichalcogenide, ACS Nano. 10 (2016) 2628–2635.

[14] C.R. Ryder, J.D. Wood, S.A. Wells, M.C. Hersam, Chemically tailoring semiconducting two-dimensional transition metal dichalcogenides and black phosphorus, ACS Nano. 10 (2016) 3900–3917.

[15] G. Gao, T. Çagin, W.A. Goddard, Energetics, structure, mechanical and vibrational properties of single-walled carbon nanotubes, Nanotechnology. 9 (1998) 184–191.

[16] Y. Sun, N. Liu, Y. Cui, Promises and challenges of nanomaterials for lithium-based rechargeable batteries, Nat. Energy. 1 (2016) 16071.

[17] F. Bonaccorso, L. Colombo, G. Yu, M. Stoller, V. Tozzini, A.C. Ferrari, R.S. Ruoff, V. Pellegrini, Graphene, related two-dimensional crystals, and hybrid systems for energy conversion and storage, Science (80). 347 (2015) 1246501.

[18] B. Anasori, M.R. Lukatskaya, Y. Gogotsi, 2D metal carbides and nitrides (MXenes) for energy storage, Nat. Rev. Mater. 2 (2017) 16098.

[19] M.R. Lukatskaya, S. Kota, Z. Lin, M.-Q. Zhao, N. Shpigel, M.D. Levi, J. Halim, P.-L. Taberna, M.W. Barsoum, P. Simon, Y. Gogotsi, Ultra-high-rate pseudocapacitive energy storage in two-dimensional transition metal carbides, Nat. Energy. 2 (2017) 17105.

[20] J. Qiao, X. Kong, Z. Hu, F. Yang, W. Ji, High-mobility transport anisotropy and linear dichroism in few-layer black phosphorus, Nat. Commun. 4 (2014) 1–7.

[21]   Y. Xia, T.S. Mathis, M. Zhao, B. Anasori, A. Dang, Z. Zhou, H. Cho, Y. Gogotsi, S. Yang, Thickness-independent capacitance of vertically aligned liquid-crystalline MXenes, Nature. 557 (2018) 409–412.

[22]   G. Centi, S. Perathoner, Catalysis by layered materials: A review, Microporous Mesoporous Mater. 107 (2008) 3–15.

[23]   G. Zhou, L. Xu, G. Hu, L. Mai, Y. Cui, Nanowires for electrochemical energy storage, Chem. Rev. 119 (2019) 11042–11109.

[24]   K. Yu, X. Pan, G. Zhang, X. Liao, X. Zhou, M. Yan, L. Xu, L. Mai, Nanowires in energy storage devices: structures, synthesis, and applications, Adv. Energy Mater. 8 (2018) 1–19.

[25]   L. Mai, X. Tian, X. Xu, L. Chang, L. Xu, Nanowire electrodes for electrochemical energy storage devices, Chem. Rev. 114 (2014) 11828–11862.

[26]   M. Nehra, N. Dilbaghi, G. Marrazza, A. Kaushik, R. Abolhassani, Y.K. Mishra, K.H. Kim, S. Kumar, 1D semiconductor nanowires for energy conversion, harvesting and storage applications, Nano Energy. 76 (2020) 104991.

[27]   F. Liu, X. Yang, Z. Qiao, L. Zhang, B. Cao, G. Duan, Highly transparent 3D NiO-Ni/Ag-nanowires/FTO micro-supercapacitor electrodes for fully transparent electronic device purpose, Electrochim. Acta. 260 (2018) 281–289.

[28]   T. Youngblood, Time to get excited about these new developments with nanowires two exciting discoveries were made in the field of nanotechnology by universities on two different continents, each by research teams specializing in different scientific disciplines. Lund, (2016) 1–5.

[29]   L. Huang, Q. Wei, R. Sun, L. Mai, Nanowire electrodes for advanced lithium batteries, Front. Energy Res. 2 (2014).

[30]   N. Mahmood, Y. Hou, Electrode nanostructures in lithium-based batteries, Adv. Sci. 1 (2014) 1–20.

[31]   J. Goodenough, K. Park, The Li-ion rechargeable battery: a perspective., J. Am. Chem. Soc. 135 4 (2013) 1167–1176.

[32]   G. Girishkumar, B. McCloskey, A.C. Luntz, S. Swanson, W. Wilcke, Lithium−air battery: Promise and challenges, J. Phys. Chem. Lett. 1 (2010) 2193–2203.

[33]   A. Manthiram, Y. Fu, S.H. Chung, C. Zu, Y.S. Su, Rechargeable lithium-sulfur batteries, Chem. Rev. 114 (2014) 11751–11787.

[34]   S. Chowdhury, Nanowire battery anodes (2010) 2015–2017, http://large.stanford.edu/courses/2010/ph240/chowdhury1/.

[35]   Y. Yang, M.T. McDowell, A. Jackson, J.J. Cha, S.S. Hong, Y. Cui, New nanostructured Li_2S/silicon rechargeable battery with high specific energy, Nano Lett. 10 (2010) 1486–1491.

[36]   Y. Cui, F.B. Prinz, C.K. Chan, C. Xie, H. Peng, H. Huang, Nanowire lithium-ion batteries as electrochemical energy storage for electric vehicles, Mater. Sci. (n.d.).

[37]   H. Kim, Y. Son, C. Park, J. Cho, H.C. Choi, Catalyst-free direct growth of a single to a few layers of graphene on a germanium nanowire for the anode material of a lithium battery, Angew. Chemie - Int. Ed. 52 (2013) 5997–6001.

[38]   H. Zhang, Y. Tian, S. Wang, J. Feng, C. Hang, C. Wang, J. Ma, X. Hu, Z. Zheng, H. Dong, Robust Cu-Au alloy nanowires flexible transparent electrode for asymmetric electrochromic energy storage device, Chem. Eng. J. 426 (2021) 131438.

[39]   G. Wang, L. Zhang, J. Zhang, A review of electrode materials for electrochemical supercapacitors, Chem. Soc. Rev. 41 (2012) 797–828.

[40]   S.J. Uke, V.P. Akhare, D.R. Bambole, A.B. Bodade, G.N. Chaudhari, Recent advancements in the cobalt oxides, manganese oxides, and their composite as an electrode material for supercapacitor: A review, Front.Mater. 4 (2017) 2–7.

[41]   C. Sealy, Nanowires promise stable supercapacitors (n.d.) 9–10.

[42]   Q. Abbas, Advances of Electrode Materials, Ref. Modul. Mater. Sci. Mater. Eng., Elsevier, 2020.

[43]   A. Ray, A. Roy, P. Sadhukhan, S.R. Chowdhury, P. Maji, S.K. Bhattachrya, S. Das, Electrochemical properties of TiO_2 -V_2O_5 nanocomposites as a high performance supercapacitors electrode material, Appl. Surf. Sci. 443 (2018) 581–591.

[44]   E. Umeshbabu, G. Ranga Rao, Vanadium pentoxide nanochains for high-performance electrochemical supercapacitors, J. Colloid Interface Sci. 472 (2016) 210–219.

[45] J. Xu, F. Zheng, C. Xi, Y. Yu, L. Chen, W. Yang, P. Hu, Q. Zhen, S. Bashir, Facile preparation of hierarchical vanadium pentoxide (V_2O_5)/titanium dioxide (TiO_2) heterojunction composite nano-arrays for high performance supercapacitor, J. Power Sources. 404 (2018) 47–55.

[46] Q. Zhang, X. Wu, Q. Zhang, F. Yang, H. Dong, J. Sui, L. Dong, One-step hydrothermal synthesis of MnO_2/graphene composite for electrochemical energy storage, J. Electroanal. Chem. 837 (2019).

[47] R. Zou, M.F. Yuen, L. Yu, J. Hu, C.S. Lee, W. Zhang, Electrochemical energy storage application and degradation analysis of carbon-coated hierarchical $NiCo_2S_4$ core-shell nanowire arrays grown directly on graphene/nickel foam, Sci. Rep. 6 (2016) 1–9.

[48] M. Winter, R.J. Brodd, What are batteries, fuel cells, and supercapacitors?, Chem. Rev. 104 (2004) 4245–4269.

[49] X. Zhang, L. Gong, K. Liu, Y. Cao, X. Xiao, W. Sun, X. Hu, Y. Gao, J. Chen, J. Zhou, Z.L. Wang, Tungsten oxide nanowires grown on carbon cloth as a flexible cold cathode, Adv. Mater. 22 (2010) 5292–5296.

[50] P.G. Bruce, B. Scrosati, J.M. Tarascon, Nanomaterials for rechargeable lithium batteries, Angew. Chem. Int. Ed. 47 (2008) 2930–2946.

[51] X. Ma, W. Luo, M. Yan, L. He, L. Mai, In situ characterization of electrochemical processes in one dimensional nanomaterials for energy storages devices, Nano Energy. 24 (2016) 165–188.

[52] X. Su, L. Yu, G. Cheng, H. Zhang, M. Sun, X. Zhang, High-performance α-MnO_2 nanowire electrode for supercapacitors, Appl. Energy. 153 (2015) 94–100.

[53] Q. Xue, Y. Tian, S. Deng, Y. Huang, M. Zhu, Z. Pei, H. Li, F. Liu, C. Zhi, LaB_6 nanowires for supercapacitors, Mater. Today Energy. 10 (2018) 28–33.

[54] H. Wang, C. Zhu, D. Chao, Q. Yan, H.J. Fan, Nonaqueous hybrid lithium-ion and sodium-ion capacitors, Adv. Mater. 29 (2017) 1–18.

[55] S. Rashidi, J.A. Esfahani, F. Hormozi, Classifications of porous materials for energy applications, Ref. Modul. Mater. Sci. Mater. Eng., Elsevier, 2020.

[56] K. Leng, F. Zhang, L. Zhang, T. Zhang, Y. Wu, Y. Lu, Y. Huang, Y. Chen, Graphene-based Li-ion hybrid supercapacitors with ultrahigh performance, Nano Res. 6 (2013) 581–592.

[57] A.G. Pandolfo, A.F. Hollenkamp, Carbon properties and their role in supercapacitors, 157 (2006) 11–27.

[58] Y. Soneda, Carbons for supercapacitors, in: Handb. Adv. Ceram. Mater. Appl. Process. Prop. Second Ed., Elsevier Inc., 2013: pp. 211–222.

[59] D. Dodoo-Arhin, R.A. Nuamah, P.K. Jain, D.O. Obada, A. Yaya, Nanostructured stannic oxide: Synthesis and characterisation for potential energy storage applications, Results Phys. 9 (2018) 1391–1402.

[60] J. Liu, J. Jiang, C. Cheng, H. Li, J. Zhang, H. Gong, H.J. Fan, Co_3O_4 nanowire@MnO_2 ultrathin nanosheet core/shell arrays: A new class of high-performance pseudocapacitive materials, Adv. Mater. 23 (2011) 2076–2081.

[61] R. Bolagam, S. Um, Hydrothermal synthesis of cobalt ruthenium sulfides as promising pseudocapacitor electrode materials, Coatings. 10 (2020).

[62] T. Yumak, D. Bragg, E.M. Sabolsky, Effect of synthesis methods on the surface and electrochemical characteristics of metal oxide/activated carbon composites for supercapacitor applications, Appl. Surf. Sci. 469 (2019).

[63] S.D. Raut, H.R. Mane, N.M. Shinde, D. Lee, S.F. Shaikh, K.H. Kim, H.J. Kim, A.M. Al-Enizi, R.S. Mane, Electrochemically grown MnO_2 nanowires for supercapacitor and electrocatalysis applications, New J. Chem. 44 (2020) 17864–17870.

[64] S.D. Perera, X. Ding, A. Bhargava, R. Hovden, A. Nelson, L.F. Kourkoutis, R.D. Robinson, Enhanced supercapacitor performance for equal Co-Mn stoichiometry in colloidal $Co_{3-x}Mn_xO_4$ nanoparticles, in additive-free electrodes, Chem. Mater. 27 (2015) 7861–7873.

[65] C.Y. Hui, C.W. Kan, C.L. Mak, K.H. Chau, Flexible energy storage system-an introductory review of textile-based flexible supercapacitors, Processes. 7 (2019) 1–26.

[66] P. Simon, Y. Gogotsi, P. Simon, Y. Gogotsi, N. Materials, Materials for electrochemical capacitors, Nature Mater., 7 (2008) 845–854.

[67] F. Sheet, Flexible nanowire devices for energy harvesting, ERC-2014-STG, Project reference: 639052.

[68] E. Zdraveva, J. Fang, B. Mijovic, T. Lin, Electrospun nanofibers, in: Struct. Prop. High-Performance Fibers, Elsevier Inc., 2017: pp. 267–300.

[69] Q. Li, Z. Hu, Z. Liu, Y. Zhao, M. Li, J. Meng, X. Tian, X. Xu, L. Mai, Recent advances in nanowire-based, flexible, freestanding electrodes for energy storage, Chem. - A Eur. J. 24 (2018) 18307–18321.

[70] K. Shang, J. Gao, X. Yin, Y. Ding, Z. Wen, An overview of flexible electrode materials/substrates for flexible electrochemical energy storage/conversion devices, Eur. J. Inorg. Chem. 2021 (2021) 606–619.

[71] S. Zhu, J. Sheng, Y. Chen, J. Ni, Y. Li, Carbon nanotubes for flexible batteries: recent progress and future perspective, Natl. Sci. Rev. 8 (2021) nwaa261.

9 Nanowires for Supercapacitors

Sanjeev Verma
Department of Chemical Engineering & Technology, Indian Institute of
Technology, Banaras Hindu University, Varanasi, India

Shivani Verma
Department of Chemistry, CBSH, G. B. Pant University of Agriculture and
Technology, Pantnagar, Uttarakhand, India

Saurabh Kumar
Department of Electronics and Communication Engineering, NIT,
Hamirpur, Himachal Pradesh, India

Bhawna Verma
Department of Chemical Engineering & Technology, Indian Institute of
Technology, Banaras Hindu University, Varanasi, India

CONTENTS

9.1   INTRODUCTION

The continued depletion of fossil fuels, as well as the resulting pollution and climate change, need the innovation of sustainable and green energies [1]. Wind, tidal, and solar energy are all neat and

DOI: 10.1201/9781003296621-9

sustainable sources of energy, but they confront obstacles such as weather variability and irregular distribution [2]. As a result, energy storage science is critical to addressing the need for expanding the usage of intermittent energies like those indicated above. Energy storage electrochemical devices like supercapacitors have lately focused on electrical energy storage applications, such as electric vehicles and portable electric gadgets due to their cost-effectiveness and high yielding [3,4]. The need for power source performance is rapidly growing due to the exponential growth of machine learning and artificial intelligence. Supercapacitor devices store ions at the electrode/electrolyte interface by generating double layers (EDLCs) and redox reactions between electrolyte/active material and may charge-discharge in under 10 seconds and cycle for almost indefinite periods [5,6].

All the energy storage electrochemical systems, active materials with a high storage capacity, good cycling life, and cheap cost are widely sought. Nanomaterials hold promise for the easy and accurate development of electrode nanostructured materials, which can improve the stability of high-capacity electrode materials significantly [7,8]. Furthermore, decreasing the material size from micro to nanoscale can considerably improve the dynamics of the materials since ionic diffusion lengths are shorter at lower scales. Smaller-sized materials allow more active areas on the surface and easier movement of ions between the electrode/electrolyte interface. Nanosized materials have gained a lot of attention in recent years because of their photonic, thermal, electronic, opto-electronics, mechanical capabilities, and electrochemical, as well as their many prospective uses. Nanowires (NWs), nanofibers, nanotubes, nanoribbons, nanorods, and nanobelts, among other 1D nanostructured materials, have made significant progress and attracted significant interest in recent decades because of their specific physical and chemical properties and potential application in LEDs, field emitters, and the active material of electrode for electrochemical energy storage systems [9,10].

Among 1D nanostructure, NWs attract minds in energy-storage-related materials because of their large surface/volume ratio and also a huge active material interface between the electrode-electrolyte, along with 1D ions' pathway for better and quick charge movement during the electrochemical volumetric contraction-expansion process [10,11]. NWs have unique qualities that make them advantageous in a variety of real-world uses. First, it is feasible to adjust NWs' radial dimensions such that their diameters are equal to or lower than their typical lengths for a variety of

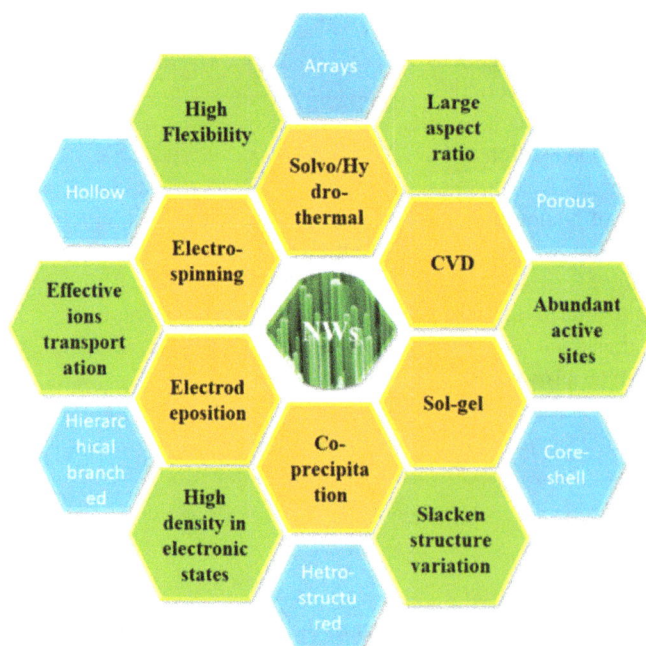

FIGURE 9.1   Different synthesis methods, properties, and structures of nanowires.

applications. This distinctive quality is what causes the observed alterations in NWs' physical properties. There are many synthesis routes of NWs, including physical and chemical approaches such as solvothermal/hydrothermal, coprecipitation, sol-gel, electrodeposition, electrospinning, chemical-vapor-deposition (CVD), and others (**Figure 9.1**) [10–13]. NW-based material electrode performance can be notably changed with changes in its structures like morphology, length, composition, diameter, crystallinity, surface defects, phase purity, etc. As a result, monitoring the shape, size, and morphology variation of NW-based materials throughout the cycling and electrochemically reactions is crucial for designing the NW electrodes with stable stability and enhanced capacitive nature [14]. A wide range of materials based on NWs and its composites' surfaces has been extensively researched for the creation of modern electrical gadgets with many functions. Mai et al. stated the phase shifting of active materials on real-time observation for the creation of an enhanced electrochemical energy storage system, in addition to increasing the nanostructure of the electrode materials [15].

In this chapter, the formation of NWs via different methods has been discussed, followed by the different challenges and advantages in the field of energy storage application. Furthermore, the current progress of NW-based material for energy storage devices mainly supercapacitors has been shown. The most important factors to consider while building NW structures, like core-shell, hierarchical branched, array, porous, and hetero-structured NWs, are to obtain high specific energy, high specific power, and extended cycle stability in energy storage devices. In the end, the issues and future perspectives for the use of NWs in energy storage systems have been compiled.

9.2 THE DESIGN PRINCIPLE OF NANOWIRES

In general, most NW building procedures are founded on the idea that in 1D, atoms develop in a certain direction. There are two types of NW manufacturing strategies: "spontaneous growth" and "space-constrained growth." In most situations, the intrinsic progress approach is denoted as the aqueous-phase action to create one plane of active materials made up of certain elements (like Ti, Mo, and so on) that are simple to develop 1D morphologies [16,17]. Auxiliary tools like templates and molds are used for those substances, which cannot develop in one plane structure, and these approaches are summed up in the space-containment growth strategy. Pure phase NWs' materials may not give optimal performance due to their limited electrical conductivity, slow kinetics, low rate capability, and capacity in energy storage electrochemical application [18]. The cycle stability of materials that experience significant changes in volume during charge injection may be reduced. To address these issues, certain hierarchical and heterogeneous NW structures have been developed to increase the energy and power density of the NW electrode materials. The conductivity of heterogeneous structures can be improved by covering the NW surface with conductive material. Assembling hierarchical nanostructures like branch/linkage in NWs can also help to improve structural as well as cyclic stability. One-step and multi-step techniques are commonly used to design hierarchical and diverse NW structures. A one-step approach is defined as a response in which distinct materials or frameworks develop in situ to generate the desired structure during only a single reaction. A preceding synthetic step is usually required in a multi-step technique to generate the bare phase nanomaterials like substrate or template under the changes will be conducted. The next sections go through the various synthesis processes in depth [19,20].

9.3 HYDROTHERMAL/SOLVOTHERMAL APPROACH

Hydrothermal/solvothermal processes occur as completely heterogeneous processes at the temperature over the steaming point of the solvent in a closed-batched chamber at higher enforcement. The major solvent in this approach (hydrothermal) is water, while the prime solvent/mineralizer in this process (solvothermal) is an organic solvent. In other words, the hydro/solvothermal technique is a type of self-growth approach that uses aqueous phase reactions to accelerate the reactions of soluble materials (precursors) by adding specified solvents at precise pressure and temperature (**Figure 9.2**). A range of

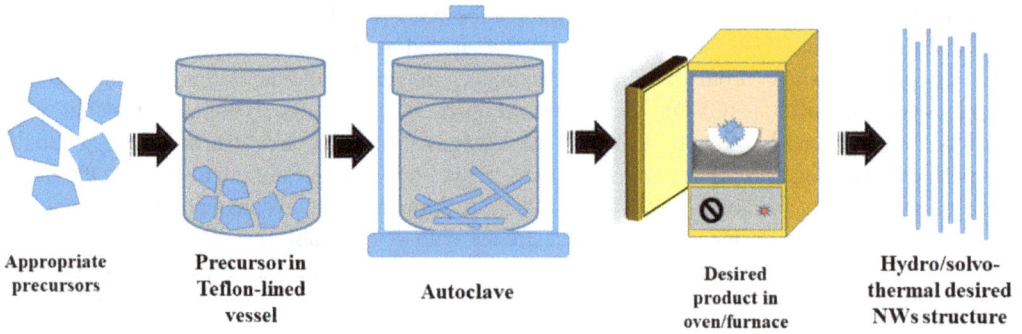

Appropriate precursors   Precursor in Teflon-lined vessel   Autoclave   Desired product in oven/furnace   Hydro/solvo-thermal desired NWs structure

FIGURE 9.2   Schematic of hydro/solvothermal approach for nanowires.

factors, such as choices of solvent, pH values, reaction time, pressure, temperature, etc., can affect the structure of the desired product. In comparison to other approaches, the hydrothermal/solvothermal method has fast kinetics, phase purity, uses less energy, and is very simple to implement [21,22].

Many NW nanostructures, such as TiO_2, Sb_2O_3, NaV_3O_8, $H_2V_3O_8$, and others, have been effectively synthesized by manipulating the factors in the synthesis process. Luo et al. employed hydrazine hydrate, Na_3SeO_3, and $Sb(CH_3COO)_3$ as precursors dissolving in a certain quantity of water and synthesized ultra-long, freestanding Sb_2Se_3 NW material [23]. The authors concluded that extending hydrothermal time helps to produce consistent and lengthy NWs. The synthesized NWs had a diameter of around 90 nm and a length approximate of micrometers after a 36-h reaction period. Li et al. used a simple one-pot hydrothermal process to make Co_3Se_4 NWs develop on Co foam. After hydrothermal salinization, the NW base on Co_3Se_4 is evenly formed on the raw foam of Co as a current precursor and collector [24].

The solvothermal approach, which involves using a solvent like organic to improve the dispersion of appropriate precursors that are poorly soluble in water, is also normally employed in the synthesis of NWs. The solvothermal method ensures that NWs are always properly crystalline. Liao et al. built NWs based on trigonal selenium using a simple solvothermal procedure with the help of CTAB (cetyltri-methyl-ammonium bromide) [25]. The formation of Se trigonal NWs is believed to be like a solid/solution/solid process. The resulting materials are well-distributed NWs with around 100 nm diameter when the solvothermal duration is 12 h. Zhou et al. used solvothermal two-phase processes to form the $Na_2Ti_3O_7$ intertwined NWs, which were uniformly enfolded in the graphene sheets [26]. The $Na_2Ti_3O_7$ graphene cited NW design has high structural stability and porosity as well as high rate capability and specific capacity. Solvo/hydro-thermal methods have a few drawbacks, consisting of the necessity for costly Teflon liners and autoclaves, the likelihood of safety hazards at the time of reactions, and the inability to analyze in situ reactions.

9.4   SOL-GEL APPROACH

Sol-gel is a technique for obtaining the desired output by leveraging the properties of hydrolysis followed by polymerization condensation of readily appropriate precursor in water or a specific solvent, subsequently drying as well as sintering, and further post-processing procedures (**Figure 9.3**). The hydrolysis process and the polymerization reaction are the two primary reactions in the sol-gel technique. This technique is a productive way to fabricate nanomaterials with a huge surface area while also reducing the amount of temperature and time required at the time of sample synthesis. The sol-gel approach has various features, including the capacity to tune the structure and the capability to manage the synthesized nanoarchitecture materials, which makes it appealing for conversion and energy storage [27].

Hou et al. used a hydrothermal procedure to make graphene-TiO_2(B) NW composites from titania gel and graphene oxide produced using the sol-gel technique [28]. The generated NW composites

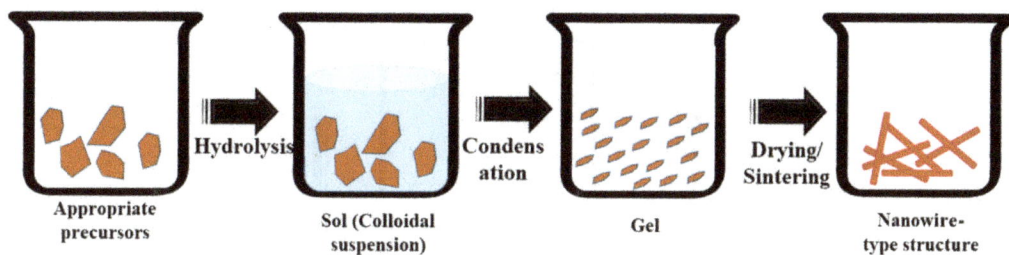

FIGURE 9.3   Schematic diagram of a sol-gel method for nanowires.

have a huge area (270 m^2/g) according to this unique process. Similarly, Wang et al. used a template with a sol-gel solution comprising citric acid and $Mn(CH_3COO)_2$ to make manganese dioxide NWs [29]. The diameter and length of MnO_2 NWs were determined by the thickness of the applied template and pore diameter and NWs based on MnO_2 have a maximum 165 F/g capacitance. Magnesium-ion electrode materials are also prepared using a similar process. An et al. described the preparation of graphene/hydrated-vanadium oxide NW nanostructure by a simple synthesis and subsequent freeze-drying of a hydrogel [30]. They discovered that the presence of water in aerogel can help increase Mg^{2+} insertion kinetics by providing charge shielding. In addition, the porous nature of aerogel might facilitate electrolyte entry into the active substance.

9.5   CO-PRECIPITATION APPROACH

Co-precipitation is usually characterized as a simple and economical "one-pot" approach for creating nanomaterials. To make homogenous composites, this approach uses numerous positive ions in a precipitation process. Co-precipitation has been used to make a variety of NW-structured materials including Fe_3O_4, $NiFe_2O_4$, Fe_3O_4-coated Ag, and indium tin oxide (ITO) NWs, throughout the last several decades. Yu et al. recently used a hydrothermal and co-precipitation technique to create Ni/Co oxide shells derived from ZIF-67 on Co_3O_4 nanoarrays [31]. The $CoCo_3O_4$@ZIF-$NiCo_2O_4$ nanoflake structure was visible, and the cores and shells were shown as a single unit. It is worthwhile to note the fact that interior nanoholes could not provide more area but reduce the charge diffusional path, giving better storage of energy. Furthermore, the completed asymmetric supercapacitor has 50.6 Wh/kg energy density and an 856 W/kg specific power density. The device has 30.2 Wh/kg energy density despite having 11.1 kW/kg power density. Although the co-precipitation approach is an easy, affordable, and high outcome, it may result in non-homogeneous distribution and agglomeration of NWs.

9.6   ELECTRODEPOSITION APPROACH

Another liquid-based process is electrodeposition, in which materials develop due to the energy given by an electric field. The use of electrodeposition to create NW materials always demands the need for a substrate (**Figure 9.4**).

Xiao et al. used a simple electrodeposition technique to create the nanosheets of MnO_2, $Co_xNi_{1-x}(OH)_2$, and nanoparticles of FeOOH placed on nanoarrays of $NiCo_2S_4$ [32]. The surface of the material is influenced by the concentration, temperature of the solution, and deposition voltage, that is necessary for getting better results. Moreover, nanoarrays of Cu have been created with the help of Cu^{2+} ions' electrodeposition in the pores of the anodized-alumina-oxide (AAO) membrane, according to Taberna et al. An additional electrodeposition method was used to cover the surface of the Cu nanoarrays with Fe_3O_4 after the AAO membrane was removed [33]. Due to the limitations of the experimental setup and the substance type, this method is not mostly used to

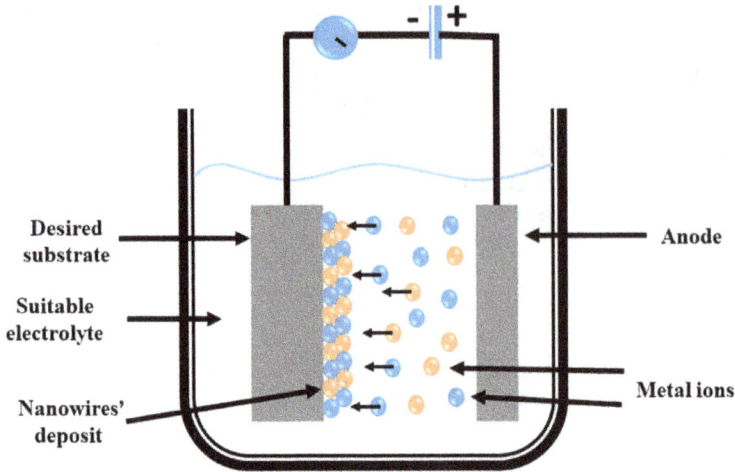

FIGURE 9.4   Pictorial diagram of electrodeposition for nanowire production.

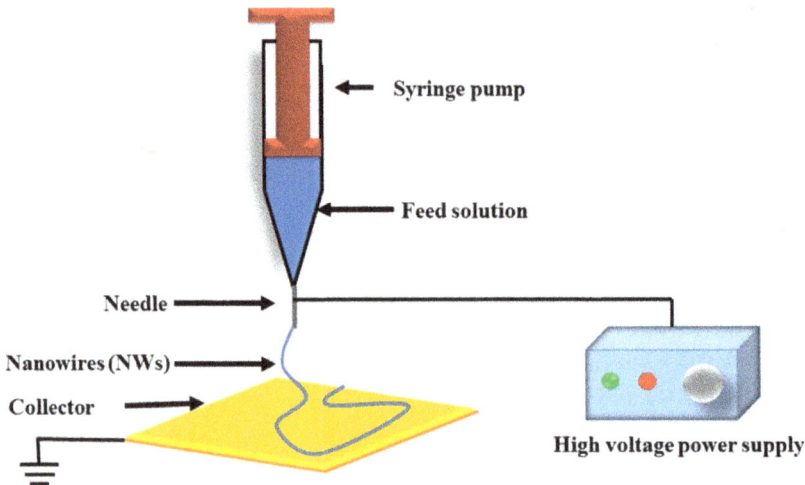

FIGURE 9.5   Schematic flow diagram of electrospinning process for nanowires.

produce nanomaterials up to this point. As a result, more efficient electrodeposition modalities must be created in the future.

9.7   ELECTROSPINNING APPROACH

For practically all inorganic materials, electrospinning is a commonly used and easy way of fabricating hierarchical, self-assembled, and mesoporous NWs. The syringe head (needle) of the electrospinning setup works as a mold to regulate the desired product in nanostructure during the electrospinning process (**Figure 9.5**). Three procedures have gotten a lot of attention to get proper structured NWs: making appropriate precursor polymer dispersion, electrospinning setup, and then sintering. All three procedures are simple to change when compared to other manufacturing methods. The two most essential qualities of raw precursor dispersion, conduction, and viscosity, which may be easily modified by modifying the polymer and inorganic component ratio, have a significant impact on the final morphology. The receiving distance for the marching procedure is normally 30 cm. The arrival of the NWs reaching the receiver takes time. As a result, influencing their behavior is simple. There are a variety of polymers with distinct thermal characteristics that may be used in the sintering process. Various morphologies may be achieved by combining different polymers. When it comes to constructing NW structures, electrospinning outperforms the other ways, and numerous sophisticated technologies of electrospinning have been produced by changing the above three main paths. The downside is that the entire process is highly complicated to determine clear causation from different variables to the desired structure [34].

Electrospinning may easily construct mesoporous-based NWs that have a substantial capacity fluctuation at the time of charge/discharge procedures using the insertion of the template into the appropriate precursor. By combining ZIF-8 with nitrogen-doped carbon nanofibers, Chen et al. created highly spongy carbon nanofibers doped by nitrogen [35]. ZIF-8 dissolved and created mesoporous during the sintering process, and Zn metal developed at the nanoscale. Due to its minimal boiling range, Zn metal dissociated from the fibers. As a result, the formed porous hierarchical structure was effectively constructed using a soft template.

To produce core-shell NWs, electrospinning based on a double coaxial nozzle is a revolutionary approach. Researchers may tailor the cone by using a dual-coaxial nozzle to deliver two distinct solutions pf polymers: the external one produces a shell, while the internal one makes the core. Zhang et al. used coaxial electrospinning to create core-shell silicon/carbon fibers [36]. The shell polymer solution was polyacrylonitrile (PAN around 15 wt%) in dimethylformamide (DMF), while the core solution of polymer was Si-PAN coupled poly-styrene. To construct the secondary structure based on NWs, modifying the sintering procedure is beneficial. Chen et al. used ZIF-67 containing polymer solution to make microtubes based on Co_3O_4/CNT [37]. The ZIF-67 is converted into cobalt nanocrystals during the sintering process, which accelerates CNTs generated on nanofibers' surfaces utilizing a pyrolytic carbon source.

9.8 CHEMICAL VAPOR DEPOSITION (CVD) APPROACH

CVD on substrates has been a popular method for producing NWs because it is both affordable and large-scale (**Figure 9.6**). The NWs were made in a quartz tube containing a furnace. Different

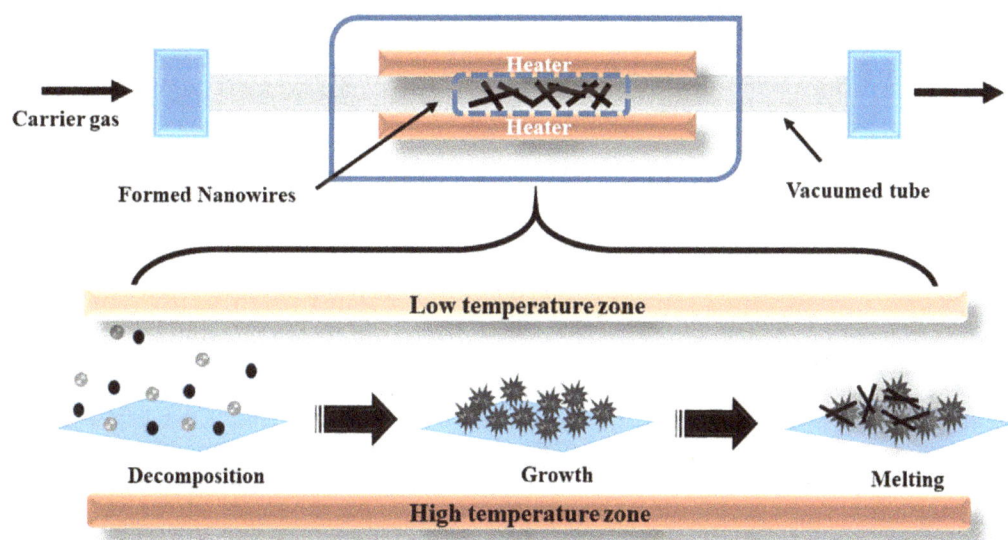

FIGURE 9.6 Flow diagram of CVD process for nanowires.

NWs, like Si-carbide, Fe_3O_4, Si, ZnO, GaN, and Ge, have been manufactured using this process throughout the last decade [34]. Wang et al. reported the successful synthesis of Ge-based NWs using Au nanocrystals as seed particles and CVD of GeH_4 (germane) at 275 °C [38]. Chan et al. used a similar method to create as-grown NWs with a 50–100 nm diameter and 20–50 μm of length [39]. Zhao et al. synthesized Cu@carbon NWs with 30 to 90 nm outer diameter and 700 nm to 10 μm length were produced with maximum yield using a CVD approach and subsequently thermal breakdown of copper(II)-acetylacetonate [40].

9.9   SUPERCAPACITORS

Our present technological world necessitates large-scale energy usage and storage. Supercapacitors have gained a lot of interest in energy storage applications. Supercapacitor offers a higher power density and better cycling durability than the battery, although having a lower energy density. By raising the capacitance and charging voltage, supercapacitors may achieve increased energy density [41]. As a consequence, the supercapacitor has bridged the energy and power density gap between batteries and regular dielectric capacitors, owing to the high power density and extended cycle life, and will display a larger part in upcoming energy storage systems. EDLCs (electric double-layer capacitors) and Faradaic (pseudocapacitors) are two different types of supercapacitors. In EDLCs, mainly physical charge buildup on the electrolyte/electrode interface during the charging and discharging operations in the EDLC. In pseudocapacitors, rapid redox processes also occur, which might lead to irreversible capacitance degradation. As a result of the redox processes, pseudocapacitors can have larger capacitive values than EDLCs but have little cycle durability because of their distinct charge storage mechanism [42,43].

9.10   CHARACTERISTICS OF NANOWIRES FOR SUPERCAPACITORS

In supercapacitors, 1D electrode active nanowire materials have been thoroughly investigated. The intrinsic 1D NWs architecture gives a rapid channel for effective ion transit, which is extremely favorable to SC rate performance and power density enhancement. NWs have a greater specific surface area, which enhances the electrolyte/electrode contact area and subsequently takes a huge amount of charges from the electrolyte, allowing quick kinetics of active material. Furthermore, agglomeration at the nanoscale level is a concern for carbon-based materials with a larger specific area according to the EDLC-based storage system, and 1D nanostructure may eliminate this type of issue and vastly boost the internal area. Faradaic NW materials might reduce the volumetric expansion at the time of cycling and hence improve electrode performance. The SCs effectively highlight the NWs' unique structural benefits [44].

9.11   CARBON-BASED NANOWIRE MATERIALS IN EDLCS

Carbon-based materials are suited to most used electrode active materials because of their remarkable cycle stability and inexpensive cost. Adsorption and buildup of electrolyte ions are all that are required for carbon materials to store energy. As a result, the uniform pore structure with a high surface area is an essential indicator for improving supercapacitor properties. The high porosity and surface area of the 1D porous carbon NWs. Their 1D shape decreases electron transport resistance while simultaneously increasing the absorbable surface of electrolyte ions. The highly common carbon NW materials are carbon nanotubes (CNTs) and carbon nanofibers (CNFs). The first carbon-derived supercapacitor active material to be studied was CNTs. Because of its porous structure, which can absorb a high amount of electrolyte ions, and the graphitic carbon wall, which is exceptionally conducting for effective electron transport, supercapacitors have high specific strength capacity and rate performance. Using the CVD technique, Frackowiak et al. produced multi-walled

CNTs with a 410 m^2/g specific area and 137 F/g of specific capacitance [45]. Using single-walled CNT as active materials, An et al. enhanced the capacitance up to 180 F/g with 6.5 Wh/kg energy density [46]. However, because of its high pricing and the restricted application of supercapacitors, the number of publications on CNTs for supercapacitor materials has decreased in recent times, and they are now used as outstanding conductive additives.

Due to their outstanding flexibility and cheap synthesis techniques, CNFs are the second most intensively investigated 1D carbon-based materials. CNFs get over the drawbacks of the CNTs' synthesis technique. Cross-linked CNFs are also favored as good electrode materials because of their huge specific surface area and structural durability. Electrospinning is a popular method for obtaining nanofibers that has been extensively researched. Chen et al. used ZIF-8 as a precursor to establishing a successful technique for synthesizing N-doped CNFs [35]. CNFs with a high capacitance of 307 F/g at 1 A/g and 10.96 Wh/kg energy density have a huge area of 418 m^2/g. Yu's team also used Te templates for NWs and carbon precursors as glucose to create a series of CNFs. After hydrothermal treatment, the mechanically strong monolith was formed, which can be readily cut into any shape. The CNFs were obtained after the Te NWs were removed, and they had a high desired output with a huge area. The capacitance was around 202 F/g after polypyrrole coating on CNFs, and the power density reached 90 kW/kg. Furthermore, this study team used another precursor bacterial cellulose to create heteroatom doping CNFs, with 7.76 Wh/kg of energy density. CNF's benefits in supercapacitors are fully exploited thanks to these enhancements [34].

9.12 PSEUDOCAPACITORS BASED ON NANOWIRE MATERIALS

Pseudocapacitor materials store ions/charges on the active material surface through a reversible Faradaic mechanism. Pseudocapacitive electrode materials have been researched and categorized as conducting polymers, metal sulfides, and metal oxides. The energy density and capacitance of pseudocapacitive materials are greater than those of carbon materials. They are, however, hampered mainly through low cycle stability and rate capability. Because NWs have a higher length, pseudocapacitive NW material electrode geometries have efficient conducting routes, and large specific areas, allowing for faster redox reactions and increased cycle stability.

9.12.1 PSEUDOCAPACITORS BASED ON METAL OXIDE NW MATERIALS

The most used pseudocapacitor material is RuO_2 due to its high capacitance value, high conductivity, and cyclic stability characteristics. Due to the high cost of RuO_2 nanomaterials, various low-cost metal oxides like Co_3O_4 and MnO_2 have been investigated. Metal oxide NW electrodes may be designed to improve the conducting channels for quick redox actions, giving improved electrochemical properties. For instance, NWs based on MnO_2 matured on carbon cloth, have 150 mF/cm^2 of areal capacitance at 1 mA/cm^2 current density and a superb 73.6% rate ability at 10 mA/cm^2, according to Yang and coworkers [47]. In a three-electrode setup, Rakhi and coworkers successfully manufactured brush-like structured Co_3O_4 NWs with 1525 F/g high capacitance at 1 A/g [48]. Furthermore, synthesizing carbon/pseudocapacitive composite materials of different metal oxides and polymers with coaxial/core-shell nanostructured can also increase the capacitance of formed NW composites. The increased surface area and conductivity are responsible for the improved electrochemical performance. Zhou and colleagues successfully built a CoO@PPy core-shell structure with high 2223 F/g specific capacitance at 1 mA/cm^2 and outstanding capacitance retention [49].

Due to their better electrochemical activity, binary metal oxides like $MnCo_2O_4$, $ZnCo_2O_4$, and $NiCo_2O_4$ have recently been favored as supercapacitor active materials in recent research. For example, Shen et al. described mesoporous NW based on $NiCo_2O_4$, which was grown onto the carbon fiber cloth and obtained a 1283 F/g high capacitance value at 1 A/g with 79% rate ability at 20 A/g and cyclic stability up to 5000 loops [50]. Transition metal sulfides, in comparison to their oxide equivalents, have higher conductivities, hence NW materials like $ZnCo_2S_4$, $NiCo_2S_4$, and CoS_2 have been produced to achieve larger capacitance values.

9.12.2 PSEUDOCAPACITORS WITH CONDUCTIVE POLYMER NWS

Pseudocapacitive materials made of conducting polymers (PPy, PANI, and PEDOT) have good conductivity and a huge capacitance value. While, within conducting polymer electrodes, limited ion mobility leads to low cycle stability and higher bulk resistance. Because of the short diffusion length in NWs, synthesizing 1D conducting NW polymers on an appropriate current collector assures optimal exploitation of active materials and minimizes total resistance by removing the usage of the desired binder. For instance, Wang et al. used a simple one-step galvanostatic deposition process to successfully create PANI NWs (50 nm thickness) on a conductive collector [51]. At 1 A/g, the formed PANI NW material had a high 950 F/g of capacitance and kept 82% of capacitance retention at 40 A/g current density.

By combining conducting polymer NW electrodes with highly conductive carbon nanomaterials, the cycle stability and capacitance values of conducting polymer-based NW materials may be significantly increased. By encapsulating PPy from 25–85% and PANI from 20–95% NWs with carbonaceous shells (5 nm), Liu et al. increased their cycling performance [52]. Yu et al. used an in situ synthesis approach to generate well-distributed PPy NWs on graphene nanosheets, with a maximum 434.7 F/g capacitance at 1 A/g and retention of 83% at 20 A/g [53].

9.13 HYBRID NW MATERIALS FOR SUPERCAPACITOR

For renewable energy storage systems, electrochemical energy storage technologies have advanced quickly. Due to their distinct characteristics, supercapacitors are seen to be the most promising of these technologies. However, their low energy density and poor cycle stability make them unsuitable for practical use. Battery-type/pseudocapacitive materials have been proposed to address these concerns. The majority of them have low-rate capabilities. As a result, high energy density hybrid supercapacitors must be built without losing cycle stability or rate capability. Dai et al. synthesized hybrid structured supercapacitor material of Ni-Co-S nanosheets with $ZnCo_2O_4$ nanowires by a simple electrochemical process [54]. And shows the 53.1 Wh/kg and 3375 W/kg as energy density and power density, respectively. Thalji et al. examined a new type of nanowire-based nanocomposite of $W_{18}O_{49}$ NWs-rGO and used it as an asymmetric supercapacitor [55]. This asymmetric assembly shows the 365.5 F/g specific capacitance at 1 A/g current density with 96.7% retention up to 12000 cycles and 28.5 Wh/kg, 751 W/kg of energy density and power density. Liu et al. reported conductive coating of PEDOT around SiC freestanding nanowires by oxidative-CVD method and illustrated 26.53 mF/cm^2 specific capacitance at 0.2 mA/cm^2 [56]. Mahmood et al. synthesized MoO_3 nanowire-based MoO_3/MXene@carboncloth electrode material for supercapacitor application [57]. This fabricated assembly displayed a high capacitance of 783.4 F/g at a scan rate of 5 mV/s and 775 F/g at a current density of 1 A/g with retention around 96.4% up to 6000 cycles.

9.14 FUTURE CHALLENGES

The challenges of various NW applications in supercapacitors have been outlined and debated. Commercial electrode materials depend upon the physical adsorption process and are yet the top option, and it is necessary to improve the cycle stability of pseudocapacitance materials. As a result, hybrid supercapacitors are now the most promising electrochemical storage devices for advanced research. The supercapacitors' small specific energy continues to hinder their long-term evolution. In recent years, electrode material development has been too attentive to its area, although a higher surface area can only provide severally benefits. The device's energy density will be considerably lowered. Future studies will focus on how to coordinate the pore distribution and specific surface area of spongy NW electrode materials with the solvated ion transport kinetics. In addition, greater attention should be made to the transport mechanism of dispersed ions in electrolytes and the modulation of the stable potential window. Ionic electrolytes with a huge stable potential window up to 5.0 V can help alleviate the reduced energy density that is a major problem in supercapacitors. NW material electrodes with a solid surface with strong conductivity can withstand higher potential voltages, and the porous cross-linked structure efficiently reduces the organic electrolytes' viscosity with sluggish ions' diffusion. As a result, in the supercapacitor electrode materials now being explored, NW materials have a few competitiveness with a wide development process and are projected to be used commercially. The main objective is to combine compatible, NW-based higher conducting electrolytes with well-effective NW-assembled anodes and cathodes to produce NW-based supercapacitor devices, keeping safety and dependability in mind.

9.15 ADVANTAGES AND DISADVANTAGES OF NANOWIRE-BASED MATERIALS

1D NW electrode materials outperform the majority of potential materials for supercapacitors because of their major advantages such as the following:

- The ion movement rate is substantially increased and the diffusion length is greatly reduced, resulting in the outstanding supercapacitor performance of NW-based materials.
- The huge length-to-diameter ratio results in a large specific area for charge absorption-desorption and a large number of active sites for Faradaic pseudocapacitance, making surface modification or loading more favorable for further improving performance.
- The cross-linked structured NWs prevent the clusters of NW materials (like sheets, spheres, etc.) and speed up electrolyte infiltration and ion absorption-desorption. As a result of these benefits, NWs in supercapacitors have high specific capacitance, strong rate capability, and cycle stability.

  Despite the numerous substantial advancements in both synthesis and use of NWs, their implementation continues to lag behind the needs. There are still a lot of drawbacks, and there is plenty of potential for further research to accelerate practical uses of NW-based energy storage systems, such as the following:
- The productivity of scalable NWs' fabrication for actual uses is still limited, and manufacturing NW-based materials with diverse architectures (like hierarchical, branching, and so on) is also expensive. As a result, large amounts manufactured with minimal cost, high-performance NWs with programmable architectures will be required.
- NWs qualities are influenced by their size and shape, an appropriate synthesis approach that allows for exact size and morphology is still needed. Analyzing strategies for the regulated fabrication of multi-phase or multi-element NWs would provide the nanomaterials with additional characteristics and broaden their application possibilities. A possible method to obtain high power and energy densities is to create innovative multi-dimensional nanomaterials for electrodes with a rapid charge movement mechanism.
- Synthesizing materials with a protective carbon coating interface and choosing appropriate electrolytes with stable potential windows are helpful, but the possibility of unwanted reactions increases at the interface. Other techniques, like hierarchy, branching, and so on, can help to mitigate the problem.
- To obtain the high energy/volumetric energy density of supercapacitor devices, NW nanomaterials with optimized volume and pore size may also enhance the mass loading of active materials.

9.16 CONCLUSION

In summary, we have analyzed the current advances in NW synthesis processes and addressed the special characteristics of NW-based nanomaterials for electrochemical energy storage in the fields of primary supercapacitors. The benefits of NW-based materials over typical other materials in resolving essential difficulties have been thoroughly explained and proved in a variety of energy storage systems. NW electrodes' specific physical and chemical characteristics, along with numerous formation approaches for controlling NW size and shapes, offer an intriguing possibility to develop superior energy storage devices for advanced research. However, by synthesizing binary and ternary hybrid nanocomposites with conducting polymers and metal oxides, extensive research on nanowire-based supercapacitor materials has progressed to a new level. As a result, we were able to create enhanced supercapacitors with a high capacitance value, strong cycle stability, and greater energy and power density. Additionally, with the correct government assistance, NW-based materials can be a more efficient nanomaterial in every aspect of research and engineering. As a result, we may conclude that a wide range of research is being conducted to improve supercapacitor performance by employing hybrid nanomaterials based on NW-based materials.

REFERENCES

[1]   S. Chu, Y. Cui, N. Liu, The path towards sustainable energy, Nat. Mater. 16 (2017) 16–22.

[2]   Z. Yang, J. Zhang, M.C.W. Kintner-Meyer, X. Lu, D. Choi, J.P. Lemmon, J. Liu, Electrochemical energy storage for green grid, Chem. Rev. 111 (2011) 3577–3613.

[3]   P.G. Bruce, B. Scrosati, J.-M. Tarascon, Nanomaterials for rechargeable lithium batteries, Angew. Chemie Int. Ed. 47 (2008) 2930–2946.

[4]   S. Verma, V.K. Pandey, B. Verma, Facile synthesis of graphene oxide-polyaniline-copper cobaltite (GO/PANI/CuCo2O4) hybrid nanocomposite for supercapacitor applications, Synth. Met. 286 (2022) 117036.

[5]   P. Simon, Y. Gogotsi, Materials for electrochemical capacitors, in: Nanosci. Technol., Co-Published with Macmillan Publishers Ltd, UK, 2009: pp. 320–329.

[6]   V.K. Pandey, S. Verma, B. Verma, Polyaniline/activated carbon/copper ferrite (PANI/AC/CuF) based ternary composite as an efficient electrode material for supercapacitor, Chem. Phys. Lett. 802 (2022) 139780.

[7]   G. Xu, W. Wang, X. Qu, Y. Yin, L. Chu, B. He, H. Wu, J. Fang, Y. Bao, L. Liang, Electrochemical properties of polyaniline in p-toluene sulfonic acid solution, Eur. Polym. J. 45 (2009) 2701–2707.

[8]   S. Verma, V.K. Pandey, B. Verma, Synthesis and supercapacitor performance studies of graphene oxide based ternary composite, Mater. Technol. (2022) 1–17.

[9]   B. Tian, T.J. Kempa, C.M. Lieber, Single nanowire photovoltaics, Chem. Soc. Rev. 38 (2009) 16–24.

[10] A.I. Hochbaum, P. Yang, Semiconductor nanowires for energy conversion, Chem. Rev. 110 (2010) 527–546.

[11] L. Mai, X. Tian, X. Xu, L. Chang, L. Xu, Nanowire electrodes for electrochemical energy storage devices, Chem. Rev. 114 (2014) 11828–11862.

[12] N.P. Dasgupta, J. Sun, C. Liu, S. Brittman, S.C. Andrews, J. Lim, H. Gao, R. Yan, P. Yang, 25th Anniversary Article: Semiconductor nanowires - synthesis, characterization, and applications, Adv. Mater. 26 (2014) 2137–2184.

[13] Y. Xia, P. Yang, Y. Sun, Y. Wu, B. Mayers, B. Gates, Y. Yin, F. Kim, H. Yan, One-dimensional nanostructures: Synthesis, characterization, and applications, Adv. Mater. 15 (2003) 353–389.

[14] Q. Wei, F. Xiong, S. Tan, L. Huang, E.H. Lan, B. Dunn, L. Mai, Porous One-Dimensional nanomaterials: Design, fabrication and applications in electrochemical energy storage, Adv. Mater. 29 (2017) 1602300.

[15] L. Mai, M. Yan, Y. Zhao, Track batteries degrading in real time, Nature. 546 (2017) 469–470.

[16] J.-Y. Liao, D. Higgins, G. Lui, V. Chabot, X. Xiao, Z. Chen, Multifunctional TiO_2–C/MnO_2 core–double-shell nanowire arrays as high-performance 3D electrodes for lithium ion batteries, Nano Lett. 13 (2013) 5467–5473.

[17]   B.L. Mai, B. Hu, W. Chen, Y. Qi, C. Lao, R. Yang, Y. Dai, Z.L. Wang, Lithiated MoO_3 nanobelts with greatly improved performance for lithium batteries, Adv. Mater. 19 (2007) 3712–3716.

[18]   Z. Cai, L. Xu, M. Yan, C. Han, L. He, K.M. Hercule, C. Niu, Z. Yuan, W. Xu, L. Qu, K. Zhao, L. Mai, Manganese oxide/carbon yolk–shell nanorod anodes for high capacity lithium batteries, Nano Lett. 15 (2015) 738–744.

[19]   Q. Wei, Q. An, D. Chen, L. Mai, S. Chen, Y. Zhao, One-pot synthesized bicontinuous hierarchical $Li_3V_2 (PO_4)_3/C$ mesoporous nanowires for high-rate and ultralong-life lithium-ion batteries, Nano Lett. 2 (2014) 1042–1048.

[20]   X. Wang, C. Niu, J. Meng, P. Hu, X. Xu, X. Wei, Novel $K_3V_2 (PO_4)_3/C$ bundled nanowires as superior sodium-ion battery electrode with ultrahigh cycling stability, Adv. Energy Mater. 2 (2015) 1–8.

[21]   M.A. Mahdi, J.J. Hassan, S.S. Ng, Z. Hassan, Growth of CdS nanosheets and nanowires through the solvothermal method, 359 (2012) 43–48.

[22]   G. Zou, H. Li, Y. Zhang, K. Xiong, Solvothermal/hydrothermal route to semiconductor nanowires, Nanotechnology 17 (2006) S313.

[23]   W. Luo, A. Calas, C. Tang, F. Li, L. Zhou, L. Mai, Ultralong Sb_2Se_3 nanowire-based free-standing membrane anode for lithium/sodium ion batteries, ACS App. Mat. Inter. 8 (2016) 35219–35226.

[24]   W. Li, X. Gao, D. Xiong, F. Wei, W. Song, J. Xu, L. Liu, Hydrothermal synthesis of monolithic Co_3Se_4 nanowire electrodes for oxygen evolution and overall water splitting with high efficiency and extraordinary catalytic stability, Adv. Energy Mater. 1602579 (2017) 1–7.

[25]   F. Liao, X. Han, Y. Zhang, H. Chen, C. Xu, CTAB-assisted solvothermal synthesis of ultralong t-selenium nanowires and bundles using glucose as green reducing agent, Mater. Lett. 214 (2018) 41–44.

[26]   Z. Zhou, H. Xiao, F. Zhang, X. Zhang, Y. Tang, Electrochimica acta solvothermal synthesis of $Na_2Ti_3O_7$ nanowires embedded in 3D graphene networks as an anode for high-performance sodium-ion batteries, Electrochim. Acta. 211 (2016) 430–436.

[27]   M. Niederberger, Nonaqueous sol–gel routes to metal oxide nanoparticles, 40 (2007) 793–800.

[28]   J. Hou, R. Wu, P. Zhao, A. Chang, G. Ji, B. Gao, Q. Zhao, Graphene – TiO_2 (B) nanowires composite material: Synthesis, characterization and application in lithium-ion batteries, 100 (2013) 173–176.

[29]   X. Wang, X. Wang, W. Huang, P.J. Sebastian, S. Gamboa, Sol–gel template synthesis of highly ordered MnO_2 nanowire arrays, J. Power Sources. 140 (2005) 211–215.

[30]   Q. An, Y. Li, H. Deog Yoo, S. Chen, Q. Ru, L. Mai, Y. Yao, Graphene decorated vanadium oxide nanowire aerogel for long-cycle-life magnesium battery cathodes, Nano Energy. 18 (2015) 265–272.

[31]   D. Yu, B. Wu, L. Ge, L. Wu, H. Wang, T. Xu, Decorating nanoporous ZIF-67-derived $NiCo_2O_4$ shells on a Co_3O_4 nanowire array core for battery-type electrodes with enhanced energy storage performance, J. Mater. Chem. A. 4 (2016) 10878–10884.

[32]   J. Xiao, L. Wan, S. Yang, F. Xiao, S. Wang, Design hierarchical electrodes with highly conductive $NiCo_2S_4$ nanotube arrays grown on carbon fiber paper for high-performance pseudocapacitors, Nano Lett. 14 (2014) 831–838.

[33]   P.L. Taberna, S. Mitra, P. Poizot, P. Simon, J.-M. Tarascon, High rate capabilities Fe_3O_4-based Cu nano-architectured electrodes for lithium-ion battery applications, Nat. Mater. 5 (2006) 567–573.

[34]   G. Zhou, L. Xu, G. Hu, L. Mai, Y. Cui, Nanowires for electrochemical energy storage, Chem. Rev. 119 (2019) 11042–11109.

[35]   L.-F. Chen, Y. Lu, L. Yu, X.W. (David) Lou, Designed formation of hollow particle-based nitrogen-doped carbon nanofibers for high-performance supercapacitors, Energy Environ. Sci. 10 (2017) 1777–1783.

[36]   H. Zhang, X. Qin, J. Wu, Y.-B. He, H. Du, B. Li, F. Kang, Electrospun core–shell silicon/carbon fibers with an internal honeycomb-like conductive carbon framework as an anode for lithium ion batteries, J. Mater. Chem. A. 3 (2015) 7112–7120.

[37]   Y.M. Chen, L. Yu, X.W. (David) Lou, Hierarchical tubular structures composed of Co_3O_4 hollow nanoparticles and carbon nanotubes for lithium storage, Angew. Chemie Int. Ed. 55 (2016) 5990–5993.

[38]   D. Wang, H. Dai, Low-temperature synthesis of single-crystal germanium nanowires by chemical vapor deposition, Angew. Chemie Int. Ed. 41 (2002) 4783–4786.

[39]   C.K. Chan, X.F. Zhang, Y. Cui, High capacity Li ion battery anodes using Ge nanowires, Nano Lett. 8 (2008) 307–309.

[40]   Y. Zhao, Y. Zhang, Y. Li, Z. Yan, A flexible chemical vapor deposition method to synthesize copper@ carbon core-shell structured nanowires and the study of their structural electrical properties, New J. Chem. 36 (2012) 1161–1169.

[41]   S. Verma, T. Das, V.K. Pandey, B. Verma, Nanoarchitectonics of GO/PANI/CoFe$_2$O$_4$ (graphene oxide/ polyaniline/cobalt ferrite) based hybrid composite and its use in fabricating symmetric supercapacitor devices, J. Mol. Struct. (2022) 133515.

[42]   T. Das, B. Verma, Polyaniline based ternary composite with enhanced electrochemical properties and its use as supercapacitor electrodes, J. Energy Storage. 26 (2019) 100975.

[43]   S. Verma, B. Verma, Graphene-Based Nanomaterial for Supercapacitor Application, in: 2022: pp. 221–244.

[44]   Z. Chen, V. Augustyn, J. Wen, Y. Zhang, M. Shen, B. Dunn, Y. Lu, High-performance supercapacitors based on intertwined CNT/V$_2$O$_5$ nanowire nanocomposites, Adv. Mater. 23 (2011) 791–795.

[45]   E. Frackowiak, K. Metenier, V. Bertagna, F. Beguin, Supercapacitor electrodes from multiwalled carbon nanotubes, Appl. Phys. Lett. 77 (2000) 2421–2423.

[46]   K.H. An, W.S. Kim, Y.S. Park, Y.C. Choi, S.M. Lee, D.C. Chung, D.J. Bae, S.C. Lim, Y.H. Lee, Supercapacitors using single-walled carbon nanotube electrodes, Adv. Mater. 13 (2001) 497–500.

[47]   P. Yang, Y. Ding, Z. Lin, Z. Chen, Y. Li, P. Qiang, M. Ebrahimi, W. Mai, C.P. Wong, Z.L. Wang, Low-cost high-performance solid-state asymmetric supercapacitors based on MnO$_2$ nanowires and Fe$_2$O$_3$ nanotubes, Nano Lett. 14 (2014) 731–736.

[48]   R.B. Rakhi, W. Chen, D. Cha, H.N. Alshareef, Substrate dependent self-organization of mesoporous cobalt oxide nanowires with remarkable pseudocapacitance, Nano Lett. 12 (2012) 2559–2567.

[49]   C. Zhou, Y. Zhang, Y. Li, J. Liu, Construction of high-capacitance 3D CoO@polypyrrole nanowire array electrode for aqueous asymmetric supercapacitor, Nano Lett. 13 (2013) 2078–2085.

[50]   L. Shen, Q. Che, H. Li, X. Zhang, Mesoporous NiCo$_2$O$_4$ nanowire arrays grown on carbon textiles as binder-free flexible electrodes for energy storage, Adv. Funct. Mater. 24 (2014) 2630–2637.

[51]   K. Wang, J. Huang, Z. Wei, Conducting polyaniline nanowire arrays for high performance supercapacitors, J. Phys. Chem. C. 114 (2010) 8062–8067.

[52]   T. Liu, L. Finn, M. Yu, H. Wang, T. Zhai, X. Lu, Y. Tong, Y. Li, Polyaniline and polypyrrole pseudocapacitor electrodes with excellent cycling stability, Nano Lett. 14 (2014) 2522–2527.

[53]   C. Yu, P. Ma, X. Zhou, A. Wang, T. Qian, S. Wu, Q. Chen, All-solid-state flexible supercapacitors based on highly dispersed polypyrrole nanowire and reduced graphene oxide composites, ACS Appl. Mater. Interfaces. 6 (2014) 17937–17943.

[54]   M. Dai, D. Zhao, H. Liu, X. Zhu, X. Wu, B. Wang, Nanohybridization of Ni-Co-S nanosheets with ZnCo$_2$O$_4$ nanowires as supercapacitor electrodes with long cycling stabilities, ACS Appl. Energy Mater. 4 (2021) 2637–2643.

[55]   M.R. Thalji, G.A.M. Ali, P. Liu, Y.L. Zhong, K.F. Chong, W$_{18}$O$_{49}$ nanowires-graphene nanocomposite for asymmetric supercapacitors employing AlCl$_3$ aqueous electrolyte, Chem. Eng. J. 409 (2021) 128216.

[56]   W. Liu, X. Li, W. Li, Y. Ye, H. Wang, P. Su, W. Yang, Y. Yang, High-performance supercapacitors based on free-standing SiC@PEDOT nanowires with robust cycling stability, J. Energy Chem. 66 (2022) 30–37.

[57]   M. Mahmood, K. Chaudhary, M. Shahid, I. Shakir, P.O. Agboola, M. Aadil, Fabrication of MoO$_3$ nanowires/MXene@CC hybrid as highly conductive and flexible electrode for next-generation supercapacitors applications, Ceram. Int. 48 (2022) 19314–19323.

10 Nanowires for Electrocatalytic Activity

A. Rebekah, A. Rajapriya, and N. Ponpandian
Department of Nanoscience and Technology, Bharathiar University,
Coimbatore, Tamil Nadu, India

CONTENTS

10.1   INTRODUCTION

Nanowires can be identified as nanostructured material that exhibits the form of a wire with a diameter varying from 10–100 nm. The prominent characteristics of nanowires are their length-to-width ratio being greater than 1000. Due to the huge variation in the ratio of length and diameter, nanowires are often regarded as quasi-one-dimensional material. Also, it is capable of allowing the desired space for quantum confinement effects, which will also be called "quantum wires" [1]. Depending on the fabrication techniques of nanowires, the properties can be suitably tuned due to the modulation in the porosity, length of the nanowires, thickness, etc. In the preparation of nanowires by epitaxial growth techniques, atomically sharp junctions can be achieved, which is an essential parameter in the field of band-engineering. In addition, this parameter enables the nanowires to be used in countless applications, including optoelectronics, quantum technologies, and electronic field at a large scale [2,3]. Nanowires can be prepared from a wide variety of materials, such as semiconductors (silicon and germanium), carbon materials, conductive noble metals (platinum, gold, and copper), and non-noble metals. It is worth mentioning that nanowires are distinct from nanotubes, because nanotubes exhibit hollow structures, whereas nanowires possess solid nanostructural characteristics. One-dimensional (1D) nanowires exhibit abundant promising properties for electrocatalysis. Nanowires possess a large surface area with high aspect ratios that is responsible for exposing numerous active sites for electrocatalytic reaction. Moreover, it effectively suppresses dissolution, aggregation, and

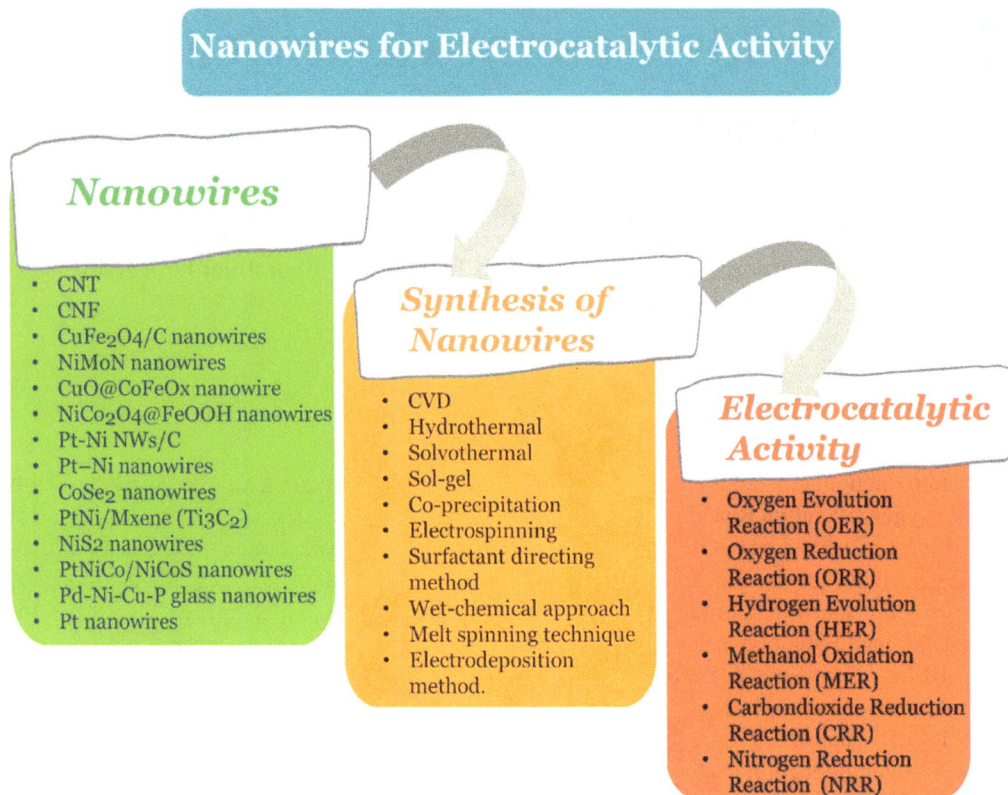

FIGURE 10.1   Flowchart of nanowires for electrocatalytic activity.

Ostwald ripening, during the electrocatalytic reactions [4]. 1D nanowires allow the flow channel along axial direction for transport of mass and electron, which thereby greatly enhances the conductivity, which is highly beneficial for elevating the electrocatalytic performance. Flowchart of nanowires for electrocatalytic activity is depicted in **Figure 10.1.** Also, the nanowires exhibit highly active and high-index atomic steps, which help in promoting maximized activity [5]. The nanowires can be easily assembled into the 3D structure, which is again an effective route for enhancing the electrocatalytic activity and stability due to the availability and utilization of large surface area utilization [6].

10.2   OXYGEN EVOLUTION REACTION (OER)

The OER is a crucial step in energy conversion devices, including water electrolyzers and metal-air batteries. OER is a multistep four electron-proton coupled processes proceed via the reaction $4OH^- \rightarrow O_2 + 2H_2O + 4e^-$ in an alkaline medium. This requires a high overpotential (η) to overcome the sluggish reaction kinetics for achieving a desired current density. So far, noble metal-based electrocatalysts, such as RuO_2 and IrO_2 have been widely accepted as active electrocatalysts for catalyzing OER. However, scarcity, high cost, and poor stability have greatly hindered it in commercializing the energy conversion systems for large-scale production of clean energy (H_2). Thus, developing economical, abundantly available, and stable electrocatalysts with superior catalytic activity in terms of low overpotential, and good withstanding capability for catalyzing OER is an urgent necessity to be identified [7].

Various cost-effective and widely acquirable electrocatalysts such as spinel oxides (mixed transition metals), perovskites, transition metal hydroxides, etc. have been identified as efficient electrocatalysts for effectively catalyzing OER. The electrocatalysts, which expose a greater number of reactive sites have been the target of intense research focus due to their significantly enhanced catalytic performances. The surface morphology and size modulation of the nanostructure is a prime key factor for enhancing the reactivity and tuning the reactive sites. Considering this aspect, 1D nanowires with self-standing features exhibiting a greater number of reactive facets are an effective approach for elevating the OER electrocatalytic activity. So, this part deals with the influencing features of nanowire-based electrocatalyst for catalyzing the evolution of oxygen in both acidic and alkaline media [8].

10.2.1 NANOWIRES IN OER

Li et al. designed $CuFe_2O_4$ spinel coupled with carbon nanowires by electrospinning technique for catalyzing OER and hydrogen peroxide reduction. The prepared nanowires exhibited more active facets created due to the embedment of $CuFe_2O_4$ into the carbon nanowires. Also, the electrocatalyst, $CuFe_2O_4$ coupled with porous carbon nanowires has enhanced the mass transport and a couple of conductive carbon nanowires have boosted the electron conductivity. Owing to the combined synergistic effect of the existence of abundant porous channels and elevated electron conducting ability, the catalyst revealed inherently superior electrocatalytic performance in catalyzing OER with the overpotential of 1.598 V vs. RHE for attaining the current density (10 mA/cm²) [9]. Chang et al. prepared bimetallic NiMoN nanowire electrocatalyst by managing the exchange of elements *insitu* and achieved various morphologies by varying the nitridation temperature for catalyzing both HER and OER. The electrocatalyst prepared at 700 °C coded as NiMoN-NF700 has revealed better performance with a low overpotential of 290 mV to achieve the current density of 50 mA/cm² in alkaline conditions and efficient in catalyzing HER in an acidic medium. The authors claimed that the low adsorption free energy of H* ($\Delta GH*$), (100) prominent facet served as the reactive site of the NiMoN nanowire for facilitating HER activity in an acidic medium. As a bifunctional electrocatalyst for overall water splitting, it required a cell voltage of 1.498 V to achieve 20 mA/cm² and revealed good stability during its prolonged electrochemical reaction [10].

Chen et al. constructed a robust and cost-effective metal $CuO@CoFeO_x$ self-reliant nanowire electrocatalyst via the solution oxidation method. The electrocatalyst has good durability with the application of a constant current density of 10 mA/cm² for a longer duration of 300 h continuously without observable degradation in 0.1 M KOH solution. When the anode catalyst $CuO@CoFeO_x$ nanowire anode was coupled with a cathode Pt-Ni foam it revealed a superior activity, which required an overpotential of 0.39 V @ 10 mA/cm² for overall water electrolysis in an alkaline solution. Li et al. synthesized core-shell structured $NiCo_2O_4@FeOOH$ nanowire arrays over Ni foam substrate by electrodeposition technique. The fabricated $NiCo_2O_4@FeOOH/NF$ electrode exhibiting core-shell nanowire arrays possess a mesoporous structure that facilitates diffusion of ions in a shorter path and enhances the electrical conductivity and electrode/electrolyte contact interface. The vertically aligned $NiCo_2O_4$ NWs ensure good electrical contact, display a high surface area, and contain a higher degree of porosity, which helps in the diffusion of electrolytes into inner active sites and for rapidly delivering the H_2 or O_2 bubbles produced during the electrochemical reaction. In addition, the synergistic effect between FeOOH and $NiCo_2O_4$ ensures a high charge transfer rate and electron transfer at the interface that is responsible for enhancing the electrocatalytic performance [10].

Wang et al. prepared Fe-doped Ni_3S_2 nanowires on Ni foam through a facile solvothermal route for overall water splitting in an alkaline medium. The electrocatalyst displayed excellent electrocatalytic activity and durability for catalyzing both OER and HER under elevated current densities. The doping of iron concomitantly alters the electronic structure and growth of Ni_3S_2, which helps in enhancing the conductivity and generating abundant active edge sites. A negligible *in situ* oxidation of nanowires in a strongly alkaline medium generated more interfaces that have significantly enhanced the reactivity and endurance of the electrocatalyst during OER and HER electrocatalysis. The incorporation of Fe species modulated the electronic structure in which the electron conductivity will be enhanced due to the rapid transfer of electrons and the morphology is altered in which abundant active edge sites will be created in the thin nanowires structure [11].

Yan et al. fabricated a homologous Ni-Co-based nanowire pair, which is obtained from the combination of $Ni_xCo_{3-x}O_4$ nanowires synthesized by hydrothermal technique and $NiCo/NiCoO_x$ hybrid nanowires. The $Ni_xCo_{3-x}O_4$ nanowires were tested towards OER catalysis and the same was converted into $NiCo/NiCoO_x$ heterostructural combination through hydrogenation and it was tested towards catalyzing HER. The $Ni_xCo_{3-x}O_4$ nanowires disclosed better OER activity with an overpotential of 335 mV @ 10 mA/cm². The $NiCo/NiCoO_x$ electrocatalyst revealed a superior HER performance with the least overpotential of 155 mV @ 10 mA/cm². The prepared OER and HER electrocatalyst were demonstrated for overall water splitting performance in which it required a cell voltage of 1.75 V to split the water molecules and excellent stability in an alkaline medium [12]. SEM images of $NiCo_2Se_4$, $NiCo_2O_4$, $NiCo_2S$, and $NiCo_2Se_4$ nanowires as high-performance bi-functional oxygen electrocatalysts are depicted in **Figure 10.2**. Czioska reported the fabrication of novel $NiFeO_x/CuO$ nanosheet/nanowire combination and investigated their efficacy in catalyzing OER. The $Cu(OH)_2$ nanowires were grown *insitu* on a copper substrate by a simple solution technique and it then converted to CuO nanowires by calcination in air. The $NiFeO_x$ nanosheets will be coated over CuO nanowires by anodic co-deposition. This approach of fabrication controls the destruction of CuO nanowires caused in acidic solutions. Electrocatalysis results revealed that the fabricated electrocatalyst requires the overpotential of only 300 mV to facilitate OER and the catalyst showed high durability with no loss in efficiency during the electrocatalytic performance [13].

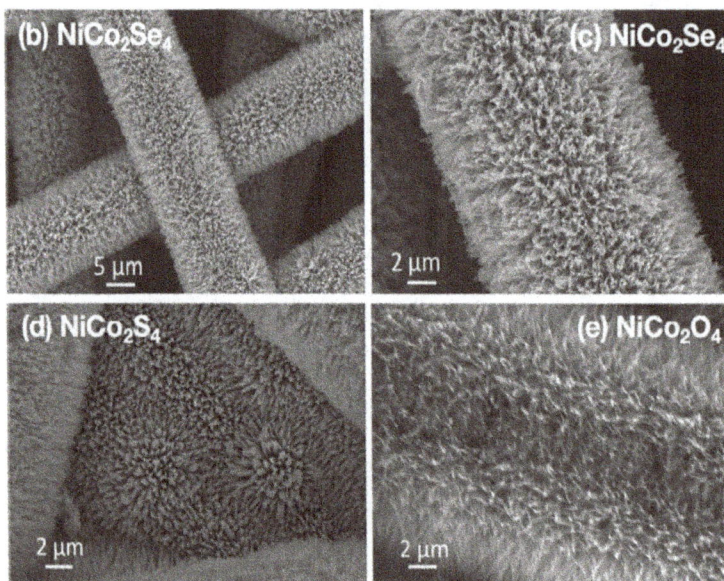

FIGURE 10.2 SEM images of $NiCo_2Se_4$, $NiCo_2O_4$, $NiCo_2S$, and $NiCo_2Se_4$ nanowires as a high-performance bi-functional oxygen electrocatalyst. Adapted from the open access journal [14]. Copyright (2020) J. Electrochem. Soc. Creative Commons Attribution 4.0 License (CC BY).

10.3 OXYGEN REDUCTION REACTION (ORR)

The development of advanced electrocatalysts with low cost with Earth-abundant materials is a key factor in the development of fuel cells and energy conversion devices. In a fuel cell, electricity is generated by electrochemical reduction of oxygen to oxidize fuel into the water. In that, cathodic reduction of oxygen plays a vital role in generating electricity and is a key problem limiting the performance of the fuel cell. However, the sluggish kinetics of ORR and the creation of intermediate products extremely hamper technological developments. As a consequence, many efforts have been promoted to the design and development of high-performance electrocatalysts. Conventionally, platinum (Pt) has been considered the best electrocatalyst for the energy conversion process, although it is still affected by various drawbacks, such as high cost, insufficient resistance to methanol, CO poisoning effects, and poor tolerance of Pt catalysts. The large-scale application of fuel cells needs the development of the high performance of efficient electrochemical catalysts. Nowadays, more potential candidates are developed for the replacement of Pt catalysts for sustainable energy conversion systems [15].

The electrocatalysts are based on transition metal chalcogenides, carbon derivatives, metal oxides, and polymer nanocomposites. In the development of the promising electrocatalysts for the ORR, the following requirements are essential: (a) abundant active sites in the electrocatalysts to improve the activity and kinetics of ORR; (b) high selectivity in ORR for four electron-transfer pathways; (c) long-term stability and better tolerance to methanol and carbon monoxide poisoning. The high performance of ORR catalysts generally has an improved surface area, hierarchical structures, functionalization, and higher electrical conductivity. However, nanostructured materials still suffer from several issues, like difficult synthesis process, lower durability, and insufficient catalytic active sites. 1D nanowires have attracted great attention due to their greater surface area, tunable chemical compositions, abundant porous nature, and variable morphologies. Mostly, 1D nanofibers offer extreme flexibility in fabricating distinctive structures that are modified with defect engineering, functional groups, heteroatom-doping, and other active materials, which are of great importance to overcoming contests in the improvement of electrocatalysts.

10.3.1 Nanowires for ORR

A study enables the large-scale synthesis of metallic nanowires by Pt-Ni-based alloys for the development of enhanced ORR catalysts. These structures enable the enhancement of catalysis due to their surface-to-volume ratio, alloy effect, and 1D morphology. The Pt-Ni-based metallic nanowires were prepared with high yield and a unique 1D structure enhanced the catalytic activity while combining with other metals. These 1D Pt alloy nanowires have a high density of atomic steps, ledges, and kinks, serving as an ORR catalyst with an excellent specific activity [16]. To explore inexpensive, highly active, and stable electrocatalysts toward ORR, a 3D hybrid carbon nanowire (i.e., Fe/P/C) network was developed. This could be because the Fe atom in Fe/N/C helps develop strong back bonding with the adsorbed O_2, which causes an increase of O–O bond distance and hence facilitates ORR. Due to their distinctive 3D interconnected network structure, many active sites (Fe–P bonds, $Fe_3C@C$ units, and P–C bonds), huge numbers of pores, high surface area, and high density of plane edges the resultant Fe/P/C networks show better ORR electrocatalytic activity in alkaline media/comparable activity in acidic media. Fe/P/C exhibits better durability and stronger methanol tolerance compared to those of 20 wt.% Pt/C.

The ORR takes place on the surface of the catalyst, and the precise control of surface components at the atomic scale has been an important factor in effectively improving the performance of catalysts. 1D Pt alloy catalysts display remarkable potential for ORR catalysis due to their high surface area, high flexibility, high conductivity, and 1D morphology, making them work more operative related to Pt electrocatalyst. So, a highly active and stable catalyst is prepared by constructing Pt-Ni

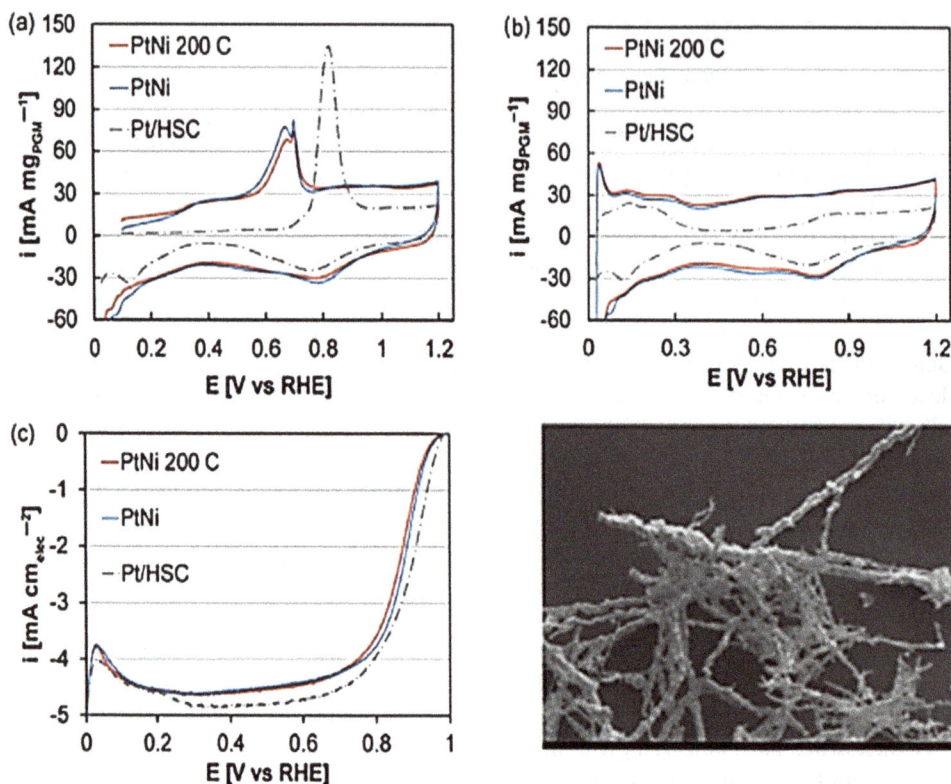

FIGURE 10.3 ORR polarization curves of Pt-Ni nanowires and their SEM image. Adapted from the open access journal [17]. Copyright (2016), J. Electrochem. Soc. Creative Commons Attribution 4.0 License (CC BY).

alloy nanowires. The prime Pt-Ni/C-NWs exhibit enhanced mass activity (1.02 A/mg) and specific activity (3.86 mA/cm^2) for ORR, which are higher than those of commercial Pt/C, respectively. ORR polarization curves of Pt-Ni nanowires and their SEM image are depicted in **Figure 10.3**. Platinum-based one-dimensional nanowires (Pt-NWs) display unique advantages, such as (a) nanowires have a large surface area with very high aspect ratios; (b) numerous catalytic sites for reaction; (c) greater conductivity; (d) the nanowire catalysts can be easily assembled into the 3D structure, which is a smart structure for electrocatalysis due to the large surface area consumption and high stability. Furthermore, silver nanowires with silver nanoparticles have been prepared to enhance the hydroxide-exchange membrane fuel cell development with variable diameters and sizes. The prepared silver-based catalysts exhibited specific mass activity and also showed alcohol tolerance towards ORR.

FePt grew over the FeNiPt nanowires by seed-mediated growth, resulting in FeNiPt/FePt core/shell nanowires. Through acid etching and thermal annealing, the surface profile of the core/shell NWs may be adjusted to have either a Pt-skeleton or Pt-skin structure while maintaining their 1D morphology. In catalyzing ORR in 0.1 m HClO$_4$, these core/shell NWs exhibit increased activity as well as greater durability [18]. However, in the acidic electrochemical environment, these early transition metals are susceptible to acid/electrochemical etching, leaving a Pt-skeleton-type structure with many low coordination sites all around Pt. Such weakly coordinated Pt locations are vulnerable to more than only the ORR's "poisoning" of the Pt surface by strong binding to oxygenated species ("O") but also to migration and dissolution, resulting in poor catalyst stability in the acidic ORR condition.

10.4   HYDROGEN EVOLUTION REACTION (HER)

The conventional electrocatalysts used for catalyzing HER in commercial electrolysis cells utilize Pt-based catalysts owing to their high catalytic activity and withstanding capability in acidic and alkaline media. However, higher cost and poor abundance are the major bottlenecks that greatly hinder the commercialization of water electrolyzers and fuel cells. One major focus towards the reduction of the overall fabrication cost of electrolyzers is to dramatically minimize the quantity of Pt used while not compromising the catalytic activity. The effective strategies for enhancing the catalytic performance are by controlling the morphology by modulating the reaction parameters such as temperature, time, pH, and alloying Pt with other metal species through interfacial inter-action engineering. When Pt nanostructures were integrated with the metal hydroxides, the edges of the hydroxides will promote or stimulate the dissociation of water molecules followed by the generation of hydrogen intermediates. The produced hydrogen intermediates will be subsequently adsorbed on the Pt surfaces and finally recombines into molecular hydrogen. This strategy will be therefore an effective route to reduce the cost of electrocatalysts also with enhanced HER activity in basic conditions [19].

The electrocatalyst surface of the electrode encountering HER electrocatalyst differs in an acidic and alkaline medium. As a first step in the acidic condition, protons will be generated from the deprotonation of water molecules. These protons bind with the electrons that emerged during elec-trochemical reactions and release adsorbed hydrogen (H_{ads}) species. This step is described as the Volmer reaction, also called the discharge reaction. Based on the quantity of adsorbed hydrogen species (H_{ads}), the next subsequent step in which the reaction proceeds is either the Heyrovsky or Tafel reaction. If the mass of H_{ads} species is less, the electrocatalyst-coated electrode surface contains abundant active sites nearer to H_{ads} sites, and they will preferably bond with a proton and an electron for the simultaneous evolution of H_2 molecules, which is denoted by the relation H_{ads} + H^+ (aq) + $e^- \rightarrow H_2$. This step is described as the Heyrovsky step. In the Tafel step, the H_{ads} species will be high and hence the two adjacent protons combine and release H_2 molecules more rapidly. Those are the electrochemical reaction steps that occur during the evolution of hydrogen molecules [20]. A few catalysts exhibiting nanowire-shaped morphology were discussed below to understand the efficiency of 1D nanowires in catalyzing HER.

10.4.1   NANOWIRES IN HER

Wang developed interface-engineered Pt-Ni NWs/C electrocatalyst synthesized through a simple post-annealing treatment in which Pt_3Ni_3 NWs/C electrocatalyst revealed comparatively excep-tional HER performance with a very low overpotential of 40 mV @ 10 mA/cm² and excellent durability than that of commercial Pt/C electrocatalyst in alkaline medium. Xie et al. prepared Pt–Ni nanowires and achieved nitrogen doping with the prepared Pt–Ni nanowires by N-induced orbital tuning. This modification greatly enhances the HER electrocatalytic activity with an ultralow overpotential of 13 mV to achieve the benchmarking current density of 10 mA/cm² and this is the best HER performance in alkaline medium among the so far reported electrocatalytic activities [21]. Cyclic voltammetry (CV), linear sweep voltammetry (LSV), and chronoamperometry (CA) of the Rh nanowires are depicted in **Figure 10.4**. Lin et al. fabricated cobalt-doped molybdenum-carbide nanowires for efficiently catalyzing HER in an alkaline medium by consequently improving the electron density around Fermi energy E_f to facilitate the HER kinetics more rapidly with low overpotential. Liu et al. developed a $CoSe_2$ nanowire array modified on carbon cloth substrate through a facile hydrothermal synthesis technique and investigated the nanoarray electrode towards electrochemical hydrogen evolution reaction. The $CoSe_2$ nanowire modified carbon cloth electrode exhibited excellent electrocatalytic activity and withstanding capability in acidic media. It required the overpotentials of 130 mV @ current densities of 10 mA/cm² with good durability for about 48 h during electrocatalytic activity [22].

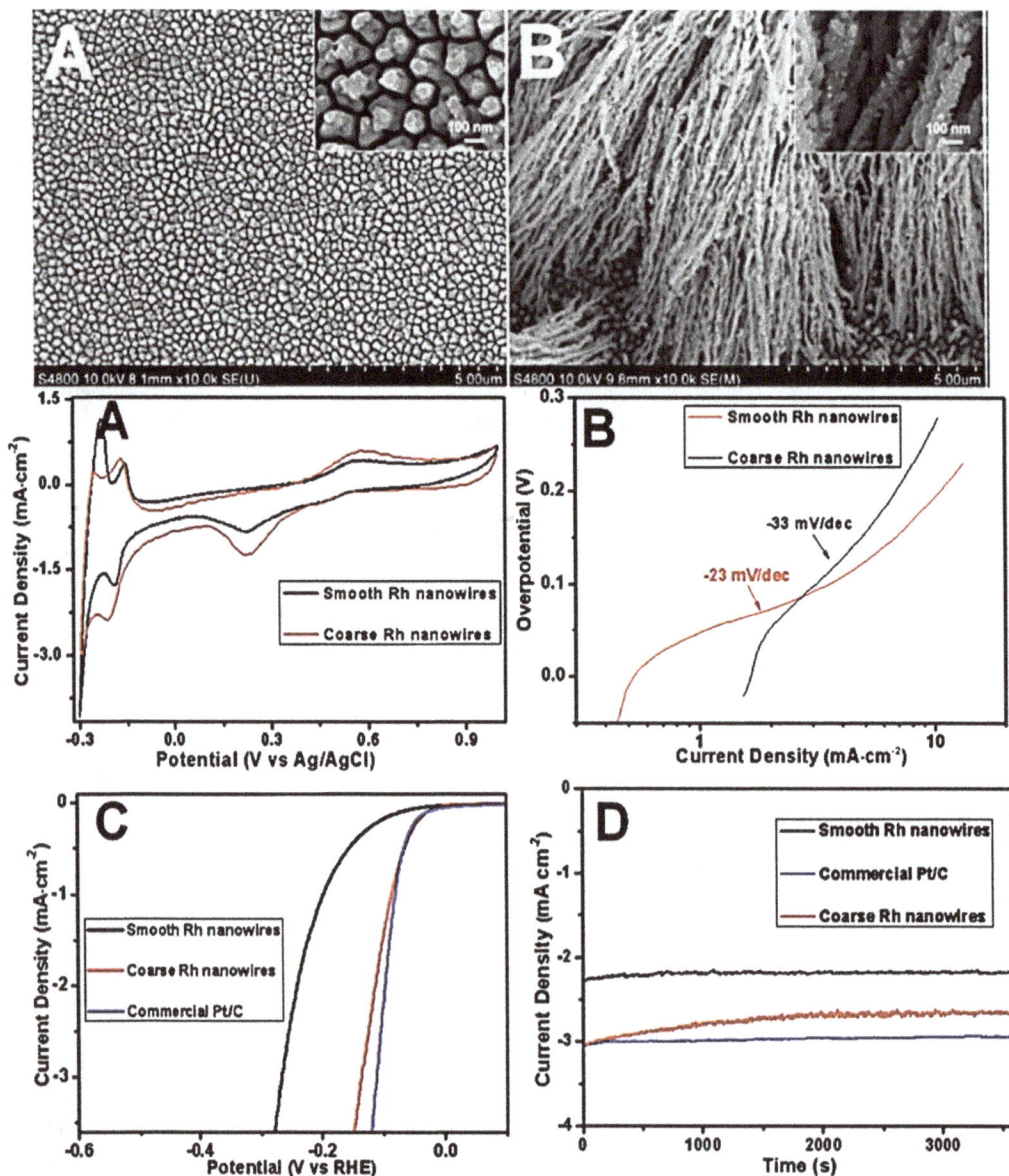

FIGURE 10.4   CV, LSV, and CA of the Rh nanowires. Adapted from the open access journal [23]. Copyright (2017), Nanomaterials. Creative Commons Attribution 4.0 License (CC BY).

Jiang et al. prepared PtNi ultrathin nanowires with varying concentrations coupled with MXene (Ti_3C_2) for catalyzing HER in both acidic and alkaline media. Theoretical calculations revealed that the coupling of MXenes with the nanowires reaches the Gibbs free energy for hydrogen adsorption close to zero through the electron transfer between them in acidic media. It also provides more active sites for dissociation of water molecules in alkaline solution, which is highly beneficial for enhancing the HER performance. Liu et al. synthesized 1D NiS_2 nanowires and investigated their efficiency in catalyzing HER. The prepared nanowires exhibited a large surface area and a higher degree of active sites which enabled them to be highly efficient towards the electrocatalytic performance of HER with long-term stability.

Peng et al. developed a homologous nanowire-based system of $NiCo_2O_4$ and $Ni_{0.33}Co_{0.67}S_2$, for catalyzing OER and HER reaction kinetics, respectively. The high surface area and crystalline structures of these nanowires enable excellent electrocatalytic activities with good stability. Zhang et al. reported a new strategy to fabricate PtNi@Pt dendritic nanowires decorated with surface-enriched amorphous tungsten oxide (WOx) for attaining excellent HER catalysis [24]. The WOx decorated PtNi@Pt dendritic largely promotes the water electrolysis process in alkaline HER and revealed outstanding HER with an overpotential of 24 mV @ 10 mA/cm^2, which is substantially superior to the commercial Pt/C. Cheng et al. fabricated CoP_3 nanowire decorated with copper phosphides nanodot structures synthesized through a simple precursor-transformation strategy for catalyzing HER. The optimized CoP_3 nanowires were decorated with copper phosphide nanodots in which the prepared catalyst-modified electrode exhibited excellent catalytic activity and long-term durability for HER in alkaline conditions, achieving a low overpotential of 49.5 mV at a benchmarking current of 10 mA/cm^2. The decoration of copper phosphide nanodots over the CoP_3 nanowire structure avoided the aggregation issues and thus took full advantage of the high HER activity.

10.5 ALCOHOL (METHANOL, ETHANOL) OXIDATION REACTION

A direct alcohol fuel cell (DAFC) is different from a proton exchange membrane (PEM) fuel cell. A fuel cell that uses methanol as a fuel is called a direct methanol fuel cell (DMFC) and a fuel cell that utilizes ethanol as the fuel for electricity production is called a direct ethanol fuel cell (DEFC). Generally, oxidation of fuels (methanol and ethanol) in the anode and the oxygen reduction kinetics occurring in the cathode is comparatively faster in the alkaline medium than in the acidic medium. Pt has been widely accepted as an efficient electrocatalyst for catalyzing the oxidation of ethanol, and methanol fuels. The development of Pt-based electrocatalyst for reaching superior mass activity and excellent durability is the crucial factor to be considered in catalytic reactions such as the methanol/ethanol oxidation reaction (MOR/EOR) in which they undergo sluggish kinetics during the electrochemical reactions. Moreover, CO poisoning of the high-cost Pt electrocatalyst occurred due to the intermediate products formed during the reaction also greatly hinder the reaction kinetics, which therefore impedes its further commercialization. Thus, it is highly beneficial for generating novel and efficient anti-poisoning Pt-based electrocatalysts with superior electrocatalytic activities to address this drawback. One of the desirable strategies is to develop multimetallic Pt-based alloys or composites with suitable size and structure [25]. In this context, 1D nanowires have gained immense attention in enhancing the sluggish kinetic reactions owing to the beneficial merits such as high conductivity, a higher degree of porosity, large specific surface area, and strong intercalation between the substrate support and electrocatalyst. Their elevated behavior in their activities and withstanding capability during the electrochemical reactions are mainly due to the structural and electronic effects. The comparison of the electrocatalyst to electrocatalytic activity in OER and HER is illustrated in **Table 10.1**.

10.5.1 Nanowires for Alcohol (Methanol, Ethanol) Oxidation Reaction

Sun et al. developed PtNiCo/NiCoS nanowires by the interfacial engineering technique and employed it for the oxidation process of ethanol and methanol fuel. The authors would have claimed that the developed PtNiCo/NiCoS nanowires delivered better activity and stability in oxidizing methanol and ethanol fuels in alkaline electrolytes. The elevated performance was due to certain factors aroused from their unique structure in which it can suppress the impact created by particle ripening, movement of electrons rapidly, and dissolution due to their interaction with carbon support [18]. Zhang et al. developed wicker-like Pt-Fe nanowires with branch-rich outfits for efficiently catalyzing methanol and ethanol oxidation reactions. The enhanced catalytic behavior is ascribed

TABLE 10.1

Comparison of the Electrocatalyst to Electrocatalytic Activity in OER and HER

Electro-catalystic nanowires	Method of synthesis	Electro-catalytic activity	Overpotential	Tafel value	Electrolyte (in molar)	Current density
NiFeO$_x$/CuO nanosheets/ nanowires	Anodic co-deposition	OER	300 mV	36 mV/dec	1 M KOH	100 mA/cm^2
Bimetallic NiMoN nanowires	Hydrothermal	HER		46 mV/dec	1.0 M KOH	10 mA/cm^2
Co-Mo$_2$C nanowires		HER	140 mV	39 mV/dec	0.5 M H$_2$SO$_4$	5.1 mA/cm^2
Porous iridium nanowires (Ir PNWs)	Hydrothermal	OER	1.65 mV	74 mV/dec	1M KOH	10 mA/cm^2
PtNi Ultrathin nanowires with MXenes (Pt3.21Ni@ Ti3C2)	solvothermal	HER	18.55 mV	13.37 mV/ dec	0.5 m H$_2$SO$_4$ solution	10 mA/cm^2
Ultra-thin PdPt bimetallic nanowires	Surfactant directing method	HER	0.8 mV	-28.8 mV/ dec	0.5 M H$_2$SO$_4$	10 mA/cm^2
WOx-surface decorated PtNi@ Pt dendritic nanowires (WOx-PtNi@Pt DNWs	Wet-chemical approach	HER	24 mV		0.1 m KOH	10 mA/cm^2
Three-dimensional NiFe/Cu$_2$O nanowires/Cu foam	Electrodeposition method	OER	215 mV	42 mV/dec	1.0 M KOH	10 mA/cm^2
Hybrid Co$_3$O$_4$-carbon porous nanowire arrays	Hydrothermal process	OER	85 mV	70 mV decade-1;61 mVdec-1	M KOH 1.0 M KOH	10 mA/cm^2

to stable interconnected networked nanowires structure, branched exteriors, and a large number of defective sites due to the alloying effect [26]. Sekol and his team fabricated Pd-Ni-Cu-P metallic glass nanowires by nanomolding technique and their efficiency in catalyzing methanol and ethanol oxidation reaction in alkaline media was investigated. These nanowires revealed superior activity with an onset potential of 300 mV for CO oxidation, which suggests the resistance against poisoning is much higher than pure Pd. The enhanced behavior is attributed to the unique properties, such as viscosity, homogeneity, and surface tension exhibited by bulk metallic glass nanowires that promote the formation of porous and high surface area electrocatalysts at low reaction temperatures [27].

Zheng et al. developed platinum-ruthenium nanotubes (PtRuNTs) and platinum-ruthenium-coated copper nanowires by galvanic displacement reaction using copper nanowires as a template. The different composition of PtRu combination is investigated towards MOR catalysis and observed that the optimum bulk Pt/Ru atomic ratio of 4 is suitable for both PtRu nanotubes and PtRu/Cu nanowires. The authors would have claimed that the combinations of platinum-ruthenium nanotubes and platinum-ruthenium-coated copper nanowires revealed outstanding MOR performance owing to the modulation of the d-band center, which leads to a weaker bonding of Pt when poisoned by the CO intermediate. This behavior resulted in improved performance in catalyzing MOR with the combination of PtRu nanotube and PtRu/Cu nanowire electrocatalysts [28].

Kim et al. fabricated Pt nanowires by electrospinning technique. The physicochemical and electrocatalytic properties of the Pt nanowires are greatly influenced by the preparatory conditions, such as rate of heating, temperature, time, and atmosphere. The annealing conditions are optimized for preventing the sintering of the Pt particles existing in the nanowires during the burning out of the carbonaceous materials in the fibers. The prepared Pt nanowires exhibited a larger electrochemical surface area of 6.2 m^2/g due to the surface roughness and the specific activity approaches the current density value of 1.41 mA/cm^2, which is comparatively higher than the specific activity of commercially used Pt black electrocatalyst. The prepared Pt nanowires can be suitably optimized for higher mass activity by suitably minimizing the diameter of nanowires or roughening their surface [29]. Li et al. prepared a series of surface oxygen-mediated ultrathin PtRuM nanowires, where M corresponds to Ni, Fe, and Co species. All the prepared electrocatalysts revealed ultrahigh electrocatalytic performance and good durability for MOR in an acidic medium due to their modulated electronic structures generated by the addition of electroactive O species. Scofield et al. synthesized thin ternary PtRuFe nanowires (NW) with different chemical compositions by a simple solution-assisted technique to examine their CO tolerance and electrochemical activity towards the methanol oxidation reaction (MOR) and formic acid oxidation reaction. Specifically, the Pt_7Ru_2Fe and Pt_7Ru_3 nanowire electrocatalyst possessed the lowest onset of formation of Ru–OH species even with a mere 10% loss of Ru and a corresponding 10% increase in Fe content, suggestive of a ligand-induced effect, likely as a result of the presence of Fe lowering the d-band center of Pt and thereby altering the electronic properties of the overall alloy [30].

10.6   NITROGEN REDUCTION REACTION (NRR)

In the chemical industry, ammonia (NH_3) is one of the most frequently manufactured compounds. It is fundamental to the development of drugs, fertilizers, and plastics and it finds new opportunities as a carbon-neutral fuel source. Ammonia fuel cells are also a perspective option for the hydrogen economy due to their increased energy density and massive hydrogen capacity. Moreover, from a thermodynamic perspective, the conversion of nitrogen (N_2) to NH_3 is a complex multistep reaction due to the high bond energy and triple bond in N_2. Nowadays, the Haber-Bosch technique is most extensively used because of the industrial-scale yield of effective NH_3 synthesis from N_2 to H_2. There is a significant need to develop a carbon-neutral synthetic method that can efficiently and selectively reduce the formidable N–N triple bond. Undoubtedly, finding better catalysts to effectively activate and hydrogenate the stable nitrogen molecules is the key to addressing this problem. The investigation of transition metal, noble metal, and non-metal systems have led to a recent acceleration in research on the electrocatalytic NRR. In recent years, significant efforts have been undertaken to convert N_2 to NH_3 by employing biocatalytic, electrocatalytic, and photocatalytic approaches in the search for effective and sustainable catalysis. The electroreduction of N_2 to NH_3 is one potential method for producing NH_3, which permits the operation of the NRR process at mild temperature and pressure using renewable electrical energy. A lot of effort has been implemented to change the NRR of electrocatalysts including size regulation, crystal engineering, ion integration, the addition of defect sites, and component modulations [31].

10.6.1   NANOWIRES FOR NRR

Despite the strong conceptual framework, only a few selections of metal nitride catalysts have been experimentally investigated for electrochemical NRR formation. Recent advancements in molybdenum or vanadium nitride materials in the forms of nanowires, nanosheets, or nanoparticles have demonstrated NRR activity with enhanced Faradaic efficiencies. Inspired by the rich NRR chemistry on Mo atoms, the demonstration of production and characterization of highly dispersed, defect-rich MnO_2 nanoparticles grafted onto the surface of a conductive carbon nanowire substrate. The greatest activity and selectivity towards NRR on a Mo-based electrocatalytic system have an

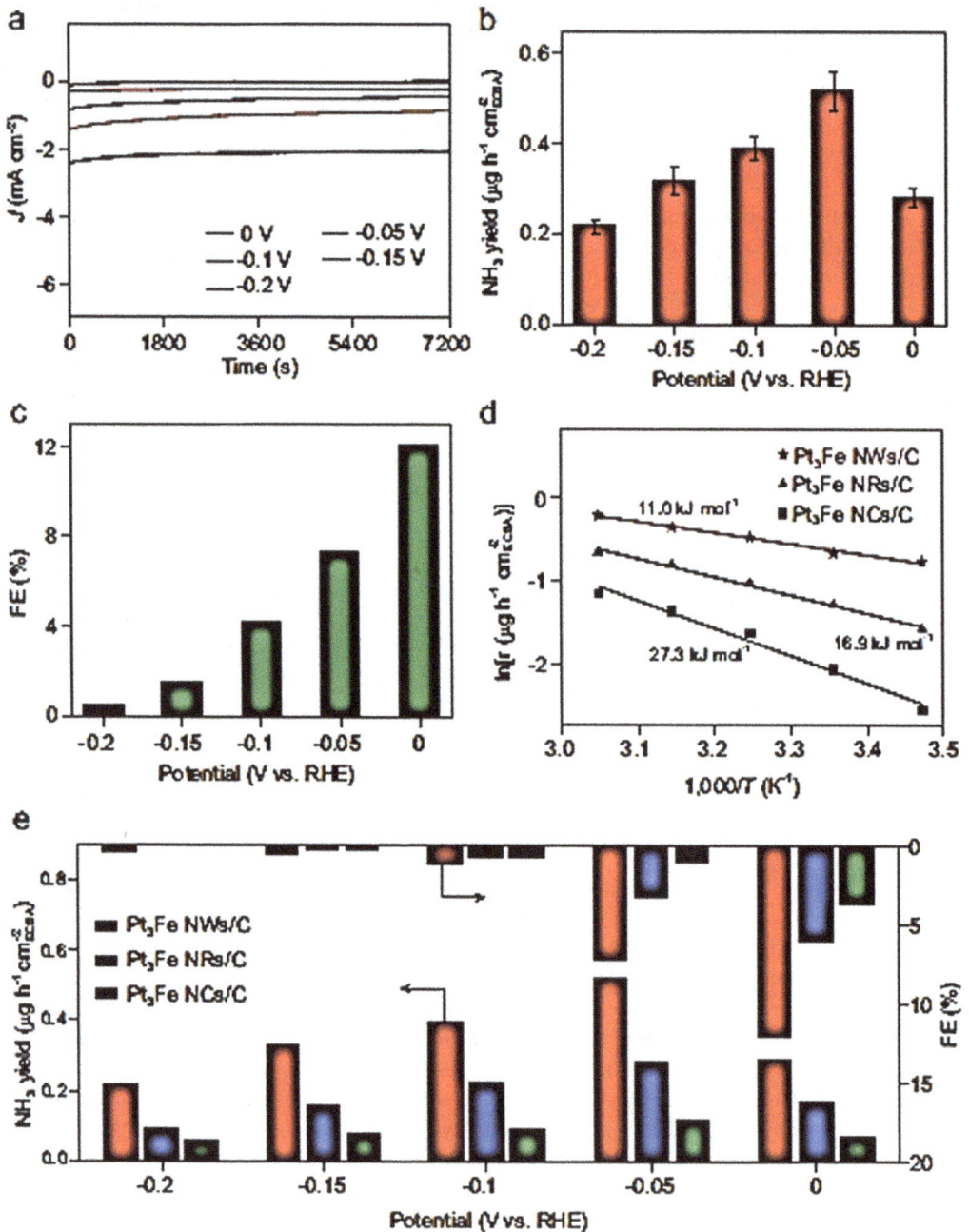

FIGURE 10.5   I-t curves, NH_3 yield, FE of Pt_3Fe NWs/C at different applied potentials, and apparent activation energy for NRR with different Pt_3Fe electrocatalysts. Adapted from the open access journal [32]. Copyright (2020), National Science Review. Creative Commons Attribution 4.0 License (CC BY).

average Faradaic efficiency of 13.3% and NH_3 yield of 21.2 µg/h.mg at 0.1 V of low applied potential. The optimization model electrocatalysts for analyzing the NRR on various exposed facets are Pt_3Fe nanocrystals with well-defined adjustable morphologies (nanowire, nanorod, and nanocube). According to thorough electrocatalytic experiments, Pt_3Fe nanocrystals exhibit shape-dependent NRR electrocatalyst [32]. I-t curves, NH_3 yield, FE of Pt_3Fe NWs/C at different applied potentials, and apparent activation energy for NRR with different Pt_3Fe electrocatalysts are depicted in **Figure 10.5**. The prepared Pt_3Fe nanowires embedded with high-index facets provide outstanding selectivity and electrocatalytic activity at 18.3 µg/h.mg yield of NH3 and 7.3% of Faradaic efficiency at an applied voltage of -0.05 V. Pt_3Fe nanowires exhibit good stability after five cycles. Ru:Pt (88:12) atomic ratio is used to create Pt-decorated Ru nanowires. The produced Ru88Pt12 nanowires exhibit a few scattered Pt atoms on the Ru lattice. They outperformed Ru76Pt24, Ru47Pt53, Ru, and Pt nanowires with a significantly improved NH_3 generation rate of 47.1 and Faradaic efficiency of 8.9 percent at -0.2V. Theoretical studies show that the presence of a Pt atom causes the d-band center of Ru atoms to shift upward, selectivity enhancing N_2 adsorption and stabilizing N_2H.

10.7   CARBON DIOXIDE REDUCTION REACTION (CO_2RR)

Carbon dioxide (CO_2), a greenhouse gas that rises as a result of the combustion of fossil fuels, is responsible for a variety of environmental hazards, including ocean acidification, global warming, and sea level rise. The simultaneous reduction of greenhouse gas emissions and use of carbon resources in CO_2 can be achieved by the conversion of CO_2 into highly valuable fuels and chemicals like formate (HCOOH) and carbon monoxide (CO). The electrochemical CO_2 reduction reaction (CO_2RR), is gaining a lot of attention due to its mild operating conditions and possible conversion efficiencies, but generating desirable products is extremely hard due to the high energy barriers and proton coupling multielectron steps.

10.7.1   NANOWIRES FOR CO_2RR

Various types of electrocatalysts are engineered for the electrochemical reduction of CO_2 such as transition metal oxides, chalcogenides, and carbon nanostructures. In particular, copper-based nanostructures, have been widely studied for the conversion of CO_2 Bismuth (Bi) has been considered an auspicious electrocatalyst for CO_2 reduction to produce formate. The fabrication of different nanostructures with different strategies, such as controlled synthesis with increased features, a combination of bimetals, and grain boundary engineering. Among all the strategies bimetallic material design may both boost single metal selectivity and catalytic activity. The research on Bi-based bimetallic electrocatalysts for CO_2RR is still in its early stages [33].

For the preparation of bimetallic compounds, copper (Cu) has fascinating properties like electron configuration, the ability to produce hydrocarbons, and higher electrical conductivity. However, individual Bi still has a poor electrocatalytic activity towards CO_2RR to formate, Bi is incorporated with Cu due to its high conductivity and the ability to act as an electron donor or acceptor. Cu nanowires (NWs) bridged with Bi nanosheet (Cu NWs-Bi NSs) arrays produced on carbon cloth. In comparison to pristine Bi NSs, the as-obtained electrocatalysts displayed a higher partial current density and improved selectivity for CO_2RR to formate [33]. Cu NWs better charge transfer and electronic effect may be responsible for the improved selectivity and increased electrocatalytic activity. The one-dimensional Ag incorporated CuO nanowires (Ag-CuO NW) constructed by the combination of nanoparticles as electrocatalyst for the CO_2RR. The prepared Ag-CuO NWs exhibit a higher electrochemical active surface area, and it shows a relatively lower overpotential [34].

However, palladium-silver (Pd-Ag) based electrocatalysts reduce the CO poisoning from the intermediates from the side reactions. Pd-Ag alloy nanowires are prepared by solution method and they exhibit various properties, such as large surface area, high aspect ratio, and high roughened surfaces. The explored Pd-Ag nanowires highly promote the CO_2RR to formate. So, 1D nanostructures provide fascinating electrocatalytic activity due to their inherent anisotropy, high flexibility, and ease to modify the phases [35].

10.8 CONCLUSION

Unquestionably, one of the biggest needs of a sustainable society is energy. The conversion and storage of renewable energy are crucial to the social and economic advancement of the modern world. However, since the start of the twentieth century, there has been an enormous increase in the use of non-renewable fossil fuels, which has led to grave concerns about energy shortages and the related carbon emission that degrade the environment. There is an urgent need for access to reliable, inexpensive, and sustainable energy that can replace fossil fuels. Energy conversion has drawn concern from all around the world and unwavering scientific interest as a fundamentally intermediate step to effectively use clean and renewable energy. Electrochemical conversion systems that can achieve efficient, diverse, and ecological friendly use of energy are given serious consideration with the rising need for portable gadgets, electric vehicles, and consumer electronics.

REFERENCES

[1] S. V. P. Vattikuti, "Heterostructured nanomaterials: latest trends in formation of inorganic heterostructures," *Synthesis of Inorganic Nanomaterials*, 1 (2018) 89–120.

[2] I. Khan, K. Saeed, and I. Khan, "Nanoparticles: Properties, applications and toxicities," *Arab. J. Chem.*, 7 (2019) 908–931.

[3] A. Rajapriya, S. Keerthana, A. Rebekah, C. Viswanathan, and N. Ponpandian, "Enriched oxygen vacancy promoted heteroatoms (B, P, N, and S) doped CeO_2: Challenging electrocatalysts for oxygen evolution reaction (OER) in alkaline medium," *Int. J. Hydrogen Energy*, 46 (2021) 37281–37293.

[4] P. Chen, K. Xu, Z. Fang, Y. Tong, J. Wu, X. Lu, and Y. Xie, "Metallic Co_4N porous nanowire arrays activated by surface oxidation as electrocatalysts for the oxygen evolution reaction," *Angew. Chemie*, 127 (49) (2015) 14923–14927.

[5] T. Wasiak and D. Janas, "Nanowires as a versatile catalytic platform for facilitating chemical transformations," *J. Alloys Compd.*, 892 (2022) 162158.

[6] M. Li, M. Lu, J. Yang, J. Xiao, L. Han, Y. Zhang, and X. Bo, "Facile design of ultrafine $CuFe_2O_4$ nanocrystallines coupled porous carbon nanowires: Highly effective electrocatalysts for hydrogen peroxide reduction and the oxygen evolution reaction," *J. Alloys Compd.*, 809 (2019) 151766.

[7] A. Rajapriya, S. Keerthana, C. Viswanathan, and N. Ponpandian, "Three dimensional integrated architecture of Sr – Fe LDH on hierarchical NiS framework as a flexible electrode for efficient energy storage and conversion applications," *J. Energy Storage*, 53 (2022) 105091.

[8] B. Chang, J. Yang, Y. Shao, L. Zhang, W. Fan, B. Huang, and X. Hao, "Bimetallic NiMoN nanowires with a preferential reactive facet: An ultraefficient bifunctional electrocatalyst for overall water splitting," *Chem Sus Chem*, 11 (18) (2018) 3198–3207.

[9] G. Chen, Y. Guo, S. Miao, H. Sun, B. Gu, W. Li, and Z. Shao, "Constructing self-standing and non-precious metal heterogeneous nanowire arrays as high-performance oxygen evolution electrocatalysts: Beyond the electronegativity effect of the substrate," *J. Power Sources*, 396 (5) (2018) 421–428.

[10] M. Li, L.Tao, X. Xiao, X. Lv, X. Jiang, M. Wang, and Y. Shen, "Core-shell structured $NiCo_2O_4$@ FeOOH nanowire arrays as bifunctional electrocatalysts for efficient overall water splitting," *Chem Cat Chem*, 10 (18) (2018) 4119–4125.

[11] X. Wang, W. Zhang, J. Zhang, and Z. Wu, "Fe-doped Ni_3S_2 nanowires with surface-restricted oxidation toward high-current-density overall water splitting," *Chem Electro Chem*, 6 (17) (2019) 4550–4559, 2019.

[12] X. Yan, K. Li, L. Lyu, F. Song, J. He, D. Niu, and X. Chen, "From water oxidation to reduction: Transformation from $Ni_xCo_{3-x}O_4$ nanowires to $NiCo/NiCoO_x$ heterostructures," *ACS Appl. Mater. Interfaces*, 8 (5) (2016) 3208–3214.

[13] S. Czioska, J. Wang, S. Zuo, X. Teng, and Z. Chen, "Hierarchically structured $NiFeO_x/CuO$ nanosheets/nanowires as an efficient electrocatalyst for the oxygen evolution reaction," *Chem Cat Chem*, 10 (5) (2018) 1005–1011, 2018.

[14] H. Sancho, Y. Zhang, L. Liu, V. G. Barevadia, S. Wu, Y. Zhang, and N. Liu, "$NiCo_2Se_4$ nanowires as a high-performance bifunctional oxygen electrocatalyst," *J. Electrochem. Soc.*, 167 (5) (2020) 056503.

[15] M. Li, T. Liu, X. Bo, M. Zhou, L. Guo, and S. Guo, "Hybrid carbon nanowire networks with Fe–P bond active site for efficient oxygen/hydrogen-based electrocatalysis," *Nano Energy*, 33 (2016) 221–228.

[16] H. Tang, Y. Su, B. Chi, J. Zhao, D. Dang, X. Tian, and G. R. Li, "Nodal PtNi nanowires with Pt skin and controllable near-surface composition for enhanced oxygen reduction electrocatalysis in fuel cells," *Chem. Eng. J.*, 418 (2021) 129322.

[17] S. M. Alia, S. Pylypenko, A. Dameron, K. C. Neyerlin, S. S. Kocha, and B. S. Pivovar, "Oxidation of platinum nickel nanowires to improve durability of oxygen-reducing electrocatalysts," *J. Electrochem. Soc.*, 163 (3) (2016) 296–301.

[18] H. Zhu, S. Zhang, D. Su, G. Jiang, and S. Sun, "Surface profile control of FeNiPt/Pt core/shell nanowires for oxygen reduction reaction," *Small*, 11 (29) (2015) 3545–3549.

[19] G. H. Gu *et al.*, "Autobifunctional mechanism of jagged Pt nanowires for hydrogen evolution kinetics via end-to-end simulation," *J. Am. Chem. Soc.*, 143 (14) (2021) 5355–5363.

[20] Q. Liu, J. Shi, J. Hu, A. M. Asiri, Y. Luo, and X. Sun, "$CoSe_2$ nanowires array as a 3D electrode for highly efficient electrochemical hydrogen evolution," *ACS Appl. Mater. Interfaces*, 7 (7) (2015) 3877–3881.

[21] Y. Xie, J. Cai, Y. Wu, Y. Zang, X. Zheng, J. Ye, and Y. Qian, "Boosting water dissociation kinetics on Pt–Ni nanowires by N-induced orbital tuning," *Adv. Mater.*, 31 (16) (2018) 1–7.

[22] H. Lin, N. Liu, Z. Shi, Y. Guo, Y. Tang, and Q. Gao, "Cobalt-doping in molybdenum-carbide nanowires toward efficient electrocatalytic hydrogen evolution," *Adv. Funct. Mater.*, 26 (31) (2016) 5590–5598.

[23] L. Zhang, L. Liu, H. Wang, H. Shen, Q. Cheng, C. Yan, and S. Park, "Electrodeposition of rhodium nanowires arrays and their morphology-dependent hydrogen evolution activity," *Nanomaterials*, 7 (5) (2017) 1–14.

[24] Y. Jiang, X. Wu, Y. Yan, S. Luo, X. Li, J. Huang, and D. Yang, "Coupling PtNi ultrathin nanowires with MXenes for boosting electrocatalytic hydrogen evolution in both acidic and alkaline solutions," *Small*, 15 (12) (2019) 1–9.

[25] Y. Sun, Y. Li, Y. Qin, L. Wang, and S. Guo, "Interfacial engineering in PtNiCo/NiCoS nanowires for enhanced electrocatalysis and electroanalysis," *Chem. - A Eur. J.*, 26 (18) (2020) 4032–4038, 2020.

[26] Y. Zhang, F. Gao, T. Song, C. Wang, C. Chen, and Y. Du, "Novel networked wicker-like PtFe nanowires with branch-rich exteriors for efficient electrocatalysis," *Nanoscale*, 11 (33) (2019) 15561–15566.

[27] R. C. Sekol *et al.*, "Pd-Ni-Cu-P metallic glass nanowires for methanol and ethanol oxidation in alkaline media," *Int. J. Hydrogen Energy*, 38 (26) (2013) 11248–11255.

[28] J. Zheng *et al.*, "Platinum-ruthenium nanotubes and platinum-ruthenium coated copper nanowires as efficient catalysts for electro-oxidation of methanol," *ACS Catal.*, 5 (3) (2015) 1468–1474.

[29] J. M. Kim *et al.*, "Preparation and characterization of Pt nanowire by electrospinning method for methanol oxidation," *Electrochim. Acta*, 55 (16) (2010) 4827–4835.

[30] M. E. Scofield, C. Koenigsmann, L. Wang, H. Liu, and S. S. Wong, "Tailoring the composition of ultrathin, ternary alloy PtRuFe nanowires for the methanol oxidation reaction and formic acid oxidation reaction," *Energy Environ. Sci.*, 8 (1) (2015) 350–363.

[31] X. Zhang, R. M. Kong, H. Du, L. Xia, and F. Qu, "Highly efficient electrochemical ammonia synthesis: Via nitrogen reduction reactions on a VN nanowire array under ambient conditions," *Chem. Commun.*, 54 (42) (2018) 5323–5325.

[32] W. Tong, B. Huang, P. Wang, Q. Shao, and X. Huang. "Exposed facet-controlled N_2 electroreduction on distinct Pt_3Fe nanostructures of nanocubes, nanorods and nanowires," *National Science Review*, 12 (2021) 1–9.

[33] L. Li, F. Cai, F. Qi, and D. K. Ma, "Cu nanowire bridged Bi nanosheet arrays for efficient electro-chemical CO_2 reduction toward formate," *J. Alloys Compd.*, 841 (2020) 1–8.

[34] C. J. Chang, S. F. Hung, C. S. Hsu, H. C. Chen, S. C. Lin, Y. F. Liao, and H. Chen, "Quantitatively unraveling the redox shuttle of spontaneous oxidation/electroreduction of CuO_x on silver nanowires using in situ x-ray absorption spectroscopy," *ACS Cent. Sci.*, 5 (12) (2019) 1998–2009.

[35] N. Han, M. Sun, Y. Zhou, J. Xu, C. Cheng, R. Zhou, and Y. Li, "Alloyed palladium–silver nanowires enabling ultrastable carbon dioxide reduction to formate," *Adv. Mater.*, 33 (4) (2021) 1–7.

11 Nanowires for Fuel Cells

Jie Li, Cheng Wang, and Yukou Du
College of Chemistry, Chemical Engineering and Materials Science,
Soochow University, Suzhou, PR China

CONTENTS

11.1 INTRODUCTION

In the past few decades, the energy demand on a global scale has significantly increased due to the rapid development of industrial society and the booming growth of the population. The excessive consumption of fossil energy and the environmental pollution caused by its use have prompted the research on renewable and green energy devices to become a hot spot. The fuel cells (FCs) have been regarded as a promising energy conversion device owing to unique advantages, including diversified fuel sources, low radiation, high specific power, and low carbon emissions [1, 2]. Regardless of these advantages of FCs over fossil energy for producing electricity, the sluggish reaction kinetics of cathodic and anodic reactions have severely limited the development of FCs. High-performance electrocatalysts can effectively improve the reaction kinetics and determine the overall energy conversion efficiency [3]. FCs electrocatalysts fall into two categories: noble metal-based catalysts, and non-noble metal-based catalysts. Of this category, noble platinum (Pt)-group (Pt, Pd, Ru, Rh, Ir, Os) catalysts have attracted remarkable attention, which originates from their inherent electronic structural properties and relatively high activity [4, 5]. Among Pt-group nanocatalysts with various geometric structures, one-dimensional (1D) nanowires (NWs) have drawn intensive attention due to their inherent structural advantages, including high aspect ratio for sufficient exposure to active sites, outstanding electron mobility for fast charge transfer, and

DOI: 10.1201/9781003296621-11

high density of unsaturated atoms for improved intrinsic activity, etc. [6–8]. Furthermore, elaborate nanostructure engineering endows the development of Pt-group NWs with unprecedented opportunities. The rational design of NWs integrates well-defined 1D nanostructure with interface, composition, size, defect, and morphology engineering to realize the controllable synthesis of excellent nanocatalysts. Regardless of the significant progress achieved in Pt-group NWs for enhanced catalysis based on diversified architecture, a detailed summary of the achievements and conundrums in this field is insufficient but essential.

In consequence, we herein focus on the characteristic NWs for boosting FCs' electrocatalysis reaction, including 1D heterogeneous, core-shell, ultrathin, defective, and networked NWs. Notably, the key issues addressed in this chapter are the reaction pathways and the mechanism of catalytic performance enhancement, namely, the influence of the electronic structure and active sites of Pt-group metal NWs on the adsorption strength of intermediates and catalytic activity. Additionally, the characterization and applications of NWs are summarized and prospected, which may offer insights into the fabrication of advanced NWs electrocatalysts for FCs.

11.2 OVERVIEW OF FUEL CELLS

FCs include two half-reactions of fuel oxidation reaction (FOR) and oxygen reduction reaction (ORR) that occur on the anodic and cathodic sides, respectively, each of which follows a different reaction mechanism and pathway depending on the natural properties of the electrode surface. The choice of advanced Pt-group NW electrocatalysts is of paramount importance for improving the sluggish reaction kinetics of FOR and ORR. Aiming to provide available guidance for the design of efficient NW catalysts, it is necessary to probe the possible reaction mechanism and summarize some property descriptors.

11.2.1 FUNDAMENTAL UNDERSTANDING OF CATHODIC REACTION IN FCS

ORR occurs at the cathodes in FCs. The sluggish kinetics of ORR is the major bottleneck restricting the development of FCs. ORR contains an incompletely reduced two-electron pathway and a fully reduced four-electron pathway, generating hydrogen peroxide and water, respectively. Complete reduction of oxygen is preferred in FCs, owing to its higher energy efficiency and avoidance of the corrosion problem of hydrogen peroxide to electrode materials. Theoretical studies suggest that an associative mechanism dominates at high oxygen coverage. The associative mechanism consists of four elementary reactions. Oxygen adsorption is followed by protonation to form *OOH (eq 11.1) (* denotes the adsorption sites of catalysts); the second proton/electron is transferred to the same O atom, and subsequently the O-O bond of *OOH dissociates to form *O and H_2O (eq 11.2); a proton/electron is transferred to the *O (eq 11.3); a proton/electron is transferred to the *O to form H_2O (eq 11.4).

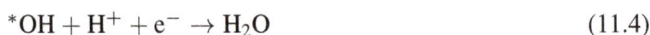

$$O_2 + H^+ + e^- \rightarrow {}^*OOH \tag{11.1}$$

$$^*OOH + H^+ + e^- \rightarrow {}^*O + H_2O \tag{11.2}$$

$$^*O + H^+ + e^- \rightarrow {}^*OH \tag{11.3}$$

$$^*OH + H^+ + e^- \rightarrow H_2O \tag{11.4}$$

11.2.2 FUNDAMENTAL UNDERSTANDING OF ANODIC REACTION IN FCS

In this section, two types of FOR and reaction mechanism are mainly introduced, namely, alcohol oxidation reaction (AOR) and hydrogen oxidation reaction (HOR). In virtue of high energy density, a wide range of fuel sources, and environment friendliness, direct alcohol fuel cells and hydrogen fuel cells have attracted wide attention, in which the energy carriers are alcohols (such as methanol, ethanol, ethylene glycol, and glycerol, etc.) and hydrogen, respectively [9, 10].

11.2.2.1 AOR

Methanol oxidation reaction (MOR) on pure Pt surface is the most classic and easy monoalcohol oxidation reaction without C-C bond cleavage. At present, a dual pathway MOR mechanism has been proposed, including the direct pathway (or non-CO pathway) and the indirect pathway (or CO pathway) [11]. MOR proceeds in a specific way heavily depending on the cleavage of C-H or O-H bond during the initial dehydrogenation step. As illustrated in eq 11.5, the indirect pathway firstly dissociates the C-H bond to form $*CH_2OH$ intermediate, followed by a stepwise dehydrogenation to form $*CO$ intermediate. Finally, $*CO$ intermediate is oxidized by $*OH$ (originated from the dissociation of water) to form CO_2 (eq 11.6 and 11.7). In direct pathway, methanol molecules initially dissociate the O-H bond to form $*CH_3O$ intermediate, and then dehydrogenates to form $*HCHO$ intermediate. Subsequently, $*HCHO$ may desorb from the catalyst surface and diffuse into the solution to generate by-product HCHO, a two-electron pathway. In addition, two parallel reactions can simultaneously occur, in which the $*HCHO$ intermediate would be further oxidized to form by-product HCOOH or complete oxidation product CO_2, which are four-electron and six-electron pathways, respectively (eq 11.8).

(i) The indirect pathway:

$$CH_3OH \rightarrow *CH_2OH \rightarrow *CHOH \rightarrow *COH \rightarrow *CO \qquad (11.5)$$

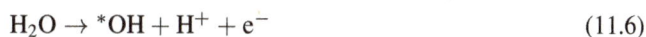

$$H_2O \rightarrow *OH + H^+ + e^- \qquad (11.6)$$

$$*CO + *OH \rightarrow *COOH \rightarrow CO_2 \qquad (11.7)$$

(ii) The direct pathway:

$$CH_3OH \rightarrow *CH_3O \rightarrow *HCHO \rightarrow *HCOOH \rightarrow CO_2 \qquad (11.8)$$
$$\text{diffuse} \downarrow \quad \text{diffuse} \downarrow$$
$$\text{HCHO} \qquad \text{HCOOH}$$

Ethanol oxidation reaction (EOR) on the pure Pt surface is another classic monoalcohol oxidation reaction, featuring 12-electron transfer and C-C bond cleavage [12]. Currently, it is well accepted that ethanol oxidation takes place through parallel pathways. At present, it is widely accepted that multiple parallel reactions occur in EOR, namely, C1 and C2 pathways. The C1 pathway produces the $*CH_{(x)}$ and $*CO$ intermediates with C-C bond cleavage, and then they are oxidized to generate final product CO_2 (CO_3^{2-} in alkaline electrolyte) (eq 11.9). As shown in eq 11.10, incomplete oxidation of ethanol generates acetic acid (acetate in alkaline electrolyte) in the C2 pathway.

(i) C1 pathway:

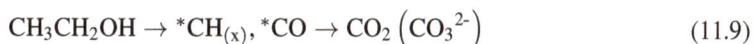

$$CH_3CH_2OH \rightarrow *CH_{(x)}, *CO \rightarrow CO_2 \left(CO_3^{2-}\right) \qquad (11.9)$$

(ii) C2 pathway:

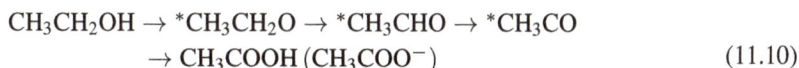

$$CH_3CH_2OH \rightarrow *CH_3CH_2O \rightarrow *CH_3CHO \rightarrow *CH_3CO$$
$$\rightarrow CH_3COOH \,(CH_3COO^-) \qquad (11.10)$$

Recently, the oxidation of polyhydric alcohol (mainly including ethylene glycol and glycerin) has gradually attracted increasing attention due to the advantages of fuels including low toxicity, low volatility, and high boiling point. The reaction mechanism of ethylene glycol oxidation reaction (EGOR) is more complicated than MOR and EOR, because of the active two hydroxyl groups and C-C bond. Researchers have proposed the possible dual pathway, similarly involving C1 and C2 pathways [13]. In the C1 pathway (eq 11.11), after the dehydrogenation process, the C-C bond of generated *HOCCHO intermediate dissociates, followed by generating *COH$_X$ and *CO intermediates, which are oxidized to HCOOH (HCOO$^-$ in the alkaline electrolyte) or CO$_2$ (CO$_3^{2-}$ in the alkaline electrolyte). As shown in eq 11.12, the products of glycolate or oxalate would generate during the C2 pathway.

(i) C1 pathway:

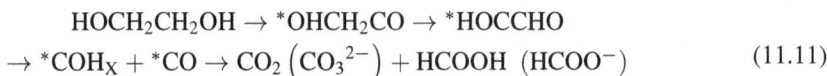

$$HOCH_2CH_2OH \rightarrow {}^*OHCH_2CO \rightarrow {}^*HOCCHO$$

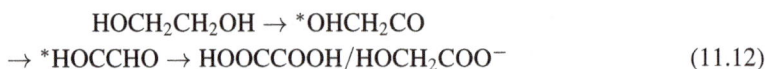

$$\rightarrow {}^*COH_X + {}^*CO \rightarrow CO_2\left(CO_3^{2-}\right) + HCOOH\ (HCOO^-) \tag{11.11}$$

(ii) C2 pathway:

$$HOCH_2CH_2OH \rightarrow {}^*OHCH_2CO$$

$$\rightarrow {}^*HOCCHO \rightarrow HOOCCOOH/HOCH_2COO^- \tag{11.12}$$

11.2.2.2 HOR

HOR occurs at the anode of the HFCs, and the hydrogen as reactant generates protons. The working electrolyte of HOR is divided into two types: acidic and alkaline. The reaction kinetics is 2 to 3 orders of magnitude faster in acid than in base. Whether in an acidic or alkaline environment, hydrogen is oxidized to water through Tafel-Volmer or Heyrovsky-Volmer mechanisms [14]. The oxidation process consists of three elementary steps. In Tafel (eq 11.13 and 11.16), hydrogen adsorbed at catalytic sites without electron transfer. The Heyrovsky reaction represents the adsorption of H$_2$ with simultaneous proton/electron transfer and generation of *H (eq 11.14 and 11.17). The Volmer reaction displays the discharge of *H and the release of a proton (eq 11.15 and 11.18). At present, it is believed that the key to improving the activity of HOR is to optimize the hydrogen binding energy (HBE) through strategies, such as heteroatom introduction, size modulation, and optimization of carriers, or to introduce oxygenophilic elements and construct heterostructures for increasing the oxhydryl binding energy (OHBE) [15].

(i) In acidic electrolyte:

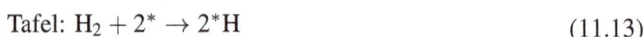

$$\text{Tafel: } H_2 + 2^* \rightarrow 2^*H \tag{11.13}$$

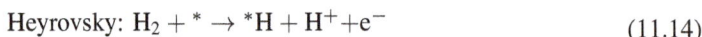

$$\text{Heyrovsky: } H_2 + {}^* \rightarrow {}^*H + H^+ + e^- \tag{11.14}$$

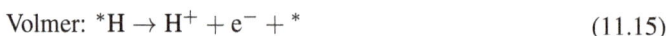

$$\text{Volmer: } {}^*H \rightarrow H^+ + e^- + {}^* \tag{11.15}$$

(ii) In alkaline electrolyte:

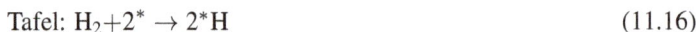

$$\text{Tafel: } H_2 + 2^* \rightarrow 2^*H \tag{11.16}$$

$$\text{Heyrovsky: } H_2 + OH^- + {}^* \rightarrow {}^*H + H_2O + e^- \tag{11.17}$$

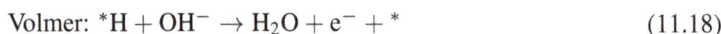

$$\text{Volmer: } {}^*H + OH^- \rightarrow H_2O + e^- + {}^* \tag{11.18}$$

11.2.3 Activity Descriptors for Cathodic and Anodic Reaction

Cathodic and anodic reactions belong to surface-related reactions, in which activation of reactants, adsorption of intermediates, and desorption of products all occur on the catalyst's surface. The phenomenon shows that the binding strength between catalyst and reactants/intermediates would directly affect the catalytic performance [16]. Hence, this relation indicates that the binding strength of reactants/intermediates on catalysts can be used to predict catalytic performance. On the other hand, the interactions between the catalyst surface and the reactants/intermediates are determined by the inherent electronic structure and physicochemical properties of the NW catalysts. Hence, it is imperative to establish a visible correlation between these interactions and the intrinsic structure of NWs catalysts, which can provide favorable guidance for the design of efficient NWs electrocatalysts. Benefiting from the development of computational quantum chemistry, the density functional theory (DFT) calculation applied in the field of electrocatalysis provides a feasible way to establish this visible correlation. Several activity descriptors based on DFT calculation have been proposed to probe the correlation among intermediate adsorption energy, reaction energy barrier, NWs electrocatalysts nanostructure, and catalytic performance. Consequently, activity descriptors are gradually playing an important role in constructing structure-function relationships and analyzing reaction pathways/mechanisms. In the following sections, we summarize some typical descriptors, looking forward to providing theoretical guidance for enhancing the catalytic performance of NWs electrocatalysts.

11.2.3.1 The D-Band Center Theory

The d-band center theory clarifies the relationship between the electronic structure of the catalyst and the adsorption energy of the adsorbed molecule on the catalyst surface [17, 18]. When the adsorbed molecules approach the surface of the catalyst, the molecular orbitals interact with the s-, p-, and d-orbitals of the catalyst. The interactions between s- and p-orbitals of the catalyst and molecular orbitals do not affect the formation of metal-adsorbate bonds, owing to weak energy variation. The interaction of molecular orbitals with d-orbitals results in the formation of filled bonding and incompletely filled antibonding orbitals. The energy level position of the d-orbital determines the energy level of the antibonding orbital, which plays a decisive role in the adsorption of the adsorbate. The position of the energy level of the antibonding orbital relative to the Fermi level determines the strength of the metal-adsorbate bond [19]. As the d-band center moves down, the energy level of the antibonding orbital is lower than that of the Fermi level, which is beneficial to weak metal-adsorbate bond and represents low adsorption energy of adsorbed molecules. For transition metals, the energy levels of the bonding orbitals are almost the same, but the energy levels of the antibonding orbitals are quite different, due to the different positions of the d-band centers. The position of the d-band center. For NWs catalysts mostly composed of transition metals could exert a distinct influence on the adsorption energy of reactants/intermediates, which in turn affects the catalytic performance.

Many studies have focused on the relationship between the d-band center and the electrocatalytic performance of metal nanomaterials. Huang's group advanced ultrathin PtGa alloy NWs via a two-step approach (**Figure 11.1a**), featuring a strong hybridization interaction between p-orbitals of Ga and d-orbitals of Pt (**Figure 11.1b, c**) [20]. The graph of projected d-density of states (PDOS) of surface atoms was used to estimate the d-band center of PtGa NWs (**Figure 11.1d**). Compared with the pure Pt slab, the PtGa slab displays the lower d-band center position, indicating the weaker adsorption strength of oxygenated species, which is beneficial to ORR catalysis. Besides, the surface valence band photoemission spectra indeed verify the downshift of the d-band center from the experimental evidence. The results of electrochemical tests show that PtGa alloy NWs exhibit higher electrocatalytic performance for ORR than pure Pt NWs, which is consistent with the prediction of the d-band center theory.

Furthermore, the d-band center is also an effective descriptor to indicate the adsorption energy of carbonaceous intermediate and predict the AOR activity. Zhu and co-workers developed a series of subnanometer Pt-Rh NWs with ultrathin features (**Figure 11.1e**) [21]. The component-optimized PtRh nanowires possess the lowest d-band center (**Figure 11.1f**), which facilitates the desorption

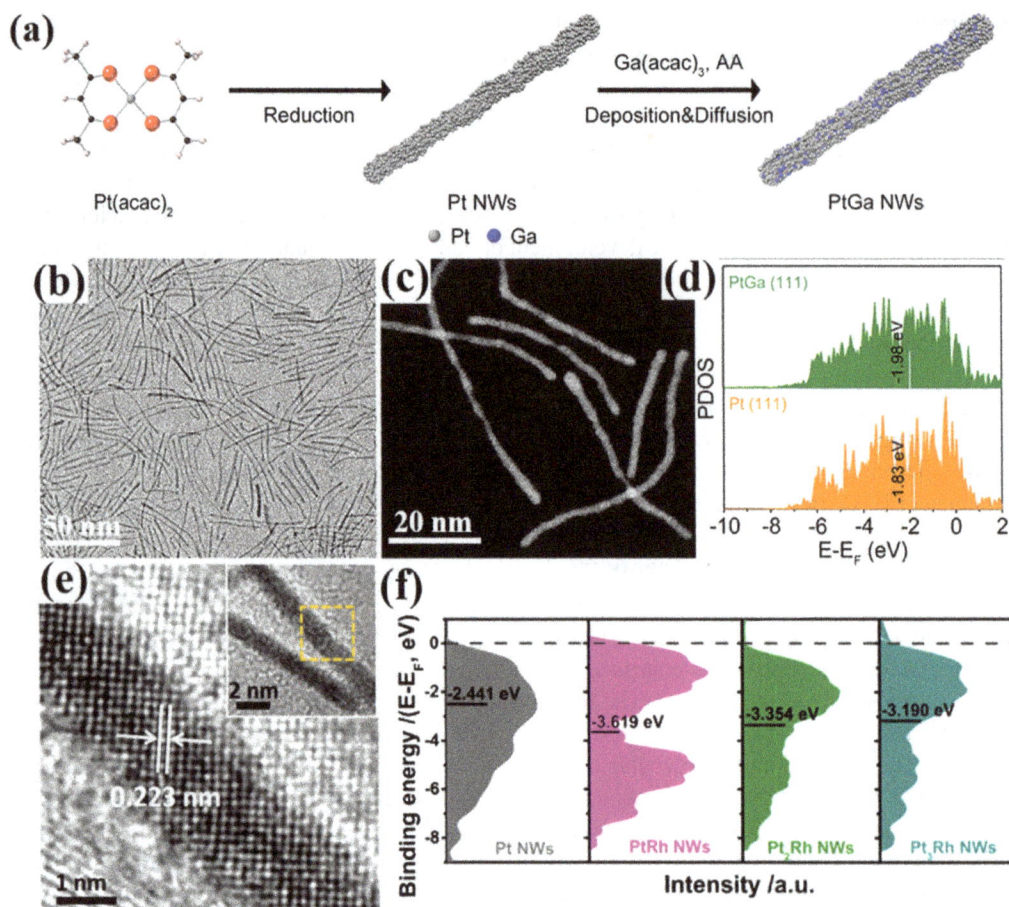

FIGURE 11.1 (a) The synthesis route and (b, c) TEM image and HAADF-STEM image of PtGa alloy NWs. (d) The plots of PDOS of surface Pt atoms for Pt and PtGa alloy slabs. (e) The HRTEM image of PtRh NWs and (f) surface valence band photoemission spectra of PtRh-based NWs. (a-d) Adapted with permission [20]. Copyright (2019), American Chemical Society. (e, f) Adapted with permission [21]. Copyright (2019), American Chemical Society.

of toxic intermediates such as *CO and $*CH_{(x)}$ in the C1 pathway and releases the active sites as soon as possible, thereby promoting the EOR reactivity and long-term stability. Accordingly, electrochemical CO stripping experiments were conducted to disclose the CO tolerance of Pt-Rh NWs. Consistent with the lowest d-band center, the PtRh NW catalyst exhibited the most negative onset potential of CO oxidation, suggesting the weakest adsorption on surface active sites.

11.2.3.2 Adsorption Free Energy of Specific Intermediate (ΔG_x)

In addition to the d-band center, the development of computational chemistry has also developed a descriptor of adsorption free energy to illustrate the relationship between the adsorption state of intermediates and catalytic behavior, which provides a good theoretical basis for the design of efficient catalysts and the improvement of catalytic performance [22]. During electrocatalysis, there is a linear proportional relationship between the adsorption energies of reaction intermediates, which allows a variable to be used to represent the adsorption energies of other intermediates. Therefore, ΔG_x is considered as one of the parameters that can reflect the intrinsic activity of the catalyst (ΔG_x represents the adsorption free energy of X).

For the cathodic ORR, the reaction intermediates include *OOH, *O, and *OH, among which the ΔG_O and ΔG_{OH} are widely used to predict the catalytic performance. Sabatier made a detailed study on the relationship between the overpotential of ORR and ΔG_{OH} on the surface of Pt-group metal catalysts (**Figure 11.2a**) [23]. In the relationship curve, the abscissa refers to ΔG_{OH} and the ordinate represents overpotential. This plot is in the shape of a volcano, and the closer to the peak, the smaller the overpotential and the higher ORR activity of catalysts. The adsorption energy of Pt is closest to the peak position, which represents the optimal adsorption energy and catalytic performance of Pt among Pt-group metals, being consistent with the experimental results. Besides, the left and right sides of the peak represent that the adsorption of *OH on the catalyst surface is too strong or too weak, which is not conducive to the catalytic process. The volcano plot provides an effective strategy to improve catalytic performance. For example, for metals (such as Pt, Pd, Ir, Rh) with too strong adsorption of oxygen-containing species, it is wise to tune the electronic structure of nanomaterials for weakening the adsorption of oxygen-containing intermediates. Lu's group revealed that the optimized $Pt_{16}Pd_5$ NWs exhibit the best ORR performance, owing to the efficient cleavage of O-O bonds and weakest adsorption of *O and *OH on $Pt_{16}Pd_5$ NWs surface among a series of catalysts of different compositions (**Figure 11.2b, c**) [24].

For the anodic HOR, it is of great significance to study the variation of HBE and OHBE in catalytic process. Zhan and co-workers investigated the application of high-entropy alloy NWs in HOR (**Figure 11.2d–g**) [25]. Compared with commercial Pt/C and PtRu/C, the adsorption strength of *H on high-entropy alloy NWs is significantly weakened, which is almost close to the ideal value of HBE (0 eV) (**Figure 11.2h**). In addition, from the perspective of OHBE, the adsorption strength of *OH on high-entropy alloy NWs is relatively suitable, which is the result of neutralizing the excessively low

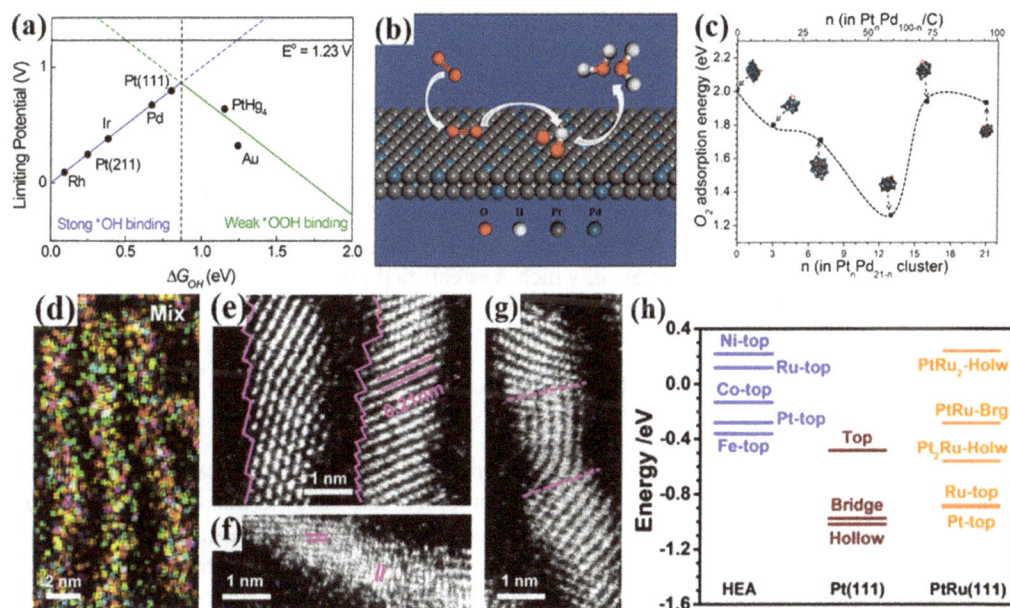

FIGURE 11.2 (a) Activity trends for a series of monometallic surfaces plotted as a function of ΔG_{OH}. (b) Catalytic synergy for the ORR on PtPd NWs. (c) Adsorption free energy of molecularly adsorbed O on Pt_nPd_{21-n} clusters (n = 1–21). (d) The elemental mapping images and (e-g) HRTEM images of high-entropy alloy NWs. (h) The HBE of high-entropy alloy NWs, Pt(111), and PtRu(111) on the hollow, bridge, and top sites, respectively. (a) Adapted with permission [23]. Copyright (2018), American Chemical Society. (b, c) Adapted with permission [24]. Copyright (2020), American Chemical Society. (d–h) Adapted with permission [25]. Copyright (2017), Nature Publishing Group.

binding energy of Pt with the high binding energy of Ru. As a result, the weak adsorption of *H and appropriate adsorption of *OH conspire to improve the electrocatalytic properties for HOR.

11.2.3.3 Free Energy Diagrams

Adsorption free energy, as a descriptor in the field of thermodynamics, cannot completely and clearly describe the rate-limiting step of elementary reaction in the whole reaction. It has been noted that adsorption energy would no longer be applicable for evaluating superior electrocatalysts with fast reaction kinetics. The emergence of free energy diagrams solves this dilemma, which can simultaneously reflect intermediate thermodynamic properties and provide information about dynamics, such as the kinetic barrier of the elementary step. The reaction rate is not only affected by the free energy change of each reaction step but also closely related to the kinetic barrier between intermediate states. Therefore, despite the superiority of adsorption energy in predicting performance, there is still a strong incentive to utilize free energy diagrams containing kinetic and thermodynamic information to describe the catalytic performance of nanomaterials.

For example, Wang's group elucidated the rate-determining step of ORR catalysis and the reasons for the enhanced activity of defective PtCu NWs (**Figure 11.3a–c**) using free energy diagrams at different potentials [26]. As the potential increases, endothermic steps in the ORR catalysis process can be seen from the free energy diagram, wherein the downhill represents the exothermic, and the uphill represents the endothermic. As can be seen from **Figure 11.3d**, the rate-determining step on the surface of PtCu (111) and PtCu (111) with abundant vacancies is the elementary reaction forming OH intermediates, the overpotential (η) could be calculated by the equation: $\eta = 1.23$ V-min[ΔG]/e. Conclusively, the overpotential on PtCu (111) with Cu vacancy is lower than that of perfect PtCu (111), indicating the improved reaction kinetics of the former.

Free energy diagrams are also of great significance in FC anode reactions with more complex reaction pathways and more diverse intermediates. Li and co-workers proposed that the adverse indirect pathway of MOR prepared on the YO_x/MoO_x-Pt NWs surface is different from that on the pure Pt surface (**Figure 11.3e–g**).[27] As shown in **Figure 11.3h**, the oxidation pathway is $CH_3OH \rightarrow {}^*CH_2OH \rightarrow {}^*CH_2O \rightarrow {}^*CHO \rightarrow {}^*CO \rightarrow {}^*COOH \rightarrow CO_2$. In the third and fourth steps, $^*CH_2OH$ is oxidized to $^*CH_2O$ and $^*CHO$ on YO_x/MoO_x-Pt NWs, rather than to $^*CHOH$ and $^*COH$ like Pt. The free energy diagram simultaneously provides kinetic information, that is, the rate-determining step of MOR on YO_x/MoO_x-Pt NWs is the final oxidation of $^*COOH$, which originates from the reduction of $^*COOH$ free energy. Consequently, the decoration of YO_x and MoO_x on Pt NWs contributes to the enhanced CO tolerance, which was certified by the reduction of the free energy barriers from CO* into COOH*.

11.3 DISTINCT NANOSTRUCTURES OF NWS FOR ADVANCED FUEL CELL CATALYSIS

1D Pt-group NWs have given rise to significant interest due to their unique flexibility, satisfactory conductivity, outstanding resistance to structure deformation, and efficient mass/electron transport. To further enhance catalytic performance, it is necessary to create novel nanocatalysts with highly open nanostructure for maximizing the active sites, which remains a captivating but challenging area. Advanced nanostructure engineering has been adopted to design and synthesize various NWs, including 1D heterogeneous, core-shell, ultrathin, defective, and networked nanowires. 1D NWs with distinctive structural merits have exhibited remarkable catalytic performance in both ORR and FOR.

11.3.1 1D Heterogeneous NWs

Among 1D NWs with different sub-nanostructure, nanomaterials with heterogeneous architecture are drawing increasing attention. The heterojunction structure is characterized by various types of

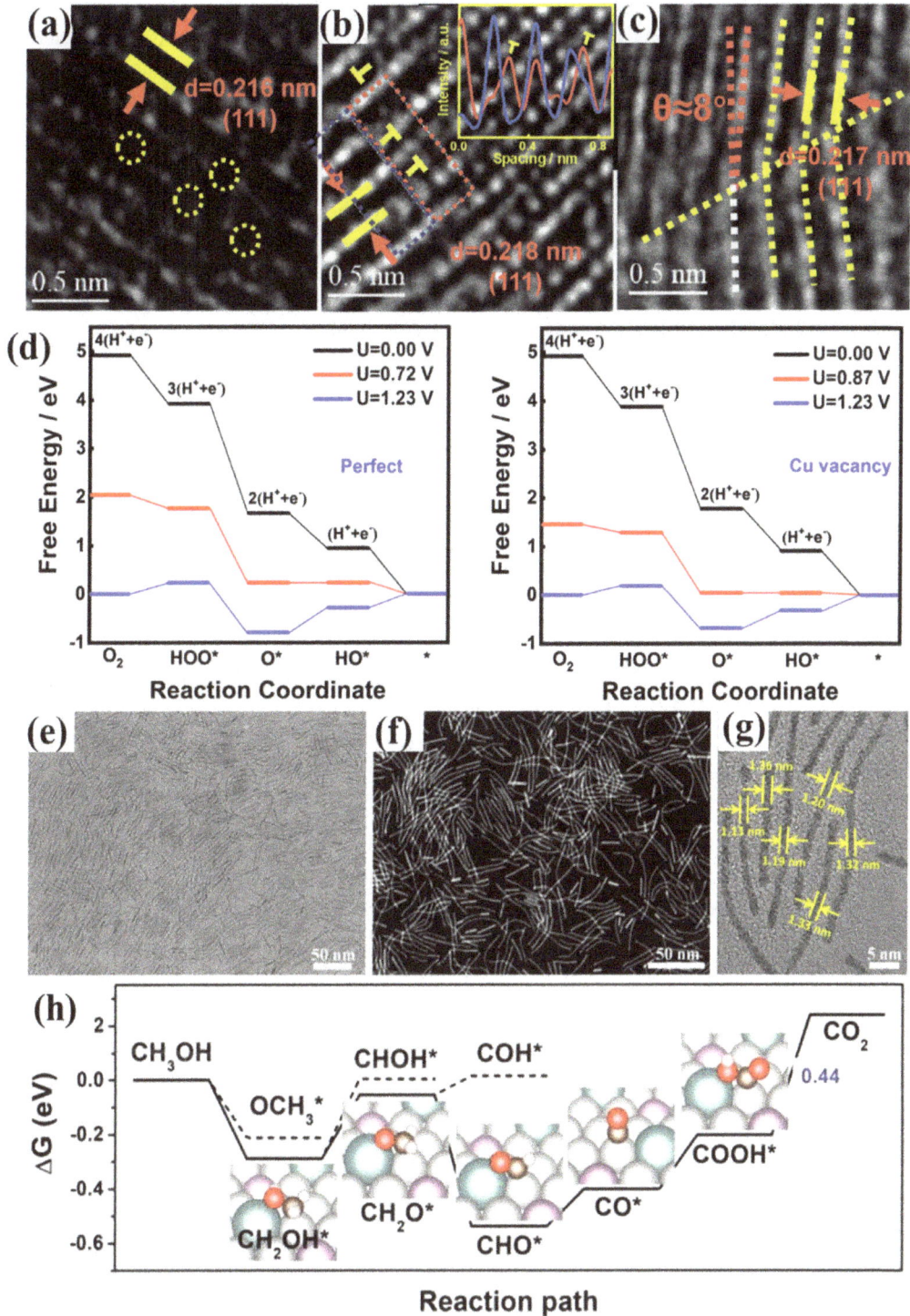

FIGURE 11.3 (a–c) Point, line defects, and plane defects on defective PtCu NWs. (d) Free energy diagrams for ORR on perfect PtCu(111) surface and defective PtCu(111) surface, respectively. (e–g) Representative TEM images and HAADF-STEM image of YO_x/MoO_x-Pt NWs, (h) Free energy diagram for MOR on YO_x/MoO_x–Pt NWs surface. (a–d) Adapted with permission [26]. Copyright (2020), Wiley-VCH Verlag GmbH & Co. KGaA Weinheim. (e–h) Adapted with permission [27]. Copyright (2021), Wiley-VCH GmbH.

interfaces, such as metal/metal, metal/oxide, metal/hydroxide, metal/sulfide interface, etc. [28-30]. Many studies have shown that the interfaces between Pt-group metal and oxide/hydroxide in heterogeneous NWs play an important role in enhancing electrocatalytic properties because they provide the channel for electron redistribution and intermediates' spillover. Interface engineering can integrate these facilitators of abundant interfaces and multicomponent architectures into heterogeneous NWs and generate exceptional catalytic performance.

The construction of interfaces is usually achieved by a two-step method that exploits differences in the chemical reactivity of components or employs post-processing techniques. Yu's group accurately manipulate the replacement reaction between Pd and Te by the "precursor solution-aging" approach [31]. Utilizing $PtCl_4^{2-}$ and Pt_mCl_n complexes as Pt precursors with different aging times, the approach leads to the formation of Pt/PtTe heteronanowires and PtTe NWs (**Figure 11.4a**), respectively, which is attributed to the tunable oxidation ability of Pt precursors. The active Te template, different concentrations of $PtCl_4^{2-}$ ions, and Pt_mCl_n complexes conspired to Pt/PtTe heterogeneous NWs (**Figure 11.4b, c**), creating a novel interface structure that served as a catalytic platform and charge transfer location. DFT calculations revealed that the superior activity and enhanced stability of Pt/PtTe heteronanowires toward ORR than PtTe nanowires stemmed from the surface charge polarization at the interface and the structural stability of the interface, respectively. Furthermore, the interfacial effect of metal/metal oxide plays a crucial role in performance improvement.

In addition, post-processing techniques have been widely used to achieve the construction of heterostructures' catalysts, such as electrochemical treatment, chemical treatment, and plasma surface treatment. Chen and co-workers reported the controllable synthesis of $PdNW@aCuO_x$ (a refers to amorphous) through a two-step hydrothermal method (**Figure 11.4d**) [32]. Cold plasma deriving from glow discharge provides a fast and practical way for the surface modification of nanomaterials, enabling the crystal phase transition of metallic nanocatalysts during plasma treatment. With the help of the ingenious application of cold plasma treatment, they realized the nanostructure transformation from the amorphous CuO_x nanolayer to crystalline CuO_x nanoparticles, synthesize $PdNW@cCuO_x$ (c refers to crystalline) heterogeneous NWs (**Figure 11.4e–i**). $Pd-cCuO_x$ interfaces endow the $PdNW@cCuO_x$ heterostructure NWs with electron-deficient $Pd^{\delta+}$, which contribute to improving catalytic activities of anode reaction, including MOR and EOR. Besides, the existence of inverse interfaces also enables $PdNW@cCuO_x$ enhanced stability, which can be ascribed to the enhanced capacity of CO removal.

11.3.2 1D CORE-SHELL NWS

The core-shell nanostructures are characterized by different compositions of the core and shell, where the interactions between core and shell exert a significant effect on the catalytic performance of the catalytically active shell. Core-shell nanomaterials constitute a crucial type of electrocatalyst. Rational selection of core and shell compositions could maximize the structural advantage of core-shell nanomaterials for promoting catalytic performance. At present, a variety of approaches have been developed to realize the preparation of core-shell catalysts, such as seed-mediated growth strategy, dealloying, and CuUPD-mediated electrodeposition. Particularly, the seed-mediated growth strategy has been demonstrated as a valid approach for the construction of 1D core-shell NWs. Chao et al. disclosed that screwlike Pd/Au–Pt NWs take a core-shell structure with atomically dispersed Pt on the surface (**Figure 11.5a–c**) [33]. The growth of Pd/Au–Pt NWs starts from the pre-preparation of Pd nanowires, followed by the reduction of Au precursor through the galvanic replacement reaction, and ends up with the reduction of Pt on the Pd/Au core-shell nanostructures. The formation of Pd/Au core-shell structure is beneficial to creating defect sites for loading the atomically dispersed Pt atoms, which act as active sites to improve the reaction kinetics and achieve high-performance hydrogen evolution reaction and ORR catalysis.

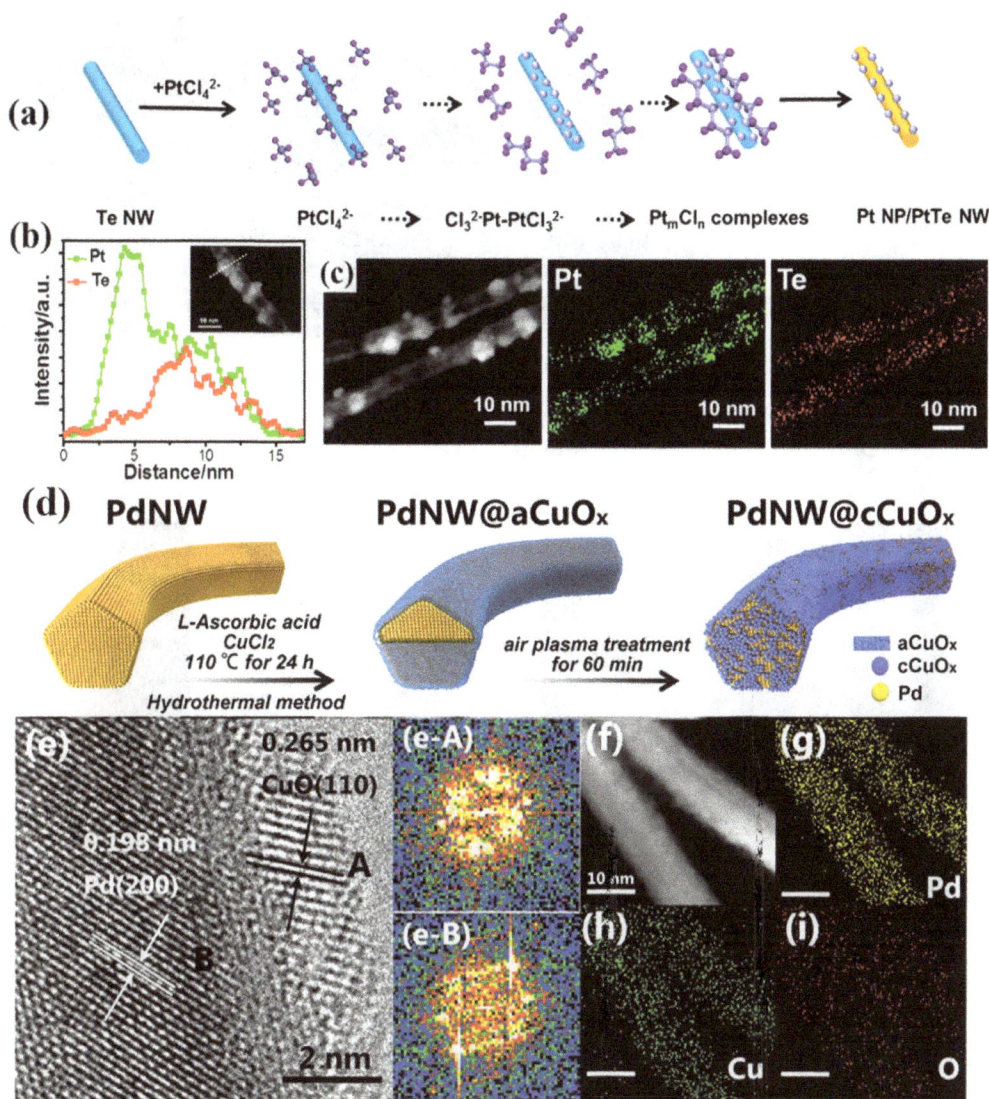

FIGURE 11.4 (a) The synthesis route, (b) line-scanning profiles, and (c) the elemental mapping images of heterogeneous Pt/PtTe NW. (d) The synthesis route, (e) HRTEM image and the corresponding FFT patterns, and (f–i) the elemental mapping images of PdNW@cCuO$_x$ heterostructures NWs. (a–c) Adapted with permission [31]. Copyright (2016), American Chemical Society. (d–i) Adapted with permission [32]. Copyright (2019), WILEY-VCH Verlag GmbH & Co. KGaA, Weinheim.

Pt-based nanomaterials with non-Pt core and limited thickness shell nanostructures are considered as one of the top-ranked cathode catalysts, which can not only effectively reduce the consumption of expensive Pt, but also optimize the Pt-O interaction by shifting the d-band center. Unfortunately, the ultrathin shell with high activity is vulnerable to dissolution and poor stability. Therefore, there is a strong impetus to exploit core/shell electrocatalysts with durable stability. Furthermore, composition engineering offers an opportunity for the construction of 1D core-shell NWs with well-tuned composition. The introduction of heteroatoms could tailor the electronic structure of the active center, and thus improve the catalytic behavior in terms of activity and stability. Based on the above considerations, Guo's group skillfully introduced inert metal Au into Pd NWs by

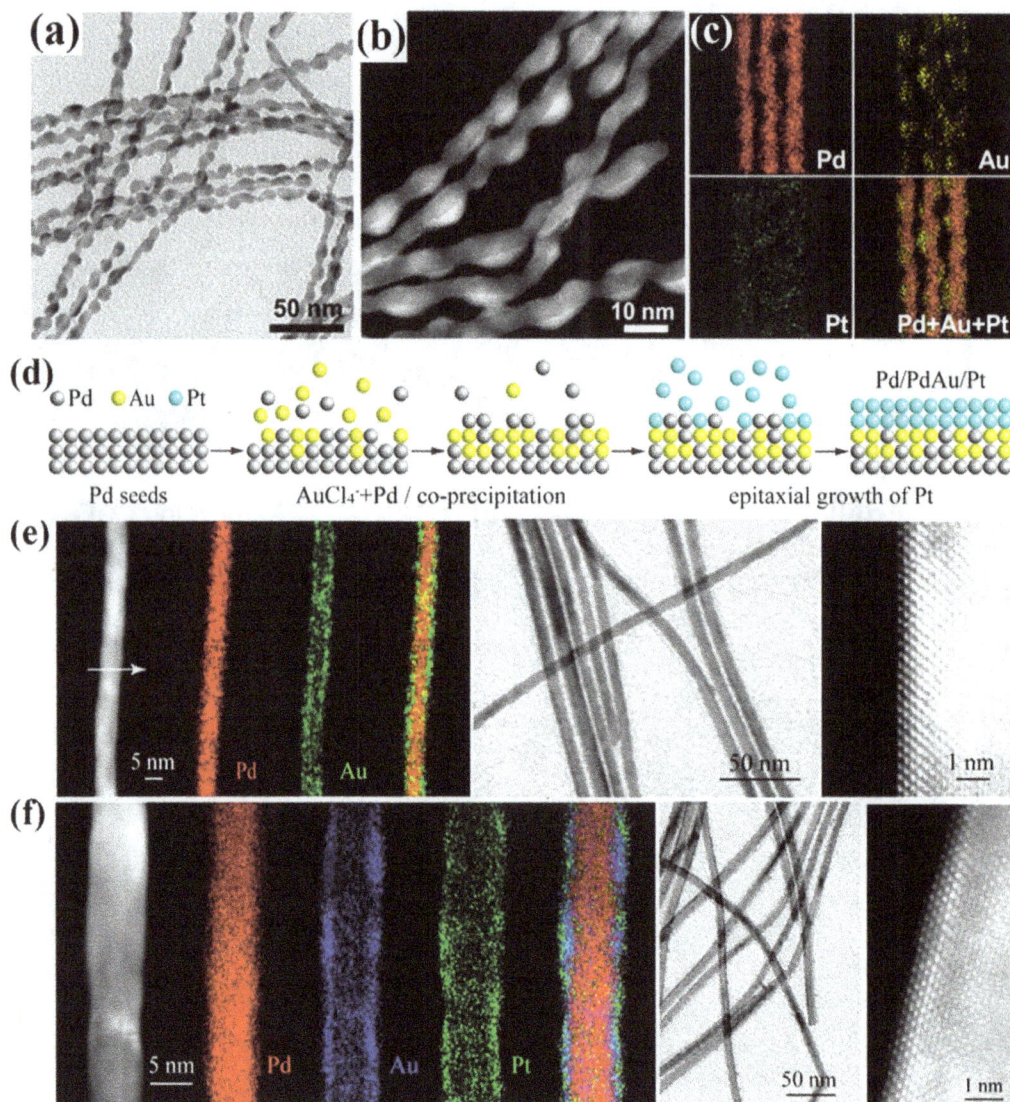

FIGURE 11.5 (a) TEM image and (b, c) the HAADF-STEM image and the corresponding elemental mapping of Pd/Au-Pt NWs. (d) The synthesis route of Pd/PdAu/Pt NWs. (e) The elemental mapping and TEM images of Pd/PdAu NWs. (f) The elemental mapping and TEM images of Pd/PdAu/Pt NWs. (a–c) Adapted with permission [33]. Copyright (2019), Wiley-VCH Verlag GmbH & Co. KGaA, Weinheim. (d–f) Adapted with permission [34]. Copyright (2020), American Chemical Society.

a displacement reaction to generate Pd/PdAu NWs (**Figure 11.5d, e**) [34]. Finally, accompanying with the deposition of Pt atoms, Pd/PdAu NWs act as seeds to induce the formation of Pd/PdAu/Pt core/shell/shell NWs (**Figure 11.5f**). DFT calculations and experiments elucidated that the PdAu interlayers offer good preservation of surface Pt active sites, originating from the electron-transfer bridge Pd and electron reservoir Au of the PdAu interlayers.

11.3.3 1D Ultrathin NWs

The limited diameter and high aspect ratio of 1D ultrathin NWs conspire to increase the Pt-group metal atomic utilization and contribute to realizing extraordinary electrocatalytic activity. However,

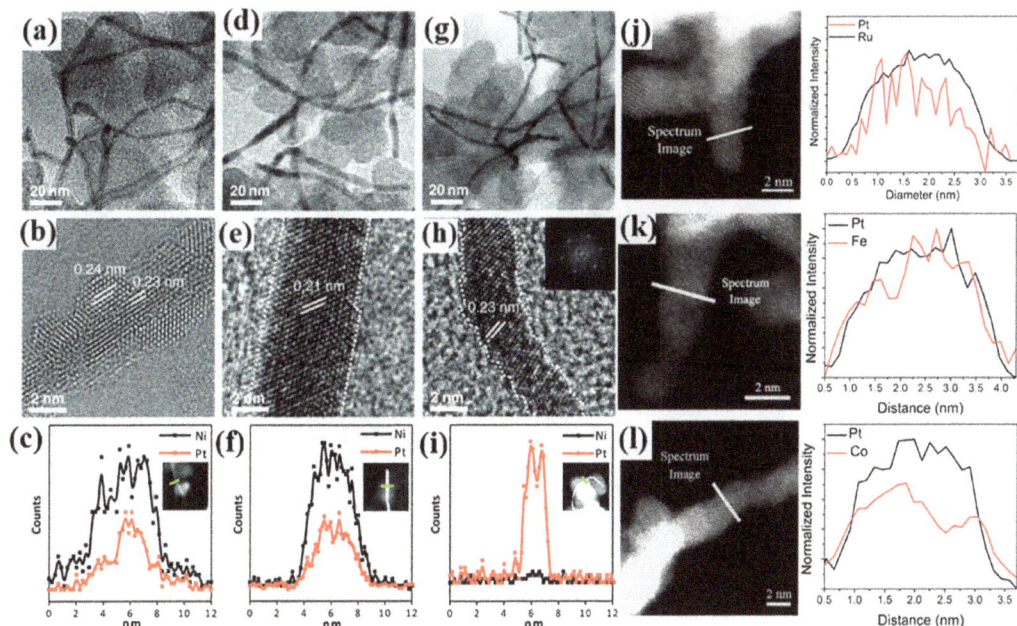

FIGURE 11.6 (a–c) Representative TEM image, HRTEM image, and line-scanning profiles of Pt/NiO core/shell NWs, respectively. (d–f) Representative TEM image, HRTEM image, line-scanning profiles of PtNi alloy NWs, respectively. (g–i) Representative TEM image, HRTEM image, line-scanning profiles of ultrafine jagged Pt NWs, respectively. (j–l) HRTEM images and corresponding line-scanning profiles of ultrathin PtRu, PtFe, PtCo) binary NWs, respectively. (a–i) Adapted with permission [35]. Copyright (2016), American Association for the Advancement of Science. (j–l) Adapted with permission [36]. Copyright (2016), American Chemical Society.

it remains a challenging field to elaborately regulate the surface structure of ultrafine NWs without the degradation of original 1D nanostructures. Innovatively, Duan's group promoted the structural evolution of ultrafine NWs from Pt/NiO core-shell NWs (**Figure 11.6a–c**) and PtNi alloy NWs (**Figure 11.6d–f**) into jagged Pt NWs (**Figure 11.6g–i**) through two-step post-processing techniques, which include thermal annealing and electrochemical dealloying [35]. Reactive molecular dynamics simulations demonstrated that high mechanical strain and the undercoordinated rhombus-rich surface of ultrafine jagged Pt NWs can decrease the binding energy of adsorbents on active sites and thus effectively contribute to the activity enhancement.

1D ultrafine NWs also have broad application prospects in HOR. A solution-based method was used to synthesize a class of ultrathin PtM (M = Fe, Co, Ru, Cu, Au) binary NWs toward HOR catalysis by Scofield and co-workers (**Figure 11.6j–l**).[36] The author confirmed the d-band center moving closer to the Pt Fermi level enable the decrease in hydrogen binding energy of PtM NWs, owing to the desirable ligand and lattice strain effects between Pt and M. Reducing HBE of ultrathin PtRu NWs is conducive to improving sluggish reaction kinetics of HOR and give rise to excellent catalytic performance.

11.3.4 1D Defective NWs

There are two basic strategies to improve catalytic performance. One is to increase the amounts of active sites, which maximizes the exposure of surface atoms by increasing the openness of nanomaterials as much as possible. The other is to increase the intrinsic activity of a single active site, which can be achieved by constructing a surface rich in unsaturated low-coordination sites or highly active defective sites. Defects represent the distortions of periodic nanostructure

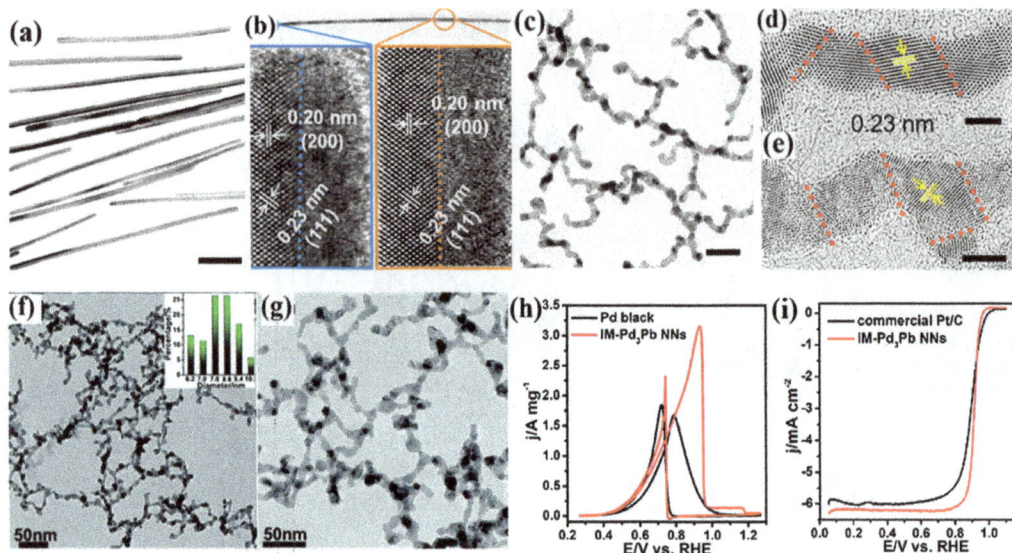

FIGURE 11.7 (a, b) TEM and HRTEM images of Pd$_4$Sn NWs. (c–e) TEM and Cs-corrected TEM images of defective Pd$_4$Sn NWs. (f, g) TEM image of networked Pd$_3$Pb NWs. (k, l) Pd mass-normalized EOR and ORR catalytic performance of networked Pd$_3$Pb NWs, respectively. (a–e) Adapted with permission [37]. Copyright (2019), American Chemical Society. (f–i) Adapted with permission [39]. Copyright (2017), Royal Society of Chemistry.

in crystals, which are capable of tuning electronic structures and optimizing the adsorption state of intermediates. Hence, the defect engineering of nanomaterials is of significant importance for boosting electrocatalysis, especially along the longitudinal of 1D NWs. Extensive efforts have been dedicated to probing advanced 1D defective NWs. Huang's group synthesized two types of similar Pd$_4$Sn NWs by the facile solvothermal method, called Pd$_4$Sn NWs (**Figure 11.7a, b**) and defective Pd$_4$Sn NWs (**Figure 11.7c–e**), differing in the absence and presence of high density of defects on the surface [37]. In contrast to Pd$_4$Sn NWs, the defective Pd$_4$Sn WNWs delivered the optimal activity and stability toward ORR and MOR. The enhanced electrocatalytic performance can be ascribed to the presence of electronic active grain boundaries composed of numerous surface vacancies/agglomerated voids, which can be beneficial to minimize the intrinsic site-to-site electron-transfer barriers.

11.3.5 1D NETWORKED NWS

Among various 1D nanostructures, networked NWs are known for their interconnected self-supporting structures and excellent electrical conductivity as well as superior stability, which contribute to eliminating the use of carbon support. Inherent structural advantages make networked NWs more resistant to deformation and dissolution than bulk metals and zero-dimensional nanostructures. At present, multitudinous networked NWs have been used for advanced electrocatalysis, such as Pt/Pd/Ir/Rh-based alloy networked NWs. Notably, our group developed a class of networked PtFe NWs with branch-rich exteriors on the rough surface [38]. The interconnected networked structure and rich steps and defects on branch-rich structure conspire to realize excellent EOR and MOR electrocatalysis. In particular, Shi et al. disclosed the synthesis of intermetallic Pd$_3$Pb nanowires network by a one-step wet chemical method (**Figure 11.7f, g**) [39]. The networked Pd$_3$Pb NWs exhibit enhanced activity and long-term stability for EOR and ORR (**Figure 11.7h, i**), owing to the self-supported nanostructures and ordered atomic structures. Besides, the designed networked construction and the defect-enriched surface enable methanol- and CO-tolerant ability as cathode and anode catalysts by means of increasing active sites, accelerating electron transfer, and utilizing geometric and electronic effects.

11.4   SUMMARY AND OUTLOOK

The chapter has outlined the research progress in Pt-group metal electrocatalysts with well-defined nanowire structures, for FCs electrocatalysis from the aspects of reaction mechanism and material characteristics for property enhancement. We mainly focused on the superiorities for the construction of 1D NWs with diversified nanostructures and the superior catalytic behaviors in anode reaction and cathode reaction. In spite of the great progress, some problems and challenges still exist. In the final section, these related challenges will be summarized and perspectives on the future development of NWs electrocatalysts for FCs catalysis are presented.

First, it is imperative to further study and realize the controllable synthesis of nanowires for practical application. Most of the existing synthesis and regulation methods, such as ligand-assisted synthesis, seed-mediated growth strategy, and post-processing technology, are cumbersome and complicated, in which too many factors affect the nucleation, growth, and structure evolution of nanocrystals. Considerable efforts have been devoted to the formation of well-defined nanowires by the trial-and-error method. The current approaches for the preparation of NWs exhibit the worseness with poor yields and high cost as well as poor reproducibility, leading to the failure of large-scale practical production. Consequently, it is necessary to develop universal and facile strategies for the synthesis of various NWs electrocatalysts.

Second, the identification and characterization of true active sites remain challenging. The electrochemical testing working environment (at a certain potential and an acidic or alkaline electrolyte) may lead to the structural evolution of NWs electrocatalysts, which results in the surface electronic structure and local coordination of the real active sites being different from the initial state. The application of advanced in-situ characterization techniques (such as in-situ XRD, in-situ Raman, and in-situ FTIR) over FCs catalysis is conducive to building an accurate structure-activity relationship and thus providing correct guidance for the improvement of catalytic performance.

Lastly, the catalytic mechanism of certain fuel cell reactions remains controversial and ambiguous. For example, there are conflicts and contradictions between HBE theory and bifunctional theory in HOR [15]. The HBE theory suggests that the improvement of HOR performance comes from the optimization of HBE [40]. On the other hand, bifunctional theory proposes that the oxyphilic element or OHBE plays an important role in the property improvement of HOR in alkaline media. Hence, it is anticipated that a combination of theoretical calculations, experimental work, and in-situ characterization could be used to elucidate the detailed mechanism of nanowire-catalyzed FCs reactions.

REFERENCES

[1]   Y. Wang, D.Y.C. Leung, J. Xuan, H. Wang, A review on unitized regenerative fuel cell technologies, part B: Unitized regenerative alkaline fuel cell, solid oxide fuel cell, and microfluidic fuel cell, *Renew. Sust. Energy Rev.* 75 (2017) 775–795.

[2]   S. Fukuzumi, Production of liquid solar fuels and their use in fuel cells, *Joule*, 1 (2017) 689–738.

[3]   Q. Shao, K. Lu, X. Huang, Platinum group nanowires for efficient electrocatalysis, *Small Methods*, 3 (2019) 1800545.

[4]   H. Xu, H. Shang, C. Wang, Y. Du, Ultrafine Pt-based nanowires for advanced catalysis, *Adv. Funct. Mater.* 30 (2020).

[5]   W. Wang, F. Lv, B. Lei, S. Wan, M. Luo, S. Guo, Tuning nanowires and nanotubes for efficient fuel-cell electrocatalysis, *Adv. Mater.* 28 (2016) 10117–10141.

[6]   J. Ge, P. Wei, G. Wu, Y. Liu, T. Yuan, Z. Li, Y. Qu, Y. Wu, H. Li, Z. Zhuang, X. Hong, Y. Li, Ultrathin palladium nanomesh for electrocatalysis, *Angew. Chem. Int. Ed.* 57 (2018) 3435–3438.

[7]   J. Li, Z. Zhou, H. Xu, C. Wang, S. Hata, Z. Dai, Y. Shiraishi, Y. Du, In situ nanopores enrichment of mesh-like palladium nanoplates for bifunctional fuel cell reactions: A joint etching strategy, *J. Colloid Interface Sci.* 611 (2022) 523–532.

[8]   C. Wang, H. Shang, H. Xu, Y. Du, Nanoboxes endow non-noble-metal-based electrocatalysts with high efficiency for overall water splitting, *J. Mater. Chem.A* 9 (2021) 857–874.

[9]   I. Staffell, D. Scamman, A. Velazquez Abad, P. Balcombe, P.E. Dodds, P. Ekins, N. Shah, K.R. Ward, The role of hydrogen and fuel cells in the global energy system, *Energy Environ. Sci.* 12 (2019) 463–491.

[10] X. Yang, Q. Wang, S. Qing, Z. Gao, X. Tong, N. Yang, Modulating electronic structure of an Au-nanorod-core–PdPt-alloy-shell catalyst for efficient alcohol electro-oxidation, *Adv. Energy Mater.* 11 (2021).

[11] Y.-W. Zhou, Y.-F. Chen, K. Jiang, Z. Liu, Z.-J. Mao, W.-Y. Zhang, W.-F. Lin, W.-B. Cai, Probing the enhanced methanol electrooxidation mechanism on platinum-metal oxide catalyst, *Appl. Catal. B: Environ.* 280 (2021).

[12] X. Lao, M. Yang, J. Chen, L.Y. Zhang, P. Guo, The ethanol oxidation reaction on bimetallic Pd_xAg_{1-x} nanosheets in alkaline media and their mechanism study, *Electrochim. Acta* 374 (2021).

[13] J. Qi, Z. An, C. Li, X. Chen, W. Li, C. Liang, Electrocatalytic selective oxidation of ethylene glycol: A concise review of catalyst development and reaction mechanism with comparison to thermocatalytic oxidation process, *Curr. Opin. Electroche.* 32 (2022).

[14] S. Lu, Z. Zhuang, Electrocatalysts for hydrogen oxidation and evolution reactions, *Sci. China Mater.* 59 (2016) 217–238.

[15] L. Su, D. Gong, Y. Jin, D. Wu, W. Luo, Recent advances in alkaline hydrogen oxidation reaction, *J. Energy Chem.* 66 (2022) 107–122.

[16] L. Li, P. Wang, Q. Shao, X. Huang, Metallic nanostructures with low dimensionality for electrochemical water splitting, *Chem. Soc. Rev.* 49 (2020) 3072–3106.

[17] B. Hammer, J.K. Norskov, Why gold is the noblest of all the metals, *Nature*, 376 (1995) 238–240.

[18] A. Nilsson, L.G.M. Pettersson, B. Hammer, T. Bligaard, C.H. Christensen, J.K. Nørskov, The electronic structure effect in heterogeneous catalysis, *Catal. Lett.* 100 (2005) 111–114.

[19] M.T.M. Koper, R.A. van Santen, Interaction of H, O and OH with metal surfaces, *J. Electroanal. Chem.* 472 (1999) 126–136.

[20] L. Gao, X. Li, Z. Yao, H. Bai, Y. Lu, C. Ma, S. Lu, Z. Peng, J. Yang, A. Pan, H. Huang, Unconventional p-d hybridization interaction in PtGa ultrathin nanowires boosts oxygen reduction electrocatalysis, *J. Am. Chem. Soc.* 141 (2019) 18083–18090.

[21] Y. Zhu, L. Bu, Q. Shao, X. Huang, Subnanometer PtRh nanowire with alleviated poisoning effect and enhanced C–C bond cleavage for ethanol oxidation electrocatalysis, *ACS Catal.* 9 (2019) 6607–6612.

[22] A. Kulkarni, S. Siahrostami, A. Patel, J.K. Norskov, Understanding catalytic activity trends in the oxygen reduction reaction, *Chem. Rev.* 118 (2018) 2302–2312.

[23] C. Xie, Z. Niu, D. Kim, M. Li, P. Yang, Surface and interface control in nanoparticle catalysis, *Chem. Rev.* 120 (2019) 1184–1249.

[24] F. Chang, Z. Bai, M. Li, M. Ren, T. Liu, L. Yang, C.J. Zhong, J. Lu, Strain-modulated platinum-palladium nanowires for oxygen reduction reaction, *Nano Lett.* 20 (2020) 2416–2422.

[25] C. Zhan, Y. Xu, L. Bu, H. Zhu, Y. Feng, T. Yang, Y. Zhang, Z. Yang, B. Huang, Q. Shao, X. Huang, Subnanometer high-entropy alloy nanowires enable remarkable hydrogen oxidation catalysis, *Nat. Commun.* 12 (2021) 6261.

[26] N. Guo, H. Xue, A. Bao, Z. Wang, J. Sun, T. Song, X. Ge, W. Zhang, K. Huang, F. He, Q. Wang, Achieving superior electrocatalytic performance by surface copper vacancy defects during electrochemical etching process, *Angew. Chem. Int. Ed.* 59 (2020) 13778–13784.

[27] M. Li, Z. Zhao, W. Zhang, M. Luo, L. Tao, Y. Sun, Z. Xia, Y. Chao, K. Yin, Q. Zhang, L. Gu, W. Yang, Y. Yu, G. Lu, S. Guo, Sub-monolayer YO_x/MoO_x on ultrathin Pt nanowires boosts alcohol oxidation electrocatalysis, *Adv. Mater.* 33 (2021) e2103762.

[28] X. Wang, M. Xie, F. Lyu, Y.M. Yiu, Z. Wang, J. Chen, L.Y. Chang, Y. Xia, Q. Zhong, M. Chu, H. Yang, T. Cheng, T.K. Sham, Q. Zhang, Bismuth oxyhydroxide-Pt inverse interface for enhanced methanol electrooxidation performance, *Nano Lett.* 20 (2020) 7751–7759.

[29] P. Wang, X. Zhang, J. Zhang, S. Wan, S. Guo, G. Lu, J. Yao, X. Huang, Precise tuning in platinum-nickel/nickel sulfide interface nanowires for synergistic hydrogen evolution catalysis, *Nat. Commun.* 8 (2017) 14580.

[30] Z. Lyu, X.G. Zhang, Y. Wang, K. Liu, C. Qiu, X. Liao, W. Yang, Z. Xie, S. Xie, Amplified interfacial effect in an atomically dispersed RuO_x-on-Pd 2D inverse nanocatalyst for high-performance oxygen reduction, *Angew. Chem. Int. Ed.* 60 (2021) 16093–16100.

[31] H.-H. Li, M.-L. Xie, C.-H. Cui, D. He, M. Gong, J. Jiang, Y.-R. Zheng, G. Chen, Y. Lei, S.-H. Yu, Surface charge polarization at the interface: enhancing the oxygen reduction via precise synthesis of heterogeneous ultrathin Pt/PtTe nanowire, *Chem. Mater.* 28 (2016) 8890–8898.

[32] Z. Chen, Y. Liu, C. Liu, J. Zhang, Y. Chen, W. Hu, Y. Deng, Engineering the metal/oxide interface of Pd nanowire@CuO_x electrocatalysts for efficient alcohol oxidation reaction, *Small*, 16 (2020) 1904964.

[33] T. Chao, Y. Zhang, Y. Hu, X. Zheng, Y. Qu, Q. Xu, X. Hong, Atomically dispersed Pt on screw-like Pd/Au core-shell nanowires for enhanced electrocatalysis, *Chem. Eur. J.* 26 (2020) 4019–4024.

[34] L. Tao, B. Huang, F. Jin, Y. Yang, M. Luo, M. Sun, Q. Liu, F. Gao, S. Guo, Atomic PdAu interlayer sandwiched into Pd/Pt core/shell nanowires achieves superstable oxygen reduction catalysis, *ACS Nano*, 14 (2020) 11570–11578.

[35] M. Li, Z. Zhao, T. Cheng, A. Fortunelli, C.-Y. Chen, R. Yu, Q. Zhang, L. Gu, V. Merinov Boris, Z. Lin, E. Zhu, T. Yu, Q. Jia, J. Guo, L. Zhang, A. Goddard William, Y. Huang, X. Duan, Ultrafine jagged platinum nanowires enable ultrahigh mass activity for the oxygen reduction reaction, *Science*, 354 (2016) 1414–1419.

[36] M.E. Scofield, Y. Zhou, S. Yue, L. Wang, D. Su, X. Tong, M.B. Vukmirovic, R.R. Adzic, S.S. Wong, Role of chemical composition in the enhanced catalytic activity of Pt-based alloyed ultrathin nanowires for the hydrogen oxidation reaction under alkaline conditions, *ACS Catal.* 6 (2016) 3895–3908.

[37] Y. Zhang, B. Huang, Q. Shao, Y. Feng, L. Xiong, Y. Peng, X. Huang, Defect engineering of palladium-tin nanowires enables efficient electrocatalysts for fuel cell reactions, *Nano Lett.* 19 (2019) 6894–6903.

[38] Y. Zhang, F. Gao, T. Song, C. Wang, C. Chen, Y. Du, Novel networked wicker-like PtFe nanowires with branch-rich exteriors for efficient electrocatalysis, *Nanoscale*, 11 (2019) 15561–15566.

[39] Q. Shi, C. Zhu, C. Bi, H. Xia, M.H. Engelhard, D. Du, Y. Lin, Intermetallic Pd_3Pb nanowire networks boost ethanol oxidation and oxygen reduction reactions with significantly improved methanol tolerance, *J. Mater. Chem. A* 5 (2017) 23952–23959.

[40] Y. Wang, G. Wang, G. Li, B. Huang, J. Pan, Q. Liu, J. Han, L. Xiao, J. Lu, L. Zhuang, Pt–Ru catalyzed hydrogen oxidation in alkaline media: Oxophilic effect or electronic effect?, *Energy Environ. Sci.* 8 (2015) 177–181.

12 Nanowires for Solar Cells

Ujjwal K. Prajapati, Endersh Soni, Mohit Solanki, and Jyoti Rani
Department of Physics, Maulana Azad National Institute of Technology
Bhopal, India

CONTENTS

12.1   INTRODUCTION

The energy demand is increasing exponentially for the use of electronics and other energy-consuming equipment. Currently, fossil fuels fulfill a major portion of demand, but they are limited and are on the verge of getting exhausted. On the other hand, the consumption of fossil fuels also pollutes the environment [1–2]. Hence, to meet the exponentially growing demand for energy worldwide, developments in renewable energy sources are much needed. Among the eco-friendly energy sources (wind power/solar cell/hydropower), a solar cell seems to be a potential candidate because easily available sunlight can be converted into electrical current with the help of solar cells. Today, most commercialized solar cells are based on silicon. Purification of silicon and fabrication technology of silicon-based solar cells is bit-expansive and hence silicon-based solar cells are not affordable for common people [3–5].

Nanotechnology has grown very rapidly over the last decade and given a new technology for manipulating material properties by changing particle size. Manipulation of the material property has enabled us to use nanomaterials like quantum dots, graphene, nanowires, and carbon nanotubes in solar cell applications [6]. Among these, nanowires look more promising because of their better electronic, and optical properties, strain relaxation effect, low cost, and new charge separation method over the traditional materials. Nanowires are the one-dimensional cylinder-shaped nanostructure having a diameter of a few nanometers and lengths up to a few micrometers. The one-dimensional nature of allowing high power, high-efficiency of harvesting electricity from heat, plasmonic effect, and geometry favors better photovoltaic performance [7]. The main aim is to use the III-V semiconductor nanowire to fabricate high-performing solar cells at a low cost for the upcoming generation. Fabricating a nanowire-based solar cell with an efficiency greater than 20% is the primary goal for researchers today. A wide range of materials comprising silicon, germanium, zinc oxide, zinc sulfide, cadmium telluride, cadmium selenide, copper oxide, titanium oxide, gallium nitride, indium gallium nitride, gallium arsenide, indium arsenide, and many polymer nanowires are strong candidates for the photovoltaic application. Some of the material combinations have achieved power conversion efficiency (PCE) near about 15%, which is below the commercialization benchmark. We need to work on these materials to solve the problems in the way of achieving higher efficiency [8–9].

This chapter summarizes different nanowires, their synthesis process, and their application for solar cells. Starting with the basics of the synthesis method, the importance of a substrate during nanowire growth, and the importance of patterning. Then, a discussion of the advantages of nanowires for solar cell applications, such as maximizing photon absorption, improving charge carrier generation, separation, and transportation is provided. Later, a review of the most promising nanowires for solar cell application is presented. Finally, we discuss the future possibilities of nanowire-based photovoltaics.

12.2   SYNTHESIS OF NANOWIRES

There are many methods to fabricate a nanowire, which include top-down and bottom-up approaches. Some of the top-down and bottom-up approaches are listed below:

- Patterned chemical etching
- Electron beam lithography
- Metal-catalyzed electroless etching
- Reactive-ion etching method
- Chemical vapor deposition
- Chemical beam epitaxy
- Hydride vapor phase epitaxy
- Laser ablation method
- Electrospinning
- Template synthesis
- Molecular beam epitaxy (MBE)

There are many more techniques besides the ones listed above used by many research groups. Top-down is a conventional subtractive technique, which involves down-sizing a bulk material into a nanostructure with the help of chemical etching or lithography. A large amount of material is lost during fabricating a nanostructure using the top-down method, hence the method is expensive. Our main focus will be on the bottom-up approach, as it is an additive technique where the minimum loss of materials takes place. The bottom-up technique is very much similar to the natural process of the self-assembling of molecules into growing and bigger structures. Nanostructures can be synthesized in a controlled manner by this technique. The yield of material through the bottom-up approach is uniform, perfectly ordered, and crystallized.

Chemical vapor deposition (CVD) is a simple technique to deposit any gaseous substance on some substrate. Nanowires can be easily synthesized using chemical vapor deposition by flowing vapor precursor near the furnace hot zone so it can react with the substrate. Generally, a metal catalyst is used to assist the reaction. The source of vapor may be a liquid, gas, or solid. An inert gas is used as a carrier to carry the precursor near the substrate. The substrate is kept in the deposition zone near the furnace so that deposition can take place easily [10–11]. Cumbul et al. have used chemical vapor deposition to synthesize Ge nanowires on a Si substrate. His group experimented with a hot

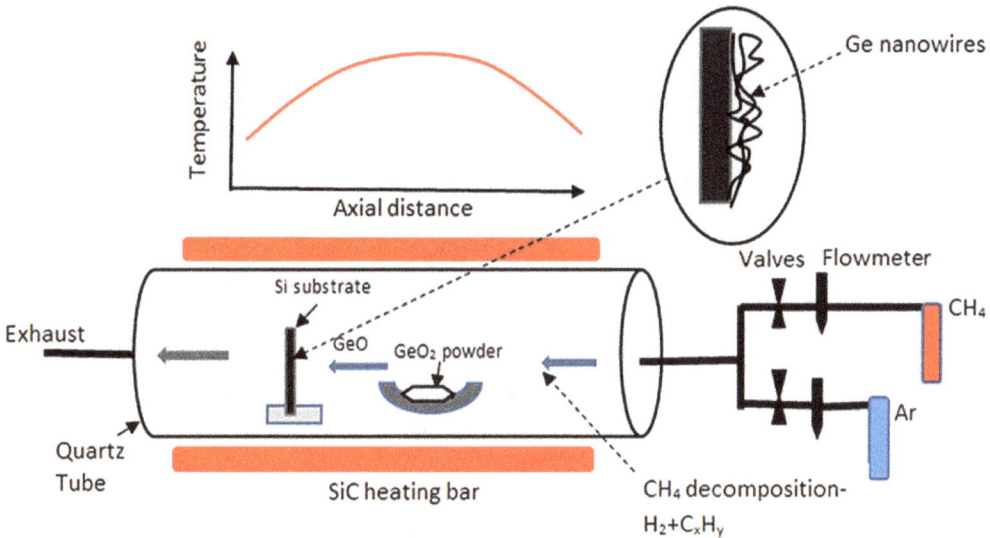

FIGURE 12.1 Experimental setup for the synthesis of the Ge nanowire by chemical vapor deposition Adapted with permission [12]. Copyright 2020, Elsevier.

quartz tube reactor at atmospheric pressure. The reactor was heated by SiC bar, CH_4, and Ar gas supplied through a flowmeter into the chamber. CH_4 was used as a reducing precursor while GeO_2 was used as a Ge precursor. A boat of alumina was used to place the Ge precursor near the deposition zone (uniform temperature zone in the furnace). Si substrate is placed vertically in an alumina boat downstream to the flow of gases as shown in **Figure 12.1**. Variation of temperature in the axis of the tube has also been represented in **Figure 12.1** [12].

Pattern chemical etching is a reduction or subtracting method for nanostructure synthesis. Chemical etchants are used to remove unwanted material from the structure. This is a top-down approach to down-size a structure into a nanoscale structure. Wang et al. have synthesized GaN nanowire arrays by pattern metal-assisted chemical etching [13]. Wang et al. have fabricated ZnO nanowire using electron beam lithography. They set up PMMA resist to spin coat on a Si substrate. Then E-beam lithography is used to develop a pattern on the PMMA resist. After patterning, metal is deposited into the pattern finally resist is left off to get the nanowire. The length of the nanowire was equal to the thickness of the PMMA resist. **Figure 12.2** shows the fabrication steps of ZnO nanowire by E-beam lithography [14].

Chemical beam epitaxy is an important technique for the semiconductor industry, especially for III-V semiconductors. An ultra-high vacuum system is needed for this technique. This technique is better than molecular beam epitaxy and metal-oxide molecular beam epitaxy. Noori et al. [15] have synthesized InAs nanowire using chemical beam epitaxy.

Electrospinning is a method generally used to synthesize polymer fibers. In this process, a high voltage is used, which is given to a capillary through which solvent of polymer is made to pass through to a metallic plate, which acts as the counter electrode or collector. As the jet moves between the electrodes, a critical field force charge carrier transport is observed due to the formation of fibers on the metallic counter-electrode via evaporation of the solvent on the fluid string between the electrodes. This method can be used for the synthesis of inorganic nanowires also by using a triphasic solute (solvent, polymer, and precursor) [16].

The template synthesis technique is a simple and affordable synthesis technique for the preparation of nanowires. It requires only basic instrumentation for metal deposition and sputtering. In this method, a host material with pores is used, which directs the growth of the nanowire. The host or template is preconfigured to obtain the desired nanofiber morphology (**Figure 12.3**).

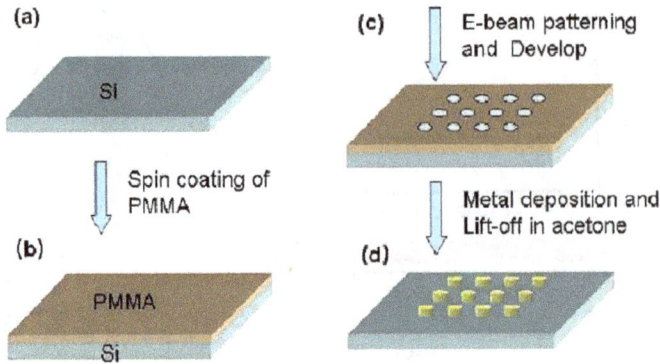

FIGURE 12.2 Steps involved in the fabrication of ZnO on a Si substrate using E-beam lithography. Adapted with permission [14]. Copyright 2007, SPIE conference preceding.

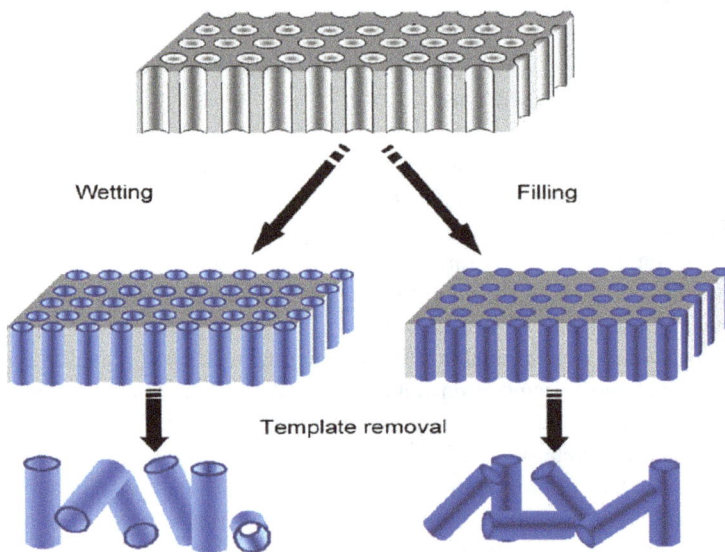

FIGURE 12.3 Steps involved in nanowire synthesis by template synthesis technique. Adapted with permission [18]. Copyright 2010, Elsevier.

12.3 ADVANTAGES OF NANOWIRE FOR SOLAR CELL APPLICATIONS

The process of converting light into electricity can be understood by the following steps (**Figure 12.4**). During this process, there are many losses and inefficiency in the process. To achieve a highly efficient solar cell, these losses need to be minimized and processes will have to be efficient. Nanowires have the potential to minimize these losses at a lower cost. The major benefits of using nanowires instead of thin film or wafers are discussed below.

12.3.1 ABSORPTION

Absorption of the photon into the solar cell absorption layer is key for the generation of excitons. Absorption losses can occur in two ways. First, when the absorption layer is too thin to absorb the incident light or does not have sufficient light trapping capabilities to absorb all incident photons.

TABLE 12.1

The Different Synthesis Techniques of Different High Aspect Ratio Nanowire Semiconductors

Semiconductor	Band Gap in Bulk (eV)	Synthesis Technique
Si	1.12	CVD Laser ablation MBE Physical transport Etching
Ge	0.97	CVD Laser ablation Template synthesis etching
GaN	3.36	CVD Laser ablation Template synthesis etching
ZnO	3.37	CVD Laser ablation Template synthesis etching MBE Carbothermal electrospinning
SnO_2	3.6	CVD Laser ablation Template synthesis etching MBE Carbothermal electrospinning

So, most of the incident light gets transmitted through the layer. The second is reflection, in this case, incident light gets reflected and cannot enter the semiconductor absorber layer. Using a material of optimum thickness and sufficient light trapping capabilities can resolve the first loss whereas using an anti-reflecting coating may address the second loss [19–20].

When there is a difference in the refractive index of the air and the semiconductor layer, the light incident to the interface gets reflected. Generally, 10–50% of light having a wavelength 300–2000 nm gets reflected. This leads to a major loss of sunlight energy. These losses can be minimized by using another layer having an intermediate refractive index between that of the air and semiconductor layer [21]. The reflexing can be eliminated by using a coating layer of optimum thickness and suitable thickness. But this technique works only for single wavelength and single angle of light incidence. In addition to a double- or triple-layer coating spectrally broadening the response, an ideal coating has successively graded refractive indices. An effective medium is formed by a tapered array of nanowires (nanocones) with tips that are smaller than the wavelength of light, with gradual changes in refractive index depending on the weighted average of the material and the air. By doing so, a great improvement has been observed in the absorption of the semiconductor material over the broad wavelength and angle of incidence.

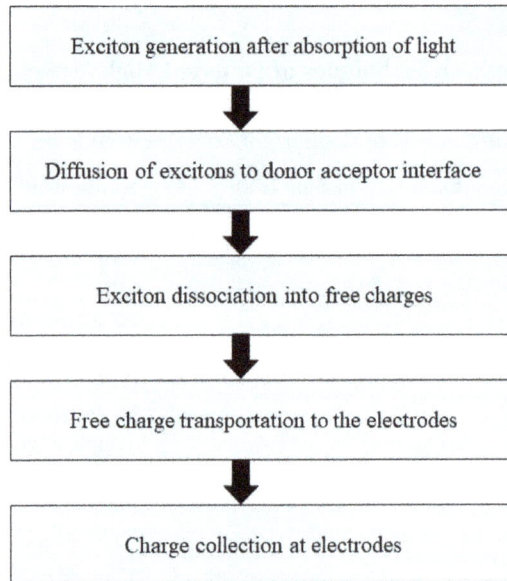

FIGURE 12.4   Steps involved in the process of converting light into electricity.

Figure 12.5 shows the comparative study of absorption of a thin film, nanowires, and nanocones. It can be seen by the plot that, absorption, when nanocones were used, was highest, followed by the nanowires and thin film [19–21]. **Figure 12.5(a)** shows absorption vs wavelength of the light of nanocone, nanowires, and thin film. The value for the plot has been generated by doing experimental measurement. **Figure 12.5(b)** shows the same but the value has been generated by simulations. **Figure 12.5(c–d)** shows a variation of absorption vs different angles of an incident of nanocone, nanowire, and thin film.

12.3.2   Exciton Generation

Exciton generation takes place as soon as the light gets absorbed in the absorber layer of the solar cell. Depending upon the binding energy of the excitons as compared to the thermal energy available at room temperature, either exciton will dissociate into free charges or remain bound as an exciton. When the thermal energy available is sufficient to overcome the binding energy, charge carriers' relaxation takes place by carrier-phonon coupling in such a way that, energy lost is in the form of heat. This loss is the maximum (~ 30–40% of the absorbed photons in most of the photovoltaics) throughout the whole process of converting light into electricity. For the higher efficiency of the device, this loss needed to be minimized. Choosing a material with optimum band gap can be a method to do so [22–23]. Alloying multiple materials together to adjust the band gap of a material is the simplest method but by doing so, other properties may be degraded. Besides this, synthesizing such material with multiple materials with arbitrary concentration as a single crystal is a challenge. Indium gallium nitride, an alloy shows a miscibility gap as a result of lattice train mismatch. Nanowires have capabilities to close this miscibility gap as they show enhanced strain relaxation, enabling absorption of a broad range of energies into thin films. Varying the radius of nanowires for the Bohr radius provides additional tuning, as quantum confinement leads to an increment in the band gap of the material. This is only possible with the material having larger Bohr radii and relatively small band gap material like germanium and lead selenide. Applying some additional carrier generation schemes such as a multiple exciton generator (MEG) can minimize heat loss. Quantum confined systems also exhibit greatly enhanced MEG yield. The mechanism is still under

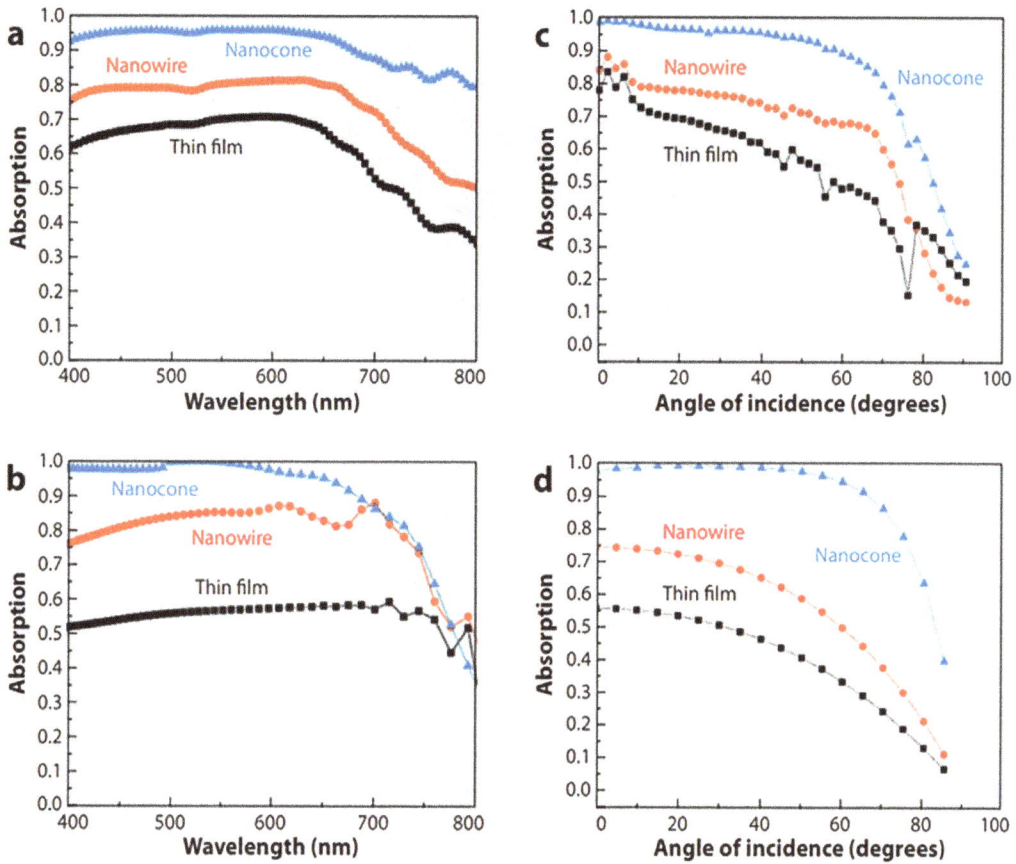

FIGURE 12.5 Anti-reflection properties of nanowires and nanocones. (a) Measured and (b) simulated values of absorption on samples with amorphous Si:H (a-Si:H) thin films, nanowire arrays, and nanocone arrays over a large range of wavelengths at normal incidence. (c) Measured and (d) calculated values of absorption on samples with a-Si:H thin films, nanowire arrays, and nanocone arrays over different angles of incidence. Adapted with permission [19]. Copyright 2009, American Chemical Society.

process but shows reduced phonon coupling, and relaxed momentum conversation, which leads to the betterment of the device [24–25].

12.3.3 CHARGE SEPARATION

Nanowire has shown a new charge separation mechanism that is more efficient than the traditional charge separation mechanism. Wu et al. [26] demonstrated that tapered nanowires can separate electrons and holes without doping using density functional theory as tapered nanowires show different degrees of quantum confinement along the length. **Figure 12.6** shows the separation between HOMO and LUMO wave functions of a tapered silicon nanowire. Some other researchers have also reported in their paper that varying strain along the length of a nanowire induces a charge in the band gap, hence the mechanism can be used to separate the charge similarly. But it is doubtful that this mechanism will influence the other parameters, such as fill factor and open circuit voltage, positively or negatively. This kind of study needs to be carried out intensively. This new charge separation mechanism still needs to be validated experimentally but it gives solid hopes to fabricate a solar cell without using any dopant with a lesser auger recombination rate [26–28].

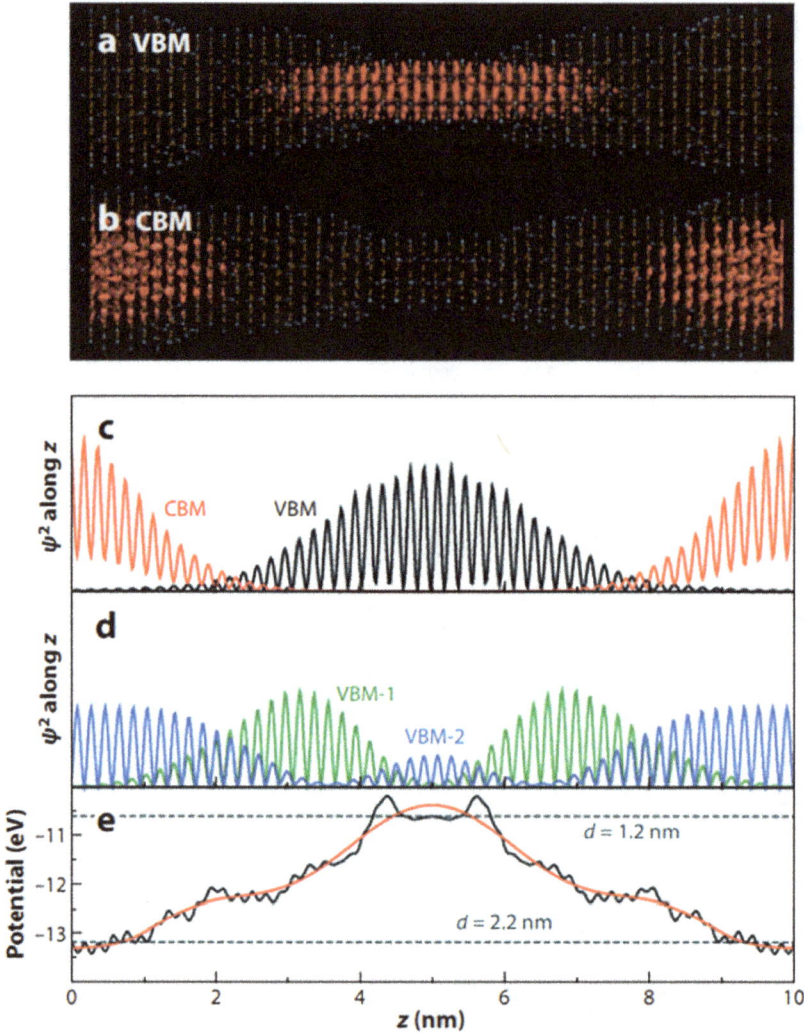

FIGURE 12.6 Charge separation using quantum confinement. (a) Iso-surfaces (red) of the square of the wave functions of the valence band maximum (VBM), (b) shows iso-surfaces (red) of the square of the wave functions of the conduction band minimum (CBM0. (c) Charge density distribution of the VBM, CBM, VBM-1, (d) represents VBM-2 states along the wire axis (integrated over the x-y plane). (e) Planar-averaged total potential V(z) along the wire axis for the tapered nanowires (solid black) and the straight-edged nanowires (dashed black). Adapted with permission [24]. Copyright 2008, American Physical Society.

12.3.4   CHARGE CARRIER COLLECTION

Once carriers are separated, they need to be collected at the charge collection electrode to produce electricity. The geometry of the nanowire has been proved advantageous for this step as it can provide rapid radial charge separation and collection at the contact electrode efficiently. Law et al. [29] have shown that the efficiency of charge collection is much higher when zinc nanowire has been used in dye-sensitized solar cells as compared to standard nanoparticles. Nanowires showed faster band conduction rather than a trap-limited diffusion transport mechanism standard nanoparticle. Hence, it has been seen that the use of nanowires as a charge collection mechanism can improve device performance. Other researchers have also used titania shells to reduce surface recombination rates or titania nanotubes to increase the total surface area to boost the efficiency of the cell. Still, dye-sensitized solar cell lags behind the nanoparticle solar cell in terms of performance due to lower dye loading in a lower surface area [29–30].

In inverted polymer bulk heterojunction solar cells formed with zinc oxide nanowires, a much shorter nanowire network showed better charge transport and a higher fill factor (FF) and overall efficiency rather than controls on planar zinc oxide films. If we take into consideration what was discussed above about light trapping, ordered nanowire arrays may be able to provide further gains in efficiency by improving absorption in both dye-sensitized and bulk heterojunction polymer solar cells [28,30]. Kayes et al. [31] reported that using a radial p-n junction causes less reduction in the cell efficiency and lifetime as compared to the planner p-n junction when the minor charge diffusion length becomes less than the film thickness.

Figure 12.7 confirms that the effect of a p-n junction on efficiency is more effective when the mismatch between the absorption and minority carrier diffusion lengths is greater. Generally, this happens more in indirect band gap materials compared to the direct band gap materials. As radial junctions are less sensitive to bulk defects, hence radial junction solar cell is advantageous when the defects are concentrated in the neutral region of the device. The device performance is decreasing in both the case of radial and planner junction when defect density is high in the depletion region.

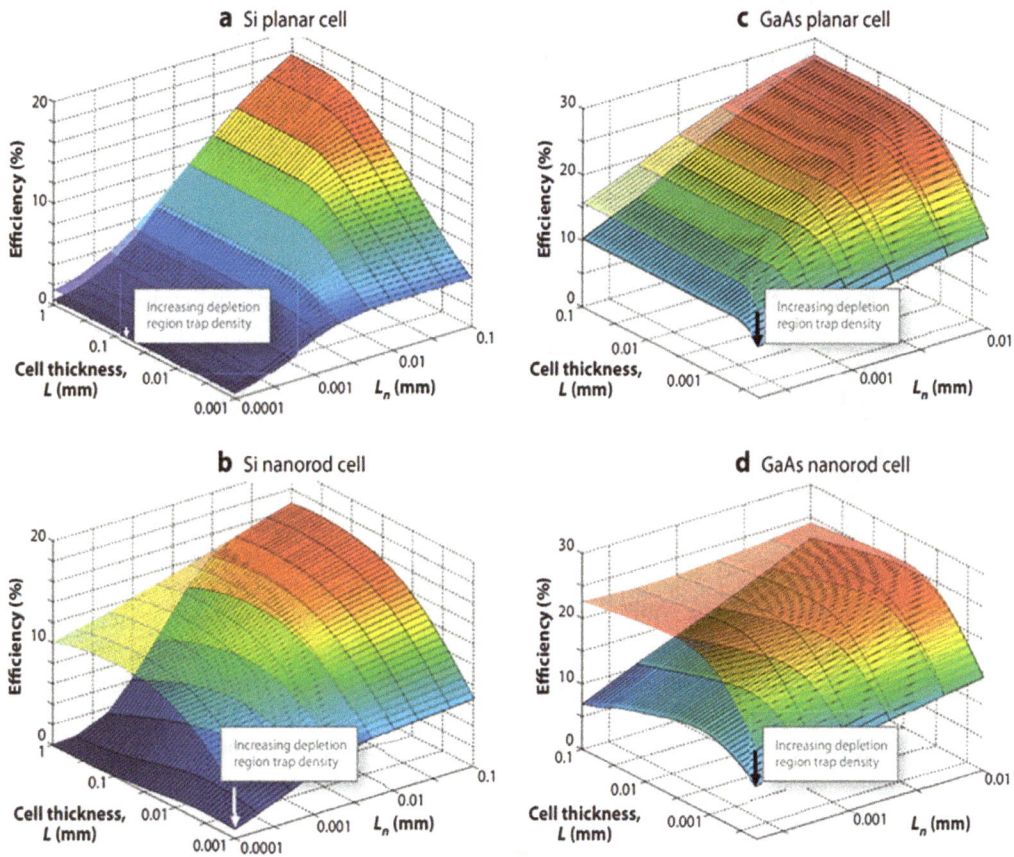

FIGURE 12.7   Plot of efficiency vs cell thickness and minor charge diffusion length. (a) A conventional planar p-n junction silicon cell, (b) a radial p-n junction nanorod silicon cell, (c) a conventional planar p-n junction gallium arsenide (GaAs) cell, and (d) a radial p-n junction nanorod GaAs cell, Adapted with permission [31]. Copyright 2005, American Institute of Physics.

To validate the benefits of using radial junction, control over diffusion length and charge extraction length of the nanowires is needed. An efficiency improvement has been seen in single nanowire photovoltaic cells when a radial junction is used. This improvement can be attributed to the improved photocurrent, which is the result of enhanced carrier collection efficiency in radial junction [32–33].

The charge separation advantage of radial junction can also be verified by the research work of Fan et al. [34], who observes a boost in the efficiency of 0.5–6% by increasing the interpenetration of cadmium sulfide nanowires into a constant-thickness cadmium telluride thin film.

12.3.5 COST

As we have discussed, there are a lot of benefits of using nanowires in solar cell applications, apart from that nanowires are economical too. As nanowires have an excellent absorption coefficient, less material is needed to capture the photons and hence, we need to expend less for less material. Other factors of nanowires, which make them very economical as compared to the traditionally crystalline semiconductors, are that band gap tuning is easy, the charge separation mechanism is efficient, there is less temperature requirement, and no need for purification. Additionally, nanowires may provide a huge cost advantage in the fabrication steps of high-performance multijunction solar cells by removing the need for an expensive substrate matched to the lattice. Many employable materials for solar cell application, such as zinc oxide, germanium, silicon, indium gallium nitride, and cadmium sulfide can be grown on a non-epitaxial substrate through the bottom-up technique, while some can be directly grown on an electrode like aluminum foil, stainless steel, or indium tin oxide-coated glass [34–35].

12.4   PROMISING NANOWIRES FOR SOLAR CELL

12.4.1   SILICON NANOWIRE

Si is one of the famous materials in the field of solar cells and is being used in about 90% of the commercialized solar cells, however, the low efficiency of these solar cells is the main drawback. Besides this, the form in which Si is used is very expensive to obtain. Traditionally, Si is used as a Si wafer, because of the high cost of manufacturing and purification of Si wafer, the cost of currently commercialized solar cells is quite high. The development of Si nanowires over the years has assured us of the future requirements of energy. A single nanowire has shown very exotic properties compared to the traditional solar cell. These nanoscale single Si nanowires can be integrated to supply electronic devices. Jia et al. [36] reported the core-shell heterojunction solar cell TCO/a-Si/Si nanowires, which are fabricated on the SiNW arrays with diameters of about 300 nm and lengths of about 900 nm and synthesized by wet chemical etching of an n-type silicon wafer. The reported solar cell with 7 mm^2 contacted area showed an open circuit voltage (V_{oc}) of 476 mV, short circuit current density (J_{sc}) of 27 mA/cm^2, fill factor (FF) of 56.2%, and power conversion efficiency (PEC) of is 7.29% at AM 1.5. Pal et al. [37] fabricated and studied Si/InP core-shell radial heterojunction solar cell nanowire array based on an average diameter of 80–150 nm with an average length of 12 μm, which was fabricated using a chemical wet etching process. The reported solar cell shows the V_{OC} ~0.56 V, J_{sc} ~ 14.26 mA/cm^2, and PEC 4.39% at room temperature and 1.5 AM illumination condition.

12.4.2   GERMANIUM (GE) NANOWIRES

Germanium shows carrier motilities similar to silicon, therefore, it could be a possible substitute for Si nanowires. Ge nanowires can be grown using bottom-up techniques like chemical vapor deposition, template methods, supercritical fluid synthesis, molecular beam epitaxy, solution phase synthesis, and thermal evaporation. Normally Ge nanowires are produced by VLS and VSS mechanisms. Kim et al. [38] reported a Ge nanopillar solar cell grown by metal-organic chemical vapor deposition. The solar cell consisted of a three-layer combination of n-type Ge, InGaP as the emitter and window layer, and an n-type GaAs ohmic layer grown through MOCVD on the Ge nanopillar templates, which behave like p-type. The optimal height of 600 nm showed PEC~ 4.55%, FF%~ 68.42, V_{OC} ~ 0.24 V, and J_{SC} ~ 27.68.

12.4.3   Gallium Nitrate (GaN) Nanowires

GaN is a semiconductor with a direct band gap of 3.4 eV at room temperature. GaN semiconductor is an active material used in various applications because of its wide direct band gap. GaN nanowires can be grown using various processes like CVD, hydride vapor phase epitaxy, metal-organic chemical vapor deposition, laser ablation, etc. GaN nanowires are used in ultraviolet (UV) emitters and high-power electronic devices and research dominating fields are solar cells, field effect transistors, light emitting diodes, nanometers, and lasers. In addition to its wide band gap and direct carrier mobility, GaN nanowire possesses high crystalline, thermal and chemical stability, and the ability to dope n-type or p-type semiconductors. Tchutchulashvili et al. [39] fabricated a heterostructure of solar cells on silicon substrate using GaN nanowires. The purposed model Si/GaN/P3HT:PC_{71}BM is an inverted hybrid heterostructure. The GaN nanowire array affects the charge transfer between the silicon and organic layer. For the morphological study of the solar, the diameter of the nanowires varied from 30 nm to 150 nm, and the effect of diameter was studied. The device showed a V_{OC} of 0.52 V and J_{SC} of 1.7 mA/cm^2 for a 50 nm diameter of the GaN nanowire.

Park et al. [40] fabricated a solar cell using III-nitride nanowire as hybrid coaxial and uniaxial InGaN/GaN, which is the multi-quantum well (MQW) nanostructure. III-nitride showed excellent physical properties due to increased surface area and thus enhanced light absorption and direct path for carrier transport suggesting potential application as an active material for photovoltaic devices. GaN nanorods/Si heterojunctions have been demonstrated to be useful for solar application by Tang et al. [41]. By using a CVD method with gold nanoparticles as a catalyst, the team synthesized p-type GaN nanorod arrays on n-type Si substrates for the fabrication of p-n heterojunction PV cells. With an AM 1.5G illumination of 100 mW/cm^2 and a short-circuit photocurrent density of J_{SC} = 7.6 mA/cm^2, the cells had an energy conversion efficiency of 2.73%. In a study [42], Li and coworkers described the hydride vapor phase epitaxy (HVPE) method for synthesizing GaN nanorod/Si heterostructures as well as their application in UV PV cells. A UV lamp with a power of 6 W and a wavelength of 365 nm was used in this experiment. According to the results, the maximum open-circuit voltage, the short-circuit current density, and the fill factor was V_{OC} = 0.7 V, J_{SC} = 146 µA/cm^2, and 0.38, respectively.

12.4.4   Zinc Oxide (ZnO) Nanowires

ZnO is a semiconductor material that has a direct wide band gap of 3.37 eV and large exciton binding energy of 60 meV at room temperature. ZnO nanowires show remarkable performance in optics, electronics, photonics devices, LEDs, solar cells, and photocatalysts. ZnO dye-sensitized nonporous material is suitable for solar cells.

Kao et al. [43] reported a dye-sensitized solar cell using ZnO nanowires. The NWs were deposited on the indium titanium oxide-coated substrate using the solution phase deposition method. It was observed that an increase in ZnO nanowire length showed an increase in absorption of N3 dye, and thus improves J_{SC} and V_{OC} of the solar cell. The length of ZnO nanowire varied from 1 µm to 4 µm. At 4 µm J_{SC} = 4,8 mA/cm^2 and V_{OC} = 0.58 V was observed, which is lower for the ZnO nanowire with a length of 3 µm (J_{SC} = 5.6 mA/cm^2 and V_{OC} = 0.62 V).

Krishna et al. [44] reported an organic dye-sensitized solar cell fabricated using solution-processed aligned ZnO nanowires as anti-reflection (AR) and electron transfer layer (ETL). ZnO nanowire as the AR/ET layer minimized the electrical and optical losses and thus enhanced the photocurrent density due to the enhanced light-coupling. A morphological study was done through the combination of AR and ET layers. N719 with the AR layer showed good performance with $V_{OC} = 0.708$ V, $J_{sc} = 4.39$ mA/cm^2, FF% = 57.5%, and PCE = 1.78%. Qiu et al. [45] reported a solution-derived 40-micron ZnO nanowire array for dye-sensitized solar cells. ZnO nanowire was fabricated successfully using the polyethlenimine (PEI)-assisted preheating hydrothermal method (PAPHT). Based on the peak shift after annealing in various atmospheres, they attributed the UV and yellow emission of ultralong ZnO nanowire arrays to absorbed hydroxyl groups. Axial electrochemical impedance spectroscopy (EIS) revealed that dye-sensitized solar cells performed better as the length of ZnO nanowire arrays increased, mostly due to the increased photocurrent and reduced recombination loss. Solar performance parameters as $J_{sc} = 4.26$ mA/cm^2, $V_{OC} = 0.69$ V, FF% = 42%, and PEC = 1.3% were observed.

12.5 CONCLUSIONS

Increasing energy demand has reminded researchers to look for alternative sources of energy. A solar cell is one of the best promising candidates on the list as it provides many advantages over other energy sources. Currently, commercialized solar cell technology is a bit expensive and has low efficiency. Nanowires are the 1D material and have properties to be utilized as solar cell materials to improve efficiency and are economical also. This chapter deals with the synthesis of nanowires and their advantages in solar cell applications. Although nanowires have the required properties to be used as solar cell material, it shows low efficiency than the existing technology. More research is needed in this field to search for the best-suited material for high-performance solar cells.

REFERENCES

1. G. Yu, J. Gao, J. C. Hummelen, F. Wudl, A. J. Heeger, Polymer photovoltaic cells: enhanced efficiencies via a network of internal donor-acceptor heterojunctions. Science 270 (1995) 1789–1791.
2. B.E. Logan, M. Elimelech, Membrane-based processes for sustainable power generation using water. Nat. Commun. 488 (2012) 313–319.
3. Y. Liu, Y. Chen, Integrated perovskite/bulk-heterojunction organic solar cells. Adv. Mater. 32 (2020) 1805843.
4. Y. Zhang, J. Wang, X. Wang, Review on probabilistic forecasting of wind power generation. Renew. Sustain. Energy Rev. 32 (2014) 255–270.
5. J. Zhao, A. Wang, M. Green, High-efficiency PERL and PERT silicon solar cells on FZ and MCZ substrates. Sol. Energy Materials Sol. Cells 65 (2001) 429–435.
6. E. Garnett, P. Yang, Light trapping in silicon nanowire solar cells. Nano Lett. 10 (2010) 1082–1087.
7. K. Wang, J.J. Chen, Z.M. Zeng, J. Tarr, W.L. Zhou, Synthesis and photovoltaic effect of vertically aligned ZnO/ZnS core/shell nanowire arrays. Appl. Phys. Lett. 96 (2010) 123105.
8. M. Green, K. Emery, Y. Hishikawa, W. Warta, E. Dunlop, Solar cell efficiency tables. Prog. Photovolt. Res. Appl. 24 (2016) 3–11.
9. B.D. Yuhas, P. Yang, Nanowire-based all-oxide solar cells. J. Am. Chem. Soc. 131 (2009) 3756–3761.
10. M. Beard, J. Luther, A. Nozik, The promise and challenge of nanostructured solar cells. Nature Nanotech. 9 (2014) 951–954.
11. K. Hillerich, K.A. Dick, M.E. Messing, K. Deppert, J. Johansson, Simultaneous growth mechanisms for Cu-seeded InP nanowires. Nano Res. 5 (2012) 297–306.
12. M.C. Altay, S. Eroglu, Chemical vapor deposition of Ge nanowires from readily available GeO$_2$ and CH$_4$ precursors. Journal of Crystal Growth. 551 (2020) 125886.
13. Q. Wang, G. Yuan, S. Zhao, W. Liu, Z. Liu, J. Wang, J. Li, Metal-assisted photochemical etching of GaN nanowires: The role of metal distribution. Electrochemistry Communications, 103, (2019) 66–71.

14. A.K. Sood, Y.R. Puri, P. Gao, C. Lao, Z.L. Wang, D.L. Polla, M.B. Soprano, Proceedings of SPIE, Volume 6556, (2007) 6556IL.
15. F.T. Mohummed Noori, Chemical beam epitaxy growth of III–V semiconductor nanowires. AIP Conference Proceedings 1569 (2013) 355.
16. H. Wu, W. Pan, D. Lin, H. Li, Electrospinning of ceramic nanofibers: Fabrication, assembly and applications. J Adv Ceram 1 (2012) 2–23.
17. J.E. Graves, M.E.A. Bowker, A. Summer, A. Greenwood, C. Ponce de León, F.C. Walsh, A new procedure for the template synthesis of metal nanowires. Electrochemistry Communications, 87 (2018) 58–62.
18. S. Barth, F. Hernandez-Ramírez, J. Holmes, A. Romano-Rodríguez. Synthesis and applications of one-dimensional semiconductors. Progress in Materials Science 55 (2010) 563–627.
19. J. Zhu, Z. Yu, G.F. Burkhard, C. Hsu, S.T. Connor, Y. Xu, Q. Wang, M. McGehee, S. Fan, Y. Cui, Optical absorption enhancement in amorphous silicon nanowire and nanocone arrays. Nano Lett. 9 (2009) 279–282 85.
20. O.L. Muskens, J.G. Rivas, R.E. Algra, E.P. Bakkers, A. Lagendijk, Design of light scattering in nanowire materials for photovoltaic applications. Nano Lett. 8 (2008) 2638–2642 86.
21. L. Tsakalakos, J. Balch, J. Fronheiser, B.A. Korevaar, O. Sulima, J. Rand, Silicon nanowire solar cells. Appl. Phys. Lett. 91 (2007) 233117.
22. T. Kuykendall, P. Ulrich, S. Aloni, P. Yang, Complete composition tunability of InGaN nanowires using a combinatorial approach. Nat. Mater. 6 (2007) 951–956 100.
23. Z. Wu, J.B. Neaton, J.C. Grossman, Quantum confinement and electronic properties of tapered silicon nanowires. Phys. Rev. Lett. 100 (2008) 246804 101.
24. M.C. Beard, A.G. Midgett, M.C. Hanna, J.M. Luther, B.K. Hughes, A.J. Nozik, Comparing multiple exciton generation in quantum dots to impact ionization in bulk semiconductors: Implications for enhancement of solar energy conversion. Nano Lett. 10 (2010) 3019–3027 102,
25. A.J. Nozik, Nanoscience and nanostructures for photovoltaics and solar fuels. Nano Lett. 10 (2010) 2735–2741.
26. Z. Wu, J. B. Neaton, J.C. Grossman, Charge separation via strain in silicon nanowires. Nano Lett. 9 (2009) 2418–2422 104.
27. T. Tiedje, E. Yablonovitch, G. Cody, B. Brooks. Limiting efficiency of silicon solar cells. IEEE Trans. Electron Devices 31 (1984) 711–716 105.
28. K. Takanezawa, K. Hirota, Q. Wei, K. Tajima, K. Hashimoto, Efficient charge collection with ZnO nanorod array in hybrid photovoltaic devices. J. Phys. Chem. C 111 (2007) 7218–7223,
29. M. Law, L.E Greene, J.C, Johnson, R. Saykally, P. Yang, Nanowire dye-synsitizd solar cell. Nat. Mater. 4(6) (2005) 455–459T.
30. K. Takanezawa, K. Tajima, K. Hashimoto, Efficiency enhancement of polymer photovoltaic devices hybridized with ZnO nanorod arrays by the introduction of a vanadium oxide buffer layer. Appl. Phys. Lett. 93 (2008) 063308.
31. B.M. Kayes, H.A. Atwater, N.S. Lewis, Comparison of the device physics principles of planar and radial p-n junction nanorod solar cells. J. Appl. Phys. 97 (2005) 114302
32. C.K. Chan, H. Peng, G. Liu, K. McIlwrath, X.F. Zhang, R.A. Huggins, Y. Cui, High-performance lithium battery anodes using silicon nanowires. Nat. Nanotechnol. 3 (2008) 31–35 108.
33. M. Green, P. Basore, N. Chang, D. Clugston, R. Egan, R. Evans, D. Hogg, S. Jarnason, M.Keevers, P. Lasswell, J. O'Sullivan, Crystalline silicon on glass (CSG) thin-film solar cell modules. Solar Energy 77 (2004) 857–863.
34. Z. Fan, H. Razavi, J. Do, A. Moriwaki, O. Ergen, et al. 2009. Three-dimensional nanopillar-array photovoltaics on low-cost and flexible substrates. Nature Mater. 8 (8) 648–653.
35. M. Law, L.E. Greene, J.C. Johnson, R. Saykally, P. Yang, Nanowire dye-sensitized solar cells. Nat. Mater. 4 (6) (2005) 455–459.
36. G. Jia, M. Steglich, I. Sill, F. Falk, Core–shell heterojunction solar cells on silicon nanowire arrays, solar energy materials and solar cells, 96 (2012) 226–230.
37. B. Pal, K. J. Sarkar, P. Banerji, Fabrication and studies on Si/InP core-shell nanowire based solar cell using etched Si nanowire arrays. Solar Energy Materials and Solar Cells, 204 (2020) 110217.
38. Y. Kim, N.D. Lam, K. Kim, W.-K. Park, J. Lee, Ge nanopillar solar cells epitaxially grown by metalorganic chemical vapor deposition. Scientific Reports, 7 (2017) 42693.

39. G. Tchutchulashvili, S. Chusnutdinow, W. Mech, K.P. Korona, A. Reszka, M. Sobanska, Z.R. Zytkiewicz, W. Sadowski, GaN nanowire array for charge transfer in hybrid GaN/P3HT:PC71BM photovoltaic heterostructure fabricated on silicon. Materials, 13 (2020) 4755.

40. J.-H. Park, R. Nandi, J.-K. Sim, D.-Y. Um, S. Kang, J.-S. Kim, C.-R. Lee, A III-nitride nanowire solar cell fabricated using a hybrid coaxial and uniaxial InGaN/GaN multi quantum well nanostructure. RSC Advances, 8 (2018) 20585–20592.

41. Y.B. Tang, Z.H. Chen, H.S. Song et al., Vertically aligned p-type single-crystalline GaN nanorod arrays on n-type Si for heterojunction photovoltaic cells. Nano Letters, 8 (2008) 4191–4195.

42. F. Li, S.H. Lee, J.H. You, T.W. Kim, K.H. Lee, J.Y. Lee, Y.H. Kwon, T.W. Kang, UV photovoltaic cells fabricated utilizing GaN nanorod/Si heterostructures. Journal of Crystal Growth, 312 (2010) 16–17, 2320–2323.

43. M.C. Kao, H.Z. Chen, S.L. Young, C.C. Lin, C.Y. Kung, Structure and photovoltaic properties of ZnO nanowire for dye-sensitized solar cells. Nanoscale Research Letters, 7 (2012) 260.

44. V.S Jonnadula, K.G. Reddy, K. Devulapally, N. Islavatha, L. Giribabu, Solution processed aligned ZnO nanowires as anti-reflection and electron transport layer in organic dye-sensitized solar cells. Optical Materials, 95 (2019) 109243.

45. J. Qiu, X. Li, F. Zhuge, X. Gan, X. Gao, W. He, S.-J. Park, H.-K. Kim, Y.-H. Hwang, Solution-derived 40 microm vertically aligned ZnO nanowire arrays as photoelectrodes in dye-sensitized solar cells. Nanotechnology. 14 21(19) (2010) 195602.

46. W. Zhang, W. Song, J. Huang, L. Huang, T. Yan, J. Ge, R. Penga, Z. Ge, Graphene: Silver nanowires composite transparent electrode based flexible organic solar cells with 13.4% efficiency. J. Mater. Chem. A, 7 (2019) 22021–22028.

13 Nanowires for Photonic Applications

Benjamin Agyei-Tuffour
Department of Materials Science and Engineering, School of Engineering
Sciences, College of Basic and Applied Sciences, University of Ghana,
Legon, Accra, Ghana
African Research Universities Alliance (ARUA) Center of Excellence in
Energy, Stellenbosch University, South Africa

Elizabeth Adzo Addae
Department of Materials Science and Engineering, School of Engineering
Sciences, College of Basic and Applied Sciences, University of Ghana,
Legon, Accra, Ghana

Reuben Amedalor
Department of Materials Science and Engineering, School of Engineering
Sciences, College of Basic and Applied Sciences, University of Ghana,
Legon, Accra, Ghana
Institute of Photonics, University of Eastern Finland, Joensuu, Finland
Faculty of Engineering and Natural Science, Photonics,
Tampere University, Tampere, Finland

Ibrahim Issah
Faculty of Engineering and Natural Science, Photonics,
Tampere University, Tampere, Finland

Joseph Asare
Department of Physics, School of Physical and Mathematical Sciences,
College of Basic and Applied Sciences, University of Ghana, Legon,
Accra, Ghana

Neill J. Goosen
African Research Universities Alliance (ARUA) Center of Excellence in
Energy, Stellenbosch University, South Africa

Kwadwo Mensah Darkwa and Daniel Nframah Ampong
Department of Materials Engineering, College of Engineering, Kwame
Nkrumah University of Science and Technology, Kumasi, Ghana

Ram K. Gupta
Department of Chemistry and National Institute for Materials
Advancement, Pittsburg State University, Pittsburg, KS, USA

DOI: 10.1201/9781003296621-13

CONTENTS

13.1   INTRODUCTION

In the past, the terms optics and photonics were often used interchangeably. In recent years, although there exist similarities between the two, the difference in the two fields is in terms of device applications. In a broader sense, optics involves guided-wave and free-space propagation and touches on topics such as diffraction, interference, photon optics, statistical optics, and imaging. Photonics, on the other hand, is understood to encompass optics and involve topics that concern the generation, transmission of light, and its interaction with matter in systems and devices [1]. The fast progress in this field is due to the unprecedented advent of new materials that meet the demands of the photonic industry. Photonic materials can be defined as metals, semiconductors, and dielectrics that possess desired optical characteristics and functionalities. Unlike metals and semiconductors, known to have relatively free mobile electrons, which enable them to conduct electricity, dielectrics are poor electrical conductors consisting of bound charges. Dielectrics are materials that are easily polarized by an externally applied electromagnetic field. They can also be described in terms of the energy-storing potential of the material to polarizability. An applied electric field causes displacement of the otherwise tightly bound charges in the dielectric material. The positive charges are displaced slightly in the direction of the applied electric field, while the negative charges are in the opposite direction. The polarizability of a dielectric is a measure of the material's electric field susceptibility or electric permittivity. Examples include glass, plastics, ceramics, and oxides of various metals. Dry air and distilled water are also good examples of dielectric materials. Dielectrics are desired photonic materials for various reasons, including their low energy dissipation in transporting electromagnetic fields and ability to guide electromagnetic beams. Photonics has evolved over the years to encompass nanophotonics. Technological advances and the quest for miniaturized photonic systems and devices, such as integrated photonic circuitry places demand on the research into multifunctional and effective photonic and nanophotonic materials and the design of nanostructures that meet the occasion (Figure 13.1). One of such materials is photonic nanowires.

Nanowires have emerged as potential materials for diverse applications in photonic systems. They are one-dimensional photonic materials with advantageously high aspect ratios. These have quantum mechanical effects on optoelectronic devices fabricated from nanowires and are called "quantum wires" [3]. Nanowires possess exceptional properties not usually observed in other bulk materials because there are quantum restrictions of electrons laterally in nanowires, leading to a variation in the energy levels between bulk materials and nanowires. Due to their unique electrical and optical

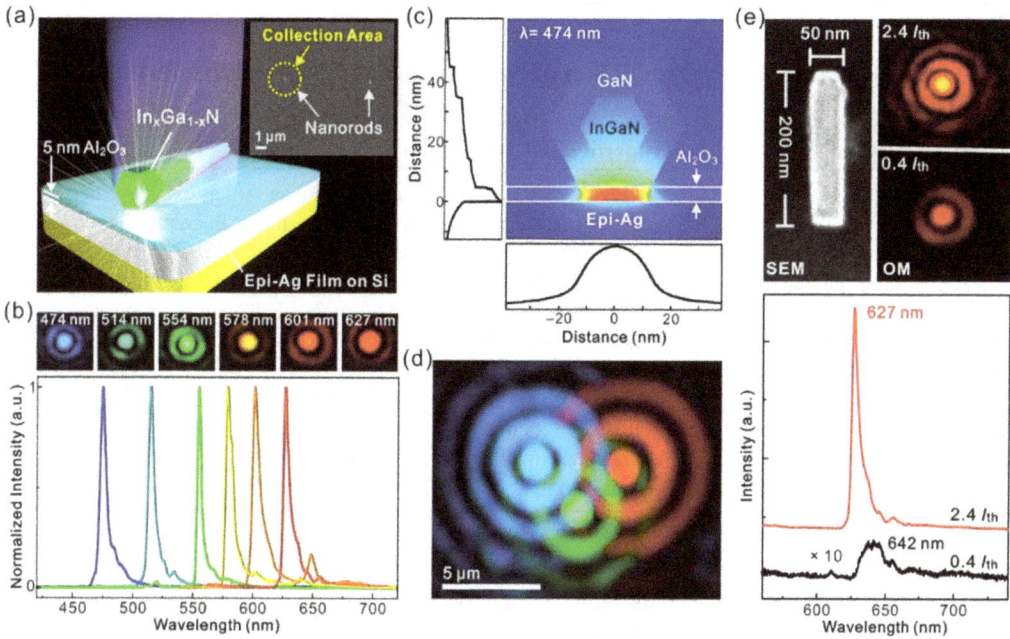

FIGURE 13.1 Representation of plasmonic nanolaser emission from single nanowire (a), ~4 nm nanowire for single-mode laser emission of all colors (b). 474 nm wavelength nanolaser stimulation and field distribution (c). Three different nanowires placed side-by-side for concurrent RGB lasing (d). SEM and optical microscope images' lasing threshold (e). Spontaneous nanolaser emission spectra (627 nm, red; 642 nm, black). Adapted with permission [2]. Copyright (2014), American Chemical Society.

characteristics, nanowires have become widely reliable applications in the field of nanotechnology, including nanophotonics and nanoelectronics. Nanowires have the potential to control or manipulate atoms, photons, electrons, phonons, and plasmons. In wave optics, nanowires are employed for waveguiding tightly confined light fields [4,5]. These characteristics make them greatly desired for various photonic technological applications. They can function as optical waveguides for light transport [6], and optical sensors in miniaturized platforms [7]. Thanks to the state of the art fabrication techniques, such as the physical drawing method [8,9], electron beam lithography, chemical growth method of glass nanowires, and electrospinning technique [10], nanowires of different mechanical, optical, and surface texture qualities can be fabricated. Among the above fabrication techniques, the physical drawing method produces nanowires of relatively better quality in terms of diameter uniformity and surface smoothness of the fabricated nanowire. Moreover, the good mechanical properties of photonic nanowires make it possible to manipulate single nanowires into different desired positions and geometries such as micro ring for optical resonance sensing and Mach-Zehnder geometry for optical interferometry. The manipulation process is possible using designed tapered fiber probes under an optical microscope [5].

13.2 SYNTHESIS OF NANOWIRES

The manufacture of a 1D nanostructure having a carefully designed phase purity, dimension, lucidity, and component presents various problems. Understanding and controlling the growth processes and nucleation at the nanoscale is critical for constructing precisely tailored nanostructures [11]. Based on quantum restriction, nanowires have a duo path, but the third is left free for electrical conduction purposes. They, therefore, find applications when tunneling transport is not needed but rather electrical conduction. Nanowires with tiny diameters are projected to have considerably different

TABLE 13.1

Very Useful Materials for the Production of Nanowires and the Respective Techniques Used [14]

Material	Method	Properties and Applications
MgO	VS	MgO with ~ 2400 ºC fusion temperature as well as extensive heat capacity. They are adopted as reinforcements in composites structures.
Cu_2O	Vapor Phase	Copper (I) oxide with the band gap of ~2eV is useful in the conversion of all other forms of (surfactant-assisted chemical energy, etc.)
SiO_2	VLS, CVD, and Laser Ablation	Optical waveguiding
Ga_2O_3	Vapor, Liquid Phase	Gallium oxide has a 4.9 eV band gap. Used for catalysis, gas sensors, etc. due to its wide band gap. It also emits in blue.
Al_2O_3	Vapor phase, template-assisted etching	Alumina has low thermal conductivity and is usually adopted as an insulation material.

characteristics compared to their bulk samples (3D crystalline ones) equivalents due to the unusual state of their electronic densities. The production of 1D nanostructures, specifically for nanowires, involves limiting material development in two directions to a few nanometers while permitting growth in the third. To successfully implement nanowire synthesis approaches in a manufacturing setting, we must guarantee that these technologies are simple to scale up and economical [12]. The reagents and derivatives of nanowire production should be safe for the environment. We should model each process parameter's impact and devise ways to manage the specifications during extensive production accurately. The nanowire synthesis procedure should be robust enough to handle noise-based parameters [13]. Table 13.1 highlights some of the significant materials that may be generated in the form of nanowires using various techniques.

13.2.1 VAPOR-LIQUID-SOLID METHOD

The most widely researched method for producing 1D nanostructures, such as whiskers, nanorods, and nanowires, is vapor-liquid-solid (VLS) phase synthesis [15]. The key to achieving regulated 1D development is to avoid supersaturation or minimize it. The gaseous reactions are used to create vapor species in a typical process (Figure 13.2). Following that, they are transferred and liquefied on lower-temperature support. A suitable catalyst is used in the VLS technique, which specifies the nanowire size and guides selectively, reactants' addition. The key processes that drive the development of nanowires have been given a detailed explanation. Lieber's group has shown that this process can be used to generate semiconductor nanowires from a variety of materials [16]. This approach, however, will most likely be confined to simple oxides. Unlike chemical-based approaches, the laser-assisted VLS method necessitates a costly experimental apparatus. The VLS synthesis of nanowires is confined to the solid-liquid system under catalysis since many of the oxides (high, T_{mp}) require high temperatures to attain eutectic points. This allows for nanowires to be produced at extremely high temperatures [17].

13.2.2 ELECTRODEPOSITION

Electrodeposition is a low-cost alternative to more time-consuming procedures like laser ablation [18]. The growth, in particular, occurs closer to equilibrium than with high-temperature vacuum deposition procedures. In this, the porous membrane is deposited with a thin metal acting as a

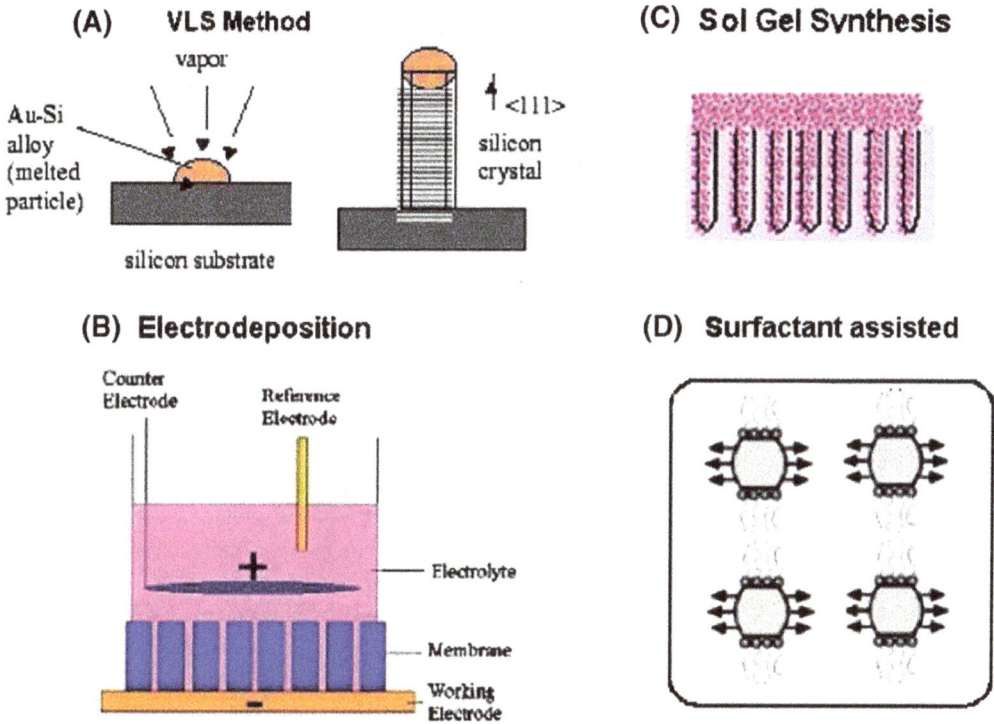

FIGURE 13.2 Synthesis methods of nanowires, including VLS (A), sol-gel (B) electrodeposition (C), and (D) surfactant-assisted. Adapted with permission [14]. Copyright (2005), Elsevier.

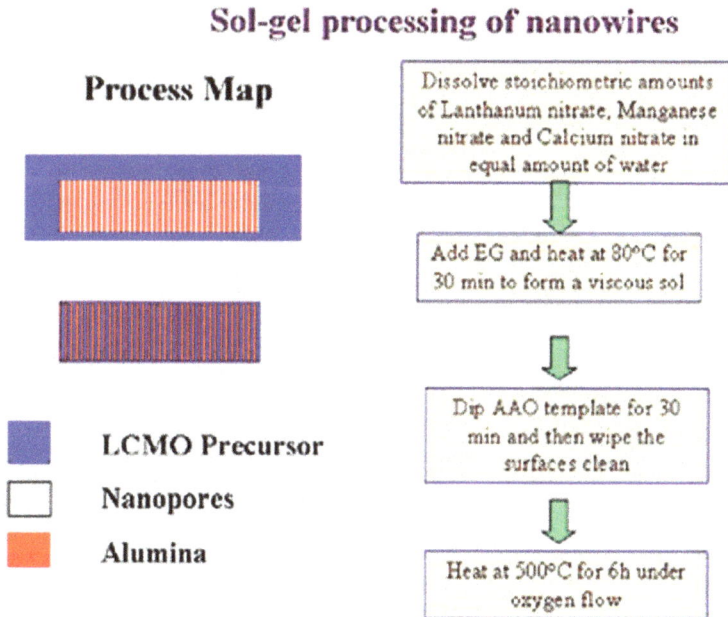

FIGURE 13.3 Schematic diagram of sol-gel processing for nanowires. Adapted with permission [14]. Copyright (2005), Elsevier.

cathode in electroplating, and it is commonly adopted to produce different classes of nanowires. This technique is easy to use and economical. The major drawback is the compositional modulation, which is not suitable for multiple oxide-based nanowires. This is because of the different ionic radii of the cations and their respective diffusivities [19].

13.2.3   NANOWIRE SYNTHESIS BY SOL-GEL TECHNIQUE

The sol-gel method has gained popularity because it produces extremely stoichiometric and this technique, combined with template-assisted, electrophoretic deposition, etc., has achieved successful outcomes of nanowires of multicomponent nature in a repeated order. The hydrolysis of a solution of precursor molecules to produce a suspension of colloidal particles (the sol), followed by the condensation of sol particles to produce a gel, is the foundation of sol-gel processing. Whereas in aqueous environments, precursors can be inorganic salts, they are metal alkoxides in organic solvents. The capacity to handle multi-component complex oxides is the most significant benefit of sol-gel processing. Controlling hydrolysis and condensation properly is critical. The formation of InP and InAs nanowires in the self-catalyzed VLS mode has been shown in Figure 13.3, and their structural features have been studied. With their tunable wavelength luminescence, a variety of MQDs can be interspersed with InP nanowires. This shows that strong growth controllability guarantees that the gain medium is homogeneous across the MQD area. The expansion is consistent with a dislocation-free InAs/InP interface despite a 3.1% lattice mismatch between InP and InAs. Both the InAs QDisks and the InP coatings exhibit few stacking defects similar to a crystalline structure of zinc-blende, according to the matching, selective electron diffraction (SED) pattern (Figure 13.4). Micro-photoluminescence is used to characterize the optical characteristics of InP/InAs MQD nanowires (PL), which are disseminated regularly over a gold-plated SiO_2/Si support for optical measurements, as shown in Figure 13.5. Overall, it can be said that nanowires are of high optical and mechanical strength and hence would be essential in fabricating photonic devices or applications. Figure 13.6 shows photoluminescence (PL) and SEM measurements of the mechanically dispersed nanowire. However, the fabrication method can affect its efficiency and hence suitable stoichiometry and morphological control, low-cost infrastructure needs, and ease of scalability needs to be applied. Chemical solution processing, such as the VLS, is much more advantageous.

13.3   NANOWIRES FOR LIGHT-EMITTING DIODES

The enormous research into semiconductor technology has proven its importance for light generation. Some good examples include the advent of semiconductor light-emitting diodes and semiconductor-based lasers. A wide range of semiconductor materials, both organic and inorganic

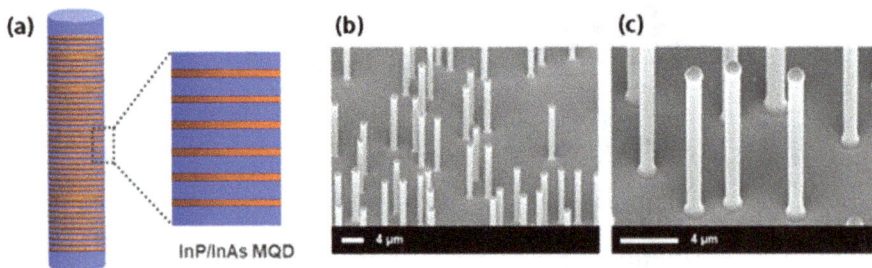

FIGURE 13.4   Representation of heterostructure nanowires (InAs/InP) (a) MQD. (b and c) Inp (111) B substrate-grown InP/InAs SEM images. Adapted from [20]. Copyright (2019) The Authors. Some rights reserved; exclusive licensee American Association for the Advancement of Science. Distributed under a Creative Commons Attribution License 4.0 (CC BY).

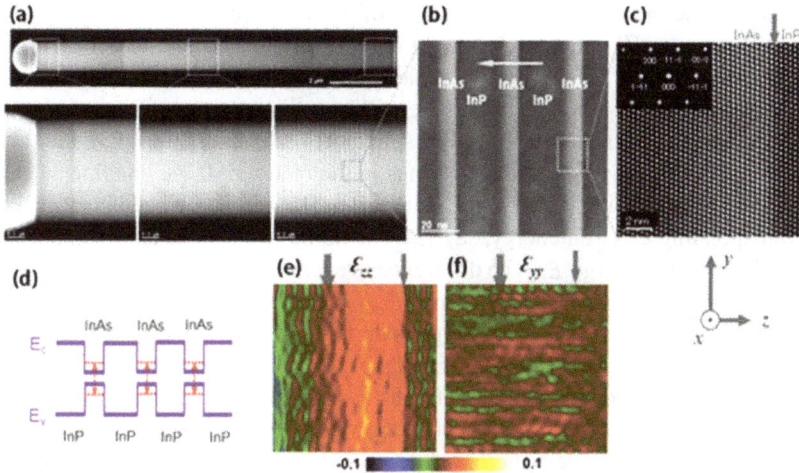

FIGURE 13.5   Transmission electron microscopy analysis and schematic representation of MQD nanowire: (a) STEM images (b and c). Band diagram of nanowires (d), Y-z direction strain mapping of MQD nanowire with its lattice spacing (e, f). Adapted from [20]. Copyright (2019) The Authors. Some rights reserved; exclusive licensee American Association for the Advancement of Science. Distributed under a Creative Commons Attribution License 4.0 (CC BY).

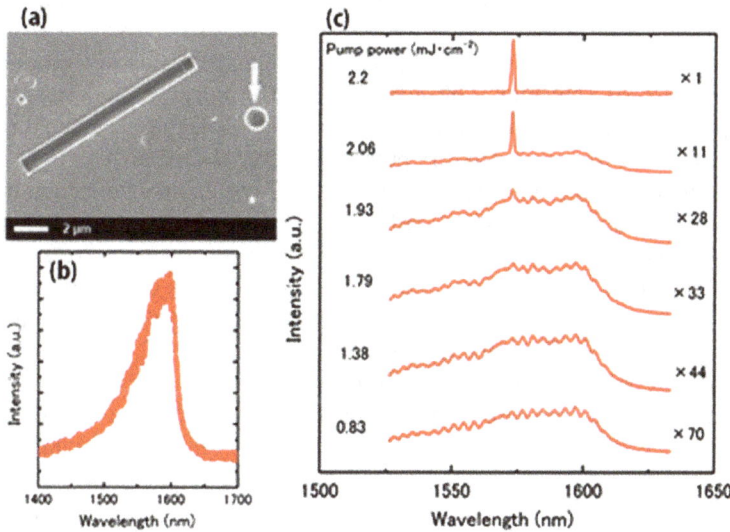

FIGURE 13.6   Photoluminescence (PL) and SEM measurements of the mechanically dispersed nanowire with varying pump laser power. (b) PL at specified laser power, (c) increasing pump power. Adapted from [20]. Copyright (2019) The Authors. Some rights reserved; exclusive licensee American Association for the Advancement of Science. Distributed under a Creative Commons Attribution License 4.0 (CC BY).

has been extensively exploited, including composite perovskites, group II-VI, and III-V elements of the periodic table. Semiconductor materials exhibit electronic properties in-between conductors (metals) and insulators (dielectrics). Although they possess relatively fewer free electrons, which aid conductions than in metals, the individual atoms of the semiconductor materials combined to form a periodic crystal lattice that enables tunability of their electrical conduction through a hole-electron pair; a process often termed doping. By doping, it means engineering the electrical conductivity of intrinsic semiconductor materials such as those of group IV (e.g., Si and Ge) by introducing a trivalent material, a group III element such as Al into the lattice structure. Electron transfer occurs resulting in the generation of an electron-hole pair, thus improving the material's conductivity. In other words, it means bandgap engineering as we will later discuss.

Semiconductor materials are characterized by split electronic bands; a result of the solution of the Schrodinger equation to the energy of the electrons in the material. If we imagine an electron transition between two such existing bands in the material, the lower energy band is termed the valence band with corresponding electron energy, say E_v. The higher energy band corresponds to the conduction band with electron energy E_c. The difference in energy between the conduction band and the valence band is the bandgap energy, often denoted E_g. That is $E_g = E_c - E_v$. This is precisely the energy potential an electron in the lower energy state needs to overcome to successfully transition to the higher energy band (excited state). The bandgap energy is different for different materials and thus, defines the electrical property of the materials. Some materials exhibit direct bandgap, while others have indirect bandgap. The former is when the valence band's maximum and the conduction band's minimum coincide with the same k, wavevector, on the energy-wavevector (E-k) dispersion diagram. The latter occurs when the maximum of the valence band and the minimum of the conduction band are at different k values on the E-k dispersion diagram [11]. Bandgap energies of Si and Ge, which exhibit direct bandgap structure at room temperature, are 1.12 eV and 0.66 eV, respectively, corresponding to wavelengths around 1100 nm and 1879 nm. In practice, these materials will reflect all energies corresponding to wavelengths above the bandgap wavelength as those are not strong enough to cause any transition to a suitable higher energy level. A lower wavelength will, however, produce enough energy for electron transitions. Bandgap energies of other semiconductor materials can be found in ref. [11].

The quest for downsizing photonic devices (e.g., light-emitting diodes, lasers modulators) for greater integration led to the emergence of semiconductor nanowires for light transmission. Owing to their high aspect ratio, tunability of the emission wavelength, and high light confinement, among others, the semiconductor nanowires are very promising material structures for light generation and transmission for attaining the goal of photonic devices and system miniaturization beyond the size of electronic devices, although the diffraction limits highly restrict this. Moreover, aside from the great optical waveguiding characteristic, high refractive index nanowires serve as the high-quality optical gain medium and optical cavities, and also the simultaneous confinement of both photons and generated electron-hole pairs results in effective photon electron-hole pair interaction, thus, making them desirable for light-emitting diodes [11].

As hinted earlier, doping is one way of controlling or tuning the bandgap of semiconductor materials. By varying the compositions of compound $GaAs_xP_{(1-x)}$ nanowire, the bandgap energy can be tuned from 1.38 eV to 2.25 eV, corresponding to the visible wavelength range of 550 nm to 900 nm. Moreover, varied compositions of $In_xGa_{(1-x)}P$ and $In_xGa_{(1-x)}N$ nanowires exhibited energy bandgaps that span from UV to near-infrared (IR) [11,21,22]. The bandgap tunability paves the way for numerous photonic devices and systems such as multicolored light-emitting diodes (LEDs), lasers, and wide-spectrum semiconductor photodetectors. Si technology amid its low efficiency of light generation is outliving its usefulness as a standalone semiconductor technology as demand for high-performing and multifunctional devices increases. The need to push Si technology further by integrating with other high-performing semiconductor devices has led to several research outputs on the integration of other semiconductor materials on Si platforms. A publication by Samuelson and the group reported the successful integration of group III-IV compounds with tunable bandgaps by epitaxial growth. These semiconductor nanowires are shown in Figure 13.7 [23].

This later set the pace for the development of high-performing semiconductor nanowires as nano light sources. As an example, Tomioka and coworkers demonstrated the integration of GaAs/AlGaAs core-multishell nanowire-based LEDs on Si substrate. The LED has seen improved light output efficiency by designing and growing a double heterostructure on Si [24]. This has the potential

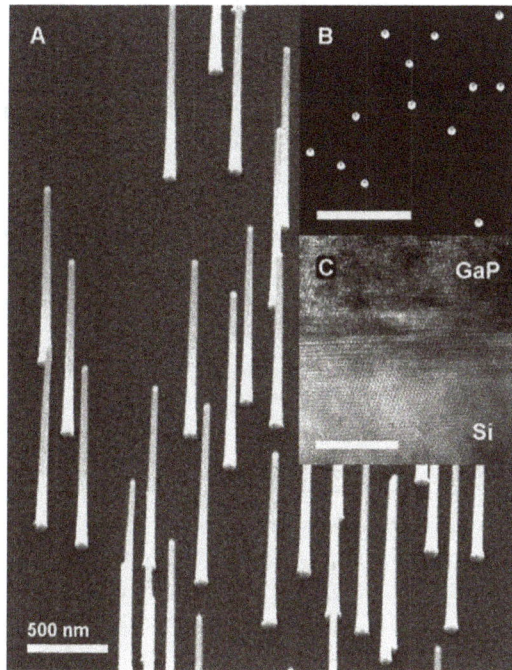

FIGURE 13.7   Nanowire growth: (a) GaP image with a scanning electron microscope. The inserts: (b) the plan view and (c) TEM image of used Si substrate. Adapted with permission [23]. Copyright (2004), American Chemical Society.

for on-chip integration through optical interconnections for high-performance applications, thus, replacing metal-based on-chip connections

Besides, many research works on efficient LEDs have been reported by many researchers. Lots of researchers have exploited ZnO for its photoluminescence (PL) around 384 nm. It also has wide luminescence, which is associated with defects at room temperature at about 600 nm wavelength electroluminescence (EL) emission of epitaxially grown n-type ZnO nanowires based on homojunction and heterojunction has also been demonstrated for LEDs. LEDs based on core-multishell nanowire have been fabricated in wide bandgap semiconductors. GaN-based LED with ZnO nanorod arrays has been reported to show improved light emission. According to the report, the incorporation of ZnO nanowire arrays has seen the rather low light extraction efficiency of conventional GaN LEDs improve to 100% at an applied current of 50 mA. Later in the same year, Lai and colleagues reported the GaN/ZnO nanowire array-based LEDs with p-type GaN and n-type ZnO p-n junction composition with both compositions grown by MOCVD and epitaxially, respectively. ZnO-based nanorod arrays UV and red LED was also demonstrated. The p-type ZnO constitutes an As$^+$ doped as-grown ZnO by the method of implantation. Embedded ZnO nanowires in polymer matrix for flexible LEDs have also been realized [25–28]. The above-cited works on bandgap tunable nanowire LEDs, constitute research efforts to move away from the low-efficiency bulk material semiconductor LEDs. Some bottlenecks persist and these drive ongoing research works in this field.

13.4   NANOWIRES FOR SUBWAVELENGTH WAVEGUIDES

Optical waveguides are components for photonic circuits in confining and transporting optical signals from inputs/transmitters to targets/receivers. The optical nanowire is one such structure for optical waveguiding that offers an advantage in guiding optical beams due to their high aspect ratio.

To qualify as a subwavelength structure, the nanowire diameter is much smaller than the wave-length of the launched light or photon beam. The dielectric and metallic (plasmonic) nanowires are desired for waveguiding purposes. The main difference between the dielectric nanowire and plasmonic nanowire waveguides lies in the confinement of propagating photons. While the dielectric nanowire propagates tightly confined photons inside the material, confinement in plasmonic nanowire waveguides occurs close to the surface.

Dielectric nanowires function as waveguides by the principle of total internal reflection of the incident light beam. Waveguiding is possible due to the refractive index contrast between the waveguiding element and the surrounding environment. The higher the contrast in the refractive index, the higher the light beam is confined in the element. For example, given that the waveguiding medium is glass $n = 1.5$) and the surrounding medium is air ($n = 1.0$), the refractive index contrast is 0.5. Mostly, single-mode propagation is preferred for photonic waveguides due to high confinement. This implies that the diameter of the nanowire waveguide must be designed to accommodate only single modes. In practice, however, the Abbe diffraction greatly influences the diameter of dielectric nanowires [5,11].

The concept of metallic (plasmonic) nanowires surfaced to break this challenge. Plasmonic nanowire functions as a waveguide by propagating light with surface plasmon polaritons (SPPs) excited on the surface of the waveguide. It is associated with ohmic and other structural losses that limit their application [11]. The next attempt at solving this challenge led to the exploitation of hybrid photonic-plasmonic waveguides, which takes advantage of the individual properties of the nanowires to minimize loss and attain optimum performance. The field is still under exploitation, with current research results yielding mixed successes. The following sections will discuss dielectric nanowire waveguides, metallic nanowire waveguides, and hybrid dielectric-plasmonic nanowire waveguides.

13.4.1 DIELECTRIC NANOWIRE WAVEGUIDES

Dielectrics nanowire waveguides are photonic waveguides that rely on bounded electrons' response to electromagnetic light [5]. The response manifests as a propagating photon beam in the dielectric material. The propagation of photons in dielectric waveguides happens by total internal reflection (TIR), governed by Snell's law. To illustrate this principle, let us consider two media with refractive indices n_1 and n_2, with $n_1 > n_2$ the propagation of generated photons in the media is governed by Snell's law, mathematically expressed as

$$n_1 \sin \theta_1 = n_2 \sin \theta_2, \quad (13.1)$$

$$\sin \theta_1 = \frac{n_2}{n_1} \sin \theta_2, \quad (13.2)$$

where θ_1 and θ_2 are the angle of incidence and refraction, respectively. The physical interpretation of the law is that the direction of photon propagation in any media is a function of the properties of the medium (the refractive index). From Equation 13.2, at a certain angle of incidence $\theta_1 = \theta_c$, which is termed the critical angle of incidence, $\theta_2 = 90°$, and the refracted light propagates at the interface between the two media n_1 and n_2. When $\theta_1 > \theta_c$, light is completely reflected into the same medium, and this condition is referred to as "total internal reflection" [29].

Let us consider the case where the high refractive medium n_2 is sandwiched between the low refractive index medium n_1; we have a waveguide where the light propagates in the high refractive index medium n_2 (Figure 13.8a). Okay, what if $\theta_1 < \theta_c$? In this condition, the photons are not resonating with the waveguiding medium and thus, leak out of the waveguide. The shapes of the transverse electric and magnetic field distributions of the allowed propagating photons in the waveguides are called modes. A mode is a unique solution of the electromagnetic wave equation obtained by solving Maxwell's equation. Other conditions such as phase matching (constructive interference),

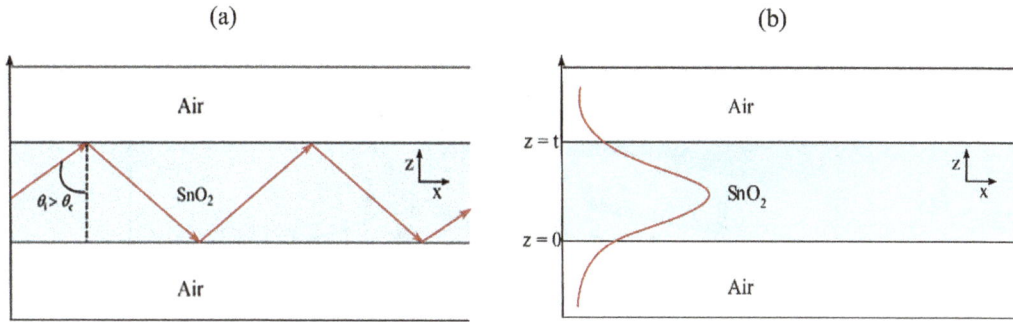

FIGURE 13.8 Waveguiding in a nanowire. (a) Ray illustration of total internal reflection. (b) Typical shape of a fundamental mode in a waveguide.

propagation-induced phase shift, and Fresnel phase shift must be satisfied [11]. Figure 13.8b shows a highly confined propagating single-mode shape in the waveguide with evanescent (decay) tails into the external medium.

For a photonic dielectric nanowire, a single-mode condition is governed by,

$$\pi \frac{d}{\lambda} \sqrt{(n_1{}^2 - n_2{}^2)} < 2.405, \tag{13.3}$$

where d and λ represent the diameter of the nanowire waveguide and wavelength of the incident photons, n_1 and n_2. This equation denotes the nanowire's refractive indices and the outside medium. The smallest attainable diameter d (Abbe diffraction limit) is proportional to $\lambda/(2n \sin \theta)$, where n is the known refractive index of the nanowire [5,11]. From the Abbe diffraction limit equation, high refractive index materials are more suited to achieving deep subwavelength waveguides. SiO_2 $n_1 = 1.5$) known for its dielectric characteristics is one of the most studied materials for nanowire waveguides. Over the years, research has geared towards materials with a relatively higher refractive index to replace SiO_2. This is necessary due to the quest for miniaturized nanowires for photonic integrated circuits and devices. Wide bandgap SnO_2 nanowire waveguide with a refractive index of 2.1 has been exploited and reported to exhibit excellent subwavelength waveguiding characteristics. It exhibits low propagation loss (1–8 dB/mm) at frequencies in the visible spectrum [30]. Moreover, tangential light coupling into the SnO_2 nanowire waveguide had been successfully realized by employing active GaN and ZnO nanowires [31].

13.4.2 Plasmonic Nanowire Waveguides

A plasmonic nanowire is highly desired for high-sensitivity optical sensing applications due to its propagation mode (metal surface modes). It offers high propagating metal surface energy and strong evanescent fields for high sensitivity applications. Its notable materials are the noble metals, gold (Au), silver (Ag), and copper (Cu), due to their unique electronic properties. However, the major challenge with a plasmonic waveguide is the propagation loss suffered due to surface roughness and crystallinity of the metals. To reduce the propagation loss, the metals are preferably chemically synthesized to improve the surface quality and increase crystallinity. In addition, these metals suffer absorption losses. Ag nanowire instead of Au and Cu is preferably exploited for plasmonic waveguides. This is because, in the visible spectrum, there are absorption losses for Cu and Au. This is because of the d-band transitions associated with Au and Cu. On the other hand, Ag is liable to propagation loss because of its chemical instability under ambient conditions than Au. Research into Ag nanowire plasmonic waveguides revealed that measured propagation length showed significant variances in the diameter. The attempts at explaining the discrepancies resulted in the discovery

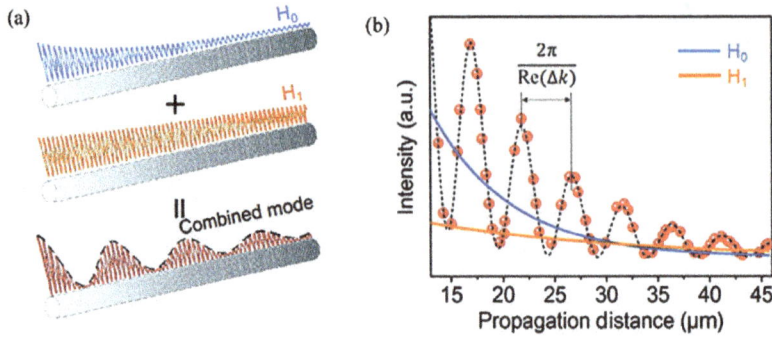

FIGURE 13.9 Nanowires made from Ag showing the surface plasmon polariton modes of the lowest order (a). Effect of propagation distance on the intensity of plasmon emissions (b). Adapted with permission [32]. Copyright (2017) Springer Nature.

of the chemically grown Ag nanowire-based surface plasmon polariton (SPP) multi-modes by employing a tapered optical fiber tip attached to a suspending Ag nanowire [11].

In a publication by Kim et al. Ag nanowire SPP modes have been decoupled and analyzed as a function of the mode excitation wavelengths and the diameter using a developed mode-interference. Because lower-order modes are more confined than the higher-order modes, they are of great interest in plasmonic nanowire waveguide applications. Figure 13.9 shows H_0 and H_1, the dual lowest-order modes, with different propagation distances and interference modes of the multi-SPP modes of the Ag nanowire. These two lowest-order modes are great for optical sensing and imaging applications [11]. With Au nanowire, excited SPP modes have been realized by employing a scanning tunneling microscopy (STM) tip. The STM tip generates low-energy electrons that excite gap plasmons. At ambient conditions, the excited gap plasmons can couple to propagating SPPs. The result complements current metal oxide semiconductor technology [32]. Despite the effort to reduce losses in plasmonic nanowire waveguides, some bending loss still rises when integrating optical elements. For example, bending loss for Ag nanowire increases with a rising bend radius due to increasing refractive index contrast [33].

Moreover, although lower-order modes are more tightly confined, thus reducing ohmic losses, they also suffer coupling losses with high nanowire diameters, thus limiting their application. The lower-order mode might not be suitable for applications that require a larger diameter for propagating light. As such, these nanowire waveguides demand a careful selection of the geometry of the optical elements to compensate for ohmic and bending losses if one ought to attain a high-performance plasmonic nanowire waveguide.

13.4.3 HYBRID: DIELECTRIC-PLASMONIC NANOWIRE WAVEGUIDES

So far, we have considered the two different photonic nanowire waveguides. We have noticed that some pertinent issues remain unresolved despite the research efforts over the years to resolve those. We have noted that the Abbe diffraction limit hinders the performance of dielectric nanowire waveguides. Also, plasmonic nanowire waveguides were noted to be characterized by ohmic losses. The idea of a hybrid photonic nanowire waveguide emerged to solve these challenges and pave the way for high-performing optical elements. The hybridization must be such that there is optical coupling between the dielectric and the plasmonic waveguides. In a study carried out, a hybrid optical dielectric (SnO_2)-metal (Ag)-dielectric (SnO_2) configuration has realized light coupling. However, the problem of crystal quality and surface roughness of Ag remain issues under investigation [11].

13.5   NANOWIRES FOR OPTICAL SENSORS

Considering the continued advancement of micro/nanotechnology and the rising need for downsized sensors with improved performance, optical sensing has the potential for high sensitivity, rapid response, and immunity to electromagnetic interference, and safety procedures in explosive or combustible environments, as well as numerous signal retrieval possibilities from fluorescence lifespan, optical intensity, spectrum, phase, and polarization [5]. Using single-mode subwavelength-diameter silica nanowires for optical sensing is suggested based on the evanescent-wave guiding capabilities of nanowire waveguides [21]. The evanescent coupling technology is used for broadband optical launching and collection because of its high efficiency (up to 90%), stability, and compactness. Subwavelength-diameter silica nanowires have recently been proved to guide light in the visible and near-infrared spectral regions [34]. Because of their exceptional diameter uniformity and atomic-level sidewall smoothness, these wires can direct light with minimal optical losses. Because light led along such nanowires releases a considerable percentage of the guided field beyond the wire as evanescent waves, it is extremely sensitive to changes in the index of the surrounding medium [34].

13.5.1   REFRACTIVE INDEX SENSOR

RI is one of the fundamental physical aspects of optical sensing (Figure 13.10). This is most typically used to determine the solute concentration in an aqueous medium. A conduit for the liquid analyte is often created in the area of the deceased. Evanescent regions that proceed into the tapering segment of the fiber, expand through channels and render an extremely responsive system to changes in liquid refractive index. The sensor may function as a sharper on-off gadget as well as in constant estimation mode using an approximated RI measurement accuracy of 5×10^{-4}. The use of an interferometric structure can considerably boost the sensitivity of nanowire sensors. The FFPI sensor, for example, shows a much better detection rate compared with FBG detectors due to its smaller spectral feature and can accurately predict an index fluctuation of 1.4×10^{-5}. Fiber dimension also affected RI sensitivity; typically, a smaller diameter provided a sharper detection [35].

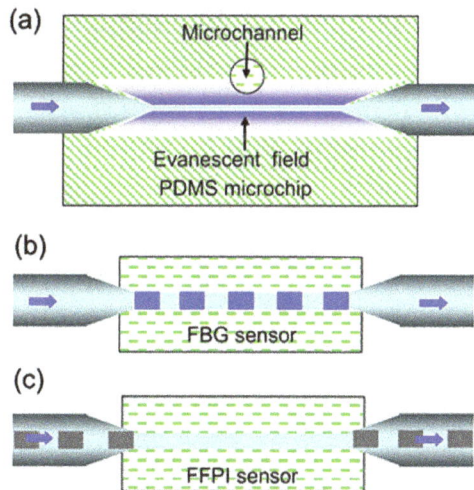

FIGURE 13.10   Schematics of nanowire RI sensor: (a) MNF sensor with liquid channel. (b) FBG unitary detector. (c) Fiber Fabry-Pérot interferometer (FFPI) sensor. Adapted from reference [36]. Copyright (2011) the authors. This article is distributed under the terms of the Creative Commons Attribution 2.0 International License.

13.5.2 Humidity Sensor

Nanowires made from polyacrylamide (PAM) have found application nanosensors for relative humidity detection, Gu et al. PAM nanowire was generated from a PAM aqueous solution with RH sensing based on RH-dependent evanescent power loss, [37]. At one end of the nanowire, a diameter of the 500 nm distal end of the tapered fiber fabricated from mode-one fiber is placed parallel and close to this polymer. Light may be effectively thrown into and taken up from the nanowire within a few micrometers of overlap because of the interaction that takes place in the nanowire against the tapered region of the fiber [38]. The sensor's rapid response can be due to the nanowire's tiny diameter and wide surface-to-volume ratio, which allows for quick diffusion or evaporation of water molecules and fast signal recovery utilizing an optical technique. RH detectors are demonstrated by Zhang et al. having gelatin plated nanowires in the subwavelength dimension ranger, as illustrated in Figure 13.11. The detection component included a nanowire of 680 nm diameter and tested at 1550 nm after coating it with 8 nm long, 80 nm wide film produced from gelatin [36]. When exposed or subjected to moisture, the variation in the gelatin layer's refractive index altered the coated fiber's mode field, converting a part of the power from guided mode to radiated mode, resulting in RH-dependent loss for optical sensing [39].

13.5.3 Chemical Sensor

Humidity and chemical gas sensors have similarities [40]. For example, because hydrogen gas lacks inherent absorption and or emissions that may be employed for light detection, an extra surface treatment is often adopted in the detection of hydrogen. Hydrogen detection, according to Villatoro et al., was achieved by coating the nanowire with a palladium sheet, and upon exposure to H_2 gas, the aggressive reaction between the palladium and the hydrogen witnessed a red shift from ~850 nm to ~1550 nm in the optical transmittance. High sensitivity and rapid reaction were produced by utilizing a thin nanowire to enhance the interaction between the coated platinum and the incident radiation. Polymer nanowires, which inherit the permeably selective character and biocompatibility of polymer materials, provide a variety of extremely appealing benefits for sensing applications as compared to glass or oxide nanowires [41]. Reports have revealed that a highly sensitive polymer single-nanowire optical sensor could also conduct quick response gas sensing. Before drawing the nanowire, solvated polymers and other doped functional components are mixed to make them suitable for optical sensing.

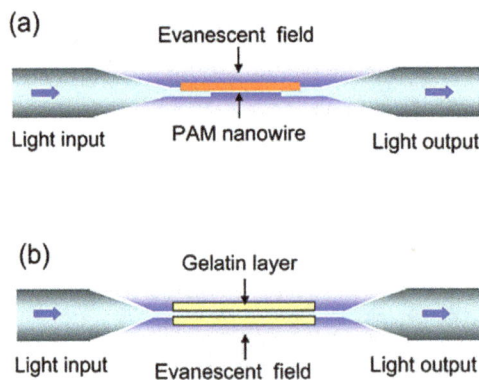

FIGURE 13.11 (a) Representation of humidity sensor made from nanowires of PAM and fitted with evanescent. (b) Optical fiber sensor coated with gelatin. Adapted from reference [36]. Copyright (2011) the authors. This article is distributed under the terms of the Creative Commons Attribution 2.0 International License.

13.6 NANOWIRES FOR FLEXIBLE PHOTONICS

Flexible photonic or electronic devices have gained advanced interest due to their new functionalities, compact size, and interesting photonic features. They have established enormous benefits compared to their rigid planar film counterparts [42], including low-cost fabrication via self-response roll-to-roll processing [43–46] and self-flexibility to conform with different materials [47]. As a result, different fabrication techniques have been explored to practically realize these flexible integrated electronic and photonic devices, which can be used for applications, such as wearable devices, flexible sensors, and rollable displays. Wearable device functionalities, in particular, require flexible materials, and as such, the need to study different optical or electronic materials needed to attain these functionalities is imperative. Similarly, flexible sensors also require unique materials to achieve their unique properties relevant for robotic skin, intelligent movement, and clothing applications [48,49].

These technologies employ the usage of one-dimensional (1D) nanowires as a result of their peculiar properties (electrical and optical). As stated earlier, nanowire properties depend on their physically confined structural parameters, such as diameters, lengths, and chemical compositions. Depending on their application, the latter categorized these nanowires into semiconducting, metallic, and dielectric. Metallic and dielectric nanowires, for instance, are suitable for plasmonic and all-optical photonics, dielectric, and metallic nanowires are preferred, and for sensing, light generation, etc., nanowires of semiconducting nature are employed [50]. Conventionally, the rigidity or flexibility of electronic or photonic devices inherently depends on the fabricated substrates. As such, soft materials that can mechanically deform (i.e., bending, folding, stretching, compressing, and twisting) without damage are key properties needed to achieve this flexibility or stretchability in optical or electronic systems.

Recently, there has been a surge in the development of flexible cell phones, which requires flexible light-emitting technologies. Though, organic light-emitting diodes (LEDs), hybrid perovskites, and colloidal quantum-dots have emerged as attractive technologies in the direction of flexible photonics applications; nonetheless, nanowires have demonstrated to be a plausible candidate for this technology due to their extraordinary chemical and mechanical stability as well as their wide bandgap tenability, and flexibility based on their selected underlying substrate [51–54]. As such, the combination of light-generating nanowires and suitable flexible substrates could play a crucial role in the flexible or stretchable electronic or photonic industry.

The approach used to realize these unique structures on flexible substrates is categorized into two (i.e., direct growth or transfer techniques). The direct growth approach is by growing the nanowires directly on a flexible substrate, while the second approach deals with using epitaxial lift-off techniques to transfer already grown nanowires on a flexible substrate. Until now, Wö Lz et al. have demonstrated the direct growth of GaN nanowires on Ti films deposited on sapphire due to Ti thermal management capability [55]. Zhao et al. also showed the possibility of growing red nanowire LEDs on Ti-coated bulk polycrystalline Mo substrate [56]. The different research works on GaN on diverse rigid substrates show the possibility of growing quality nanowires on metallic substrates.

However, for techniques such as roll-to-roll processing, substrates with flexible and stretchable properties are required and the prospect of using thin metal foils is among the state-of-the-art in flexible photonics and electronics applications. Per se, [57] reported on self-assembled AlGaN nanowires grown directly on Ta and Ti foils using the molecular beam epitaxy (MBE) technique as shown in Figure 13.12. Figure 13.12(a) shows a photograph of the GaN nanowires grown on a flexible Ti foil. Similarly Figures 13.12(b–c) show the scanning electron microscope (SEM) images of the nanowires grown on multiple grains of the Ti foil with its corresponding zoomed image of the Ti grain section. Figure 12(d) also represents nanowires grown on a single-crystalline Si substrate with its c-axis roughly perpendicular to the substrate. Figure 12(e) also portrays nanowires grown on a Ta foil. The nanowires are uniformly tilted to the surface normal, and they show approximately similar heights and radii on the Ti or the Ta foil as compared to the single-crystalline case, which is perpendicular.

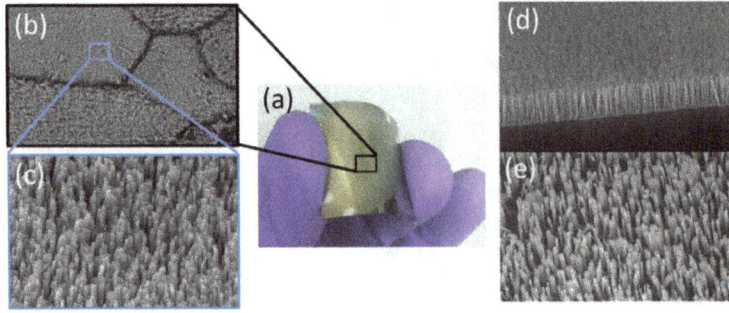

FIGURE 13.12 (a). Ti-foil flexible GaN nanowires (b-e), SEM of GaN nanowires on (b, c) different grains in the Ti foils, (d) Si single crystal substrate, and (e) Ta foil. Adapted with permission [57]. Copyright (2016) AIP Publishing.

Moreover, different characterizations of the fabricated samples such as photoluminescence (PL), electroluminescence (EL), and time-resolved lifetime measurements, were carried out. The sample showed a dominant neutral donor bound A exciton recombination at ~358 nm. The EL spectra demonstrate ultraviolet (UV) emission at ~350nm for the samples fabricated on the Ti foils depicting a good diode characteristic. Nevertheless, the time-resolved lifetime values were slightly different for the two substrates utilized in the work. Primarily, two metals were used in the latter example for the direct growth approach; however, a wider variety of metal foils compatible with a nanowire growth temperature and mechanical property could be instigated in these technologies.

In contrast to direct growth, the state-of-the-art in flexible photonics relies enormously on the transfer approach, which utilizes epitaxial lift-off of nanowires from an underlying substrate to another with investigations on their compatibility properties as well as the mechanical transferable property of the nanowires. In recent times, epitaxial lift-off processes, such as the transfer of individual nanowires on a plastic substrate, have been demonstrated [58]. Chung et al. [59] further illustrated flexible InGaN/GaN micro-rod LEDs by developing them on graphene-coated Si substrate, taken after by an exchange onto Ag-coated polymer substrate utilizing lift-off preparing procedures.

InGaN/GaN nanowires produce an LED display in blue when fabricated in the core/shell format [60]. This occurred in a large area and was highly flexible as illustrated in Figure 13.16. They exchanged InGaN nanowire clusters by implanting them in a flexible film, taken after by confining those nanowires from the previous sapphire substrate. Figure 13.13(a) represents the SEM image of the fabricated single nanowire showing an enlarged image in an artificial colors' system produced by MOCVD, which is comprised of a pristine n+GaN base and the InGaN/GaN core/shell region. The nanowires were mechanically immersed in a polymer (PDMS) and taken off from their original substrate. By adopting a nanowire network made from Ag, composite membranes are produced when coupled with transparent and conductive electrodes. Figure 13.13(b) illustrates a relatively broad EL emission spectra of the LED for different voltage biases, which could be relevant in rollable displays and Figure 13.13(c) shows the brightening of the LED under current injection. The bright blue emission of the device shows good performance even under mechanical tension, illustrating the potency of using such technologies in optoelectronic instruments such as light-emitting systems with optimally flexible tendencies.

13.7 CONCLUSION AND FUTURE REMARKS

In this chapter, nanowires, which are one-dimensional nanoscale structural elements have been examined. They offer excellent synthetic growth control over their dimensions, compositions, crystal

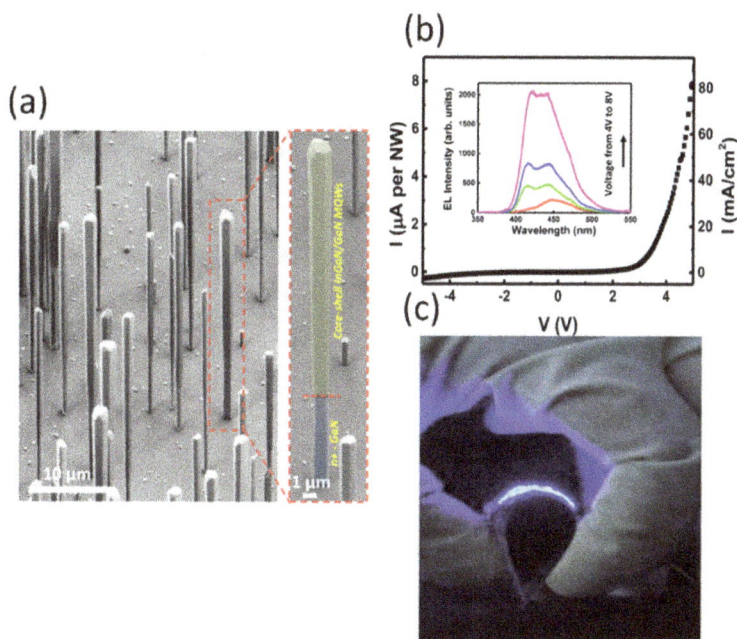

FIGURE 13.13 (a) SEM image of nanowires of the fabricated sample showing in artificial colors the bare n+-GaN (base) and the core/shell using the metal-organic chemical vapor deposition (MOCVD). (b) The I–V characteristic of the flexible LED normalized to the number of contacted nanowires (left scale) or the device surface (right scale). The inset presents EL spectra at various applied biases from 4 to 8 V. (c) Picture of blue emission of LED at 7.0 mm diameter curvature with further illustration on the vertical nanoarrays' mechanical properties. Adapted with permission [60]. Copyright (2015) American Chemical Society.

structures, and associated compositional and structural complexities, making them great candidates for downsizing photonics and optical devices and systems. The avalanche of nanowire-based high-performance miniaturized photonic devices and systems, including optical waveguides in-circuit integration platforms, LEDs, lasers, photodetectors, and flexible photonic devices, are proofs of their superior optical and electrical characteristics. Based on their compositions and electrical/optical characteristics, these nanowires may be classified as semiconducting, plasmonic, dielectric, or hybrid of any of these forms. Each type of nanowire exhibits unique optical and electrical properties in its generation, transmission, and optical confinement of optical waves and electrical signals. As optical waveguides, dielectric waveguides route tightly confined light within the nanowire; while plasmonic waveguides function with tightly confined localized surface plasmon polaritons on the metal surface and allow for enhancement of electromagnetic fields. This allowed for the advent of nanowire optical sensors. Effective interaction of photons and electron-hole pairs, and the good optical cavity functionality coupled with bandgap tunability of semiconductor nanowires paved the way for multicolored LEDs, and lasers. At length, a flexible substrate with the capability to bear extensive response when embedded with nanowire arrays and their application in flexible photonics can be enormous.

REFERENCES

[1] B. Culshaw, J.M. López-Higuera, Fundamentals of photonics, in: Optochemical Nanosensors, 2016: pp. 3–34.

[2] Y.-J. Lu, C.-Y. Wang, J. Kim, H.-Y. Chen, M.-Y. Lu, Y.-C. Chen, W.-H. Chang, L.-J. Chen, M.I. Stockman, C.-K. Shih, S. Gwo, All-color plasmonic nanolasers with ultralow thresholds: Autotuning mechanism for single-mode lasing, Nano Lett. 14 (2014) 4381–4388.

[3]   M. Gupta, P.K. Dhawan, S.K. Verma, R.R. Yadav, M. Gupta, P.K. Dhawan, S.K. Verma, R.R. Yadav, Diameter dependent ultrasonic characterization of InAs semiconductor nanowires, Open J. Acoust. 5 (2015) 218–225.

[4]   L.N. Quan, J. Kang, C.Z. Ning, P. Yang, Nanowires for photonics, Chem. Rev. 119 (2019) 9153–9169.

[5]   X. Guo, Y. Ying, L. Tong, Photonic nanowires: From subwavelength waveguides to optical sensors, Acc. Chem. Res. 47 (2014) 656–666.

[6]   L. Tong, J. Lou, E. Mazur, Single-mode guiding properties of subwavelength-diameter silica and silicon wire waveguides, Opt. Express. 12 (2004) 1025.

[7]   D. Sirbuly, A. Tao, M. Law, R. Fan, P.Y.-A. Materials, Multifunctional nanowire evanescent wave optical sensors, Wiley Online Libr. 19 (2007) 61–66.

[8]   L. Tong, E. Mazur, Glass nanofibers for micro- and nano-scale photonic devices, J. Non. Cryst. Solids. 354 (2008) 1240–1244.

[9]   L. Tong, L. Hu, J. Zhang, J. Qiu, Q. Yang, J. Lou, Y. Shen, J. He, Z. Ye, Photonic nanowires directly drawn from bulk glasses, Opt. Express. 14 (2006) 82.

[10] X.Lu, C. Wang, Y. Wei, One-dimensional composite nanomaterials: Synthesis by electrospinning and their applications, Small. 5 (2009) 2349–2370.

[11] L.N. Quan, J. Kang, C.Z. Ning, P. Yang, Nanowires for photonics, Chem. Rev. 119 (2019) 9153–9169.

[12] J. Hu, T.W. Odom, C.M. Lieber, Chemistry and physics in one dimension: Synthesis and properties of nanowires and nanotubes, Acc. Chem. Res. 32 (1999) 435–445.

[13] Y. Xia, P. Yang, Chemistry and physics of nanowires, Adv. Mater. 15 (2003) 351–352.

[14] K.S. Shankar, A.K. Raychaudhuri, Fabrication of nanowires of multicomponent oxides: Review of recent advances, Mater. Sci. Eng. C. 25 (2005) 738–751.

[15] P. Klason, K. Magnusson, O. Nur, Q.X. Zhao, Q.U. Wahab, M. Willander, Synthesis and structural and optical properties of ZnO micro- and nanostructures grown by the vapour-liquid-solid method, Phys. Scr. T. T126 (2006) 53–56.

[16] T. Arjmand, M. Legallais, T.T.T. Nguyen, P. Serre, M. Vallejo-Perez, F. Morisot, B. Salem, C. Ternon, Functional devices from bottom-up silicon nanowires: A review, Nanomater. 2022, Vol. 12, Page 1043. 12 (2022) 1043.

[17] J.C. Cardoso, M. Valnice, B. Zanoni, A.A. Ullah, H. Munir, M. Shahid, C. Balasubramanian, V.P. Godbole, V.K. Rohatgi, A.K. Das, S. V Bhoraskar, Synthesis of nanowires and nanoparticles, Inst. Phys. Publ. Nanotechnol. 15 (2004) 370–373.

[18] E. Barrigón, M. Heurlin, Z. Bi, B. Monemar, L. Samuelson, Synthesis and applications of III-V nanowires, Chem. Rev. 119 (2019) 9170–9220.

[19] L.D. Zhang, G.W. Meng, F. Phillipp, Synthesis and characterization of nanowires and nanocables, Mater. Sci. Eng. A. 286 (2000) 34–38.

[20] Z. Guoqiang, T. Masato, T. Kouta, T. Takehiko, N. Masaya, G. Hideki, Telecom-band lasing in single InP/InAs heterostructure nanowires at room temperature, Sci. Adv. 5 (2022) eaat8896.

[21] L. Gagliano, A. Belabbes, M. Albani, S. Assali, M.A. Verheijen, L. Miglio, F. Bechstedt, J.E.M. Haverkort, E.P.A.M. Bakkers, Pseudodirect to direct compositional crossover in wurtzite GaP/In$_x$Ga$_{1-x}$P core-shell nanowires, Nano Lett. 16 (2016) 7930–7936.

[22] T. Kuykendall, P. Ulrich, S. Aloni, P. Yang, Complete composition tunability of InGaN nanowires using a combinatorial approach, Nat. Mater. 6 (2007) 951–956.

[23] T. Mårtensson, C.P.T. Svensson, B.A. Wacaser, M.W. Larsson, W. Seifert, K. Deppert, A. Gustafsson, L.R. Wallenberg, L. Samuelson, Epitaxial III-V nanowires on silicon, Nano Lett. 4 (2004) 1987–1990.

[24] K. Tomioka, J. Motohisa, S. Hara, K. Hiruma, T. Fukui, GaAs/AlGaAs core multishell nanowire-based light-emitting diodes on Si, Nano Lett. 10 (2010) 1639–1644.

[25] J.-H. Lim, C.-K. Kang, K.-K. Kim, I.-K. Park, D.-K. Hwang, S.-J. Park, UV electroluminescence emission from ZnO light-emitting diodes grown by high-temperature radiofrequency sputtering, Wiley Online Libr. 18 (2006) 2720–2724.

[26] X. Fang, J. Li, D. Zhao, D. Shen, B. Li, X. Wang, Phosphorus-doped p-type ZnO nanorods and ZnO nanorod p-n homojunction LED fabricated by hydrothermal method, J. Phys. Chem. C. 113 (2009) 21208–21212.

[27] O. Lupan, T. Pauporté, B.V.-A. Materials, undefined 2010, Low-voltage UV-electroluminescence from ZnO-nanowire array/p-GaN light-emitting diodes, Wiley Online Libr. 22 (2010) 3298–3302.

[28] E. Lai, W. Kim, P. Yang, Vertical nanowire array-based light emitting diodes, Nano Res. 1 (2008) 123–128.

[29] J.O. Grepstad, J. Skaar, Total internal reflection and evanescent gain, Opt. Express. 19 (2011) 21404.

[30] M. Law, D.J. Sirbuly, J.C. Johnson, J. Goldberger, R.J. Saykally, P. Yang, Nanoribbon waveguides for subwavelength photonics integration, Science (80) 305 (2004) 1269–1273.

[31] S. Kim, R. Yan, Recent developments in photonic, plasmonic and hybrid nanowire waveguides, J. Mater. Chem. C. 6 (2018) 11795–11816.

[32] S. Kim, S. Bailey, M. Liu, R. Yan, Decoupling co-existing surface plasmon polariton (SPP) modes in a nanowire plasmonic waveguide for quantitative mode analysis, Nano Res. 10 (2017) 2395–2404.

[33] W. Wang, Q. Yang, F. Fan, H. Xu, Z.L. Wang, Light propagation in curved silver nanowire plasmonic waveguides, Nano Lett. 11 (2011) 1603–1608.

[34] L. Tong, E.M.-O.C. and M. II, Subwavelength-diameter silica wires for microscale optical components, Spiedigitallibrary.Org. 5723 (2005) 105.

[35] W. Liang, Y. Huang, Y. Xu, R.K. Lee, A. Yariv, Highly sensitive fiber Bragg grating refractive index sensors, Appl. Phys. Lett. 86 (2005) 1–3.

[36] L. Zhang, J. Lou, L. Tong, Micro/nanofiber optical sensors, Photonic Sensors. 1 (2011) 31–42.

[37] F. Gu, L. Zhang, H. Zeng, Polymer micro/nanofibre waveguides for optical sensing applications, Optoelectron. Mater. Devices, InTech, 2015.

[38] S. Sikarwar, B.C. Yadav, Opto-electronic humidity sensor: A review, Sensors Actuators, A Phys. 233 (2015) 54–70.

[39] G. Wypych, Physical properties of fillers and filled materials, Handb. Fill. (2021) 339–407.

[40] K.S. Johnson, J.A. Needoba, S.C. Riser, W.J. Showers, Chemical sensor networks for the aquatic environment, Chem. Rev. 107 (2007) 623–640.

[41] H. Liu, J. Kameoka, D.A. Czaplewski, H.G. Craighead, Polymeric nanowire chemical sensor, Nano Lett. 4 (2004) 671–675.

[42] R.H. Reuss, B.R. Chalamala, A. Moussessian, M.G. Kane, A. Kumar, D.C. Zhang, J.A. Rogers, M. Hatalis, D. Temple, G. Moddel, B.J. Eliasson, M.J. Estes, J. Kunze, E.S. Handy, E.S. Harmon, D.B. Salzman, J.M. Woodall, M.A. Alam, J.Y. Murthy, S.C. Jacobsen, M. Olivier, D. Markus, P.M. Campbell, E. Snow, Macroelectronics: Perspectives on technology and applications, Proc. IEEE. 93 (2005) 1239–1256.

[43] A.V. Shneidman, K.P. Becker, M.A. Lukas, N. Torgerson, C. Wang, O. Reshef, M.J. Burek, K. Paul, J. McLellan, M. Lončar, All-polymer integrated optical resonators by roll-to-roll nanoimprint lithography, ACS Photonics. 5 (2018) 1839–1845.

[44] R. Bruck, P. Muellner, N. Kataeva, A. Koeck, S. Trassl, V. Rinnerbauer, K. Schmidegg, R. Hainberger, Flexible thin-film polymer waveguides fabricated in an industrial roll-to-roll process, Appl. Opt. 52 (2013) 4510–4514.

[45] J.G. Ok, H. Seok Youn, M. Kyu Kwak, K.T. Lee, Y. Jae Shin, L. Jay Guo, A. Greenwald, Y. Liu, Continuous and scalable fabrication of flexible metamaterial films via roll-to-roll nanoimprint process for broadband plasmonic infrared filters, Appl. Phys. Lett. 101 (2012) 223102.

[46] H.L. Liang, M.M. Bay, R. Vadrucci, C.H. Barty-King, J. Peng, J.J. Baumberg, M.F.L. De Volder, S. Vignolini, Roll-to-roll fabrication of touch-responsive cellulose photonic laminates, Nat. Commun. 9 (2018) 4632.

[47] Y. Liu, M. Pharr, G.S.-A. nano, Lab-on-skin: a review of flexible and stretchable electronics for wearable health monitoring, ACS Publ. 11 (2017) 9614–9635.

[48] A. Palczynska, A. Prisacaru, P.J. Gromala, B. Han, D. Mayer, T. Melz, Towards prognostics and health monitoring: The potential of fault detection by piezoresistive silicon stress sensor, Microelectron. Reliab. 74 (2017) 165–172.

[49] S. Gong, W. Schwalb, Y. Wang, Y. Chen, Y. Tang, J. Si, B. Shirinzadeh, W. Cheng, A wearable and highly sensitive pressure sensor with ultrathin gold nanowires, Nat. Commun. 5 (2014) 3132.

[50] N.P. Dasgupta, J. Sun, C. Liu, S. Brittman, S.C. Andrews, J. Lim, H. Gao, R. Yan, P. Yang, N.P. Dasgupta, J. Sun, C. Liu, S. Brittman, S.C. Andrews, J. Lim, H. Gao, R. Yan, P. Yang, 25th anniversary article: Semiconductor nanowires – synthesis, characterization, and applications, Adv. Mater. 26 (2014) 2137–2184.

[51] X. Zhang, B. Xu, J. Zhang, Y. Gao, Y. Zheng, K. Wang, X.W. Sun, X.L. Zhang, B. Xu, K. Wang, X.W. Sun, Y. Gao, Y.J. Zheng, J.B. Zhang, All-inorganic perovskite nanocrystals for high-efficiency light emitting diodes: Dual-phase $CsPbBr_3$-$CsPb_2Br_5$ composites, Wiley Online Libr. 26 (2016) 4595–4600.

[52] B.R. Sutherland, E.H. Sargent, Perovskite photonic sources, Nat. Photonics. 10 (2016) 295–302.

[53] X. Li, F. Cao, D. Yu, J. Chen, Z. Sun, Y. Shen, Y. Zhu, L. Wang, Y. Wei, Y. Wu, H. Zeng, All inorganic halide perovskites nanosystem: Synthesis, structural features, optical properties and optoelectronic applications, Small. 13 (2017) 1603996.

[54] Y. Shirasaki, G.J. Supran, M.G. Bawendi, V. Bulović, Emergence of colloidal quantum-dot light-emitting technologies, Nat. Photonics. 7 (2013) 13–23.

[55] M. Wőlz, C. Hauswald, T. Flissikowski, T. Gotschke, S. Fernández-Garrido, O. Brandt, H.T. Grahn, L. Geelhaar, H. Riechert, Epitaxial growth of GaN nanowires with high structural perfection on a metallic TiN film, ACS Publ. 15 (2015) 3743–3747.

[56] C. Zhao, T.K. Ng, N. Wei, A. Prabaswara, M.S. Alias, B. Janjua, C. Shen, B.S. Ooi, Facile formation of high-quality InGaN/GaN quantum-disks-in-nanowires on bulk-metal substrates for high-power light-emitters, Nano Lett. 16 (2016) 1056–1063.

[57] B.J. May, A.T.M.G. Sarwar, R.C. Myers, Nanowire LEDs grown directly on flexible metal foil, Appl. Phys. Lett. 108 (2016) 141103.

[58] M.C. McAlpine, R.S. Friedman, S. Jin, K.H. Lin, W.U. Wang, C.M. Lieber, High-performance nanowire electronics and photonics on glass and plastic substrates, Nano Lett. 3 (2003) 1531–1535.

[59] K. Chung, H. Beak, Y. Tchoe, H. Oh, H. Yoo, M. Kim, G.C. Yi, Growth and characterizations of GaN micro-rods on graphene films for flexible light emitting diodes, APL Mater. 2 (2014) 092512.

[60] X. Dai, A. Messanvi, H. Zhang, C. Durand, J. Eymery, C. Bougerol, F.H. Julien, M. Tchernycheva, Flexible light-emitting diodes based on vertical nitride nanowires, Nano Lett. 15 (2015) 6958–6964.

14 Nanowires for Thermoelectrics

Ming Tan
College of Science, Henan Agricultural University, Zhengzhou, China

CONTENTS

14.1 INTRODUCTION

With the rapid development of the global economy at present, people's demand for energy is increasing, which has greatly affected the social and political fields. But fossil energy, as the world's main energy consumption resource (**Figure 14.1**), is non-renewable energy. While fossil energy resources are decreasing, they also cause serious pollution to the environment, which makes people have to find other ways to solve the environmental problem. Thermoelectric (TE) conversion, using the thermoelectric effect, is a very quiet and portable form of converting heat into electricity directly. The TE effect can be used to convert waste heat into electricity, which is an environmentally friendly way of producing electricity. A large amount of waste heat comes from household heat, automobile exhaust, industrial production, and other processes. According to the US Department of Energy, 60% of the energy used by human beings is discharged in the form of waste heat. The waste heat is transferred to electricity by the TE effect, however, it will have a profound impact on our society. The efficiency of TE material is evaluated by the dimensionless figure of merit ZT, which is expressed as $ZT=S^2\sigma T/\kappa$[1]. Here, κ is the thermal conductivity; σ is the electrical conductivity; S is the Seebeck coefficient; T was the temperature. Before the 1990s, TE materials were limited due to their low energy conversion efficiency. A bottleneck is encountered in the development of TE materials. In 1993, Dresselhaus *et al.* theoretically predicted that the TE properties of low-dimensional materials would be significantly improved by the quantum limit effect [2,3]. Thus the development of TE fields is in a new stage.

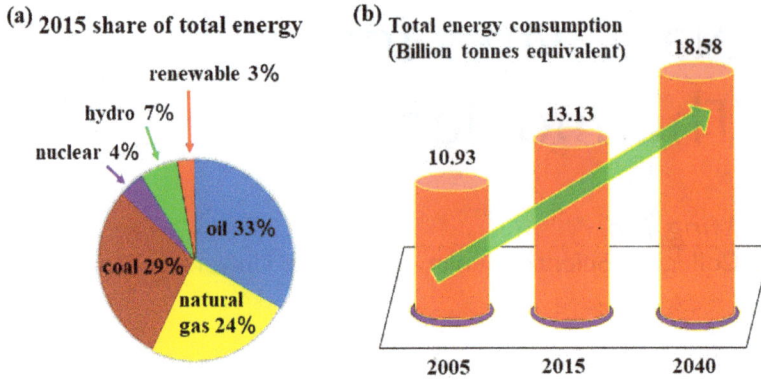

FIGURE 14.1   Energy problems. (a) Share of the total energy in 2015. (b) Statistic and estimated energy consumption by the source of 2005, 2015, 2040.

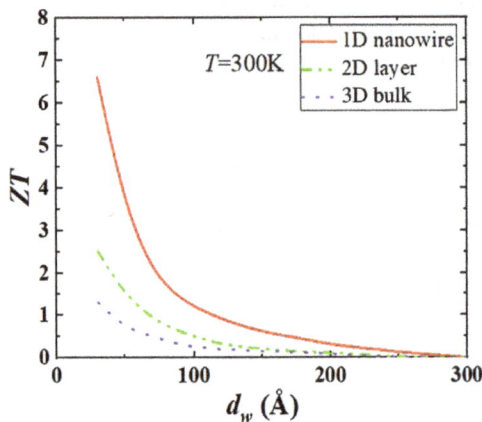

FIGURE 14.2   Dependence of ZT for 1D, 2D, and 3D TE materials on quantum well width d_w.

Because of its special structure, low-dimensional TE materials have unique TE properties, which are greatly different from corresponding bulk materials. Quantum theory calculation showed that as the material size decreases, and thermal conductivity of materials greatly reduces [2,3]. The figure-of-merit ZT of nanosuperlattice will achieve tremendous enhancement compared to the bulk material. The quantum state density near the Fermi level of nanowires (NWs) has greatly improved more than superlattices. Therefore, NWs have more excellent TE properties. **Figure 14.2** shows the relationship between the ZT value of one-dimensional (1D), two-dimensional (2D), three-dimensional (3D) thermoelectric materials, and the quantum well width d_w [4]. It can be seen that the ZT value for the lower dimension of materials is greater.

14.2   METHODS OF SYNTHESIS OF NWS

Due to the smaller size of NWs, the material has a stronger quantum confined effect, which leads to enhanced phonon scattering. At the same time, the density of states near the Fermi level of one-dimensional structure is much higher than that of two- and three-dimensional structures. So the TE performance of NWs is better. There are many ways to prepare NWs, as shown in **Figure 14.3**. The preparation methods of NWs mainly include a solvothermal method, physical vapor deposition, chemical vapor deposition, template method, laser cauterization, self-assembly technology,

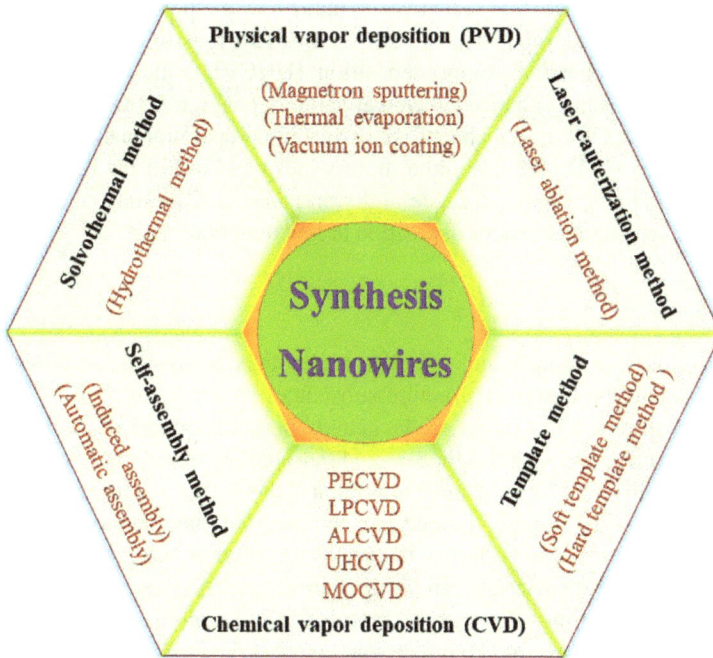

FIGURE 14.3 Some methods of synthesis of NWs.

electrochemical method, microwave-assisted method, *etc.* The following are several synthesis methods and their research progress.

14.2.1 SOLVOTHERMAL METHOD

The solvothermal method is one of the most important methods for the synthesis of NWs. It refers to a multiphase reaction process in which insoluble or undissolved substances are dissolved and recrystallized by heating a certain solvent as a reaction medium (water is used as a solvent in the hydrothermal method) in a closed container. And the temperature and pressure are generated in a closed container through heating. The principle of the solvothermal method is that the precursor is fully dissolved in the reaction system and reaches a certain supersaturation, to nucleate and crystallize and precipitate products under a special condition of high temperature and high pressure solvothermal. The morphology, crystal structure, and crystallinity of the product can be controlled by selecting and regulating the reaction medium, surfactant, temperature, time, pH value, and other parameters. At present, there is much research on the synthesis of NWs by a solvothermal method. Zhang *et al.* [5] and Yu *et al.* [6] prepared Bi_3S NWs through a simple one-step hydrothermal reaction, *etc.*

14.2.2 CHEMICAL VAPOR DEPOSITION

Chemical vapor deposition (CVD) is a new technology that is developed to prepare inorganic materials in recent decades. It refers to the introduction of the reactant vapors of raw materials into a reaction chamber under conditions far higher than the critical reaction temperature. Through a chemical reaction, the vapor atoms condense into crystal nucleation and grow up continuously, and gather to form solid sediments. The morphology, size, and crystal structure of products can be controlled by adjusting air pressure, gas flow rate, reaction temperature, and other conditions in the chemical vapor deposition process. In recent years, many new methods are improved and optimized based on traditional chemical vapor deposition. Such as low-pressure chemical vapor deposition (LPCVD), ultra-vacuum chemical vapor deposition (UHCVD), metal-organic chemical vapor deposition (MOCVD), plasma-enhanced chemical vapor deposition (PECVD), atomic layer chemical vapor deposition (ALCVD), *etc.* This provides specific preparation methods for different kinds of materials and makes CVD here more and more widespread use in the research field. Through bottom-up growth, O'Donnell *et al.* [7] successfully prepared Si NWs solar cells on glass substrates using plasma-enhanced chemical vapor deposition at low temperature, *etc.*

14.2.3 TEMPLATE METHOD

The template method is a simple and effective technology to synthesize NWs. The so-called template method, using the designed structure substances as templates, can induce and limit the nucleation and growth process of target products and obtain nanomaterials with a designed structure. The method can effectively control the size and morphology of products, and solve the problem of dispersion and stability, of which the most important is the design and utilization of templates [8]. The template method can be roughly divided into the hard template method and the soft template method. The hard template method refers to the use of the inner or outer surface of the preprepared rigid template so that the raw materials can occur chemical reaction and growth along the template by controlling the reaction time. The specific morphology of nanomaterials can be obtained after removing the template. The usual hard templates are nanoporous structure materials, such as porous aluminum oxide, carbon nanotubes, mesoporous silicon, and polymer esters [9]. The soft template method can prepare nanomaterials that are guided and controlled by ordered and special structures when amphiphilic molecules reach a certain concentration [10]. Soft templates mainly include surfactants, biomacromolecules, *etc.* Many one-dimensional nanomaterials have been developed by the template method.

14.2.4 LASER CAUTERIZATION

Laser cauterization, also known as laser ablation, is a method of synthesizing nanomaterials. The surface of the target materials is irradiated by a high-energy laser, which is rapidly heated up, evaporated, and produced a plasma effect. Then the plasmas cool and crystallize and grow into nanostructure materials. Generally speaking, the target material has a low reflectivity of a quasi-molecular laser. The short-wavelength laser is used to bombard the target material to make the material melt into plasmas. The plasmas were sputtered out from the surface of the material, which can prevent the condensation and aggregation of particles and obtain thin nanomaterials [11]. In this case, the role of the laser is mainly to provide a heat source, so that the target material can be evaporated quickly. The temperature required for the preparation of NWs is much higher than that of nanoparticles.

14.2.5 SELF-ASSEMBLY METHOD

The self-assembly method is a new technique in recent decades. The basic units, such as atoms, molecules, and clusters, can spontaneously connect and form the ordered aggregations through molecular recognition and weak non-covalent bond force under a thermodynamic equilibrium state. The aggregations possess thermodynamic stability and definite structure and peculiar properties [12]. When the self-assembly process occurs, single basic structural units will automatically combine to form a structure with an ordered or complex shape without external forces [13]. There are two ways of self-assembly processes: one is the automatic assembly of tiny particles to form one-dimensional nanomaterials under special conditions; the other is to use macromolecules and react with reactants, inducing the growth of one-dimensional nanostructures [13].

14.2.6 PHYSICAL VAPOR DEPOSITION

The surface of the material source (solid or liquid) is gasified into gaseous atoms or molecules or partially ionized into ions under vacuum conditions by physical vapor deposition (PVD) technology. And the thin film is deposited on a substrate surface under a low-pressure gas or plasma. The PVD technology is mainly divided into three categories: vacuum evaporation coating, vacuum sputtering coating, and vacuum ion coating. The technology has the advantages of a simple process, no pollution, uniform and compact film formation, and strong binding force with a matrix. For example, Tan *et al.* have developed Cu and Bi_2Te_3-based nanowire arrays on a large scale using magnetron sputtering and thermal evaporation deposition, respectively [14,15].

14.3 DEVELOPMENT OF TE NW MATERIALS

Although the TE effect was discovered in the 1820s, since then, the slow progress has remained for a century in the field. At that time, the research direction was mainly focused on metals, and the Seebeck coefficient of metals was only a few μV/K. After some time, the Seebeck coefficient of semiconductors was found to be much larger than that of metals. And Goldsmid's and Ioffe's research based on Bi_2Te_3 and PbTe TE materials led to a small upsurge in the TE field. The *ZT* value of the PbTe material was ~0.5 in the 1950s. A high *ZT*~1 in the Bi_2Te_3 material was obtained after 10 years [16]. These researches have also brought about the development of thermoelectricity in practical applications. Since then, however, the better properties of TE materials have not emerged, resulting in slow progress in TE research.

From the 1950s to the 1990s, the *ZT* value remained at about 1.0. To break the limit of *ZT*, the density of state (DOS) and phonon transport of materials are adjusted. One of the main methods is as follows: low-dimensional structures are used to make the unique density of state of materials, which is different from those of the bulk materials. In the 1990s, nanostructure and low-dimensional materials attracted much attention. The density of electronic states is greatly modulated by quantum confinement in these materials. Therefore, the quantum confinement effect of one-dimensional materials will be a very promising way to improve TE properties.

The study of one-dimensional nanomaterials is a hot spot in the TE materials field. Lots of one-dimensional nanomaterials including NWs, nanorods, nanotubes, and nanoribbons have been successfully synthesized. In 2002, Sander *et al.* prepared Bi_2Te_3 NWs by the porous anodic alumina oxide template using an electrodeposition method. The obtained NWs have high yield, high density, high aspect ratio, and are evenly distributed parallel to the template [17]. Jang *et al.* synthesized PbTe NWs with diameters of 100–150 nm by chemical vapor deposition. A single nanowire was used to establish a field-effect transistor, which proved that the PbTe nanowire was *n*-type material. And the electrical transport parameters of the NWs were also calculated by a one-dimensional electron transport model. Its Seebeck coefficient is $S = -72$ μV/K at room temperature [18]. Zhang *et al.* produced Bi_2Te_3 NWs by facile solution phase method in two steps. First, ultrafine Te nanowire with a diameter of 5 nm was synthesized, then Bi precursor solution was added. After contact between Bi atoms and Te nanowire, they diffused with each other and reacted to form an *n*-type Te-rich Bi_2Te_3 nanowire. The diameter of the Bi_2Te_3 NWs is about 8 nm. The *n*-type Bi_2Te_3 NWs retained their original nanostructures after spark plasma sintering (SPS). A large number of nanoscale interfaces in the bulk materials enhanced phonon scattering and reduced the thermal conductivity. Compared with the best commercial *n*-type $Bi_2Te_{2.7}Se_{0.3}$ bulk material, the *ZT* value (0.96) of the nanowire SPS material is higher than 13% [19], as shown in **Figure 14.4**. Boukai *et al.* prepared Si nanowire arrays by superlattice nanowire pattern transfer process (SNAP). The thermoelectric performance of Si NWs is approximately 100 times as high as that of bulk Si materials, and the *ZT* can reach ~1.0 at 200 K [20]. Mehta *et al.* found that the conductivity of Sb_2Se_3 NWs was at least a 10^4-fold improvement over that of bulk Sb_2Se_3 materials using the microwave synthesis method [21]. Mannam *et al.* showed that BiSbTe NWs with (015) preferred orientation were prepared by electrodeposition

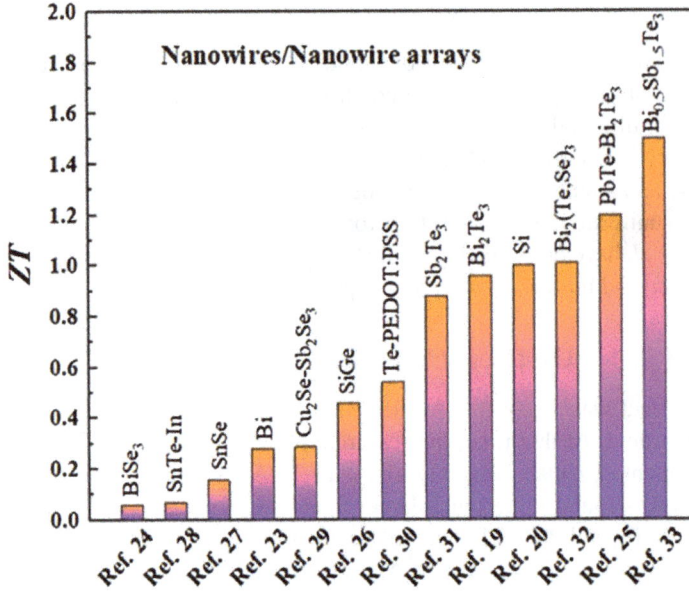

FIGURE 14.4 *ZT* of TE NW materials.

and template-assisted method. And a very high TE coefficient $S = -630\ \mu V/K$ was achieved in the BiSbTe NWs [22]. Kim *et al.* have investigated the individual single-crystalline Bi NWs, which were grown by the on-film formation of NW technology. The maximum *ZT* value is 0.28 for the Bi nanowire with a diameter of 109 nm [23]. Dedi *et al.* reported that the topological insulators' Bi_2Se_3 NWs were successfully prepared by the stress-induced growth method. The *ZT* value of Bi_2Se_3 NWs with a diameter of 200 nm is about 0.06 at room temperature [24]. Fang *et al.* demonstrated the modulated telluride nanowire heterostructures, which contain PbTe and Bi_2Te_3 via the solution-phase synthesis. That is, the PbTe-Bi_2Te_3 "barbell" nanowire heterostructures were developed by the solution-phase one-pot three-step reaction. The largely improved TE figure of merit of 1.2 at 620 K was achieved, owing to a greatly reduced thermal conductivity [25]. Lee *et al.* presented a large improvement in *ZT* values of ~0.46 for SiGe NWs at 450 K. The *ZT* enhancement is mainly attributed to reduced thermal conductivity. The low thermal conductivity of SiGe NWs was obtained by alloying Si with Ge and thin nanowire size effect. This kind of alloying and nanowire structure can effectively scatter a broad range of phonons and limit phonons' transport, respectively [26]. Hernandez *et al.* developed the SnSe NWs by the vapor-liquid-solid (VLS) process for the first time. The *ZT* of 130 nm diameter nanowire SnSe material is ~0.156 at 370 K. An obvious increase in *ZT* values was observed with decreasing nanowire diameter [27]. Li *et al.* reported the first growth of In-doped SnTe NWs using the chemical vapor deposition method. The *ZT* of the SnTe nanowire doped by ~1% of indium is 0.07, which is higher than that of the undoped SnTe nanowire. The *ZT* value of the indium-doped nanowire was enhanced by improving thermopower and suppressing thermal conductivity [28]. Kim *et al.* have synthesized Cu_2Se and Sb_2Se_3 NWs, respectively, using hydrothermal reactions and water-evaporation-induced self-assembly. The nanowire mixture containing 30% Cu_2Se and 70% Sb_2Se_3 NWs exhibited a maximal *ZT* value of 0.288 at 473 K [29]. Yang *et al.* systematically investigated the single core/shell nanowire of Te-PEDOT: PSS. The *ZT* value increases with decreasing nanowire diameter, and the hybrid nanowire with a diameter of 42 nm is 0.54 at 400 K. This work provided valuable insights into the decoupling of charge and heat transport, which indicated electrical transport through the organic shell and thermal transport by the inorganic core [30]. Tan *et al.* demonstrated that the *p*-type Sb_2Te_3 nanowire arrays have been self-assembled using the simple vacuum thermal evaporation method. A TE dimensionless $ZT = 0.88$ in the nanowire arrays was achieved at room temperature. The ZT value is likely to rise even higher with NW diameter reduction and composition optimization [31]. They further reported that the ternary compound n-type $Bi_2(Te,Se)_3$ nanowire array has been successfully prepared by the thermal co-evaporation technique. The $Bi_2(Te,Se)_3$ nanowire array with a ZT value of 1.01 was obtained at room temperature [32]. In 2020, Tan *et al.* used a thermal evaporation deposition to prepare $Bi_{0.5}Sb_{1.5}Te_3$ nanowire array film. Subsequently, the crystal-amorphous hybrid $Bi_{0.5}Sb_{1.5}Te_3$ nanowire array films were fabricated by post-annealing treatment. An optimized hybrid amorphous-crystalline $Bi_{0.5}Sb_{1.5}Te_3$ possessed an ultrahigh ZT value ~1.5 at room temperature, which indicates that crystal-amorphous compositing can be used as a new methodology to achieve ultrahigh properties in thermoelectric materials [33]. Theoretical and experimental studies have shown that the nanowire structure of thermoelectric materials can greatly improve the ZT value of materials. Therefore, this is an important way to achieve a breakthrough in the performance of thermoelectric materials and practical application.

14.4 DEVELOPMENT OF TE POLYMER-BASED NANOCOMPOSITES WITH INORGANIC NWS

In recent years, an explosion of growth has occurred in the smart electronics field due to the rising need for wearable devices. The development of wearable devices with small sizes and flexible characteristics and lightweight is a tendency so that the comfortability of wearing on the human body can be improved [34]. Therefore, a flexible lightweight power supply is desired for these ultraportable feature electronic products. The energy conversion TE polymer-based composites and devices have been extensively developed recently.

N-type TE Ag_2Te nanowire and poly (methyl methacrylate) composites were fabricated using hot pressing. The monoclinic Ag_2Te NWs were synthesized by a topotactic reaction of Te NWs. Then the surfaces of poly (methyl methacrylate) beads were uniformly coated by Ag_2Te NWs. The nanocomposites were formed using the hot compaction technique. Finally, the nanocomposite containing 50 wt% Ag_2Te NWs has achieved a ZT of 5×10^{-3} at room temperature [35], as shown in **Figure 14.5**. The composites based on carbon nanotubes (CNTs) and polymer were prepared by

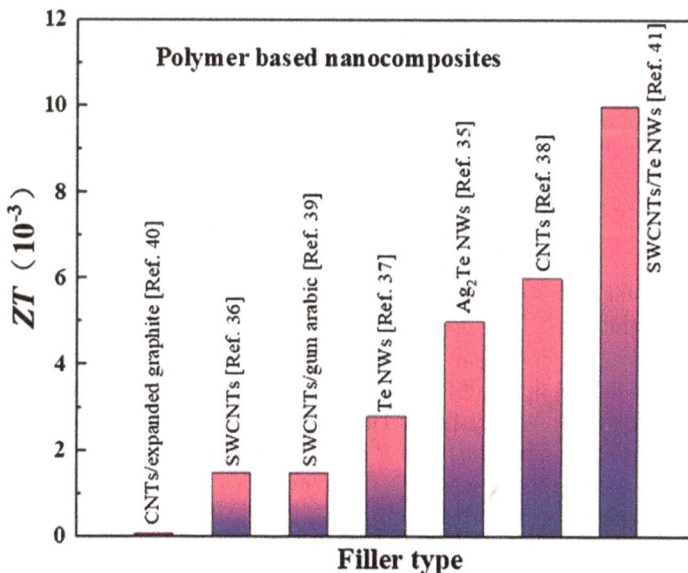

FIGURE 14.5   *ZT* of TE polymer-based nanocomposite materials.

over-coating the dense and dispersed CNTs solution. The ZT value along the in-plane direction of composites is 1.5×10^{-3} at room temperature. The results indicate that the performance of devices can be improved by controlling the orientation of CNTs in composites [36]. The composites based on tellurium NWs and thermoplastic polymer were fabricated by a simple hot compaction method. The thermoelectric ZT value of the composite with 50 wt% tellurium NWs content was $\sim 2.8 \times 10^{-3}$ at room temperature. The aligning NWs' TE materials are segregated in a polymer beads' matrix, which benefits considerable improvement in the properties of polymer-based composites [37]. The carbon nanotubes were a segregated network in the polymer-based composites. This network of CNTs in the composite can dramatically increase the electrical conductivity while remaining relatively insensitive to the thermal conductivity. This study shows that the composite is thermally disconnected but electrically connected. The composite with a CNT concentration of 20 wt % exhibits $ZT > 6 \times 10^{-3}$ at room temperature [38]. The single-walled carbon nanotubes (SWCNTs) and the gum arabic were dispersed into the polymer. The transport properties of SWCNT-filled nanocomposites were investigated. The surface of nanotubes creates an interfacial barrier, which alters the thermal and electrical transport behaviors. The nanocomposites have achieved a ZT of 1.5×10^{-3} at room temperature. It provides an excellent opportunity for increasing the TE figure of merit [39]. It is the most effective to enhance the properties of the conductive polymer composites using the eGR-CNT hybrid fillers. The ZT of the composite reaches 7×10^{-5} at room temperature, which is attributed to the ability of the CNT network [40]. A TE nanocomposite was composed of polymethyl methacrylate microbeads and tellurium NWs and single-walled carbon nanotubes using hot compaction via solvent casting. The nanocomposite with 48 wt% Te NWs and 2 wt% SWCNTs exhibited a high ZT of 1.0×10^{-2} at room temperature. This is because heterostructured interfaces were formed between the SWCNTs and the Te NWs in the nanocomposite [41].

14.5   STRUCTURE AND MEASUREMENT OF NWS AND NW DEVICES

The previous results have shown that NW structure could selectively scatter phonons more than carriers due to unique NW structure (**Figure 14.6a-h**), resulting in an enhanced ZT value of Bi_2Te_3-based alloys [32,42]. The thin $Bi_2(Te,Se)_3$ NWs with a diameter of ~ 20 nm were densely grown perpendicular to the substrate, along their preferential growth direction, using a simple thermal co-evaporation method (**Figure 14.6g,h**). The change of the Fermi level of the $Bi_2(Te,Se)_3$ has been induced by unique nanowire array structuring, which could favorably influence the phonon and carrier transport, leading to greatly improving an in-plane ZT value [32]. To overcome further a problem of phonon and carrier transport along the in-plane direction, the $(Sb,Bi)_2Te_3$ nanowire array with a tilted angle of 45, 60, and 90° have been fabricated by a simple thermal evaporation technique, as shown in **Figure 14.6a-f**. There exists a large number of interfaces in the nanowire array. Many interspaces between NWs are also found, however, the adjacent NWs have very good contact with each other, guaranteeing carriers' transport in the NW film. The results show that the carrier mobility of tilted growth NWs with a diameter of < 20 nm increases as the tilted angle of NWs decreases along the in-plane direction [42]. Ref. [43] shows that the hybrid Te-PVP NWs are uniform surface morphologies. The core-shell structure of the Te crystalline core is coated by the conducting polymer PEDOT: PSS shell. A typical way for nanowire measurement is the four-point contacts' method described in detail in Ref. [20]. When a temperature difference (ΔT) is applied to the sample, generating an open voltage (ΔV). The Seebeck coefficients of the nanowire material can be calculated by $S = \Delta V / \Delta T$. The resistance R of the nanowire can be obtained by measuring the I-V characteristics. According to the known NW dimensions (area A, length L), the electrical conductivities of the nanowire can be estimated by $\sigma = L/(A \cdot R)$. Finally, the thermal conductivities of the nanowire material can be determined by measuring the temperature difference between the hot side and cold side of the sample. Similar methods, such as the two-point technique, have also been widely employed in many kinds of research. Besides, the thermal conductivity of materials can be

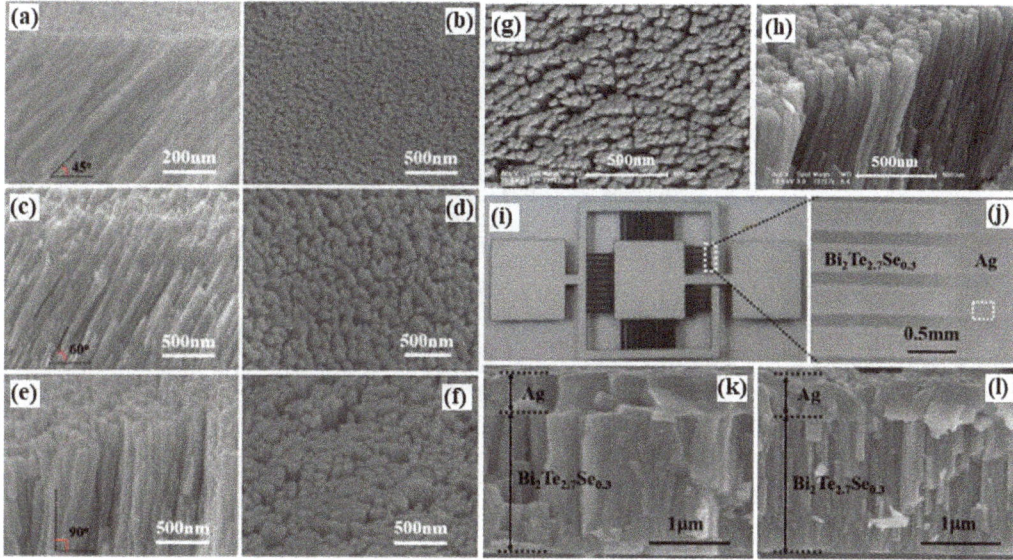

FIGURE 14.6 Structure and measurement of NWs and NW devices, (a-f) SEM images of p-(Sb,Bi)$_2$Te$_3$ nanowire arrays. Adapted with permission [42]. Copyright (2018), Elsevier; (g, h) SEM images of ordered n-Bi$_2$(Te,Se)$_3$ nanowire arrays. Adapted with permission [32]. Copyright (2014), Elsevier; (i-l) SEM image of the electrode-related TE leg and photograph of planar Bi$_2$Te$_{2.7}$Se$_{0.3}$ nanowire array device. Adapted with permission [45]. Copyright (2014), Elsevier.

measured by the 3ω method, such as the thermal conductivity of bulk crystal and thin film materials [44]. The size of the 38 n-legs parallel devices is 22 × 10 mm^2 in area, as shown in **Figure 14.6i-l**. The 38 n-Bi$_2$Te$_{2.7}$Se$_{0.3}$ nanowire legs were grown to a thickness of 2 μm using the mask. Then the Ag electrode was sputtered onto the substrate with the TE legs using the electrode mask. The ordered TE NW structure and layered electrode structure were introduced into the planar micro-devices, synergistically improving the performance of devices [45]. The size of the 4 n-legs' parallel generators is 8 × 22 mm^2, as shown in Ref. [46]. The 8 × 3 mm^2 n-InAs nanowire legs were performed on the substrate using the mask. The device is further sealed into a flexible PDMS enclosure. This is a promising route for the introduction of semiconductor NWs into flexible thermoelectrics electronics.

14.6   TE EFFECT OF NANOWIRE DEVICES

The TE effect describes the physical phenomenon of direct conversion between heat and electric energy, including the Seebeck effect, Peltier effect, and Thomson effect. Two different metals form a closed loop that generates an electric current when there is a temperature difference between the two joints. This phenomenon is known as the Seebeck effect. That is, the carriers in the metals or semiconductors can be free to move. A temperature difference is applied to both ends of the material, and the carriers in the hot end will diffuse to the cold end. The positive and negative charges will respectively accumulate at both ends, resulting in a potential difference. When a current flows through the junction of two different conductors, it will lead to a temperature change at the junction. The phenomenon is known as the Peltier effect. Thomson effect refers to the endothermic or exothermic phenomenon when a current is applied to a uniform conductor under a temperature gradient.

The TE NWs are not used to obtain the high superposition voltage by connecting all NWs in series. It is impossible to connect each nanowire in series due to the limitations of IC technology. If a connection is in series, the internal resistance of the device will become greatly large, leading

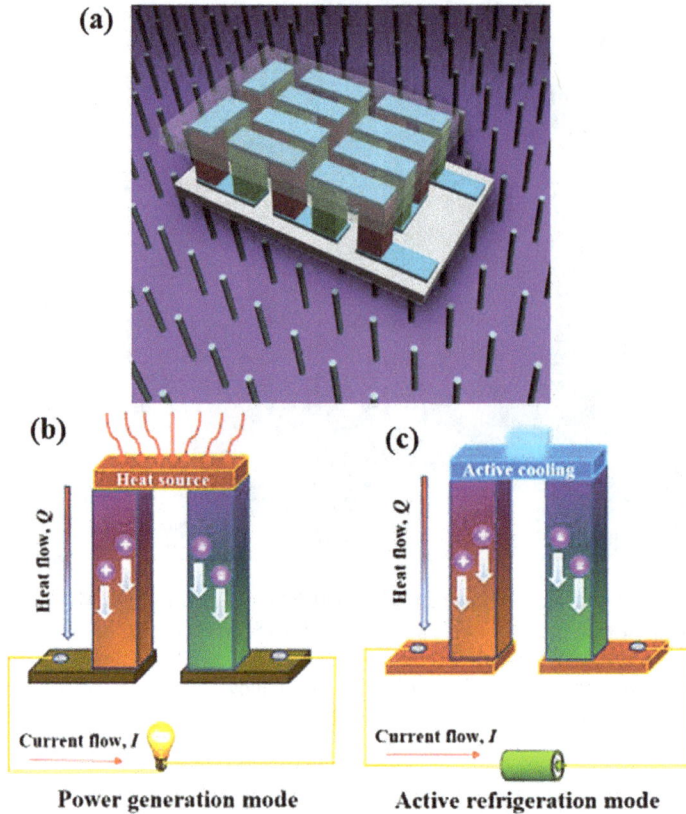

FIGURE 14.7   Schematic diagram of (a) TE NW device and (b) generator and (c) cooler.

to reduction of the TE conversion efficiency. However, it must be connected in series to obtain high output voltage. To solve this problem, a unique structure needs to be designed. That is, all NWs are firstly connected in parallel in each micro-area zone. Then each micro-area zone is connected in series, as shown in **Figure 14.7a**.

In general, TE devices are composed of an *n*-type and a *p*-type semiconductor component, which are connected by conductors (such as copper) to form thermocouples. When a temperature difference is applied to the device, the temperature difference drives the carriers (electrons or holes) in the material to spread from the hot end to the cold end, resulting in a current in the circuit, which is a TE generator (**Figure 14.7b**). When a current (voltage) in the direction shown in **Figure 14.7c** is applied to the device, the carriers move down from above and converge to the bottom end. At the same time, the heat also diffuses from the top junction to the bottom, causing the top junction to produce a cooling effect. The hot and cold ends are switched accordingly when the current is reversed. The TE generator used in practical applications is realized by connecting a large number of such thermocouples. The thermocouples are connected electrically in series and thermally in parallel. The device with a heat source at the top and a heat sink at the bottom can generate voltage by the temperature difference.

14.7   DEVELOPMENT OF TE NW DEVICES

TE power generation and refrigeration are two forms of TE devices in TE energy conversion. The existence of temperature difference is the most basic working condition of TE devices. Therefore,

TABLE 14.1

Performance of NW Devices

Micro-devices	ΔT (K)	Power/Power Density	Refs.
$Bi_2Te_{2.7}Se_{0.3}$ NWs	65	16.02 μW	[45]
InAs NWs	5	0.44 nW	[46]
Si NWs	56	29.3 μW	[47]
SiGe NWs	~80	45.2 μW/cm^2	[48]
Si NWs-Bi_2Te_3 NPs	~160	~1 nW	[49]
Bi_2Te_3 NWs-PEDOT: PSS	~60	130 μW	[50]

TE materials must have good electrical conductivity and poor thermal conductivity, resulting in high TE conversion efficiency. TE conversion efficiency is defined as the ratio of the output power to the absorbed heat of a TE unit. The output power is larger and the absorbed heat is less, so the TE conversion efficiency is higher in the devices.

Tan *et al.* have designed and fabricated a 38 *n*-$Bi_2Te_{2.7}Se_{0.3}$ NWs array leg micro-device using a magnetron sputtering and mask-assisted deposition method. The maximum power of the 38 leg device was up to 16.02 μW for a temperature difference of 65 K (see **Table 14.1**). The introduction of such ordered NWs into planar micro-devices is therefore an effective approach to improve device performance [45]. Koskinen *et al.* have reported that a thermoelectric InAs NW generator with four 8 × 3 mm^2 legs could produce a maximum power of 0.44 nW at a temperature difference of 5 K, which indicates a promising method for the development of flexible TE NWs applications [46]. Curtin *et al.* have presented that a 50 μm × 50 μm Si NW-based device could generate a maximum power of 29.3 μW at a temperature difference of 56 K. The power generation would be increased using higher packing densities of thin Si NWs [47]. Noyan *et al.* have manufactured a planar TE micro-generator (μTEG) with a power density of about 45 μW/cm^2 at a temperature difference of ~80 K. The μTEG performance was significantly increased using the heat exchanger [48]. Hsin *et al.* have demonstrated that the output power of the nanocomposite based on Si NWs and Bi_2Te_3 nanoparticles reached a maximum of ~1 nW at a temperature difference of about 160 K. This approach has provided an applicable route to developing high-performance TE nanocomposites at micro-scale for future thermal cooling and scavenging in electronics [49]. Thongkham *et al.* have demonstrated the TE nanofilms' application based on hybrid NWs of topological Bi_2Te_3 within the conductive poly (3,4-ethylenedioxythiophene): polystyrenesulfonate (PEDOT: PSS) matrix. The device exhibited an unprecedented output power of 130 μW at a temperature difference of ~60 K. The findings provide a route for the enhancement of TE generations based on the topological NWs [50].

14.8 CONCLUSIONS

TE NW materials and micro-devices have achieved excellent results. However, the various promising *p*-type and *n*-type thermoelectric NW materials still have problems. It is hoped that the output power of TE devices can be improved by enhancing the properties of NW materials in the future. It is expected to be promoted and used in more applications. It is not difficult to see that the TE properties of some NW materials have been able to meet the requirements for low-temperature power generation devices. But there is still a considerable gap in the medium- and high-temperature NW materials. Therefore, the key to research is how to improve the thermal stability and TE properties of NWs for the corresponding temperature. It is believed that more and more research resources are devoted to TE technology. All of these considerations can be systematically advanced and resolved. The TE technology is also expected to use in the development and breakthrough in the power generation and the active refrigeration fields.

The most studied Bi_2Te_3-based NWs attain the best *ZT* of ~1.5 via the optimization of nanowire structure and preferential orientation. The design and fabrication of suitable nanowire structures by defect engineering are indispensable for obtaining TE nanowire materials in the further, greatly enhanced *ZT* values. TE nanowire materials can be fabricated as flexible micro-devices. In general, the cross-plane devices are favorable for cooling applications, while the in-plane devices are beneficial for power generation. It is a key to improving the contact electrical and thermal resistances of micro-devices. To obtain high output voltage and conversion efficiency, a unique structure device needs to be designed. All NWs are firstly connected in parallel in each micro-area zones. Then each micro-area zone is connected in series in micro-devices.

In recent years, an explosion of growth has occurred in the smart electronics field due to the rising need for wearable devices. TE organic and inorganic hybrid nanocomposites are very important for wearable electronic devices. However, the TE properties of organic-inorganic hybrid materials need to be improved at present. It is an effective way to enhance the TE properties of inorganic-organic nanocomposites using the inorganic nanowire structure.

REFERENCES

[1]   M. Tan, X. L. Shi, W. D. Liu, M. Li, Y. L. Wang, H. Li, Y. Deng, Z. G. Chen. Synergistic texturing and Bi/Sb-Te anti-site doping secure high thermoelectric performance in $Bi_{0.5}Sb_{1.5}Te_3$-based thin films. Advanced Energy Materials, 11, 2021, 2102578.

[2]   L. D. Hicks, M. S. Dresselhaus. Effect of quantum-well structures on the thermoelectric figure of merit. Physical Review B, 47, 1993, 12727–12731.

[3]   L. D. Hicks, M. S. Dresselhaus. Thermoelectric figure of merit of a one-dimensional conductor. Physical Review B, 47, 1993, 16631–16634.

[4]   M. S. Dresselhaus, G. Dresselhaus, X. Sun, Z. Zhang, S. B. Cronin, T. Koga. Low-dimensional thermoelectric materials. Physics of the Solid State, 41, 1999, 679–682.

[5]   H. Zhang, Y. Ji, X. Ma, J. Xu, D. Yang. Long Bi_2S_3 nanowires prepared by a simple hydrothermal method. Nanotechnology, 14, 2003, 974.

[6]   S. H. Yua, Y. T. Qian, Y. H. Zhang. A solvothermal decomposition process for fabrication and particle sizes control of Bi_2S_3 nanowires. Journal of Materials Research, 14, 1999, 4157–4162.

[7]   B. O'Donnell, L. Yu, M. Foldyna, P. R. Cabarrocas. Silicon nanowire solar cells grown by PECVD. Journal of Non-Crystalline Solids, 358, 2012, 2299–2302.

[8]   S. K. Chakarvarti, J. Vetter. Template synthesis-a membrane based technology for generation of nano-/micro-materials: a review. Radiation Measurements, 29, 1998, 149–159.

[9]   J. C. Hulteen, C. R. Martin. A general template-based method for the preparation of nanomaterials. Journal of Materials Chemistry, 7, 1997, 1075–1087.

[10]   E. Gazit. Use of biomolecular templates for the fabrication of metal nanowires. Febs Journal, 274, 2007, 317–322.

[11]   A. M. Morales, C. M. Lieber. A laser ablation method for the synthesis of crystalline semiconductor nanowires. Science, 279, 1998, 208–211.

[12]   G. M. Whitesides, J. P. Mathias, C. T. Seto. Molecular self-assembly and nanochemistry: A chemical strategy for the synthesis of nanostructures. Science, 254, 1991, 1312–1319.

[13]   G. M. Whitesides. Self-assembling materials. Sci. Am., 273, 1995, 114–117.

[14]   M. Tan, X. Z. Wang, Y. M. Hao, Y. Deng, Strikingly enhanced cooling performance for a micro-cooler using unique Cu nanowire array with high electrical conductivity and fast heat transfer behavior. Chem. Phys. Lett., 678, 2017, 40–45.

[15]   M. Tan, Y. M. Hao, Y. Deng, D.L. Yan, H. Li, Z. H. Wu, Tilt-structure and high-performance of hierarchical Bi1.5Sb0.5Te3 nanopillar arrays. Scientific Reports, 8, 2018, 6384.

[16]   J. P. Heremans, M. S. Dresselhaus, L. E. Bell, D. T. Morelli. When thermoelectrics reached the nanoscale. Nature Nanotechnology 8, 2013, 471.

[17]   M. S. Sander, A. L. Prieto, R. Gronsky, T. Sands, A. M. Stacy. Fabrication of high-density, high aspect ratio, large-area bismuth telluride nanowire arrays by electrodeposition into porous anodic alumina templates. Advanced Materials, 14, 2002, 665–667.

[18] S. Y. Jang, H. S. Kim, J. Park, M. Jung, J. Kim, S. H. Lee, J. W. Roh, W. Lee. Transport properties of single-crystalline n-type semiconducting PbTe nanowires. Nanotechnology, 20, 2009, 415204.

[19] G. Q. Zhang, B. Kirk, L. A. Jauregui, H. Yang, X. Xu, Y. P. Chen, Y. Wu. Rational synthesis of ultrathin n-type Bi_2Te_3 nanowires with enhanced thermoelectric properties. Nano. Lett., 12, 2012, 56–60.

[20] A. I. Boukai, Y. Bunimovich, J. Tahir-Kheli, J. K. Yu, W. A. Goddard, J. R. Heath. Silicon nanowires as efficient thermoelectric materials. Nature, 451, 2008, 168–171.

[21] R. J. Mehta, C. Karthik, W. Jiang, B. Singh, Y. F. Shi, R. W. Siegel, T. B. Tasciuc, G. Ramanath. High electrical conductivity antimony selenide nanocrystals and assemblies. Nano Lett., 10, 2010, 4417–4422.

[22] R. S. Mannam, D. Davis. High Seebeck coefficient BiSbTe nanowires. Electrochemical and Solid State Letters, 13, 2010, 15–17.

[23] J. Kim, S. Lee, Y. M. Brovman, P. Kim, W. Lee. Diameter-dependent thermoelectric figure of merit in single-crystalline Bi nanowires. Nanoscale, 7, 2015, 5053.

[24] Dedi, P. C. Lee, P. C. Wei, Y. Y. Chen. Thermoelectric characteristics of a single-crystalline topological insulator Bi_2Se_3 nanowire. Nanomaterials, 11, 2021, 819.

[25] H. Fang, T. Feng, H. Yang, X. Ruan, Y. Wu. Synthesis and thermoelectric properties of compositional modulated lead telluride-bismuth telluride nanowire heterostructures. Nano Lett., 13, 2013, 2058–2063.

[26] E. Lee, L. Yin, Y. Lee, J. Lee, S. Lee, J. Lee, S. Cha, D. Whang, G. S. Hwang, K. Hippalgaonkar, A. Majumdar, C. Yu, B. Choi, J. Kim, K. Kim. Large thermoelectric figure-of-merits from SiGe nanowires by simultaneously measuring electrical and thermal transport properties. Nano Lett., 12, 2012, 2918–2923.

[27] J. Hernandez, A. Ruiz, L. F. Fonseca, M. Pettes, M. Jose-Yacaman, A. Benitez. Thermoelectric properties of SnSe nanowires with different diameters. Scientific Reports, 8, 2018, 11966.

[28] Z. Li, E. Xu, Y. Losovyj, N. Li, A. Chen, B. Swartzentruber, N. Sinitsyn, J. Yoo, Q. Jia, S. Zhang. Surface oxidation and thermoelectric properties of indium-doped tin telluride nanowires. Nanoscale, 9, 2017, 13014.

[29] M. Kim, D. Park, J. Kim. Synergistically enhanced thermoelectric performance by optimizing the composite ratio between hydrothermal Sb_2Se_3 and self-assembled beta-Cu_2Se nanowires. Cryst Eng Comm., 23, 2021, 2880–2888.

[30] L. Yang, M. P. Gordon, A. K. Menon, A. Bruefach, K. Haas, M. C. Scott, R. S. Prasher, J. J. Urban. Decoupling electron and phonon transport in single-nanowire hybrid materials for high-performance thermoelectrics. Sci. Adv., 7, 2021, eabe6000.

[31] M. Tan, Y. Deng, Y. Wang. Unique hierarchical structure and high thermoelectric properties of antimony telluride pillar arrays. J. Nanopart. Res., 14, 2012, 1204.

[32] M. Tan, Y. Deng, Y. Wang. Ordered structure and high thermoelectric properties of $Bi_2(Te,Se)_3$ nanowire array. Nano Energy, 3, 2014, 144–151.

[33] M. Tan, W. D. Liu, X. L. Shi, J. Shang, H. Li, X. B. Liu, L. Z. Kou, M. Dargusch, Y. Deng, Z. G. Chen. In situ crystal-amorphous compositing inducing ultrahigh thermoelectric performance of p-type $Bi_{0.5}Sb_{1.5}Te_3$ hybrid thin films. Nano Energy, 78, 2020, 105379.

[34] L. Huang, S. Lin, Z. Xu, H. Zhou, J. Duan, B. Hu, J. Zhou. Fiber-based energy conversion devices for human-body energy harvesting. Adv. Mater., 32, 2020, 1902034.

[35] S. Kim, S. Ryu, Y. Kwon, H. Lim, K. Park, Y. Song, Y. Choa. Synthesis and thermoelectric characterization of high density Ag_2Te nanowire/PMMA nanocomposites. Materials Chemistry and Physics, 190, 2017, 187–193.

[36] K. Suemori, T. Kamata. Thermoelectric characterization in out-of-plane direction of thick carbon nanotube-polystyrene composites fabricated by the solution process. Synth. Met., 227, 2017, 177–181.

[37] S. Kim, Y. I. Lee, S. H. Ryu, T. Y. Hwang, Y. Song, S. Seo, B. Yoo, J. H. Lim, H. B. Cho, N. V. Myung, Y. H. Choa. Synthesis and thermoelectric characterization of bulk-type tellurium nanowire/polymer nanocomposites. J. Mater. Sci., 52, 2017, 12724–12733.

[38] C. Yu, Y. S. Kim, D. Kim, J. C. Grunlan. Thermoelectric behavior of segregated-network polymer nanocomposites. Nano Lett., 8, 2008, 4428–4432.

[39] Y. S. Kim, D. Kim, K. J. Martin, C. Yu, J. C. Grunlan. Influence of stabilizer concentration on transport behavior and thermopower of CNT-filled latex-based composites. Macromol. Mater. Eng., 295, 2010, 431–436.

[40] Z. Antar, J. F. Feller, H. Noël, P. Glouannec, K. Elleuch. Thermoelectric behavior of melt processed carbon nanotube/graphite/poly (lactic acid) conductive biopolymer nanocomposites (CPC). Mater. Lett., 67, 2012, 210–214.

[41] S. Kima, Y. Songb, S. Ryu, T. Hwang, Y. Lee, J. Lim, J. Lee, K. Leea, Y. Choa. Thermoelectric behavior of bulk-type functionalized-SWCNT incorporated Te nanowire/PMMA hybrid nanocomposites with a segregated structure. Synthetic Metals, 254, 2019, 56–62.

[42] M. Tan, Y. M. Hao, Y. Deng, J. Y. Chen. High thermoelectric properties of $(Sb,Bi)_2Te_3$ nanowire arrays by tilt-structure engineering. Appl. Surf. Sci., 443, 2018, 11–17.

[43] M. P. Gordon, K. Haas, E. Zaia, A. K. Menon, L. Yang, A. Bruefach, M. D. Galluzzo, M. C. Scott, R. S. Prasher, A. Sahu, J. J. Urban. Understanding diameter and length effects in a solution-processable tellurium-poly (3,4-ethylenedioxythiophene) polystyrene sulfonate hybrid thermoelectric nanowire mesh. Adv. Electron. Mater., 7, 2021, 2000904.

[44] Y. Li, G. Wang, M. Akbari-Saatlu, M. Procek, H. H. Radamson. Si and SiGe nanowire for micro-thermoelectric generator: A review of the current state of the art. Frontiers in Materials, 8, 2021, 611078.

[45] M. Tan, Y. Deng, Y. M. Hao. Synergistic effect between ordered $Bi_2Te_{2.7}Se_{0.3}$ pillar array and layered Ag electrode for remarkably enhancing thermoelectric device performance. Energy, 77, 2014, 591–596.

[46] T. Koskinen, V. Khayrudinov, F. Emadi, H. Jiang, T. Haggren, H. Lipsanen, I. Tittonen. Thermoelectric characteristics of InAs nanowire networks directly grown on flexible plastic substrates. ACS Appl. Energy Mater., 4, 2021, 14727–14734.

[47] B. M. Curtin, E. W. Fang, J. E. Bowers. Highly ordered vertical silicon nanowire array composite thin films for thermoelectric devices. Journal of Electronic Materials, 41, 2012, 887–894.

[48] I. Noyan, M. Dolcet, M. Salleras, A. Stranz, C. Calaza, G. Gadea, M. Pacios, A. Morata, A. Tarancon, L. Fonseca. All-silicon thermoelectric micro/nanogenerator including a heat exchanger for harvesting applications. Journal of Power Sources, 413, 2019, 125–133.

[49] C. Hsinn, Y. Tsai. Power conversion of hybrid Bi_2Te_3/Si thermoelectric nanocomposites. Nano Energy, 11, 2015, 647–653.

[50] W. Thongkham, C. Lertsatitthanakorn, K. Jiramitmongkon, K. Tantisantisom, T. Boonkoom, M. Jitpukdee, K. Sinthiptharakoon, A. Klamchuen, M. Liangruksa, P. Khanchaitit. Self-assembled three-dimensional Bi_2Te_3 nanowire-PEDOT: PSS hybrid nanofilm network for ubiquitous thermoelectrics. ACS Appl. Mater. Interfaces, 11, 2019, 6624–6633.

15 Nanowires for Piezotronics and Piezo-Phototronics

Magdalene A. Asare
Department of Chemistry and National Institute for Materials
Advancement, Pittsburg State University, Pittsburg, KS, USA

Felipe de Souza
National Institute for Materials Advancement, Pittsburg State University,
Pittsburg, KS, USA

Ram K. Gupta
Department of Chemistry and National Institute for Materials
Advancement, Pittsburg State University, Pittsburg, KS, USA

CONTENTS

15.1 INTRODUCTION

The study of the interaction of devices with their environment is a crucial subject that has stipulated lots of research. An in-depth study of the sensory nature of devices to blood pressures, UV radiations, and harmful chemicals among many others is currently investigated. As such, scientists have

DOI: 10.1201/9781003296621-15

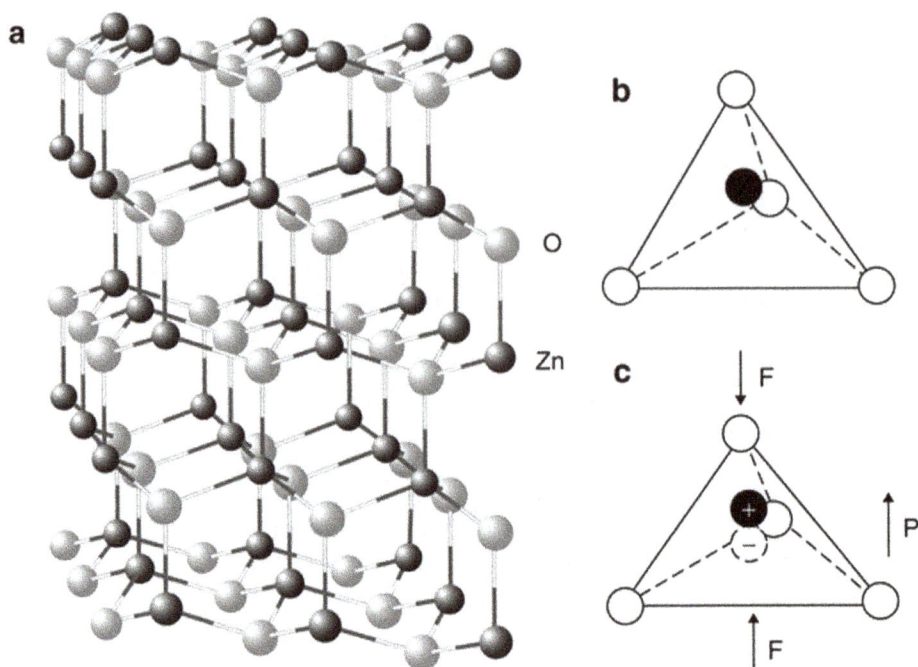

FIGURE 15.1 Structure of wurtzite ZnO. (b) A tetrahedral network between a neutral Zn and oxygen structure. (c) The relative displacement of the center of positive and negative charges by external force causes polarization and a piezoelectric effect. Adapted with permission [3]. Copyright (2007), John Wiley and Sons.

estimated that in the future, consumers will look beyond the fast-performing nature of devices and will require personal, flexible, and portable technology with multifaceted sensors and self-powering abilities [1,2]. Piezoelectricity, a main player in this area, is a phenomenon observed when a material generates electrical potential under the influence of mechanical change or deforms due to a voltage drop. The coupling effect of electrical polarization and mechanics in piezoelectricity is observed in non-centrosymmetric crystalline materials. Here, an applied stress disrupts the center of positive and negative charges relative to each other. And even though the overall structure of the material stays neutral when no force is applied for example in ZnO, the difference in their charge-center displacement under an external force induces a reversible dipole moment (Figure 15.1). The polarization resulting from the mechanical deformation and vice-versa is what is termed piezoelectricity [3].

Diverse crystalline materials like quartz, Rochelle salt, barium titanate, and polyvinylidene fluoride show this behavior, but the perovskite structure of lead zirconate titanate (PZT) is the most typical piezoelectric material utilized in actuators, electromechanical sensors, and energy devices. Despite the bulk or film form of ceramic PZT, it has a high piezoelectric voltage and dielectric constant in addition to its delicate and sensitive response to alternating voltages. However, based on their insulating nature, they have a high resistance, which is not satisfactory in some electronic devices. Third-generation semiconductors on the other hand are observed to have piezoelectric properties due to their asymmetry. Examples of these include ZnO, GaN, InN, and ZnS. These wurtzite materials have been explored for electronics, sensors, actuators, and transducers [4–6]. Piezotronics and piezo-phototronics, are more recent terms discovered and introduced by Wang and his co-workers as a link for the connection of biomechanical stimuli with the functioning of electronics [1].

The piezotronics effect is seen when a material has both piezoelectric polarization and semiconductor properties, which can facilitate the interaction between stimuli and devices. Piezotronics was initially discovered by Wang's team in ZnO nanowires (NW) with n-type conductivity. Here, an external strain stimulated a piezoelectric polarization, which eventually tuned the conductivity of the nanowire. ZnO is the most extensively studied and used nanowire because it is easy to synthe-size on a large scale, environmentally friendly, and can be grown on different substrates with diverse shapes. In addition, it has good piezoelectric and semiconducting properties [1,7]. As piezotronics was primarily discovered for strain sensing applications and lots of research on ZnO nanowires has been done, other researchers are diving into diverse nanowires and sensory applications that can detect pressure, temperature, and humidity just to mention a few [8,9]. Case in point, scientists have discovered that the adsorption of chemicals onto semiconductor materials serves as a gate voltage that controls the conductivity of the device and has taken advantage of this phenomenon for highly sensitive chemical sensory applications [10].

Piezotronics is seen in a heterojunction system, hence, when a piezoelectric material encounters an external pressure, polarization occurs at the junction between the materials, which alters the interfacial band composition [4]. In 2010, piezo-phototronics, which is slightly different from piezotronics, was coined. Piezo-phototronics occurs in materials with concurrent properties of piezoelectric, semiconductor, and photoexcitation [11]. In piezo-phototronics, mechanical inputs from the inner-crystals work as a gate to control the generation, diffusion, separation, and recom-bination of photocarriers for several applications [12]. Overall, the discovery and manipulation of piezoelectricity, semiconductors, and photoexcitation have brought forth coupling and intersecting phenomena with useful biomedical applications, wearable electronics, robotics, and other human-device applications. It is worth mentioning that one-dimensional nanowires and nanobelts play an important role and are more feasible for piezotronics and piezo-phototronics applications because they have a high tolerance to a large mechanical strain. Potential candidates researched include ZnO, GaN, InN, and doped PZT materials. This chapter explores the synthesis and the mechanism of nanowires in piezotronics and piezo-phototronics. An emphasis is placed on the applications of several nanowires for novel piezoelectric functions.

15.2 MECHANISM OF PIEZOTRONICS AND PIEZO-PHOTOTRONICS

15.2.1 MECHANISM OF PIEZOTRONICS

Piezotronics is a phenomenon observed in a heterojunction system, typically between a piezo-electric semiconductor and a metal. A heterojunction, which signifies the interface between two dissimilar materials, is important because interactions that occur between two different materials are vital for the mechanism of piezotronics [13]. A basic NW-based piezotronics transistor can be made of a metal-semiconductor-metal (MSM) structure like Ag-ZnO-Ag, Pt-ZnO-Pt, or Au-ZnO-Au (Figure 15.2). Generally, the deformation of piezoelectric material as a result of an external force results in the observation of a polarization effect between the junction. Here, the interfacial band structure of the piezoelectric material is altered and modified, which further classifies the piezotronics effect as a change in the interfacial structure. Figure 15.2a shows a typical ZnO wire with two gold electrodes on the ends with a crystallographic c-direction pointing one way to the left. A characteristic of ZnO and some piezotronics nanowires is the presence of a Schottky barrier, which signifies a rectifying junction of metal and semiconductor. In other words, this formation permits only the flow of current in one direction, and in this instance, electron transport is allowed from the ZnO to the Au. After the application of stress (Figure 15.2b), polarization is induced in the piezotronics material with positive charges and negative charges on the left and right, respectively [4]. Even though the charges could eventually disappear, they can be maintained in the nanowire due to some levels of doping and the finite charge-screening depth present at the interface [13]. Hence, when the material is in a stable state, remnant piezoelectric charges can still be observed at the inter-face. The Schottky barrier height on the left $e\phi_{SB1}$ is reduced and that on the right $e\phi_{SB2}$ is increased due to the electrostatic field from the positive and negative charges at the interface, respectively [4].

When a ZnO nanowire is subjected to a strain, two types of effects that control the carrier trans-portation processing are observed. As described above in Figure 15.2, a piezoelectric effect can be seen from the polarization of the crystals in the nanowire, which is asymmetric and can occur in

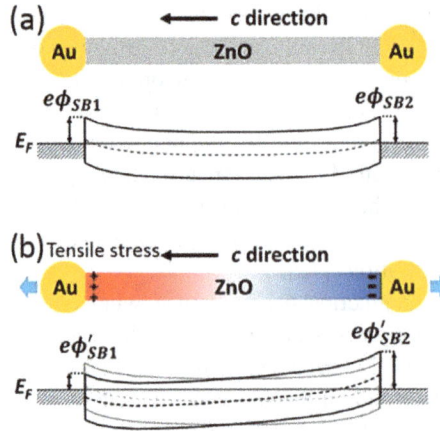

FIGURE 15.2 Illustration of Schottky interface between ZnO and Au. (a) Band diagram with equal Schottky height. (b) Band diagram with reduced barrier height eϕ_{SB1} and increase Schottky barrier height eϕ_{SB2} due to polarization under applied tensile stress. Adapted from reference [4]. Copyright (2015), MDPI. This is an open access article distributed under the Creative Commons Attribution License.

semiconductors and other wurtzite structures, such as ZnO, GaN, and PbS. Here, polarization or a change in the barrier height in an n-type semiconductor and a metal, for example, may change the Schottky interface into an Ohmic contact and vice-versa. It is worth noting that changes in the height barrier depend on the type of semiconductor and doping density present [14]. Alternatively, the piezoresistive effect, which is the other type of tuning effect on the carrier processing transportation, has a symmetric effect on the nanowires without polarization on both ends. Piezoresistive effects can be seen in semiconductors, such as GaAs, Si, and other wurtzite structures [15,16]. In another illustration (Figure 15.3), the mechanism of piezotronics is observed at the junction between a metal and semiconductor. Applying a strain on the semiconductor depletes the major electrons in the material. As projected, there is an increase in the Schottky barrier height as a result of the negative polarization charges (Figure 15.3a) and a reduced barrier height (Figure 15.3b) due to positive charges and owing to a lesser depletion at the junction. In a p-n junction (Figure 15.3c), a positive polarization at the n-type side increases and decreases the depletion on the p- and n-type sides, respectively. This expands and causes the depletion region to lean towards the p-type side and eventually causes a downward bending in the local band structure. In effect, the deformation allows for the trapping of electrons through a charge channel. The opposite is observed where negative charges are induced in the n-type region (Figure 15.3d). Here, the depletion width on the p-type area decreases, causing an expanded depletion region and shift towards the n-type side. The continuous increase in an external strain leads to an upward bend in the local band [2].

15.2.2 MECHANISM OF PIEZO-PHOTOTRONICS

The unique coupling effect of piezoelectricity, semiconductor, and the optical process is termed as the piezo-phototronics effect. In a typical piezo-phototronics device, there is the generation, separation, and recombination of charge carriers present. To explain the mechanism of piezo-phototronics, a Schottky-contacted photodetector, which has a similar mechanism to other optoelectronics is described (Figure 15.4). When no external strain is applied, the presence of incident photons generates electron-hole pairs, however, the electrons and holes are split up by the electric field at the Schottky interface. This generates a photocurrent, which is affected by an effective separation and transport of both carriers and the barrier properties. If positive polarization charges are induced at the reversely biased Schottky contact under an applied strain (Figure 15.4a), there

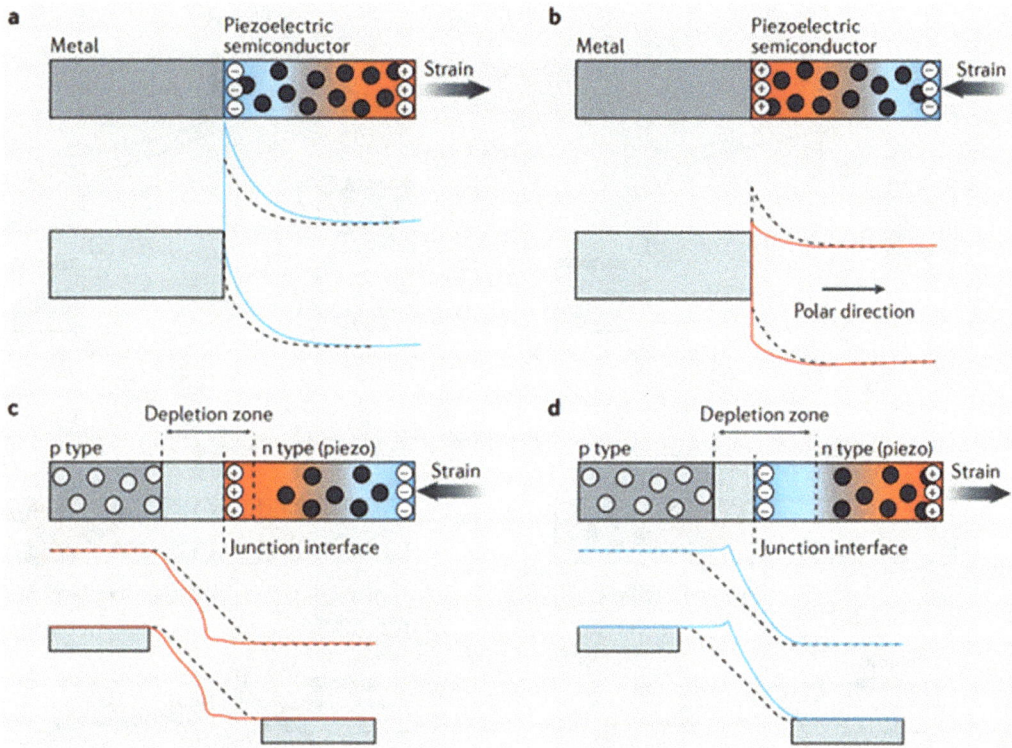

FIGURE 15.3 Image showing the mechanism of the piezotronics effect under an applied strain. (a) Increase in Schottky barrier height due to negative polarization. (b) Piezotronics with a reduced Schottky barrier height as a result of positive induced charges at the junction. (c) Presence of a downward bend in the local band structure of the piezotronics material. (d) Further increase in strain leads to an upward bend in the local band structure. Adapted with permission [2]. Copyright (2016), Springer Nature.

is a smaller driving force for the separation and redistribution of holes and electrons. The energy barrier for the electron transport at the positive interface is increased and the total photoinduced carriers in the device are reduced. Alternatively, if negative polarization charges are induced on the Schottky contact, the band structure facilitates the separation, transport, and collection of the holes and electrons (Figure 15.4b). During a band-to-band optical absorption, there is a photogeneration of the electron-hole pair at the p-n junction leading to a photoresponse. When positive charges are formed on the n-type area (Figure 15.4c), the depletion towards the p-type side increases the effective series resistance and reduces the current of the device. The creation of a charge channel in the conduction band and the downward bend of the valence band on the n-type region leads to poor separation of the electron-hole pairs. On the reverse, when negative charges are induced at the n-type region (Figure 15.4d), the upward bending of the band restrains electron-hole recombination and enhances the separation of photogeneration carriers [2].

15.3 SYNTHESIS OF NANOWIRES

Nanowires play a crucial role as building blocks for nanosensors, nanogenerators, and other piezo-electric applications. With regards to that, the secret to high-performing devices depends on nanowires whose composition, geometry, uniformity, and density can be controlled at the time of synthesis [17]. Several tunable techniques such as wet methods, chemical and physical deposition, molecular beam epitaxy, sputtering, electrospinning, and top-down fabrication exist for the synthesis of nanowires [18].

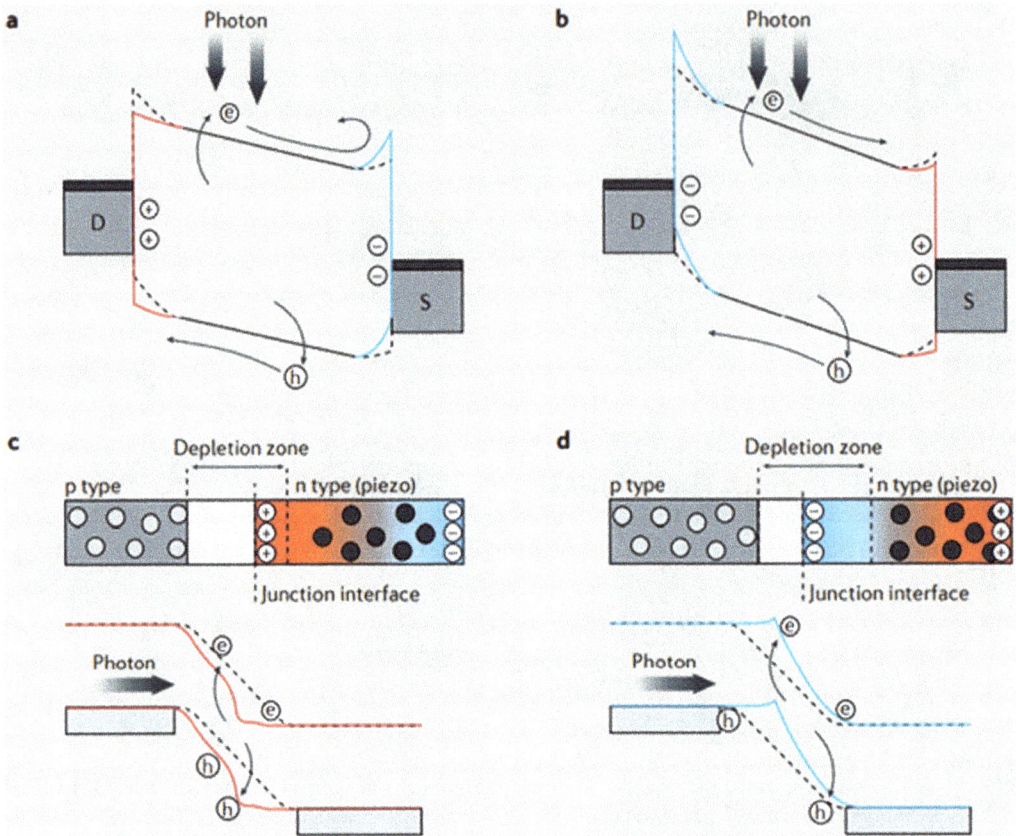

FIGURE 15.4 Illustration of the piezo-phototronics mechanism. (a) Generation of positive polarization at the Schottky contact. (b) Diagram with negative polarization at the heterojunction. (c) Downward bending due to charge channel formation. (d) Upward bending can permit an effective electron separation. Adapted with permission [2]. Copyright (2016), Springer Nature.

These techniques have diverse mechanisms with advantages and drawbacks that are discussed below. In addition, these techniques will largely touch on the fabrication of ZnO, which is the most researched NW as well as other emerging and useful nanowires for piezotronics and piezo-phototronics applications.

15.3.1 WET CHEMICAL APPROACH

This method is known as a fast-synthetic route because of its low cost and high yield. It can also be facilitated at low temperatures, it's compatible with flexible organic substrates, and can be adopted without the need for metal catalysts [19,20]. This approach usually requires a liquid phase and incorporates one or more metal salts or oxides. Here, crystallization or precipitation can be used to form nanocrystals. The size of the nanomaterials can also be altered by controlling the nucleation and agglomeration process of the nanowires in an appropriate chemical reaction system. Wet chemical synthesis includes sol-gel, hydrothermal, and electrodeposition among the others, which operate on a similar principle [14,21].

15.3.1.1 Hydrothermal Synthesis of Nanowires

This is one of the popularly used liquid phase methods for the synthesis of nanomaterials. Hydrothermal synthesis offers a wide range of temperatures and pressures in the fabrication of nanomaterials such that the integrity of products that are usually unstable under extreme conditions can be maintained [22]. This approach is performed in a hermetic vessel, usually an autoclave in which the formation of the nanomaterial involves the dissolution and crystallization according to the solubility of the chemical. In much detail, the reactants dissolve into ions, which are then separated as a result of temperature differences following a series of transformations before crystallizing into nanostructured forms [23]. In the case of oxides, the solubility of hydroxides in water needs to be higher than that of oxides for the oxides to precipitate [14]. Due to the advantages of the hydrothermal method, it has been adopted in the synthesis of thin films, bioceramics, mixed metal oxides, and simple oxides, such as ZnO [23]. In the fabrication of ZnO nanowires, zinc nitrate, and hexamethylenetetramine (HMTA) are typically employed as precursors. And based on equation 15.1, Zn^{2+} ions are obtained from zinc nitrate and OH^- ions for the crystal growth from HMTA.

$$Zn^{2+} + 2OH^- \leftrightarrow Zn(OH)_2 \rightarrow ZnO + H_2O \qquad (15.1)$$

Ou et al. experimented with a hydrothermal approach for the synthesis of aligned ZnO nanowires for piezoelectric energy harvesting applications as shown in Figure 15.5. Here, a polycarbonate (PC) template was used with the successive growth of the ZnO NW in a Teflon-sealed autoclave. The aligned nanowires were obtained after the burning of the PC template as shown in the SEM images [24] (Figure 15.6).

15.3.1.2   Electrochemical Deposition

This is a robust and wet technique that can be used for the synthesis of large and evenly distributed nanomaterials at a low cost. Electrochemical deposition is mainly driven by an electric field, which is advantageous for some non-spontaneous reactions [18]. Additionally, it has been observed that the electric force encourages a better alignment and stronger adhesion of the nanowires to substrates, especially in the case of ZnO [25]. Generally, electrochemical deposition is done in a standard three-electrode device and it requires a substrate on which the nanowire will be attached. Typically, the cathode and anodes are portioned parallel to each other while Ag/AgCl and Pt are used as reference

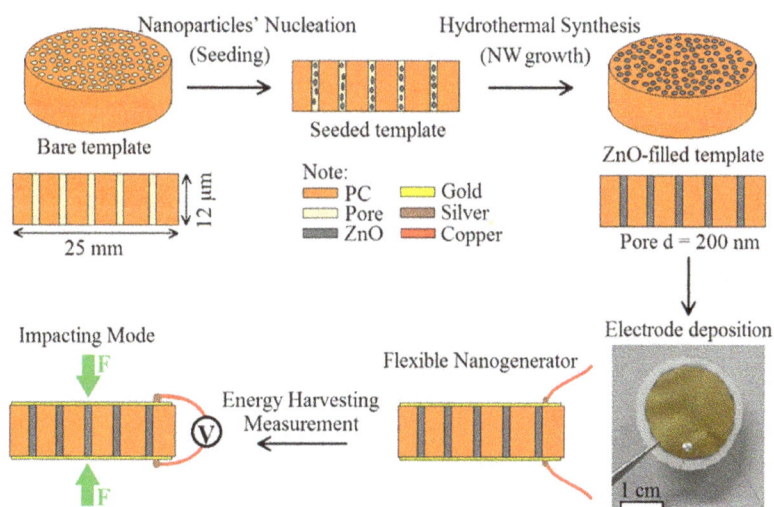

FIGURE 15.5   Hydrothermal synthesis of ZnO nanowires using PC template for piezoelectric energy harvesting application. Adapted from reference [24]. Copyright (2016), American Chemical Society. This is an open access article distributed under the Creative Commons Attribution License.

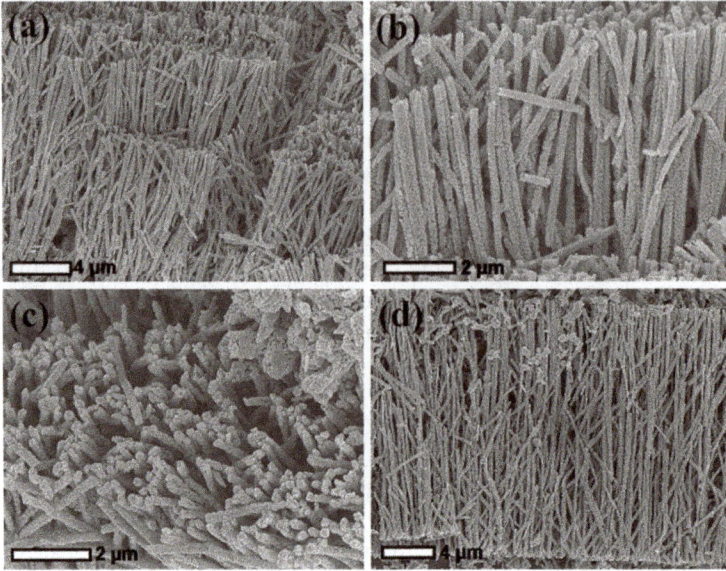

FIGURE 15.6 SEM images of aligned ZnO nanowires at various magnifications. Adapted from reference [24]. Copyright (2016), American Chemical Society. This is an open access article distributed under the Creative Commons Attribution License.

and counter electrodes, respectively. And to reduce any discrepancies, a constant current or voltage is applied for a constant reaction rate or driving force in the system [18]. As a result of the low cost and fast nature of this method, it can be scalable to meet industrial demands. It has been used in the fabrication of several nanowires including Bi_2Te_3 [26], ZnS [27], and ZnO on different substrates like Si, Cu, and conductive polymers [14].

15.3.2 VAPOR DEPOSITION METHODS

This method usually occurs in a vapor phase involving either a chemical or physical process. Several examples that fall under this technique include chemical vapor deposition, physical vapor deposition, and pulsed vapor deposition just to mention a few. In general, vapor deposition methods demand high temperatures, cost, and complex processing, however, they are advantageous for the synthesis of nanowires like GaN that are not feasibly fabricated with wet chemical methods. In addition, varying the deposition technique can be used to manipulate the desired nanowires. In the fabrication of ZnO nanowires with chemical vapor deposition, an oxide powder is heated in a furnace and through an argon gas medium, the evaporated oxide is deposited on the appropriate substrate [14]. In research by Puglisi and the team, Si nanowires were generated using chemical vapor deposition and high pressure was used in the synthesis of controlled nanowires [28]. Similarly, physical vapor deposition (PVD) was adopted by Tigli et al. [29] for the synthesis of ZnO nanowires. Here, the zinc powder was sublimated under a high-temperature zone and deposited onto the substrate at a lower temperature zone. In another experiment, a gold-assisted vapor-liquid-solid in a metal-organic chemical vapor deposition system was used to fabricate a core-shell InGaP nanowire. Microscopic imaging of the nanowires gave information about the varied size and morphology that could be improved in other syntheses. As shown in the SEM image (Figure 15.7), there were size differences in the top and bottom of the InGaP nanowire [30]. The reduction in the overall uniformity of the NW sizes distinguishes bottom-up methods from top to bottom techniques that are currently available.

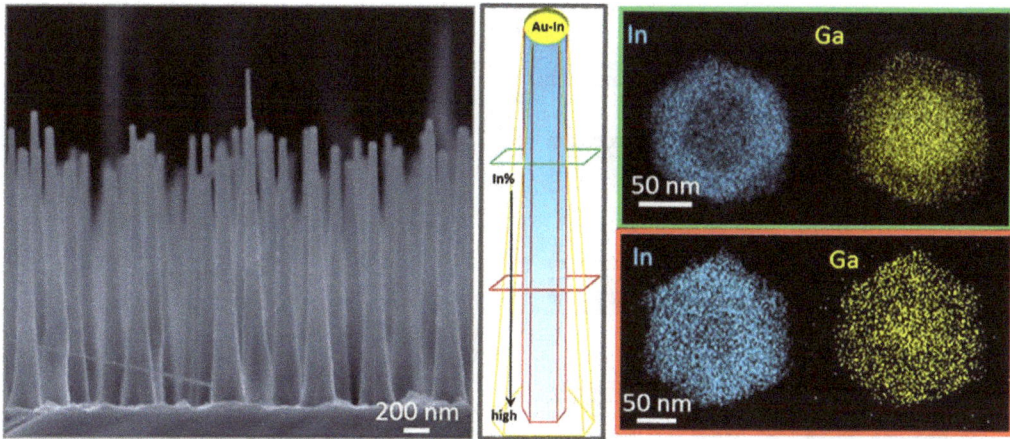

FIGURE 15.7 SEM image of InGaP nanowires synthesized using vapor-liquid-solid in a metal-organic chemical vapor deposition system. Adapted with permission [30]. Copyright (2019), American Chemical Society.

15.3.3 TOP TO BOTTOM METHOD

This method is an emerging technique geared for specific sizes of nanowire and nanomaterials. Top-down fabrication is more advantageous than bottom-up approaches because, on an industrial scale application, they can generate nanowires with a more uniform height, diameter, and distribution. However, this method is expensive and involves very complicated techniques [14].

15.4 APPLICATIONS OF NANOWIRES FOR PIEZOTRONICS AND PIEZO-PHOTOTRONICS

15.4.1 PIEZOTRONICS NANOWIRES FOR STRAIN SENSING

The ability to detect strains plays a significant role in the manufacturing of flexible wearable devices, micro/nanoelectromechanical systems (MEMS/NEMS), human-machine interaction, and health monitoring among many others. To detect such minute levels of strains in either the environment or from humans, several nanowires, nanotubes, and in general, nanomaterials have been investigated [31–33]. There is a broadened research on piezotronics devices for strain sensing applications due to their inherent response to strain through charge movements across their semiconductor-piezoelectric heterojunction [34]. In addition, researchers have studied that a higher Schottky barrier decreases the signal level and a lower Schottky barrier reduces the sensitivity of the sensor owing to the increased Ohmic transport. Hence, an effective way to manipulate the Schottky contact could potentially optimize the performance of Schottky-contact devices [35,36]. The piezotronics effect seen in wurtzite materials and some nanowires have shown promising solutions due to the tunability of their Schottky barrier height at the metal-semiconductor (M-S) contact with the application of a strain [37,38]. Hence, nanowires from nickel [39] and ZnO [31], have been synthesized for strain sensing applications.

In an experiment by Zhang et al., sensitive and flexible sensors were made from ZnO nanowires. Here, a hydrothermal synthesis was adopted in the fabrication of the vertically aligned ZnO NW with the utilization of photolithography and metallization in the production of the strain sensing device. As shown in Figure 15.8, chromium (Cr) and ZnO seed layers were deposited on a polyethylene terephthalate (PET) substrate. Photolithography was used to open a channel in the layer for the hydrothermal synthesis of ZnO nanowires. Polydimethylsiloxane (PDMS) was spin-coated for a protective layer and the tips of the nanowires were exposed through dry etching. A gold layer

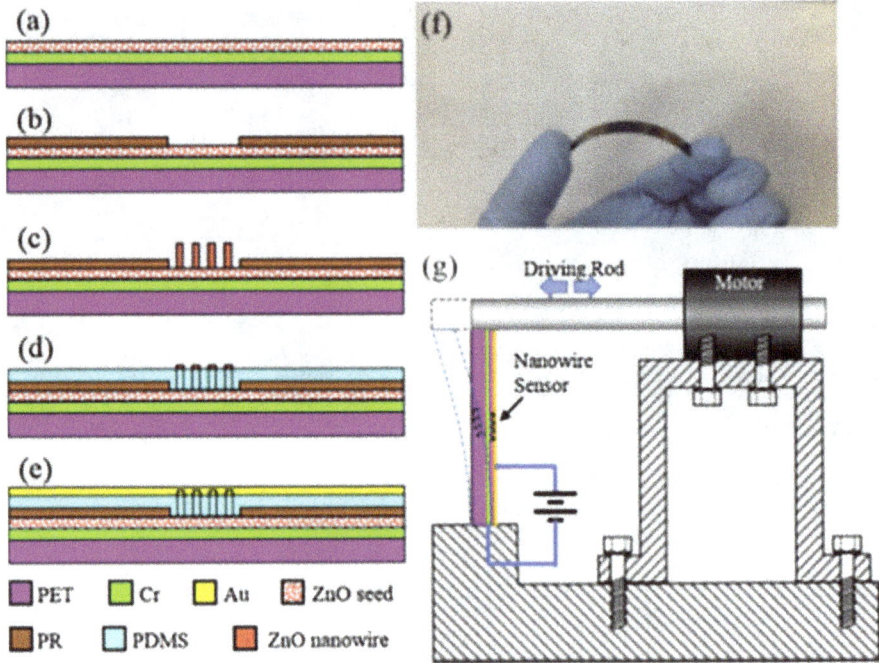

FIGURE 15.8   (a-e) Sequence for the fabrication of ZnO nanowires in the strain sensor with photolithography, hydrothermal synthesis, and dry etching. (f) Image of the synthesized strain senor device. (g) Illustration of the process of testing the strain characteristics of the device. Adapted with permission [40]. Copyright (2014), Elsevier.

was then deposited on top of the device as an electrode. In the strain testing, one end of the sensor was firmly fixed in a sample holder and the other was connected to a moving rod stimulated by a programmable linear motor. Under a controlled temperature and environment, a current amplifier and functional generator were used to monitor the current and sinusoidal bias voltage, respectively, through the wire. The device had a contact formed between the Cr electrode and the continuous layer of the ZnO seed. In addition, the graphs (Figure 15.9a) from the strain test exhibit typical current-voltage (I-V) characteristics. The non-linear and rectifying behavior is a result of the Schottky contact between the ZnO nanowire and gold electrode. Also, the dramatic change in the curve is due to the response of the device to external stress. A high gauge factor was recorded (Figure 15.9b) with a gradual increase in strain. The gauge factor, which is the ratio of the change in current with respect to a strain, is an important parameter for strain testing [4]. From this experiment, Zhang et al. [40] concluded that the high sensitivity of the fabricated strain sensor is due to the metal-semiconductor-metal (MSM) effect that generated a piezoelectric and semiconducting effect. Consequently, the conductivity of the device was remarkably tuned by the change in the ZnO/Au Schottky barrier due to a strain-induced piezoelectric potential. Final applications of the vertically aligned ZnO nanowires could be in medical, civil, and wearable devices [40].

15.4.2   PIEZOTRONICS NANOWIRES FOR CHEMICAL SENSING

Nanomaterials have been investigated for chemical sensing applications owing to their high surface area to volume ratio. In addition, the adsorption of chemicals on their surface affects their carrier charge density and current flow, which is important to determine the performance of the device [4,36]. In typical Ohmic-contact devices, smaller nanowires with low contact resistance are required for high sensitivity, however, it can be challenging to fabricate. Schottky contact devices, on the

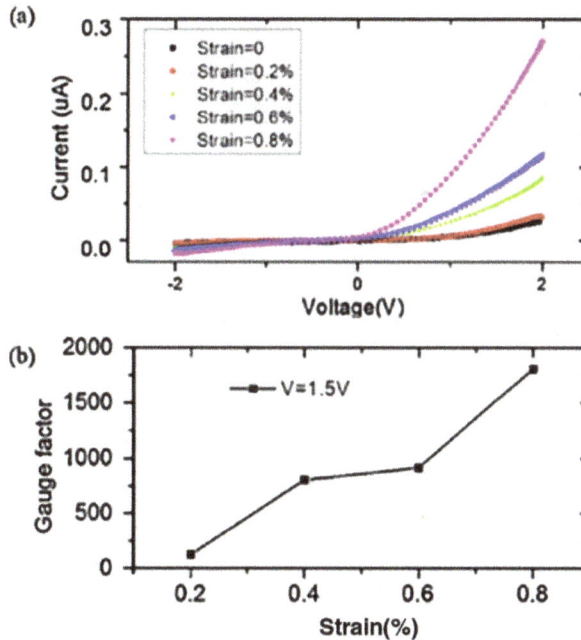

FIGURE 15.9 Results from the electrochemical testing of the fabricated strain sensor device. (a) Graph showing the typical non-linear I-V characteristics due to the Schottky barrier between the ZnO nanowire and gold electrode. (b) Graph displaying the increase in the gauge factor. Adapted with permission [40]. Copyright (2014), Elsevier.

other hand, exhibit higher sensitivity due to the easy tunability of the Schottky barrier height (SBH) at the heterojunction [36]. In effect, piezotronics nanowires have found uses in biosensors, gas sensors, and detecting other chemicals. The detection of gases is very vital in agriculture, healthcare, and in environmental monitoring purposes [41].

15.4.2.1 Nanowires for Gas Sensors

Most importantly, the sensitive detection of odorless and colorless gases like NO_2 and H_2 is crucial to avoid the likelihood of poisoning or an explosion since the normal human senses cannot naturally identify those. Zhou et al. [42] enhanced piezotronics sensor characteristics by detecting NO_2 and H_2 gases at room temperature using ZnO micro/nanowires. The researchers used an MSM structure, and took advantage of the properties of the Schottky-contact of the ZnO nanosensor such as their quick response and high sensitivity as compared to typical Ohmic-contact. The vapor-liquid-solid method was used in the fabrication of the ZnO micro/nanowires. At room temperature, Pd nanoparticles were synthesized on the device using radio frequency sputtering. The overall nanosensor was constructed on a flexible polystyrene substrate and further tested for H_2/N_2 sensing ability. When hydrogen gas is present in the chamber of the synthesized nanosensor, the adsorbed hydrogen molecules react with the pre-existing oxygen and release trapped electrons, resulting in a dissociation of the Pd nanoparticles in the device. With respect to that, injected electrons into the ZnO conduction band cause a build-up of charge carriers on the surface of the nanowires. This further increases the conductivity and output current of the synthesized nanosensors. On the other hand, when different amounts of NO_2 are tested in the device, the NO_2 molecules react with the oxygen due to their oxidizing nature. And they become negatively charged by removing electrons from the ZnO conduction band. Following, electrons in the ZnO NW are reduced, which leads to a depletion layer and reduces the conductivity and output current of the ZnO gas sensor. Depending on the crystal orientation of the ZnO and other factors, when an external strain is applied, positive polarization charges are induced at the M-S contact, which reduces the Schottky barrier height. This working mechanism enabled the scientists to observe and develop an approach to increase the sensitivity, boost the detection resolution and improve the overall function of the H_2/NO_2 gas sensor [42].

15.4.2.2 Nanowires for Glucose Detection

Glucose is another important chemical to be monitored in the food and health industry. A variety of sensors using field-effect transistors (FET) and typical NW-based glucose sensors are on the market [43]. However, due to their Ohmic-based contact mechanism, their overall output signal can be restricted. In addition, their sensitivity is more effective using very small-size nanowires and this causes challenges in their fabrications. Scientists have therefore investigated piezotronics nanowires for glucose sensing owing to their advantageous Schottky barrier contact. In research by Yu et al., a self-powered glucose sensor with enhanced performance from ZnO NW was fabricated and tested. The scientists were drawn to the Schottky-contact sensor due to the ease in their synthesis with moderate sizes without compromising their high sensitivity. In another aspect, these glucose sensors require a source of power for functioning. Batteries are typically used, which need frequent replacement. Hence, a self-powered glucose monitoring system that manipulates the piezotronics effect was developed [44]. High thermal evaporation was used to synthesize the ZnO NW and through surface functionalizing with glucose oxidase (GOx), the sensor was manufactured by bonding the ZnO NW laterally onto a polymer substrate (Figure 15.10a). Following this, a silver paste was used at both ends of the NW to serve as a source or drain electrode. GOx was introduced onto the ZnO NW and incubated in a fume hood for 2 hours for drying. After several repetitions of the above, the device was rinsed to remove any GOx that was not strongly adhered to the nanowire. A compressive strain was applied to the fabricated device as shown in Figure 15.10b. The strain response of the device was first tested in DI water instead of glucose. Here, one end of the device was free while the other was fixed on a manipulation holder. With the application of different compressive strains, the I-V graph displayed typical piezotronics characteristics. This is attributed to the piezo-charges generated at the interfacial region such that positive and negative charges reduced and increased the barrier height at the Schottky contact, respectively, leading to the manipulation of the charge transport system. In effect, increased output signals were recorded and at a bias voltage of 1.8 V, the current increased from 0.15 µA to more than 25 µA. The device was then tested with an increasing concentration of glucose under a strain-free condition at a fixed bias of 2 V. With the

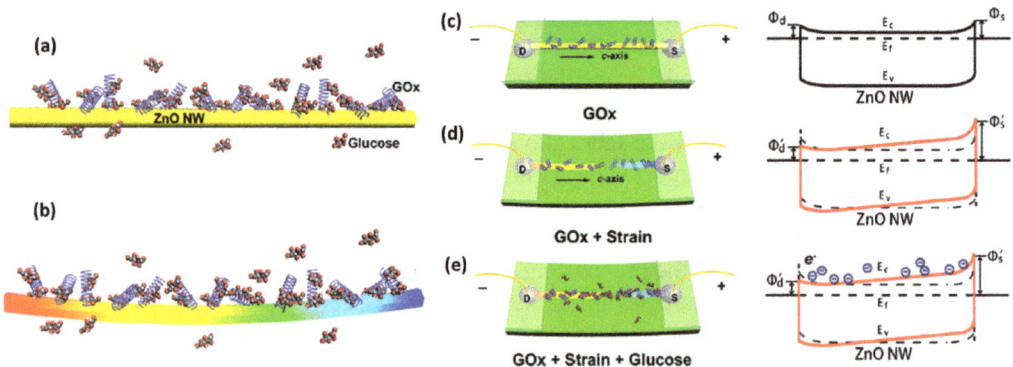

FIGURE 15.10   (a) Strain-free glucose sensor. (b) Image of glucose sensing device after applying a strain. (c-d) Energy diagram for a synthesized glucose device. (c) Under normal conditions. (d) Under strain. (e) Under the influence of strain and different glucose concentrations. Adapted with permission [44]. Copyright (2013), John Wiley and Sons.

gradual increase of glucose, there was a high output signal recorded. Each stepwise increase in glucose resulted in a significant current change that lasted for more than 10 s. And the ZnO NW sensor could detect as low as 0.49 g/L. The researchers also found that the device could detect about 0.09 to 1.5 g/L of glucose, which will be practical for human blood glucose monitoring since it has a range of 0.8 to 1.2 g/L. With several tests of the ZnO glucose sensor under strain and different glucose amounts, it was found that both parameters had significant effects on the overall signal outputs owing to the piezotronics effect. A theoretical model with energy diagrams was used to explain the piezotronics effect observed in the fabricated device. In Figure 15.10c, the Schottky barrier height remains unchanged under no strain or glucose environment. Figure 15.10d illustrates the energy band diagram in the presence of a strain. It can be observed that in response to the applied strain, piezo-charges were generated leading to a reduction and increase in the Schottky barrier height in the case of positively or negatively induced charges, respectively. This advanced to a high output signal recorded. A similar trend is seen in Figure 15.10e, however, a higher output signal is recorded. This is because when the glucose reacts with GOx on the device, gluconic acid and H_2O_2 are produced and H_2O_2 has been investigated to increase the electron transfer on the surface of the ZnO NW, leading to an additional output recorded [44].

15.4.2.3 Nanowires for Target Proteins

In another chemical sensing experiment, a high-performance ZnO NW for protein detection was synthesized and improved by the piezotronics effect. The target protein was Immunoglobulin (IgG) and the identification of such biomolecules has clinical, industrial, and environmental applications. A temperature evaporation method was adopted to synthesize the ZnO NW, which was later bonded to a polymeric substrate. In the study, an MSM structure was utilized, and gold nanoparticles-anti-Immunoglobulin G conjugates were functionalized into the sensor due to their receptive nature to the target protein. The utmost performance of the ZnO sensor was tested under diverse strains and IgG concentrations. It was found that both variables influenced the chemical sensing of the device. Notable from the experiment, the piezotronics effect increased the resolution of the protein sensor by ten times, which significantly enhanced the sensitivity and detection range of the ZnO NW, especially under a compressive strain [45].

15.4.3 PIEZOTRONICS NANOWIRES FOR HUMIDITY DETECTION

In the investigation of other applications of nanowires for piezotronics, ZnO NW was introduced as a building block by a research team in the synthesis of a relative humidity sensor in ambient temperature. The accurate monitoring of humidity is vital in medicine, electronics, national defense, and our everyday activities. Vapor-liquid-solid (VLS) method at 960 °C was used to grow the ZnO nanowires and SEM images were used for the morphology studies. Thereafter, the nanowires were selectively dispersed on PET or polystyrene (PS) substrates while both ends were fixed as electrodes using silver paste. The fabricated sensor was closed in a humidity chamber with one end tightly fixed in a clamp and the other loose. An external strain was applied at the loose end to introduce the piezotronics effect for further analysis of the humidity sensor. In the theoretical model devised to explain the effect of piezotronics on the relative humidity sensor, an energy band diagram of the strain-free sensor with the c-axis directed from the source to the drain is represented. In the presence of humidity, water molecules are adsorbed onto the ZnO surface with water dissociation happening at free surface oxygen sites. With respect to that, there was a formation of two bridging hydroxyl groups, further decreasing surface carrier concentrations, leading to the formation of an electron depletion layer and reducing the overall conductance of the device. In effect, the more the relative humidity present in the chamber, the lower the output signal due to reduced electron carrier density in the nanowire. However, under strain, polarization induces piezo-charges at the local M-S contacts at the ends of the device. And even though studies have found that the charges could be partially screened in a humid atmosphere, there is a directional movement of the free electrons. Hence, the Schottky barrier height at the drain is reduced due to the induced positive charges and was increased because of negative charges and this explains aspects of the piezotronics effect. The observed trend of the experiment established that the piezotronics effect introduced with an external strain in a humid condition significantly increased the sensitivity and sensing resolution of the fabricated ZnO NW humidity sensor [35].

15.4.4 PIEZOTRONICS NANOWIRES FOR TEMPERATURE SENSING

In one of the first studies of piezotronics for temperature sensing applications, Xue et al. [46] investigated the potential of ZnO nanowire film for a temperature sensor. Properties of a bimetallic thermometer such as an easy synthesis, robustness against magnetic interference, etc., were combined with the piezotronics effect in the fabrication of an ultra-sensitive temperature sensor. A wet chemical deposition was used to synthesize the ZnO nanowire film, which was placed on a PET substrate using photoresist. In addition, the two main processes; piezoresistive and piezotronics effects that are unique characteristics of piezoelectric semiconductor films were studied. The piezoresistive effect is a volume effect and is generally seen to change linearly to certain amounts of strain. Conversely, the piezotronics effect, which is nonlinear, is an interface effect due to the polarization of charges at the metal-semiconductor junction. To test this, a three-dimensional mechanical set-up with a displacement resolution of 1 μm was adopted to create a compressive or tensile strain through bending at the free end of the device. Based on the unchanged I-V characteristics under increasing strains to about ± 0.55%, there was no significant difference in the conductivity of the ZnO NW sensor. This ruled out the possibility of the piezoresistive effect and confirmed that of the piezotronics effect, which was tested under an external strain. Here, current moving through the device increased under a compressive strain and reduced with tensile strain. The I-V characteristics were indicative of the changes in the Schottky barrier height. Moving on, the response of the sensor to temperature changes was conspicuous from 10 °C to 110 °C at a fixed bias of 6 V. The signal response reduced drastically from a few microamperes to nanoamperes owing to the piezoelectric polarization charges induced at the contact interface and thermal disturbance of the temperature sensor. Overall, it could be inferred from the results that the fabricated ZnO NW film based nanosensor had a great response to temperature changes, which was even more heightened due to the piezotronics effect [46].

15.4.5 PIEZO-PHOTOTRONICS NANOWIRES FOR PHOTODETECTORS

Piezo-phototronics as previously discussed, is an emerging field that incorporates properties from the three-way coupling of piezoelectricity, photoexcitation, and semiconductor in a material to improve its performance for optoelectronic devices. The fundamentals of piezo-phototronics involve the generation, separation, transport, and recombination of charges at the interface in piezoelectric-semiconductors. Nanowires and other materials that exhibit the piezo-phototronics effect are ZnO, CdS, and GaN [47–49]. With the tunability of piezopotential through an external force, the effects of piezo-phototronics are gradually being studied in its combination with flexible optoelectronics for innovative devices. Examples of recent advancements are portrayed in UV photodetectors [50], photocells [51], light-emitting diodes [52], and electrochemical processes [53]. It is worth noting that diverse research has studied some aspects of the three-way coupling effect in photodetectors, however, there is still a lot that is uncovered about piezo-phototronics.

In an experiment by Liu et al. [54], the coupling effect of photoexcitation and piezoelectricity in photodetectors was analyzed based on a single-Schottky and double-Schottky based metal-semiconductor-metal set-up. Based on the results from photodetectors from CdS nanowires for visible light and ZnO NW for UV light, some theoretical results were verified. For instance, it outlined that piezo-phototronics detection necessitates a charge barrier, photoexcitation has an effect on I-V characteristics due to the generation of charges and the piezoelectric effect manipulates photodetection at the ends of nanowires under the influence of an applied strain [48]. In a more practical application, a research team introduced and investigated the effect of piezo-phototronics on an enhanced UV photodetector using carbon fiber and ZnO-CdS double-shell nanowire. Despite the use of ZnO nanowires in broad applications in optoelectronics, electrochemical and electronics, their pristine form cannot absorb wavelengths below the band gap. And so, in UV ranges, it restricts their sensitivity for photovoltaics and other wide photodetection applications.

To resolve this issue while maintaining the piezo-phototronics characteristics of the ZnO nanowires, a composite structure with carbon fibers, cadmium sulfide (CF/ZnO-CdS) nanowire was devised. A two-step hydrothermal synthesis was used in the fabrication of the device and SEM was used for characterization. The unique combination in the synthesized double-shell nanowire resulted in an enlarged surface area for UV/visible light absorption, enlarged channels for light-generated electron transport, and ultrahigh photoresponse (Figure 15.11). The CF/ZnO-CdS wire under blue light around 480 nm wavelength and defined dimensions was tested. The I-V characteristics in the

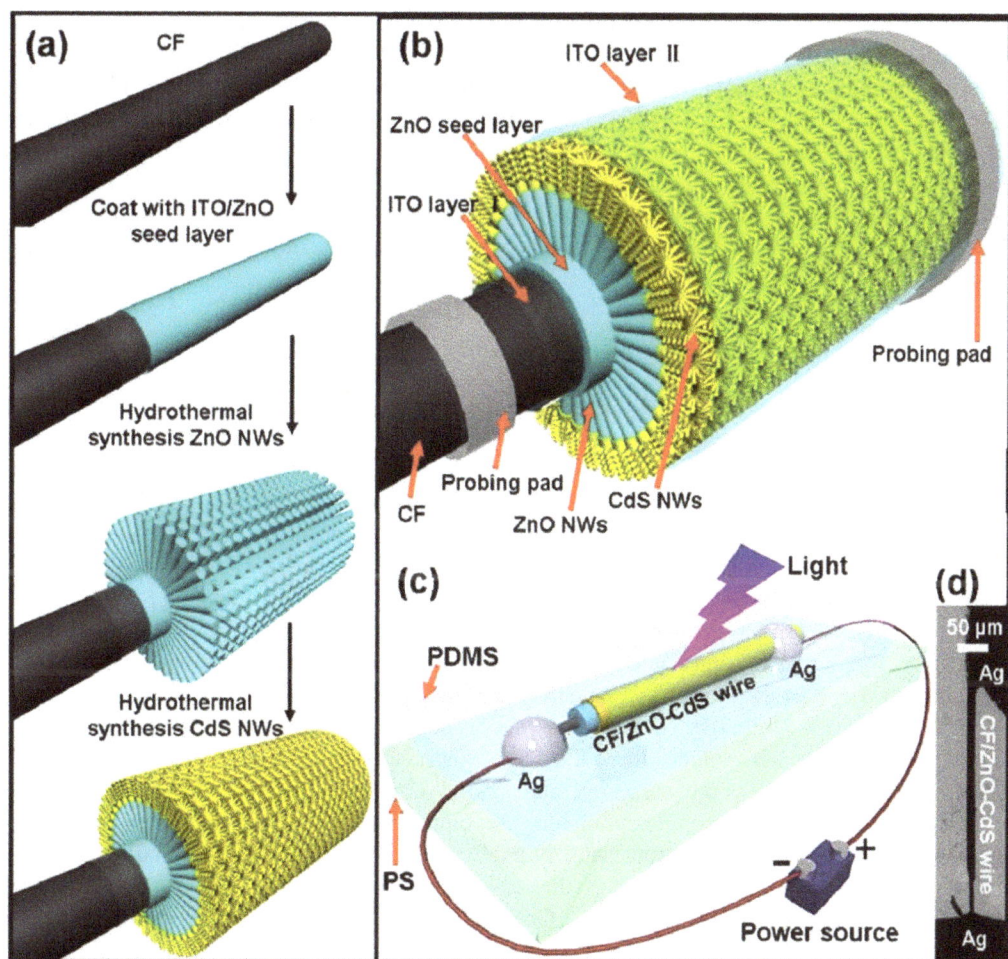

FIGURE 15.11 (a-c) Illustration of a two-step hydrothermal method for the synthesis of ZnO nanowires for the fabrication of a UV sensor device. Adapted with permission [54]. Copyright (2013), American Chemical Society.

dark and under light from 8.99 × 10^{-8} to 7.19 × 10–4 W/cm^2 were studied. At an applied voltage of 2.0 V, the absolute current rose from 3.91 μA observed in the dark to 12.9 μA at the highest intensity tested. The current sharply increases from one state to another with a response faster than 0.2 s with an on or off light irradiation. The photocurrent increased with optical power at high power intensities without any saturation. The photon responsivity with respect to excitation intensity was studied. Overall, the good photocurrent generation and fast response proved the potential of CF/ZnO-CdS wire in photodetection. Continuing, the researchers introduced an external strain and recorded an enhanced UV detector. This was attributed to the electron-hole pair generation, transport, and recombination at the junction, referencing the piezo-phototronics effect and piezopotential by tuning the ZnO-CdS heterojunction [54].

15.4.6 PIEZO-PHOTOTRONICSNANOWIRES IN SOLAR CELLS

Solar cells can be manufactured with diverse materials and approaches. However, silicon-based solar cells are the most popular and widely used in various sectors. In contrast, silicon-based solar cells are challenged with light absorptions, electron-hole pair separation, interface recombination, and surface barrier limitations. For instance, antireflective coatings and light traps have been seeking to minimize the inhibition and increase the magnitude of photoexcited carriers in the absorption layer of these silicon solar cells. As seen in previous studies, piezo-phototronics could be introduced as a solution. In an experiment by Zhu et al. [55], ZnO NW was incorporated into a silicon-based solar cell through a hydrothermal method. Implantations of ions and gold sputtering were performed before the synthesis of the nanowires as shown in Figure 15.12. Based on the static compressive strain tests performed, piezoelectric charges generated at the n-doped interface of the Si-ZnO improved the band structure and electron gas trapped, which enhanced the separation of light-induced carriers. Most significantly, the piezo-phototronics effect and strain that were introduced increased the efficiency of the solar cell from 8.97% to 9.51% [55].

15.5 CONCLUSION AND FUTURE REMARKS

Piezotronics and piezo-phototronics are unique emerging phenomena that have intrigued an in-depth study into their mechanisms and applications. Piezotronics occurs because of the coupling of properties from piezoelectricity and semiconductor. It can be seen through a piezoresistive process that involves a volume effect that is linear or a piezotronics process that occurs with a change in the interface. The effects of piezotronics are typically observed at a heterojunction and show non-linear current characteristics. When a heterojunction forms at the interface between a metal-semiconductor, a Schottky contact that allows the flow of current through a rectifying effect is formed. When an external strain is introduced, polarization occurs at the junction, which induces piezo-charges that alter the interfacial carrier composition. In typical cases, positive charges are induced on one end of the device and negative charges on the other. Positive charges have been observed to reduce the Schottky barrier height (SBH) and negative charges the converse effect at the SBH. This manipulation under the influence of a strain that was initially discovered in ZnO nanowires has sparked lots of other research interests and found many applications. Piezotronics can be used in lots of human-device interactions for pressures and chemical sensors, actuators, robots, medical implants, and other numerous sensory applications. Nanomaterials, especially nanowires have heavily contributed to piezotronics due to their non-centrosymmetric crystalline nature. Examples of these nanowires include ZnO, GaN, InN, and doped PZT. However, ZnO has been the most studied due to its easy fabrication in several shapes on a large scale and its less toxicity. Several methods have been used to synthesize nanowires that include wet approaches like hydrothermal and electrodeposition. As well as non-wet methods such as chemical or vapor deposition. It

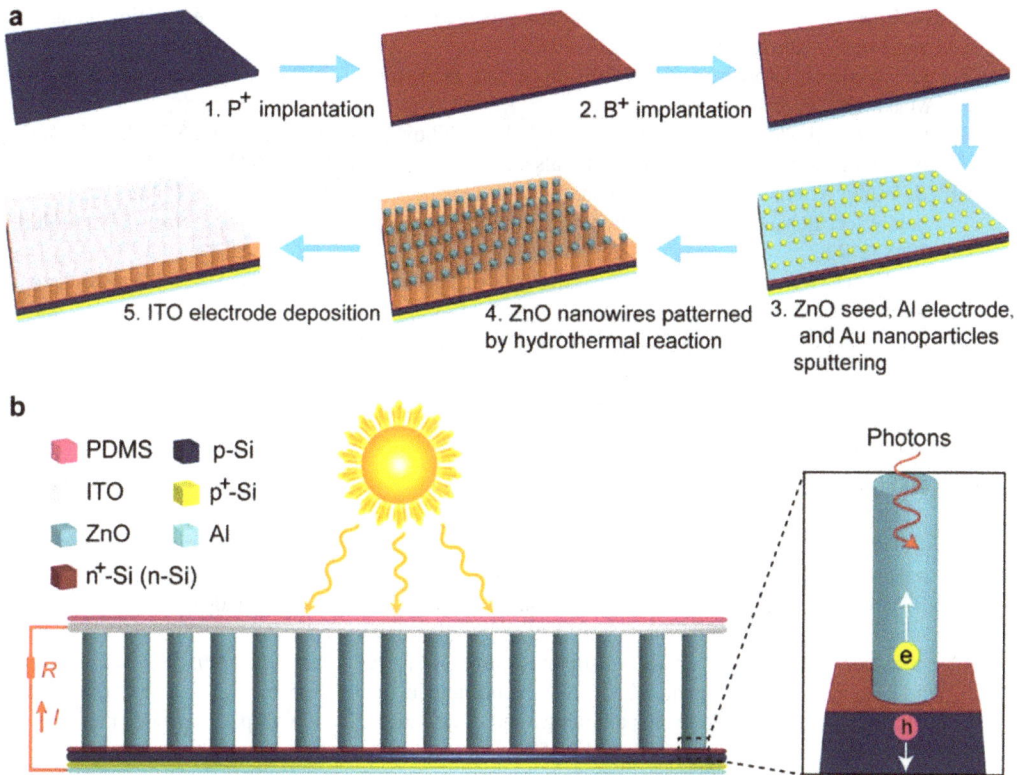

FIGURE 15.12 (a) Schematic for the fabrication of solar cells with the hydrothermal synthesis of ZnO nanowires. (b) Image of the silicon-based solar cell under the light. Adapted with permission [55]. Copyright (2017), American Chemical Society.

is also worth noting that the above synthesis, which typically involves a bottom to top approach can have issues of non-uniformity in the nanowire structure. Piezo-phototronics, which involves the three-way coupling of piezoelectricity, photoexcitation, and semiconductor has been investigated for optoelectronic devices. They can be used in UV detectors, electrochemical applications, solar cells, and human-device applications. Overall, piezotronics and piezo-phototronics have ignited an interesting field in science, especially for some nanoscale applications in devices. It is worth noting that even though current research is progressing in this field, there is still a lot to be uncovered. And more importantly, looking into the mechanisms of nanowires besides ZnO will be an interesting path to follow.

REFERENCES

[1] Z.L. Wang, Progress in piezotronics and piezo-phototronics, Adv. Mater. 24 (2012) 4632–4646.
[2] W. Wu, Z.L. Wang, Piezotronics and piezo-phototronics for adaptive electronics and optoelectronics, Nat. Rev. Mater. 1 (2016) 16031.
[3] Z.L. Wang, Nanopiezotronics, Adv. Mater. 19 (2007) 889–892.
[4] K. Jenkins, V. Nguyen, R. Zhu, R. Yang, Piezotronic effect: An emerging mechanism for sensing applications, Sensors (Switzerland). 15 (2015) 22914–22940.
[5] Z.L. Wang, Piezopotential gated nanowire devices: Piezotronics and piezo-phototronics, Nano Today. 5 (2010) 540–552.

[6] X. Chen, S. Xu, N. Yao, Y. Shi, 1.6 v nanogenerator for mechanical energy harvesting using PZT nanofibers, Nano Lett. 10 (2010) 2133–2137.

[7] L. Vayssieres, Growth of arrayed nanorods and nanowires of ZnO from aqueous solutions, Adv. Mater. 15 (2003) 464–466.

[8] Ü. Özgür, Y.I. Alivov, C. Liu, A. Teke, M.A. Reshchikov, S. Doğan, V. Avrutin, S.J. Cho, H. Morko, A comprehensive review of ZnO materials and devices, J. Appl. Phys. 98 (2005) 1–103.

[9] X. Wang, X. Wang, J. Zhou, J. Zhou, J. Song, J. Song, J. Liu, J. Liu, N. Xu, N. Xu, Piezoelectric field effect transistor and nanoforce sensor based on a single ZnO nanowire, Nano Lett. 6 (2006) 2768–2772.

[10] Y. Cui, Q. Wei, For highly sensitive and selective detection of biological and chemical species, Science (80) 293 (2001) 1289–1292.

[11] Z.L. Wang, W. Wu, C. Falconi, Piezotronics and piezo-phototronics with third-generation semiconductors, MRS Bull. 43 (2018) 922–927.

[12] C. Pan, L. Dong, G. Zhu, S. Niu, R. Yu, Q. Yang, Y. Liu, Z.L. Wang, High-resolution electroluminescent imaging of pressure distribution using a piezoelectric nanowire LED array, Nat. Photonics. 7 (2013) 752–758.

[13] J. Shi, M.B. Starr, X. Wang, Band structure engineering at heterojunction interfaces via the piezotronic effect, Adv. Mater. 24 (2012) 4683–4691.

[14] C. Pan, J. Zhai, Z.L. Wang, Piezotronics and piezo-phototronics of third generation semiconductor nanowires, Chem. Rev. 119 (2019) 9303–9359.

[15] B. Wei, K. Zheng, Y. Ji, Y. Zhang, Z. Zhang, X. Han, Size-dependent bandgap modulation of zno nanowires by tensile strain, Nano Lett. 12 (2012) 4595–4599.

[16] Y. Kanda, Piezoresistance effect of silicon, Sensors Actuators A. Phys. 28 (1991) 83–91.

[17] T.D. Nguyen, J.M. Nagarah, Y. Qi, S.S. Nonnenmann, A.V. Morozov, S. Li, C.B. Arnold, M.C. McAlpine, Wafer-scale nanopatterning and translation into high-performance piezoelectric nanowires, Nano Lett. 10 (2010) 4595–4599.

[18] S. Xu, Z.L. Wang, One-dimensional ZnO nanostructures: Solution growth and functional properties, Nano Res. 4 (2011) 1013–1098.

[19] S. Xu, Y. Wei, M. Kirkham, J. Liu, W. Mai, D. Davidovic, R.L. Snyder, L.W. Zhong, Patterned growth of vertically aligned ZnO nanowire arrays on inorganic substrates at low temperature without catalyst, J. Am. Chem. Soc. 130 (2008) 14958–14959.

[20] D.P. Singh, A.K. Ojha, O.N. Srivastava, Synthesis of different $Cu(OH)_2$ and CuO (nanowires, rectangles, seed-, belt-, and sheetlike) nanostructures by simple wet chemical route, J. Phys. Chem. C. 113 (2009) 3409–3418.

[21] L. Sztaberek, H. Mabey, W. Beatrez, C. Lore, A.C. Santulli, C. Koenigsmann, Sol-gel synthesis of ruthenium oxide nanowires to enhance methanol oxidation in supported platinum nanoparticle catalysts, ACS Omega. 4 (2019) 14226–14233.

[22] Y.X. Gan, A.H. Jayatissa, Z. Yu, X. Chen, M. Li, Hydrothermal synthesis of nanomaterials, J. Nanomater. 2020 (2020) 8917013.

[23] G. Yang, S.J. Park, Conventional and microwave hydrothermal synthesis and application of functional materials: A review, Materials (Basel). 12 (2019) 1177.

[24] C. Ou, P.E. Sanchez-Jimenez, A. Datta, F.L. Boughey, R.A. Whiter, S.L. Sahonta, S. Kar-Narayan, Template-assisted hydrothermal growth of aligned zinc oxide nanowires for piezoelectric energy harvesting applications, ACS Appl. Mater. Interfaces. 8 (2016) 13678–13683.

[25] L. Yu, G. Zhang, S. Li, Z. Xi, D. Guo, Fabrication of arrays of zinc oxide nanorods and nanotubes in aqueous solution under an external voltage, J. Cryst. Growth. 299 (2007) 184–188.

[26] C. V. Manzano, M. Martin-Gonzalez, Electrodeposition of v-vi nanowires and their thermoelectric properties, Front. Chem. 7 (2019) 1–22.

[27] X.J. Xu, G.T. Fei, W.H. Yu, X.W. Wang, L. Chen, L. De Zhang, Preparation and formation mechanism of ZnS semiconductor nanowires made by the electrochemical deposition method, Nanotechnology. 17 (2006) 426–429.

[28] R.A. Puglisi, C. Bongiorno, S. Caccamo, E. Fazio, G. Mannino, F. Neri, S. Scalese, D. Spucches, A. La Magna, Chemical vapor deposition growth of silicon nanowires with diameter smaller than 5 nm, ACS Omega. 4 (2019) 17967–17971.

[29] O. Tigli, J. Juhala, ZnO nanowire growth by physical vapor deposition, Proc. IEEE Conf. Nanotechnol. (2011) 608–611.

[30] H. Gao, W. Sun, Q. Sun, H.H. Tan, C. Jagadish, J. Zou, Compositional varied core-shell InGaP nanowires grown by metal-organic chemical vapor deposition, Nano Lett. 19 (2019) 3782–3788.

[31] Y. Yang, J.J. Qi, Y.S. Gu, X.Q. Wang, Y. Zhang, Piezotronic strain sensor based on single bridged ZnO wires, Phys. Status Solidi - Rapid Res. Lett. 3 (2009) 269–271.

[32] P. Dharap, Z. Li, S. Nagarajaiah, E. V. Barrera, Nanotube film based on single-wall carbon nanotubes for strain sensing, Nanotechnology. 15 (2004) 379–382.

[33] N. Hu, Y. Karube, M. Arai, T. Watanabe, C. Yan, Y. Li, Y. Liu, H. Fukunaga, Investigation on sensitivity of a polymer/carbon nanotube composite strain sensor, Carbon N. Y. 48 (2010) 680–687.

[34] J. Zhou, Y. Gu, P. Fei, W. Mai, Y. Gao, R. Yang, G. Bao, Z.L. Wang, Flexible piezotronic strain sensor, Nano Lett. 8 (2008) 3035–3040.

[35] G. Hu, R. Zhou, R. Yu, L. Dong, C. Pan, Z.L. Wang, Piezotronic effect enhanced Schottky-contact ZnO micro/nanowire humidity sensors, Nano Res. 7 (2014) 1083–1091.

[36] C. Pan, R. Yu, S. Niu, G. Zhu, Z.L. Wang, Piezotronic effect on the sensitivity and signal level of schottky contacted proactive micro/nanowire nanosensors, ACS Nano. 7 (2013) 1803–1810.

[37] Y. Zhang, Y. Liu, Z.L. Wang, Fundamental theory of piezotronics, Adv. Mater. 23 (2011) 3004–3013.

[38] W. Wu, C. Pan, Y. Zhang, X. Wen, Z.L. Wang, Piezotronics and piezo-phototronics - From single nanodevices to array of devices and then to integrated functional system, Nano Today. 8 (2013) 619–642.

[39] S. Wang, K. Chen, M. Wang, H. Li, G. Chen, J. Liu, L. Xu, Y. Jian, C. Meng, X. Zheng, S. Liu, C. Yin, Z. Wang, P. Du, S. Qu, C.W. Leung, Controllable synthesis of nickel nanowires and its application in high sensitivity, stretchable strain sensor for body motion sensing, J. Mater. Chem. C. 6 (2018) 4737–4745.

[40] W. Zhang, R. Zhu, V. Nguyen, R. Yang, Highly sensitive and flexible strain sensors based on vertical zinc oxide nanowire arrays, Sensors Actuators, A Phys. 205 (2014) 164–169.

[41] A.F. Smith, X. Liu, T.L. Woodard, T. Fu, T. Emrick, J.M. Jiménez, D.R. Lovley, J. Yao, Bioelectronic protein nanowire sensors for ammonia detection, Nano Res. 13 (2020) 1479–1484.

[42] R. Zhou, G. Hu, R. Yu, C. Pan, Z.L. Wang, Piezotronic effect enhanced detection of flammable/toxic gases by ZnO micro/nanowire sensors, Nano Energy. 12 (2015) 588–596.

[43] J. Wang, Electrochemical glucose biosensors, Electrochem. Sensors, Biosens. Their Biomed. Appl. (2008) 57–69.

[44] R. Yu, C. Pan, J. Chen, G. Zhu, Z.L. Wang, Enhanced performance of a ZnO nanowire-based self-powered glucose sensor by piezotronic effect, Adv. Funct. Mater. 23 (2013) 5868–5874.

[45] Z.L.W. Ruomeng Yu, Caofeng Pan, High performance of ZnO nanowire protein sensors enhanced by piezotronic effect, Energy. (2012) 3–10.

[46] F. Xue, L. Zhang, W. Tang, C. Zhang, W. Du, Z.L. Wang, Piezotronic effect on ZnO nanowire film based temperature sensor, ACS Appl. Mater. Interfaces. 6 (2014) 5955–5961.

[47] Y. Purusothaman, N.R. Alluri, A. Chandrasekhar, V. Venkateswaran, S.J. Kim, Piezophototronic gated optofluidic logic computations empowering intrinsic reconfigurable switches, Nat. Commun. 10 (2019) 1–9.

[48] Y. Liu, Q. Yang, Y. Zhang, Z. Yang, Z.L. Wang, Nanowire piezo-phototronic photodetector: Theory and experimental design, Adv. Mater. 24 (2012) 1410–1417.

[49] Z. Wang, R. Yu, X. Wen, Y. Liu, C. Pan, W. Wu, Z.L. Wang, Optimizing performance of silicon-based p-n junction photodetectors by the piezo-phototronic effect, ACS Nano. 8 (2014) 12866–12873.

[50] M. Peng, Y. Liu, A. Yu, Y. Zhang, C. Liu, J. Liu, W. Wu, K. Zhang, X. Shi, J. Kou, J. Zhai, Z.L. Wang, Flexible self-powered gan ultraviolet photoswitch with piezo-phototronic effect enhanced on/off ratio, ACS Nano. 10 (2016) 1572–1579.

[51] Y. Hu, Y. Zhang, Y. Chang, R.L. Snyder, Z.L. Wang, Optimizing the power output of a ZnO photocell by piezopotential, ACS Nano. 4 (2010) 4962.

[52] Q. Yang, W. Wang, S. Xu, Z.L. Wang, Enhancing light emission of ZnO microwire-based diodes by piezo-phototronic effect, Nano Lett. 11 (2011) 4012–4017.

[53] J. Shi, M.B. Starr, H. Xiang, Y. Hara, M.A. Anderson, J. Seo, Z. Ma, X. Wang, Interface engin-
      eering by piezoelectric potential in ZnO-based photoelectrochemical anode, Small. (2011)
      5587–5593.

[54] F. Zhang, S. Niu, W. Guo, G. Zhu, Y. Liu, X. Zhang, Z.L. Wang, Piezo-phototronic effect enhanced
      visible/UV photodetector of a carbon-fiber/ZnO-CdS double-shell microwire, ACS Nano. 7 (2013)
      4537–4544.

[55] L. Zhu, L. Wang, C. Pan, L. Chen, F. Xue, B. Chen, L. Yang, L. Su, Z.L. Wang, Enhancing the
      efficiency of silicon-based solar cells by the piezo-phototronic effect, ACS Nano. 11 (2017)
      1894–1900.

16 Silver Nanowire-based Capacitive Type Pressure and Strain Sensors for Human Motion Monitoring

Fevzihan Basarir, Hamidreza Daghigh Shirazi,
Zahra Madani, and Jaana Vapaavuori
Department of Chemistry and Materials Science, Aalto University,
Finland

CONTENTS

DOI: 10.1201/9781003296621-16

16.1   INTRODUCTION

Recently, significant research effort has been dedicated to human motion detection and monitoring that enables tracking and recording a subject's real-time body movement [1]. Quantitative evaluation of joint and muscle movement is precious for patient rehabilitation and early diagnosis of aging-related diseases [2]. Skin mountable flexible and stretchable sensors have played an important role in monitoring and investigating human motion and particularly, utilization of pressure and strain sensors have been successfully demonstrated [3,4]. Depending on the working principle, pressure and strain sensors could be classified into piezoelectric, piezoresistive, and capacitive sensors. Compared to the others, capacitive sensors provide higher accuracy and linearity, faster response time, less power consumption and hysteresis, and less dependence on external conditions (temperature and humidity) [5].

Conventional flexible capacitive sensors include a parallel-plate capacitor, which comprises a dielectric layer sandwiched in-between two conformable electrodes. Electrically conductive electrodes and dielectric layer are the fundamental parts of flexible capacitive pressure and strain sensors. The incorporation of electrically conductive materials, including carbon nanotubes, graphene, Mxenes, and silver nanowires (AgNWs) into a flexible polymer matrix has been successfully implemented to fabricate the flexible and stretchable electrodes [6]. Among these, AgNWs have been preferred because of their high inherent conductivity, excellent flexibility, and resistance to oxidation [7]. For the construction of sensors, typically, AgNWs are coated on a flexible substrate to fabricate the electrodes and a flexible dielectric layer is sandwiched between the electrodes. Concerning the pressure sensor, applied pressure results in a dielectric layer thickness decrease, which provides a change in the capacitance of the entire system. However, in the strain sensors, uniaxial stretching of the sensors renders a change in the active area of the sensor and a decrease in thickness, which in turn leads to capacitance variation [5].

This chapter aims to summarize how it is possible to construct AgNW-based pressure and strain sensors for human motion monitoring applications. Additionally, we demonstrate the sensor properties and performance of such devices.

16.2   SYNTHESIS METHODS OF SILVER NANOWIRES

Studying different methods to fabricate AgNWs is of great importance due to the high dependence of pressure and strain sensors on the cost, morphology, and size of nanowires. AgNWs can be obtained by using different synthesis methods, including polyol, solvothermal, hydrothermal, and electrochemical methods. The latter, however, has lost its popularity due to irregularity in the resulting morphology, low aspect ratio, and low yield.

In the polyol method, AgNWs are obtained by using ethylene glycol as the solvent/reducing agent and reducing $AgNO_3$ [8]. In the presence of poly(vinylpyrrolidone) (PVP) capping agent, the AgNWs are achieved via unidirectional growth. The process is often performed at high temperatures to dissolve $AgNO_3$ and PVP. The process also results in byproduct nanoparticles, which are removed by a centrifugation process at the end. The process can be precisely tailored by adjusting the process parameters, such as $AgNO_3$ concentration, PVP molecular weight and concentration, and temperature [9]. It is proposed that the growth of five-fold multiple-twinned particles (MTPs) forms the final AgNWs. PVP facilitates the addition of ions to {111} faces and passivates the {100} faces. Therefore, silver nanorods are formed by crystallization of Ag atoms on MTPs, and their further growth leads to the formation of AgNWs, with typical diameters of

30–60 nm and lengths of up to 50 μm. **Figure 16.1** demonstrates MTPs, as well as the standard morphology and crystallinity of AgNWs.

The solvothermal synthesis method is introduced as an alternative to the polyol method since it offers a facile cost-effective synthesis process with simple tuning capability [12]. In this method, $AgNO_3$, PVP, metal salts, and ethylene glycol are typically mixed, transferred to an autoclave, and ultimately heated in an oven. The molecular weight of PVP, type of solvent/salt, and reducing agent are the influential parameters affecting the final morphology of the AgNWs. A hydrothermal method is considered a green process to obtain AgNW, due to substituting ethylene glycol with water. A mild reducing agent (e.g., glucose) is used, and similar to the solvothermal method, the process is done in an autoclave [13]. This method is a relatively new and comprehensive research on the process parameters is still required.

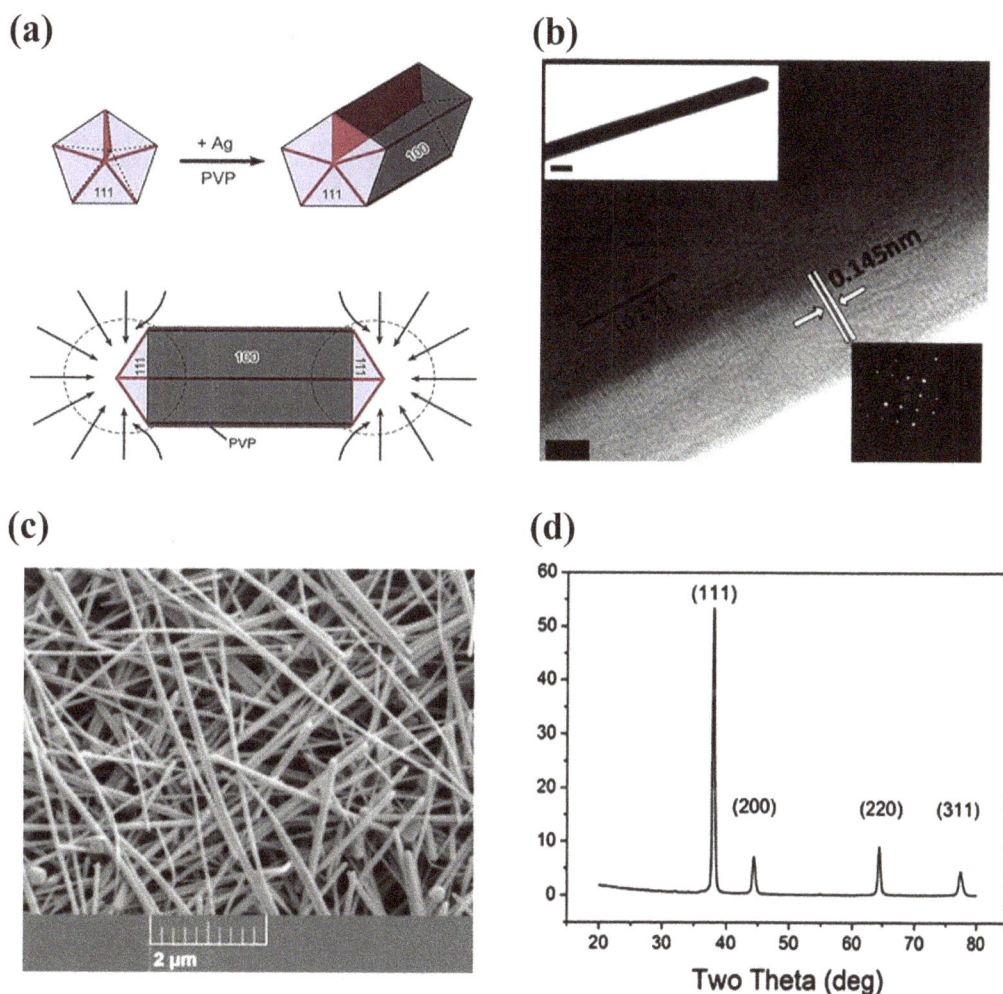

FIGURE 16.1 (a) Schematic illustration of the AgNW growth mechanism (adapted with permission [10]. Copyright 2003, American Chemical Society), (b) TEM image and electron diffraction pattern of AgNWs synthesized with polyol method (adapted with permission [11]. Copyright 2012, American Chemical Society), (c) SEM image of AgNWs synthesized with polyol method (adapted with permission [9]. Copyright 2017, The Electrochemical Society) and (d) X-ray diffraction pattern of AgNWs synthesized with polyol method (adapted with permission [11]. Copyright 2012, American Chemical Society).

In other attempts to produce AgNWs, porous anodic aluminum oxide and hexagonal mesoporous silica have been employed in template-directed synthesis approaches [14]. Despite enabling control over the morphology of AgNWs, these techniques suffer from limitations in large-scale fabrication, requiring template removal post-treatment, and being time-consuming.

Fabrication of AgNWs is also feasible by decomposing $AgNO_3$ precursors via a UV irradiation method. This approach is also time-consuming and difficult to scale up. Thus, further methods based on inducing laser photoactuation or photothermal heating/welding have been introduced to obtain AgNWs from $AgNO_3$ by using templates in a photoreduction technique [14]. This method also suffers from scaling-up limitations and time-consuming processes. All in all, the polyol method has been widely used in pressure sensor applications, since it benefits from high yield, facile process, and low cost.

16.3 FLEXIBLE CAPACITIVE PRESSURE SENSORS

Typically, flexible capacitive pressure sensors comprise sandwiching a flexible dielectric layer between two flexible electrodes. In this section, we will investigate the components of AgNW-based flexible capacitive pressure sensors as well as the sensing mechanism of the sensor.

16.3.1 PRINCIPLE OF CAPACITIVE SENSING

The capacitive sensor usually consists of two parallel-plate electrodes with a dielectric interlayer sandwiched in-between. The capacitance of the structure can be calculated by

$$C = \frac{\varepsilon\,A}{d} \tag{16.1}$$

where ε, A, and d are the dielectric permittivity of the interlayer material, effective electrode area, and the distance between the two plate electrodes, respectively [5]. It is apparent from Equation 16.1 that the capacitance change could be simply achieved by a change in either ε, A, or d. However, since the applied pressure mainly influences the ε and d, the capacitance change of the sensors emerges from the change in ε and d. The schematic working mechanism is briefly demonstrated in **Figure 16.2**.

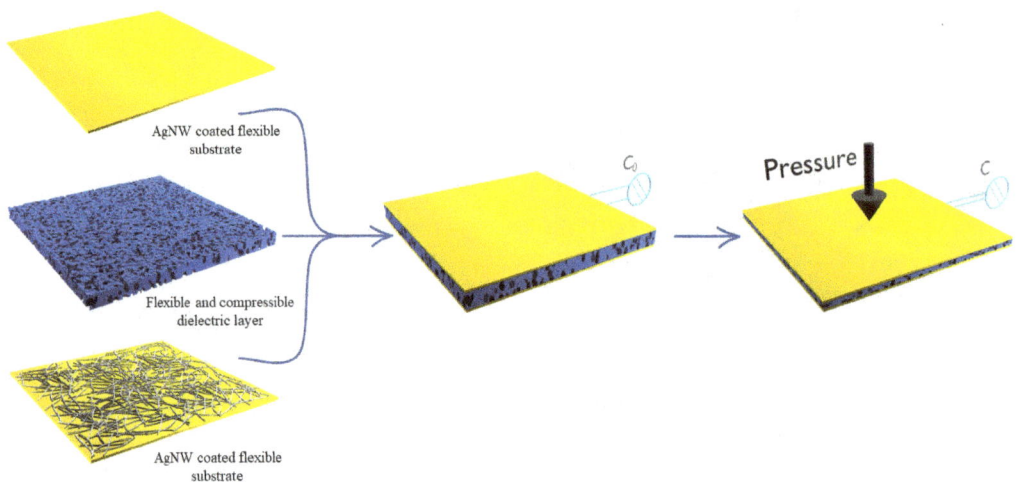

FIGURE 16.2   Schematic demonstration of capacitive pressure sensor working mechanism.

16.3.2   AgNW-Based Electrode Formation

To fabricate pressure and strain sensors, embedding, coating, or patterning the AgNWs on the flexible and/or stretchable substrate is the key step.

16.3.2.1   Plain Electrodes

In plain electrodes, mainly the formation of an interconnected network is sought. This is, for instance, achieved by annealing an AgNW-coated glass and coating it with PDMS, which results in a flexible PDMS/AgNW electrode [15], as shown in **Figure 16.3a**.

16.3.2.2   Microstructured Electrodes

The presence of microstructures on electrodes, such as wrinkles, increases sensitivity by causing localized stress [7]. These microstructures can be achieved in the form of wrinkles and microarrays, for instance, by stretching a multilayer with mismatching mechanical properties or replicating wide varieties of existing microstructures to the substrate by lithography and **Figure 16.3b** depicts these microstructures.

On the other hand, biomimicry is an intriguing approach to replicate the existing biological microstructures onto the electrodes. A renowned example of this approach is the replication of lotus leaf microstructures, followed by coating an ultrathin layer of AgNWs (**Figure 16.3c**) [16]. The skeleton of leaves contains interconnected microstructures, which can be used for coating AgNWs to reach a conductive network. Biomimicry of different biological surfaces has shown successful results, e.g., in the case of rubber tree leaves [17], however, the current information on natural surfaces is quite limited and requires extensive exploration.

16.3.2.3   Porous Electrode

Porous electrodes can be simply formed with inherently connected conductive and dielectric layers by using, e.g., sacrificial space-holders [18], porous knitted wool fabric [19], and electrospinning of nanofiber membranes [20]. A profound approach to achieving such sensors has been proposed by depositing AgNWs on the opposite side of a melamine sponge and having the internal medium as the dielectric layer [21]. The porosity of these electrodes enables production of breathable sensors with potential wearable applications. **Figure 16.3d** shows the porous electrodes fabricated using the abovementioned methods.

16.3.2.4   Yarn-Based Electrode

The next generation of electronic textiles (e-textiles) requires flexible, lightweight, and wearable sensors. Using AgNWs in yarns can lead to desired electrical properties, realizing these requirements. Sensors with polyurethane fiber twined polyester yarn coated with AgNWs, along with Ecoflex, PDMS, and Dragon skin dielectric materials have already been successfully fabricated (**Figure 16.3e**) [22]. Furthermore, fabrication of hierarchical porous structure of wet-spun bacterial cellulose/AgNWs with PDMS as the dielectric layer has been reported, showing notably high radial compressibility and thus a large capacitive change [23]. Perpendicular assembly of fibers on the substrates leads to the formation of sensing pixels.

16.3.2.5   Patterned Electrodes

Patterning of AgNWs enables the possibility to fabricate capacitive pressure sensor arrays. Several methods have been employed to produce arrays in different length scales, including screen printing [24], parylene-C stencil patterning [25], lift-off micropatterning [26], selective light irradiation [27], and selective spray coating of Mxene/AgNW ink. Due to the presence of flexible polymeric substrates, harsh conditions are avoided via these approaches. Although the sensor arrays provide more information as compared to a single sensor, in most cases, the performances of the fabricated

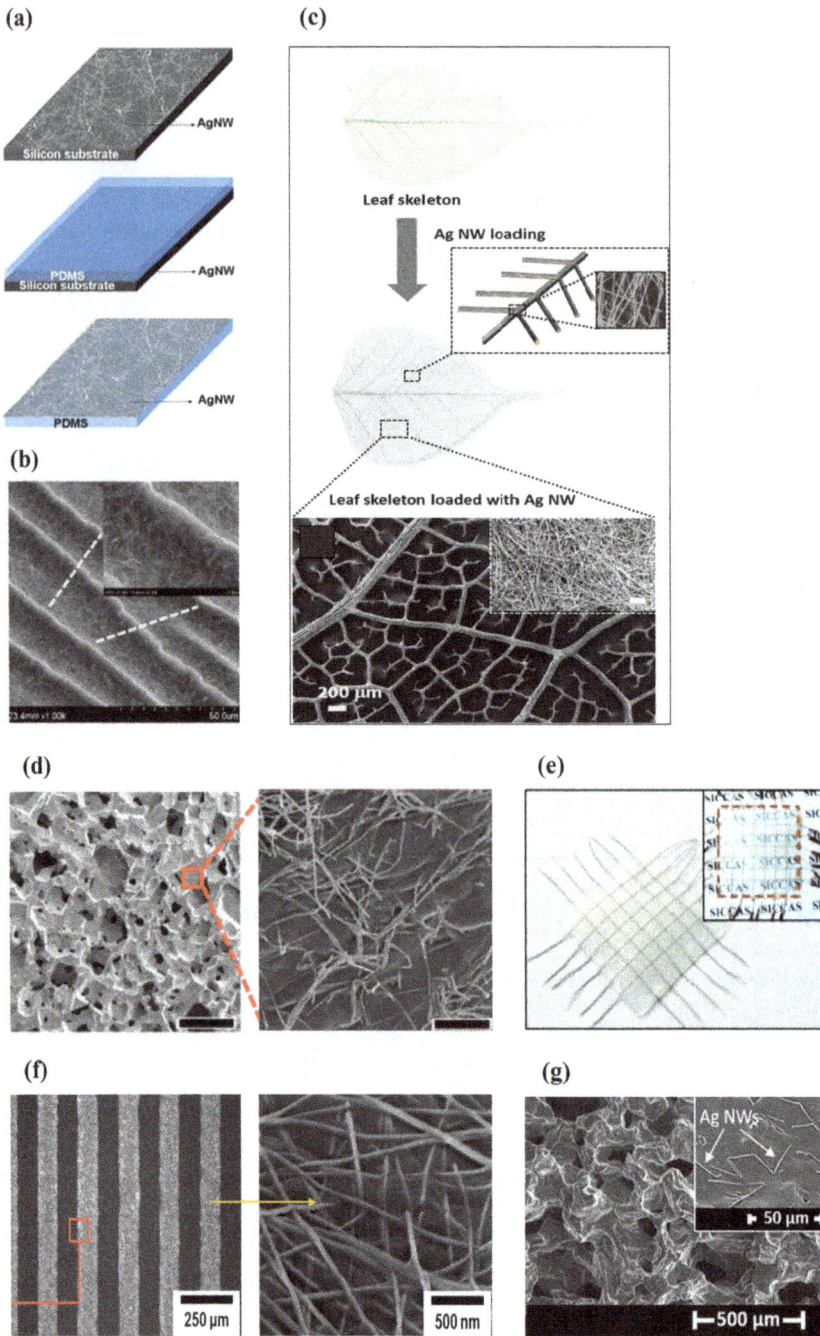

FIGURE 16.3 (a) Schematic plain electrode fabrication process via transferring of the AgNWs to PDMS substrate (adapted with permission [15]. Copyright 2015, IEEE), (b) AgNW embedded buckle like microstructured electrode (adapted with permission [7]. Copyright 2015, Royal Society of Chemistry), (c) AgNW coated Bodhi tree leaf electrode (adapted with permission [16]. Copyright 2020, Wiley-VCH GmbH), (d) porous PDMS/AgNW electrode fabricated from sugar template (adapted with permission [18]. Copyright 2020, American Chemical Society), (e) photograph of the pressure sensor array with AgNW coated yarns (adapted with permission [22]. Copyright 2017, Royal Society of Chemistry), (f) SEM images of the AgNW lines prepared by selective light irradiation (adapted with permission [27]. Copyright 2017, The Royal Society of Chemistry), and (g) SEM image porous Ag NW/PVA nanocomposite dielectric layer (adapted with permission [31]. Copyright 2021, Wiley-VCH GmbH).

electrodes were mainly tested as a proof of concept. Thus, a comprehensive investigation of the sensor performance, along with optimization of patterning techniques, is still required. A typical AgNW pattern is demonstrated in **Figure 16.3f**.

16.3.3 AgNW-Based Dielectric Formation

The capacitance change of pressure sensors is highly dependent on the compressibility and response of the dielectric layer upon externally induced thickness change. For this means, pressure sensors were prepared by using different methods, such as electro-spun core-shell AgNWs/TPU nanofiber dielectric in between gold/PDMS [28] and CNT/TPU electrodes [29], and also inserting AgNWs/PDMS microstructured layer in between ITO/PET electrodes [30]. Furthermore, since sensitivity is highly influenced by dielectric constant and porosity, porous PVA/AgNW dielectric layers were introduced as shown in **Figure 16.3g** [31]. All in all, the great potential of tunable porous materials, such as flexible sponges, aerogels, and foams should yet be investigated.

16.3.4 Performance of the Pressure Sensors

16.3.4.1 Sensitivity

Sensitivity is the most important parameter of a flexible pressure sensor, which identifies the range of available applications. The sensitivity of a capacitive sensor is typically calculated with Equation 16.2

$$S = \frac{(C/C_0)-1}{\Delta P} = \frac{\Delta C}{\Delta P} \tag{16.2}$$

where S, C, C_0, and ΔP represent the sensitivity, capacitance after applying pressure, initial capacitance, and pressure change, respectively.

As described in Equation 16.2 and illustrated in **Figure 16.4a**, the sensitivity of a pressure sensor depends on relative capacitance change with the applied external pressure. It is worthy to note that sensitivity is higher if the capacitance change is higher with the applied external pressure [20]. Five different pressure regimes are typically categorized in sensor performance characterization: (1) ultralow pressure (<1 Pa), (2) subtle pressure (1 Pa–1 kPa), (3) low pressure (1 kPa–10 kPa), (4) medium pressure (10–100 kPa), and (5) high pressure (>100 kPa) range [32]. Sensitivity in one or more regimes might be important, which depends on the selected application. Notably, achieving high sensitivity in low and medium pressure regimes is very crucial for human motion monitoring.

16.3.4.2 Response Time

Response time is another significant performance indicator for the pressure sensors, which demonstrates the sensor's response rate to reach the steady-state capacitance value after applying pressure, as shown in **Figure 16.4b** [20]. It is highly crucial to note that faster response, at millisecond level, is highly needed in human motion monitoring applications.

16.3.4.3 Dynamic Durability

Dynamic durability is considered an important indicator for flexible pressure sensor performance. It is related to the sensor's electrical and mechanical integrity to pressure loading/unloading cycles. **Figure 16.4c** demonstrates a cycling test of the pressure sensor over 10,000 loading-unloading cycles under the pressure of 15 kPa. It is notable that selecting the right pressure and loading/unloading frequency is very important because they have a great influence on the fatigue or irreversible deformation of the sensor. However, the selected pressure should be in line with the human body pressure generated by the body's motion activities. On the other hand, bendability is an important performance criterion in sensor performance and long-term dynamic durability.

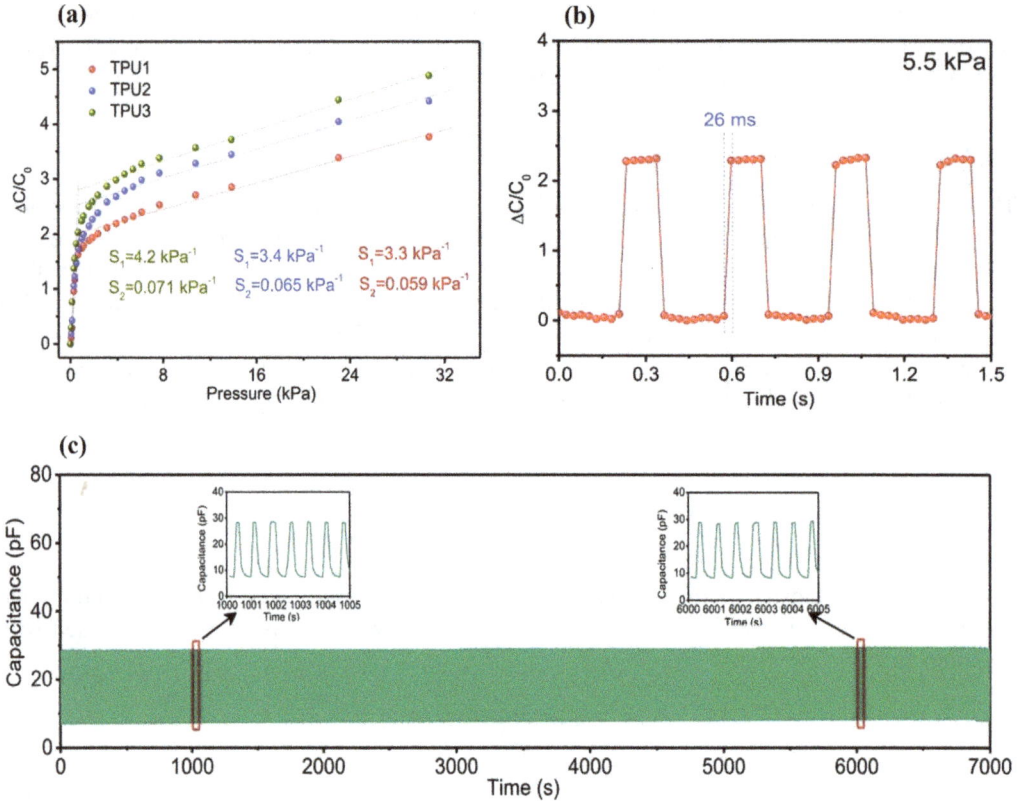

FIGURE 16.4 Performance of flexible pressure sensors. (a) Capacitance change with the applied pressure and the related sensitivities within specific pressure regimes, (b) demonstration of a response time of pressure sensor under an applied pressure of 5.5 kPa, and (c) cycling test of pressure sensor over 10,000 loading-unloading cycles at a frequency of 1.43 Hz under the pressure of 15 kPa (a-c, adapted with permission [20]. Copyright 2017, Wiley-VCH GmbH).

16.3.4.4 Limit of Detection

The minimum pressure that can be recognized by the pressure sensor might be described as a limit of detection (LOD). The sensors with higher sensitivity have lower LOD values and the AgNW-based sensors have LOD typically in the range of 1–3 Pa.

16.4 FLEXIBLE CAPACITIVE STRAIN SENSORS

Stretchable capacitors could be utilized as strain sensors owing to their ability under large strain. The critical elements of the stretchable strain sensors are the stretchable electrodes and dielectric layers, which should be mechanically and electrically reliable over large strain ranges. Upon applying strain, due to the mechanical deformation, the geometry of the capacitor sensor changes, which in turn leads to capacitance change [4,33]. The working principle of the strain sensor is schematically illustrated in **Figure 16.5**.

16.4.1 PRINCIPLE OF SENSING

Stretching capacitive-type sensors results in a change in capacitive area and a reduction in the thickness of a dielectric layer, which gives rise to the capacitance. The initial capacitance value of the strain sensor is calculated with Equation 16.3.

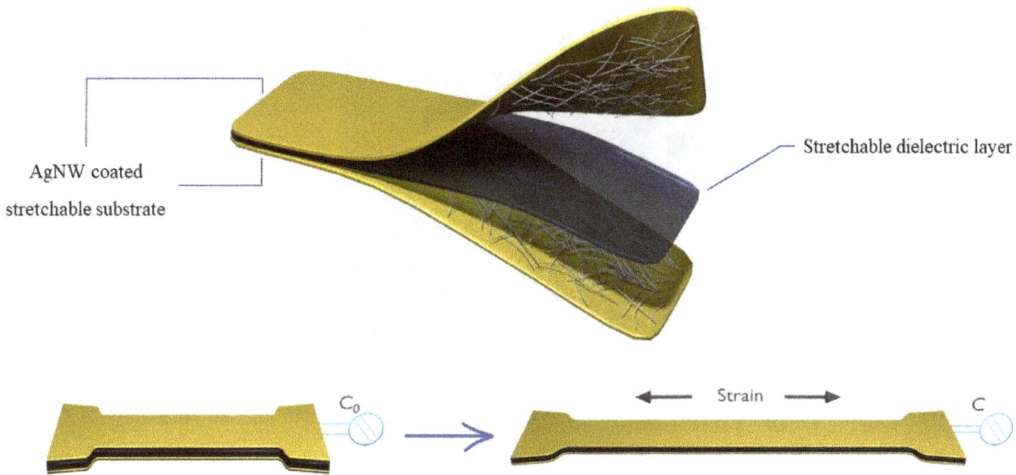

FIGURE 16.5 Schematic demonstration of capacitive strain sensor working mechanism.

$$C_0 = \varepsilon_0 \varepsilon_r \, l_0 \, w_0 / d_0 \qquad (16.3)$$

where $\varepsilon_0, \varepsilon_r, l_0, w_0$, and d_0 symbolize the dielectric constant of vacuum, the relative permittivity of dielectric media, initial length and width of the capacitor, and thickness of the dielectric layer, respectively.

Stretching the sensor results in a change in the geometry of the sensors and the deformation of the sensor is mainly influenced by Poisson's ratio of the dielectric layer and flexible substrate. For example, if the sensor is stretched to a specified strain ε, the geometry of the sensor including capacitor length, capacitor width, and dielectric layer thickness will change to $(1+\varepsilon)l_0$, $(1-v_{electrode})w_0$, and $(1-v_{dielectric})d_0$, respectively.

By assuming that Poisson's ratio for both stretchable electrodes and dielectric layer are similar, capacitance under stretching can be calculated with Equation 16.4.

$$C = \varepsilon_0 \varepsilon_r \, (1+\varepsilon)l_0 \, (1-v_{electrode})w_0 / (1-v_{dielectric})d_0 = \varepsilon_0 \varepsilon_r \, l_0 \, w_0 / d_0 = (1+\varepsilon)C_0 \qquad (16.4)$$

Based on the equation, it is obvious that there is a linear relationship between the applied strain and the capacitance of the sensors.

16.4.2 AGNW-BASED ELECTRODE FORMATION

16.4.2.1 Plain Electrode

Exploring new ways of directly coating or transferring the AgNWs in the form of an interconnected network to a stretchable substrate is a crucial element for the fabrication of plain electrodes. An important example is the stretchable PDMS/AgNW electrode fabricated by Xu et al. via coating of the silicon wafer with AgNWs and subsequently annealing them, followed by coating with liquid PDMS rubber and curing it, which in turn led to the transfer of the AgNWs onto the PDMS surface [34]. The schematic demonstration of the fabrication process and the transfer of the AgNWs to the PDMS substrate is shown in **Figure 16.6a**. A similar approach was also utilized by Hu et al. to transfer the AgNWs to a stretchable polyurethane (PU) substrate [35] to fabricate strain sensor.

16.4.2.2 Patterned Electrode Arrays

To obtain more information from the sensor, placing the electrodes in array form in the sensor is essential. Therefore, various approaches have been exhibited in the literature for the patterning of

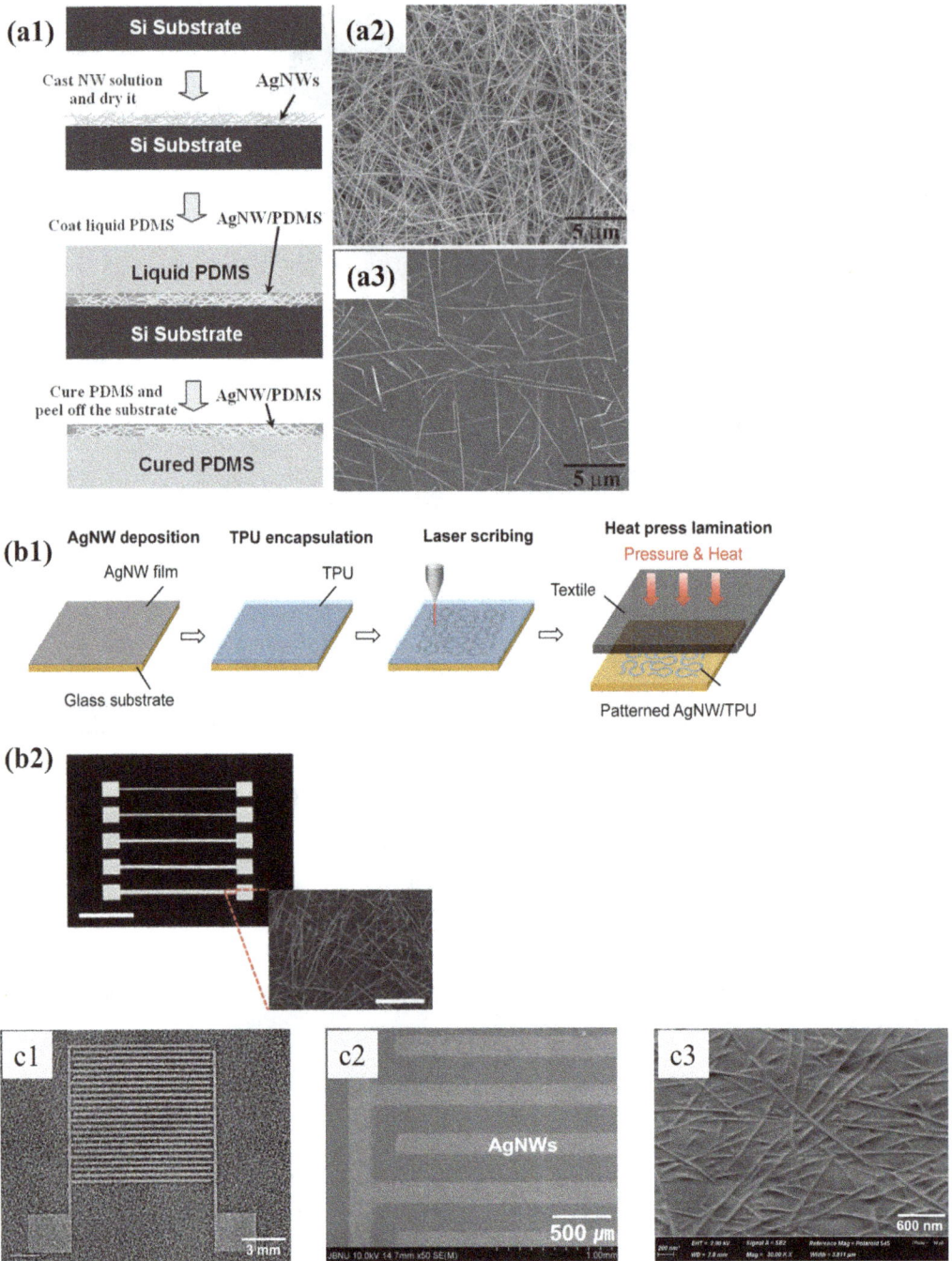

FIGURE 16.6   (a1) Schematic demonstration of the plain electrode fabrication, (a2) SEM image of the AgNW film on the silicon substrate, (a3) SEM image of the AgNW film transferred to flexible PDMS substrate (a1-a3, adapted with permission [34]. Copyright 2012, Wiley-VCH GmbH), (b1) schematic demonstration of patterned electrode arrays, (b2) AgNW patterns with different line widths (scale bar: 5 mm) and the inset shows the close-up SEM image of the AgNW pattern (scale bar: 5 μm) (b1-b2, adapted with permission [38]. Copyright 2019, American Chemical Society), (c1) optical micrograph of a tandem compound AgNW pattern impregnated under the surface of 184 PDMS, (c2) SEM image of the AgNW pattern impregnated under the surface of the 184 PDMS, and (c3) close-up SEM image of the AgNW pattern (c1-c3, adapted with permission [40]. Copyright 2021, Wiley-VCH GmbH).

AgNWs to fabricate stretchable capacitive strain sensor arrays. For instance, Yao *et al.* demonstrated the utilization of screen printing for the preparation of AgNW patterns on a PDMS substrate that has a linewidth and a spacing of 2 mm [24]. Besides, AgNW patterns were achieved on a PDMS [36] and Ecoflex substrate [37] via drop casting of AgNW solution through a polyimide mask. On the other hand, a mechanically and electrically robust stretchable strain sensor integrated with textile was fabricated via laser scribing and heat press lamination [38]. The schematic fabrication process and the ultimate AgNW patterns are shown in **Figure 16.6b**.

16.4.2.3 Interdigitated Electrodes

The parallel-plate-type capacitive strain sensors discussed above have a confirmation difficulty for human skin and additionally, the capacitance can be influenced both by the strain and pressure simultaneously [39]. Interdigitated capacitive strain sensors (ICSSs) showed better performance in terms of sensitivity, linearity, and hysteresis [40–42]. The ICSSs comprise in-plane electrodes which are parallel to their substrates (**Figure 16.6c1**). For instance, Ju *et al.* demonstrated the fabrication of AgNW-based interdigitated electrodes via patterning of AgNWs on a Kapton™ film by coating a UV curable polymer and selective UV light irradiation through a photomask, followed by transferring the electrode patterns to a PDMS substrate [40]. The ultimate ICSS structure is demonstrated in **Figure 16.6c**. Similarly, interdigitated electrodes were prepared on a stretchable polyurethane urea (PUU) substrate via patterning the AgNWs on a Kapton™ film by spin coating of the AgNWs and patterning via photolithography, which was followed by transferring to PUU substrate [41]. Consequently, Kim *et al.* introduced the capillary force lithography (CFL) technique to form the interdigitated electrodes [39], which includes filling the hydrogel solution in the PDMS stamp channels placed on AgNW coated substrate by capillary force channels, followed by UV curing and developing the patterns.

16.4.3 PERFORMANCE OF THE STRAIN SENSORS

A stretchable strain sensor's performance is evaluated by numerous parameters, including sensitivity, stretchability, linearity, response time, hysteresis, and dynamic durability. In this section, we focus on the performance parameters of stretchable strain sensors in detail.

16.4.3.1 Sensitivity

The sensitivity or gauge factor (GF) of stretchable strain sensors can be calculated by GF = $(\Delta C/C_0)/\varepsilon$, which could be evaluated by the slope of the relative change of the capacitance, versus applied strain [33], as demonstrated in **Figure 16.7a**. However, the theoretical gauge factor is limited (GF ≤ 1) as it can be determined from Equation 16.5.

$$GF = (\Delta C/C_0)/\varepsilon = ((1 + \varepsilon)C_0 - C_0)/\varepsilon C_0 = 1 \qquad (16.5)$$

Therefore, the closer the gauge factor is to one, the sensor exhibits better performance. **Figure 16.7a** demonstrates the performance of an AgNW-based strain sensor with a calculated GF of 0.96.

16.4.3.2 Stretchability

Stretchability is also an important indicator of sensor performance, and it can be defined as the maximum strain that can be achieved with the sensor without losing the mechanical integrity and electrical response [4,43]. It is worthy to note that the stretchability of the substrate and dielectric layer should comply with each other to maintain the performance of the sensor. Utilization of AgNWs can have a great impact on the stretchability of strain sensors resulting from a high aspect ratio, and easy and effective percolation network. It is important to keep in mind that stretchable sensors must withstand high levels of strain (e.g., 50% or more) to track human motion [5]. **Figure 16.7b** shows the stretchability of an AgNW-based strain sensor up to 50% strain.

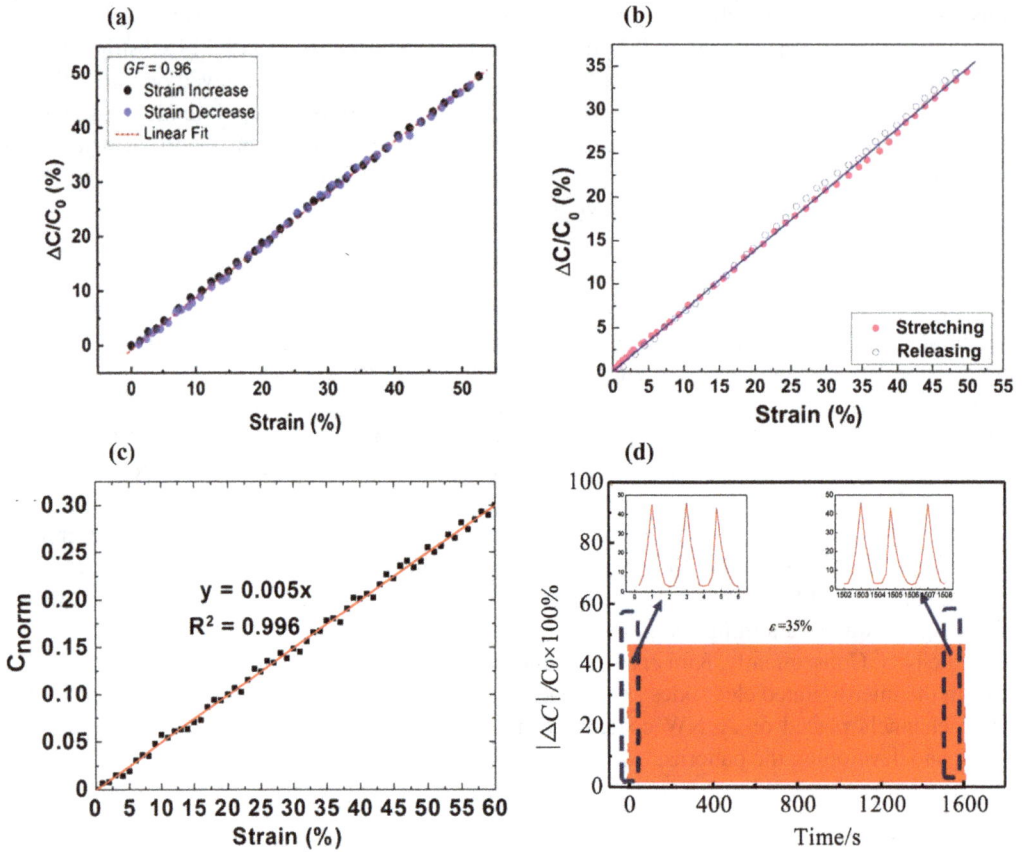

FIGURE 16.7 Performance of the stretchable strain sensors. (a) Textile-integrated capacitive strain sensor during reversible straining (adapted with permission [38]. Copyright 2019, American Chemical Society), (b) relative capacitance change versus applied strain for stretching and releasing (adapted with permission [24]. Copyright 2014, The Royal Society of Chemistry), (c) linearity of the strain sensor (adapted with permission [35]. Copyright 2013, American Institute of Physics), and (d) long-term stable response of the stretchable sensor to cyclic strain loads (adapted with permission [44]. Copyright 2022, IOP Publishing Ltd).

16.4.3.3 Linearity

Linearity is another performance indicator for a stretchable strain sensor and as shown in **Figure 16.7c**, it is evaluated by the coefficient of determination (denoted as R^2) obtained with a simple linear regression analysis. If the sensor is regarded as linear, it has a higher R^2 value. It was reported that capacitive-type sensors provide excellent linearity with low sensitivity [4].

16.4.3.4 Response Time

Response time indicates the time needed for the sensor to reach the steady-state capacitance value and it is usually quantified experimentally from the capacitance vs. time graph [4].

16.4.3.5 Dynamic Durability

Dynamic durability indicates the sensor's long-term stability to stretching and releasing cycles without losing the mechanical and electrical properties [43], as demonstrated in **Figure 16.7d**. Highly durable sensors are required for wearable sensor applications.

16.4.3.6   Hysteresis

Hysteresis depicts the dependence of the state of a sensor on its history and it becomes important when the sensor is subjected to dynamic starching/releasing cycles [43]. If the sensor is believed to have large hysteresis, it will face irreversible sensor performance. Hysteresis is mainly caused by the viscoelastic nature of the substrate and/or dielectric layer.

16.5   HUMAN MOTION MONITORING APPLICATION

Human motion monitoring is very crucial for tracking patient rehabilitation or early diagnosis of aging-related diseases [1,2]. Recent advances in functional materials, artificial intelligence, and Internet of Things (IoT) technology enabled the development of wearable sensors for human motion monitoring applications. Flexible pressure sensors and or stretchable strain sensors attached to the human body can monitor physiological parameters, such as hand motion (finger movement), joint movement (knee, wrist, and scapular movement), and subtle human motion (eye blinking, pulse, and respiration). For instance, Cheng *et al.* demonstrated the utilization of yarn based flexible pressure sensor that was conformably mounted on the back of the human hand, as shown in **Figure 16.8a** [22]. The sensor successfully provided real-time electrical signal change upon bending the palm. Additionally, Kim *et al.* fabricated interdigitated capacitive strain sensors (ICSSs) and applied them to body motion sensing including finger and wrist bending (**Figure 16.8b, c**) [39]. Upon bending and releasing the finger and wrist, capacitance changes of 10% and 6% were recorded, respectively. The results proved that ICSSs could be used as wearable strain sensors for human motion monitoring. In addition, Yao *et al.* integrated strain sensor arrays on textile via laser scribing and heat press lamination [38]. As illustrated in **Figure 16.8d**, the strain sensor was placed on the elbow to monitor the elbow bending at different bending speeds. It was shown that the measured strain is linearly proportional to the bending angle of the elbow. Real-time information about bending angle, strain and angular velocity could be obtained with the mentioned textile integrated strain sensors. Moreover, Mo *et al.* employed a novel microstructured dielectric layer for developing a flexible pressure sensor [45]. The flexible sensor was mounted on the human arm and used for monitoring the arm bending, as shown in **Figure 16.8e**. Real-time capacitance change was observed with bending and releasing the arm with the sensor. In another work, Shuai *et al.* prepared a flexible pressure sensor with microstructured electrodes [46]. The sensor was mounted on the human neck to monitor the voice vibrations during talking (**Figure 16.8f**) and capacitance measurements reflected the outstanding capability of the sensor to detect the human voice. Besides that, Yang *et al.* developed a smart mask with a breathable pressure sensor based on nanofiber membranes [20], as illustrated in **Figure 16.8g**. The pressure on the mask caused by the breath was successfully monitored in real-time with the sensor. Lately, Chhetry *et al.* demonstrated a highly sensitive flexible pressure sensor based on randomly distributed microstructured iontronic film [47]. The sensor was utilized for monitoring physiological signals from eye blinking and the human wrist, as shown in **Figure 16.8h, i**. The capacitance measurements provided not only the wrist pulse rate of a healthy volunteer but also the profile of the open/close movement of an eye.

16.6   CONCLUSIONS

Herein, we have summarized the recent progress of AgNW-based flexible capacitive pressure sensors and stretchable strain sensors and their possible implementations in human motion monitoring. Synthesis methods of the AgNWs were briefly introduced and compared. Owing to their facile synthesis, processability, outstanding electrical property, and flexibility, AgNWs have been widely utilized as an electrode material as well as a filler material in the dielectric layer. The structure, sensing mechanism, and performance indicator of the pressure and strain sensor were discussed in detail. The impact of electrode structure on the sensor performance was demonstrated. Sensor

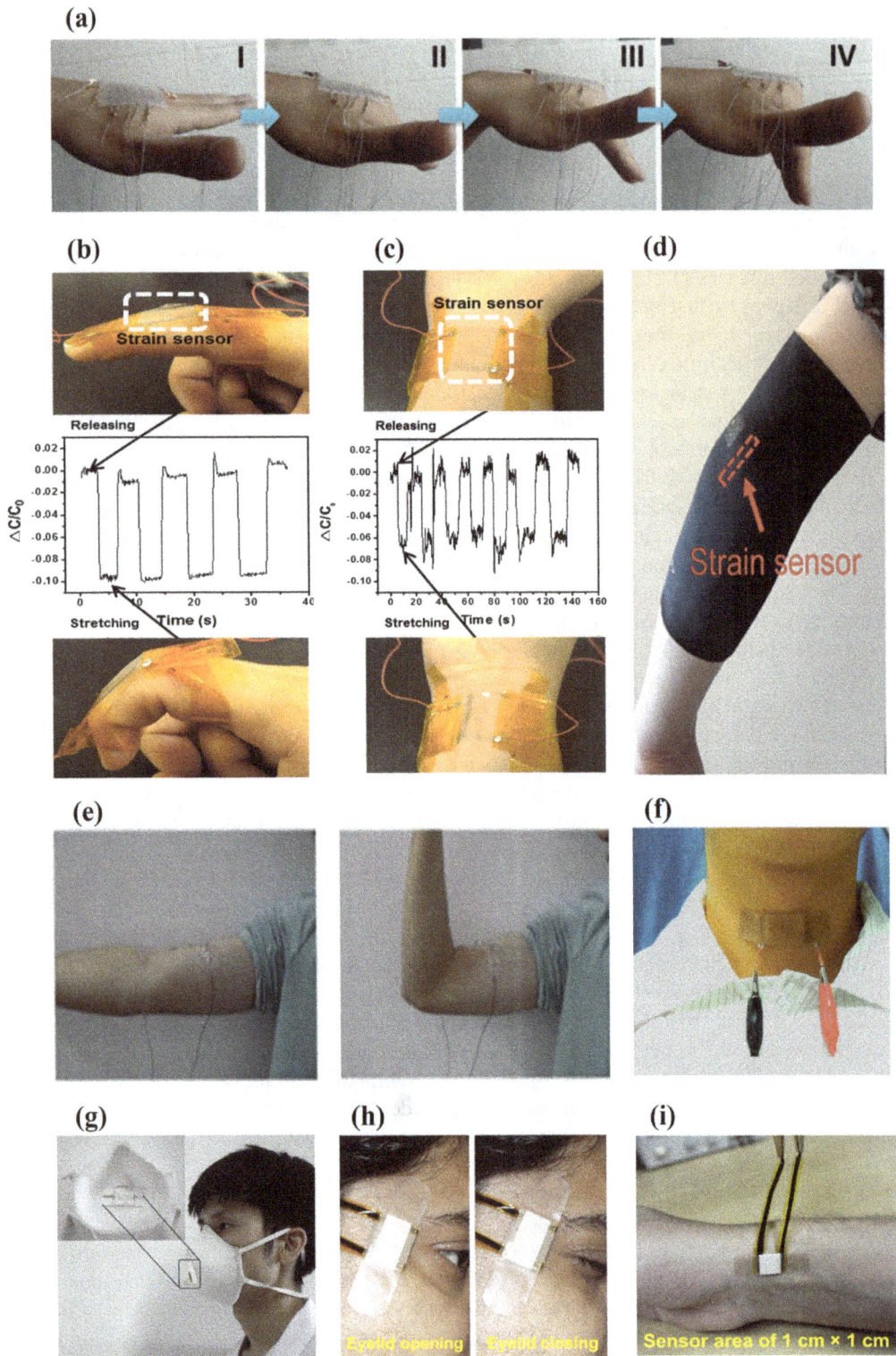

FIGURE 16.8   Utilization of pressure and strain sensors in human motion monitoring. (a) Pressure sensor monitoring palm bending with different angles (adapted with permission [22]. Copyright 2017, Royal Society of Chemistry), (b) strain sensor for finger bending monitoring and related capacitance change upon bending

FIGURE 16.8   (Continued)

and releasing, (c) strain sensor for wrist bending monitoring and related capacitance change upon bending and releasing (b-c, adapted with permission [39]. Copyright 2017, American Chemical Society), (d) textile integrated strain sensor for elbow bending tracking (adapted with permission [38]. Copyright 2019, American Chemical Society), (e) pressure sensor mounted on human arm to detect arm bending (adapted with permission [45]. Copyright 2021, IOP Publishing), (f) pressure sensor placed on the human neck to detect the vocal cords' vibration (adapted with permission [46]. Copyright 2017, American Chemical Society), (g) pressure sensor attached to the mask for respiration monitoring (adapted with permission [20]. Copyright 2017, WILEY-VCH), (h) pressure sensor tracking muscle movement during eyelid opening and closing (adapted with permission [47]. Copyright 2018, American Chemical Society), and (i) pressure sensor mounted on the radial artery for real-time monitoring of the wrist pulse (adapted with permission [47]. Copyright 2018, American Chemical Society).

performance indicators, including, sensitivity, response time, dynamic durability, the limit of detection, hysteresis, linearity, and stretchability were explained. Finally, we have demonstrated that both the flexible pressure and stretchable strain sensors are very promising in human motion monitoring physiological parameters, including hand motion (finger movement), joint movement (knee, wrist, and scapular movement), and subtle human motion (eye blinking, pulse, and respiration).

REFERENCES

1. S. Z. Homayounfar; T. L. Andrew, Wearable Sensors for Monitoring Human Motion: A Review on Mechanisms, Materials, and Challenges. *SLAS Technol.* **2020**, *25*, 9–24.
2. J. Wang; C. Lu; K. Zhang, Textile-Based Strain Sensor for Human Motion Detection. *ENERGY Environ. Mater.* **2020**, *3*, 80–100.
3. R. B. Mishra; N. El-Atab; A. M. Hussain; M. M. Hussain, Recent Progress on Flexible Capacitive Pressure Sensors: From Design and Materials to Applications. *Adv. Mater. Technol.* **2021**, *6*, 2001023.
4. H. Souri; H. Banerjee; A. Jusufi; N. Radacsi; A. A. Stokes; I. Park; M. Sitti; M. Amjadi, Wearable and Stretchable Strain Sensors: Materials, Sensing Mechanisms, and Applications. *Adv. Intell. Syst.* **2020**, *2*, 2000039.
5. J. Qin; L. -J. Yin; Y. -N. Hao; S. -L. Zhong; D. -L. Zhang; K. Bi; Y. -X. Zhang; Y. Zhao; Z.-M. Dang, Flexible and Stretchable Capacitive Sensors with Different Microstructures. *Adv. Mater.* **2021**, *33*, 2008267.
6. N. Wen; L. Zhang; D. Jiang; Z. Wu; B. Li; C. Sun; Z. Guo, Emerging Flexible Sensors Based on Nanomaterials: Recent Status and Applications. *J. Mater. Chem. A* **2020**, *8*, 25499–25527.
7. Y. Joo; J. Byun; N. Seong; J. Ha; H. Kim; S. Kim; T. Kim; H. Im; D. Kim; Y. Hong, Silver Nanowire-Embedded PDMS with a Multiscale Structure for a Highly Sensitive and Robust Flexible Pressure Sensor. *Nanoscale* **2015**, *7*, 6208–6215.
8. K. E. Korte; S. E. Skrabalak; Y. Xia, Rapid Synthesis of Silver Nanowires Through a CuCl- or CuCl2-Mediated Polyol Process. *J. Mater. Chem.* **2008**, *18*, 437–441.
9. S. Hemmati; D. P. Barkey, Parametric Study, Sensitivity Analysis, and Optimization of Polyol Synthesis of Silver Nanowires. *ECS J. Solid State Sci. Technol.* **2017**, *6*, P132–P137.
10. Y. Sun; B. Mayers; T. Herricks; Y. Xia, Polyol Synthesis of Uniform Silver Nanowires: A Plausible Growth Mechanism and the Supporting Evidence. *Nano Lett.* **2003**, *3*, 955–960.
11. J. H. Lee; P. Lee; D. Lee; S. S. Lee; S. H. Ko, Large-Scale Synthesis and Characterization of Very Long Silver Nanowires via Successive Multistep Growth. *Cryst. Growth Des.* **2012**, *12*, 5598–5605.
12. B. Liu; H. Yan; S. Chen; Y. Guan; G. Wu; R. Jin; L. Li, Stable and Controllable Synthesis of Silver Nanowires for Transparent Conducting Film. *Nanoscale Res. Lett.* **2017**, *12*, 212.
13. B. Bari; J. Lee; T. Jang; P. Won; S. H. Ko; K. Alamgir; M. Arshad; L. J. Guo, Simple Hydrothermal Synthesis of Very-Long and Thin Silver Nanowires and their Application in High Quality Transparent Electrodes. *J. Mater. Chem. A* **2016**, *4*, 11365–11371.

14. F. Basarir; F. S. Irani; A. Kosemen; B. T. Camic; F. Oytun; B. Tunaboylu; H. J. Shin; K. Y. Nam; H. Choi, Recent Progresses on Solution-Processed Silver Nanowire Based Transparent Conducting Electrodes for Organic Solar Cells. *Mater. Today Chem.* **2017**, *3*, 60–72.

15. S. Chen; X. Guo, Improving the Sensitivity of Elastic Capacitive Pressure Sensors Using Silver Nanowire Mesh Electrodes. *IEEE Trans. Nanotechnol.* **2015**, *14*, 619–623.

16. A. Elsayes; V. Sharma; K. Yiannacou; A. Koivikko; A. Rasheed; V. Sariola, Plant-Based Biodegradable Capacitive Tactile Pressure Sensor Using Flexible and Transparent Leaf Skeletons as Electrodes and Flower Petal as Dielectric Layer. *Adv. Sustain. Syst.* **2020**, *4*, 2000056.

17. Koivikko; V. Lampinen; K. Yiannacou; V. Sharma; V. Sariola, Biodegradable, Flexible and Transparent Tactile Pressure Sensor Based on Rubber Leaf Skeletons. *IEEE Sens. J.* **2021**, *1*, 11241–11247.

18. S. Peng; S. Wu; Y. Yu; B. Xia; N. H. Lovell; C. H. Wang, Multimodal Capacitive and Piezoresistive Sensor for Simultaneous Measurement of Multiple Forces. *ACS Appl. Mater. Interfaces* **2020**, *12*, 22179–22190.

19. A. Gurarslan; B. Özdemir; İ. H. Bayat; M. B. Yelten; G. Karabulut Kurt, Silver Nanowire Coated Knitted Wool Fabrics for Wearable Electronic Applications. *J. Eng. Fiber. Fabr.* **2019**, *14*, 1–8.

20. W. Yang; N. -W. Li; S. Zhao; Z. Yuan; J. Wang; X. Du; B. Wang; R. Cao; X. Li; W. Xu, A Breathable and Screen-Printed Pressure Sensor Based on Nanofiber Membranes for Electronic Skins. *Adv. Mater. Technol.* **2018**, *3*, 1700241.

21. M. O. Cicek; D. Doganay; M. B. Durukan; M. C. Gorur; H. E. Unalan, Seamless Monolithic Design for Foam Based, Flexible, Parallel Plate Capacitive Sensors. *Adv. Mater. Technol.* **2021**, *6*, 2001168.

22. Y. Cheng; R. Wang; H. Zhai; J. Sun, Stretchable Electronic Skin Based on Silver Nanowire Composite Fiber Electrodes for Sensing Pressure, Proximity, and Multidirectional Strain. *Nanoscale* **2017**, *9*, 3834–3842.

23. F. Guan; Y. Xie; H. Wu; Y. Meng; Y. Shi; M. Gao; Z. Zhang; S. Chen; Y. Chen; H. Wang, Silver Nanowire–Bacterial Cellulose Composite Fiber-Based Sensor for Highly Sensitive Detection of Pressure and Proximity. *ACS Nano* **2020**, *14*, 15428–15439.

24. S. Yao; Y. Zhu, Wearable Multifunctional Sensors using Printed Stretchable Conductors Made of Silver Nanowires. *Nanoscale* **2014**, *6*, 2345–2352.

25. Y. Kim; N. Chou; S. Kim, Highly Sensitive Capacitive Tactile Sensor Based on Silver Nanowire using Parylene-C Stencil Patterning Method. *Proceedings of the 2015 IEEE SENSORS* **2015**, 1–3.

26. D.-W. Jeong; N.-S. Jang; K.-H. Kim; J.-M. Kim, A Stretchable Sensor Platform Based on Simple and Scalable Lift-off Micropatterning of Metal Nanowire Network. *RSC Adv.* **2016**, *6*, 74418–74425.

27. C.J. Han; B.-G. Park; M. Suk Oh; S.-B. Jung; J.-W. Kim, Photo-induced Fabrication of Ag Nanowire Circuitry for Invisible, Ultrathin, Conformable Pressure Sensors. *J. Mater. Chem. C* **2017**, *5*, 9986–9994.

28. S. Zhao; W. Ran; D. Wang; R. Yin; Y. Yan; K. Jiang; Z. Lou; G. Shen, 3D Dielectric Layer Enabled Highly Sensitive Capacitive Pressure Sensors for Wearable Electronics. *ACS Appl. Mater. Interfaces* **2020**, *12*, 32023–32030.

29. J. Wang; Y. Lou; B. Wang; Q. Sun; M. Zhou; X. Li, Highly Sensitive, Breathable, and Flexible Pressure Sensor Based on Electrospun Membrane with Assistance of AgNW/TPU as Composite Dielectric Layer. *Sensors* **2020**, *20*, 2459

30. R. Shi; Z. Lou; S. Chen; G. Shen, Flexible and Transparent Capacitive Pressure Sensor with Patterned Microstructured Composite Rubber Dielectric for Wearable Touch Keyboard Application. *Sci. China Mater.* **2018**, *61*, 1587–1595.

31. P. Wang; G. Li; J. Liu; Z. Hou; C. Meng; S. Guo; C. Liu; S. Fan, Tailorable Capacitive Tactile Sensor Based on Stretchable and Dissolvable Porous Silver Nanowire/Polyvinyl Alcohol Nanocomposite Hydrogel for Wearable Human Motion Detection. *Adv. Mater. Interfaces* **2021**, *8*, 2100998.

32. R. Li; Q. Zhou; Y. Bi; S. Cao; X. Xia; A. Yang; S. Li; X. Xiao, Research Progress of Flexible Capacitive Pressure Sensor for Sensitivity Enhancement Approaches. *Sensors Actuators A Phys.* **2021**, *321*, 112425.

33. J. Park; I. You; S. Shin; U. Jeong, Material Approaches to Stretchable Strain Sensors. *ChemPhysChem* **2015**, *16*, 1155–1163.

34. F. Xu; Y. Zhu, Highly Conductive and Stretchable Silver Nanowire Conductors. *Adv. Mater.* **2012**, *24*, 5117–5122.

35. W. Hu; X. Niu; R. Zhao; Q. Pei, Elastomeric Transparent Capacitive Sensors Based on an Interpenetrating Composite of Silver Nanowires and Polyurethane. *Appl. Phys. Lett.* **2013**, *102*, 83303.

36. S. Yao; J.S. Lee; K. James; J. Miller; V. Narasimhan; A.J. Dickerson; X. Zhu; Y. Zhu, Silver Nanowire Strain Sensors for Wearable Body Motion Tracking. *Proceedings of the 2015 IEEE SENSORS* **2015**, 1–4.

37. S. Yao; L. Vargas; X. Hu; Y. Zhu, A Novel Finger Kinematic Tracking Method Based on Skin-Like Wearable Strain Sensors. *IEEE Sens. J.* **2018**, *18*, 3010–3015.

38. S. Yao; J. Yang; F. R. Poblete; X. Hu; Y. Zhu, Multifunctional Electronic Textiles Using Silver Nanowire Composites. *ACS Appl. Mater. Interfaces* **2019**, *11*, 31028–31037.

39. S.-R. Kim; J.-H. Kim; J.-W. Park, Wearable and Transparent Capacitive Strain Sensor with High Sensitivity Based on Patterned Ag Nanowire Networks. *ACS Appl. Mater. Interfaces* **2017**, *9*, 26407–26416.

40. Y.H. Ju; C.J. Han; K.-S. Kim; J.-W. Kim, UV-Curable Adhesive Tape-Assisted Patterning of Metal Nanowires for Ultrasimple Fabrication of Stretchable Pressure Sensor. *Adv. Mater. Technol.* **2021**, *6*, 2100776.

41. B. You; C. J. Han; Y. Kim; B.-K. Ju; J.-W. Kim, A Wearable Piezocapacitive Pressure Sensor with a Single Layer of Silver Nanowire-based Elastomeric Composite Electrodes. *J. Mater. Chem. A* **2016**, *4*, 10435–10443.

42. S. Jun; B.-K. Ju; J.-W. Kim, Ultra-Facile Fabrication of Stretchable and Transparent Capacitive Sensor Employing Photo-Assisted Patterning of Silver Nanowire Networks. *Adv. Mater. Technol.* **2016**, *1*, 1600062.

43. M. Amjadi; K.-U. Kyung; I. Park; M. Sitti, Stretchable, Skin-Mountable, and Wearable Strain Sensors and Their Potential Applications: A Review. *Adv. Funct. Mater.* **2016**, *26*, 1678–1698.

44. Z. Tu; Z. Xia; W. Luo; P. Huang; J. Lin, Structural Design of Flexible Interdigital Capacitor based upon 3D Printing and Spraying Process. *Smart Mater. Struct.* **2022**, *31*, 45005.

45. L. Mo; X. Meng; J. Zhao; Y. Pan; Z. Sun; Z. Guo; W. Wang; Z. Peng; C. Shang; S. Han, Full Printed Flexible Pressure Sensor based on Microcapsule Controllable Structure and Composite Dielectrics. *Flex. Print. Electron.* **2021**, *6*, 14001.

46. X. Shuai; P. Zhu; W. Zeng; Y. Hu; X. Liang; Y. Zhang; R. Sun; C. Wong, Highly Sensitive Flexible Pressure Sensor Based on Silver Nanowires-Embedded Polydimethylsiloxane Electrode with Microarray Structure. *ACS Appl. Mater. Interfaces* **2017**, *9*, 26314–26324.

47. A. Chhetry; J. Kim; H. Yoon; J.Y. Park, Ultrasensitive Interfacial Capacitive Pressure Sensor Based on a Randomly Distributed Microstructured Iontronic Film for Wearable Applications. *ACS Appl. Mater. Interfaces* **2019**, *11*, 3438–3449.

17 Nanowires for Flexible Electrochemical Energy Devices

Tapas Das, Sanjeev Verma, Vikas Kumar Pandey, and
Bhawna Verma

Department of Chemical Engineering & Technology, Indian Institute of
Technology, Banaras Hindu University, Varanasi, India

CONTENTS

17.1 INTRODUCTION

In addition to becoming a dominant societal path in the modern world for a variety of constant uses, the rapid expansion of portable and flexible electronics also demonstrates enormous relevance for enduring profitable hikes and also improvement of human life quality. Flexible electrochemical storage devices, smart sensors, flexible LEDs, flexible energy harvesters, and flexible transistors are just a few of the advanced electronic components that have allured significant regard in ongoing decades because of their high elastic nature and good performance [1,2]. Flexible electronics demonstrate many distinctive advantages in comparison to conventional indium tin oxides (ITO) and Si-based hard electric frameworks, such as lightweight, high portability, and good flexibility, which will fulfill the main significance for the upcoming decades. Numerous structural and fabrication designing strategies have been described to enhance the behavior defined by the structure to achieve highly efficient, straightforward, affordable, and portable flexible electric devices [3]. It is significant to highlight that the fields of nanotechnology and nanoscience have developed, with an active concentration on both applied and basic research.

In comparison to their bulk material, low-dimensional nanomaterials have garnered a lot of attention because they have greater ion diffusion and physical/chemical size-dependent nature that

DOI: 10.1201/9781003296621-17

have a wide range of possible applications in improving the behavior of flexible electronic gadgets [4,5]. The distinctive two-dimensional (2D) confinement surfaces, optoelectronics/electrical capabilities, and high surface/volume ratio of one-dimensional (1D) nanofibers, like nanowires (NWs), have piqued widespread attention among the numerous nanostructures. Their high capacity and remarkable flexibility for the diffusion of quantum ions like photons, phonons, and electrons, are also responsible for building flexible blocks in flexible device applications [6]. The performance of a variety of flexible NW-based devices, including field-effect transistors, flexible displays, flexible electrochemical gadgets, flexible botanic sensors, and flexible photodetectors, is admirable and equivalent to thin films. But one of the most important methods for enhancing the performance of flexible electronics based on NWs would be to integrate each NW into clearly desired surfaces over sizable pores. This is in addition to NW synthesis and device construction.

The unique spectroscopic, electrical, magnetic, and optical capabilities of formed NW structures can be explored for future flexible device applications by first creating chances for indirect as well as direct connections with its environment. The qualities resulting from assembled/arranged NWs are caused by the development of collective properties as well as the random enhancement in surface to volume ratio of nanowires. Additionally, assembled NW-based flexible electronics outperform conventional electronics in terms of material usage, durability, and versatility in a variety of uses like healthcare sensors, RFID cards for communication, energy batteries, photovoltaics, displays, actuators, and defensive line sensors, types of products. Normally, there are lots of challenges to arrange nanowires in effective and dense nanosurfaces, however novel bottom/up approaches allow us to form flexible dense nanoarrays with remarkable precision [7,8].

In this chapter, we provided a general review of the many flexible energy storage electrochemical devices, including supercapacitors, lithium-ion batteries (LIBs), sodium-ion batteries (SIBs), and metal-air batteries (**Figure 17.1**). An in-depth discussion will also be given on various flexible NW composites made of carbon, polymer, organic, and inorganic materials. To support the claim that

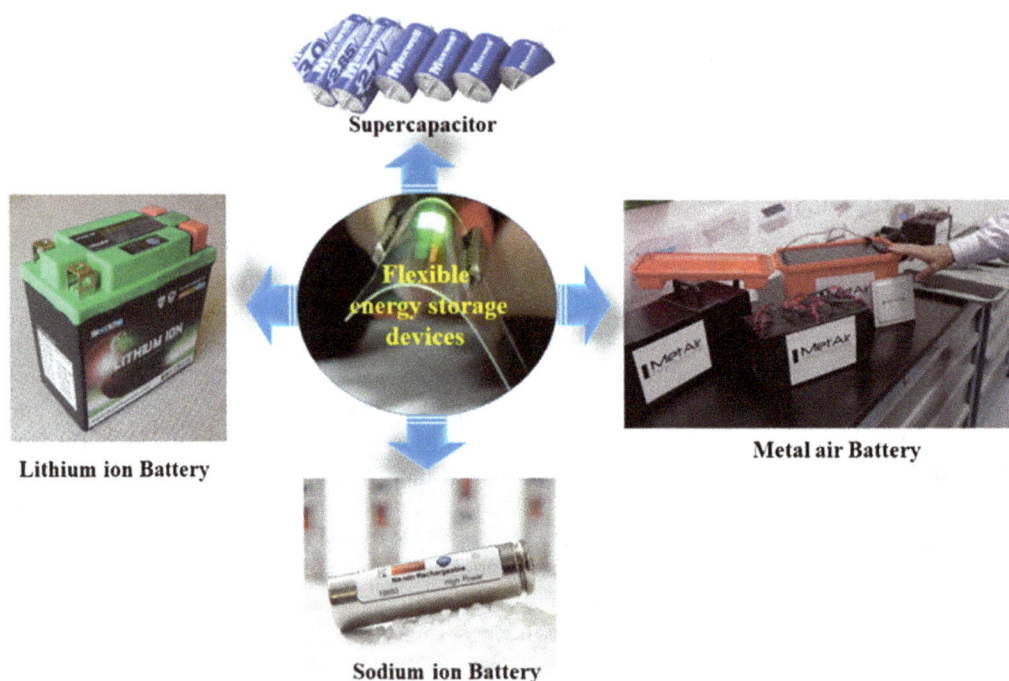

FIGURE 17.1   Different flexible energy storage devices.

employing interlinkage and organization in well-defined structures is the most successful approach for freestanding electronics devices with good controllability and remarkable performance than disarray type structures, and also highlights and emphasizes the current prospects in the field of research. Finally, the direction framework assembly of NWs toward flexible electronic devices has positive effects on various prospective applications.

17.2 FLEXIBLE NANOWIRES FOR ENERGY STORAGE DEVICES

Flexible materials with sizes down to the nanoscale give the materials several distinctive features that set them apart from bulk macroscale materials. In terms of providing the necessary energy and power for wearable electronics to operate, efficient, flexible electrochemical energy storage devices are excellent prospects to serve as a significant component. Recently, the research community has made tremendous efforts toward developing flexible electrochemical energy storage by redesigning its active components, current collectors, separators, and designs to improve performance even in a deformed state. These flexible energy storage devices are far better than their conventional, non-flexible counterparts. For effective systems, there has been a constant improvement in device design and material properties since engineering is the key to creating adaptable electrochemical energy storage devices, flexible nanowires (FNWs) being amongst the most widely investigated nanomaterials, provide a range of benefits for energy storage applications for example facilitating better contact between electrolyte and electrode and making it simpler for electrolytes to access the electrochemically active locations, which increases capacity and material consumption; FNWs offer an uninterrupted path for electron transmission, lowering the resistance to charge transfer, which boosts conductivity and electrode dynamics; since diffusion time is directly proportional to diffusion length, FNWs help shorten the ion transport pathway, which can increase electrodes' rate capability; the comparatively tiny structure of FNWs allows for volume fluctuations and structural deterioration caused by electrochemical reactions, as a result of frequent lithiation and delithiation processes; a high length-to-diameter ratio of FNWs helps create three-dimensional models that can stand alone without a binder; using FNWs as the building blocks, it is simple to put up sophisticated designs for various applications, including branched, hierarchical, and other types [9–11].

Despite the considerable benefits of FNWs outlined above, there are still several obstacles to their use in electrochemical energy storage, including the following; FNWs are readily accumulated with frequent charging and discharging, causing augmented internal impedance and capacity decay; larger accessible surface areas of FNWs lead to more interface side reactions, which diminish cycle life and Coulombic efficiency; FNWs' compact densities are typically lower than those of their micrometer-sized particles, resulting in lower volumetric energy density; because surface scattering shortens the electron mean free path, which results in FNWs having a lower electrical conductivity than their bulk analogs; it is difficult to synthesize FNWs with tunable size and shape, and much work needs to be done to achieve large-scale, low-cost manufacturing of FNWs [12,13].

17.2.1 SUPERCAPACITORS

Modern society's reliance on technology necessitates extensive energy use and storage. Batteries and electrochemical supercapacitors (SCs) have garnered much attention as promising electrochemical energy storage devices [14]. It is evident that even while SCs have a greater power density than the battery, despite having a much lower energy density, they have a much longer cycling life of up to 100000 cycles. Moreover, by raising the capacitance (C) and charging voltage (V), SCs can produce a more significant energy density (E), according to the formula $E = 0.5CV^2$. Due to their higher specific power density and extended lifecycles, the power and energy density gap between batteries and conventional dielectric capacitors has been narrowed by SCs. They will consequently

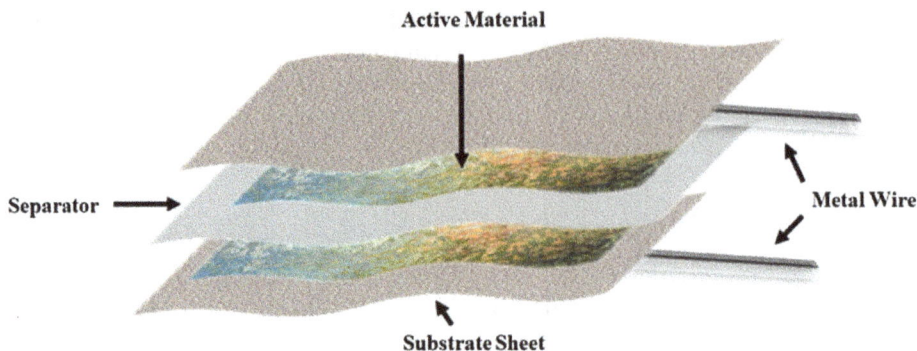

FIGURE 17.2   Schematic representation of flexible supercapacitor.

play a more significant part in developing new energy storage technologies. Electric double-layer capacitors (EDLCs) and pseudocapacitors (Faradaic SCs) are two types of SCs. In EDLCs, only physical electrostatic charge accumulation takes place at the electrode-electrolyte interface during charging and discharging operations without any physical changes. Pseudocapacitors also experience quick redox reactions, which might result in irreversible capacitance degradation. As a result, due to their differing methods for storing charges, pseudocapacitors have a shorter cycle life than EDLC materials but can have a larger specific capacitance.

In SCs, 1D FNW electrode materials have been thoroughly investigated (**Figure 17.2**). The 1D morphology of the FNWs naturally offers a quick route for adequate electron transportation, which is particularly advantageous for raising the power density and rate performance of supercapacitors. Due to their larger surface area and higher electrode-electrolyte contact, FNWs have a greater length-to-diameter ratio. This causes a significant number of electrolyte ions to be adsorbed on the electrode surface, which speeds up the electrode reaction kinetics. Additionally, aggregation at the nanoscale level impacts EDLC-based carbon materials with high specific surface areas. The 1D FNW structure may effectively address this issue and significantly increase the specific surface area. Pseudocapacitance-based flexible nanowire materials could lessen volumetric fluctuations during cycling and improve the electrode's stability. The SCs effectively illustrate the distinctive structural advantages of the NWs. Frackowiak et al. reported a specific capacitance of 137 F/g from MWCNTs prepared by chemical vapor deposition (CVD) with 410 m^2/g [15]. Chen et al. synthesized N-doped CNFs using ZIF-8 precursor with a high specific capacitance of 307 F/g at 1 A/gwith 10.96 Wh/kgenergy density [16]. Recently, dual metal oxides like $MnCo_2O_4$, $ZnCo_2O_4$, $CuCo_2O_4$, and $NiCo_2O_4$ have also been investigated due to their superior electrochemical performance compared to mono metal oxides [17]. Shen et al. synthesized mesoporous $NiCo_2O_4$ nanowire array grown on carbon cloth and incurred a high specific capacitance of 1283 F/gat 1 A/gand a stable cycling performance up to 5000 cycles [18]. Also, owing to their synergistic effects, a high specific capacitance of 2244 F/gwas obtained by Kong et al. with $NiCo_2O_4$@PPy NW [19]. Li et al. fabricated nickel foam electrodes with a Ni-Co sulfide nanowire array and obtained high gravimetric and areal capacitance of 2415 F/gand 6.0 F/cm^2 [20].

17.2.2   LITHIUM-ION BATTERIES (LIBS)

The development of autonomous vehicles has significantly led to the growth of electric vehicles. Electric vehicles provide several advantages, like zero-air pollution and less noise and vibration. However, electric vehicles have limitations, such as long charging times, cost, mileage, limited lifetime, and safety. These issues led to the development of new generation power systems with lithium-ion battery packs [21].

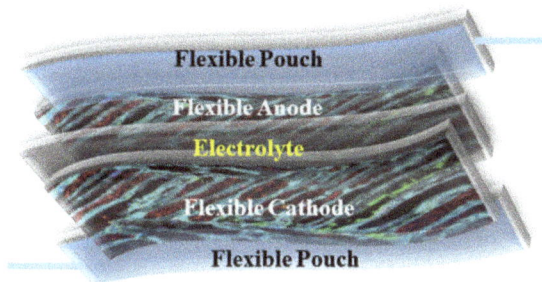

FIGURE 17.3 Pictorial diagram of flexible lithium-ion battery.

The widely used LIBs consist of a non-aqueous liquid electrolyte, graphite anode, and layered $LiCoO_2$ cathode (**Figure 17.3**). To achieve energy storage, the $LiCoO_2$ host discharges Li ions while charging, crosses the electrolytes and separator, and then intercalates into the graphite layers. At the same time, the external circuit facilitates the flow of electrons. In contrast, Li ions are efficiently and reversibly removed from the electrodes in a perfect scenario through intercalation and deintercalation. However, lithium-consuming side reactions occur throughout the charge/discharge process, resulting in steady capacity decay and Coulombic efficiency of less than 100%. Therefore, the foundation and key to developing improved batteries are creating a strong and chemically stable solid electrolyte interface. $LiCoO_2$/graphite system-based LIBs deliver 3.7 V output voltage and about 150–200 Wh/kg of energy density. The energy density for $LiCoO_2$/graphite-based LIBs is between 150 and 200 Wh/kg, with an output voltage of 3.7 V. Based on the intercalation process, the currently employed electrode materials are getting close to their capacity limit and still fall well short of what will be required for future applications. It is imperative to develop better electrode materials to address the rapidly rising need for high energy density rechargeable batteries for smart grids, portable electronic devices, and electric cars. Therefore, tremendous efforts have been devoted to developing battery systems or electrode materials with high power and energy density with long cycle lives.

Previous studies have shown that the electrode materials' nanostructures significantly impact the electrochemical performance of LIBs. Reducing particle size is one of the most effective strategies to relieve internal mechanical stresses and trim the Li-ion diffusion length, which improves capacity retention and cycling life. The FNW electrode has several advantages over bulk materials for use in LIBs, such as the following: by more accurately adapting volume changes experienced by electrodes during lithiation and delithiation, NWs with diameters in the nanometer range prevent structural fracturing that typically happens in micron-sized or bulk materials. Zhu et al. synthesized core-shell $LiFePO_4$/C nanowires and claimed a battery capacity of 169 mAh/g at 0.1 C and 93 mAh/g at 10 C with 86% of capacitance retention after 100 cycles at 1C [22]. Chain-like $LiCoO_2$ nanowire arrays reported by Xia et al. extracted a battery capacity of 135 mAh/g at 0.1 C with a retention of 90% after 50 cycles in the potential range of 3.0–4.2 V when used as the cathode [23]. Mesoporous nanowires of $Li_3V_2(PO_4)_3$/C used as the cathode and synthesized by the hydrothermal method showed 120 mAh/g reversible capacity at 1C with initial Coulombic efficiency of 85% as reported by Ding et al. between 3.0–4.3 V [24]. Electrospun $LiFe_{0.5}Mn_{0.5}PO_4$/C nanowire-based cathode showed 125 mAh/g at 0.5 C and 102 mAh/g at 4C between 3.0–4.9 V potential range as reported by von Hagen et al. [25].

17.2.3 SODIUM-ION BATTERIES (SIBS)

In some specific applications, non-lithium battery systems with the appropriate qualities can replace flexible LIBs in terms of functionality, price, and safety. Because of their appealing qualities of natural abundance and inexpensive cost, as well as their resemblance in the operating principles to that of LIBs, SIBs have drawn the most attention in recent years. Additionally, SIBs have many technical advantages over LIBs, including more electrolyte options with a wider electrochemical window, the ability to produce liquid electrolytes with high ionic conductivities while using less salt, quick charging, and excellent working behaviors over a wider temperature range of 40 to 80 °C. These amazing features have sparked the rapid development of SIBs as prospective options for electrical energy storage technology. Therefore, the creation of high-performance flexible SIBs is anticipated to hold significant potential for the development of flexible electronic gadgets of the future [26].

Because sodium is widely available and reasonably priced, SIBs are currently being developed as a feasible alternative to LIBs. It is challenging for the built-in SIBs to outperform the LIBs in terms of energy density, specific capacity, or rate capability since Na is three times heavier than Li and has a lower standard electrochemical potential than Li. The examination of cathode materials in SIBs, such as polyanionic compounds and layered transition metal oxides, has so far dominated active research, with anode materials receiving very little attention.

Rechargeable batteries are appealing for next-generation devices and electric vehicle applications. Due to their broad availability, inexpensive cost, and physical similarities to lithium, ambient temperature SIBs are promising alternatives for extensive grid-based energy storage devices. However, the Na ion has a lower reversible capacity and a lower rate capability than the Li-ion because it has a larger ionic radius and diffuses more slowly during charge/discharge cycles. Therefore, improving the performance of electrode materials and creating appropriate electrode nanoarchitecture, are the main concerns to address the continuously increasing energy and power densities of Na-ion-based storage devices. Due to its proficient electrolyte access to the active material, smaller ion diffusion length, and capability to tolerate volume fluctuations during the electrochemical reactions, 1D FNW structures stand out among other nanostructures as viable candidates for the Na-ion-based electrodes. Su et al. fabricated bilayer V_2O_5 nanowire-based cathodes via a hydrothermal route, which exhibited 235.7 mAh/g at 40 mA/g in 1–4 V [27]. Further, the Fe-VO$_x$ nanobelt-based cathode demonstrated 196 mAh/g at 50 mA/g from 1–4 V potential as incurred by Wei et al. [28]. Jin et al. reported 120 mAh/g from NaVPO$_4$F/C nanofibers prepared by electrospinning between the 2.6–4.5 V range [29]. Dong et al. claimed a battery performance of 158.7 mAh/g at 200 mA/g from Na$_{1.25}$V$_3$O$_8$ hierarchical zigzag NWs used as the cathode and prepared by the hydrothermal method [30].

17.2.4 METAL-AIR BATTERIES

Future generations of EVs may benefit from metal-air batteries like aluminum, magnesium, zinc, and lithium since they take O_2 from the environment air, which is considered the prime reactant of this battery (**Figure 17.4**). The weight and space for ion storage are reduced by using air oxygen as a reactant. The lithium-air battery exhibits a larger potential specific density than all metal-air electrochemical batteries [31].

An air-type cathode and metal anode, both submerged in the appropriate electrolyte with a proper separator, are used to make metal-air electrochemical batteries. The metal anode occupies a large portion of the cell volume in metal-air batteries, which has the highest charge-to-mass ratio, these batteries have high energy densities. Similar to metal-ion batteries, metal ions move between the cathode and anode at the charging and discharging mechanism, but the primary distinction is the electrochemical processes occurring at the air cathode surface. Two different classes of metal-air electrochemical batteries, like aqueous and non-aqueous types, depending on the electrolyte's composition.

$$A \leftrightarrow A^{n+} + ne^- \text{ (anode)}$$

$$O_2 + 4e^- + 2H_2O \leftrightarrow 4OH^- \text{ (cathode)}$$

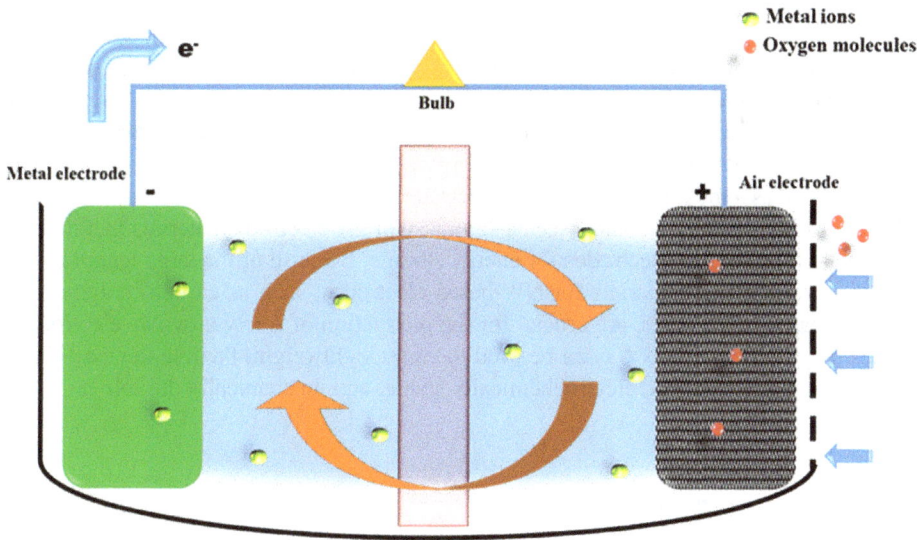

FIGURE 17.4   Schematic diagram of metal-air battery.

The metal (A = Li, Mg, Al, Fe, and Zn) oxidizes at the anode surface during the discharge process when the electrolyte is aqueous, and O_2 breaks at the cathode surface (considered as ORR, oxygen-reduction reaction). During charging, the mechanism is biased, with anodic metal plating and O_2 evaporating at the cathode surface (also known as OER, oxygen-evolution reaction).

In a non-aqueous type electrolyte, during the discharging process, the metal atom (A = K, Na, and Li) also oxidizes at the anode surface to generate A^+ metal ions, and these A^+ metal ions then migrate across the non-electrolyte where they are combined with O_2 atoms to create metal peroxides/superoxides.

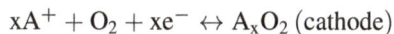

$$A \leftrightarrow A^+ + e^- \text{ (anode)}$$

$$xA^+ + O_2 + xe^- \leftrightarrow A_xO_2 \text{ (cathode)}$$

Non-aqueous metal-air batteries also charge through an anodic metal plating surface and OER at the cathode surface, just like aqueous metal-air electrochemical batteries do. The porous structure and configuration of the air-type cathode and inherent catalytic activity of OER/ORR put air-type cathode, are usually thought to have a major role in the overall nature of metal-air electrochemical batteries [32]. Air cathodes have more functions in metal-air batteries than metal anodes, including allowing oxygen to pass, storing discharge ions (particularly in non-aqueous type metal-air electrochemical batteries), and conducting OER/ORR. Furthermore, the surfaces of these discharge ions in non-aqueous metal-air batteries, cathode ion diffusion nature, and cathode wettability are all impacted by the air cathode electrode structures. Therefore, the air cathodes will be largely responsible for the metal-air batteries' capacity, cyclability, rate capability, and cycle trip efficiency. As a result, ongoing efforts are being made to develop and create perfect cathodes with improved architecture and cutting-edge materials to produce metal-air batteries with desired performance [33,34].

Zhao et al. synthesized a highly active FeNi@NCNT/CC NW-array-based air cathode for highly flexible solid-state rechargeable Zn-air batteries, which demonstrated a high discharge voltage (~ 1.0 V at 2 mA/cm^2) and low charge voltage ($\sim$1.65 V at 2 mA/cm^2) [35]. Yin et al. prepared NiS$_2$/CoS$_2$ -Oporous NWs for Zn-air portable batteries with a high open circuit voltage (1.49 V) and stable rechargeable performance for more than 30 h at 3 and 5 mA/cm^2 current densities [36].

17.3 FLEXIBLE NANOWIRE-BASED COMPOSITE

Flexible electrodes, such as substrates and composite active material electrodes with good electrical conductivity, are considered superior materials to use for creating flexible energy storage devices since they guarantee quick charge/discharge capabilities. According to current developments, there are two ways to create flexible electrodes: (a) incorporate composite active materials in flexible thin sheets (also known as a movable freestanding electrode), and (b) affix composite active materials to movable substrates. Recent years have seen a summary of the electrode materials that may be used to build flexible freestanding electrodes for energy storage. Because of the large length to diameter ratio of nanowires (NWs), materials for NW-based electrodes, such as carbon, polymer, organic, and inorganic-based composites, are crucial for the production of freely movable electrode sheets. Composites based on flexible NWs can be used to create lightweight, flexible devices. Each component must be highly efficient, electrochemically stable, and mechanically durable in research on flexible power sources.

17.3.1 CARBON NANOWIRE-BASED COMPOSITE

Since carbon compounds have a low capacity, they are known to have a poor energy density. Carbon nanowire (CNW) materials have been mostly used for synthesizing carbon-based composite nanowire structured materials, where the interlinkage of one-dimensional carbon materials forms a 3D flexible base to support the composite materials. To create high-performance flexible electrode materials, carbon nanowire networks are a considerable advantage. The network provides a solid foundation that effectively cushions the volumetric changes of high capacitive materials that are supported, thereby increasing the cyclic performance of electrode materials. Additionally, the linkage offers quick and proper electron movement routes that significantly increase the electrode's electrical conductivity and boost the kinetics of electrochemical reactions. The network, which creates effective ion transport channels, enables the individual dispersion of active chemicals. A large proportion of elements onto carbon enhances the capacitive nature but accords the electrical conductivity. This is how the capacitive nature of flexible carbon-based electrodes and electrical conductivity typically show a common relationship. To achieve good stable stability and capacity, the active composite materials might be largely presented in the pores of three-dimensional carbon structure, with strong coupling effects, and a high interlinkage area between carbon and active composite material.

For instance, flexible LIB high-performance cathode materials were created by growing $LiMn_2O_4$ in a 3D CNT framework, where bonds between the $LiMn_2O_4$ and carbon improved cycling and mechanical stabilities [37]. Guo et al. synthesized NVP@C/CC composite by taking high mass loading of NVP@C in carbon [38]. (NVP=$Na_3V_2(PO_4)_3$), where it not only coated the carbon fibers' surfaces but also filled the spaces in between them. The flexible cathode demonstrated great power/energy density and exceptional cycling stability using 3.5 mg/cm^2 high mass loading and a good coupling effect between CC and NVP@C. Additionally, by filtering $LiCoO_2$ and $Li_4Ti_5O_{12}$ electrodes into the CNT network's pores, a significant mass loading (5 mg/cm^2) was achieved, creating a robust interface and enhancing mechanical stability [39].

The Qian group successfully manufactured free movable hierarchical composite material of P/CNTs@rGO with outstanding stability and good capacity for LIBs as well as SIBs [40]. At high P content (70 wt percent), the composite P/CNTs@rGO along with the 3D interlinked CNT@rGO framework nearly consisting of red P usually reduced the volumetric changes of red P. As a result, LIBs have 2189.1 mAh/g gravimetric capacity while SIBs have 2112 mAh/g. These numbers are getting close to the red P's theoretical specific capacity. The P/CNT/rGO electrode has a low fading capacity rate of around 0.08 percent in each cycle till 500 cycle tests of stability.

Guo et al. created $CNT/Na_{0.44}MnO_2$ (NMO) and $CNT/NaTi_2(PO_4)_3$@C (NTPO@C) composites for SIB electrodes [41]. The NMO or NTPO@C active composite materials were highly injected into an aligned CNTF-based surface using a wet twisting approach. The system has a good electrochemical nature, including high power as well as energy density, superb cycle stability, high rate, and large flexibility with good safety for all wearable sources. The formed fiber-shaped SIBs continued to function well when the researchers added Na$^+$ containing aqueous electrolytes to the cell-culture media, indicating potential uses in implanted electrical devices.

17.3.2 POLYMER NANOWIRE-BASED COMPOSITE

Conducting polymer (CP) materials, particularly in supercapacitors, are frequently employed in energy storage to contribute pseudocapacitance. Examples of these materials include PPy, PANI, and PEDOT [42]. Building flexible energy-storage devices with CPs is favorable because of their high electrical conductivity, huge tensile strength, good flexibility, quick Faradaic action, and lightweight. Pure CPs, on the other hand, typically have poor cycle stability because of the rupture brought on by the significant volume fluctuations between doping and undoping surfaces during charge and discharge cycles. As a result, creating carbon/CP nanocomposites is a useful technique for raising the cyclic performance, improving electrode's conductivity, and enhancing specific areas to aid in charge movement. For example, using rGO-F/PANI electrodes (foam-reduced graphene oxide films with polyaniline nanoarrays) in which PANI is inserted on pores of rGO-F high energy/power densities and flexible, lightweight symmetric supercapacitors were produced [43]. Using hydrogel of rGO/PANI as electrode materials with PVA/H$_2$SO$_4$ as the gel electrolyte/separator, formed a hydrogel-based supercapacitor [44]. The combined gel fiber-based supercapacitor benefited from the all-gel construction when stable under full stretching and given 8.80 mWh/cm^3 volumetric energy density at 30.77 mW/cm^3 power density. Wei et al. used a different approach for creating a highly flexible integrated gel-based supercapacitor, using PVA/H$_2$SO$_4$ gel with PANI nanoarrays grown on both electrodes of the flexible supercapacitor [45]. The CP-active material-carbon framework matrix nanocomposites are anticipated to exhibit good electrochemical performance due to continuous electron/ion transport since CPs may adhere tightly to active material surfaces because of their flexible nature and abundant polymer functional groups.

17.3.3 ORGANIC NANOWIRE-BASED COMPOSITE

Given the qualities of organic materials, such as their tunable redox capabilities, flexibility, lightweight, environmental friendliness, and low cost, organic nanowire material electrodes are also appropriate for freestanding electrochemical energy storage. For the next generation of LIBs, organic carbonyl compounds hold the most promise [46]. Low solubility and conductivity of materials in the appropriate electrolytes are mainly two significant obstacles, even though these compounds can theoretically give a large capacity. One approach to addressing these challenges is carbonaceous material composition. To create hybrid electrodes, for instance, PI was combined with SCNTs, and the resulting devices demonstrated promising electrochemical capabilities in flexible LIBs [47]. A flexible hybrid electrode made of few-walled CNTs that have been dopamine-coated through self-polymerization was created, according to Li et al. [48]. The freestanding composite electrodes had 133 mAh/g gravimetric capacity for LIBs and approximately 109 mAh/g for SIBs, where polydopamine-fueled redox processes and CNTs contributed to EDLC. For LIBs, to create a foldable flexible organic-based cathode, Xu et al. created AAQ (aminoanthraquinone) NWs inserted in a 3D graphene structure [49]. The composite electrode benefits from a one-dimensional AAQ nanostructure, which reduces the charge movement path, encourages electron diffusion, and also the 3D graphene framework, that offers a proper conducting link and guards against the dissolution of AAQ. These two components work together to give the composite electrode good flexibility, high-rate capability, remarkable cycling, and superior reversibility.

17.3.4 Inorganic Nanowire-Based Composite

Freestanding flexible electrodes for different electrochemical charge storage systems have been made out of inorganic-based NW material with varied hierarchical structures and rich compositions. Although inorganic materials are typically rigid, if inorganic NW materials are long and adequate to form in densely twisted structure, inorganic NW-based membranes can exhibit significant flexibility. According to Zhou et al., it is simple to adjust the length of the NWs by adjusting the hydrothermal reaction duration [50]. For example, the Mai group produced extremely long flexible Sb_2Se_3 NW materials for electrodes with exceptional elasticity for LIB and SIB electrodes, which demonstrated their remarkable electrochemical nature [51]. Moreover, using the liquid exfoliation technique with Li^+ ion inserted, the above teammate also created a 3D hierarchic $H_2V_3O_8$ hydrogel nanostructure made up of interconnected NRs and self-coiled nanoscrolls [52]. The 3D $CNT/H_2V_3O_8$ hydrogel hybrid nanoarrays offered 3D ion and electron diffusion with CNTs to create the freestanding electrode. As a result, it demonstrated a significantly improved cycling stability and rate, at high mass charging. Because of its highly twisted topologies with various contact places among distinct NWs, ultra-long NW constructed network electrodes were employed to build freestanding flexible electrodes with specialized functionality.

A few techniques, like the addition of oxygen-deficient sites and hetero atom doping, have also been shown to be successful means of improving electron and ion transport [53]. Developing inorganic-based NWs on freely movable substrates, including carbonaceous and Ti foil substrates, was another method used to create flexible electrodes. For instance, firstly developing ZNCONW nanoarrays on the surface of CNTF and then building $Ni(OH)_2$ nanostructures on NWs, 3D hierarchic uniform aligned ZNCO (zinc nickel cobalt oxide) at $Ni(OH)_2$ nanoarrays were produced [54]. The flexible electrode provided an extremely high 2847.5 F/cm^3 (10.678 F/cm^2) specific capacitance at 1 mA/cm^2 current density, making use of the special hetero nanostructure and the synergic effects of the uniformed ZNCONW nanoarrays and $Ni(OH)_2$ nanostructures. A solid-state wire-shaped supercapacitor was assembled with CNTF/ZNCO@$Ni(OH)_2$ NW, considered as a core electrode, outer electrode VN@C, and the desired electrolyte (PVA/KOH gel). This resulted in extremely high electrochemical performances, having 94.67 F/cm^3 (573.75 mF/cm^2) high specific capacitance and 33.66 mWh/cm^3 (204.02 mWh/cm^2) exceptional energy density. Notably, the inorganic-based NW array materials on carbon wires have also served as a useful arena for the development of the materials, enabling the creation of hierarchic heterogeneous nanostructures for a variety of applications.

17.4 SUMMARY AND OUTLOOK

In this chapter, we reviewed FNW-based electrodes for electrochemical charge storage in devices (like SCs, LIBs, SIBs, and metal-air batteries). The benefits of the FNW electrode materials over the traditional materials addressing the primary concerns have been thoroughly explained and illustrated in different energy storage electrochemical systems. Even though there have been several significant developments in the synthesis and use of FNWs, their implementation is still lagging what is needed. There are still many obstacles to overcome. There is a lot of potential for their future development to advance real-world uses of NW material-based energy storage electrochemical devices, like the following: (i) the manufacturing of FNWs with different architectures (such as branching, core-shell, hierarchical, etc.) is still expensive. The desired products of NWs are still insufficient for easy synthesis in real-world uses. Therefore, it will be essential to fabricate NWs on a wide scale with cheap cost, appropriate performance, and controllable architectures. (ii) Because FNWs' qualities are correlated with their form and size, it is imperative to precisely control their size and morphology to create an effective synthesis technique. Developing methods for the multi-phase flexible nanowires or multi-element-controlled synthesis of NWs might result in new properties, a greater variety of materials, and a wider range of applications. The synthesis of new 3D flexible materials with a quick electron and charge transport mechanism is one potential method for achieving high power with huge-energy densities. (iii) Designing and logically assembling compelled FNW structures is another area that merits research since they might offer greater flexibility for modifying the electrochemical performance than single structured electrode materials. Coaxial, tri-axial, hierarchical heterostructures, core-shell, multi-shell, and NW arrays are nanostructured designs that can further improve electrochemical performance. (iv) It is essential to design shielding interface materials (like carbon-coating) or choose appropriate electrolytes with stable potential to create a realizable solid-state electrolyte interface layer. This is mainly because of the NW structure's expanding outer surface area, which raises the probability of other unfavorable side reactions occurring at the electrode-electrolyte interface. This effect can be lessened by employing hierarchical, branching, or secondary particle-based systems. (v) The spacing and porosity between FNWs facilitate augmented specific areas for synthesized electrode materials' storage/ deposition and electrolyte penetration, enabling fast kinetics and stability. NW-based 3D hosts/current collectors, either in porous or arrays nanostructures, can even be used for flexible energy storage materials. (vi) The active material content/mass loading of FNW electrode materials with customized volume and pore size volume can be maximized to provide high volumetric energy density in SCs/batteries. The assembly of thin FNW framework materials provides a new prototype for transparent electrochemical energy storage devices in addition to a thick electrode. FNWs' hierarchical porous structure will help the Li ions move towards the active regions and leave space for simultaneously incorporating active components.

REFERENCES

[1]   K. Song, J. Noh, T. Jun, Y. Jung, H.-Y. Kang, J. Moon, Fully Flexible Solution-Deposited ZnO Thin-Film Transistors, Advanced Materials. 22 (2010) 4308–4312.

[2]   N. Kahn, O. Lavie, M. Paz, Y. Segev, H. Haick, Dynamic Nanoparticle-Based Flexible Sensors: Diagnosis of Ovarian Carcinoma from Exhaled Breath, Nano Letters. 15 (2015) 7023–7028.

[3]   Z. Yu, Q. Zhang, L. Li, Q. Chen, X. Niu, J. Liu, Q. Pei, Highly Flexible Silver Nanowire Electrodes for Shape-Memory Polymer Light-Emitting Diodes, Advanced Materials. 23 (2011) 664–668.

[4]   H. Wu, Y. Huang, F. Xu, Y. Duan, Z. Yin, Energy Harvesters for Wearable and Stretchable Electronics: From Flexibility to Stretchability, Advanced Materials. 28 (2016) 9881–9919.

[5]   W. Liu, M.-S. Song, B. Kong, Y. Cui, Flexible and Stretchable Energy Storage: Recent Advances and Future Perspectives, Advanced Materials. 29 (2017) 1603436.

[6]   Z. Liu, J. Xu, D. Chen, G. Shen, Flexible Electronics Based on Inorganic Nanowires, Chemical Society Reviews. 44 (2015) 161–192.

[7]   J.-L. Wang, J.-W. Liu, B.-Z. Lu, Y.-R. Lu, J. Ge, Z.-Y. Wu, Z.-H. Wang, M.N. Arshad, S.-H. Yu, Recycling Nanowire Templates for Multiplex Templating Synthesis: A Green and Sustainable Strategy, Chemistry - A European Journal. 21 (2015) 4935–4939.

[8]   Z. He, M. Hassan, H.-X. Ju, R. Wang, J.-L. Wang, J.-F. Chen, J.-F. Zhu, J.-W. Liu, S.-H. Yu, Stability and Protection of Nanowire Devices in Air, Nano Research. 11 (2018) 3353–3361.

[9]   C. Cheng, H.J. Fan, Branched Nanowires: Synthesis and Energy Applications, Nano Today. 7 (2012) 327–343.

[10] N.P. Dasgupta, J. Sun, C. Liu, S. Brittman, S.C. Andrews, J. Lim, H. Gao, R. Yan, P. Yang, 25th Anniversary Article: Semiconductor Nanowires - Synthesis, Characterization, and Applications, Advanced Materials. 26 (2014) 2137–2184.

[11] P.G. Bruce, B. Scrosati, J.-M. Tarascon, Nanomaterials for Rechargeable Lithium Batteries, Angewandte Chemie International Edition. 47 (2008) 2930–2946.

[12] L. Mai, X. Tian, X. Xu, L. Chang, L. Xu, Nanowire Electrodes for Electrochemical Energy Storage Devices, Chemical Reviews. 114 (2014) 11828–11862.

[13] Q. Zhang, E. Uchaker, S.L. Candelaria, G. Cao, Nanomaterials for Energy Conversion and Storage, Chemical Society Reviews. 42 (2013) 3127.

[14] V.K. Pandey, S. Verma, B. Verma, Polyaniline/Activated Carbon/Copper Ferrite (PANI/AC/CuF) Based Ternary Composite as an Efficient Electrode Material for Supercapacitor, Chemical Physics Letters. 802 (2022) 139780.

[15] E. Frackowiak, K. Metenier, V. Bertagna, F. Beguin, Supercapacitor Electrodes from Multiwalled Carbon Nanotubes, Applied Physics Letters. 77 (2000) 2421–2423.

[16] L.-F. Chen, Y. Lu, L. Yu, X.W. (David) Lou, Designed Formation of Hollow Particle-Based Nitrogen-Doped Carbon Nanofibers for High-Performance Supercapacitors, Energy & Environmental Science. 10 (2017) 1777–1783.

[17] Y. Zhang, L. Li, H. Su, W. Huang, X. Dong, Binary Metal Oxide: Advanced Energy Storage Materials in Supercapacitors, Journal of Materials Chemistry A. 3 (2015) 43–59.

[18] L. Shen, Q. Che, H. Li, X. Zhang, Mesoporous $NiCo_2O_4$ Nanowire Arrays Grown on Carbon Textiles as Binder-Free Flexible Electrodes for Energy Storage, Advanced Functional Materials. 24 (2014) 2630–2637.

[19] D. Kong, W. Ren, C. Cheng, Y. Wang, Z. Huang, H.Y. Yang, Three-Dimensional $NiCo_2O_4$ @ Polypyrrole Coaxial Nanowire Arrays on Carbon Textiles for High-Performance Flexible Asymmetric Solid-State Supercapacitor, ACS Applied Materials & Interfaces. 7 (2015) 21334–21346.

[20] Y. Li, L. Cao, L. Qiao, M. Zhou, Y. Yang, P. Xiao, Y. Zhang, Ni–Co Sulfide Nanowires on Nickel Foam with Ultrahigh Capacitance for Asymmetric Supercapacitors, J. Mater. Chem. A. 2 (2014) 6540–6548.

[21] Z. Yin, S. Cho, D.-J. You, Y. Ahn, J. Yoo, Y.S. Kim, Copper Nanowire/Multi-Walled Carbon Nanotube Composites as All-Nanowire Flexible Electrode for Fast-Charging/Discharging Lithium-Ion Battery, Nano Research. 11 (2018) 769–779.

[22] C. Zhu, Y. Yu, L. Gu, K. Weichert, J. Maier, Electrospinning of Highly Electroactive Carbon-Coated Single-Crystalline $LiFePO_4$ Nanowires, Angewandte Chemie International Edition. 50 (2011) 6278–6282.

[23] H. Xia, Y. Wan, W. Assenmacher, W. Mader, G. Yuan, L. Lu, Facile Synthesis of Chain-Like $LiCoO_2$ Nanowire Arrays as Three-Dimensional Cathode for Microbatteries, NPG Asia Materials. 6 (2014) e126–e126.

[24] Y.-L. Ding, J. Xie, G.-S. Cao, T.-J. Zhu, H.-M. Yu, X.-B. Zhao, Single-Crystalline LiMn2O4 Nanotubes Synthesized Via Template-Engaged Reaction as Cathodes for High-Power Lithium Ion Batteries, Advanced Functional Materials. 21 (2011) 348–355.

[25] R. von Hagen, H. Lorrmann, K.-C. Möller, S. Mathur, Electrospun $LiFe_{1-y}MnyPO_4$/C Nanofiber Composites as Self-Supporting Cathodes in Li-Ion Batteries, Advanced Energy Materials. 2 (2012) 553–559.

[26] L. Zhao, Z. Qu, Advanced Flexible Electrode Materials and Structural Designs for Sodium Ion Batteries, Journal of Energy Chemistry. 71 (2022) 108–128.

[27] D. Su, G. Wang, Single-Crystalline Bilayered V_2O_5 Nanobelts for High-Capacity Sodium-Ion Batteries, ACS Nano. 7 (2013) 11218–11226.

[28] Q. Wei, Z. Jiang, S. Tan, Q. Li, L. Huang, M. Yan, L. Zhou, Q. An, L. Mai, Lattice Breathing Inhibited Layered Vanadium Oxide Ultrathin Nanobelts for Enhanced Sodium Storage, ACS Applied Materials & Interfaces. 7 (2015) 18211–18217.

[29] T. Jin, Y. Liu, Y. Li, K. Cao, X. Wang, L. Jiao, Electrospun $NaVPO_4$ F/C Nanofibers as Self-Standing Cathode Material for Ultralong Cycle Life Na-Ion Batteries, Advanced Energy Materials. 7 (2017) 1700087.

[30] Y. Dong, S. Li, K. Zhao, C. Han, W. Chen, B. Wang, L. Wang, B. Xu, Q. Wei, L. Zhang, X. Xu, L. Mai, Hierarchical zigzag $Na_{1.25}V_3O_8$ Nanowires with Topotactically Encoded Superior Performance for Sodium-Ion Battery Cathodes, Energy & Environmental Science. 8 (2015) 1267–1275.

[31] D. Zhou, X. Qiu, F. Liang, S. Cao, Y. Yao, X. Huang, W. Ma, B. Yang, Y. Dai, Comparison of the Effects of $FePO_4$ and $FePO_4 \cdot 2H_2O$ as Precursors on the Electrochemical Performances of $LiFePO_4$/ C, Ceramics International. 43 (2017) 13254–13263.

[32] Z. Ma, X. Yuan, L. Li, Z.-F. Ma, D.P. Wilkinson, L. Zhang, J. Zhang, A Review of Cathode Materials and Structures for Rechargeable Lithium–Air Batteries, Energy & Environmental Science. 8 (2015) 2144–2198.

[33] D.U. Lee, P. Xu, Z.P. Cano, A.G. Kashkooli, M.G. Park, Z. Chen, Recent Progress and Perspectives on Bi-Functional Oxygen Electrocatalysts for Advanced Rechargeable Metal–Air Batteries, Journal of Materials Chemistry A. 4 (2016) 7107–7134.

[34] F. Liang, X. Qiu, Q. Zhang, Y. Kang, A. Koo, K. Hayashi, K. Chen, D. Xue, K.N. Hui, H. Yadegari, X. Sun, A Liquid Anode for Rechargeable Sodium-Air Batteries with Low Voltage Gap and High Safety, Nano Energy. 49 (2018) 574–579.

[35] X. Zhao, S.C. Abbas, Y. Huang, J. Lv, M. Wu, Y. Wang, Robust and Highly Active FeNi@NCNT Nanowire Arrays as Integrated Air Electrode for Flexible Solid-State Rechargeable Zn-Air Batteries, Advanced Materials Interfaces. 5 (2018) 1701448.

[36] J. Yin, Y. Li, F. Lv, M. Lu, K. Sun, W. Wang, L. Wang, F. Cheng, Y. Li, P. Xi, S. Guo, Oxygen Vacancies Dominated NiS_2/CoS_2 Interface Porous Nanowires for Portable Zn-Air Batteries Driven Water Splitting Devices, Advanced Materials. 29 (2017) 1704681.

[37] T. Gu, Z. Cao, B. Wei, All-Manganese-Based Binder-Free Stretchable Lithium-Ion Batteries, Advanced Energy Materials. 7 (2017) 1700369.

[38] D. Guo, J. Qin, Z. Yin, J. Bai, Y.-K. Sun, M. Cao, Achieving High Mass Loading of $Na_3V_2(PO_4)_3$@ Carbon on Carbon Cloth by Constructing Three-Dimensional Network between Carbon Fibers for Ultralong Cycle-Life and Ultrahigh Rate Sodium-Ion Batteries, Nano Energy. 45 (2018) 136–147.

[39] J.W. Hu, Z.P. Wu, S.W. Zhong, W.B. Zhang, S. Suresh, A. Mehta, N. Koratkar, Folding Insensitive, High Energy Density Lithium-Ion Battery Featuring Carbon Nanotube Current Collectors, Carbon N Y. 87 (2015) 292–298.

[40] J. Zhou, Z. Jiang, S. Niu, S. Zhu, J. Zhou, Y. Zhu, J. Liang, D. Han, K. Xu, L. Zhu, X. Liu, G. Wang, Y. Qian, Self-Standing Hierarchical P/CNTs@rGO with Unprecedented Capacity and Stability for Lithium and Sodium Storage, Chem. 4 (2018) 372–385.

[41] Z. Guo, Y. Zhao, Y. Ding, X. Dong, L. Chen, J. Cao, C. Wang, Y. Xia, H. Peng, Y. Wang, Multi-Functional Flexible Aqueous Sodium-Ion Batteries with High Safety, Chem. 3 (2017) 348–362.

[42] S. Verma, T. Das, V.K. Pandey, B. Verma, Nanoarchitectonics of GO/PANI/$CoFe_2O_4$ (Graphene Oxide/polyaniline/Cobalt Ferrite) Based Hybrid Composite and its Use in Fabricating Symmetric Supercapacitor Devices, Journal of Molecular Structure. 1266 (2022) 133515.

[43] P. Yu, X. Zhao, Z. Huang, Y. Li, Q. Zhang, Free-Standing Three-Dimensional Graphene and Polyaniline Nanowire Arrays Hybrid Foams for High-Performance Flexible and Lightweight Supercapacitors, J. Mater. Chem. A. 2 (2014) 14413–14420.

[44] P. Li, Z. Jin, L. Peng, F. Zhao, D. Xiao, Y. Jin, G. Yu, Stretchable All-Gel-State Fiber-Shaped Supercapacitors Enabled by Macromolecularly Interconnected 3D Graphene/Nanostructured Conductive Polymer Hydrogels, Advanced Materials. 30 (2018) 1800124.

[45] K. Wang, X. Zhang, C. Li, X. Sun, Q. Meng, Y. Ma, Z. Wei, Chemically Crosslinked Hydrogel Film Leads to Integrated Flexible Supercapacitors with Superior Performance, Advanced Materials. 27 (2015) 7451–7457.

[46] C. Luo, R. Huang, R. Kevorkyants, M. Pavanello, H. He, C. Wang, Self-Assembled Organic Nanowires for High Power Density Lithium Ion Batteries, Nano Letters. 14 (2014) 1596–1602.

[47] H.P. Wu, Q. Yang, Q.H. Meng, A. Ahmad, M. Zhang, L.Y. Zhu, Y.G. Liu, Z.X. Wei, A Polyimide Derivative Containing Different Carbonyl Groups for Flexible Lithium Ion Batteries, Journal of Materials Chemistry A. 4 (2016) 2115–2121.

[48] T. Liu, K.C. Kim, B. Lee, Z. Chen, S. Noda, S.S. Jang, S.W. Lee, Self-Polymerized Dopamine as an Organic Cathode for Li- and Na-ion Batteries, Energy & Environmental Science. 10 (2017) 205–215.

[49] G. Yang, F. Bu, Y. Huang, Y. Zhang, I. Shakir, Y. Xu, In Situ Growth and Wrapping of Aminoanthraquinone Nanowires in 3 D Graphene Framework as Foldable Organic Cathode for Lithium-Ion Batteries, ChemSusChem. 10 (2017) 3419–3426.

[50] B. Yao, L. Huang, J. Zhang, X. Gao, J. Wu, Y. Cheng, X. Xiao, B. Wang, Y. Li, J. Zhou, Flexible Transparent Molybdenum Trioxide Nanopaper for Energy Storage, Advanced Materials. 28 (2016) 6353–6358.

[51] W. Luo, A. Calas, C. Tang, F. Li, L. Zhou, L. Mai, Ultralong Sb_2Se_3 Nanowire-Based Free-Standing Membrane Anode for Lithium/Sodium Ion Batteries, ACS Applied Materials & Interfaces. 8 (2016) 35219–35226.

[52]   Y. Dai, Q. Li, S. Tan, Q. Wei, Y. Pan, X. Tian, K. Zhao, X. Xu, Q. An, L. Mai, Q. Zhang, Nanoribbons and Nanoscrolls Intertwined Three-Dimensional Vanadium Oxide Hydrogels for High-Rate Lithium Storage at High Mass Loading Level, Nano Energy. 40 (2017) 73–81.

[53]   T. Zhai, X. Lu, Y. Ling, M. Yu, G. Wang, T. Liu, C. Liang, Y. Tong, Y. Li, A New Benchmark Capacitance for Supercapacitor Anodes by Mixed-Valence Sulfur-Doped V_6O_{13-x}, Advanced Materials. 26 (2014) 5869–5875.

[54]   Q. Zhang, W. Xu, J. Sun, Z. Pan, J. Zhao, X. Wang, J. Zhang, P. Man, J. Guo, Z. Zhou, B. He, Z. Zhang, Q. Li, Y. Zhang, L. Xu, Y. Yao, Constructing Ultrahigh-Capacity Zinc–Nickel–Cobalt Oxide@ $Ni(OH)_2$ Core–Shell Nanowire Arrays for High-Performance Coaxial Fiber-Shaped Asymmetric Supercapacitors, Nano Letters. 17 (2017) 7552–7560.

18 Nanowires for Flexible Electronics

Chohdi Amri
Department of Physics, College of Science, United Arab Emirates University, Sheik Khalifa Bin Zayed Street, Al Ain, Abu Dhabi, United Arab Emirates
Photovoltaic Laboratory, Energy Research and Technology Center, Soliman Tourist Route, Borj Cerdria, Hammam-Lif, Tunisia

Abderrahmane Hamdi
Université Polytechnique, Hauts-de-France, INSA Hauts-de-France, CERAMATHS - Laboratoire de Matériaux Céramiques et de Mathématiques, Valenciennes, France

Adel Najar
Department of Physics, College of Science, United Arab Emirates University, Sheik Khalifa Bin Zayed Street, Al Ain, Abu Dhabi, United Arab Emirates

CONTENTS

DOI: 10.1201/9781003296621-18

18.1 INTRODUCTION

Nanowires (NWs) are one dimension (1-D) size-reduced cylindrical shaped materials grown or grooved in a specific direction, with diameters in the range of 1–100 nanometers (nm) and lengths ranging from a few nm to hundred micrometers (μm). This structure was originally elaborated by Wagner and Ellis in 1964 using the vapor-liquid-solid (VLS) mechanism catalyzed by Au nanoparticles to grow silicon-NWs (Si-NWs) [1]. Following that, many investigative works have been carried out regarding the development of new elaboration techniques [2] and the study of the interesting properties of the NWs [3] making them promising materials and building blocks for electronic devices. NWs are integrated into a wide variety of applications, such as transistors, photovoltaic cells, touch screens, pressure sensors, electric nanogenerators, batteries, etc. Over the past few years, the demand for smaller, lighter, and wirelessly connected high technology devices increased significantly, replacing the bulky, heavy, and wired electronics. In addition, the flexibility of these devices becomes a critical property in the development of emerging electronic devices. As it is well known, the flexibility of the device means to comply with bending to some degree for many cycles without losing its function or its initial properties [4].

Many types of flexible substrates were used in electronics, including metallic and inorganic semiconductor flexible substrates with a thickness of less than 100 μm. However, low-cost, transparent, lightweight, and flexible polymers such as polydimethylsiloxane (PDMS), polyimides (PIs), polyethylene naphthalates (PEN), Polyethylene terephthalate (PET), and poly(3,4-ethylenedioxythiophene): poly(styrene sulfonate) (PEDOT:PSS) are widely used in flexible electronics [5]. A pairing between NWs and flexible substrates makes them a promising material for flexible electronic applications. Plenty of articles, review papers, and book chapters have been published to summarize and discuss the effectiveness of applying NWs in flexible electronics. By way of example, Liu et al. and later Han et al. summarized the recent progress in the assembly method and the large-scale and high-frequency flexible electronics of inorganic semiconductor NWs [6,7]. However, several interesting applications, such as touch screens, batteries, nanogenerators based on NWs were not discussed. However, the authors discussed only the inorganic NW and neglected the importance of organic NWs in flexible electronics.

This chapter offers an in-depth study of the application of metallic, inorganic semiconductors, and organic NWs in flexible electronics. At first, the recent progress in NW properties is summarized. Then, a brief discussion of the flexible substrates suitable for electronic applications is highlighted focusing on the enhancement of their physical properties. In the last part of this chapter, we deeply discuss the recent progress of the NWs in flexible electronics, including transistors, touch screen, chemical and pressure sensors, piezoelectric nanogenerators, photovoltaic cells, batteries, and supercapacitors.

18.2 NWS FOR FLEXIBLE ELECTRONICS

18.2.1 Properties of NWs

Since the discovery of the NW structure, research works have moved from the field of scientific curiosity to that of the promising nano-object suitable for sophisticated electronic devices. When the NW dimensions (diameter or length) are reduced to a few nm or their sidewalls are coated with a thin porous layer [8], interesting physical properties, absent in the bulk and macroscopic material, emerge such as follows:

(i)   The large surface-to-volume ratio giving the NWs distinct structural and distinguished chemical reactivity suitable for sensing applications.
(ii) The quantum confinement effect and the modification of the exciton properties making the NWs a suitable 2-D material for photonic and optoelectronic applications.
(iii) The low total reflectivity values of the NWs nearly about 1% allowing an efficient light trapping due to the multiple scattering between the sidewalls and the gradual increase in refractive index with depth.

These outstanding properties allow good integrity of NWs in various fields, in particular electronic devices.

18.2.2 Different Types of NWs

Metallic-NWs (M-NWs), inorganic semiconductor-NWs, perovskite-NWs, and organic NWs have been widely applied in flexible electronics. In this section, we briefly introduce the different NW properties making them promising materials suitable for flexible electronic applications.

18.2.2.1 M-NWs

M-NWs have many excellent features, such as simple elaboration techniques, high transparency, and low sheet resistance. A wide variety of M-NWs has been used in flexible electronics, among them Au-NWs, Ag-NWs, Cu-NWs, etc. However, there are some drawbacks limiting the application of M-NWs in flexible electronics where the most common is the weak electrode-substrate bonding strength due to the poor van der Waals forces between the M-NWs and the substrate surface leading to the delamination or the fall-off from the substrates when bent or compressed, resulting in the device failure and less durability.

18.2.2.2 Semiconductor and Semiconductor Oxide NWs

Semiconductor and semiconductor oxide NWs, such as Si-NWs, GaN-NWs, ZnO-NWs, and MoS_2-NWs are good candidates for flexible electronics [9–13]. The elaboration techniques of these NWs are similar to M-NWs and made by transferring or growing the NWs onto flexible substrates. Recently, some efforts have been carried out to groove NWs into flexible semiconductor substrates. Wang et al. have fabricated large-area ultrathin and highly flexible monocrystalline Si films in a free-standing form by KOH etching solution [14]. Based on that, Omar et al. grooved Si-NWs in a flexible Si substrate using metal-assisted chemical etching resulting in Si-NWs with the same properties as the starting material and strong bonding forces [15].

18.2.2.3 Perovskite-NWs

Perovskite-NWs such as $CH_3NH_3PbI_3$-NWs [16], $BaTiO_3$-NWs [17], $LaCo_{0.8}Fe_{0.2}O_3$-NWs [18] with excellent elastic property and mechanical robustness demonstrate their effectiveness in flexible electronics. However, the stability of perovskite-NWs is a serious drawback limiting their applications in flexible electronics since they strongly interact with moisture. Rajbari et al. demonstrated that super-hydrophilic $SrMnO_3$-NWs can be turned into a water repellent upon modification by higher energy nitrogen ion beam irradiation [19]. This nitrogen treatment was recommended to be applied to other perovskite-NWs aiming to enhance their stability against environmental change.

18.2.2.4 Organic NWs

Organic-NWs such as PEDOT-NWs, [20], carbon-NWs [21], and P3HT-NWs [22] were also applied in flexible electronics [23]. However, these materials suffer from many drawbacks such as low mobility (~10 cm²/Vs) leading to poor performance of the devices especially when the high speed is required, weak mismatched interfaces between organic NWs and metal electrodes, poor heat tolerance, and eventual organic pollution.

18.3 FLEXIBLE SUBSTRATES

Substrates in flexible electronics must satisfy specific requirements other than the mechanical elasticity, such as high optical transmission, chemical stability at high temperatures, environmental resistance, and large area stability to avoid current leakage [24], etc.

By reducing the thickness of rigid metallic or inorganic semiconductor substrates to less than 100 μm they become flexible. However, pairing between these ultra-thin flexible substrates and NWs is not thoroughly reported in the literature since they were replaced by polymer substrates. Even that, by using flexible Ti mesh, Liu et al. fabricated TiO_2-NWs as a photoanode of dye-sensitized solar cells using a hydrothermal and spin-coating method [12]. In addition, Omar et al. grooved Si-NWs in flexible Si substrates for photovoltaic application [15].

Despite these efforts, polymer substrates remain the most used in flexible electronics due to their low cost, lightweight, and distinguished ultra-flexibility as reported in the yield strength and Young's modulus diagram [25]. However, the polymer flexible substrates suffer mainly from the low chemical stability at high temperatures and poor conductivity. Many investigative works have been carried out to boost their physical properties and to make them the leading candidates for flexible electronics. For example, PDMS suffers from low thermal stability above 200 °C and hydrophobicity resulting in a poor adhesion between the substrate and the deposit thin film. To boost their performances, Dhanabalan et al. reported that treating PDMS with the combination of O_2 plasma and sodium dodecyl sulfate enhances its thermal stability to withstand up to 400 °C and boosts the bonding force between the deposited material and the substrate [26]. Furthermore, Hwang et al. fabricated a porous PDMS by embedding sugar particles in its volume to boost its deformability with voids after dissolving the sugar particles [27]. The modified PDMS substrate shows excellent performances in flexible capacitive pressure sensors tested on finger-attached sensors (gloves), respiration monitoring system (face mask), and sensor array to measure the pressure distribution for objects with irregular shapes.

Polymer performances can be also enhanced by embedding NWs within their volume [28] or by depositing them on their surface to enhance chemical stability [29], electrical conductivity [30], thermal conductivity [31], and even flexibility [32]. In the next sections, we discuss in detail the pairing between polymer flexible substrates and NWs for their applications in flexible electronics.

18.4 APPLICATIONS OF NWS FOR THE FLEXIBLE ELECTRONICS

Within the last decades, the demand for smaller, lighter, and wirelessly connected high technology devices increased significantly, replacing the bulky, heavy, and wired electronics. For that, the scientific community faced a serious challenge in the development and miniaturization of electronic devices. NWs coating flexible substrates have attracted great attention for their application in next-generation technology. In this section, we review the different applications of NWs in flexible electronics focusing on the recent advances and challenges.

18.4.1 TRANSISTORS

Nowadays, field effect transistors (FETs) have received great attention as they are the fundamental components in electronics. However, the most used transistors are based on two-dimensional (2-D) materials representing some limitations, such as the degradation of their electrical properties due to the polycrystalline structure of the thin film and the deterioration of the device under a few hundred of bending cycles. To overcome these limitations, NWs have been widely investigated as the building blocks for flexible FETs. Si-NWs with high sensitivity, large dynamic ranges, and fast responses can be used in FETs as a single Si-NW [33], Si-NW multi-parallel channels [34], and Si-nanonets (Si-NNs) [35]. However, due to the small area in the cross-section of single Si-NWs, the drain current (I_{ds}) of single Si-NW FETs is still limited, which causes difficulty in applying to various flexible electronic devices.

Organic-NWs have been incorporated within the channel of flexible organic electrochemical transistors (OECTs) as an efficient electronic device suitable for the detection of some ions. Wang et al. developed cotton fiber (CF) based OECTs functionalized with PEDOT-NWs/multi-walled carbon nanotube (MWCNT) channel as a transducer to detect an ultralow concentration of potassium ions (K^+) in the human body [20]. **Figure 18.1** shows the elaboration procedure of the OECT, which starts with the immersion of the CF in MWCNT dispersion until the complete coverage of the

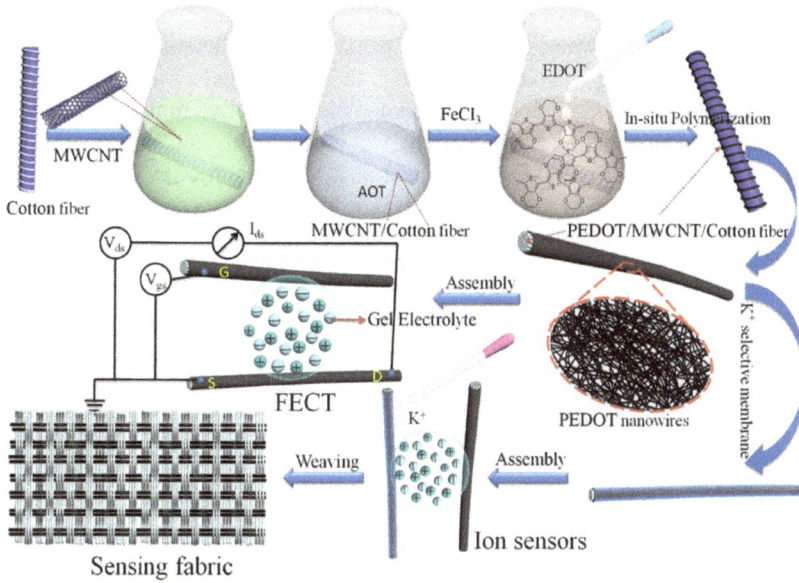

FIGURE 18.1 Elaboration procedure of the OECT. Adapted with permission [20]. Copyright 2022, Elsevier.

surface. Then, PEDOT-NWs were deposited on the treated CF by an inverse micro-emulsion method to be ready for the K^+ detection after a final coverage with the ion-selective membrane solution. The OECTs based on PEDOT-NWs show a fast response and good reproducibility detection limit of 1 nM. Since the normal range of K^+ level is 3.5–5.5 mM in the human plasma and 2–8 mM in the sweat, this OECT is sensitive enough to detect K^+ in the human body. The significant increase in sensitivity is attributed to the enhanced surface area on the channel associated with strong interactions between PEDOT-NWs/MWCNT and the active sites. Furthermore, no matter the degree of bending, the OECT based on PEDOT-NWs shows a good and stable characteristic under a bending radius of 2 to 0.5 cm. These results confirm the effectiveness of applying the PEDOT-NWs in flexible OECTs and the possibility of integrating them into the fabric by hand weaving or weaving machine.

18.4.2 Sensing Applications

Human-oriented sensing technologies such as pressure and chemical sensors are becoming omnipresent and part of everyday life. However, these sensors need deep transformations for the design and the fabrication of size-reduced and flexible devices. NWs are promising platforms for flexible sensors due to their superior sensing performance, long-term stability, ultra-low power consumption, and facile integration in flexible electronics. In this section, we report the recent advances of the NW in flexible sensors in particular pressure and gas sensors.

18.4.2.1 Flexible Pressure Sensors

The incorporation of M-NWs with flexible substrates as an electrode for pressure sensors with high sensing capability, flexibility, and lightweight has attracted great attention for a wide field of applications, including artificial skin (e-skin), wearable devices, humanoid robotics, and smart prosthetics. Furthermore, the biodegradability of the flexible pressure sensor is a parameter that should be seriously considered during the manufacture of the devices due to the increasing awareness of environmental sustainability.

Song et al. produced a biodegradable flexible capacitive sensor composed of a degradable glycerol/chitosan dielectric layer with micro-protrusions and two degradable Cu-NW/chitosan electrodes [36]. The elaboration procedure of the biodegradable flexible pressure sensor starts by mixing acetic acid, chitosan powder, glycerin, and pentane under a magnetic stirrer to be deposited onto sandpaper by spin coating technique. After that, the formed dielectric layer was removed, and finally, a Cu-NWs/chitosan flexible degradable composite conductive film was stuck on both sides of the glycerin/chitosan elastic film. The sensitivity was considerably enhanced with well-embedded Cu-NWs in the chitosan layer reducing the contact area between the Cu-NW and the air and significantly improving the oxidation resistance of the Cu-NWs/chitosan composite film. **Figure 18.2** shows the real-time applications of the degradable sensor under airflow impact, artery pulse of the

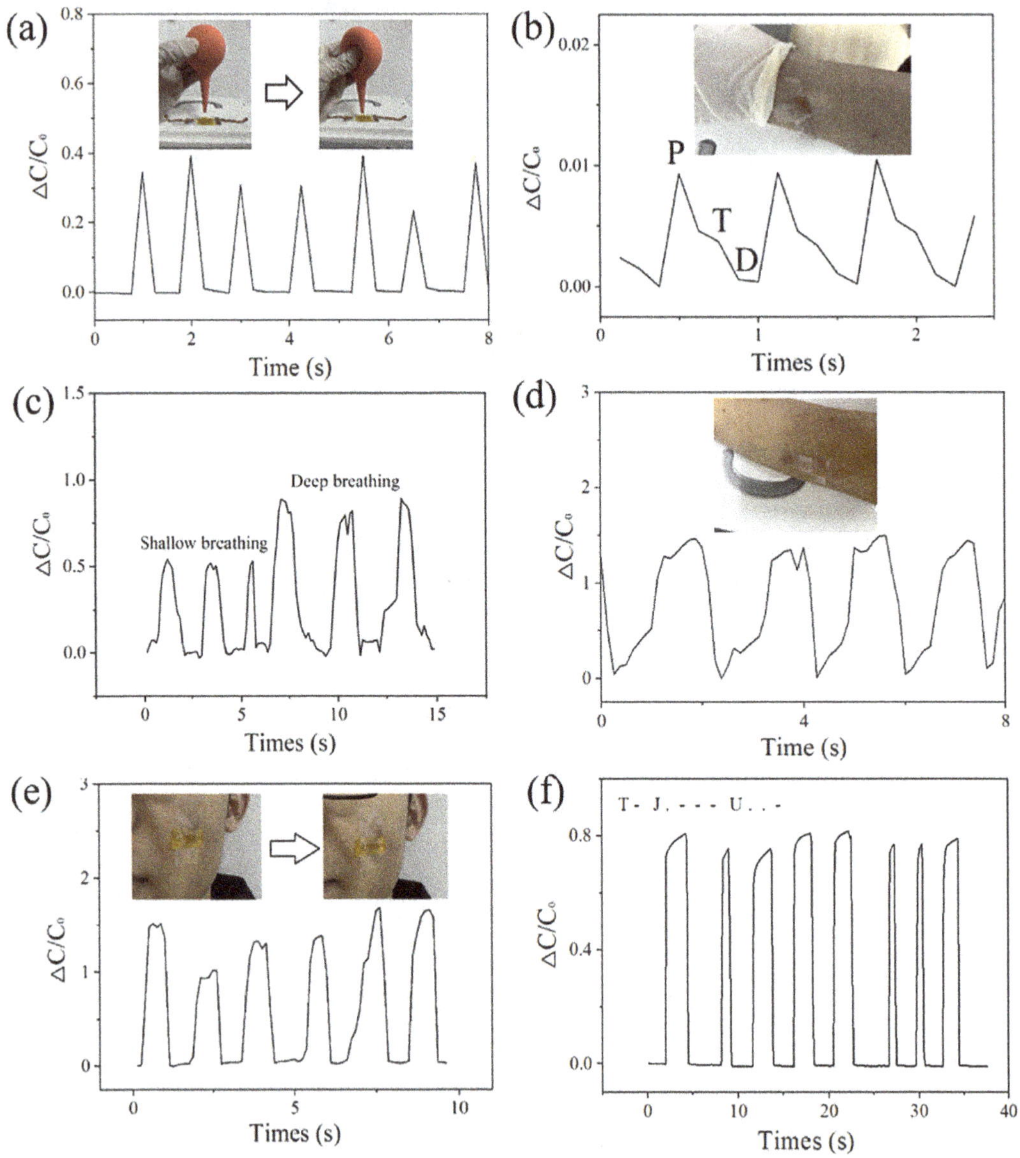

FIGURE 18.2 Capacitive responses of the real-time applications of the degradable sensor (a) under airflow impact, (b) human wrist pulse signals, (c) deep and shallow breathing, (d) muscle changes in the arm, (e) smiling. (f) Send the letters "T", "J" and "U" in Morse code. Adapted with permission [36]. Copyright 2022, Elsevier.

human wrist, deep and shallow breathing, arm muscle changes, and face smile. The capacitance of the sensor is recorded to be proportional to the air blown due to its high sensitivity. By comparison with a professional instrument, the sensor shows a rate of about 75 beats per minute after being installed on the volunteer's wrist to detect the radial artery pulse wave and the abdomen to monitor shallow and deep breathing and clearly distinguish the characteristic peaks in the pulse wave waveform. For the electronic skin application, the pressure sensor can detect the movement of the human arm muscles and micro-expression of the face while smiling. This flexible sensor exhibits excellent sensitivity of 1.7 kPa^{-1}, a fast response time of about 180 ms, and good repeatability, which can be reused more than 1700 times. The biodegradability test demonstrates that the sensor is completely dissolved in urban tap water containing 0.02% acetic acid and snail enzyme within 5 days under natural conditions. All these previous results confirm the effectiveness of embedding the Cu-NWs' network in the chitosan film to form the biodegradable, ultra-flexible, pressure sensor.

18.4.2.2 Flexible Gas Sensors

Many harmful gases were used to test the sensors such as H_2, H_2S, NH_3, CO, volatile organic compounds, and oxidative gases NO, NO_2, and CO_2. Among these gases, H_2 is characterized by its huge security risk (explosive limit ranging from approximately 4.1%–74.8% volume in the air) and it being colorless and odorless. Thus, the development of highly sensitive flexible H_2 gas sensors is a must for early warning and vigilant control in production, transport, storage, and usage. Due to their high surface area, unidirectional charge transport, and excellent reactivity with the environment, NWs grown or deposited on flexible substrates are the main candidates for sensing applications.

Pradeep et al. report the use of ZnO decorated SnO_2-NWs on overhead projector (OHP) substrate for H_2 sensing [37]. The flexible H_2 sensor shows almost the same response under different curvatures: flat condition (0°) of the substrate, flexible (curved) such as 45°, 60° because of the NW length and flexibility. However, a drop in the sensing response at a 90° angle was attributed to cracks in the film. In addition, the performance of the flexible sensor is limited by the high minimum gas detection level (1000–3000 ppm) and the low operating temperature of around 100 C. On the other hand, Luo et al. reported the elaboration of a flexible H_2 sensor based on ultra-long free-standing MoO_3-NW paper [38]. **Figure 18.3** shows the characterization of MoO_3-NWs using SEM, TEM, and SAED patterns as well as the digital photo of the corresponding NW paper. As exhibited in **Figure 18.3(a–c)** MoO_3-NWs display good dispersion, uniform diameter (~300 nm), long length (~1 mm), and high crystallinity. The digital photo of the self-assembly flexible MoO_3-NW paper shown in **Figure 18.3(d)** reveals the excellent flexibility of the sensor making a cylindrical shape. The NWs' paper sensor shows excellent performances: a low minimum gas detection level of about 100 ppm, fast response, and recovery time is about 3.0 s and 2.7 s, respectively, and good selectivity toward H_2 against other reduce gas, such as C_2H_5OH, CO, and CH_4, and outstanding stability and reliability over 5.6 years. These interesting results are obtained thanks to the 1-D structure of NW coated with a porous thin layer and the high specific surface of MoO_3-NW paper making them suitable H_2 sensors that could be integrated into flexible electronics.

18.4.3 Touch Screens

Touch screens are very commonplace in our daily lives and are one of the most used interfaces in electronic devices. The bulky touchscreens use indium tin oxide (ITO) as a transparent and conductive electrode. However, ITO is an expensive and brittle substrate that can be replaced by flexible and abundant organic substrates modified with NWs aiming to elaborate flexible touch screens with better brightness, longer battery life per charge, and reduced size.

Among all M-NWs, Ag-NWs are the most used in touch screen applications. De Guzman et al. reported the elaboration of Ag-NWs on cellulose acetate flexible substrate by electroless deposition using $CuCl_2$ as a stabilizer for transparent conductive electrodes (TCEs) of the resistive touch

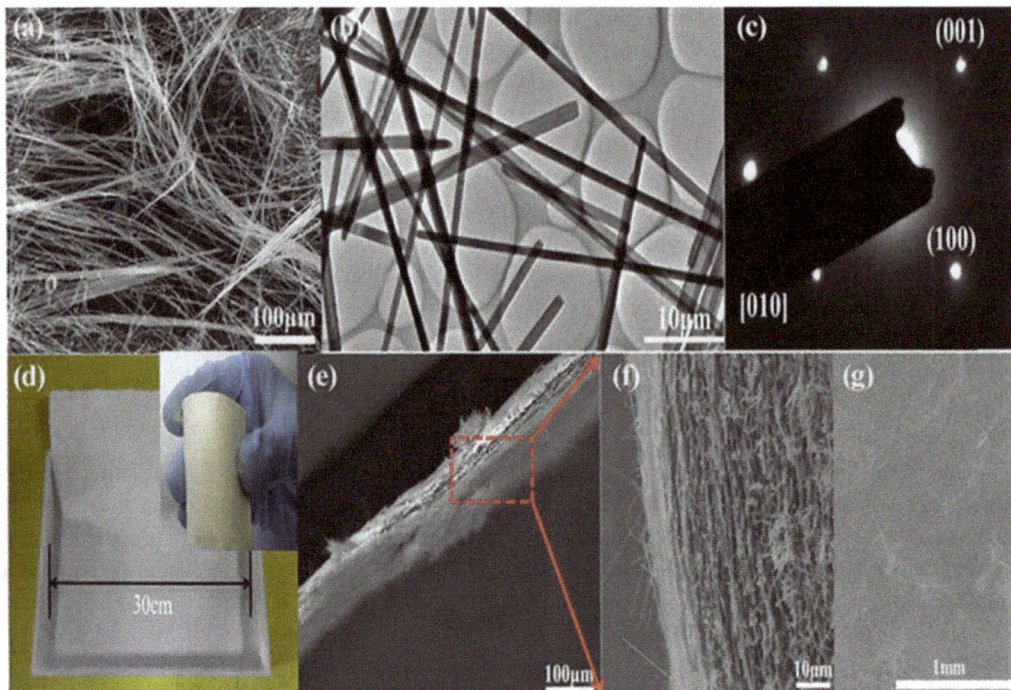

FIGURE 18.3 (a–c) SEM and TEM images of MoO$_3$-NWs (d) digital photo of the corresponding NW paper (e–g) cross-section and surface of MoO$_3$-NWs' paper. Adapted with permission [38]. Copyright 2017, Elsevier.

screens [39]. The high-quality and ultra-long Ag-NWs of the TCEs have an average diameter and length of about 96.3 nm and 46.3 µm, respectively, and are obtained only in the presence of the stabilizer with an optimum amount of about 1.6 µM. The TCEs show low sheet resistances in the range of 10–150 Ω/sq at high optical transmittances (~90%) and good sensitivity upon the application of vertical and horizontal pressure. However, the use of Ag-NWs for the touch screen suffers from the weak bonding strength with the polymer substrates leading to the delamination or the NW falling off when the touch screen is bent resulting in its rapid degradation.

According to the Ag-S bond theory, the 5 s-orbital lone electrons of Ag atoms can be introduced to the electron hybrid orbitals of sulfur atoms and form an Ag-S bond with a strong bonding force energy of 217 kJ/mol [40]. Inspired by this theory, Yu et al. developed tough and strong electrode-substrate bonds by using Ag-NWs as a metal conductive electrode and thiol-modified nanofibrillated cellulose nanopaper (NFC-HS) as substrates [41]. Therefore, with Ag-containing Ag-NW network electrodes and S-containing NFC-HS nanopaper substrates, tough TCEs for touch screens with strong Ag-S bonds are produced. By comparing with Ag-NWs/PET electrode, the resistance of the Ag-NWs/NFC-HS electrode increases nearly five times after 10,000 bending cycles with a radius of 5 mm. In contrast, the resistance of Ag-NWs/PET electrode increases dramatically by a factor of 28 after only 1100 bending cycles. **Figure 18.4** shows the schematic illustration and the digital photos of the flexible touch screen and its performances under different conditions. **Figure 18.4(a)** depicts the schematic illustration of the flexible touch screen where the two TCEs are made by Ag-NWs/NFC-HS and the spacer is formed by a commercial UV-cured adhesive printed onto the bottom surface electrode. As shown in **Figure 18.4(b, c)** TCE exhibits excellent flexibility and transparency as a logo printed under the device is observed due to the excellent transparency of Ag-NWs and strong bonds between Ag-NWs and the NFC-HS substrates. Furthermore, the touch screen shows a stable high sensitivity, high reliability, and durability even after 500 bending cycles in flat and bent conditions as depicted in **Figure 18.4(d–f).**

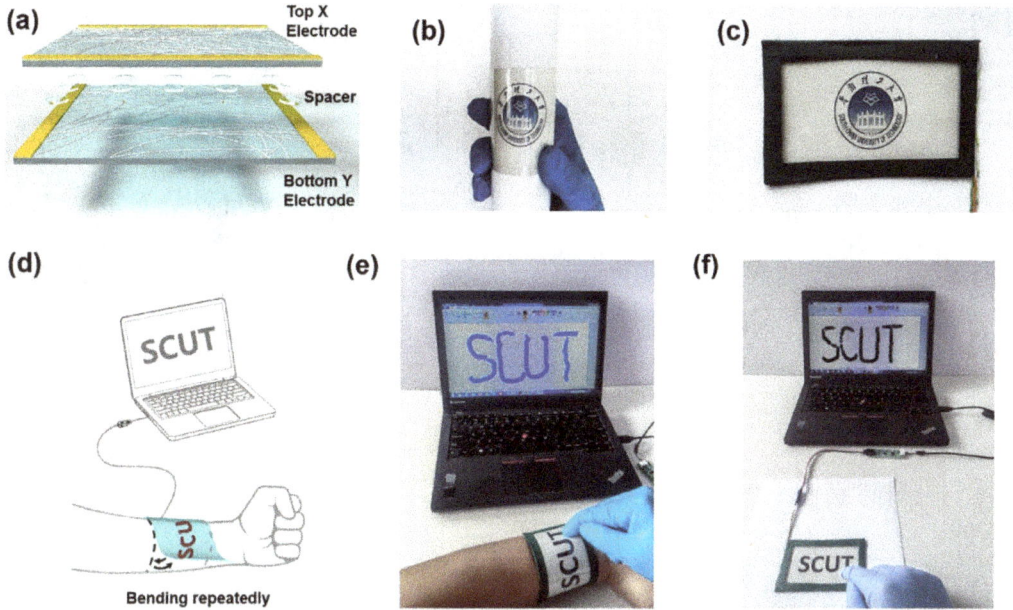

FIGURE 18.4 (a) Schematic illustration and digital photos of the flexible touch screen showing its excellent (b) flexibility, (c) transparency. (d) Schematic illustration of the touch screen exhibiting its sensitivity and reliability after 500 bending cycles under (e) bent and (f) flat conditions. Adapted with permission [41]. Copyright 2021, Elsevier.

18.4.4 PIEZOELECTRIC NANOGENERATORS (PNGs)

Nanogenerators capable of converting mechanical energy abundant in the environment into electricity using piezoelectric NWs have received the attention of the scientific community due to their significant superior piezoelectric properties and high sensitivity to small forces compared to bulk and thin layers [42].

ZnO-NWs are highly used materials for the elaboration of flexible PNGs due to their eco-friendly elaboration techniques, high force-to-displacement sensitivities, and high piezoelectric response. Chelu et al. studied the effect of the substrate on the piezoelectricity of ZnO-NWs [43]. The flexible metallic substrate is Ti foil while the rigid ones are $Pt/Ti/SiO_2/Si$ and $Au/Ti/SiO_2/Si$. The ZnO-NW grown on flexible Ti substrate and covered by PMMA shows a piezoelectric coefficient d_{33} equal to 26 pC/N at 110 Hz, which is four times greater than the piezoelectric coefficient of the rigid ones. These interesting results are attributed to the mechanical deformation enhancement of the entire structure assured by the flexibility of the Ti foil. However, doping the ZnO-NWs with different impurities from group V (P, As, Sb) seems more interesting to enhance the piezoelectric voltage by acting as acceptor impurities for intrinsically n-type ZnO-NWs by which reverse leakage current through NWs has been minimized [44]. In addition, several issues related to ZnO polarity are still open and deserve particular attention [45].

Wang et al. reported the integration of Ag-NWs for the elaboration of flexible PNGs by embedding the Sm-doped $Pb(Mg_{1/3}Nb_{2/3})O_3$–$PbTiO_3$ (Sm-PMN-PT) piezoelectric ceramics into polyvinylidene fluoride-trifluoroethylene (P(VDF-TrFE)) film [46]. Ag-NWs are selected due to their excellent conductivity and used to enhance the output electrical performance of the device and improve the polarization of ceramic particles by increasing the applied electric field. **Figure 18.5** shows digital photos of piezoelectric film doped and undoped with Ag-NWs, a schematic illustration of PNGs and commercial LEDs, and the output voltages generated by the mechanical human stimulus. The

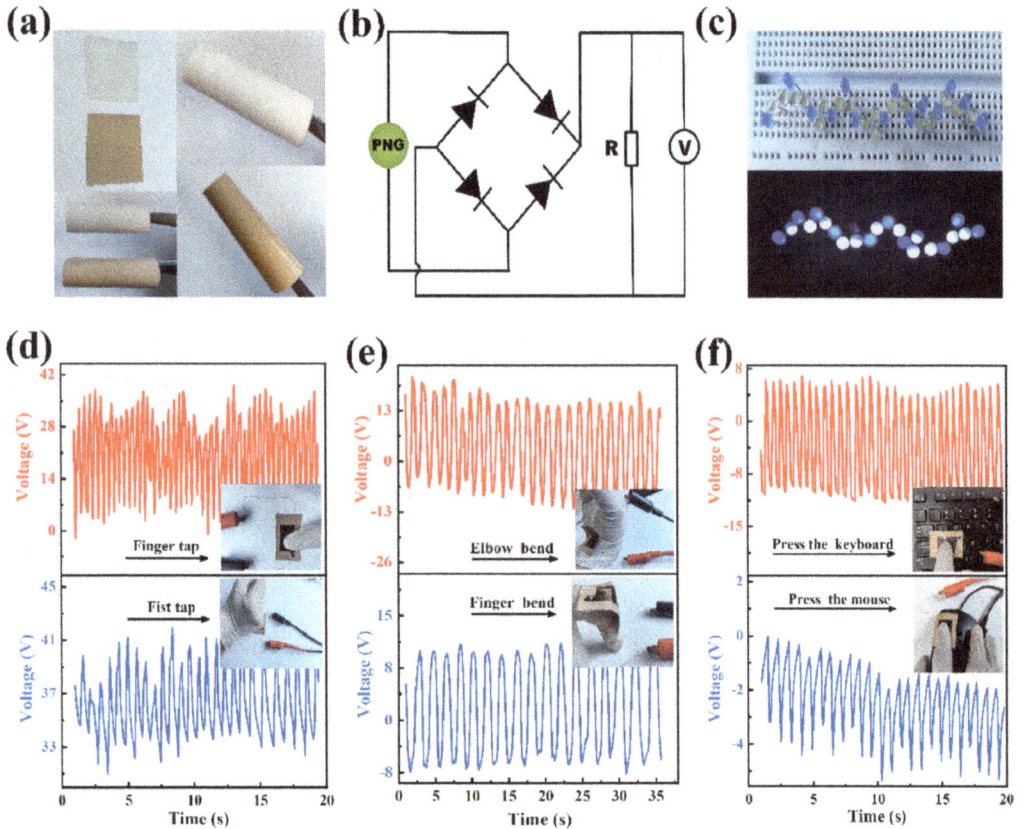

FIGURE 18.5 Digital photos of (a) piezoelectric film doped and undoped with Ag-NWs, and schematic illustration of (b) PNG circuit and (c) commercial LEDs. Output voltages generated by (d) finger and fist tap, (e) elbow and finger bend, (f) pressing the keyboard and pressing the mouse. Adapted with permission [46]. Copyright 2021, Elsevier.

output voltage and current of the PNG doped Ag-NWs increase by 1.6 times compared with the PNG without Ag-NWs to reach 83.5 V and 1.2 µA at a frequency of 1 Hz and an amplitude of 4 mm. In addition, it exhibits a maximum instantaneous power density of 7.48 µW/cm^2 at a load resistance of 70 MΩ. In addition, good mechanical stability up to 6000 bending–releasing cycles are obtained after aging for two months. As shown in **Figure 18.5**, the PNGs were tested in the real application under elbow, finger bending operating mode, and by pressing the keyboard and mouse and show outstanding results. These PNGs based on Ag-NWs exhibit an effective way to harvest and convert the mechanical energy and can be integrated into real applications.

18.4.5 Photovoltaic Cells

The performance of the flexible photovoltaic cells highly benefits from the unique properties of NWs such as the excellent anti-reflecting property (absorb nearly 99% of the incident light), the light-harvesting capacity, the long exciton diffusion lengths, the super-hydrophobic property, and the facile fabrication processes.

Further than grooving the Si-NWs in ultra-thin Si substrate to produce flexible photovoltaic cells [15], Peksu et al. reported the transfer of Si-NWs from the mother Si wafer to different foreign flexible substrates toward the fabrication of third-generation solar cells [47]. The elaboration process consists in placing Si-NWs grooved in n-type substrate onto the carrier substrate covered with the PMMA layer. The Si-NWs are embedded into the PMMA under the effect of a vertical force followed by heat treatment at 230 °C to create a viscous PMMA layer. Finally, lateral forces directed to the edge of the mother substrate are applied to separate the Si-NWs from it and embed them into the PMMA layer coating the carrier substrate. As an application of the transferred Si-NWs, a third-generation solar cell composed of soda-lime-glass/Ag-paste/n–Si-NWs/CZTS/ITO/Ag was elaborated with outstanding power conversion efficiency.

Perovskite solar cells have attracted a large interest due to their impressive photovoltaic conversion efficiency that can reach 25.5%, high flexibility, and compatibility with flexible electronic products [48]. However, the perovskite cells suffer from short-term stability and the poor conductivity of the TCE, etc. A wide variety of flexible TCEs has been investigated, including transparent conductive oxides, conductive polymers, carbon nanomaterial (nanotubes, graphene) metallic nanostructure (NWs, meshes, ultra-thin films) [49]. Flexible TCEs based on Ag-NWs are promising electrodes for flexible electronics due to their high conductivity, good optical transmittance, low sheet resistance, and good stability to repeated tensile bending by making bridges in the flexible substrate volume. Ag-NWs can be embedded into ultra-thin flexible polymer films [50] or deposited onto their surfaces [51] to improve the TCE conductivity, and the interface binding force and decrease the sheet resistance after repetitive flooding cycles. Kang et al. deposit Ag-NWs on the top of PET substrates by spray method and passivated with the ZnO layer to overcome the low interface binding force, the poor conductivity, and low stability of the TCE in the perovskite solar cells [51]. The sheet resistance of the non-treated Ag-NWs dramatically increased from 28 Ω/sq to 574.2 Ω/sq after only 300 bending cycles due to the low bending forces. Whereas the Ag-NWs-ZnO TCE maintains an average sheet resistance of 44.3 Ω/sq after 10000 bending cycles and remains stable even after 270 days of air storage due to the fixation of Ag-NW during bending and the improvement of the bending forces by filling the voids between the NWs with ZnO [52]. Furthermore, the surface roughness and the suppression of Ag migration of the flexible TCE were confirmed with the decrease in the contact angle of water from 81° to 35° and the SIMS results, respectively. Based on all these previous results, TCE based on Ag-NWs passivated with the ZnO layer enhances the stability of the solar cell and prevents the quick degradation of J_{sc} and FF.

18.4.6 FLEXIBLE ENERGY STORAGE DEVICES

The rapidly growing market of flexible electronics requires the development of an energy/power supply with high capacity, long life, and facile integration with flexible electronics. Batteries are devices that store chemical energy and convert it into electrical energy through an oxidation/reduction reaction of the active materials. Due to their high aspect ratio, fast charge transfer pathways, and stable frameworks, considerable progress has been made to apply NWs in flexible battery manufacture. Among the newly emerging batteries, flexible lithium-ion batteries (f-LIB) based on NWs are good candidates for a wide variety of applications.

Storan et al. elaborated Si-NWs on stainless steel (SS) substrate and carbon cloth (CC) catalyzed with Sn nanoparticles flexible substrates for their use in f-LIB anodes [9]. Si-NWs/CC anode exhibits high charge and discharge capacities greater than 2.2 mAh/cm^2, stable cycling capacity retention of 80% without any Si-NW delamination after 200 cycles, and excellent rate capability performance due to the good adherence between Si-NWs and CC. However, Si-NWs/SS anode shows low charge and discharge capacities of 0.3 mAh/cm^2. By using a similar experimental procedure, Imtiaz et al. reported the growth of Si-NWs on a flexible, fire-resistant, and mechanically robust stainless-steel fiber cloth (SSFC) [53]. Thanks to the high surface area of the SSFC, a dense Si-NW network completely wraps the SSFC and gives a buffer space for expansion/contraction during Li-cycling. The Si-NWs/SSFC anode displays a stable performance for 500 cycles with an average coulombic efficiency of >99.5% and a stable areal capacity of ≈2 mAh/cm^2 at 0.2 C after 200 cycles with

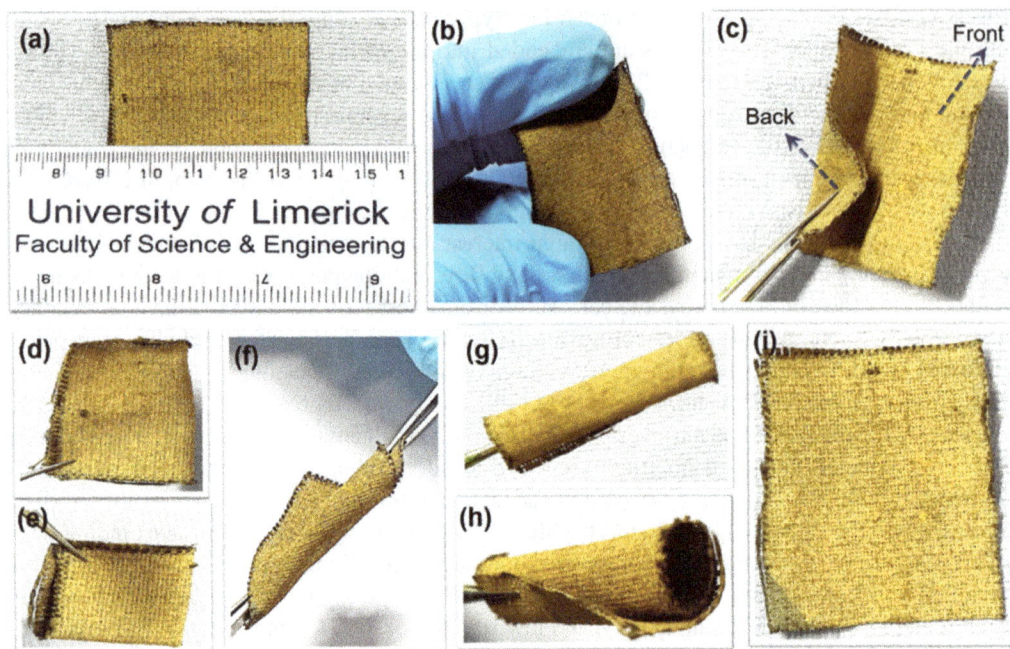

FIGURE 18.6 Digital photos of Si-NWs/SSFC anode showing (a-c) its front and back sides and under (d) single folding, (e) multiple folding, (f) twisting, rolling (g) top and (h) side view, and (i) after releasing state. Figure 18.6 is adapted with permission [53]. Copyright The Authors, some rights reserved; exclusive licensee John Wiley and Sons. Distributed under a Creative Commons Attribution License 4.0 (CC BY).

a mass loading of 1.32 mg/cm². **Figure 18.6(a–c)** shows digital photos of Si-NWs/SSFC anodes with an area of 5× 5 cm² and its front and back views. It is seen that Si-NWs/SSFC anodes exhibit outstanding mechanical robustness without any deformation or delimitation after single or multiple folding, twisting, rolling top, and side view, and even after releasing state as shown in **Figure 18.6(d–i)**. After extended cycling, SEM and TEM images show the restructuring of the Si-NWs into a more stable porous network.

On the other hand, some efforts have been reported to elaborate battery-supercapacitor hybrid (BSH) devices combining the performances of the batteries and the supercapacitors exhibiting fast charge-discharge rates, longer life cycles, and high-power density of the supercapacitors. Wan et al. developed a reasonable core-shell heterostructure consisting of 1-D $CoFe_2Se_4$-NWs working as the cores and CoNi-CH-NWs as the shells grown on CC [54]. This architecture combines the high conductivity of $CoFe_2Se_4$ required for the rapid electron transfer and the high electroactivity of CoNi-CH required for multiple redox reactions aiming to enhance the mechanical and cyclic stability and increase the specific capacity. The $CoFe_2Se_4$/CoNi-CH/CC electrode demonstrated an excellent specific capacity of 218.6 mAh/g, 65.5% of capacity retention at 20 A/g and cyclic stability after 20,000 cycles. By assembling $CoFe_2Se_4$/CoNi-CH and porous carbon as an anode, a maximum energy density of 67.3 Wh/kg at 765.9 W/kg and superior cycling stability (85.4% over 20,000 cycles) are obtained.

18.5 CONCLUSION

The research activities on 1-D material integrated into flexible electronics continues to attract the interest of researchers all over the world. In this chapter, we first presented a brief description of the properties of NWs that exhibit unique mechanical, electrical, thermal, and optical characteristics.

To manufacture these NWs, two techniques have been developed, which can be classified into top-down and bottom-up methods. Combining the superior physical/chemical properties of NWs and their unique geometric structure, many types of flexible devices have been designed. Therefore, an in-depth study of the application of different types of NWs, such as metallic, inorganic semiconductors, perovskite, and organic NWs in flexible electronics was presented in the last part of this chapter. Applications such as transistors, touch screens, chemical and pressure sensors, piezoelectric nanogenerators, photovoltaic cells, batteries, and supercapacitors were presented. We believe that the future of NWs is still bright and needs further deep studies, especially the stability of NWs under real applications and environmental conditions.

REFERENCES

[1]   R.S. Wagner, W.C. Ellis, Vapor-liquid-solid mechanism of single crystal growth, Applied Physics Letters, 4 (1964) 89–90.

[2]   A. Najar, M. Jouiad, Synthesis of InGaN nanowires via metal-assisted photochemical electroless etching for solar cell application, Solar Energy Materials and Solar Cells, 180 (2018) 243–246.

[3]   C. Amri, H. Ezzaouia, R. Ouertani, Photoluminescence origin of lightly doped silicon nanowires treated with acid vapor etching, Chinese Journal of Physics, 63 (2020) 325–336.

[4]   M. Hassan, G. Abbas, N. Li, A. Afzal, Z. Haider, S. Ahmed, X. Xu, C. Pan, Z. Peng, Significance of flexible substrates for wearable and implantable devices: Recent advances and perspectives, Advanced Materials Technologies, 7 (2022) 2100773.

[5]   X. Guo, J. Li, F. Wang, J. Zhang, J. Zhang, Y. Shi, L. Pan, Application of conductive polymer hydrogels in flexible electronics, Journal of Polymer Science, (2022).

[6]   X. Liu, Y.-Z. Long, L. Liao, X. Duan, Z. Fan, Large-scale integration of semiconductor nanowires for high-performance flexible electronics, ACS Nano, 6 (2012) 1888–1900.

[7]   N. Han, J.C. Ho, Integrating semiconductor nanowires for high performance flexible electronic circuits, in: Flexible Electronics, World Scientific, 2016: pp. 117–165.

[8]   C. Amri, R. Ouertani, A. Hamdi, R. Chtourou, H. Ezzaouia, Effect of porous layer engineered with acid vapor etching on optical properties of solid silicon nanowire arrays, Materials & Design, 111 (2016) 394–404.

[9]   D. Storan, S.A. Ahad, R. Forde, S. Kilian, T.E. Adegoke, T. Kennedy, H. Geaney, K.M. Ryan, Silicon nanowire growth on carbon cloth for flexible Li-ion battery anodes, Materials Today Energy, 27 (2022) 101030.

[10] T. Duan, C. Liao, T. Chen, N. Yu, Y. Liu, H. Yin, Z.-J. Xiong, M.-Q. Zhu, Single crystalline nitrogen-doped InP nanowires for low-voltage field-effect transistors and photodetectors on rigid silicon and flexible mica substrates, Nano Energy, 15 (2015) 293–302.

[11] L. Zhu, Y. Xiang, Y. Liu, K. Geng, R. Yao, B. Li, Comparison of piezoelectric responses of flexible tactile sensors based on hydrothermally-grown ZnO nanorods on ZnO seed layers with different thicknesses, Sensors and Actuators A: Physical, 341 (2022) 113552.

[12] W. Liu, H. Zhang, H. Wang, M. Zhang, M. Guo, Titanium mesh supported TiO_2 nanowire arrays/ upconversion luminescence Er^{3+}-Yb^{3+} codoped TiO_2 nanoparticles novel composites for flexible dye-sensitized solar cells, Applied Surface Science, 422 (2017) 304–315.

[13] B. Ranjan, G.K. Sharma, D. Kaur, Tuning growth of MoS_2 nanowires over NiTiCu nanostructured array for flexible supercapacitive electrodes with enhanced Li-ion storage, Applied Physics Letters, 118 (2021) 223902.

[14] S. Wang, B.D. Weil, Y. Li, K.X. Wang, E. Garnett, S. Fan, Y. Cui, Large-area free-standing ultrathin single-crystal silicon as processable materials, Nano Letters, 13 (2013) 4393–4398.

[15] H.D. Omar, Md.R. Hashim, M.Z. Pakhuruddin, Surface morphological and optical properties of flexible black silicon fabricated by metal-assisted chemical etching, Optics & Laser Technology, 136 (2021) 106765.

[16] D. Wu, H. Zhou, Z. Song, M. Zheng, R. Liu, X. Pan, H. Wan, J. Zhang, H. Wang, X. Li, H. Zeng, Welding perovskite nanowires for stable, sensitive, flexible photodetectors, ACS Nano, 14 (2020) 2777–2787.

[17] M. Zhang, T. Gao, J. Wang, J. Liao, Y. Qiu, H. Xue, Z. Shi, Z. Xiong, L. Chen, Single BaTiO$_3$ nanowires-polymer fiber based nanogenerator, Nano Energy, 11 (2015) 510–517.

[18] J.G. Kim, Y. Kim, Y. Noh, S. Lee, Y. Kim, W.B. Kim, Bifunctional hybrid catalysts with perovskite LaCo$_{08}$Fe$_{02}$O$_3$ nanowires and reduced graphene oxide sheets for an efficient Li–O$_2$ battery cathode, ACS Applied Materials & Interfaces, 10 (2018) 5429–5439.

[19] M.K. Rajbhar, P. Das, B. Satpati, W. Möller, N. Ramgir, S. Chatterjee, Moisture repelling perovskite nanowires for higher stability in energy applications, Applied Surface Science, 527 (2020) 146683.

[20] Y. Wang, Y. Wang, R. Zhu, Y. Tao, Y. Chen, Q. Liu, X. Liu, D. Wang, Woven fiber organic electrochemical transistors based on multiwalled carbon nanotube functionalized PEDOT nanowires for nondestructive detection of potassium ions, Materials Science and Engineering: B, 278 (2022) 115657.

[21] J. Cui, Y. Zhang, Z. Cao, D. Yu, Y. Wang, J. Liu, J. Zhang, Y. Zhang, Y. Wu, Nanoporous carbon nanowires derived from one-dimensional metal-organic framework core-shell hybrids for enhanced electrochemical energy storage, Applied Surface Science, 576 (2022) 151800.

[22] K. Kim, J.W. Lee, S.H. Lee, Y.B. Lee, E.H. Cho, H.-S. Noh, S.G. Jo, J. Joo, Nanoscale optical and photoresponsive electrical properties of P3HT and PCBM composite nanowires, Organic Electronics, 12 (2011) 1695–1700.

[23] S.-Y. Min, T.-S. Kim, Y. Lee, H. Cho, W. Xu, T.-W. Lee, Organic nanowire fabrication and device applications, Small, 11 (2015) 45–62.

[24] X. Li, P. Li, Z. Wu, D. Luo, H.-Y. Yu, Z.-H. Lu, Review and perspective of materials for flexible solar cells, Materials Reports: Energy, 1 (2021) 100001.

[25] J. Peng, G.J. Snyder, A figure of merit for flexibility, Science (1979), 366 (2019) 690–691.

[26] S.S. Dhanabalan, T. Arun, G. Periyasamy, D. N, C. N, S.R. Avaninathan, M.F. Carrasco, Surface engineering of high-temperature PDMS substrate for flexible optoelectronic applications, Chemical Physics Letters, 800 (2022) 139692.

[27] J. Hwang, Y. Kim, H. Yang, J.H. Oh, Fabrication of hierarchically porous structured PDMS composites and their application as a flexible capacitive pressure sensor, Composites Part B: Engineering, 211 (2021) 108607.

[28] L. Ju, Z. Wang, K. Sun, H. Peng, R. Fan, F. Qin, Weak epsilon-negative silver nanowires/polyimide metacomposites with extremely low losses, Composites Part A: Applied Science and Manufacturing, 153 (2022) 106755.

[29] J. Bang, S. Coskun, K.R. Pyun, D. Doganay, S. Tunca, S. Koylan, D. Kim, H.E. Unalan, S.H. Ko, Advances in protective layer-coating on metal nanowires with enhanced stability and their applications, Applied Materials Today, 22 (2021) 100909.

[30] C.F. Guo, Z. Ren, Flexible transparent conductors based on metal nanowire networks, Materials Today, 18 (2015) 143–154.

[31] W. Dai, J. Yu, Z. Liu, Y. Wang, Y. Song, J. Lyu, H. Bai, K. Nishimura, N. Jiang, Enhanced thermal conductivity and retained electrical insulation for polyimide composites with SiC nanowires grown on graphene hybrid fillers, Composites Part A: Applied Science and Manufacturing, 76 (2015) 73–81.

[32] Y. Zhou, L. Zhao, Z. Song, C. Chang, L. Yang, S. Yu, Foldable and highly flexible transparent conductive electrode based on PDMS/Ag NWs/PEDOT: PSS, Opt Mater (Amst), 126 (2022) 112175.

[33] T. Cohen-Karni, B.P. Timko, L.E. Weiss, C.M. Lieber, Flexible electrical recording from cells using nanowire transistor arrays, Proceedings of the National Academy of Sciences, 106 (2009) 7309–7313.

[34] D.H. Kim, S.J. Lee, S.H. Lee, J.-M. Myoung, Electrical properties of flexible multi-channel Si nanowire field-effect transistors depending on the number of Si nanowires, Chemical Communications, 52 (2016) 6938–6941.

[35] T.T.T. Nguyen, T. Cazimajou, M. Legallais, T. Arjmand, V.H. Nguyen, M. Mouis, B. Salem, E. Robin, C. Ternon, Monolithic fabrication of nano-to-millimeter scale integrated transistors based on transparent and flexible silicon nanonets, Nano Futures, 3 (2019) 025002.

[36] Z. Song, Z. Liu, L. Zhao, C. Chang, W. An, H. Zheng, S. Yu, Biodegradable and flexible capacitive pressure sensor for electronic skins, Organic Electronics, 106 (2022) 106539.

[37] N.Pradeep, U. Venkatraman, A.N. Grace, Flexible hydrogen gas sensor: ZnO decorated SnO$_2$ nanowire on over head projector (OHP) sheet substrate, Materials Today: Proceedings, 45 (2021) 4073–4080.

[38] X. Luo, K. You, Y. Hu, S. Yang, X. Pan, Z. Wang, W. Chen, H. Gu, Rapid hydrogen sensing response and aging of α-MoO₃ nanowires paper sensor, International Journal of Hydrogen Energy, 42 (2017) 8399–8405.

[39] N. de Guzman, M. Ramos, M.D. Balela, Improvements in the electroless deposition of Ag nanowires in hot ethylene glycol for resistive touchscreen device, Materials Research Bulletin, 106 (2018) 446–454.

[40] X.-F. Pan, H.-L. Gao, Y. Su, Y.-D. Wu, X.-Y. Wang, J.-Z. Xue, T. He, Y. Lu, J.-W. Liu, S.-H. Yu, Strong and stiff Ag nanowire-chitosan composite films reinforced by Ag–S covalent bonds, Nano Research, 11 (2018) 410–419.

[41] H. Yu, Y. Tian, M. Dirican, D. Fang, C. Yan, J. Xie, D. Jia, Y. Liu, C. Li, M. Cui, H. Liu, G. Chen, X. Zhang, J. Tao, Flexible, transparent and tough silver nanowire/nanocellulose electrodes for flexible touch screen panels, Carbohydrate Polymers, 273 (2021) 118539.

[42] Y. Wu, Y. Ma, H. Zheng, S. Ramakrishna, Piezoelectric materials for flexible and wearable electronics: A review, Materials & Design, 211 (2021) 110164.

[43] M. Chelu, H. Stroescu, M. Anastasescu, J.M. Calderon-Moreno, S. Preda, M. Stoica, Z. Fogarassy, P. Petrik, M. Gheorghe, C. Parvulescu, C. Brasoveanu, A. Dinescu, C. Moldovan, M. Gartner, High-quality PMMA/ZnO NWs piezoelectric coating on rigid and flexible metallic substrates, Applied Surface Science, 529 (2020) 147135.

[44] M. Ahmad, M.K. Ahmad, N. Nafarizal, C.F. Soon, N.M.A.N. Ismail, A.B. Suriani, A. Mohamed, M.H. Mamat, Effects of As, P and Sb on the output voltage generation of ZnO nanowires based nanogenerator: Mitigation of screening effect by using surface modified ZnO nanowires, Vacuum, 202 (2022) 111130.

[45] V. Consonni, A.M. Lord, Polarity in ZnO nanowires: A critical issue for piezotronic and piezoelectric devices, Nano Energy, 83 (2021) 105789.

[46] Wang, H. Sun, H. Guo, H. Sui, Q. Wu, X. Liu, D. Huang, High performance piezoelectric nanogenerator with silver nanowires embedded in polymer matrix for mechanical energy harvesting, Ceramics International, 47 (2021) 35096–35104.

[47] E. Peksu, O. Guller, M. Parlak, M.S. Islam, H. Karaagac, Towards the fabrication of third generation solar cells on amorphous, flexible and transparent substrates with well-ordered and disordered Si-nanowires/pillars, Physica E: Low-Dimensional Systems and Nanostructures, 124 (2020) 114382.

[48] M. Alla, V. Manjunath, N. Chawki, D. Singh, S.C. Yadav, M. Rouchdi, F. Boubker, Optimized CH3NH3PbI3-XClX based perovskite solar cell with theoretical efficiency exceeding 30%, Opt Mater (Amst), 124 (2022) 112044.

[49] Y. Xu, Z. Lin, W. Wei, Y. Hao, S. Liu, J. Ouyang, J. Chang, Recent progress of electrode materials for flexible perovskite solar cells, Nano-Micro Letters, 14 (2022) 117.

[50] R. Miao, P. Li, W. Zhang, X. Feng, L. Qian, J. Fang, W. Song, W. Wang, Highly foldable perovskite solar cells using embedded polyimide/silver nanowires Conductive Substrates, Advanced Materials Interfaces, 9 (2022) 2101669.

[51] J. Kang, K. Han, X. Sun, L. Zhang, R. Huang, I. Ismail, Z. Wang, C. Ding, W. Zha, F. Li, Q. Luo, Y. Li, J. Lin, C.-Q. Ma, Suppression of Ag migration by low-temperature sol-gel zinc oxide in the Ag nanowires transparent electrode-based flexible perovskite solar cells, Organic Electronics, 82 (2020) 105714.

[52] K. Han, M. Xie, L. Zhang, L. Yan, J. Wei, G. Ji, Q. Luo, J. Lin, Y. Hao, C.-Q. Ma, Fully solution processed semi-transparent perovskite solar cells with spray-coated silver nanowires/ZnO composite top electrode, Solar Energy Materials and Solar Cells, 185 (2018) 399–405.

[53] S. Imtiaz, I.S. Amiinu, D. Storan, N. Kapuria, H. Geaney, T. Kennedy, K.M. Ryan, Dense silicon nanowire networks grown on a stainless-steel fiber cloth: A flexible and robust anode for lithium-ion batteries, Advanced Materials, 33 (2021) 2105917.

[54] L. Wan, T. Jiang, Y. Zhang, J. Chen, M. Xie, C. Du, 1D-on-1D core–shell cobalt iron selenide @ cobalt nickel carbonate hydroxide hybrid nanowire arrays as advanced battery-type supercapacitor electrode, Journal of Colloid and Interface Science, 621 (2022) 149–159.

19 Nanowires for Heavy Metal Removal

N. Chidhambaram and S. Jasmine Jecintha Kay
Department of Physics, Rajah Serfoji Government College
(Autonomous), Thanjavur, Tamil Nadu, India

Arun Thirumurugan
Sede Vallenar, Universidad de Atacama, Costanera, Vallenar, Chile

V. Anusooya
Department of Electronics and Communication Engineering, Amrita
College of Engineering and Technology, Nagercoil, Kanyakumari District,
Tamil Nadu, India

CONTENTS

19.1 INTRODUCTION

Heavy metals containing wastewaters are increasingly being released into the environment, particularly in developing countries. This is due to various rapid developments in industries, such as mining operations, metal plating facilities, tanneries, fertilizer industries, battery recycling, paper industries, pesticide industries, and agricultural operations. Heavy metal ions have emerged as one of the most significant threats to the natural environment, and it is believed that human beings may be at risk of being exposed to them even at low concentrations. Heavy metals have atomic weights, which range from 63.5 to 200.6, and their specific gravities are more than 5. They, in contrast to

DOI: 10.1201/9781003296621-19

organic pollutants, do not degrade and instead tend to accumulate in living organisms [1]. Many heavy metal ions are carcinogenic or poisonous. Toxic heavy metals such as copper [1], zinc [2], nickel [3], mercury [4], cadmium [5], lead [6], and chromium [7] are indispensable in industrial wastewater. Further, these are venomous above a particular limit known as the permissible limit. The limit is set per the pollutant type. The Environmental Protection Agency (EPA) establishes maximum contamination levels (MCLs) for each pollutant that can be present in drinking water without causing harm to safeguard people and the environment, these harmful heavy metals should be eliminated from wastewater [8]. For the past decade, major efforts have been made around the world to improve the synthesis of porous nanostructured materials and their functional capabilities in the removal of harmful heavy metals and organic contaminants. Fe-based [9], WO_x-based [10], TiO_2 [11], MnO_2 [12], Nb_2O_5 [13], SnO_2 [14], carbon-based [15] nanowires play an important part in removing hazardous heavy metal ions from effluents as an effective adsorbent. This chapter provides additional information on the many methods that can be used to synthesize nanowires, as well as the sources of heavy metal ions, the harmful consequences they have, and how nanowires can be used to remove heavy metal ions.

19.2   NANOWIRES' SYNTHESIS

In sense, the properties of nanowires vary from that of their bulk counterparts due to the ability to disperse strain at the surfaces. It is beneficial to make use of these properties, the synthesis methods that can be controlled are obligatory, still, the synthetic mechanism alters on basis of anisotropic structure. There are two different synthesis processes that one might choose from. Both a top-down and a bottom-up approach can be taken. The top-down approach is analogous to carving a block of stone, whereas the bottom-up strategy involves attempting to construct nanodevices out of atomic or molecule components.

19.2.1   Top-Down Approach

The most prevalent ways for producing nanowires underneath top-down approaches are lithographic techniques.

19.2.1.1   Lithography

The term "lithography" was first coined by Thomson, who defined lithography as photoengraving, which is the technique of transferring a pattern onto a reactive polymer film (also known as a resist), which is then utilized to copy the design onto an underlying thin film or conductor [16]. Nanowire fabrication makes the most common use of optical lithography, electron beam lithography, and other technologies including X-ray lithography.

19.2.1.2   Optical Lithography

For decades, optical lithography, often known as "photolithography," has been the industrial work-horse in the semiconductor sector [17]. **Figure 19.1** illustrates the schematic representation of silicon nanowires' fabrication using optical lithography.

Giovanni Pennelli reported the electrical illustration of FET devices based on a single silicon nanowire with a triangular section, in which the lateral oxidation of (111) planes leads to a triangle segment, starting from a trapezoidal segment obtained via silicon anisotropic etching. In this case, the starting material is not in any way smaller, and it can also be understood by the application of cutting-edge optical lithography. Also, he concluded that it is probable to construct, in a consistent and replicable means, Si NWs less than 20 nm, and extensive numerous micrometers (larger than 320) via optical lithography [18].

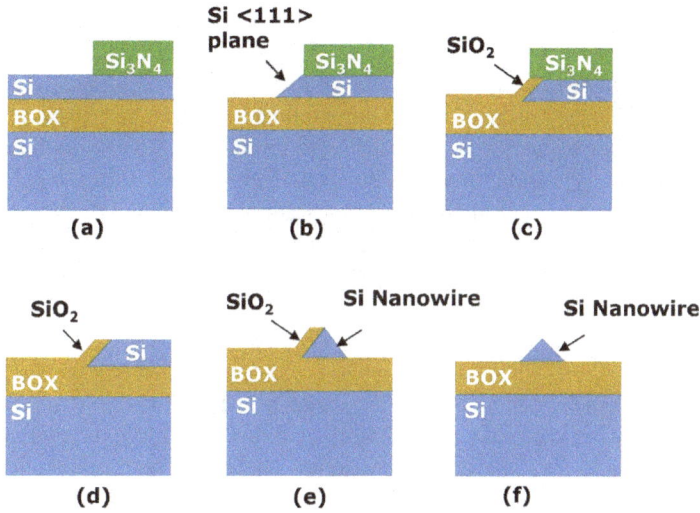

FIGURE 19.1 Schematic representation of silicon nanowires using optical lithography with oxidation and anisotropic etching processes. Adapted with permission [17]. Copyright 2012, AIP Publishing.

19.2.1.3 Electron Beam Lithography (EBL)

The advancements made in the scanning electron microscope in the early 1960s paved the way for the development of a method known as electron beam lithography, which is a well-known form of technology. Nor et al. fabricated silicon nanowires using the SEM-based EBL method, and also demonstrated a silicon nanowire array for pH sensing [19].

19.2.1.4 X-Ray Lithography

Through the use of X-rays, a geometric pattern is transferred from a mask to a light-sensitive chemical photoresist, which is simply referred to as "resist," on the substrate. It was decided to develop a brand-new method called LIGA to enhance lithographic procedures. The LIGA process stands for lithographie, galvanoformung, and abformung (lithography, electroforming, and molding), which combines X-ray lithography with electroplating, and is the best top-down strategy for the formation of 3D features with a high aspect ratio [20]. Djenizian et al. provide an example of Ni nanowires produced by the LIGA process [21].

19.2.2 BOTTOM-UP APPROACH

More often applied bottom-up approaches are, e.g., vapor-phase synthesis, template-based synthesis, and other chemical methods.

19.2.2.1 Vapor-Liquid-Solid Growth

The vapor phase synthesis technique is commonly utilized in nanostructure synthesis. Vapor-liquid-solid (VLS) and vapor-solid (VS) growth are two forms of vapor-phase methods. The VLS method is a simpler and less expensive way of generating nanowires as compared to other vapor-phase procedures. The VLS approach has seen a great deal of application, most notably in the production of 1D nanowires. In a VLS process, nanosized droplets of liquid metal are utilized in the role of catalysts and the gaseous reactants have an interaction with the nanosized liquid, which enables the nucleation and development of single-crystalline structures. Au, Cu, Ni, Sn, and other metal catalysts are used in the VLS process [22].

19.2.2.2 Template-Assisted Synthesis

The template-assisted growth technique is a relatively simple and instinctive method for growing nanowires. The many different kinds of templates that can be utilized to make nanowires are seen in **Figure 19.2**. To create the nanowire, the cylindrical pores or voids in the starting material are first filled with a material of choice that conforms to the structure and morphology of the pore. For nanowire production, typical templates include anodic alumina (Al_2O_3), mica sheets, ion track etched polymers, and nanochannel glass. The porous aluminum oxide (anodic aluminum oxide – AAO) membrane is another template for nanowire synthesis that has superior stability and chemical inertia to the polycarbonate membrane [23].

19.3 HEAVY METAL ION REMOVAL

Heavy metals, unlike organic pollutants, are not capable of biodegradation and tend to accrue in living organisms that are toxic or oncogenic [25]. Even at low levels of existence, heavy metals can cause toxicity [26]. **Figure 19.3** shows the general mechanism of Pb(II) adsorption through CNT. The common heavy metal contaminants of industrial pollutants are zinc (Zn), copper (Cu), cadmium (Cd), chromium (Cr), nickel (Ni), lead (Pb), arsenic (Ar), and mercury (Hg).

19.3.1 TOXIC EFFECTS AND SOURCES OF CONTAMINATION OF HEAVY METALS IN WASTEWATER

Zinc (Zn): Zinc is an essential nutrient that helps keep many different biochemical processes running smoothly as well as the biological activities of living tissue [2]. Refineries, brass plating, the production of wood pulp, plumbing, the production of ground and newsprint paper, cosmetics, steelworks with galvanizing lines, and zinc and brass metalwork are some of the industrial sources of zinc. Zinc

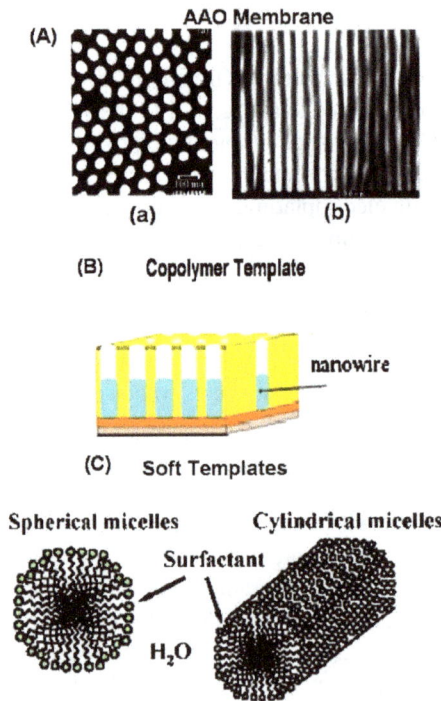

FIGURE 19.2 Schematic image of templates used for nanowire preparations. Adapted with permission [24]. Copyright 2005, Elsevier.

sample A

| Collected the MWCNTs with Pb(II) adsorption in the form of (MWCNTs-COO)₂Pb, (MWCNTs-O)₂Pb, PbO, Pb(OH)₂, PbCO₃ |

(1) adding Na_2SO_4 solution to sample A
(2) filtration

deposition C

| Solution | | MWCNTs-COONa, MWCNTs-ONa and MWCNTs with PbSO₄, Pb(OH)₂, PbCO₃, PbO |

filtrate B

(1) adding HNO_3 solution to deposition C
(2) filtration

| Pb(NO₃)₂ solution | | MWCNTs-COOH, MWCNTs-OH, PbSO₄ |

filtrate D   deposition E

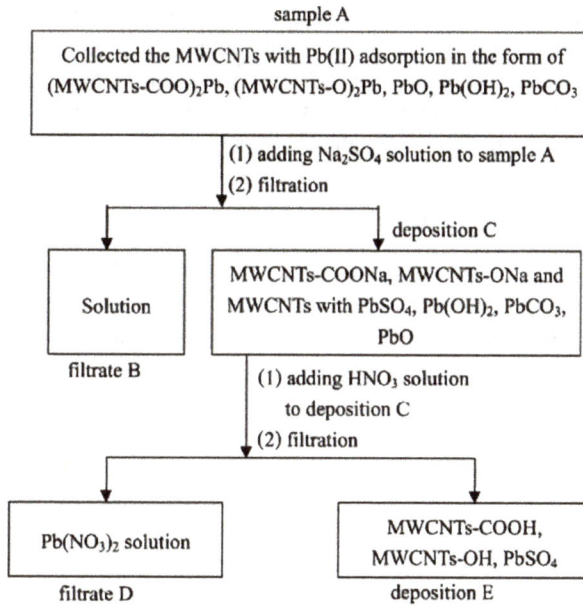

FIGURE 19.3 Schematic illustration of Pb(II) adsorption process. Adapted with permission [15]. Copyright 2007, Elsevier.

can be released into the environment by the remobilization or entrainment of sediment, the effects of agricultural practices, the disruption of groundwater flow, or any combination of these sources.

Copper (Cu): Copper can be found in rather high concentrations in some drinking water supplies [1]. Copper can be released into the environment in many different ways, the most common of which are through the mining sector, metal pipelines, the chemical business, and the pesticides' industry.

Cadmium (Cd): Cadmium is decidedly noxious and contemplated as one of the foremost precedence contaminants in drinking water. This element is a heavy metal that forms naturally in deposits that also contain other elements. The United States Environmental Protection Agency classifies cadmium as a "human carcinogen," and its biological half-life in human bodies ranges from 10 to 35 years. [5].

Chromium (Cr): The oxidation states of chromium that exist naturally are found to be in the range of 2+, 3+, and 6+, with the trivalent Cr(III) and hexavalent Cr(VI), being the firmest forms in nature. Cr(III) is a vital component in human bodies and far less noxious than Cr(VI) [7]. In contrast, Cr(VI) is extremely toxic [27] and can be found in a variety of industrial fluids.

Nickel (Ni): Nickel is a poisonous heavy metal ion that does not degrade and is classified as a "human carcinogen." It may be found in wastewater. The fabrication of batteries, the creation of certain alloys, printing, electroplating, and silver refineries are only a few of the industrial activities that are responsible for the largest amount of nickel contamination in the world's water supplies [3].

Lead (Pb): Lead is a perilous heavy metal that is non-recyclable and readily accumulates in the human body. Drinking water, which contains significant amounts of lead, is the main source of lead in the human body. It can enter the body through the digestive tract and lungs, then spread throughout the body by the blood. Sources of lead ions are PVC pipe manufacturing, agriculture processes, recycled PVC pipes, lead paints, jewelry, and lead batteries [6].

Arsenic (Ar): The presence of arsenic in the water originates from natural deposits and anthropogenic activity. Anthropogenic activities, such as refineries, mining, industrial manufacturing, metal-workings, wood conservation, and the use of pesticides inject arsenic into the aquatic environment [6].

Mercury (Hg): Mercury can cause severe harm to the central nervous system and it is a neuro-toxin [4]. The few potential sources of mercury contamination are coal burnings, ignition of solid waste, and volcanic emissions. **Table 19.1** lists the heavy metal ions, their maximum contamination level (MCL) (mg/L), WHO provisional guideline value (mg/L), related health issues, applicable treatment techniques, and the rank of heavy metals as per CERCLA.

19.3.2 REMOVAL MECHANISM OF HEAVY METAL IONS BASED ON NANOWIRES

Heavy metals are generally removed from contaminated water through chemical and physical treatment methods. The removal of such ions can be accomplished in a variety of ways, such as by chemical precipitation, ion exchange, adsorption, membrane filtration, electrochemical treatment methods, and many more. The adsorption technique is the one that is used for nanowires the most frequently, according to the results of this research [32]. In recent years, adsorption has gained popularity as a unit operation for removing contaminants from effluents, owing to the ability of the process to produce a high-quality treated effluent that complies with environmental emission regulations, as well as its ease of greater benefits, and cost-effectiveness if the used adsorbent is reusable. Static adsorption assays are conducted to assess the adsorption isotherms and kinetics of heavy metals under a variety of operating circumstances, including pH, temperature, and heavy metal ion starting concentrations.

In general, nanowires with O_2 functional groups enhance adsorbent-adsorbate interactions by forming electrostatic attraction, surface complexes, ion exchange, and cation-bonding, which results in a faster sorption rate and greater adsorption ability [33]. Also from the literature, it is evident that nanowires other than O_2 functional groups also show good adsorbent properties towards heavy metal ions.

(a)   Adsorptive Materials

Adsorptive materials are used in this technique to entrap metal ions by physical or chemical exchanges. The adsorbents are then filtered out of the solution and go through a rejuvenation process.

Activated Carbon Adsorbents

Activated carbons are mostly equipped from agronomy by-products when K_2CO_3 is used. Its utility stems mostly from its huge micropore and mesopore contents, as well as the high surface area that results. Many studies are investigating the use of activated carbon to remove heavy metals. When the adsorbent concentration is 0.25 g, studies reveal that carbon produced at 900 °C is more effective at removing Ni(II) from a solution [34].

Carbon Nanotubes' Adsorbents

Carbon nanotubes (CNTs) are well-known for their outstanding characteristics and uses. CNTs have been shown to have a high potential for eliminating heavy metal ions from wastewater. Single-walled CNTs (SWCNTs), as well as multi-walled CNTs (MWCNTs), are the two variants of carbon nanotubes. Metal ion adsorption capabilities of raw CNTs are quite poor, but after oxidation with HNO_3, NaClO, and $KMnO_4$ solutions, they dramatically increase [35].

Bioadsorbants

Bioaccumulation of heavy metals from wastewater is a fairly unique technology that is a very successful approach to heavy metal contamination removal. Weak heavy metal effluent is particularly well suited to biosorption techniques. Biosorbents can be prepared from three sources: (1) non-living biomass; (2) algal organic matter; and (3) microbial biomass. Metal biosorption utilizing algal biomass has been the subject of numerous studies. Iron-oxide nanowires, developed by bio-film debris via bacteria developed in water pipelines, were studied by Losic et al. to illustrate their

TABLE 19.1

Heavy Metal Ions, Maximum Contamination Level, WHO Guideline Value, Related Health Issues, Treatment Techniques, and the Rank of Heavy Metals

Heavy Metals	Maximum Contamination Level (MCL) (mg/L)	WHO Guideline Value (mg/L)	Related Health Issues	Applicable Treatment Techniques	The Rank of Heavy Metals as per CERCLA	References
Hg	0.002	0.006	Impaired neurologic development, immune system, lungs, kidneys, skin and eyes, increases salivation, hypotonia, and hypertension	Chemical precipitation, adsorption, ion exchange method, and membrane process	03	[1]
Zn	5.0	3	Stomach cramps, skin irritations, vomiting, nausea, and anemia	Chemical precipitation, adsorption, ion exchange method, and membrane method	74	[8]
Cr	0.2	0.05	Skin irritation, lung cancer, vomiting, liver and kidney damage, and fibrosis	Adsorption, membrane process, ion exchange method, and coagulation process	18	[13]
Cu	1.3	2	Vomiting, cramps, increased blood pressure, kidney and liver disorder, respiratory problems, or even death	Chemical precipitation. Ion exchange method, reverse osmosis, and membrane method	133	[14]
Cd	0.005	0.003	Kidney dysfunction, overexposure can cause even death, and skeletal damage	Chemical precipitation, ion exchange method, membrane process, and coagulation method	08	[28]
Ni	0.1	0.07	Gastrointestinal distress, skin dermatitis, chest pain, nausea, vomiting, headache, and breathing complications	Chemical precipitation, coagulation. Membrane method, ion exchange method, and adsorption process	-	[29]
Pb	0.15	0.01	Anemia, insomnia, headache, neurodegenerative diseases, dizziness, irritability, weakness of muscles, hallucination, and renal damages	Chemical precipitation method, adsorption, membrane method, and ion exchange method	02	[30]
Ar	0.05	0.01	Hemopoietic, dermatologic, skin cancer, anorexia, brown pigmentation, diabetes, hyper-pigmentation, and skin cancer	Ion exchange method, coagulation method, reverse osmosis process, and filtration methods	01	[31]

applicability for arsenic removal from fluids. The chemical composition and characterization of the unique nanowire architectures reveal desirable features for heavy metal adsorption from water [31].

Certain parameters need to be explored to fully comprehend the adsorption mechanism. They are as follows.

(i)   Determination of Adsorption Capacity

The following equations can be used to compute the adsorption capacity of metal ions (adsorption percentage and Qe (mg/g)):

$$\text{Adsorption } (\%) = C_oC_e/C_o \times 100\% \tag{19.1}$$

$$Q_e = C_oC_e/m \times V \tag{19.2}$$

where V denotes the volume of the solution and C_o and C_e signify the initial and equilibrium metal ion concentrations, respectively, and m is the mass of an adsorbent.

(ii)   Adsorption Isotherm

An adsorption isotherm, such as the Langmuir and Freundlich isotherms, describes the fraction of adsorbate molecules partitioned between liquid and solid phases at equilibrium [36]. The Langmuir isotherm is based on monolayer adsorption on a homogeneous surface with a finite number of saturable adsorption sites. When all of the sites are saturated with adsorbates, the maximum adsorption capacity is reached. The Langmuir isotherm model's linear form is as follows:

$$C_e/q_e = 1/K_Lq_m + C_e/q_m \tag{19.3}$$

where q_m is the maximal adsorption capacity (mg/g), and K_L is the Langmuir constant related to the energy of adsorption. q_m and K_L were calculated using the slope and intercept of plots of C_e/q_e versus C_e at various temperatures.

The Freundlich isotherm applies to both mono-layer (chemisorption) and multi-layer adsorption (physisorption) and is based on the assumption that the adsorbate adsorbs onto the heterogeneous surface of an adsorbent. The linear form of the Freundlich equation is

$$\log q_e = \log K_F + 1/n \log C_e \tag{19.4}$$

The Freundlich isotherm constants K_F and n are related to adsorption capacity and intensity, respectively, and C_e signifies equilibrium concentration (mg/L). The intercept and slope of a plot of log q_e versus log C_e yield the Freundlich isotherm constants K_F and n.

(iii)   Adsorption Process Parameters

Effect of pH

The adsorbent's surface charge, as well as the degree of ionization and speciation of the adsorbates, are affected by the pH of the solution. Kandah et al. developed carbon nanotubes for the removal of nickel ions, and they noticed that as the pH rises, the adsorption uptake for carbon nanotubes increases due to the increase in electrostatic attractive interactions between OH- and Ni^{2+} ions [29]. The extent of protonation that the functional groups undergo, and hence the electrostatic repulsions between the protonated groups and the heavy metal ions, appear to be the key causes that tend to control the pH dependency. This is most common in systems with a low pH and large availability of protons in the medium. The competitive adsorption of ions with positively charged metal species, which leads to ineffective adsorption, is another phenomenon that may exist in lower pH conditions. Metal ion hydroxylated complexes may develop at higher pH levels, reducing action by inhibiting surface compounds. A fascinating conclusion drawn is that the best adsorption occurs at a moderate pH range [37].

Effect of Contact Time

After the active sites are occupied, the effectiveness of removal appears to improve steadily at first with an increase in contact duration, and subsequently, it slows down. Mahmoud et al. investigated the use of titanate nanowires to remove lead ions from effluent and offered an experimental report showing that absorption increased with time up to 45 minutes, for 27, 40, and 60 mg/L reaching 53.7, 79.6, and 119.76 mg/g, respectively. This could be due to the adsorbent TiO_2 nanowires having an empty adsorption site. As time passed, the removal of Pb(II) ions bound became less efficient. This could be because the sorption sites on TiO_2 nanowires were saturated to their maximal absorption capability [30]. The contact time also plays a critical role in adsorption mechanisms.

Effect of Initial Ion Concentration

The number of metal ions accessible for adsorption is minimal at low concentrations, resulting in a decreased adsorption efficiency. The number of ions accessible for adsorption increases as it rises, resulting in a higher absorption rate. However, if the number of saturation sites reaches a particular level, they can no longer absorb any more ions. The work of Sun et al. demonstrates this with an initial Cr^{6+} concentration of 21.17 mg/L, the $K_2Cr_2O_7$ solution drops substantially for the first 20 minutes, then gradually reduces before stabilizing [7]. **Table 19.2** enlists the various adsorption parameters of different forms of nanowires.

(b)   Reuse of Adsorbent (Nanowires)

The desorption percentage of the desorption process was calculated using,

$$\text{Desorption } (\%) = C_{des}/C_o \times 100 \tag{19.5}$$

TABLE 19.2
Various Adsorption Parameters of Different Types of Nanowires

Nanowires	pH	Adsorption Time (t)	C_o (mg/L)	C_e (mg/L)	Q_m (mg/g)	Target Pollutant	References
Chitosan-iron nanowires	5	400 min	58.74	-	113.2	Cr(VI)	[7]
$K_2W_4O_{13}$	Above 5.3	-	100	-	228.83	Pb^{2+}	[10]
TiO_2- nanowires	3	24 hrs	40	0.15	79.7	Pb^{2+}	[30]
Bacterial iron oxide	7	120 min	10–50	-	50%	As(II) As(V)	[31]
WO_x/C	5	1 min	10–270	0	1224.7	Pb^{2+}	[33]
$\alpha - Fe_2O_3$	2	60 min	10	-	-	Cr(VI)	[38]
$Fe@Fe_2O_3$	6.1	60 min	8	-	177	Cr(VI)	[39]
Fe-Mn binary oxide-nanowires	7	75 min	200	10	171	As(III)	[40]
WO_x–EDA	5–6	20 hrs	10	0.22	925 610	Pb^{2+} $UO_2{}^{2+}$	[41]
MnO_2-nanowires	5	15 min 30 min	10–50	-	3.730 2.972	Pb^{2+} Cu^{2+}	[42]
$Na_5V_{12}O_{32}$-nanowires	1	6 hrs	187	2.1	2170	Pb^{2+}	[43]
Carbonaceous nanowires	5–11	2 hrs	10	-	85.6	Cd^{2+}	[44]

where C_{des} denotes the concentration after desorption and C_o denotes the concentration before desorption. To ensure that the adsorbents could be reused, nanowires were thoroughly cleaned with a NaOH + NaCl solution to remove heavy metal ions in preparation for the next experiment [32]. This is one of the most important long-term applications.

19.4   NANOWIRES FOR HEAVY METAL REMOVAL

In general, 1D nanostructures outperform their larger counterparts in terms of characteristics and functioning. When compared to bulk materials, small structures with enormous surface areas and nanoscale quantum confinement effects exhibit unique chemical, optical, and electrical capabilities. It is possible to create nanowire and nanotube structures with small diameters and lengths. The physicochemical properties of the nanowires and their efficiency were utilized for the removal of heavy metal ions.

19.4.1   IRON-BASED NANOWIRES

In recent times, iron mineral samples, such as goethite, magnetite, hematite, and laterite, have recently been utilized to remove heavy metal ions from drinking water as simple, inexpensive, and easily available materials. Because of their availability and high affinity for heavy metal ions, iron minerals are one of the most investigated natural materials, particularly for arsenic and chromium adsorption [7,9]. Iron nanowires have different forms. The various types of iron-based nanowires and their heavy metal ion removal efficiencies are listed in **Table 19.3**.

TABLE 19.3
Iron-Based Nanowires and their Heavy Metal Ion Removal Efficiencies

Nanowire Types	Removal of Heavy Metal Ions	References
Bacterial iron oxide nanowires from biofilm waste	Removal of arsenic contaminants.	[31]
Chitosan-iron nanowires	Chitosan-iron nanowires were synthesized in porous anodic alumina (PAA), with different adsorption capacities of 113.2 and 123.95 mg/g for Cr(VI).	[7,45]
α–Fe_2O_3	α-Fe_2O_3 nanosheets and nanowires were utilized as Cr(VI) adsorbents, and 10 mg/L of Cr(VI) was adsorbed in 60 minutes on all samples, although samples with denser nanostructures had a higher adsorption capacity.	[38,46]
	α-Fe_2O_3 nanowires showed a maximal adsorption capacity of 99.98 and 100% for the removal of As(III) and As(V) for the concentration of 10 mg/L, respectively.	
Fe@Fe_2O_3	Removal of 100% anoxic Cr(VI) in 60 minutes using core-shell Fe@Fe_2O_3 nanowires at pH of 6.1.	[39]
Manganese-based nanowires: (1) Fe-Mn binary oxide nanowires. (2) $MnFe_2O_4$ and Mn-doped Fe_3O_4	The absorption capacity for As(III) of about 171 mg/g is attained with a 1:3 ratio of Fe: Mn and permits magnetic separation from water at pH 7.0. The porous nanostructures have remarkable removal capacities for heavy metal ions like Pb(II) and Cr (VI). For Pb(II) and Cr(VI), the highest adsorption capacities of $MnFe_2O_4$ and Mn/Fe_3O_4 are 131 and 67.3 mg/g and 131 and 73.9 mg/g, respectively.	[40] [47]

19.4.2 TUNGSTATE-BASED NANOWIRES

Tungsten oxides (WO_{3-x}, $0 < x < 1$), tungsten bronzes (M_xWO_3, where M is an alkali metal), tungsten oxide hydrates ($WO_3 \cdot nH_2O$), and other tungsten-based materials have recently emerged to have outstanding adsorption capability. The tungsten-based family also includes potassium tungsten oxides, known as $K_2O \cdot nWO_3$. The $K_2W_3O_{10}$ (n = 3) and $K_2W_4O_{13}$ (n = 4) species were primarily investigated because of their ease of preparation and unique features [48]. Electronic attraction allowed Pb^{2+} to trap non-bonding electrons from the W=O, O–H groups of $K_2W_4O_{13}$. Furthermore, by ion exchange, Pb^{2+} might replace K^+ to locate the tunnels for a higher electro-negativity (2.33 > 0.82). Both reactions would result in Pb^{2+} being removed by $K_2W_4O_{13}$. Wei et al. fabricated $K_2W_4O_{13}$ and observed that it had a high absorption capacity (228.83 mg/g), a fast kinetic (141.67 mg/g in 30 min), good acid resistance (75 percent elimination at pH = 2), and out-standing reusability when it came to Pb^{2+} adsorption (over 95 percent removal after five cycles) [10]. Xu et al. developed oxygen vacancy-rich WO_x/C inorganic-organic hybrid nanowire systems and stated high adsorption capacities of 1225 and 1188 mg/g for the removal of Pb^{2+} and Methylene Blue, respectively [33]. Li et al. synthesized inorganic-organic hybrid WO_x-ethylenediamine (WO_x–EDA) nanowires and showed excellent adsorption capabilities of 925 and 610 mg/g for Pb^{2+} and UO_2^{2+}, respectively [41].

19.4.3 TITANATE-BASED NANOWIRES (TiO_2)

Titanate nanowires have gotten a lot of attention because of their high ion exchange capacity and ease of production. Nanosized titanium dioxide (TiO_2) was discovered to be a superior adsorbent for heavy metal ions in wastewater, notably, lead ions, when compared to activated carbon (the most prevalent and commercial type of carbon adsorbents), microparticles TiO_2, and multi-wall carbon nanotubes. Titanium dioxide (TiO_2) has recently emerged as a strong contender for water treatment [11]. Youssef et al. conducted studies using TiO_2 nanowires to remove heavy metal (Pb^{2+}, Cu^{2+}, Fe^{3+}, Cd^{2+}, and Zn^{2+}) deposits from polluted water and observed the adsorption capacity to be 97.06, 75.24, 79.77, 64.89, and 35.18%, respectively [49]. Yin et al. synthesized TiO_2 nanowires for the efficient removal of U(VI) and noted an extraordinary absorption capacity of 410 mg/g at 328 K within one hour, also having optimal pH 4–7 [50].

19.4.4 MANGANESE OXIDE-BASED NANOWIRES (MnO_2)

Manganese oxide (MnO_2) has gained much interest in this field due to its environmental compati-bility, easy availability, and low cost. MnO_2 is a natural adsorbent of contaminants from water and soil, and usually, has a large surface area in its nanoscaled form and a high affinity for metal ions (e.g., lead, copper, zinc, cadmium, etc.) due to its natural negative surface charge [51]. Claros et al. prepared MnO_2 nanowires with diameters of 20–100 nm as adsorbents of Pb(II) and Cu(II) ions and their result shows quick adsorption rates and improved affinity towards Pb(II) compared to Cu(II), since the adsorption capacity (3.730 [mg/g]) and the elimination percentage of 99.99% at equilib-rium for Pb(II), which are greater than that of Cu(II), 2.972 [mg/g] and 75.48% [42].

19.4.5 SODIUM VANADATE ($Na_5V_{12}O_{32}$) BASED NANOWIRES

The layered cation-exchange inorganic materials, as a relatively new cation-exchange material, are distinguished by their tunable interlayer gaps. Such compounds include vanadium-based materials, such as vanadates of $M_xV_3O_8$ where M is an alkali metal, alkaline earth metal, or transition metal ion. $M_xV_3O_8$ has two fundamental units in its crystal structure: a VO_6 octahedron and a VO_5 tri-angular double cone that are joined by a common angle to form a pleated (V_3O_8) ion layer with

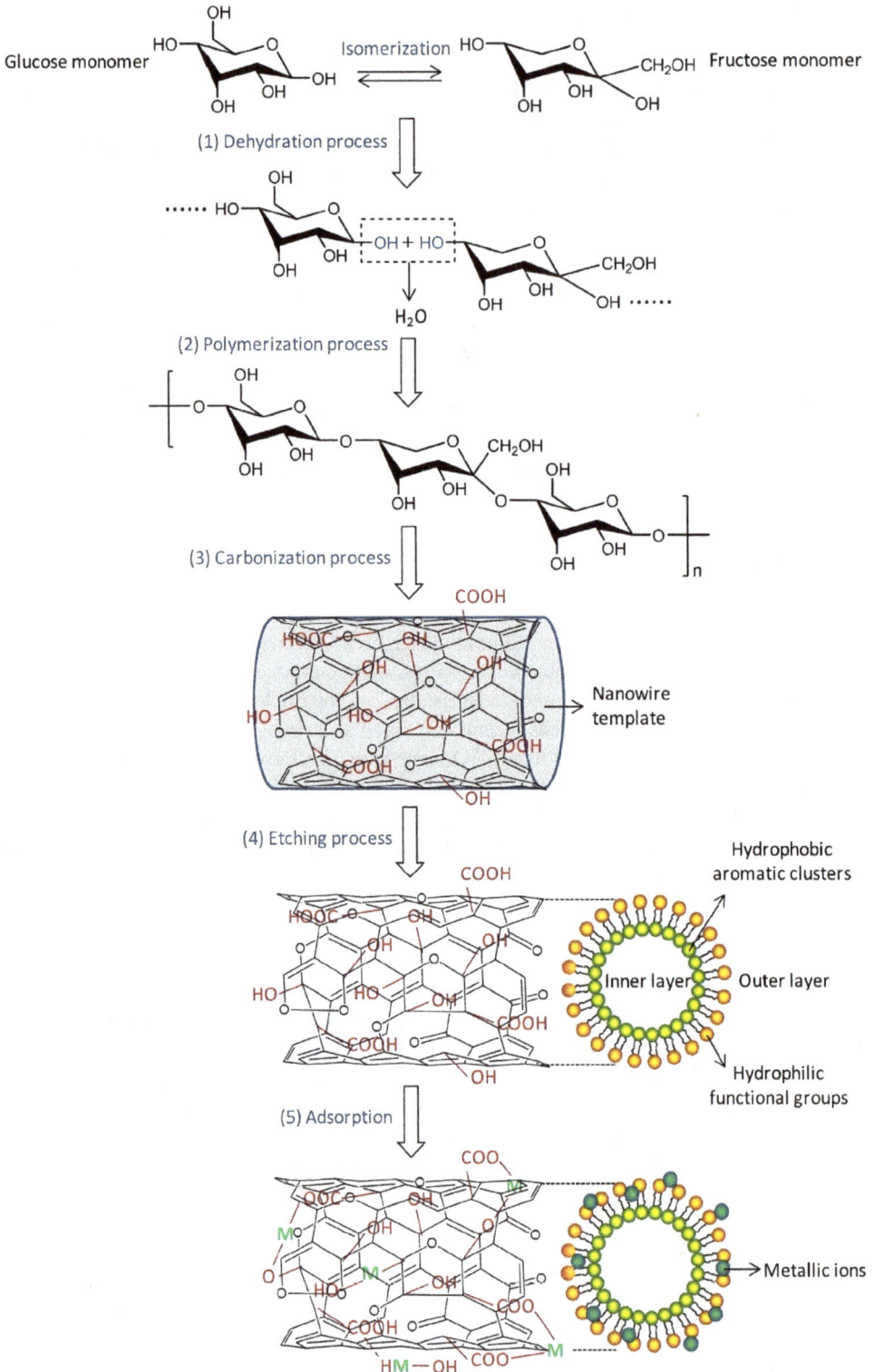

FIGURE 19.4 Synthesis of carbonaceous nanowires and their use in heavy metals' removal. Adapted with permission [44]. Copyright 2013, Elsevier.

cations between layers. The interlayer contains sodium ions from sodium vanadate ($Na_{1.25}V_3O_8$ or $Na_5V_{12}O_{32}$), with a maximum interlayer spacing of 1.16 nm. Fang et al. used hydrothermal synthesis to generate $Na_5V_{12}O_{32}$ nanowires on Ti foil, then conducted an extensive investigation on the cation exchange and determined that it has the maximum capacity for Pb^{2+} elimination. The hue of the $Na_5V_{12}O_{32}$ nanowires changed from green to gray. After two hours, the Pb^{2+} removal rate reaches 93.6 percent, and the Pb^{2+} content drops from 187.7 mg/L to 2.1 mg/L after six hours [43].

19.4.6 TIN-BASED NANOWIRES

To remove U(VI) from wastewater, Lei and colleagues created surface oxidized tin disulfide nanosheets on tellurium nanowires ($Te@O-SnS_2$) as greatly efficient photocatalysts. $Te@O-SnS_2$ showed a maximum extraction capacity of 704.8 mg/g when the initial concentration of U(VI) in the sample was 200 mg/L [52].

19.4.7 CARBON-BASED NANOWIRES

It has been demonstrated that carbon nanotubes have a considerable deal of promise for eliminating heavy metal ions from wastewater. The thermodynamics and kinetics of the adsorption capabilities of carbon nanotubes for the removal of heavy metals have been the subject of examination by a large number of researchers. This is because of their diminutive size, uniform pore distribution, and significant specific surface area of CNT. The creation of carbon-based nanowires and their application in heavy metal ion removal methods are both depicted in **Figure 19.4**.

At various pH levels and temperatures, Kuo et al. investigated the optimum adsorption of Cd^{2+} onto as-grown and customized CNTs [28]. Li et al. studied that the copper removal effectiveness of CNTs immobilized by calcium alginate is high, reaching 69.9% even at a pH of 2 [53]. Du et al. used a hydrothermal approach to synthesize niobium oxide (Nb_2O_5) nanowire-deposited carbon fiber (CF) for Cr(VI) removal/adsorption capacity of the Nb_2O_5/carbon fiber, reporting a maximum Cr(VI) adsorption capacity of 115 mg/g for the Nb_2O_5 nanowire/CF sample [13]. Luo et al. designed and analyzed ZrO_2 embedded carbon nanowires to remove arsenic (As) from water, and their findings show that at 40 °C, the highest adsorption capacity values of As(III) and As(V) on the ZCNs are 28.61 and 106.57 mg/g, respectively. These capacities are much higher than those of pure ZrO_2 (2.56 mg/g for As(III) and 3.65 mg/g for As(V)) [54].

19.5 CONCLUSIONS AND FUTURE PERSPECTIVES

The most harmful contaminants are heavy metal ions (HMIs). An updated summary of general chemical and physical approaches that have recently been used to treat and purify water using nanowires to remove heavy metal ions such as Zn, Cu, Cd, Cr, Ni, Pb, Ar, and Hg. It is also discussed how useful nanowires are as adsorbents, how well they work to get rid of heavy metal ions, and whether or not they can be reused. Nanowires have a huge specific surface area, which enables them to offer plentiful and rich active sites in addition to having robust adsorption capabilities. There are substantial latent chances for the treatment and purification of wastewater based on the exploitation of nanowires.

ACKNOWLEDGMENTS

A.T acknowledges ANID-SA 77210070 for the financial support and Universidad da Atacama for their support.

REFERENCES

[1] F. Fu, Q. Wang, Removal of heavy metal ions from wastewaters: A review, Journal of Environmental Management. 92 (2011) 407–418.

[2] E. Senturk, The treatment of zinc-cyanide electroplating rinse water using an electrocoagulation process, Water Science and Technology. 68 (2013) 2220–2227.

[3] S. Yang, J. Li, D. Shao, J. Hu, X. Wang, Adsorption of Ni(II) on oxidized multi-walled carbon nanotubes: Effect of contact time, pH, foreign ions and PAA, Journal of Hazardous Materials. 166 (2009) 109–116.

[4] C. Namasivayam, K. Kadirvelu, Uptake of mercury (II) from wastewater by activated carbon from an unwanted agricultural solid by-product: Coirpith, Carbon. 37 (1999) 79–84.

[5] M. Arias, M.T. Barral, J.C. Mejuto, Enhancement of copper and cadmium adsorption on kaolin by the presence of humic acids, Chemosphere. 48 (2002) 1081–1088.

[6] L.A. Malik, A. Bashir, A. Qureashi, A.H. Pandith, Detection and removal of heavy metal ions: A review, Environmental Chemistry Letters. 17 (2019) 1495–1521.

[7] L. Sun, L. Zhang, C. Liang, Z. Yuan, Y. Zhang, W. Xu, J. Zhang, Y. Chen, Chitosan modified Fe0 nanowires in porous anodic alumina and their application for the removal of hexavalent chromium from water, Journal of Materials Chemistry. 21 (2011) 5877–5880.

[8] Ihsanullah, A. Abbas, A.M. Al-Amer, T. Laoui, M.J. Al-Marri, M.S. Nasser, M. Khraisheh, M.A. Atieh, Heavy metal removal from aqueous solution by advanced carbon nanotubes: Critical review of adsorption applications, Separation and Purification Technology. 157 (2016) 141–161.

[9] D. Mohan, C.U. Pittman, Arsenic removal from water/wastewater using adsorbents – A critical review, Journal of Hazardous Materials. 142 (2007) 1–53.

[10] Q.-S. Huang, W. Wu, W. Wei, L. Song, J. Sun, B.-J. Ni, Highly-efficient Pb_{2+} removal from water by novel $K_2W_4O_{13}$ nanowires: Performance, mechanisms and DFT calculation, Chemical Engineering Journal. 381 (2020) 122632.

[11] X. Zhao, Q. Jia, N. Song, W. Zhou, Y. Li, Adsorption of Pb(II) from an aqueous solution by titanium dioxide/carbon nanotube nanocomposites: kinetics, thermodynamics, and isotherms, Journal of Chemical & Engineering Data. 55 (2010) 4428–4433.

[12] Y. Du, G. Zheng, J. Wang, L. Wang, J. Wu, H. Dai, MnO_2 nanowires in situ grown on diatomite: Highly efficient absorbents for the removal of Cr(VI) and As(V), Microporous and Mesoporous Materials. 200 (2014) 27–34.

[13] Y. Du, S. Zhang, J. Wang, J. Wu, H. Dai, Nb_2O_5 nanowires in-situ grown on carbon fiber: A high-efficiency material for the photocatalytic reduction of Cr(VI), Journal of Environmental Sciences. 66 (2018) 358–367.

[14] R. Alizadeh, R.K. Kazemi, M.R. Rezaei, Ultrafast removal of heavy metals by tin oxide nanowires as new adsorbents in solid-phase extraction technique, International Journal of Environmental Science and Technology. 15 (2018) 1641–1648.

[15] H. Wang, A. Zhou, F. Peng, H. Yu, J. Yang, Mechanism study on adsorption of acidified multiwalled carbon nanotubes to Pb(II), Journal of Colloid and Interface Science. 316 (2007) 277–283.

[16] L.F. Thompson, An introduction to lithography (1983): pp. 1–13.

[17] N.F. Za'Bah, K.S.K. Kwa, L. Bowen, B. Mendis, A. O'Neill, Top-down fabrication of single crystal silicon nanowire using optical lithography, Journal of Applied Physics. 112 (2012) 024309.

[18] G. Pennelli, Top down fabrication of long silicon nanowire devices by means of lateral oxidation, Microelectronic Engineering. 86 (2009) 2139–2143.

[19] M.M.N., U. Hashim, A. Ruslinda, M.K. Arshad, M.H.A. Baharin, Fabrication of silicon nanowires array using e-beam lithography integrated with microfluidic channel for pH sensing, Current Nanoscience. 11 (2015) 239–244.

[20] A. Wisitsoraat, S. Mongpraneet, R. Phatthanakun, N. Chomnawang, D. Phokharatkul, V. Patthanasettakul, A. Tuantranont, Low-cost and high-resolution x-ray lithography utilizing a lift-off sputtered lead film mask on a Mylar substrate, Journal of Micromechanics and Microengineering. 20 (2010) 075026.

[21] T. Djenizian, I. Hanzu, M. Eyraud, L. Santinacci, Electrochemical fabrication of tin nanowires: A short review, Comptes Rendus Chimie. 11 (2008) 995–1003.

[22] Y. Zhang, M.K. Ram, E.K. Stefanakos, D.Y. Goswami, Synthesis, Characterization, and applications of ZnO nanowires, Journal of Nanomaterials. 2012 (2012) 1–22.

[23] W. Lee, S.-J. Park, Porous anodic aluminum oxide: Anodization and templated synthesis of functional nanostructures, Chemical Reviews. 114 (2014) 7487–7556.

[24] K.S. Shankar, A.K. Raychaudhuri, Fabrication of nanowires of multicomponent oxides: Review of recent advances, Materials Science and Engineering C. 25 (2005) 738–751.

[25] N.K. Srivastava, C.B. Majumder, Novel biofiltration methods for the treatment of heavy metals from industrial wastewater, Journal of Hazardous Materials. 151 (2008) 1–8.

[26] V.K. Gupta, T.A. Saleh, Sorption of pollutants by porous carbon, carbon nanotubes and fullerene - An overview, Environmental Science and Pollution Research. 20 (2013) 2828–2843.

[27] L. Khezami, R. Capart, Removal of chromium(VI) from aqueous solution by activated carbons: Kinetic and equilibrium studies, Journal of Hazardous Materials. 123 (2005) 223–231.

[28] C.-Y. Kuo, H.-Y. Lin, Adsorption of aqueous cadmium (II) onto modified multi-walled carbon nanotubes following microwave/chemical treatment, Desalination. 249 (2009) 792–796.

[29] M.I. Kandah, J.-L. Meunier, Removal of nickel ions from water by multi-walled carbon nanotubes, Journal of Hazardous Materials. 146 (2007) 283–288.

[30] M.G.A. Saleh, A.A. Badawy, A.F. Ghanem, Using of titanate nanowires in removal of lead ions from waste water and its biological activity, Inorganic Chemistry Communications. 108 (2019) 107508.

[31] I. Andjelkovic, S. Azari, M. Erkelens, P. Forward, M.F. Lambert, D. Losic, Bacterial iron-oxide nanowires from biofilm waste as a new adsorbent for the removal of arsenic from water, RSC Advances. 7 (2017) 3941–3948.

[32] H. Wu, C. Wang, J. Kwon, Y. Choi, J. Lee, Synthesis of 2D and 3D hierarchical β-FeOOH nanoparticles consisted of ultrathin nanowires for efficient hexavalent chromium removal, Applied Surface Science. 543 (2021) 148823.

[33] S. Zhang, H. Yang, H. Huang, H. Gao, X. Wang, R. Cao, J. Li, X. Xu, X. Wang, Unexpected ultrafast and high adsorption capacity of oxygen vacancy-rich WOx/C nanowire networks for aqueous Pb^{2+} and methylene blue removal, Journal of Materials Chemistry A. 5 (2017) 15913–15922.

[34] S. Erdoğan, Y. Önal, C. Akmil-Başar, S. Bilmez-Erdemoğlu, Ç. Sarıcı-Özdemir, E. Köseoğlu, G. İçduygu, Optimization of nickel adsorption from aqueous solution by using activated carbon prepared from waste apricot by chemical activation, Applied Surface Science. 252 (2005) 1324–1331.

[35] T.W. Odom, J.L. Huang, P. Kim, C.M. Lieber, Atomic structure and electronic properties of single-walled carbon nanotubes, Nature. 391 (1998) 62–64.

[36] K. Pillay, E.M. Cukrowska, N.J. Coville, Multi-walled carbon nanotubes as adsorbents for the removal of parts per billion levels of hexavalent chromium from aqueous solution, Journal of Hazardous Materials. 166 (2009) 1067–1075.

[37] S. Singh, D. Kapoor, S. Khasnabis, J. Singh, P.C. Ramamurthy, Mechanism and kinetics of adsorption and removal of heavy metals from wastewater using nanomaterials, Environmental Chemistry Letters. 19 (2021) 2351–2381.

[38] F. Budiman, N. Bashirom, W.K. Tan, K.A. Razak, A. Matsuda, Z. Lockman, Rapid nanosheets and nanowires formation by thermal oxidation of iron in water vapour and their applications as Cr(VI) adsorbent, Applied Surface Science. 380 (2016) 172–177.

[39] Y. Mu, H. Wu, Z. Ai, Negative impact of oxygen molecular activation on Cr(VI) removal with core-shell $Fe@Fe_2O_3$ nanowires, Journal of Hazardous Materials. 298 (2015) 1–10.

[40] H.-J. Cui, J.-K. Cai, H. Zhao, B. Yuan, C.-L. Ai, M.-L. Fu, Fabrication of magnetic porous Fe–Mn binary oxide nanowires with superior capability for removal of As(III) from water, Journal of Hazardous Materials. 279 (2014) 26–31.

[41] W. Li, F. Xia, J. Qu, P. Li, D. Chen, Z. Chen, Y. Yu, Y. Lu, R.A. Caruso, W. Song, Versatile inorganic-organic hybrid WOx-ethylenediamine nanowires: Synthesis, mechanism and application in heavy metal ion adsorption and catalysis, Nano Research. 7 (2014) 903–916.

[42] M. Claros, J. Kuta, O. El-Dahshan, J. Michalička, Y.P. Jimenez, S. Vallejos, Hydrothermally synthesized MnO_2 nanowires and their application in lead (II) and copper (II) batch adsorption, Journal of Molecular Liquids. 325 (2021) 115203.

[43] D. Fang, X. Xu, R. Bao, R. Wan, F. Yang, J. Yi, T. Zeng, O. Ruzimuradov, Layered $Na_5V_{12}O_{32}$ nanowires with enhanced properties of removing Pb^{2+} in aqueous solution by chemical reaction, Journal of Environmental Chemical Engineering. 9 (2021) 104765.

[44] Y.-J. Zhong, S.-J. You, X.-H. Wang, X. Zhou, Y. Gan, N.-Q. Ren, Synthesis of carbonaceous nanowire membrane for removing heavy metal ions and high water flux, Chemical Engineering Journal. 226 (2013) 217–226.

[45] L. Sun, Z. Yuan, W. Gong, Z. Xu, J. Lu, Q. Zou, Y. Wu, G. Su, H. Wang, Removal of trace Cr(VI) from water using chitosan-iron nanowires in porous anodic alumina, Science China Chemistry. 59 (2016) 383–386.

[46] Y. Du, H. Fan, L. Wang, J. Wang, J. Wu, H. Dai, α-Fe_2O_3 nanowires deposited diatomite: Highly efficient absorbents for the removal of arsenic, Journal of Materials Chemistry A. 1 (2013) 7729–7737.

[47] H.-J. Cui, J.-W. Shi, B. Yuan, M.-L. Fu, Synthesis of porous magnetic ferrite nanowires containing Mn and their application in water treatment, Journal of Materials Chemistry A. 1 (2013) 5902.

[48] A.K. Nayak, S. Lee, Y.I. Choi, H.J. Yoon, Y. Sohn, D. Pradhan, Crystal Phase and size-controlled synthesis of tungsten trioxide hydrate nanoplates at room temperature: Enhanced Cr(VI) photoreduction and methylene blue adsorption properties, ACS Sustainable Chemistry & Engineering. 5 (2017) 2741–2750.

[49] A.M. Youssef, F.M. Malhat, Selective removal of heavy metals from drinking water using titanium dioxide nanowire, Macromolecular Symposia. 337 (2014) 96–101.

[50] L. Yin, P. Wang, T. Wen, S. Yu, X. Wang, T. Hayat, A. Alsaedi, X. Wang, Synthesis of layered titanate nanowires at low temperature and their application in efficient removal of U(VI), Environmental Pollution. 226 (2017) 125–134.

[51] J.G. Speight, Redox transformations, in: Reaction Mechanisms in Environmental Engineering, Elsevier, 2018: pp. 231–267.

[52] J. Lei, H. Liu, F. Wen, X. Jiang, C. Yuan, Q. Chen, J.A. Liu, X. Cui, F. Yang, W. Zhu, R. He, Tellurium nanowires wrapped by surface oxidized tin disulfide nanosheets achieves efficient photocatalytic reduction of U(VI), Chemical Engineering Journal. 426 (2021) 130756.

[53] Y. Li, F. Liu, B. Xia, Q. Du, P. Zhang, D. Wang, Z. Wang, Y. Xia, Removal of copper from aqueous solution by carbon nanotube/calcium alginate composites, Journal of Hazardous Materials. 177 (2010) 876–880.

[54] J. Luo, X. Luo, C. Hu, J.C. Crittenden, J. Qu, Zirconia (ZrO_2) embedded in carbon nanowires via electrospinning for efficient arsenic removal from water combined with DFT studies, ACS Applied Materials and Interfaces. 8 (2016) 18912–18921.

20 Nanowires for Organic Waste Removal

S. Sivaselvam, T. Sangavi, and N. Ponpandian
Department of Nanoscience and Technology, Bharathiar University,
Coimbatore, India

CONTENTS

20.1 INTRODUCTION

In recent decades, the impact of environmental pollution on public health has become a growing global concern. According to the World Health Organization (WHO), one-third of all diseases that afflict humanity arises due to long-term exposure to pollution. Scientists have identified several chemical pollutants that are toxic, environmentally persistent, prone to long-term accumulation and migration, precipitation and sedimentation, and are thought to have serious consequences for human health, wildlife, and marine life, even though their release is far from the source. A wholesome eco-sphere and uncontaminated drinking water are the intrinsic obligations for sustaining a healthy life. Clean and safe water is a vital factor of a prosperous society and economy. The cumulative usage of the water primes the affluence of wastewater on a large scale. Almost 1.2 billion people lack access to clean drinking water, and millions of children have died from water-borne illnesses. The water resources are exacerbated mainly due to urbanization, escalating industrialization, and a swift upsurge in population along with the deteriorating quality of water. Subsequently, providing reliable and clean water, especially drinking water in developing countries has become an enormous valu-ation and a key operation in the aim of humanitarian.

The discharge of organic pollutants from the industries to the water bodies is a major concern. Some organic pollutants like industrial waste dyes, pharmaceutical waste, and insecticides are highly reactive and reported to cause adverse health effects like immune suppression, decreased cognitive function, and induction of cancer [1]. Their long-term effects include the development of diabetes, cardiovascular disease, and disruption of thyroid function [2]. The prevailing water treatment technologies like precipitation, membrane separation, and flocculation are not efficient and are cost-effective in the large-scale removal of organic pollutants like pesticides and pharma-ceutical wastes [3]. Recent advancements in nanotechnology have opened up new doors of oppor-tunity for treating wastewater. Today's water treatment technologies are efficient and mobile. Highly efficient, flexible, and multipurpose procedures are designed using nanotechnology to give more efficient water treatment options that depend on massive infrastructure. Different nanostructured materials like nanotubes, nanospheres, and nanowires are used to treat wastewater. The nanowires' nanostructure has the advantage of uniform structure with a high surface area and more active sites are present in them.

This chapter aims to critically review organic waste and its treatment methods using nanowires. Further, it primarily focuses on the developments of nanowires in wastewater remediation. In add-ition, it covers the detailed mechanism involved in different types of water treatments.

20.2   ORGANIC WASTES

Organic wastes, in addition to heavy metals in water, have become one of the most important envir-onmental problems as they are subject to environmental recycling and sustainability. Removal of organic contaminants in water is a world concern. Organic waste is categorized as follows.

20.2.1   Pharmaceutical Waste

Many previous reports confirmed the presence of pharmaceutical residues in water bodies [3,4]. Drug residues like ciprofloxacin, furosemide, hydrochlorothiazide, furosemide, bezafibrate, raniti-dine, and ibuprofen are known to present in the wastewater from ng/L to μg/L concentration. These contaminants are not biodegradable and their presence in μg/L will induce serious health effects on human and marine life.

20.2.2   Pesticides

Chemical pesticides contribute to a greater global food supply in the face of rising demand and population growth. However, the used pesticides will eventually turn into a source of the pollutant over a course of time. The insecticides induce serious health effects like damage to the reproductive and neuron system, induce cancer, and affect the function of organs. Some popular pest toxins include argon chloride, N-methyl carbonate, organophosphate, and pentachlorophenol.

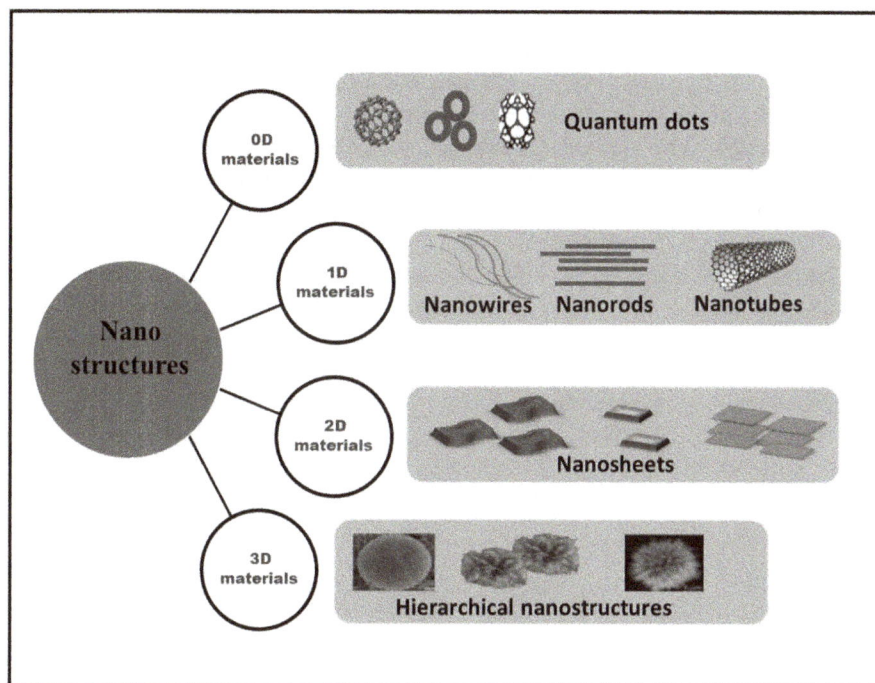

FIGURE 20.1   Types of nanostructures depend on their dimension.

20.2.3 ORGANIC DYES

Organic dyes are synthetic aromatic chemicals often used in textile and paper industries. They are known to induce a carcinogenic and mutagenic effect in humans and they are generally categorized into anionic and cationic dyes. Some common organic pollutants are methylene blue, methyl orange, crystal violet, acid red, and rhodamine-B.

20.3 NANOMATERIAL

Nanomaterials have been carefully studied and used to purify water since their discovery. The effectiveness of nanomaterials in purifying water from organic water pollutants is largely based on their physical properties. For example, 0D quantum dots show a surface plasmon resonance (SPR) effect to enhance the catalytic activity by adsorption of visible photons, the 1D nanostructures, such as nanotubes, nanorods, and nanobands show a large surface area, and one-dimensional flow of electrons and 2D nanosheets show a large surface area and more active sites to improve the photocatalytic efficiency. **Figure 20.1** depicts the types of nanomaterials depending on their dimension.

Recently, nanowires gained popularity in wastewater treatment because of their large surface area, porous nature, and quantum confinement along the axis of nanostructures. The nanowires are generally <100 nm in diameter with a length of 1–100 μm. Metal-based and semiconductor nanomaterials like Fe_2O_3, TiO_2, Ag, MnO_2, GaN, ZnO, and CuO are actively used to remove organic pollutants. The adsorption, photocatalysis, and Fenton reaction are the major treatment method used to remove organic pollutants from water.

20.4 ADSORPTION

The adsorption process has garnered a lot of interest in the field of environmental remediation for its ability to remove a wide range of contaminants in the air and water. In the water treatment methods to remove organic pollutants, the adsorption process is mainly preferred over other treatment methods because of its simplicity, low cost, and easy operation. The main advantage of the adsorption process is not forming any toxic by-products during the reaction. Adsorption is a phenomenon that takes place in the interface of the solid-liquid or solid-gas. The adsorption can be either reversible or irreversible. The solute that is retained (on the solid surface) as a result of adsorption processes is referred to as an adsorbate, whereas the solid on which the solute is held is referred to as an adsorbent. Once the adsorbent's active site has been filled by the adsorbate the state is known as equilibrium adsorption. The term "under-saturated adsorption" refers to a situation in which the amount of adsorption that takes place is lower than the amount of adsorption that would be required for equilibrium. On the other side, adsorption is referred to as being over-saturated if the quantity of adsorption that occurs is more than the amount of absorption that occurs during equilibrium. When adsorbate molecules adsorb onto the surface of an adsorbent, the adsorbent changes either its surface area or its volume as a direct result of this process. Following the completion of the equilibrium adsorption, the amount of adsorbate that has been adsorbed onto the adsorbent (q_e) is calculated using the following formula:

$$q_e = V\,(C_o - C_e)\,/W \tag{20.1}$$

where V is the solution volume (L), W is the weight of the adsorbent used (g), C_e and C_o are the adsorbate concentration at equilibrium and initial time, respectively. The adsorption of adsorbate and adsorbent can be physical, chemical, and charge-based interactions. The adsorbent with a large surface area with a greater number of active sites (surface reactivity) leads to high adsorption capacity.

20.4.1 Adsorption Kinetics Models

Adsorption kinetics is a mathematical model to evaluate the efficiency and rate limiting phases in the adsorption process. The kinetics models are also used to study the adsorption mechanism in the process. The most commonly used adsorption kinetics are as follows:

(1) Pseudo-first-order kinetics
(2) Pseudo-second-order kinetics
(3) Intraparticle diffusion model
(4) Film diffusion model

For analyzing the whole adsorption process, the pseudo-first-order and pseudo-second-order models are used, whilst the intraparticle and film diffusion models are employed to evaluate the limiting stages in adsorption. If the adsorption kinetics is corresponding to the pseudo-first-order model, the mechanism of the adsorption process is physio-sorption in the solid-liquid interface. The most accurate model is determined by the correlation coefficient (R^2) value. If the R^2 value is close to 1, it fits well with the model. If the adsorption kinetics is corresponding to the pseudo-second-order model, the mechanism of the adsorption process is chemisorption at the solid-liquid interface. The term "chemisorption" refers to the process in which electrons are exchanged or shared between the adsorbent and adsorbate.

The intra-particle diffusion model assumes that the adsorbate diffused into the adsorbent pores on its surface during the adsorption process. The linear fitting in the intra-particle diffusion model is plotted against qt and $t^{1/2}$. This model is used to identify the diffusion mechanism in the adsorption process. The film diffusion model is used to predict the rate-limiting phases in the adsorption process. This model takes into account both the physical and chemical interactions that take place between adsorbate and adsorbent.

20.4.2 Adsorption Isotherm Models

Adsorption isotherm models are used to explain the equilibrium performance of adsorbents while the temperature remains the same. It also depends upon the solution pH, concentration of adsorbent, and adsorbate. The commonly used two isotherm models are as follows:

(1) Langmuir model
(2) Freundlich model

In most cases, these models are used to investigate whether the adsorption process involves mono-layer or multilayer adsorption.

The Langmuir isotherm assumes that the adsorbate uniformly adsorbs as a monolayer on the surface of the adsorbent. If the experimental data fit well with the Langmuir model it means the adsorbent and adsorbate have no interaction and the adsorption occurs only over the surface of the adsorbent. The Freundlich model works well when attempting to describe the multilayer or non-uniform adsorption of adsorbate that occurs on the surface of the adsorbent. Mostly it describes the multilayer adsorption process. **Figure 20.2** depicts the Langmuir and Freundlich isotherm model in the adsorption process.

20.4.3 Iron-Based Nanowires in Adsorption

The iron-based nanomaterials are widely used in the adsorption process due to their intrinsic magnetic property that allows easy recovery of the nanomaterial after the adsorption process. Hao et al. fabricated $Fe@Fe_2O_3$ core-shell nanowire by co-precipitation method using $NaBH_4$ as a reducing agent in the atmospheric condition. The adsorption of humic acid by $Fe@Fe_2O_3$ nanowires was investigated in the presence of free molecular air (Oxic) and lack of molecular O_2 (Anoxic). The removal of humic acid by $Fe@Fe_2O_3$ nanowires was monitored by photoluminescent spectroscopy. In the anoxic condition by using $Fe@Fe_2O_3$ nanowires as adsorbents maximum adsorption of the humic acid was found to be 65.5 mg/g. The adsorption of humic acid by $Fe@Fe_2O_3$ nanowires obeys the Langmuir isotherm model and pseudo-second-order kinetics with the R^2 value of 0.98. The fitting of pseudo-second order indicates that the chemisorption process is governed by the

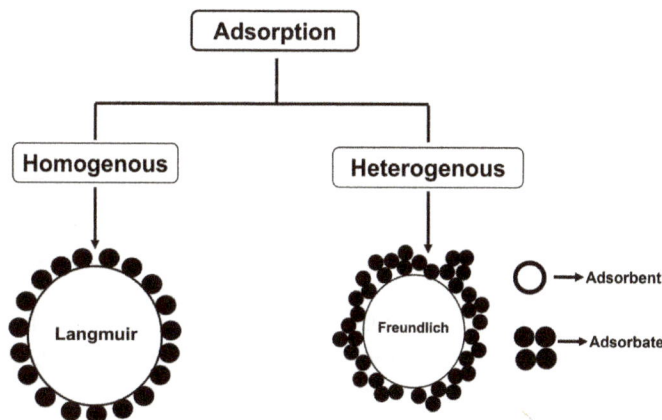

FIGURE 20.2 Graphical representation of Langmuir (homogenous) and Freundlich (heterogenous) isotherm model in the adsorption process.

adsorption process. It is interesting to note that the removal of humic acid was higher in the presence of molecular oxygen. While adding scavengers in the oxic adsorption process, the removal efficiency decreased significantly, which denotes the production of reactive oxygen species (ROS). In oxic conditions, the $Fe@Fe_2O_3$ nanowires exhibit adsorption and oxidation of humic acid. This subsequent oxidation of humic acids potentially increases the oxic removal rate to 2.5 times that of the anoxic removal rate [5].

In another study by Hao et al., they fabricated $MnFe_2O_4$ and Mn-doped Fe_2O_4 nanowires by hydrothermal method and evaluated their adsorption property using the organic pollutant Congo red. Both the prepared $MnFe_2O_4$ and Mn-doped Fe_2O_4 nanowires are super-paramagnetic with the magnetization saturation of 45.9 and 45.4 emu/g respectively. The BET surface area was also found to be 37.8 and 45.4 m^2/g for $MnFe_2O_4$ and Mn-doped Fe_2O_4 nanowires respectively with a diameter range of 100–300 nm. The maximum adsorption capacity of Congo red on $MnFe_2O_4$ and Mn-doped Fe_2O_4 nanowires is found to be 180 mg/g [6]. The adsorption experimental data were well fitted with the Langmuir isotherm model.

20.4.4 COPPER-BASED NANOWIRES IN ADSORPTION

In a study by Ghaedi et al., copper nanowires were prepared and loaded on the activated carbon (CuNW-AC) and used for the adsorption of organic pollutant malachite green. The prepared Cu nanowires were uniformly bounded on the activated carbon with ~81 nm in diameter with a length of 1–100 μm. The CuNW-AC showed a higher surface area of 689.29 m^2/g. To predict the percentage of malachite green dye removal by Cu-NWs-AC, a three-layer artificial neural network (ANN) model was used. The experimental data on adsorption of malachite green by CuNW-AC follows the Langmuir isotherm model. The adsorption follows the pseudo-second-order model in combination with the intra-particle diffusion model. The maximum adsorption capacity of malachite green over CuNW-AC is 434.8 mg/g. The ANN model generated results that agreed well with the experimental data [7]. In another study, the copper nanowires were prepared by the co-precipitation method and decorated on the activated carbon (AC@Cu-NWs). The average crystalline size of the prepared Cu-NWs is found to be ~17.48 nm using the Debye-Scherer formula. The SEM analysis confirmed that the formed Cu nanowires are aggregated to form a flower-like shape. The prepared AC@Cu-NWs showed good adsorption towards the methylene blue with a maximum adsorption capacity of 141.7 mg/g. The adsorption process follows the Langmuir isotherm model. The kinetic parameter showed that the adsorption of malachite green on AC@Cu-NWs is a physio-sorption process [8].

20.4.5 MANGANESE-BASED NANOWIRES IN ADSORPTION

Han et al. fabricated Mn_3O_4 nanowires using the hydrothermal method. Potassium permanganate ($KMnO_4$) is used as a precursor for Mn_3O_4 nanowire synthesis. The TEM analysis showed the formation of Mn_3O_4 nanowires. The Mn_3O_4 nanowires were used to adsorb four different organic pollutants methylene blue, rhodamine-B, methyl orange, and gentine violet. The effect of pH on these dye removals showed that maximum removal of dyes was observed at pH-2 with a removal efficiency of more than 97%. The Mn_3O_4 nanowires showed good adsorption efficiency even after five reusability tests with 95% removal and then it gradually decreased. After five cycles, morphological changes like agglomeration were observed in the structure of Mn_3O_4 nanowires and this may be the reason for the decrease in removal efficiency [9]. Dastjerdi et al. synthesized manganese oxide nanowires and coated them with hydrophobic polymer polydimethylsiloxane (PDMS), and used them for the adsorption of dissolved toluene in the groundwater. The synthesis of Mn_3O_4 nanowires was performed by hydrothermal method and the PDMS coating was performed using the vapor deposition method (VDM) at 250 ^{0}C. The adsorption of toluene by Mn_3O_4 nanowires follows the Freundlich isotherm model [10].

20.4.6 Titanate Nanowires in Adsorption

Huang et al. synthesized titanate nanowires by hydrothermal method and studied the adsorption of methylene blue. By varying the reaction time during the synthesis process (0.5–6 h) different nanostructures were obtained. The titanate nanowires were formed at the temperature and reaction time of 240 °C for 6 h, respectively. The formed titanate nanowires are in the diameter range of 40–250 nm. The titanate nanowires showed a BET surface area of 30 m^2/g. When compared with other morphologies obtained like nanoflower, and nanotube, the nanowires showed considerably fewer adsorption properties. The adsorption of methylene blue by titanate nanowires fits well with the Langmuir adsorption isotherm and pseudo-second-order kinetic model [11]. In a work by Hassan et al., potassium titanate nanowires ($K_2Ti_8O_{17}$) were prepared by hydrothermal method and used for the removal of methylene blue. During the preparation method, various concentrations of KOH (1.6–2.0 M) were added at the reaction temperature of 320 °C and a time of 80 min. The formed $K_2Ti_8O_{17}$ at all the KOH concentrations showed nanowire formation. The BET analysis of $K_2Ti_8O_{17}$ prepared at different concentrations of KOH like 1.6, 1.8, and 2 M showed a surface area of 135.33, 151.64, and 143.97 m^2/g, respectively. The adsorption process follows the Freundlich isotherm model and pseudo-second-order kinetics with the maximum adsorption capacity of 208.8 mg/g at neutral pH with a contact time of 21 min [12].

20.4.7 Tungsten Nanowires in Adsorption

Zhao et al. synthesized tungsten trioxide nanowires (WO_3 NWs) via a one-step hydrothermal method. Using a tubular furnace, the thermal treatment was given to the prepared WO_3 NW samples at 300 and 500 °C and the adsorption property was compared to the organic pollutant methylene blue and methyl orange. The post thermal treatment of WO_3 NWs at 0, 300, and 500 °C showed the BET surface area of 864.1, 82.62, and 28.41 m^2/g, respectively. The TEM analysis showed that the prepared WO_3 NWs are in the diameter range of ~10–30 nm. The WO_3 NWs without any post-treatment showed a high adsorption capacity of 156.25 and 54.23 mg/g for methylene blue and methyl orange, respectively. The adsorption process follows the Langmuir isotherm and pseudo-second-order kinetics model. The mechanism for the higher adsorption property was due to the electrostatic attraction between adsorbent and adsorbate at acidic pH. About 88% of the dye solution is adsorbed to the WO_3 NWs within 70 min of the contact time [13].

In a work by Shang et al., superfast efficient adsorbent WO_3 NWs were prepared by hydrothermal method. The concentration of sodium tungstate was varied between 0.5, 1, and 2 mM during synthesis. The formation of WO_3 NWs using 0.5, 1, and 2 mM of sodium tungstate showed nanowire morphology with the diameter range of 593, 527, and 497 nm, respectively. The WO_3 NWs showed high adsorption properties towards methylene blue with about 94% and ~100% removal within 5 and 6 min, respectively. The point of zero charges of the prepared WO_3 NWs is in the range of pH-4. Above this pH (pH 4–7), WO_3 NWs showed a negative charge and at the same pH, the methylene blue showed a positive charge. This electrostatic interaction plays a major role in the quicker adsorption of methylene blue. The adsorption process follows the Langmuir adsorption and pseudo-second-order model. The maximum adsorption capacity was found to be 547.32 mg/g. The thermodynamic study revealed that the adsorption process is exothermic and spontaneous [14].

20.4.8 Other Nanowires in the Adsorption Process

Xing et al. fabricated lanthanum oxide (La_2O_3) cerium nanowires decorated with nickel nanoparticles using the electrospinning method. To remove the polymer used in electrospinning, the formed La_2O_3-Ni nanowires were subjected to heat treatment at 800 °C. The SEM analysis showed that the formed La$_2$O$_3$-Ni nanowires have a smooth and continuous surface with a diameter of less than 150 nm. The TEM analysis showed that the Ni nanoparticle on the La$_2$O$_3$ is evenly and uniformly distributed along the nanowire with a particle size of 20 nm. The La$_2$O$_3$ nanowires with 6% loaded Ni nanoparticles exhibit a higher BET surface area of 143.3 m^2/g. The La$_2$O$_3$-Ni nanowires exhibit rapid removal (99%) of rhodamine-B dye within 10 sec. The adsorption follows Langmuir isotherm with an adsorption capacity of 36.1 mg/g. The mechanism behind the rapid adsorption is due to the interaction of Ni with La$_2$O$_3$. When Ni comes into contact with La$_2$O$_3$, electrons from La$_2$O$_3$ transfer to the Ni surface to neutralize the Fermi level and leaves a depletion layer on the interface of a particle with a positive surface charge. The positive layer attracts the electron from the dye and results in rapid adsorption [15]. Xi et al. fabricated silver nanowires (Ag NWs) wrapped with polyvinyl alcohol (PVA) and used them for continuous oil separation from water. The silver nanowires wrapped with PVA were prepared using a hydrothermal method and the formed product was freeze-dried to hydrogel and immersed in PDMS/hexane mixture to get superhydrophobicity. The diameter of the PVA-Ag NWs is 1–1.5 µm with a surface area of 44.6 m^2/g. The wettability of the PVA-Ag NWs was 156.6 ±2° and showed good adsorption properties against various oils and solvents dissolved in water. The adsorbed oil and solvents are easily recovered from PVA-Ag NWs by a pump and reused [16].

20.5 PHOTOCATALYSIS

Photocatalysis is a chemical reaction that uses light energy like UV and visible light to produce reactive hydroxyl radicals to degrade pollutants. Photocatalysis has emerged as a promising green method for the full mineralization of potentially harmful organic compounds at room temperature into harmless by-products such as water, carbon dioxide, and simple mineral acids. In photocatalysis, the photocatalyst used mainly is semiconductor-based nanomaterial because of their wide bandgap. Semiconductor nanomaterials like titanium dioxide (TiO$_2$), zinc oxide (ZnO), cadmium sulfide (CdS), and zinc sulfide (ZnS) are already used as a photocatalyst. The quantum-size effect, which raises the bandgap energy with a size reduction, is the primary benefit of nanosized semiconductor particles. This phenomenon is what makes nanosized particles so advantageous. When photons with energies at or above the bandgap encounter the catalyst surface, valence electrons get excited and reach the conduction band. Excited NPs proceed through the process of charge separation and produce a free radical that oxidizes organic molecules on the surface of the semiconductor. The graphical representation of free radical production in photocatalysis is shown in **Figure 20.3**. Different parameters like pH, temperature, contact time, the effect of pollutant concentration, and nanocatalyst concentration affect the photocatalysis process.

20.5.1 TiO$_2$ Nanocatalyst in Photocatalysis

The simple preparation method, easy recyclability, stable in both acid and alkaline environment makes titanium dioxide (TiO$_2$) an effective photocatalyst material for the degradation of organic pollutants by making use of the UV spectrum. Al-Hajji et al. prepared H$_2$Ti$_3$O$_7$ nanowires by hydrothermal method and varied the calcination temperature from 400–900 °C. The XRD pattern of the prepared H$_2$Ti$_3$O$_7$ revealed that when calcinated at 400 and 500 °C showed a monoclinic structure of TiO$_2$. When the calcination temperature was increased from 600 °C the anatase phase contributes majorly to some monoclinic phases of TiO$_2$. At the calcination temperature of 700 and 800 °C the brookite and anatase phase was observed. At 900 °C both brookite and rutile structures appeared. The prepared nanowires were up to a few µm in length with 50–200 nm in diameter. The calcination temperature from 400–900 °C has no significant changes in the morphology of the nanowires. The formed TiO^2 nanowires at 400 °C showed a higher BET surface area of 24.37 m^2/g when compared to other materials. The bandgap of the prepared nanomaterials is in the range of 2.36–2.24 eV. The

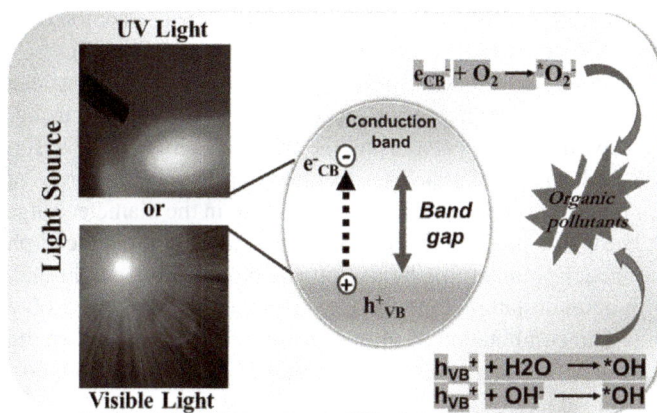

FIGURE 20.3 The mechanism of free radical production in photocatalysis reaction.

formed TiO$_2$ nanowires were used to degrade the organic pollutant resorcinol. The TiO$_2$ nanowire formed at 800 °C showed maximum removal efficiency of 99% in 180 min at neutral pH. The experimental data fit well with first-order kinetics and showed above 95% removal efficiency even after five successful runs [17]. The bandgap of TiO$_2$ nanowires is limited within the range of the UV region (3.1–3.4 eV). Much research has been promised for the application of TiO$_2$ particles in the photodegradation of pollutants in the visible region. Cheng et al. prepared TiO$_2$ nanowires on the surface of reduced graphene oxide (TiO$_2$ NWs@rGO). The prepared TiO$_2$ NWs@rGO showed a length of 10 µm with a diameter of 17 nm with more anatase phase TiO$_2$. The prepared TiO$_2$ NWs@rGO was used to remove waste engine oil and showed good removal efficiency of 86% COD removal within 30 min due to the high surface area (182 m^2/g). When the light energy is provided, the degradation efficiency rapidly increases to 98.6% removal with the final COD value of 2 mg/L from 145 mg/L. The EPR spectra analysis showed that the O$_2^-$ is the active radical species formed during the reaction [18].

20.5.2 ZnO Nanowires in Photocatalysis

Liu et al. prepared ZnO nanowire by a hydrothermal method having uniform size with a diameter of 90 nm and 700 nm in length. The ZnO nanowires exhibit a bandgap of 3.37 eV, which induces free radical formation in the UV region. The ZnO nanowire exhibits above 95% degradation of methyl orange (10 mg/L) within 150 min. The effect of pH showed a good catalytic activity from pH 6–10. The higher catalytic activity of ZnO nanowires in the wide pH range is due to the isoelectric point of ZnO nanowires (pH-9.0) and the electrostatic attraction between the and methyl orange. The ZnO nanowires showed good catalytic efficiency in the first two recycles and the efficiency decreases from the third cycle [19].

To improve the efficiency of ZnO nanowires, Hassani et al. fabricated a novel graphitic carbon coated silver and zinc oxide nanowire (g-C$_3$N$_4$/Ag/ZnO NWs) through the dip-coating method. The formed g-C$_3$N$_4$/Ag/ZnO NWs show a nanowire structure of ZnO with a diameter of ~50 nm. The photoluminescence analysis showed that the emission peak of the final composite was low when compared to g-C$_3$N$_4$ and ZnO nanowires. This indicated that the recombination of electron (e$^-$) and hole (h$^+$) in the final composite is very low and this is due to the electron trapping ability of Ag nanoparticle. The g-C$_3$N$_4$/Ag/ZnO NWs showed good photocatalytic properties against Direct Orange 26 dye (DO26) with a degradation efficiency of 94% within 120 min. The g-C$_3$N$_4$/Ag/ZnO NWs showed maximum degradation efficiency at pH-6 and the degradation process follows pseudo-first-order kinetics. The scavenger analysis confirmed the production of OH⁻ and h⁺ radicles in the g-C₃N₄/Ag/ZnO NWs composite during the degradation process [20].

Hong et al. fabricated copper sulfide/zinc oxide nanowires on the stainless steel mess (CuS/ZnS nanowires). The stainless-steel-supported ZnO nanowires were prepared using the hydrothermal method. The CuS nanoparticle was coated using the successive ionic layer adsorption and reaction (SILAR) method. The advantage of stainless-steel mess is the easy recovery of catalyst material after the degradation process. The CuS/ZnS nanowires were in the diameter range of 100 nm with a bandgap value of 2.39 eV. The prepared CuS/ZnS nanowires showed the piezo-photocatalytic property. It degrades the model pollutant methylene blue using both solar light and ultrasonic waves. When irradiated with a xenon lamp alone the CuS/ZnS nanowires showed 60% degradation efficiency within 20 min. The combination of ultra-sonication and solar light irradiation enhanced the degradation efficiency of MB (5 mg/L) up to 99% within 20 min. The degradation process obeys the first-order kinetic model [21].

20.5.3 SiO₂ NANOCATALYST IN PHOTOCATALYSIS

Ghosh et al. prepared silver nanoparticle decorated Si nanowires (Ag@Si-NWs). The photodegradation efficiency was evaluated using the model dye methylene blue (MB). When compared with Si-NWs, the Ag decorated Si-NWs showed enhanced degradation efficiency with nearly 99% removal within 200 min. The heteroatom composite Si NWs/AuPd showed 100% degradation in 150 min. The rapid degradation is explained by the difference in Fermi level between Si and metal particles. This Fermi level difference results in a Schottky barrier between the interface of Si and metal nanoparticles and efficiently separates the electron-hole recombination during the degradation process [22].

20.5.4 CuO NANOWIRES IN PHOTOCATALYSIS

Scuderi et al. synthesized CuO and Cu₂O nanowires by a thermal process in the presence of an oxygen atmosphere. The formed CuO and Cu₂O nanowires are with a length of 10 μm and 80 nm in diameter. In the degradation process of methyl orange, the CuO nanowires showed higher efficiency when compared to Cu₂O nanowires. The higher degradation was due to the electrostatic attraction of positively charged CuO and negatively charged MO dyes. When studying the degradation of methylene blue, the efficiency was low because of the electrostatic repulsion [23]. To increase the degradation efficiency of CuO nanowires, Iqbal et al. synthesized manganese doped copper oxide nanowires (Mn-CuO NWs) using the chemical precipitation method. After doping Mn⁺, the bandgap of CuO NWs reduced from 2.6 to 2.3 eV. The degradation efficiency of 2% Mn-doped CuO NWs for rhodamine-B was about 87% in 6 h. It is interesting to note that the rhodamine-B was completely mineralized without forming any secondary impurities [24].

20.5.5 OTHER NANOWIRES' CATALYST IN PHOTOCATALYSIS

Jung et al. synthesized gallium nitride nanowire (GaN NWs) using nickel as a catalyst in the chemical vapor deposition method. The prepared GaN NWs showed a length of 4–6 μm with a diameter of 20–50 nm. The GaN NWs showed better photocatalytic activity against an organic dye when compared to other GaN nanostructures. The GaN NWs showed enhanced degradation in the pH range of 2–5, even when compared with TiO₂ and ZnO nanowires. It also showed good degradation efficiency even after many recyclability in acidic solutions [25]. Ma et al. fabricated MnO₂ nanowires/carbon hollow fiber (MnO₂/Ac NWs) using a hydrothermal method. The formed MnO₂/Ac NWs showed a particle diameter of 50 nm. The MnO₂/Ac NWs showed enhanced degradation of methylene blue with 99.8% in 4 h. The formation of MnO₂ NWs on activated carbon help in preventing the aggregation and increased the active site of the material. When compared with pure

MnO_2 NWs, the activated carbon composite enhanced the degradation performance of the catalyst [26].

Liu et al. synthesized cadmium sulfide nanowire/reduced graphene oxide composite (CdS NWs/rGO) by hydrothermal method [27]. The CdS nanowires are coated to rGO via electrostatic interaction by functionalizing CdS with cationic APTES followed by the hydrothermal method. The prepared CdS NWs/rGO exhibits a diameter of 50–100 nm. The obtained bandgap of CdS NWs/rGO was about 2.3 eV. For the photocatalysis of organic nitro compounds like 4-nitroaniline (4-NA), 2-nitroaniline (2-NA), 4-nitropheneol (4-NP), bromo-4-nitrobenzene (4-NB), 4-nitroanisole (4-NAS), and 4-nitrotolune (4-NT). The xenon lamp with UV CUT filter (>420 nm) was used to initiate the photocatalysis. The degradation efficiency of about 97%, 95%, 90%, 87% 82%, and 76% was obtained for 4-NA, 2-NA, 4-NP, 4-NB, 4-NAS, and 4-NT in 8, 8, 8, 2, 4, and 4 min, respectively. The mechanism behind the rapid removal of nitro compounds are due to the presence of rGO in the composite. The rGO rapidly adsorbs the nitro compounds and shiftily transport the produced radical to the pollutant.

20.6 FENTON REACTION FOR DEGRADATION OF ORGANIC POLLUTANTS

The Fenton reaction was first observed by the British chemist Henry John Horstman Fenton. In the Fenton reaction, metal nanoparticles in the presence of a reducing agent like H_2O_2 induce the formation of free radicals like OH^- and $^-O_2$ that oxidize the pollutants. The Fenton reaction requires certain optimum conditions to reach the higher degradation efficiency like pH, temperature, catalyst, and H_2O_2 concentration. During the Fenton reaction using Fe material, both oxidation and reduction take place. When compared to oxidation the reduction process is thermodynamically low, this is due to the low solubility of Fe^{3+} ions. To overcome this low pH is preferable for the easy solubility of Fe^{3+} ions. The increase in temperature increases the OH^- radical production. **Figure 20.4** depicts the Fenton reaction process for the degradation of organic pollutants.

20.6.1 IRON-BASED NANOMATERIALS IN FENTON REACTION

The iron-based nanomaterials are generally applied for organic pollutant degradation, because of the abundant nature of iron and its safer nature from an environmental point of view. Luo et al.

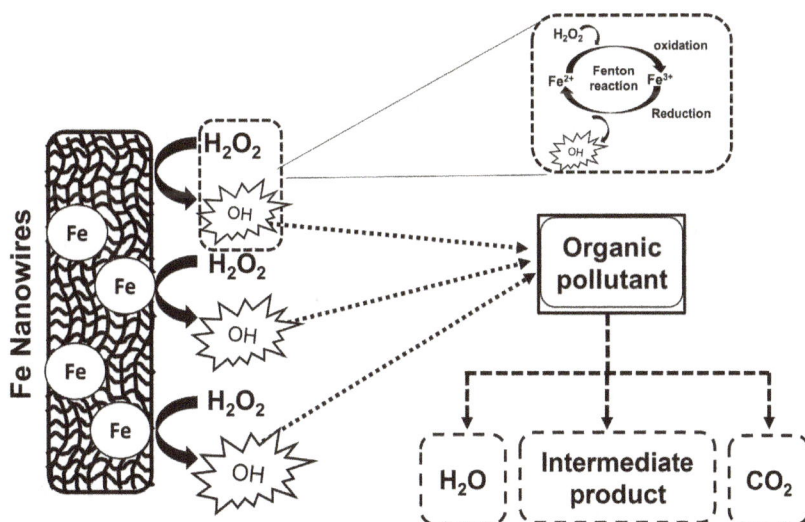

FIGURE 20.4 Schematic representation of •OH radical production in the Fenton reaction.

synthesized $Fe@Fe_2O_3$ nanowires using the chemical reduction method. The $Fe@Fe_2O_3$ nanowires were tested against two different organic pollutants like rhodamine-B and p-nitrophenol. The Fenton reaction was initiated by H_2O_2 and hydroxylamine (HA). The reaction solution was maintained at pH-3. While using $Fe@Fe_2O_3$/HA, $Fe@Fe_2O_3$/H_2O_2, and $Fe@Fe_2O_3$/H_2O_2/HA as a catalyst the degradation of rhodamine-B was found to be 8.5, 28.5, and 97.1% respectively. Likewise, the degradation of p-nitrophenol also varies with 6.5, 27.8, and 91% degradation efficiency for $Fe@Fe_2O_3$/HA, $Fe@Fe_2O_3$/H_2O_2, and $Fe@Fe_2O_3$/H_2O_2/HA as the catalyst, respectively. By varying the HA concentration, 1 mM of HA showed higher degradation with 98% removal in 2 min when compared to 0.01 mM of HA with 78% degradation in 60 min. The production of OH^- and $^-O_2$ are the main radical species produced during the Fenton reaction. Over 90% of rhodamine-B was degraded even after eight successive runs [28].

20.6.2 COPPER-BASED NANOWIRES IN FENTON REACTION

Kong et al. synthesized CuO nanowires on the graphitic carbon nitride nanosheet (CuO NWs/gC$_3$N$_4$) [29]. The H_2O_2 was used as a Fenton reaction initiator and studied the effect of various H_2O_2 (20–80 mM) concentrations against three model dyes, such as methylene blue (MB), methyl orange (MO), and rhodamine-B (Rho-B). The CuO NWs/gC$_3$N$_4$ showed enhanced catalytic activity with above 90% degradation for MB, MO, and rho-B dyes in the time of 20, 60, and 50 min, respectively. The optimum condition of H_2O_2 (0.05g) and pH-5 showed better degradation efficiency. The OH^- and $^-O_2$ are the radical species formed during the reaction and all the three model dyes were completely mineralized into CO_2 and H_2O.

20.6.3 MnO$_2$ NANOWIRES IN FENTON REACTION

Ramesh et al. synthesized reduced graphene oxide coated manganese oxide nanowire (rGO/MnO$_2$) using the hydrothermal method [30]. The prepared rGO/MnO$_2$ nanowires showed a BET surface area of 58.2 m^2/g. The pristine rGO/MnO$_2$ nanowires showed 63% degradation of reactive black 5 (RB5) in 60 min. With the addition of H_2O_2, the degradation efficiency increased to 84%. Further, with the assistance of ultrasonic waves in addition to H_2O_2 the degradation efficiency enhanced to 95%. These data showed that the ultrasonic wave combined Fenton reaction is efficient for the removal of organic wastes.

20.6.4 ZnO NANOWIRES IN FENTON REACTION

Lin et al. fabricated nickel nanoparticle-loaded ZnO nanowires (Ni@ZnO NWs) on the stainless steel substrate using hydrothermal and electrodepositing methods [31]. The Ni@ZnO NWs exhibit a diameter range of about 57 nm. The pristine Ni@ZnO NWs showed negligible degradation ability of 4-nitro phenol. After the addition of NaBH$_4$, the degradation efficiency was significantly enhanced by 95% within 30 min.

20.7 CONCLUSION

In this chapter, we discussed the most recent developments in research and technology about various nanowires used in organic waste removal from wastewater. We particularly focused on the application of nanowires in the adsorption, photocatalysis, and Fenton reaction for organic waste removal. The nanowires showed better removal and degradation property owing to their nanometer size. Fabricated nanowires with metal, semiconductor nanomaterial, or any substrate showed enhanced removal efficiency towards organic waste. In the future, more work should be focused on the development of nanowire composite with synergistic properties combining adsorption, photo-catalysis, and Fenton degradation.

REFERENCES

[1] D.O. Carpenter, Health effects of persistent organic pollutants: the challenge for the Pacific Basin and for the world, Rev. Environ. Health. 26 (2011) 61–69.

[2] J. Dębska, A. Kot-Wasik, J. Namieśnik, Fate and analysis of pharmaceutical residues in the aquatic environment, Crit. Rev. Anal. Chem. 34 (2004) 51–67.

[3] S. Sivaselvam, P. Premasudha, C. Viswanathan, N. Ponpandian, Enhanced removal of emerging pharmaceutical contaminant ciprofloxacin and pathogen inactivation using morphologically tuned MgO nanostructures, J. Environ. Chem. Eng. 8 (2020) 104256.

[4] S. Sivaselvam, R. Selvakumar, C. Viswanathan, N. Ponpandian, Rapid one-pot synthesis of PAM-GO-Ag nanocomposite hydrogel by gamma-ray irradiation for remediation of environment pollutants and pathogen inactivation, Chemosphere. 275 (2021) 130061.

[5] H. Wu, Z. Ai, L. Zhang, Anoxic and oxic removal of humic acids with $Fe@Fe_2O_3$ core–shell nanowires: A comparative study, Water Res. 52 (2014) 92–100.

[6] H. Cui, J. Shi, B. Yuan, M. Fu, Synthesis of porous magnetic ferrite nanowires containing Mn and their application in water treatment, J. Mater. Chem. A. 1 (2013) 5902.

[7] M. Ghaedi, E. Shojaeipour, A.M. Ghaedi, R. Sahraei, Isotherm and kinetics study of malachite green adsorption onto copper nanowires loaded on activated carbon: Artificial neural network modeling and genetic algorithm optimization, Spectrochim. Acta Part A Mol. Biomol. Spectrosc. 142 (2015) 135–149.

[8] S.K. Lakkaboyana, S. Khantong, N.K. Asmel, A. Yuzir, W.Z. Wan Yaacob, Synthesis of copper oxide nanowires-activated carbon (AC@CuO-NWs) and applied for removal methylene blue from aqueous solution: Kinetics, isotherms, and thermodynamics, J. Inorg. Organomet. Polym. Mater. 29 (2019) 1658–1668.

[9] R. Han, Y. Zhang, Y. Xie, Application of Mn_3O_4 nanowires in the dye waste water treatment at room temperature, Sep. Purif. Technol. 234 (2020) 116119.

[10] M.H. Tavakoli Dastjerdi, G. Habibagahi, A. Ghahramani, A. Karimi-Jashni, S. Zeinali, Removal of dissolved toluene in underground water with nanowires of manganese oxide, Adsorpt. Sci. Technol. 36 (2018) 393–407.

[11] J. Huang, Y. Cao, Z. Liu, Z. Deng, W. Wang, Application of titanate nanoflowers for dye removal: A comparative study with titanate nanotubes and nanowires, Chem. Eng. J. 191 (2012) 38–44.

[12] Q.U. Hassan, D. Yang, J.-P. Zhou, Controlled fabrication of $K_2Ti_8O_{17}$ nanowires for highly efficient and ultrafast adsorption toward methylene blue, ACS Appl. Mater. Interfaces. 11 (2019) 45531–45545.

[13] Z. Zhao, N. Ping, J. Di, H. Zheng, Highly selective adsorption of organic dyes onto tungsten trioxide nanowires, Res. Chem. Intermed. 42 (2016) 5639–5651.

[14] Y. Shang, Y. Cui, R. Shi, P. Yang, J. Wang, Y. Wang, Regenerated WO2.72 nanowires with superb fast and selective adsorption for cationic dye: Kinetics, isotherm, thermodynamics, mechanism, J. Hazard. Mater. 379 (2019) 120834.

[15] Y. Xing, J. Cheng, J. Wu, M. Zhang, X. Li, W. Pan, Direct electrospinned La_2O_3 nanowires decorated with metal particles: Novel 1D adsorbents for rapid removal of dyes in wastewater, J. Mater. Sci. Technol. 45 (2020) 84–91.

[16] G. Xi, T. Liu, C. Ma, Q. Yuan, W. Xin, J. Lu, M. Ma, Superhydrophobic, compressible, and reusable polyvinyl alcohol-wrapped silver nanowire composite sponge for continuous oil-water separation, Colloids Surfaces A Physicochem. Eng. Asp. 583 (2019) 124028.

[17] L.A. Al-Hajji, A.A. Ismail, A. Al-Hazza, S.A. Ahmed, M. Alsaidi, F. Almutawa, A. Bumajdad, Impact of calcination of hydrothermally synthesized TiO_2 nanowires on their photocatalytic efficiency, J. Mol. Struct. 1200 (2020) 127153.

[18] D. Cheng, Y. Li, L. Yang, S. Luo, L. Yang, X. Luo, Y. Luo, T. Li, J. Gao, D.D. Dionysiou, One–step reductive synthesis of Ti^{3+} self–doped elongated anatase TiO_2 nanowires combined with reduced graphene oxide for adsorbing and degrading waste engine oil, J. Hazard. Mater. 378 (2019) 120752.

[19] X. Liu, W. Huang, G. Huang, F. Fu, H. Cheng, W. Guo, J. Li, H. Wu, Synthesis of bilayer ZnO nanowire arrays: Morphology evolution, optical properties and photocatalytic performance, Ceram. Int. 41 (2015) 11710–11718.

[20] A. Hassani, M. Faraji, P. Eghbali, Facile fabrication of mpg-C_3N_4/Ag/ZnO nanowires/Zn photocatalyst plates for photodegradation of dye pollutant, J. Photochem. Photobiol. A Chem. 400 (2020) 112665.

[21] D. Hong, W. Zang, X. Guo, Y. Fu, H. He, J. Sun, L. Xing, B. Liu, X. Xue, High piezo-photocatalytic efficiency of CuS/ZnO nanowires using both solar and mechanical energy for degrading organic dye, ACS Appl. Mater. Interfaces. 8 (2016) 21302–21314.

[22] R. Ghosh, J. Ghosh, R. Das, L.P.L. Mawlong, K.K. Paul, P.K. Giri, Multifunctional Ag nanoparticle decorated Si nanowires for sensing, photocatalysis and light emission applications, J. Colloid Interface Sci. 532 (2018) 464–473.

[23] V. Scuderi, G. Amiard, S. Boninelli, S. Scalese, M. Miritello, P.M. Sberna, G. Impellizzeri, V. Privitera, Photocatalytic activity of CuO and Cu_2O nanowires, Mater. Sci. Semicond. Process. 42 (2016) 89–93.

[24] M. Iqbal, A.A. Thebo, A.H. Shah, A. Iqbal, K.H. Thebo, S. Phulpoto, M.A. Mohsin, Influence of Mn-doping on the photocatalytic and solar cell efficiency of CuO nanowires, Inorg. Chem. Commun. 76 (2017) 71–76.

[25] Jung, H.S., Hong, Y.J., Li, Y., Cho, J., Kim, Y.J. and Yi, G.C., 2008. Photocatalysis using GaN nanowires, ACS Nano, 2(4), 637–642.

[26] L. Ma, D. Li, L. Wang, X. Ma, In situ hydrothermal synthesis of α-MnO_2 nanowire/activated carbon hollow fibers from cotton stalk composite: dual-effect cyclic visible light photocatalysis perform-ance, Cellulose. 27 (2020) 8937–8948.

[27] S. Liu, Z. Chen, N. Zhang, Z. Tang, Y. Xu, An efficient self-assembly of cds nanowires − reduced graphene oxide nanocomposites for selective reduction of nitro organics under visible light irradi-ation, J. Phys. Chem. C 2013, 117, 16, 8251–8261.

[28] H. Luo, Y. Zhao, D. He, Q. Ji, Y. Cheng, D. Zhang, X. Pan, Hydroxylamine-facilitated degradation of rhodamine B (RhB) and p-nitrophenol (PNP) as catalyzed by Fe@Fe_2O_3 core-shell nanowires, J. Mol. Liq. 282 (2019) 13–22.

[29] X. Kong, A. Chen, L. Chen, L. Feng, W. Wang, J. Li, Q. Du, W. Sun, J. Zhang, Enhanced Fenton-like catalytic performance of freestanding CuO nanowires by coating with g-C_3N_4 nanosheets, Sep. Purif. Technol. 272 (2021) 118850.

[30] M. Ramesh, M.P. Rao, F. Rossignol, H.S. Nagaraja, rGO/MnO_2 nanowires for ultrasonic-combined Fenton assisted efficient degradation of Reactive Black 5, Water Sci. Technol. 76 (2017) 1652–1665.

[31] C. Lin, G. Luo, H. Zhou, A. Feng, L. Zeng, Q. Li, Ni nanoparticles-loaded ZnO nanowire as an effi-cient and stable catalyst for reduction of 4-nitrophenol, EcoMat. 4 (2022) 1–13.

21 Nanowires for Gas Sensors

Fatima Ezahra Annanouch, Shuja Bashir Malik, and Eduard Llobet
MINOS-EMaS, Universitat Rovira i Virgili, Avinguda dels Països Catalans 26, Tarragona, Catalunya, Spain

Frank Güell
MINOS-EMaS, Universitat Rovira i Virgili, Avinguda dels Països Catalans 26, Tarragona, Catalunya, Spain
ENFOCAT-IN2UB, Universitat de Barcelona, Carrer Martí i Franquès 1, Barcelona, Catalunya, Spain

CONTENTS

21.1 INTRODUCTION

Gas sensors are devices that generate an electrical signal upon interaction with gaseous species present in their surroundings. According to Liu and co-workers [1], there are two different classes of gas sensors. The first class is based on semiconductor materials that have their electrical resistivity dependent on the concentration of a surrounding gaseous species. Such devices are frequently referred to as chemiresistors, as the output signal is an electrical resistance variation upon the interaction with chemical species. The second class is based on the variation of other physical properties, such as optical, acoustic, or even the material heat capacity upon interaction with the target gas. According to the Market Size Report from Grand View Research, the gas sensor market is projected to grow from US$ 1 billion in 2022 to US$ 4.1 billion by 2030 at a rate of 8.9% per year, which demonstrates the high demand for new gas sensing technologies [2]. In recent years, with the advances in nanotechnology, tremendous efforts have been made to produce semiconductor materials in various forms to improve gas detection performances. They are dimensionally categorized, i.e., depending on the number of component dimensions of the material at the nanoscale. For instance, zero-dimensional (0D), one-dimensional (1D), two-dimensional (2D), and three-dimensional (3D) nanostructured materials. Nanowires belong to the 1D group with a length up to a few millimeters, but their thickness and width are only within a limited range between 1 and 200 nm [3,4]. Materials with this engineering, possess new and unique physical and chemical properties that do not exist in their bulk form. They have a high surface-to-volume ratio, which increases the active sites and enhances the sensitivity as well as the selectivity of the sensor. Moreover, they possess high crystallinity, which reinforces their long-term stability. Next, they are well known for their electronic confinement effect, which leads to a good charge transfer between the material and adsorbed gaseous species and superior gas detection properties.

DOI: 10.1201/9781003296621-21

This chapter provides, first an overview regarding metal oxide nanowire-based gas sensors, second a discussion on conducting polymer nanowire-based gas sensors, third a report on metallic nanowire-based gas sensors, and finally silicon nanowire-based gas sensors presented. It should be noted that for most of the mentioned sections, we have discussed the synthesis method, effect of the nanowire diameter on the sensor properties, gas sensing performance, and mechanisms.

21.2   METAL OXIDE NANOWIRE-BASED GAS SENSORS

Metal oxide semiconductors (MOS) are commonly used as chemiresistors, because they exhibit a band-gap in a range that yields considerable electrical resistance changes upon surface redox reaction with common gases [5,6]. A renewed interest in MOS has emerged in recent years due to significant improvements in its growth techniques, which have led to the fabrication of these materials at low dimension geometries, such as nanoparticles, nanorods, nanotubes, nanowires, etc. It is well known that these different morphologies have different natures; therefore, research has been carried out with the objective of better understanding the relationship between the morphologies and the gas sensing properties of the materials. Stoycheva and co-workers [7] fabricated micromachined gas sensors based on WO_3 nanowires directly integrated into the sensor substrate, via aerosol-assisted CVD. In this work, the authors prepared three different morphologies (**Figure 21.1**) of WO_3: nanoparticles, aligned, and non-aligned nanowires, and studied their effect on the gas sensing

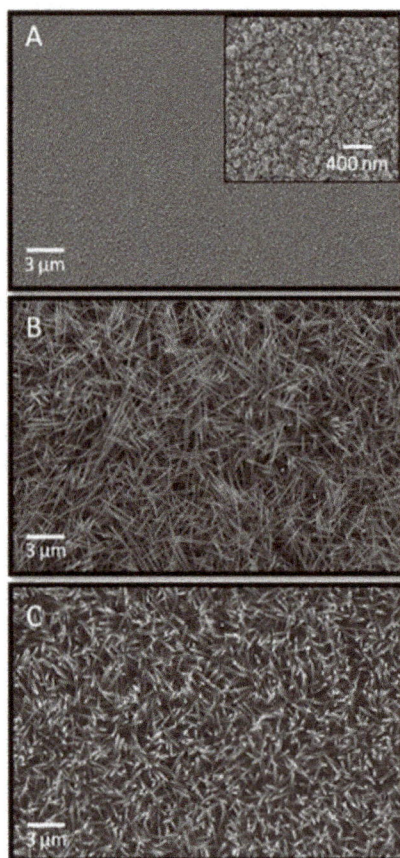

FIGURE 21.1   Field-emission scanning electron microscopy (FESEM) images of (A) WO_3 nanoparticles grown at 350 °C (inset displays a higher magnification image), (B) non-aligned nanowires grown at 450°C, and (C) quasi-aligned nanowires grown at 500°C. Adapted with permission [7]. Copyright 2014, Elsevier.

properties. Gas sensing results revealed improved sensing properties, towards various toxic gases, for sensors based on films composed of nanowires rather than those of nanoparticles, with the best results recorded from non-aligned nanowires.

Similarly, Shaalan and co-workers [8] studied the influence of morphology and structure geometry on the NO_2 gas-sensing characteristics of SnO_2 nanostructures synthesized via a thermal evaporation method. They performed three different forms: microwires, nanowires, and nanoparticles and they found that the optimal gas sensing results were registered from SnO_2 nanowires. Indeed, gas detection reactions take place at surface-active sites through gas adsorption, charge transfer, and desorption. Consequently, the number of surface-active sites will influence the gas response, which is attributed to the improved oxygen adsorption due to the increase in surface-active sites. The increment of surface-active sites can be achieved by increasing the specific surface area. This causes more gas molecules to participate in the oxidation and reduction reaction.

Besides the morphology, the uniform size, reproducible shape, perfect crystal structure, homogeneous stoichiometry, etc. all can strongly affect the physical and chemical properties of MOS nanowire-based gas sensors. Therefore, the choice of their synthesis technique is very important. Overall, various methods have been employed for successfully producing MOS nanowires. They can be classified into two categories: direct physical deposition methods and direct chemical deposition methods. The first category is in line with a top-down approach, it includes thermal evaporation, molecular beam epitaxy, sputtering, electron beam lithography, template-based synthesis, and so forth, and it generally needs etching and lithography of bulk materials to form nanowires. Khai and co-workers [9] reported on the structural, optical, and gas sensing properties of vertically well-aligned ZnO nanowires grown on graphene/Si substrate by a thermal evaporation method. Also, Hieu and co-workers [10] employed a facile thermal evaporation method to produce large-area SnO_2 nanowires. Additionally, Zhang and co-workers [11] synthesized a ZnO nanowire array using plasma-assisted molecular beam epitaxy.

Regarding the second category, it is based on the bottom-up approach, and consists of chemical vapor deposition (CVD), aerosol-assisted CVD (AACVD), hydrothermal, solvothermal, sol-gel, etc. Herein, the methods are based on the assembly of chemically synthesized building blocks, such as atoms, molecules, and/or clusters. Kien and co-workers [12] fabricated low-temperature prototype H_2 sensors using Pd-decorated SnO_2 nanowires for exhaled breath applications. In this work, the authors synthesized SnO_2 nanowires using the CVD technique. Roso and co-workers [13] demonstrated the growth of ZnO nanowires over sapphire substrates in different orientations and with different gas sensing responses by CVD. Annanouch and co-workers [14] reported a localized growth of WO_3 nanowires into microelectromechanical system (MEMS)-based gas sensor substrate, using a cold-wall AACVD.

Two major drawbacks of MOS gas sensors can be cited: (1) they often operate at high temperatures (up to 400°C) to increase the carrier density available for the surface redox reaction, and (2) low selectivity, as reactive gaseous species will tend to consume/generate oxygen ions adsorbed in the MOS surface. Advances in nanotechnology allowed a significant reduction of the operating temperatures via the integration of these materials into MEMS platforms, allowing high-temperature operation and low-power consumption [15,16], fitting the fabrication requirements of portable gas monitoring devices. Behera and co-workers [17] made MEMS-based ZnO nanowires for acetone detection. The MEMS platforms incorporated an on-chip Ni microheater and an interdigitated electrode structure. Besides, the nanowires were obtained from thermal oxidation of sputtered Zn film. As a result, the sensor showed the ability to detect 5 ppm of acetone at a moderately low operating temperature of 100°C. Additionally, at 100 ppm of acetone, it exhibited a high sensitivity of 26.3% with a power consumption of only 36 mW. Recently, Hsueh and co-workers [18] presented vertical CuO nanowires/MEMS for a NO_2 gas sensor that is produced by sputtering. Compared to the literature, the fabricated sensor showed high sensitivity (50.1%) and very good stability, toward a low concentration of NO_2 (500 ppb) at a moderately low operating temperature of 119°C.

Another strategy to reduce the operating temperature of MOS nanowire-based gas sensors consisted of their modification with additives, doping, or their combination with 2D nanomaterials for instance reduced graphene nanosheets. Annanouch and co-workers [19] functionalized WO_3 nanowires with PdO nanoparticles to enhance their three S properties (sensitivity, selectivity, and stability) towards H_2, as well as reduce their operating temperature. The authors used the AACVD method to directly synthesize and deposit the material at the MEMS platform. The results indicated that by functionalizing WO_3 nanowires with PdO nanoparticles, the response of the sensor towards 500 ppm H_2 has been increased 755 times compared to bare WO_3 nanowires. Moreover, the working temperature has shifted from 250°C for pure WO_3 nanowires to 150°C for PdO/WO_3. Quin and co-workers [20] fabricated room temperature NO_2 gas sensors using Ti-added non-stoichiometric tungsten oxide nanowires. In this work, the $W_{18}O_{49}$ nanowires were synthesized via the solvothermal method while the Ti additives were introduced via two different processes: physical impregnation and chemical synthesis. They found that by adding Ti to $W_{18}O_{49}$ nanowire-based gas sensors, the working temperature decreased from 150°C to room temperature and the sensitivity toward NO_2 increased. Bang and co-workers [21] realized a low-temperature and selective NO_2 sensor based on Bi_2O_3-branched SnO_2 nanowires. The results demonstrated the superior gas sensing properties of a modified SnO_2 nanowire gas sensor relative to pristine and branched SnO_2 gas sensors, and it worked at near room temperature (50°C). Moreover, Anasthasiya and co-workers [22] demonstrated that rGO decorated with ZnO nanowires has a much higher and faster response towards NH_3 detection at room temperature than pure ZnO nanowires. rGO nanosheets decorated with Ni-doped ZnO nanowires were also tested for H_2 detection, exhibiting approximately 30% response operating at 150°C [23]. More recently, rGO decorated with Cu-doped ZnO nanocomposites was used to detect H_2S at room temperature in concentrations lower than 250 ppm [24].

Regarding the poor selectivity issue of MOS, many researchers have overcome this shortcoming by functionalizing their host matrix nanowires with metal/metal oxide nanoparticles. For instance, Pd/PdO or Pt nanoparticles were used to enhance the sensitivity as well as the selectivity of MOS nanowires towards H_2. Kim and co-workers [25] presented Pd functionalization on ZnO nanowires for enhanced sensitivity and selectivity to hydrogen gas. Also, Meng and co-workers [26] obtained an ultra-fast response and high selectivity hydrogen gas sensor based on Pd/SnO_2 nanoparticles. CuO is a good additive to enhance the sensitivity and the selectivity of the sensors against H_2S. Annanouch and co-workers [27] fabricated a selective H_2S gas sensor based on CuO nanoparticles decorated with WO_3 nanowires. Navarrete and co-workers [28] decorated WO_3 nanowires with NiO nanoparticles in order to monitor the presence of H_2S in the environment.

Another pathway to suppress the lack of selectivity is the development of multi-sensors arrays and computing techniques to access the sensing properties of each MOS nanowire sensing layer, the so-called electronic noses (e-noses). These e-noses have found potential applications in smart farming, smart cities, biomedical diagnosis, environmental monitoring, industrial safety management, etc. [29].

21.3   CONDUCTING POLYMER NANOWIRE-BASED GAS SENSORS

During the last few decades, conducting polymer nanowires have triggered the attention of many researchers and scientists thanks to their excellent electrical properties, environmental stability, flexibility, and their facile synthesis via chemical and electrochemical pathways [30-32]. Compared to metal oxide nanomaterials, conducting polymers can operate at room temperature and can be synthesized at the desired location, making them suitable candidates for real applications. In the literature, many polymers have been used for gas sensing, but the most studied in the engineering of 1D nanowires are polyaniline (PANi), polypyrrole (PPy), and poly(3,4-ethylene-deoxythiophene)-poly(styrene sulfonate) (PEDOT:PSS). Also, conducting polymer nanowires can be adjusted or copolymerized with different monomers as reported by Jiang and co-workers [33]. They fabricated a chemiresistive gas sensor array composed of well-ordered conducting polymer nanowires for a distinct separation between ten volatile organic compounds (VOCs), including ketones, alcohols, alkanes, aromatics, and amines. The nanowires were made from PEDOT:PSS using nanoscale soft lithography. After that, they have functionalized via a simple plasma-assisted functionalization to introduce four different functional groups such as 3-aminopropyl groups, 3-bromopropyl groups, perfluorooctyl groups, and octadecyl groups to produce four different sensors BPTS, PFDTS, APTES, and OTES, respectively. Next, a sensor array of bare and functionalized PEDOT:PSS nanowires was employed as an e-nose. From the gas sensing measurement results, all the sensors showed positive responses, at room temperature, towards several VOCs with concentrations ranging from 150 to 2000 ppm. For instance, the APTES functionalized PEDOT:PSS sensor showed better performances than those of thin film devices. The detection limits for this sensor for acetone, ethanol, hexane, p-xylene, and isopropamide were calculated to be 49.88, 47.03, 139.85, 95.32, and 59.08 ppm, respectively. Moreover, the sensor exhibited very fast responses; response time was ranging from 15–20 s and the recovery time was between 50–60 s. Herein, the interaction between VOCs and the fabricated polymer nanowires was ascribed to the variation in the swelling level.

Hassanzadeh and co-workers [34] employed PPy, which is highly reported for gas sensing applications. They fabricated a dense and elongated network of PPy nanowires via a chemical polymerization route using anodic alumina membranes with a pore diameter of 200 nm and thickness of 60 μm. To ensure the pore filling and the continuity of the nanowires, they optimized the polymerization time and applied an appropriate oxidant. According to their report, they found that the optimum polymerization time is 3 h, and the oxidant ($FeCl_3$): pyrrole molar ratio that gives the highest surface area is 1:1 $FeCl_3$: pyrrole with each reagent concentration of 0.2 M. To test the gas sensing performances of the PPy nanowires, the sensors were subjected to different concentrations of H_2 (1300–12000 ppm), at room temperature. The results revealed the interaction of the H_2 molecules with the polymer by a decrease in the electrical resistance of the PPy nanowires. Moreover, the sensors were able to differentiate between the different studied concentrations while at concentrations higher than 12000 ppm the sensors started to be saturated due to a decrease in the active sites at the polymer.

Hernandez and co-workers [35] synthesized PPy nanowires employing the same technique as reported in the previous work [34], but with a slight modification on the alumina membrane pores. They applied a thin layer of SiO_2 to the inner walls of the alumina pores before initiating the chemical polymerization of the pyrrole. As a result, the nanowires were very smooth at their surface compared to the ones obtained with unmodified alumina membrane pores. The authors related this difference to the presence of anionic sites on silica that played an important role in the preferential growth of PPy along pore walls and led to the continuous filling of the pores. The diameter of the nanowires was 300 nm while the length was between 50 to 60 μm. To fabricate the gas sensor, a single PPy nanowire was deposited between two gold microelectrodes, and a series of measurements were performed to study the semiconducting properties of the nanowire. The temperature-dependent current-voltage characteristic demonstrated the ohmic contact between the electrodes and the nanowire. Moreover, by increasing the temperature, the resistance of the PPy nanowire decrease, indicating its semiconducting behavior. Besides, the sensor was exposed to NH_3 (40–300 ppm) and after each exposure, the sensor recovered its baseline resistance by purging with argon. It was observed that by increasing the NH_3 concentration the sensor response increased, while after being exposed to high concentrations, the sensor was not able to reach its baseline resistance due to some irreversible binding between the NH_3 molecules and the PPy.

Al-Mashat and co-workers [36] have developed a room temperature H_2 gas sensor based on PPy nanowires directly deposited on gold interdigitated electrodes with a gap of 50 μm. The nanowires were obtained via a template-free electropolymerization of pyrrole monomer in a three-electrode electrochemical cell. After the analysis of the obtained samples with FESEM and X-ray photoelectron spectroscopy (XPS) techniques, the results confirmed the presence of a thick layer of PPy

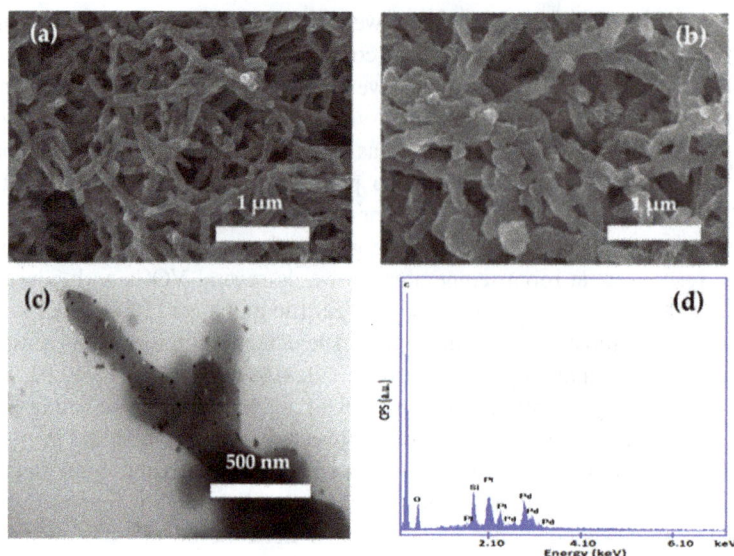

FIGURE 21.2   FESEM images of (a) PANi nanowires, (b) PANi/Pd, (c) TEM images of PANi/Pd, (d) typical
EDS spectrum of PANi/Pd. Adapted with permission [37]. Copyright 2017, Elsevier.

nanowires with a diameter ranging between 40 to 90 nm. Next, the fabricated sensors were exposed
to various H_2 concentrations starting from 600 to 10000 ppm at room temperature. The gas sensors
showed different baseline resistances of 13287, 562, and 9430 Ω, which were related to the number
of nanowires that crossed the interdigitated electrodes. It was reported that the optimization of the
electropolymerization conditions yields uniform nanowire coverage. Moreover, the sensors were
able to differentiate between all the studied H_2 concentrations, showing a decrease in the resistance upon exposure to H_2 and an increase in the resistance upon exposure to the air. Besides, the
calculated detection limit was found to be 12 ppm. In addition, the sensors showed very fast and
reversible responses with a response time of 54 s towards 10000 ppm of H_2.

Hien and co-workers [37] elaborated PANi nanowires decorated with Pd-nanoparticle for room
temperature NH_3 gas sensors. In this study, the PANi nanowires were directly electrodeposited on
interdigitated Pt electrodes via the cyclic voltammetry technique. Besides, the Pd nanoparticles were
added to the host matrix through the drop-casting method. The morphology, structure, and composition of the obtained nanowires were analyzed, and the results revealed the presence of a thick
layer of long nanowires with a diameter ranging from 50 to 100 nm decorated with well-dispersed
Pd nanoparticles (**Figure 21.2**). Six sensors were fabricated by using different Pd content (0, 0.1,
0.5, 1.0, 2.0, and 3.0 wt.%) and were tested at room temperature against various concentrations of
NH_3: 10, 25, 50, 100, 250, 500 ppm. The results revealed that sensors with 2.0 wt.% of Pd exhibited
the highest response of 21.9 towards 500 ppm of NH_3, which is eight times higher than bare PANi
nanowires. Moreover, the authors have discussed the gas sensing mechanism and they proposed that
the increase in the gas sensing performances could be due to the π-electron transfer from the PANi
nanowires to the Pd nanoparticles, which produce different work function values of PANi nanowires
and the Pd nanoparticles.

21.4   METALLIC NANOWIRE-BASED GAS SENSORS

Metals such as Pd, Pt, Au, Ag, and so forth, have been widely employed as functional additives,
to promote the gas sensing properties of semiconducting nanostructures, via chemical and/or electronic sensitization. However, their direct use as an independent sensing element was less explored,

despite their many advantages for instance: fast detection time, reversibility, room temperature operation, etc. In this section, we report on the most recent works related to metallic nanowires: their
synthesis and their gas sensing results.

Shobin and co-workers [38] reported Ag nanowire-based fiber optic sensors for room temperature
NH_3 vapor detection. The nanowires were prepared via a polyol process followed by a purification
step through a 300 nm pore membrane vacuum filter. The measured average diameter and length of
the nanowires were 60 nm and 7 μm, respectively. To fabricate the gas sensor, the nanowires were
dispersed in ethanol and coated to the cladding modified fiber via the deep coating technique. The
authors found that for concentrations ranging from 0 to 500 ppm of NH_3 with a step of 50 ppm,
the best response was obtained at a wavelength of 706 nm, at room temperature. Additionally, at
500 ppm of NH_3, the response and recovery times were 70 and 100 min, respectively, which could
be improved by using a carrier gas, such as nitrogen or argon. Tran and co-workers [39] have used
Ag nanowires as small bridges between graphene nanosheets to enhance their electrical properties
towards NH_3. In this work, the nanowires were synthesized similarly to [38] and subsequently they
were spray-coated to the deposited graphene nanosheets, onto the sensor substrate. Contingent on
the obtained results, the sensitivity of Ag-graphene composite, towards NH_3, was eight times higher
than that of pure graphene nanosheets, at room temperature, which highlights the important role of
the Ag nanowires.

According to the literature, metallic nanowires were typically produced through a top-down
approach, resulting in thick nanowires with diameters ranging from 20 nm to several hundreds of
nanometers. To obtain ultrathin nanowires (diameter near 5 nm) with properties (i.e., high surface-
to-volume ratio, quantum confinement effect, high crystallinity) analogous to those of ideal building
blocks materials, a bottom-up chemical approach is an excellent pathway to synthesize such
materials. However, some challenges discouraged researchers from pursuing this approach, such
as the need for strong binding ligands that impede charge transfer within the nanowire networks.
Ding and co-workers [40] reported on highly sensitive chemical detection with tunable sensitivity
and selectivity from ultrathin Pt nanowires. Herein, the authors used wet chemical synthesis to
obtain Pt nanowires, which do not require a strong binding ligand. As a result, the nanowires have a
diameter of 3 nm and were easily transferred to the sensor substrate via a drop-casting method. At
0.5% H_2 concentration, the response of the studied Pt nanowires was two times higher than the ones
prepared via a top-down method. Moreover, the authors studied the effect of the nanowire diameter on the sensor response. The results showed that when the diameter of the nanowire approached
its mean free path, the conductivity increases, and it shows remarkable gas sensing performances.
Therefore, at 0.5% H_2 concentration, the fabricated sensors showed a response of 261.4%, which is
much higher than the maximum response (3.5%) observed from larger nanowires. Kumar and co-
workers [41] fabricated ultrathin Pd nanowires with a diameter ≤ 5 nm, via colloidal solvothermal
synthesis (**Figure 21.3**). Before preparing a sensor, they applied an ultraviolet ozone treatment to
the as-synthesized Pd nanowires, to remove the binding ligands attached to their surface. This helps
to expose the active sites of the Pd nanowires to H_2 molecules, makes their dissociation very fast,
and decreases the resistance between narrow nanowires. Regarding gas sensing results, the sensors
showed a response of 1.7 towards 1% of H_2 in the air at room temperature, with very fast response
and recovery times of 3.4 and 11 s, respectively. Furthermore, by exposing them to decreasing
H_2 concentrations (from 1 to 0.02%) followed by increasing H_2 concentrations (0.02 to 1%), the
sensors exhibited hysteresis-free behavior. It is worth noting that hysteresis in the sensor response is
a shortcoming that can arise from mechanical deformation in the Pd lattice upon fast adsorption and
desorption of the H_2 molecules. Therefore, the authors related the high gas sensing performances
observed from Pd nanowires to their ultrathin diameter, their good interconnectivity, and the catalytic facet enriched surface.

A biological template-assembly method is an alternative route for synthesizing metals in the
form of nanowires. Moon and co-workers [42] used a viral template assembly to fabricate room

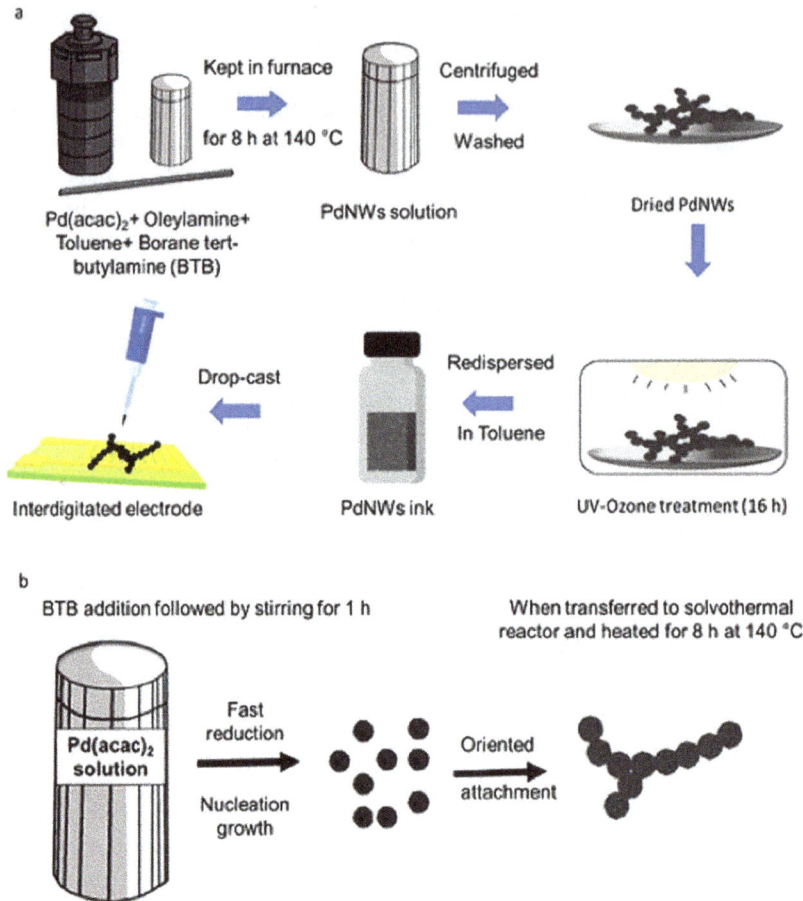

FIGURE 21.3 (a) Schematic synthesis of ultrathin Pd nanowires and fabrication of an H_2 sensor from them; (b) synthesis mechanism of ultrathin Pd nanowires. Adapted with permission from Reference [41], Copyright 2022, American Chemical Society.

temperature gold nanowire-based H_2S gas sensor. To this end, the authors employed gold binding M13 bacteriophage to assembly gold nanoparticles into a linear shape, to serve as seeds for subsequent nanowire formation using an electroless technique. The width, the connectivity, and the surface coverage of Au nanowires were adjusted by varying the phage concentration as well as the electroless deposition time. It was found that an increase in these two parameters resulted in a decrease in the sensor resistance since the yield and the number of nanowires that bridge the electrodes are increasing. The sensors were tested against H_2S at room temperature. The studied concentrations were in the range of 25 to 500 ppb. As a result, the sensor resistance increased upon exposure to the target gas, while it decreased upon purge with air, which is consistent with previously reported works. This increase in the resistance was produced by the hindrance of the charge transport between the nearby nanowires caused by the adsorption of H_2S molecules at the Au surface. Moreover, the sensors showed excellent gas sensing performance at room temperature; the sensitivity was 654%/ppm, the theoretical lowest detection limit of 2 ppb, and 70% recovery within 9 min for 0.025 ppm. By eliminating the biological template, the sensors were tested under the same conditions as before and a decrease in the sensor performances was observed, confirming the importance of the biological template in the gas sensing properties. Al-Hinai and co-workers [43] also reported DNA-templated Pd nanowires and their use as H_2 gas sensors. They found that the nanowires were highly resistive compared to bulk Pd metal since they are an assembly of small Pd grains with a diameter ≤ 2 nm. Regardless of this high resistivity, Pd nanowires were able to detect H_2 with enough sensitivity at room temperature.

21.5 SILICON NANOWIRE-BASED GAS SENSORS

Silicon (Si) nanowire-based sensors demonstrate a substantial advantage owing to their standard processing techniques and their compatibility with conventional complementary MOS (CMOS) technologies [44]. Leiber and co-workers reported amine- and oxide-functionalized Si nanowire sensors [45]. Due to this pioneering study, Si nanowires of various dimensions have gained increased interest as gas sensors, and several synthesis techniques have been suggested to enhance device functionality and sensitivity to a particular gas.

Also, Si nanowire-based sensors show excellent gas sensing properties at room temperature [46]. Song and co-workers [46] demonstrated a step-guided in-plane solid-liquid-solid (IPSLS) growth of orderly Si nanowire array-based (<80 nm) field effect sensor for detection of NH_3 using In as the catalyst. The research group adopted the IPSLS technique to overcome the challenges reported with the Si nanowire sensors fabricated by electron beam lithography (EBL), patterning, or vapor-liquid-solid (VLS) technique. The resistor-type gas sensors fabricated by the VLS technique are usually bulky and suffer from limited sensitivity due to a lack of amplification function when compared to field-effect transistor (FET) sensors. On the other hand, sensors fabricated via EBL are expensive, which hinders their scalability. The Si nanowire-FET sensors demonstrated by the group can be turned into a subthreshold detection regime to achieve field-effect sensitivity, selectivity to NH_3 at room temperature, and fast response/recovery time scales (T_{res}/T_{rec}) of 20 s for 100 ppb NH_3. The sensors show excellent repeatability and show stability over 180 days, thus promising scalable and flexible sensors for flexible, wearable electronics, and the IoT. The incorporation of the In catalyst ion atoms into Si nanowires gives rise to effective p-type doping of the Si nanowires. This arises p-type FET transfer properties of Si nanowires. When exposed to NH_3, the charge transfer properties of the Si nanowire-FET sensors shift negatively with respect to the initial state in the air. This indicates that there is a gradual accumulation of positive charge on the sidewall of Si nanowire channels on the exposed channel surface with an increase in the NH_3 adsorption. The NH_3 adsorption on the p-type Si nanowire surface donates electrons to the Si nanowire channels, leaving the positive charges on the sidewall, causing the transfer curve to shift negatively and suppressed channel currents. As a result, the Si nanowire-FET will experience a negative shift in the transfer curve and lower/suppressed channel currents, especially in the desired subthreshold region. This is similar to imposing an additional positive gating. This mechanism was reported by Gao and co-workers [47]. Kondratev and co-workers [48] also reported a Si nanowire-based sensor for NH_3 sensing fabricated via simple drop-casting of Si nanowires over the substrate with interdigitated Au electrodes. The group employed electrochemical impedance spectroscopy (EIS) to acquire impedance spectra over a broad range of frequencies. The nanowires are 10–12 µm long and 350–400 nm thick at the base and monotonously tapering towards the tip. The sensors showed a decrease in the resistance when exposed to water and NH_3 vapors. This is due to the injection of electrons in the nanowires from the adsorbates. It is found that the Schottky barrier between metal pads and Si nanowires governs the sensor resistance rather than the bulk conductivity of the nanowires. The impedance of the sensors exposed to the NH_3 aqueous solution was investigated to demonstrate the response of the increase in the NH_3 content. It was found that the electrolyte conductivity controls the resistance of the sensors. The sensors can detect small NH_3 contents of up to 8 ppm with a detection limit of 80 ppb and sensitivity of 0.1% ppb^{-1} in liquid solutions both in direct and indirect manners. Qin and co-workers [49] fabricated Si nanowire-based sensors for ethanol detection by constructing densely arranged particles of zeolite imidazole frameworks-8 (ZIF-8) and nano-Ag on the surface of Si nanowires.

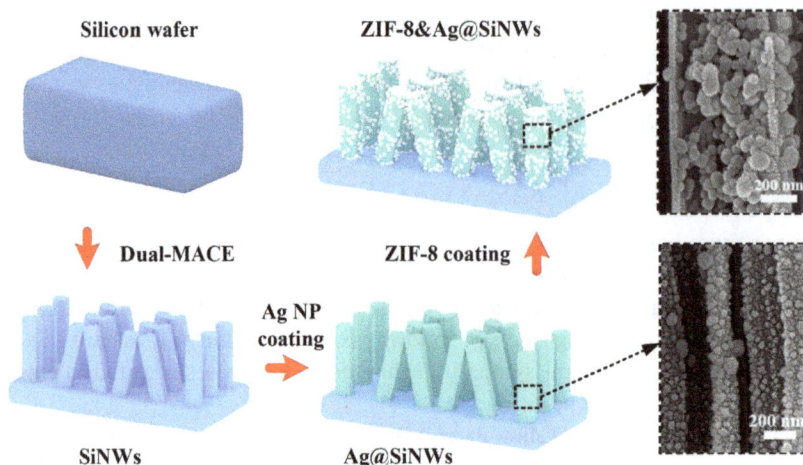

FIGURE 21.4 Illustration of the fabrication process of ZIF-8, Ag@Si nanowires, and ZIF-8&Ag@Si nanowires. Adapted with permission [49]. Copyright 2021, Elsevier.

The Si nanowires were fabricated by the dual metal-assisted chemical etching (dual-MACE) technique. The process is schematically illustrated in **Figure 21.4**. The ZIF-8 particles are spherical with a mean particle size of about 85 nm. The bare Si nanowires have a length of about 50 μm. The sensors work by combining Si nanowires and ZIF-8 with heterojunction modulation, which can lower ZIF-8 working temperature while improving the response of Si nanowires to VOCs. The introduction of nano-Ag modification leads to the formation of p-n heterojunctions between Si nanowires and p-type ZIF-8. The sensors respond highly to 125 ppb of ethanol at 20°C under 25% relative humidity. The sensors can detect ultra-low concentrations of ethanol (5.45 ppb) with excellent stability in humid conditions. When the sensors are exposed to ethanol, the resistances of the sensors decrease. This is inconsistent with the response feature of p-type semiconductor sensors and results from the Ag modification, which alters the surface energy band structure of Si nanowires.

Shehada and co-workers [50] reported an ultra-sensitive, molecularly modified Si nanowire FET for gastric cancer diagnosis from exhaled volatolome. Two main methods to enhance the selectivity and sensitivity of the Si nanowire-based gas sensors for VOC detection are lock-and-key recognition, and the second method uses cross-reactive sensor arrays to impart varying affinities from different combinations of gaseous species and accomplish the necessary selectivity using sophisticated pattern recognition algorithms. The sensors reported by Shehada and co-workers [50] are a hybrid of the two methods, which demonstrate the selectivity of the lock-and-key approach to the ability to cross-reactive sensor arrays for complex fluids. The suggested method is based on a single Si nanowire FET that has been molecularly modified. This Si nanowire FET provides a variety of independent properties, each of which reacts differently to various VOCs at the low ppb concentration level. However, the combined output of these features can be processed using straightforward pattern recognition techniques to allow for the recognition of intricate gas mixtures. Since it is based on a single sensor, this method ultimately needs only minimal hardware, which enhances the device's potential to be miniaturized, makes it easier to operate, and makes it easier to integrate into existing Si-based technologies. For the very sensitive detection of gastric cancer VOCs in both laboratory and actual clinical situations, a molecularly modified Si nanowire FET coated with trichloro-(phenethyl)silane (TPS) was fabricated. The sensor has a detection limit down to 5 ppb. This sensor was selective to VOCs linked with gastric cancer in exhaled breath and to distinguish them from environmental VOCs. Only gastric cancer-specific biomarkers, such as 2-propenenitrile, 6-methyl-5-heptene-2-one, and furfuraldehyde, are detected by the sensor; VOCs unrelated to cancer, such as 2-ethyl-1-hexanol and nonanal, do not trigger a response. The sensor's 87% sensitivity, 81% specificity, and 83% accuracy allowed it to discriminate between patients suffering from stomach cancer and healthy controls. More importantly, the sensor's ability to distinguish between early-stage (I–II) and advanced-stage (III–IV) cancer with a sensitivity of 92% and specificity of 67% suggests that it could be used to diagnose diseases at their early stages.

21.6 CONCLUSIONS

A significant amount of work has been carried out using nanowires for gas sensing applications. Here we have provided an updated account of these research efforts. Therefore, first, we showed an overview regarding metal oxide nanowire-based gas sensors, second, we discussed conducting polymer nanowire-based gas sensors, third we reported metallic nanowire-based gas sensors, and finally, we presented silicon nanowire-based gas sensors. We have discussed the synthesis method, effect of the nanowire diameter on the sensor properties, gas sensing performance, and mechanisms in the above-mentioned sections.

REFERENCES

[1] X. Liu, S. Cheng, H. Liu, S. Hu, D. Zhang, H. Ning, A survey on gas sensing technology, Sensors. 12 (2012) 9635–9665.

[2] Gas Sensor Market Size, Share & Trends Analysis Report By Product, By Type, By Technology, By End Use, By Region, And Segment Forecasts, Report ID: 978-1-68038-083-5.

[3] R.S. Devan, R.A. Patil, J. Lin, Y. Ma, One-dimensional metal-oxide nanostructures: recent developments in synthesis, characterization, and applications, Adv. Funct. Mater. 22 (2012) 3326–3370.

[4] S. Vallejos, P. Umek, T. Stoycheva, F. Annanouch, E. Llobet, X. Correig, P. De Marco, C. Bittencourt, C. Blackman, Single-step deposition of au- and pt-nanoparticle-functionalized tungsten oxide nanoneedles synthesized via aerosol-assisted CVD, and used for fabrication of selective gas microsensor arrays, Adv. Funct. Mater. 23 (2013) 1313–1322.

[5] A. Dey, Semiconductor metal oxide gas sensors: A review, Mater. Sci. Eng. B. 229 (2018) 206–217.

[6] K. Grossmann, U. Weimar, N. Barsan, Semiconducting metal oxides based gas sensors, in: Semicond. Semimetals, Elsevier, 2013: pp. 261–282.

[7] T. Stoycheva, F.E. Annanouch, I. Gràcia, E. Llobet, C. Blackman, X. Correig, S. Vallejos, Micromachined gas sensors based on tungsten oxide nanoneedles directly integrated via aerosol assisted CVD, Sens. Act. B 198 (2014) 210–218.

[8] N.M. Shaalan, T. Yamazaki, T. Kikuta, Influence of morphology and structure geometry on NO_2 gas-sensing characteristics of SnO_2 nanostructures synthesized via a thermal evaporation method, Sens. Act. B 153 (2011) 11–16.

[9] T. Van Khai, V.M. Thanh, T. Dai Lam, Structural, optical and gas sensing properties of vertically well-aligned ZnO nanowires grown on graphene/Si substrate by thermal evaporation method, Mater. Charact. 141 (2018) 296–317.

[10] N. Van Hieu, N.D. Khoang, N.T. Minh, T. Trung, N.D. Chien, A facile thermal evaporation route for large-area synthesis of tin oxide nanowires: characterizations and their use for liquid petroleum gas sensor, Curr. Appl. Phys. 10 (2010) 636–641.

[11] Z.-Y. Zheng, T.-X. Chen, L. Cao, Y.-Y. Han, F.-Q. Xu, Structure and optical properties of ZnO nanowire arrays grown by plasma-assisted molecular beam epitaxy, Wuji Cailiao Xuebao, J. Inorg. Mater. 27 (2012) 301–304.

[12] K. Nguyen, C.M. Hung, T.M. Ngoc, D.T.T. Le, D.H. Nguyen, D.N. Van, H.N. Van, Low-temperature prototype hydrogen sensors using Pd-decorated SnO_2 nanowires for exhaled breath applications, Sens. Act. B 253 (2017) 156–163.

[13] S. Roso, F. Güell, P. R. Martínez-Alanis, A. Urakawa, E. Llobet, Synthesis of ZnO nanowires and impacts of their orientation and defects on gas sensing properties, Sens. Act. B 230 (2016) 109–114 .

[14] F.E. Annanouch, I. Gràcia, E. Figueras, E. Llobet, C. Cané, S. Vallejos, Localized aerosol-assisted CVD of nanomaterials for the fabrication of monolithic gas sensor microarrays, Sens. Act. B 216 (2015) 374–383.

[15] J. Zhu, X. Liu, Q. Shi, T. He, Z. Sun, X. Guo, W. Liu, O. Bin Sulaiman, B. Dong, C. Lee, Development trends and perspectives of future sensors and MEMS/NEMS, Micromach. 11 (2019) 7.

[16] M. Mahdi, Ultra-low power MEMS micro-heater device, Microsyst. Technol. 27 (2021) 2913–2917.

[17] B. Behera, S. Chandra, A MEMS based acetone sensor incorporating ZnO nanowires synthesized by wet oxidation of Zn film, J. Micromech. Microengin. 25 (2014) 15007.

[18] T.-J. Hsueh, P.-S. Li, S.-Y. Fang, C.-L. Hsu, A vertical CuO-NWS/MEMS NO_2 gas sensor that is produced by sputtering, Sens. Act. B 355 (2022) 131260.

[19] F.E. Annanouch, Z. Haddi, M. Ling, F. Di Maggio, S. Vallejos, T. Vilic, Y. Zhu, T. Shujah, P. Umek, C. Bittencourt, C. Blackman, E. Llobet, Aerosol-assisted CVD-grown PdO nanoparticle-decorated tungsten oxide nanoneedles extremely sensitive and selective to hydrogen, ACS Appl. Mater. Inter. 8 (2016) 10413–10421.

[20] Y. Qin, X. Sun, X. Li, M. Hu, Room temperature NO_2-sensing properties of Ti-added nonstoichiometric tungsten oxide nanowires, Sens. Act. B 162 (2012) 244–250.

[21] J.H. Bang, A. Mirzaei, S. Han, H.Y. Lee, K.Y. Shin, S.S. Kim, H.W. Kim, Realization of low-temperature and selective NO_2 sensing of SnO_2 nanowires via synergistic effects of Pt decoration and Bi_2O_3 branching, Ceram. Int. 47 (2021) 5099–5111.

[22] R.K. Kampara, P.K. Rai, B.G. Jeyaprakash, Highly sensitive graphene oxide functionalized ZnO nanowires for ammonia vapour detection at ambient temperature, Sens. Act. B 255 (2018) 1064–1071.

[23] H. Abdollahi, M. Samkan, M.M. Hashemi, Fabrication of rGO nano-sheets wrapped on Ni doped ZnO nanowire p–n heterostructures for hydrogen gas sensing, New J. Chem. 43 (2019) 19253–19264.

[24] P.S. Shewale, K.-S. Yun, Synthesis and characterization of Cu-doped ZnO/RGO nanocomposites for room-temperature H2S gas sensor, J. Allo. Comp. 837 (2020) 155527.

[25] J.-H. Kim, A. Mirzaei, H.W. Kim, S.S. Kim, Pd functionalization on ZnO nanowires for enhanced sensitivity and selectivity to hydrogen gas, Sens. Act. B 297 (2019) 126693.

[26] X. Meng, M. Bi, Q. Xiao, W. Gao, Ultra-fast response and highly selectivity hydrogen gas sensor based on Pd/SnO_2 nanoparticles, Int. J. Hydro. Energy. 47 (2022) 3157–3169.

[27] F.E. Annanouch, Z. Haddi, S. Vallejos, P. Umek, P. Guttmann, C. Bittencourt, E. Llobet, Aerosol-assisted CVD-grown WO_3 nanoneedles decorated with copper oxide nanoparticles for the selective and humidity-resilient detection of H2S, ACS Appl. Mater. Inter. 7 (2015) 6842–6851.

[28] E. Navarrete, C. Bittencourt, P. Umek, E. Llobet, AACVD and gas sensing properties of nickel oxide nanoparticle decorated tungsten oxide nanowires, J. Mater. Chem. C. 6 (2018) 5181–5192.

[29] S. Jeong, J. Kim, J. Lee, Rational design of semiconductor-based chemiresistors and their libraries for next-generation artificial olfaction, Adv. Mater. 32 (2020) 2002075.

[30] X. Chen, C.K.Y. Wong, C.A. Yuan, G. Zhang, Nanowire-based gas sensors, Sens. Act. B 177 (2013) 178–195.

[31] Fratoddi, I. Venditti, C. Cametti, M.V. Russo, Chemiresistive polyaniline-based gas sensors: A mini review, Sens. Act. B 220 (2015) 534–548.

[32] E. Song, J.-W. Choi, Conducting polyaniline nanowire and its applications in chemiresistive sensing, Nanomat. 3 (2013) 498–523.

[33] Y. Jiang, N. Tang, C. Zhou, Z. Han, H. Qu, X. Duan, A chemiresistive sensor array from conductive polymer nanowires fabricated by nanoscale soft lithography, Nanosc. 10 (2018) 20578–20586.

[34] N. Hassanzadeh, H. Omidvar, S.H. Tabaian, Chemical synthesis of high density and long polypyrrole nanowire arrays using alumina membrane and their hydrogen sensing properties, Superlat. Microstruct. 51 (2012) 314–323.

[35] S.C. Hernandez, D. Chaudhuri, W. Chen, N.V. Myung, A. Mulchandani, Single polypyrrole nanowire ammonia gas sensor, Electroanal. An Int. J. Devoted to Fundam. Pract. Asp. Electroanal. 19 (2007) 2125–2130.

[36] L. Al-Mashat, C. Debiemme-Chouvy, S. Borensztajn, W. Wlodarski, Electropolymerized polypyrrole nanowires for hydrogen gas sensing, J. Phys. Chem. C. 116 (2012) 13388–13394.

[37] H.T. Hien, H.T. Giang, N. Van Hieu, T. Trung, C. Van Tuan, Elaboration of Pd-nanoparticle decorated polyaniline films for room temperature NH_3 gas sensors, Sens. Act. B 249 (2017) 348–356.

[38] L.R. Shobin, D. Sastikumar, S. Manivannan, Glycerol mediated synthesis of silver nanowires for room temperature ammonia vapor sensing, Sens. Act. A 214 (2014) 74–80.

[39] Q.T. Tran, T.M.H. Huynh, D.T. Tong, N.D. Nguyen, Synthesis and application of graphene–silver nanowires composite for ammonia gas sensing, Adv. Nat. Sci. Nanosci. Nanotechnol. 4 (2013) 45012.

[40] M. Ding, Y. Liu, G. Wang, Z. Zhao, A. Yin, Q. He, Y. Huang, X. Duan, Highly sensitive chemical detection with tunable sensitivity and selectivity from ultrathin platinum nanowires, Small 13 (2017) 1602969.

[41] A. Kumar, T. Thundat, M.T. Swihart, Ultrathin palladium nanowires for fast and hysteresis-free H_2 sensing, ACS Appl. Nano Mater. 5 (2022) 5895–5905.

[42] C.H. Moon, M. Zhang, N. V Myung, E.D. Haberer, Highly sensitive hydrogen sulfide (H_2S) gas sensors from viral-templated nanocrystalline gold nanowires, Nanotech. 25 (2014) 135205.

[43] M.N. Al-Hinai, R. Hassanien, N.G. Wright, A.B. Horsfall, A. Houlton, B.R. Horrocks, Networks of DNA-templated palladium nanowires: structural and electrical characterisation and their use as hydrogen gas sensors, Farad. Discuss. 164 (2013) 71–91.

[44] A. Mukherjee, Y. Rosenwaks, Recent advances in silicon FET devices for gas and volatile organic compound sensing, Chemosensors. 9 (2021) 260.

[45] Y. Cui, Q. Wei, H. Park, C.M. Lieber, Nanowire nanosensors for highly sensitive and selective detection of biological and chemical species, Science 293 (2001) 1289–1292.

[46] X. Song, R. Hu, S. Xu, Z. Liu, J. Wang, Y. Shi, J. Xu, K. Chen, L. Yu, Highly sensitive ammonia gas detection at room temperature by integratable silicon nanowire field-effect sensors, ACS Appl. Mater. Inter. 13 (2021) 14377–14384.

[47] X.P.A. Gao, G. Zheng, C.M. Lieber, Subthreshold regime has the optimal sensitivity for nanowire FET biosensors, Nano Lett. 10 (2010) 547–552.

[48] V.M. Kondratev, I.A. Morozov, E.A. Vyacheslavova, D.A. Kirilenko, A. Kuznetsov, S.A. Kadinskaya, S.S. Nalimova, V.A. Moshnikov, A.S. Gudovskikh, A.D. Bolshakov, Silicon nanowire-based room-temperature multi-environment ammonia detection, ACS Appl. Nano Mater. (2022) 9940–9949.

[49] Y. Qin, X. Wang, J. Zang, Ultrasensitive ethanol sensor based on nano-Ag&ZIF-8 co-modified Si NWs with enhanced moisture resistance, Sens. Act. B 340 (2021) 129959.

[50] N. Shehada, G. Bronstrup, K. Funka, S. Christiansen, M. Leja, H. Haick, Ultrasensitive silicon nanowire for real-world gas sensing: noninvasive diagnosis of cancer from breath volatolome, Nano Lett. 15 (2015) 1288–1295.

22 Metal Oxides Wired on Carbon Nanofibers for Sensors to Detect Biomolecules

Ziyu Yin and Jianjun Wei

The Department of Nanoscience, Joint School of Nanoscience and Nanoengineering, University of North Carolina at Greensboro, Greensboro, North Carolina, USA

CONTENTS

22.1 INTRODUCTION

22.1.1 FLEXIBLE SENSORS

As the demand for personal healthcare and human activity tracking, wearable flexible devices in medical and industrial markets have garnered substantial expansion. Solid data are indicating that wearable sensor users are close to 900 million by the end of 2021; furthermore, the size of the expanded market is approximately $75 billion in 2025 [1,2]. Thanks to tremendous achievements of materials science, nanotechnology, artificial intelligence, and the Internet-of-Things (IoT), a promising era where wearable flexible devices are blended into every aspects of our daily life has been adventured [3,4]. Flexible sensors with exceptional features, such as lightweight, good compatibility, great stretchability, and flexibility, have been favored in developing the next generation of healthcare devices by which real-time tracking information on the state of the human body or surrounding environment could be realized [5]. Generally, flexible sensors, consisting of functional material-based sensing modules and flexible supporting substrates, which exhibit excellent deformable capability, are capable of detecting stimuli from biophysical/biochemical signals and/or subtle movements with conspicuous current/voltage signals. The sensors generate attainable conformal intimacy with the human-machine interface and can be stretched, compressed, and twisted but cause a negligible impact on their functionalities [6]. Notwithstanding, the commercial applications of flexible sensors that can meet great characteristics and functionalities in mass production are still in their infancy. Additionally, the state of human physiological conditions and the surrounding environment is in dynamic kinetics, hence multifunctional flexible sensors towards reliable target analysis would give a new revenue to a forthcoming era of omnipotent sensors. Apart from multifunctionalities, sensing components play vital roles in sensitivity, flexibility, and compact size as well. To achieve attainable flexible sensors with ideal properties and functions, unveiling the sensing mechanism of those materials would lay the foundation of material and structural design for eliminating current obstacles of traditional sensing components, making breakthroughs in new design principles and engineering methods [7].

22.1.2 Nanowires

Thanks to the great advancements of nanofabrication and nanomaterials' synthesis technology, nanowires (NWs) have been widely exploited in flexible sensors and shown great potentiality in either working as various intelligent terminals in sensing modules or offering plentiful morphologies and combined forms. Owing to their intrinsic high-aspect-ratio structural features and quantum size effects, they are prone to a high possibility of surface functionalization for excellent sensing performance [8]. Besides, superior mechanical tolerance with great flexibility and stretchability has endowed them to construct suitable modules to prevent crack delamination. In virtue of those merits, a rational design of NWs as sensing components could greatly be attributed to the sensing performance of flexible sensors. Besides, a synergy of material composition and structural architecture would play a key role in the performance of NW-based sensors as well. To shed light on their excellent applications in flexible sensors, sundry NWs with prominent geometric structures, interconnections, and smart integration, have been exploited in biophysical/chemical sensors with conspicuous performance [9,10]. Carbon nanofiber-based network, as one of the primary and basic prototypes of simple NWs, comprised of substance with homogeneous composition, endows great potentiality in integrating flexible sensors with features of high volume-to-surface ratio, chemical friendly surface functionalization [8,11]. The large specific areas offer the great ability to release and attach functional groups, to absorb molecules, ions, catalytic parts, and nanoparticles, rendering enhancement of signals for detection of target objects and external stimuli. Besides, integration of ordered nanostructures with regular alignment by constructing interconnections between NWs would not only maintain their intrinsic superiorities of characteristics but also generate spectacular mechanical and electrical properties [12]. In contrast to the random arrangement and orientation of NWs with easy fluctuations of sensing properties, the well-ordered and aligned arrangement of NWs as functional components in sensors enables high electron transportation capability and great growth repeatability, thus offering high spatial resolution, rapid response, reliable and excellent device performance [13,14].

22.1.3 Fabrication of WA-ECNFs and Surface Functionalization

Generally, approaches to synthesizing NWs are highly categorized into a top-down method and a bottom-up approach [15]. To achieve mass-production with high-quality NWs at low cost and high efficiency, electrospinning, as one of the methods belonging to the bottom-up approach, uses electric force to draw charged polymer threads or melt polymer into nanofibers with subsequent carbonization, which produces large-scale materials with defined surface area and morphology as well [16]. It is a simple, efficient, and scalable method, enabling a smart, advanced design for various geometric morphology of NW networks. According to our previous research [17–24], a customized substrate collector was homemade by a cylinder welded with four steel poles to realize the well-aligned fibers under a specific rotating rate during the electrospinning (**Figure 22.1A**). Following subsequent stabilization and carbonization, the well-aligned electrospun carbon nanofibers (WA-ECNFs)

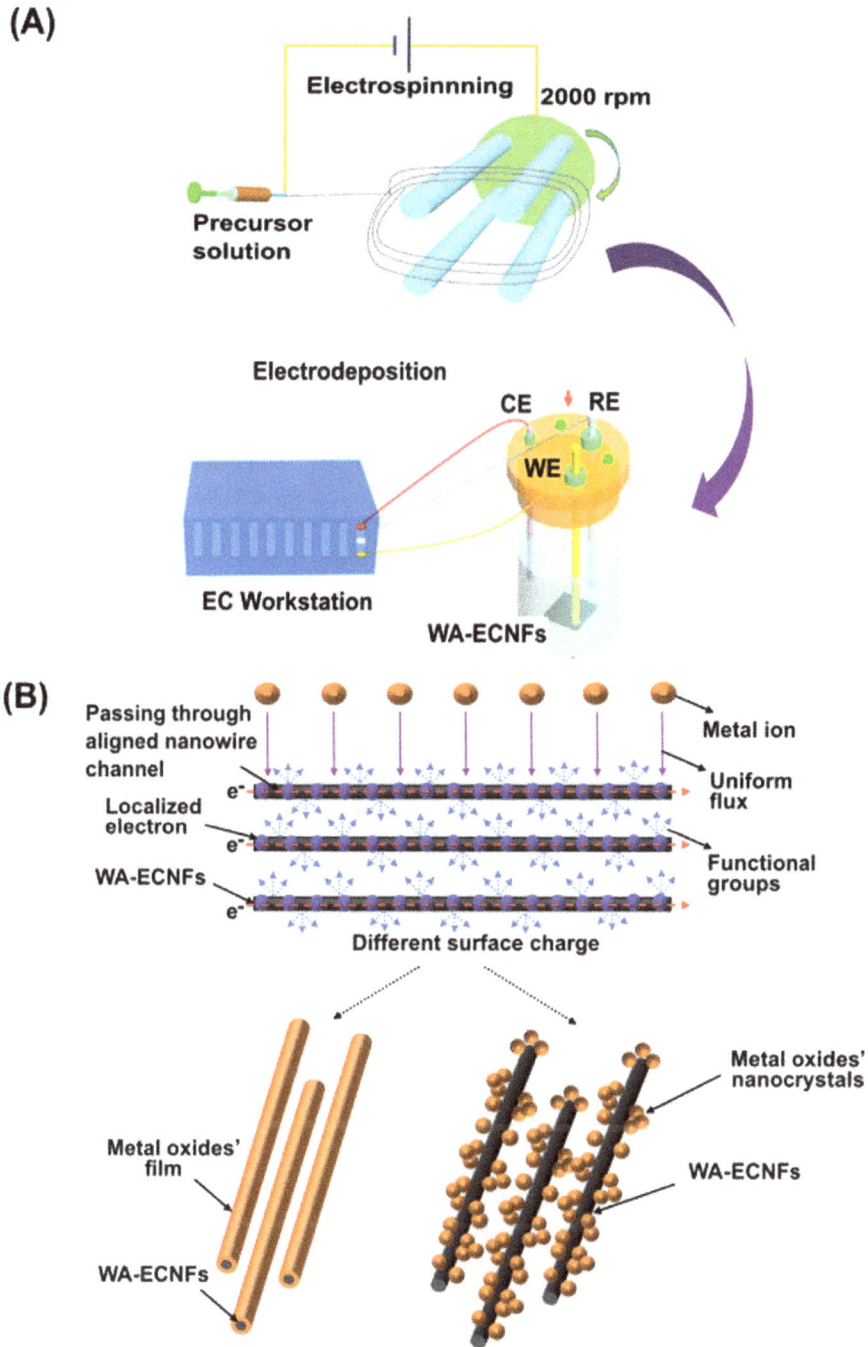

FIGURE 22.1 (A) Schematic illustration of electrospinning technique for the synthesis of WA-ECNFs followed with the low-current electrodeposition of metal oxides; (B) schematic depiction of nucleation and growth of metal oxides on the WA-ECNFs via metal ion uniform flux on the well-aligned orientation NWs channel, as well as surface functionalization for different structures and morphology of functional nanomaterials coated on NWs.

with high mechanical strength were obtained. Those well-aligned arrangements and interconnected structures can promote charges to transport and provide direct pathways along the fibers, facilitating rapid response characteristics for sensing. For instance, Han et al. [12], assembled highly aligned carbon NWs under an electric field for NO_2 sensing with superb sensitivity, owing to the great transportation of hole carriers in the aligned carbon NWs channels. Furthermore, the sensing performance can be further optimized by integrating with chemically modified functional moieties, including metal/metal oxide nanomaterials.

Inspired by both theoretical and experimental studies of revealed tunable electronic structures that affect catalytic performance in the presence of heteroatom dopants in carbon frameworks [25], surface functionalization chemistry has been extensively applied to carbon-based nanomaterials by doping heteroatom elements, e.g., N, O, P, S, and B [26]. The promoted electrocatalytic activity of those heteroatom-doped carbon-based materials could be derived from changes in charge and spin density of carbon adjacent to heteroatoms [27,28]. Besides, the catalytic performance of those materials is significantly affected by the types and numbers of heteroatoms, levels of doping, and bonds constructed between carbon and heteroatoms. Understanding and uncovering the sensing mechanism associated with the dopants is essential for providing new design principles and engineering methods to satisfy specific sensing requirements. Apart from the doping of the materials, when they are used to detect analytes (e.g., ions, biochemical molecules, and biomacromolecules), the sensing performance can be optimized by forming heterojunctions via introducing other active sites onto the surface of the NWs.

There are multiple approaches to the surface functionalization of carbon nanofibers. Specifically, a controllable low-current electrodeposition method is introduced here to obtain tunable metal oxide @ WA-ECNF hybrids for biomolecule detection. Electrodeposition, a low-cost method for coating additional functional heterojunctions on carbon-based NWs, is a process of depositing materials on a conducting surface under an electric current (**Figure 22.1A**). The WA-ECNFs without surface activation are prone to hydrophobicity due to π-π* stacking, herein, the surface is functionalized with enriched epoxy, hydroxyl, and carboxyl functional groups after acid treatment, making the surface more hydrophilic for anchoring metal ions to nucleate and grow. In addition, different surface treatments generate various charge distributions over the carbon framework, thus, creating versatile morphologies of heterojunction on the surface of carbon nanofibers (**Figure 22.1B**). Heteroatom doping, rendering charge exchange adjacent carbon, can modulate electronic structure characteristics and surface charge distribution. It further enables tailoring the nucleation and growth of heterojunctions on the surface in various dimensional morphologies. Taking nitrogen-doping as an example [23], extra electron density from graphitic nitrogen, pyrrolic nitrogen, and lone pair on pyridinic nitrogen act as Lewis's base sites strongly coupling to acidic metal ions in solution. Meanwhile, the N-dopants play as binding sites for electrochemical nucleation and growth of the metal oxide in a Frank-van der Merwe mode [29,30]. Given the difference in electronegativity between nitrogen and oxygen, the morphologies of metal oxides anchored on the surface of WA-ECNFs exhibited significant versatility.

22.2 NANOWIRE-BASED SENSORS AND DETECTION OF BIOMOLECULES USING THE HYBRID METAL OXIDE @ WA-ECNF SENSORS

Thanks to the advanced development of materials science and nanofabrication, as well as great attention in healthcare and biomedicine, flexible sensors have made encouraging achievements in monitoring physiological signals of human health, particularly, integrated with medical and wearable fields to flexible devices for point-of-care diagnosis and real-time tracking of human activity. Herein, flexible NW-based sensors have been extensively investigated in virtue of their intrinsic characteristics and macroscopic structural design for various external stimuli, including physical and chemical targets. Those recognizable signals in the body and/or surrounding environment can be arising from mechanical forces (e.g., strain and pressure), physical and physiological stimuli (e.g., gas, temperature, humidity, electrocardiogram, electrooculography, electroencephalograms, and electromyography, etc.), and biochemical targets.

Typically, the sensors acquiring physical information is mainly operating with various transduction mechanisms, such as capacitance, triboelectricity, piezoelectricity, and piezo resistivity of conductive materials in the flexible sensing modules, which works by measuring the resistance changes upon external stimuli causing deformations. Taking strain sensors as an example, external forces deform the geometry of sensing components, where electrical signals are transduced and collected in response to quantitative strain. Tang et al. [13] explored a flexible strain sensor by integrating highly aligned NWs by taking advantage of a high surface-to-volume ratio and great mechanical robustness, obtaining excellent sensitivity to even subtle external stimuli during human motion of fingers, wrist, and throat swallowing. It is noteworthy that the well-aligned arrangement of NWs creates less affected charge carrier transfer paths from applied strain, ensuring uniformity in the sensing components and reliable device-to-device performance. In contrast to strain sensors, pressure sensors detect mechanical forces longitudinally, where NWs are normally conjugated with elastic polymers/hydrogels as functional components for constructing flexible piezo-resistant pressure sensors [31,32]. Apart from mechanical force detection, gas sensors, as promising tools for monitoring the environment or detecting exhaled gas for personalized diagnosis, have also been well exploited with the increased threat from air pollution. Doan and coworkers [33] implemented highly stretchable gas sensors onto the fabric of real T-shirts, whose functional nanostructures consist of multidimensional reduced graphene oxide, zinc oxide nanorods, silver nanowires, and palladium nanoparticles for detecting multiple gases. Herein, hybrid nanomaterials offer a synergistic effect for better performance of the fabric gas sensor, where AgNWs work as an electrical bridge to prevent broken rGO under tensile strain, besides PdNPs and AgNWs facilitate oxygen adsorption and catalysis on the ZnO nanorods, generating oxygen ion species for NO_2 molecules' reduction. Besides, NW-based flexible sensors have been investigated in monitoring humidity [34], electrophysiological parameters [35], temperature [36], and so on, because the high aspect ratio of NWs offers more active sites for adsorption of target molecules, high electron transport efficiency promoting rapid response and great mechanical tolerance for conformally adorning flexible and stretchable sensors.

Furthermore, NW-based flexible sensors for biochemical signal monitoring are also significantly crucial for systematic health tracking, such as metabolites (e.g., glucose, lactate, and dopamine), and electrolytes (e.g., sodium, potassium, calcium, and chloride) [6,37]. Given those indispensable chemical and biological biomarkers are essentially significant for human-being daily lives, real-time monitoring of these signature molecules with reliable accuracy and conformable wearability has attracted tremendous attention. Glucose, as one of the typical biochemical targets, has been extensively explored and associated with human health, which leads to great progress in flexible glucose sensors [38–40]. Ideally, flexible sensors for biomarker detection are noninvasive and desirable, attached to the skin with great biocompatibility and endurant capability. Versatile flexible sensors detecting biomarkers in sweat, saliva, tear, and urine have achieved great progress [14,37]. Of note, biomarkers present in those biofluids are concentrated much lower compared to those in the human blood. Hence, superior functional nanomaterials as sensing components in flexible sensors would give insights into the next generation of flexible wearable glucose sensors. This chapter describes the catalytic performance of functional nanomaterials on the WA-ECNFs and unveils the underlying mechanism of glucose oxidation, offering a great deal of inspiration for next-generation sensors.

Yin and coworkers [23] investigated mingled MnO_2 and Co_3O_4 binary nanostructures on WA-ECNFs for rapid glucose electrooxidation and sensing. Scanning electron microscopic (SEM) images demonstrate super aligned arrangement of the carbon nanofibers, which provide more efficient electron pathway transmission for both heterojunction coating and electrocatalysis. Furthermore, a high aspect ratio of nanowires provides a large specific area for exposing more functional active sites of the hybrid nanomaterials, and the mesoporous nanostructures of hybrid components promote target adsorption. In virtue of such morphologic properties, the MnO_2/Co_3O_4@WA-ECNFs exhibited excellent electrooxidation of glucose which was recorded by electrical signals (**Figure 22.2A**). A high-intensity redox reaction pair with small overpotential from the hybrid nanocomposites of mingled MnO_2 and Co_3O_4, accompanied by a negatively shifting oxidation peak compared to the only MnO_2 or Co_3O_4 coated electrodes (**Figure 22.2B**), suggesting faster kinetics of the redox reaction. This can be attributed to the enhanced number of active sites and readily electron transport channels in the mingled binary nanostructures. Herein, the MnO_2/Co_3O_4@WA-ECNFs deliver improved sensing performance since less activation energy is required for glucose electrooxidation by analysis of the Tafel slope (**Figure 22.2C**). Overall, such a great glucose electrooxidation performance can be ascribed to a synergistic effect of both rational architectures and functionalities of materials. The binary metal oxides covert to MnOOH and CoOOH in the alkaline condition, and the redox reactions could be greatly promoted to the forward oxidative reactions (CoOOH $\rightarrow CoO_2$, MnOOH $\rightarrow MnO_2$) with the presence of glucose, thus the oxidative peak increases (**Figure 22.2D**). On the one hand, the well-aligned orientation and mesoporous morphology enhance the mobility of electrons through the whole body of MnO_2/Co_3O_4@WA-ECNFs, extending to the interconnected electron-transfer channels, and promoting the diffusion and adsorption the analytes to the catalytic active sites followed by electrocatalytic oxidation. Moreover, multiple valence states of the hybrid nanomaterials on the nanowires could provide more redox reactions, whose various energy differences between conduction band edges and conductivity in the presence of ECNFs ensure substantial charge transfer, in turn, enhance the sensitivity of the nonenzymatic sensors from MnO_2-Co_3O_4/WA-ECNFs. As aforementioned synergy in glucose oxidation mechanism, the MnO_2-Co_3O_4/WA-ECNFs electrode has shown superb performance regarding fast signal response within 5 seconds, a wide detection range from 5 µM to 10.9 mM, an excellent sensitivity of 1159 µA/mM.cm^2, and a detection limit of 0.3 µM (S/N = 3) with great selectivity, reproducibility, and stability.

By taking advantage of multifarious architectures and morphology, and incorporating multiple functional materials, nanowire-based flexible sensors have greatly enhanced their sensing performances. As aforementioned, the introduction of heteroatom dopants in the carbon framework generates pockety charge distribution and modulates electronic structure characteristics, thus, further affecting catalytic performance. For instance, heteroatoms (e.g., N, O, P, S, and B) doping can break the electroneutrality of carbon and boost their catalytic activities [28,41–43]. To get a better configuration of a nanowire-based sensing system for flexible modules, a combination of heteroatom dopants with functional materials is composed of sensing components for a great deal of inspiration in flexible sensors. Recently, we demonstrated closely packed three-dimensional Co_3O_4 nanograins on nitrogen-doped WA-ECNFs, which exhibited great sensitivity and selectivity towards high performance for dopamine (DA) detection [24], with the lowest limit of detection as low as 9 nM over a wide dynamic range of detecting concentration from 0.01 to 100 µM (**Figure 22.3A**). Moreover, the sensor provides the capability of *in-situ* real-time detection of dopamine secreted from living dopaminergic cells, i.e., pheochromocytoma (PC12), with extracellular stimulation by K$^+$. High potassium ion stimulation could depolarize the PC12 cell membrane through opening voltage-sensitive Ca^{2+} channels to induce Ca^{2+} influx accompanied by an increased level of intracellular Ca^{2+}, thus, inducing the release of DA, followed by intimate detection by the Co_3O_4 decorated nitrogen-doped WA-ECNF (Co_3O_4@WA-NECNF) electrodes (**Figure 22.3B**). Such a superior sensing performance can be attributed to the well-designed architectural scaffold, which provides a high surface area for uniform dispersion of cobalt oxide nanostructures at the WA-ECNFs, facilitating the electrochemical dopamine oxidation and high conductivity for the nanohybrid electrode. Specifically, the introduction of heteroatoms with nitrogen dopants and oxygen-containing functional groups in the carbon skeleton plays a crucial role in the formation of three-dimensional nanostructures, thus, generating more active sites for target molecules' adsorption, ion/charge transport, and diffusion for electrocatalysis. Conspicuously, nitrogen atoms and nitrogen-containing functional

(A)

(B)

(C)

(D)

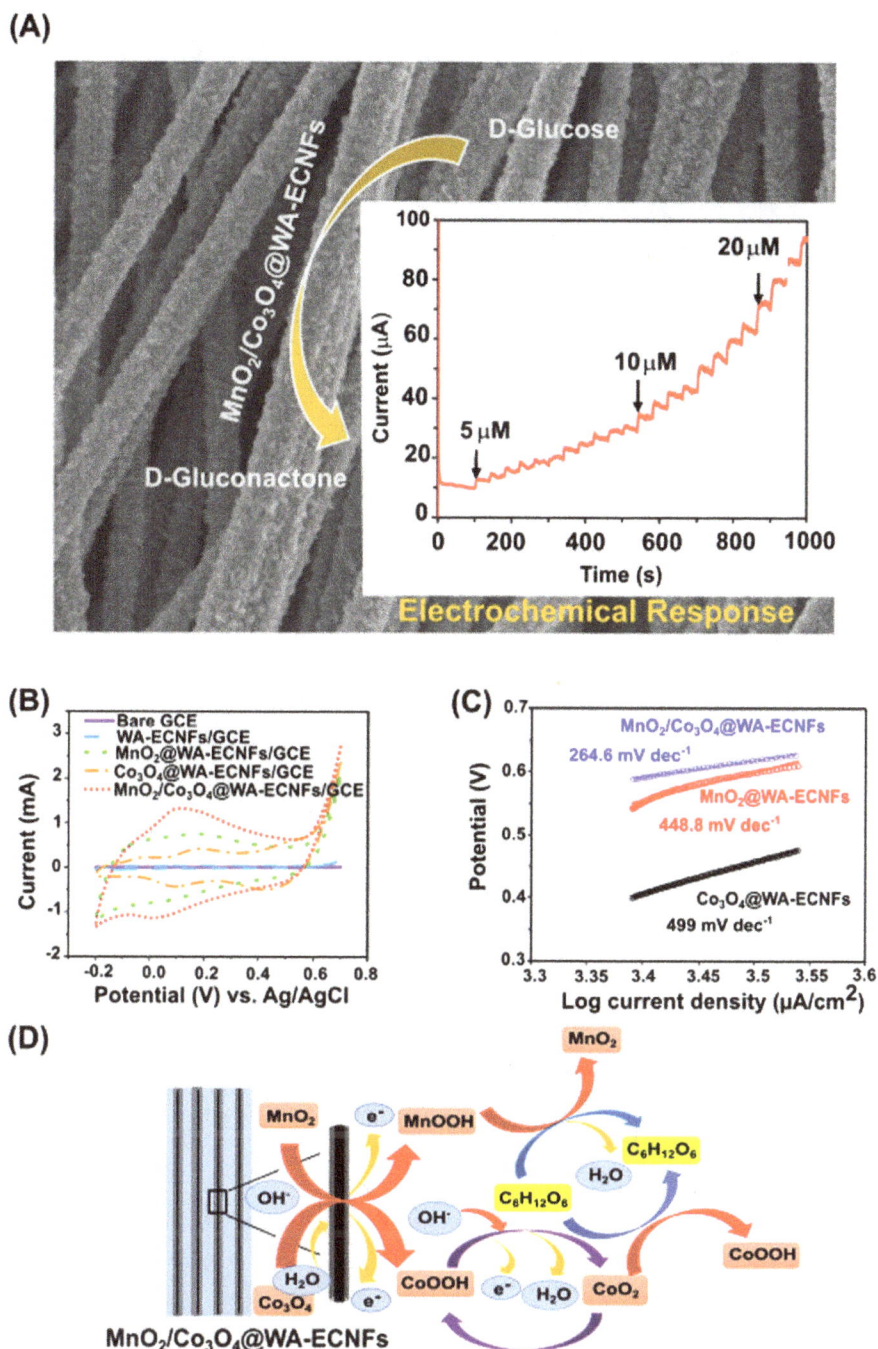

FIGURE 22.2 (A) Illustration of glucose oxidation on the hybrid system of MnO_2/Co_3O_4@WA-ECNFs recorded by electrochemical signals. (B) Cyclic voltammetric (CV) curves of different modifications of electrodes for electrochemical characterization. (C) Tafel slope diagram of various electrodes with 1 mM glucose. (D) Schematic illustration of the proposed electrocatalytic mechanisms for glucose oxidation at the MnO_2/Co_3O_4@WA-ECNF hybrid structure. Adapted with permission from [23]. Copyright (2021), American Chemical Society.

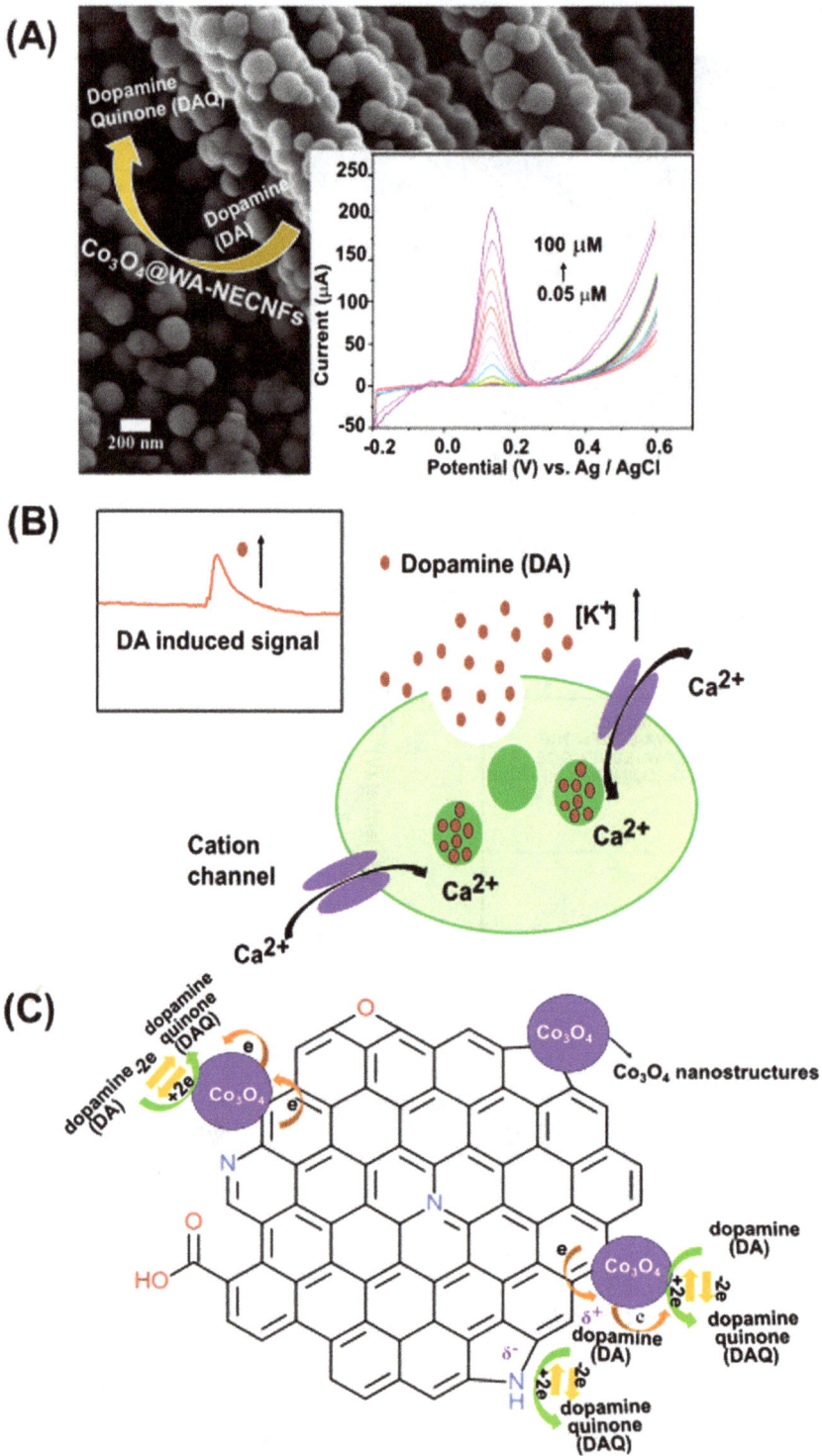

FIGURE 22.3 (A) Depiction of dopamine oxidation on the Co_3O_4 nanograins' coating on nitrogen-doped WA-ECNFs. (B) Illustration of detecting dopamine secreted by PC 12 cells via high potassium stimulation. (C) Schematic illustration of electrocatalytic reactions of dopamine on the nanohybrid exhibited great sensitivity and selectivity. Adapted with permission from [24]. Copyright (2022), Elsevier.

groups share high electronegativity, which offers strong coupling to anchor cobalt ions for the *in-situ* nucleation and growth of cobalt oxide nanomaterials, working together on the enhancement of the electrocatalytic nature (**Figure 22.3C**). Herein, by rational construction of surface chemistry of nanowires, it is capable of making precisely designed architectures for great promise in realizing flexible sensors.

In brief, nanowires coated with heterojunctions as functional materials in the flexible sensing modes have been concretely analyzed in this account, demonstrating the coupling relationship among structural arrangements and orientations, and revealing the multiple functionalities and underlying sensing mechanisms. We hope this chapter may inspire rational and optimal micro- to the macrostructural design of morphological and structural characteristic parameters affecting the sensing performance.

22.3 SUMMARY AND PERSPECTIVES

Despite great progress achieved in versatile NWs by tuning the morphology and/or composite modules, transferring the laboratory products into commercialization has yet to be met. Particularly, it is important to realize mass production of high-quality and optimized NWs in cost-effective fabrication and integration of flexible devices. Along with the great demand for flexible sensors with high accuracy and repeatability, optimization of the sensing components in the flexible platform is still a long path ahead. In this chapter, we present the WA-ECNF-based platform as a direction to NW-based flexible sensors, a promise to promoting the well-designed next generation of sensors.

The macrostructural geometry and morphology in the precise placement of NWs will provide great promise in the application of flexible sensors. Those well-arranged NWs usually show superb conductivity, excellent stability, and repeatability. Conspicuously, the alignment and ordered structures ensure charge transport efficiency, thus, facilitating signal sensitivity and rapid response of the flexible sensors. Besides, simple NWs with well-designed architectures not only enable great sensing performance in flexible platforms but also render the diversified engineering nanostructures for versatility. NWs in the forms of a hollow-structure with interconnected hierarchy, porous/mesoporous structure, core-shell/coated morphology, kinked, array, and/or bundles as building blocks could be manipulated and arranged for better controlled and tailored flexible sensors.

Apart from multifarious architectures and morphologies, introducing other functional nano/microstructured materials onto the surface of NWs could modify and optimize the sensing performance in terms of catalytic properties with enhanced sensitivity, superior conductive ability with fast response time, stability, and reproducibility. Besides, heteroatom doping of NWs may induce pockety charge distribution, forming dipoles, and further improving the response of sensors. Specifically, for the ECNFs and electrodeposition of nanostructured metal oxides, the introduction of heteroatoms in the carbon framework gives precise control of the size and morphologies of the nanostructured metal oxides for better manipulating sensing responses and modifying different electronegativity of heteroatoms' adjacent carbon elements in the catalytic process.

Machine learning would provide a great deal of prediction in properties and the discovery of functional materials. In contrast to conventional enumeration approaches from materials to properties, machine learning starts with desirable functionalities and looks for an ideal materials structure [44]. Machine-learning-enabled technology has been explored to power the IoT for flexible gas sensing [45], which could be able to provide great insights and motivation into the wireless-powered wearable devices. In addition, unveiling the rigorous sensing mechanism of nanowire-based flexible sensors and tackling the limitations of conventional sensing functions can make breakthroughs in those critical parameters, improving the performance of NW-based sensors. Furthermore, by modeling and analyzing sensing performance relevant to parameters, machine learning could foster new and advanced sensors from "big data" analytics and create powerful sensing networks, even connecting various fields, including biomedical diagnosis, global health, environmental sensing, etc. [46].

Last but not the least, macrostructural design and integration of flexible devices are also crucial factors for the deployment of flexible sensors from bench-scale production to large-scale
applications. For instance, additive manufacturing holds great promise in customization and on-
demand fabrication and has been greatly applied in producing high-quality nanowires, even
integrated flexible sensors in a manner of low cost and high automation. In contrast to conventional manufacturing of flexible sensors with layer-by-layer coating/depositing, 3D printing enables
excellent molding capabilities and convenient structural design realizability, thus, giving flexible
sensors previously unattainable geometric structures and functions. Consequently, researchers combine multidisciplinary knowledge of materials science and engineering to manipulate those NW-
based flexible sensors that exhibit great functions and high feasibility [47]. Besides, multifunctional
sensing modules are a trend in flexible sensors. Given the dynamics of the human physiological
system accompanying complexity in monitoring human status and external stimuli, flexible sensors
with a single modal, which can only perform under specific conditions responding to a sole kind
of stimulus, may not satisfy human demand for simultaneously monitoring multiple stimuli. To
this end, multifunctional sensing platforms implemented with versatile modes for detecting two or
more external stimuli to meet the practical requirements for a variety of information acquisition are
increasingly studied in both industry and academia [48]. Given this perspective, advanced manufacturing methods of integrating multifunctional sensing components to flexible systems with a relatively small change in size and manufacturing complexity are their future developing trend.

22.4   CONCLUSION

In summary, in virtue of the outstanding superiorities of volume-to-space ratio, mechanical tolerance,
ease of functionalization, flexibility, and stretchability, NWs are adopted for implementing flexible
sensors. The structural geometries and morphologies, rational nanostructures of sensing modules,
sensing mechanisms, and promising applications of NW-based flexible sensors are described in this
chapter. Specifically, metal oxides wired on aligned ECNFs as a flexible sensor platform to detect
small biomolecules via electrocatalysis are well demonstrated. In perspective, an ideal era of flexible sensors for practical applications is still in the early stage. Multidisciplinary research, including
materials science, micro-/nanofabrication, and electrical engineering, demands closely cooperate
for developing flexible sensors to operate highly conformable and deformable, self-powered, and
deployable, even realizing the transplantation. Furthermore, with growing attention to personalized
healthcare, smart information storage, and virtual reality, significant achievements of artificial intelligence and machine learning have advanced the Internet of Everything. Herein, it is crucial to think
of how to combine those flexible sensors with collected signals with other electronic components
and soft algorithms to achieve real-world applications in the human-machine interface, including
data acquisition and transmission systems, artificial sensory neutron systems, artificial-intelligent-
assisted systems, etc.

REFERENCES

[1]   T. Kim, Q. Yi, E. Hoang, R. Esfandyarpour, A 3D printed wearable bioelectronic patch for multi-
sensing and in situ sweat electrolyte monitoring, Advanced Materials Technologies. 6 (2021),
2001021.
[2]   W. Gao, H. Ota, D. Kiriya, K. Takei, A. Javey, Flexible electronics toward wearable sensing, Accounts
of Chemical Research. 52 (2019), 523–533.
[3]   W. di Li, K. Ke, J. Jia, J.H. Pu, X. Zhao, R.Y. Bao, Z.Y. Liu, L. Bai, K. Zhang, M.B. Yang, W. Yang,
Recent advances in multiresponsive flexible sensors towards e-skin: A delicate design for versatile
sensing, Small. 18 (2022), 2103734.
[4]   Q. Shi, B. Dong, T. He, Z. Sun, J. Zhu, Z. Zhang, C. Lee, Progress in wearable electronics/photonics—
Moving toward the era of artificial intelligence and internet of things, InfoMat. 2 (2020), 1131–1162.

[5]   Q. Fu, C. Cui, L. Meng, S. Hao, R. Dai, J. Yang, Emerging cellulose-derived materials: A promising platform for the design of flexible wearable sensors toward health and environment monitoring, Materials Chemistry Frontiers. 5 (2021), 2051–2091.

[6]   F. Yuan, Y. Xia, Q. Lu, Q. Xu, Y. Shu, X. Hu, Recent advances in inorganic functional nanomaterials based flexible electrochemical sensors, Talanta. 244 (2022), 123419.

[7]   J. Zhang, T. Sun, Y. Chen, L. Liu, H. Zhao, C. Zhang, X. Meng, D. Wang, Z. Hu, H. Zhang, B. Li, S. Niu, Z. Han, L. Ren, Q. Lin, Nanowires in flexible sensors: Structure is becoming a key in controlling the sensing performance, Advanced Materials Technologies. (2022) 2200163.

[8]   D. Wang, L. Wang, G. Shen, Nanofiber/nanowires-based flexible and stretchable sensors, Journal of Semiconductors. 41 (2020), 041605.

[9]   K. Wang, L.W. Yap, S. Gong, R. Wang, S.J. Wang, W. Cheng, Nanowire-based soft wearable human–machine interfaces for future virtual and augmented reality applications, Advanced Functional Materials. 31 (2021), 2008347.

[10] S.T. Han, H. Peng, Q. Sun, S. Venkatesh, K.S. Chung, S.C. Lau, Y. Zhou, V.A.L. Roy, An overview of the development of flexible sensors, Advanced Materials. 29 (2017), 1700375.

[11] A. Sabah, G.L. Hornyak, Nanofibers and nanowires, in: Nanobotany, Springer International Publishing, 2018: pp. 67–82.

[12] K.H. Han, H. Kang, G.Y. Lee, H.J. Lee, H.M. Jin, S.K. Cha, T. Yun, J.H. Kim, G.G. Yang, H.J. Choi, Y.K. Ko, H.T. Jung, S.O. Kim, Highly aligned carbon nanowire array by e-field directed assembly of PAN-containing block copolymers, ACS Applied Materials and Interfaces. 12 (2020) 58113–58121.

[13] N. Tang, C. Zhou, D. Qu, Y. Fang, Y. Zheng, W. Hu, K. Jin, W. Wu, X. Duan, H. Haick, A highly aligned nanowire-based strain sensor for ultrasensitive monitoring of subtle human motion, Small. 16 (2020), 2001363.

[14] Q. Zhai, L.W. Yap, R. Wang, S. Gong, Z. Guo, Y. Liu, Q. Lyu, J. Wang, G.P. Simon, W. Cheng, Vertically aligned gold nanowires as stretchable and wearable epidermal ion-selective electrode for noninvasive multiplexed sweat analysis, Analytical Chemistry. 92 (2020), 4647–4655.

[15] C. Jia, Z. Lin, Y. Huang, X. Duan, Nanowire electronics: From nanoscale to macroscale, Chemical Reviews. 119 (2019), 9074–9135.

[16] J. Luo, X. Luo, C. Hu, J.C. Crittenden, J. Qu, Zirconia (ZrO_2) embedded in carbon nanowires via electrospinning for efficient arsenic removal from water combined with DFT studies, ACS Applied Materials and Interfaces. 8 (2016), 18912–18921.

[17] Z. Zeng, W. Zhang, Y. Liu, P. Lu, J. Wei, Uniformly electrodeposited α-MnO_2 film on super-aligned electrospun carbon nanofibers for a bifunctional catalyst design in oxygen reduction reaction, Electrochimica Acta. 256 (2017), 232–240.

[18] Z. Zeng, T. Zhang, Y. Liu, W. Zhang, Z. Yin, Z. Ji, J. Wei, Magnetic field-enhanced 4-electron pathway for well-aligned Co_3O_4/electrospun carbon nanofibers in the oxygen reduction reaction, Chem Sus Chem. 11 (2018), 580–588.

[19] Z. Zeng, Y. Liu, W. Zhang, H. Chevva, J. Wei, Improved supercapacitor performance of MnO_2-electrospun carbon nanofibers electrodes by mT magnetic field, Journal of Power Sources. 358 (2017), 22–28.

[20] Y. Liu, Z. Zeng, B. Bloom, D.H. Waldeck, J. Wei, Stable low-current electrodeposition of α-MnO_2 on superaligned electrospun carbon nanofibers for high-performance energy storage, Small. 14 (2018), 1703237.

[21] Y. Liu, Z. Zeng, R.K. Sharma, S. Gbewonyo, K. Allado, L. Zhang, J. Wei, A bi-functional configuration for a metal-oxide film supercapacitor, Journal of Power Sources. 409 (2019), 1–5.

[22] K. Allado, M. Liu, A. Jayapalan, D. Arvapalli, K. Nowlin, J. Wei, Binary MnO_2/Co_3O_4 metal oxides wrapped on superaligned electrospun carbon nanofibers as binder free supercapacitor electrodes, Energy and Fuels. 35 (2021), 8396–8405.

[23] Z. Yin, K. Allado, A.T. Sheardy, Z. Ji, D. Arvapalli, M. Liu, P. He, X. Zeng, J. Wei, Mingled MnO_2 and Co_3O_4 binary nanostructures on well-aligned electrospun carbon nanofibers for nonenzymatic glucose oxidation and sensing, Crystal Growth and Design. 21 (2021), 1527–1539.

[24] Z. Yin, Z. Ji, B.P. Bloom, A. Jayapalan, M. Liu, X. Zeng, D.H. Waldeck, J. Wei, Manipulating cobalt oxide on N-doped aligned electrospun carbon nanofibers towards instant electrochemical detection of dopamine secreted by living cells, Applied Surface Science. 577 (2022), 151912.

[25] J. Zhu, S. Mu, Defect engineering in carbon-based electrocatalysts: insight into intrinsic carbon defects, Advanced Functional Materials. 30 (2020), 2001097.

[26] S. Liu, H. Yang, X. Su, J. Ding, Q. Mao, Y. Huang, T. Zhang, B. Liu, Rational design of carbon-based metal-free catalysts for electrochemical carbon dioxide reduction: A review, Journal of Energy Chemistry. 36 (2019), 95–105.

[27] S. Fu, C. Zhu, J. Song, M.H. Engelhard, X. Li, P. Zhang, H. Xia, D. Du, Y. Lin, Template-directed synthesis of nitrogen- and sulfur-codoped carbon nanowire aerogels with enhanced electrocatalytic performance for oxygen reduction, Nano Research. 10 (2017), 1888–1895.

[28] L. Zhou, J. Meng, P. Li, Z. Tao, L. Mai, J. Chen, Ultrasmall cobalt nanoparticles supported on nitrogen-doped porous carbon nanowires for hydrogen evolution from ammonia borane, Materials Horizons. 4 (2017), 268–273.

[29] X. Xia, L. Figueroa-Cosme, J. Tao, H.C. Peng, G. Niu, Y. Zhu, Y. Xia, Facile synthesis of iridium nanocrystals with well-controlled facets using seed-mediated growth, J Am Chem Soc. 136 (2014), 10878–10881.

[30] N. Logeshwaran, I.R. Panneerselvam, S. Ramakrishnan, R.S. Kumar, A.R. Kim, Y. Wang, D.J. Yoo, Quasihexagonal platinum nanodendrites decorated over CoS_2-N-doped reduced graphene oxide for electro-oxidation of C1-, C2-, and C3-type alcohols, Advanced Science. 9 (2022), 2105344.

[31] X. Li, Y.J. Fan, H.Y. Li, J.W. Cao, Y.C. Xiao, Y. Wang, F. Liang, H.L. Wang, Y. Jiang, Z.L. Wang, G. Zhu, Ultracomfortable hierarchical nanonetwork for highly sensitive pressure sensor, ACS Nano. 14 (2020), 9605–9612.

[32] Y. Li, D. Han, C. Jiang, E. Xie, W. Han, A facile realization scheme for tactile sensing with a structured silver nanowire-PDMS composite, Advanced Materials Technologies. 4 (2019), 1800504.

[33] T.H.P. Doan, Q.T.H. Ta, A. Sreedhar, N.T. Hang, W. Yang, J.S. Noh, Highly deformable fabric gas sensors integrating multidimensional functional nanostructures, ACS Sensors. 5 (2020), 2255–2262.

[34] J. Wu, C. Yin, J. Zhou, H. Li, Y. Liu, Y. Shen, S. Garner, Y. Fu, H. Duan, Ultrathin glass-based flexible, transparent, and ultrasensitive surface acoustic wave humidity sensor with ZnO nanowires and graphene quantum dots, ACS Applied Materials and Interfaces. 12 (2020), 39817–39825.

[35] Y. Qiao, Y. Wang, J. Jian, M. Li, G. Jiang, X. Li, G. Deng, S. Ji, Y. Wei, Y. Pang, Q. Wu, H. Tian, Y. Yang, X. Wu, T.L. Ren, Multifunctional and high-performance electronic skin based on silver nanowires bridging graphene, Carbon N Y. 156 (2020), 253–260.

[36] L. Zhang, X.L. Shi, Y.L. Yang, Z.G. Chen, Flexible thermoelectric materials and devices: From materials to applications, Materials Today. 46 (2021), 62–108. https://doi.org/10.1016/j.mat tod.2021.02.016.

[37] Q. Zhang, D. Jiang, C. Xu, Y. Ge, X. Liu, Q. Wei, L. Huang, X. Ren, C. Wang, Y. Wang, Wearable electrochemical biosensor based on molecularly imprinted Ag nanowires for noninvasive monitoring lactate in human sweat, Sensors and Actuators, B: Chemical. 320 (2020), 128325.

[38] L. Manjakkal, L. Yin, A. Nathan, J. Wang, R. Dahiya, Energy autonomous sweat-based wearable systems, Advanced Materials. 33 (2021), 2100899.

[39] H. Fujita, K. Yamagishi, W. Zhou, Y. Tahara, S.Y. Huang, M. Hashimoto, T. Fujie, Design and fabrication of a flexible glucose sensing platform toward rapid battery-free detection of hyperglycaemia, Journal of Materials Chemistry C. 9 (2021), 7336–7344.

[40] S. Zhang, J. Zeng, C. Wang, L. Feng, Z. Song, W. Zhao, Q. Wang, C. Liu, The application of wearable glucose sensors in point-of-care testing, Frontiers in Bioengineering and Biotechnology. 9 (2021), 774210.

[41] M. Li, T. Liu, X. Bo, M. Zhou, L. Guo, S. Guo, Hybrid carbon nanowire networks with Fe–P bond active site for efficient oxygen/hydrogen-based electrocatalysis, Nano Energy. 33 (2017), 221–228.

[42] J. Chen, Q. Chen, J. Xu, C.P. Wong, Hybridizing Fe_3O_4 nanocrystals with nitrogen-doped carbon nanowires for high-performance supercapacitors, RSC Advances. 7 (2017), 48039–48046.

[43] T. Huang, Y. Chen, J.M. Lee, A microribbon hybrid structure of CoO_x-MoC encapsulated in N-doped carbon nanowire derived from MOF as efficient oxygen evolution electrocatalysts, Small. 13 (2017), 1702753.

[44] M. Wang, T. Wang, Y. Luo, K. He, L. Pan, Z. Li, Z. Cui, Z. Liu, J. Tu, X. Chen, Fusing stretchable sensing technology with machine learning for human–machine interfaces, Advanced Functional Materials. 31 (2021), 2008807.

[45]   J. Zhu, M. Cho, Y. Li, T. He, J. Ahn, J. Park, T.L. Ren, C. Lee, I. Park, Machine learning-enabled textile-based graphene gas sensing with energy harvesting-assisted IoT application, Nano Energy. 86 (2021), 106035.

[46]   Z. Ballard, C. Brown, A.M. Madni, A. Ozcan, Machine learning and computation-enabled intelligent sensor design, Nature Machine Intelligence. 3 (2021), 556–565.

[47]   Y. Du, R. Wang, M. Zeng, S. Xu, M. Saeidi-Javash, W. Wu, Y. Zhang, Hybrid printing of wearable piezoelectric sensors, Nano Energy. 90 (2021), 106522.

[48]   Y. Guo, X. Wei, S. Gao, W. Yue, Y. Li, G. Shen, Recent advances in carbon material-based multifunctional sensors and their applications in electronic skin systems, Advanced Functional Materials. 31 (2021), 2104288.

23 Nanowires for Biosensors

Vy Anh Tran
Institute of Applied Technology and Sustainable Development and
Faculty of Environmental and Food Engineering, Nguyen Tat Thanh
University, Ho Chi Minh City, Vietnam

Thu-Thao Thi Vo
Department of Food Science and Biotechnology, Gachon University,
Sujeong-gu, Seongnam-si, Republic of Korea

Giang N. L. Vo
Faculty of Pharmacy, University of Medicine and Pharmacy,
Ho Chi Minh City, Vietnam

CONTENTS

23.1 INTRODUCTION

Synthesized nanomaterials including carbon nanotubes, nanoparticles, quantum dots, and nanowires (NWs) have already produced advancements in a variety of sectors, including biological sensors, during the last decade [1, 2]. The large surface area-to-volume ratio of nanoparticles significantly boosts sensitivity relative to macro-sized materials [3, 4]. Nanowires are structures with a thickness or width of tens of nanometers or less with an unrestricted length. The importance of quantum mechanical phenomena at these scales gave rise to the name "quantum wires." There are several distinct types of nanowires, such as superconducting (YBCO), metallic (Ag, Pt, Ni, Au), semiconducting InP, SiNWs, GaN, and insulating (TiO_2, SiO_2) [5, 6].

One of the most important parts of nanotechnology advancements is the creation of nanobiosensors. A biosensor is a device that detects a specific biological material by using biorecognition components to create a signal for subsequent analysis [6-8]. Nanobiosensors make use of nanoparticles to create the recognition element or transducers. Nanobiosensors are well-known and frequently used in the biomedical area. Biomedical sensors' principal function is to

DOI: 10.1201/9781003296621-23

identify and classify chemical and biological elements, with applications ranging from illness diagnosis to drug development [8-10]. Those fundamental biological activities may be easily accomplished using nanomaterials (nanotubes, nanowires, and nanoparticles) with different magnetic, electrical characteristics, and optical [11, 12].

To identify tagged disease markers and other biological species, for example, semiconductor crystalline nanoparticles were utilized, and colloidal gold has been used in optical imaging and electromagnetism [13]. Furthermore, semiconductor nanowires allowed it to identify different species electrically and without labeling. These nanowires were made using semiconductor materials, and their surfaces may be easily changed to make them biological and chemical group sensitive. However, nanotubes, which may potentially be used in biological sensors, are created by combining semiconductor materials and metals, necessitating further purification. Furthermore, the techniques for binding diverse analytes to nanotubes are not well established. As a result, nanostructures generated using nanowires are the best alternative for biological sensors with excellent sensitivity, homogeneity, repeatability, and scalability while requiring a very easy manufacturing procedure.

This book chapter discusses current developments in nanowire-based biosensors. We briefly address the development of efficient sensing strategies for coronavirus (COVID-19) detection, agricultural applications, biomarker detection, and disease diagnostics utilizing nanowire-based biosensors. This study, in particular, provides an in-depth overview of device design methods for certain biomedical sensing tasks using nanowire-based biosensors, as well as accompanying approaches. Thus, unlike evaluations that focus on a specific application or a single material type of nanowire arrays, this book chapter covers multiple areas of biosensing applications and diverse nanowire materials. Finally, the future possibilities of practical biomedical platforms are examined in terms of the development and application directions of nanowire-based biosensor arrays.

23.2 CLASSIFICATION, WORKING PRINCIPLE, AND DENATURATION AND FABRICATION METHODS

23.2.1 CLASSIFICATION OF NANOWIRE

Nanowires, which have a length-to-width ratio of 1000 or greater and a nano range diameter of 10–200 nm, are presently being employed for cancer biosensing. Fluorescent and magnetic nanowires are frequently used in fluorescence imaging and magnetic resonance imaging (MRI) due to their enormous surface area. Because of their capacity to boost surface-enhanced Raman spectroscopy (SERS), silver nanoparticles coupled with silicon nanowires have received interest. Deoxyribo nucleic acid (DNA) mutations are discovered using molecular beacon probes made of gold nanoparticles and silicon nanowires. These nanowire-linked nanoparticles are extremely sensitive, detecting DNA even at very low concentrations. Nanowires' tube-like form makes them simpler to enter and locate target cells. The ability of silicon nanowire to identify dysregulated signaling pathways in chronic lymphocytic leukemia was studied [14].

Nanowires are nanostructures with sizes in the nanometer range. Nanowires are classified as superconducting, metallic, semiconducting, or insulating [15]. There are several types of synthetic nanowires, including the following:

a. Noncatalytic development: nanowires can also be created without the use of a catalyst, resulting in pure nanowires and reducing the number of technical stages. The most basic way for producing nanowires is to heat metals, which is done by Joule heating metal wires in the air using batteries.
b. Liquid-phase synthesized: in this process, nanowires are grown in solution; in comparison with other types, liquid-phase synthesized may create a huge number of nanowires; this is a very flexible technique for creating gold, silver, platinum, and lead nanowires.
c. Vapor–liquid–solid approach: this is a typical method for producing nanowires with diameters ranging from a few hundred nanometers to hundreds of micrometers.

d. DNA-template metallic nanowire manufacture: one new topic in metallic nanowire synthesis is the use of DNA strands as scaffolds. This approach was utilized to create metallic nanowires for usage in electrical components and biosensors. This technique converts DNA strands into metallic nanowires that can be sensed electronically.

23.2.2 WORKING PRINCIPLE OF NANOWIRE FOR BIOSENSOR

Nanowires are those solid materials having the shape of a wire, which have diameters ranging between 1–100 nm, and lateral dimensions up to the range of micrometer (μm) [16] (**Figure 23.1**). NWs are one-dimensional nanostructures that can be manufactured with a variety of architectural and chemical compositions, such as metallic (e.g., Ag, Pt, Ni, and, Au), semiconducting polymers (e.g., Si, GaN, and InP), and metal oxides (e.g., SiO_2, ZnO, Fe_2O_3, and TiO_2) [17]. NWs exhibit extraordinary characteristics that are described. The electrical properties of NWs are due to sensitive to surface scatterings, a minority, and the majority carried in the nanowire can be easily manipulated during charge capturing and electrical conduction scattering. Due to the dielectric constant, nanowires absorb light differently from ~250 μm in wavelength, making NWs also suited for photovoltaic applications [18]. Nanowires are appropriate for thermoelectric applications. The scattering of phonons, which reduces heat transport, is fundamental to the formation of the thermal conductivities of NWs [19]. Nanowires have high energy anisotropy and strong magnetic capabilities, which are ideal for electromagnetic fields [20]. Furthermore, due to their large effective areas and surfaces, nanowires are extraordinarily valuable as a potential approach for chemical and biosensors [21] and the remarkable catalytic characteristics of it [22].

23.2.3 NANOWIRE DENATURATION AND FABRICATION METHODS

For the manufacture of nanowire sensors, both bottom-up and top-down approaches are used. Bottom-up techniques are routinely employed to generate Si wafers with high-quality nanowires. Although most nanowires are cylindrical, existing bottom-up procedures may produce round, square, and triangular variants by altering their cross-sectional shape. The manufacturing process begins with the growth of Si nanowires by chemical vapor deposition (CVD). Through the vapor–liquid–solid (VLS) process, Si nanowires may be produced catalytically in the CVD procedure. The Si nanowires suspended in ethanol solution are then placed on a silicon substrate. After spin-coating

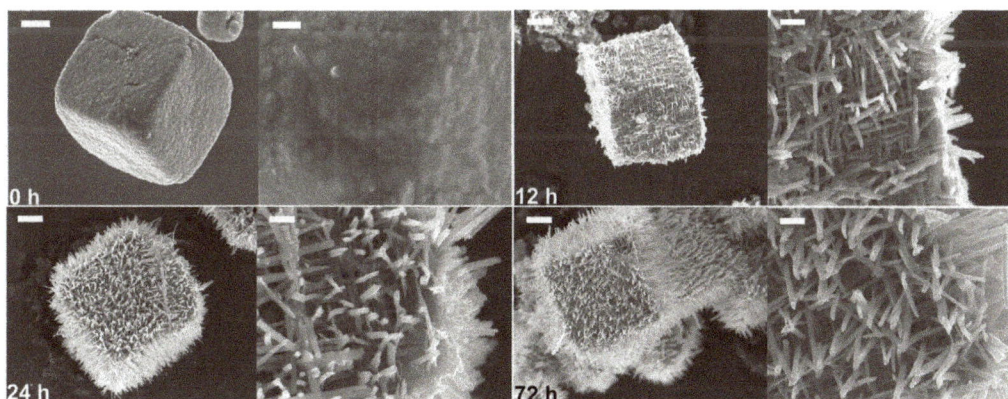

FIGURE 23.1 Scanning electron microscopy (SEM) images of $Nb_3O_7(OH)$ cube morphologies changing in hydrothermal treatment over time. Adapted with permission [16]. Copyright The Authors, some rights reserved; exclusive licensee [Royal Society of Chemistry]. Distributed under a Creative Commons Attribution License 3.0 (CC BY) https://creativecommons.org/licenses/by/3.0.

a photoresist onto the substrate with Si nanowires deposited, the metal electrodes are designed using the lift-off technique.

Bottom-up manufacturing is completed by passivation and surface modification with receptor binding. Introducing the nanowire surface to air or an oxygen atmosphere quickly creates an isolated layer. The bottom-up technique has the disadvantage of isotropic nanowires, which results in poor homogeneity and a low yield rate of nanowire sensors. Additional alignment stages during manufacturing, such as blown-bubble, Langmuir–Blodgett, microfluidic flow, electric field, and contact printing, are necessary to increase the orientation. The typical CMOS fabrication process, however, is incompatible with those alignment approaches, making mass manufacture of aligned nanowire sensors challenging [23].

Top-down approaches, as opposed to bottom-up methods, can produce aligned nanowires that are compatible with traditional CMOS technology using fabrication methods on a silicon-on-insulator (SOI) wafer or a single-crystalline silicon (SCS) wafer. The manufacturing method begins with low-density boron or phosphorous doping on the top Si layer of an SOI wafer. Following that, substantial density doping is performed on the patterning region, followed by reactive ion etching to generate a micrometer-sized source and drain electrodes (RIE). The nanometer-sized Si nanowires are then created using electron-beam (E-beam) lithography, and the metal contact leads are created using thermal evaporation. Finally, the manufacturing process concludes with passivation and surface modification, as with the bottom-up approach. While top-down approaches produce nanowires that are very compatible with CMOS technology, bottom-up methods produce nanowires with a greater diameter. Despite differences in nanowire orientation and drain and source materials, these devices have the same design [24].

Another top-down silicon nanowire production approach is based on a bulk SCS wafer. In general, fabrication is carried out using a combination of traditional semiconductor manufacturing methods such as silicon dry etching, photolithography, thermal oxidation, and anisotropic wet etching. The thermal wet oxidation period controls the width of the silicon nanowires in these top-down production processes. It produces ordered silicon nanowires as a result. After photolithographically defining the space and line, the dielectric etching procedure is used to remove the silicon dioxide from the top layer. The photoresists are then removed, and cleaning operations are carried out. The silicon deep RIE etching method is then used to define the rectangular-structured columns that house the silicon nanowires. Following that, anisotropic silicon wet etching with tetramethylammonium hydroxide solution was performed [25].

The thermal oxide growth procedure is also used in top-down (111) silicon nanowire production. The line and space patterns are then defined using photolithography and dielectric etching methods. The height of the silicon nanowire is defined using the basic deep silicon RIE method. The height and breadth of (111) silicon nanowire formations are set individually, resulting in a variety of shapes such as wires and ribbons. The sides of the whole structure are then passivated using a thermal oxidation method. To expose the bottom surface of the structure, a plasma-enhanced anisotropic dielectric dry etching procedure is used, followed by a secondary deep silicon RIE process to manufacture the sacrificial layer. Following that, TMAH performed an anisotropic silicon wet etching procedure. Following that, an anisotropic silicon wet etching technique using TMAH solution is used to show the (111) planes. After removing the passivated silicon dioxide layers, the arranged rectangular nanowire structures are created using thermal oxidation [26].

The fluid exchange system is a fundamental element of nanowire sensors. The fluid exchange system transports analytes or fluids to or near the nanowire sensor's surface. Channel devices made of polydimethylsiloxane (PDMS) are commonly utilized to feed analytes to nanowire sensors. Because of its great biocompatibility, PDMS is also useful in biosensing applications. The PDMS-based microfluidic fluid exchange technology, on the other hand, has several drawbacks. Because fluid flow in microscale channels is laminar, particles inside the microfluidic channels have a tough time approaching the surface of the nanowire sensors. Furthermore, because PDMS may absorb biomolecules, the sensor's sensitivity may be reduced. To circumvent the constraints of PDMS microfluidic channels, an acrylic chamber as a fluid exchange device was designed.

23.3 APPLICATION OF NANOWIRE FOR BIOSENSOR

23.3.1 NANOWIRE BIOSENSOR FOR CORONAVIRUS (COVID-19) DETECTION

The lack of precise and speedy detection techniques has made the continued spread of coronavirus (COVID-19) a global threat. The coronavirus, also popularly known as SARS-CoV-2, is severe and has a significant impact on human interactions, the world is gripped by 6,325,785 deaths globally and ~540 million positive cases (https://covid19.who.int/ accessed on June 28, 2022). Scientists from all over the world have attempted several methods to diagnose the sickness. The detection and analysis approach for this threatening virus initially started with various virus detection methods [27, 28], such as, (a) nucleic acid detection and amplification [29], (b) isothermal amplification technologies [30], (c) immunoassays [31], (d) DNA sequencing [32], (e) mass spectrometric methods [33], and (f) microelectronics and microfluidics-based techniques [34]. However, the drawbacks of these methods include low accuracy and sensitivity, time-consuming sample preparation and purification, complex instrumentation, laboratory setup, and maintenance, as well as the need for highly trained and skilled workers and inapplicability for rapid on-site analysis [35]. As a result, there is a pressing need to create a viral biosensor that is affordable, portable, and effective and can also overcome the obstacles presented by established diagnostic tests. For biosensors to be specific for SARS-CoV-2, they had to be able to recognize their targets, which could be viral ribonucleic acid (RNA), proteins, or human immunoglobulins. They also had to recognize their targets in a certain way, such as by detecting the binding of an antibody to an antigen through nucleic acid, aptamers, antibodies, or receptors, and they had to have a signal intensification and transduction system [36]. There are various types of biosensors that could be applied: electrochemical biosensors, electronic biosensors, physical biosensors, optical biosensors, and thermal biosensors [35, 37].

The efforts against coronavirus (COVID-19) and other upcoming viral outbreaks must incorporate nanotechnology and nanoscience-based techniques [38]. Among nanodetection methods that were developed as an alternative [39], nanowires are one type of sensor utilized [40] (**Figure 23.2**). The device had a silicon nanowire biosensor and could find suspensions of coronavirus-like particles and specific virus antibodies [41]. Based on a nanomaterial, another nanowire field-effect transistor (FET) biosensor was used to find COVID-19. This biosensor was also good in terms of selectivity, sensitivity, and low detection limit [3]. The combination of graphene-coated FETs in the identification of COVID-19 protein in various media, including clinical samples, phosphate-buffered saline, and culture medium, is extremely valuable. The nanowire graphene-based field-effect transistor was introduced as a biosensor for the rapid detection and analysis of viruses [35, 42, 43]. Due to the

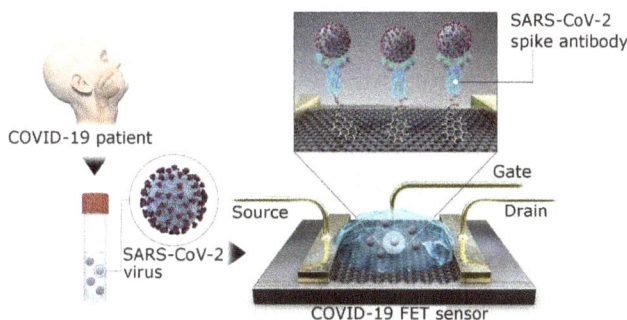

FIGURE 23.2 Schematic illustration of the biosensor for coronavirus detection. Adapted with permission [40]. Copyright The Authors, some rights reserved; exclusive licensee [Frontiers]. Distributed under a Creative Commons Attribution License 4.0 (CC BY) https://creativecommons.org/licenses/by/4.0.

high conductivity of graphene, which plays a major role in sensor response time, a very rapid reaction time was observed. This parameter determines the sensor's capacity for rapid detection, hence facilitating the rapid identification of viruses. Another nanowire was developed based on ZnO-NW, it was reported to detect different antibodies specifically CR3022 by the spike of glycoprotein S1 of SARS-CoV-2 in human serum. ZnO-NW morphologies alter the overall performance of the sensor [44]. The underlying concept of these experiments is that conductance is discrete when a virus particle attaches to an antibody receptor on a nanowire device. Simultaneous optical and electrical measurements with fluorescently labeled viruses provided conclusive evidence that the discrete conductance changes observed in these investigations were attributable to the detection of a single viral binding and subsequently unbinding [45].

23.3.2 NANOWIRE BIOSENSOR FOR AGRICULTURAL APPLICATIONS

Nowadays, it is feasible to create agricultural systems based on sensors to monitor and ensure agricultural land's wellbeing. The plant's growth may be optimized by continuously monitoring soil and environmental parameters with nanosensors (**Figure 23.3**). This is where the advantages of nanowire sensors and the potential benefits of nanowire sensors were recognized [46].

Agricultural crops are severely impacted by environmental degradation and the frequency of extreme climate disasters [48], such as the crop damage caused by freezing disasters. Various thermal management equipment and materials are available for altering the temperature in freezing disasters. Among the various approaches, the fabrication of agricultural films integrated with varied components is widely applied in agricultural production. Degradable cellulose films also have been gaining growing interest in agricultural applications. They have sufficient chemical stability and

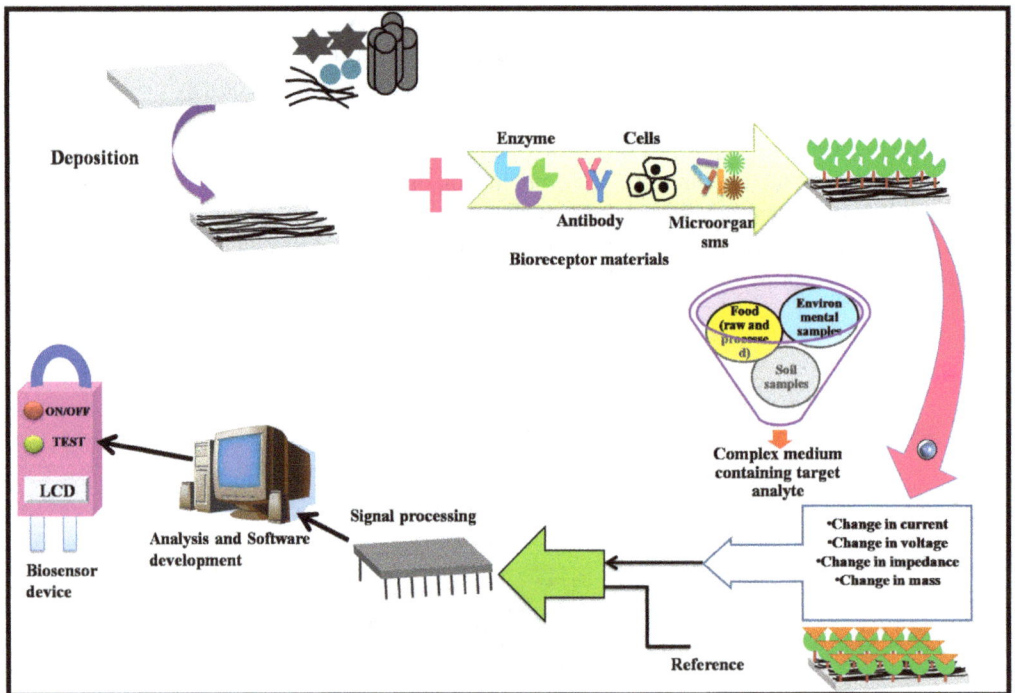

FIGURE 23.3 Schematic showing the development of biosensors for agricultural applications. Adapted with permission [47]. Copyright 2019, Elsevier. Distributed under a Creative Commons Attribution License 4.0 (CC BY) https://creativecommons.org/licenses/by/4.0.

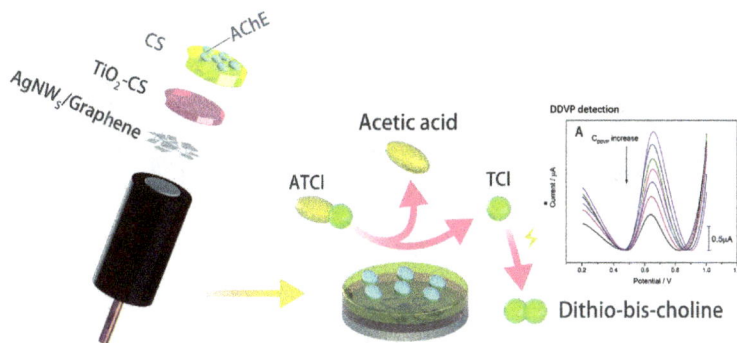

FIGURE 23.4 Structure and mechanism of the AChE biosensor in response to ATCl. Adapted with permission [57]. Distributed under a Creative Commons Attribution License 3.0 (CC BY) https://creativecommons.org/licenses/by/3.0/.

outstanding mechanical qualities [49]. Unfortunately, the utilization of biodegradable cellulose films is limited by their insulating characteristics, particularly during strong cold waves. Efforts have been made to develop functional cellulose films by using the flexible AgNW/cellulose hybrid films. The AgNWs were uniformly scattered in the cellulose grid, providing a highly conductive layer capable of efficiently harvesting heat from Joule heating [50].

Pesticides significantly contributed to agricultural productivity by improving total yield and protecting crops from pests. However, pesticide residues in numerous areas of the ecosystem have created grave worries about their use, which will outweigh their overall benefits [51]. Children and pregnant women could be especially vulnerable to the toxicity of pesticides [52]. Organophosphorus pesticides are toxic to humans and most animals because they cause permanent damage to the nervous system. They generate an accumulation of acetylcholine (ATCl) in humans by inhibiting the activity of acetylcholinesterase (AChE) and overstimulating synaptic receptors [53]. Liquid chromatography, mass spectrometry, electrochemistry, fluorometry, and colorimetry are all approaches for detecting pesticide residues, however, immobilization of acetylcholinesterase with TiO_2–CS composite film has been authorized as a more effective strategy [54]. Different low-dimensional materials have been tested and achieved certain improvements for the signal magnification, including Au NCs–CeO_2 NWs [55], AuNPs–PPy nanowires [56], and AChE/CS/TiO_2–CS/Gra/AgNWs [57]. Biosensors are a diverse class of techniques that overcome the disadvantages of conventional pesticide detection methods, which are costly, time-consuming, and need complex instruments. Developing such monitoring technologies would enable prompt decontamination decisions, preventing further damage caused by pesticides' toxicity [57] (**Figure 23.4**).

23.3.3 NANOWIRE BIOSENSOR FOR DISEASE DIAGNOSTICS

Hepatitis E virus (HEV) is among the most common causes of acute viral hepatitis across the world. Graphene quantum dots and gold-embedded polyaniline nanowires were used to create a pulse-triggered ultrasensitive electrochemical sensor utilizing a surface polymerization and subsequent self-assembly technique (**Figure 23.5a**). When compared to other traditional electrochemical sensors, due to the expanded surface of the virus particle and the length of the antibody-conjugated polyaniline chain, applying an external electrical pulse during the viral accumulation phase increases sensitivity to HEV. In cell culture supernatant and a series of fecal specimen samples taken from a G7 HEV-infected monkey, the sensor was utilized to identify HEV genotypes such as G1, G3, G7, and ferret HEV. The sensitivity matches that of real-time quantitative reverse transcription-polymerase chain reaction (RT-qPCR). These findings imply that the suggested sensor has the

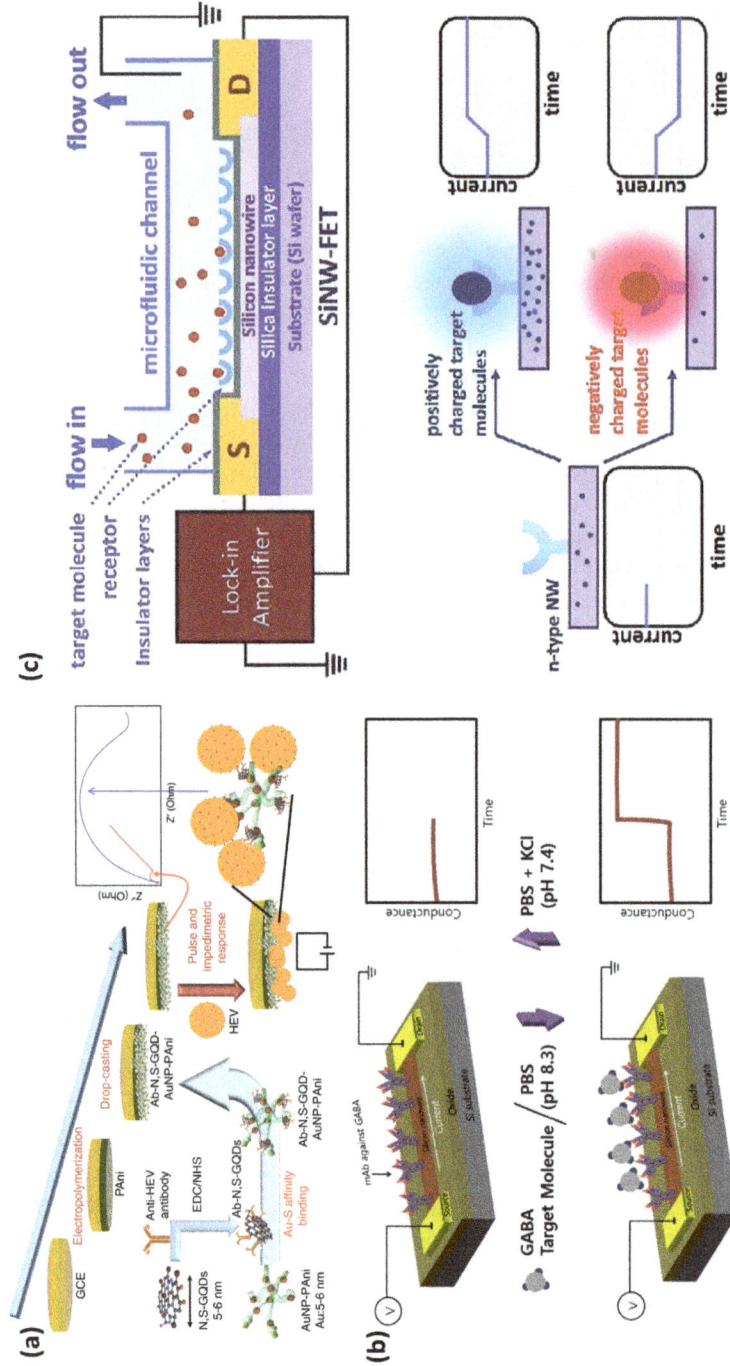

FIGURE 23.5 (a) The sensor electrode loaded with Ab-N, S-GQDs@AuNP-PAni nanocomposite and its pulse-induced impedimetric sensing of HEV are depicted schematically. (b) Schematic representation of the GABA detecting device based on silicon nanowires and field-effect transistors. Because of the immunoreaction, when GABA was supplied to this system, the conductance increased. (c) The working mechanism and concept of the SiNW-FET biosensor are illustrated schematically. A SiNW-FET is composed of a single SiNW connected to the source (S) and drain (D) electrodes on a Si wafer. The SiNW-FET device's PDMS channel is utilized to deliver the sample. A lock-in amplifier normally records the electrical signal in the presence of a water gate electrode (for example, an Ag/AgCl reference electrode) that is instantly placed into the buffer solution. A SiNW-FET biosensor recognizes particular targets by using receptor molecules mounted on the SiNW(s). When positively charged targets attach to an n-type SiNW-FET, holes accumulate in the SiNW, increasing electrical conductivity. Negatively charged targets, on the other hand, deplete charge carriers, lowering SiNW conductivity. Adapted with permission [58], [59], [60]. Copyright 2019, 2014; Nature, Springer, Elsevier; respectively. Distributed under a Creative Commons Attribution License 4.0 (CC BY) https://creativecommons.org/licenses/by/4.0/.

potential to pave the way for the development of robust, high-performance sensing approaches for HEV detection [58].

γ-aminobutyric acid (GABA) is a central nervous system (CNS) inhibitory neurotransmitter that serves as a significant biomarker for neurological illnesses such as Parkinson's disease and meningitis (**Figure 23.5b**). As a result, accurate measurement of the GABA molecule has emerged as an essential topic for the effective diagnosis and treatment of neurological illnesses. However, no reports of extremely sensitive biosensor systems capable of analyzing a wide variety of GABA molecules in a timely way have been made. To detect GABA molecules, a silicon nanowire field-effect transistor (FET) device-based immunosensor was designed. Electron-beam lithography was used to create zig-zag-shaped silicon nanowires, and the electrical properties of a p-type FET device were confirmed using a semiconductor analyzer. The fluorescent signal measurement was used to identify the best immobilizing condition of the antibody against the GABA molecule. Through immunoreactions, various concentrations of GABA ranging from 970 fM to 9.7 M were sensitively detected by conductance change on silicon nanowire-based platforms. Furthermore, because of the ease of miniaturization and the label-free system, we believe that the proposed device system has the potential to be used as an implantable biosensor to detect neurotransmitters in the brain, potentially opening up new avenues in the diagnosis and therapy of neurological disorders [59].

Silicon nanowire field-effect transistors (SiNW-FETs) have demonstrated significant potential as biosensors in very sensitive, selective, real-time, and label-free measurements. While various papers have reported uses of SiNW-FETs for biological species detection, less attention was given to summarizing the conjugating procedures involved in coupling organic bioreceptors with the inorganic transducer and strategies for enhancing device sensitivity (**Figure 23.5c**). This article attempts to summarize the different organic immobilization methods and discuss various sensitivity enhancing strategies, such as (I) minimizing nonspecific binding, (II) probe alignment, (III) signal enhancement by charge reporter, (IV) novel architecture structures, and (V) sensing in the sub-threshold regime [60].

23.3.4 NANOWIRE BIOSENSOR FOR BIOMARKERS

MnO_2 nanowires electrochemically reduced graphene-oxide-modified glassy carbon electrodes (MnO_2 NWs-ErGO/GCE) were used to create a promising sensing system for the ultra-sensitive detecting of dopamine (DA). MnO_2 NWs-ErGO/GCEs with a high electrochemical active area and a low charge transfer resistance (Rct) were presented (**Figure 23.6a**). As a consequence, the MnO_2 NWs-ErGO response peak current is approximately 13 times more than the bare GCE, exhibiting extraordinary electrocatalytic activity toward DA. According to electrochemical kinetics, DA oxidation is a quasi-reversible process with one electron and two protons. On the MnO_2 NWs-ErGO/GCE, three linear ranges (0.01−0.10, 0.10−1.0, and 1.0−80 μM) were found, with a low detection limit of 1.0 nM. Furthermore, even in the presence of 100-fold ascorbic acid (AA) and uric acid (UA), the response current remained nearly unchanged, indicating that MnO_2 NWs-ErGO had strong selectivity for DA. Finally, the MnO_2 NWs-ErGO/GCEs were used to identify DA with good accuracy and recovery in injectable solutions and human blood serum samples [61].

Using a combination of electrospinning and thermal annealing, Samie et al. developed a hybrid architecture comprising CeO_2-Au nanofibers (CeO_2-AuNFs) and single-crystalline RuO_2 nanowires (RuO_2 NWs) (**Figure 23.6b**). Electrospinning was used to create the CeO_2-Au hybrid nanofibers, which were subsequently annealed in air. On electrospun CeO_2-AuNFs, amorphous $Ru(OH)_3$ precursors were effectively transformed into highly single-crystalline RuO_2NWs at low temperatures. A unique RuO_2NWs-CeO_2-AuNFs hybrid architecture was coupled with graphite oxide (GO) and functionalized multi-walled carbon nanotubes (f-MWCNTs) to produce a sensitive electrochemical technique with acceptable characteristics for the simultaneous detection of serotonin (5-HT), dopamine (DA), and ascorbic acid (AA). Under optimal working conditions, linear calibration curves in

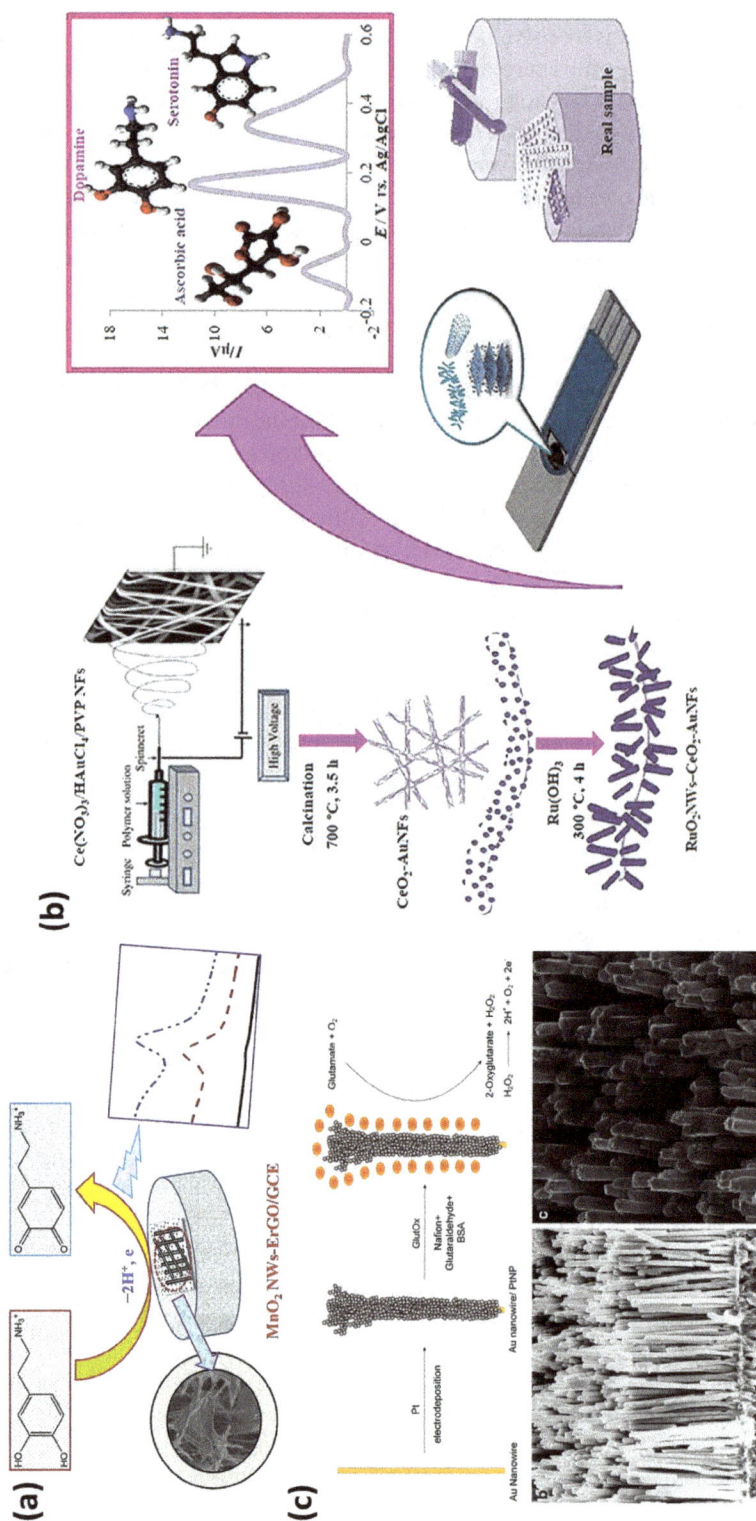

FIGURE 23.6 (a) Electrochemically reduced graphene oxide chemically modified electrode made of MnO_2 nanowires (MnO_2 NWs-ErGO/GCE); (b) hybrid architecture comprising CeO_2-Au nanofibers (CeO_2-AuNFs) and single-crystalline RuO_2 nanowires (RuO_2 NWs); (c) schematic representation of stepwise preparation of the GlutOx/PtNP/NAEs electrodes. Adapted with permission [61], [62], [63]. Copyright 2019, 2019, 2013; Elsevier, Elsevier, Elsevier; respectively. Distributed under a Creative Commons Attribution License 4.0 (CC BY) https://creativecommons.org/licenses/by/4.0/.

the ranges of 0.01–150, 0.01–120, and 0.5–100 μM were achieved, with detection limits of 2.4, 2.8, and 160 nM for 5-HT, DA, and AA, respectively. The suggested electrochemical sensor eliminates overpotentials and resolves oxidation peak potential overlaps. This sensor was successfully used to determine 5-HT, DA, and AA levels in biological fluids and pharmaceutical samples [62].

An enzyme-free electrochemical sensor platform based on vertically aligned nickel nanowire arrays (NiNAE) and Pt-coated nickel nanowire arrays (Pt/NiNAE) has been developed to detect glutamate (**Figure 23.6c**). The morphology of Ni electrodes was studied using SEM combined with energy dispersive X-ray (SEM-EDX), X-ray diffraction, and TEM. The catalytic activity of the NiNAE and the Pt/NiNAE for glutamate was measured using cyclic voltammetry (CV) and amperometry. Both NiNAE and Pt/NiNAE electrodes demonstrated significantly enhanced electrocatalytic activity toward these glutamates when compared to planar Ni electrodes, as well as a higher catalytic activity when compared to other metallic nanostructures' electrodes such as gold nanowire (AuNAE) and Pt coated gold nanowire (Pt/AuNAE). NiNAE and Pt/NiNAE have been reported to have the sensitivity of 65 and 96 A mM^{-1} cm^{-2}, respectively, which is roughly 6 to 9 times greater than the state of the glutamate sensor. Under ideal detection circumstances, both NiNAE and Pt/NiNAE sensors displayed linear behavior for glutamate detection in concentrations up to 8 mM, with limits of detection of 68 and 83 μM, respectively [63].

23.4 PERSPECTIVE APPLICATIONS

Nanowires' unique qualities, such as mechanical stability, reduced weight, current augmentation capability, potential reduction, and biocompatibility with biological and chemical species, place them among the most effective detection techniques. Sensors' true worth rests in their detection limit range and ultra-sensitivity. Their superior electrical, thermal, magnetic, and optical qualities contribute to their versatility. The overall qualities of biosensors will increase their popularity among researchers and in the industry in the future years. The bulk of the applications covered in the article is geared toward pushing the future limits of nanowire-based biosensors to detect a few molecules, if not a single molecule, in a given volume of biological fluid [64].

23.5 CONCLUSIONS

Nanowire materials have become key components in bioanalytical instruments because they improve sensitivity and detection limits down to single molecules detection. The unique features of such nanoobjects also provide alternatives to traditional transduction methods. Furthermore, combining several nanowire materials, each with unique properties, to improve the performance of biosensors is a well-accepted method. Due to a large number of various nanowires, each with its unique features, only a few examples could be described here while stressing the main benefits of such materials. The usage of the nanowire in biosensor applications is discussed and reviewed in depth.

ACKNOWLEDGMENT

Vy Anh Tran was funded by the Postdoctoral Scholarship Program of Vingroup Innovation Foundation (VINIF), code VINIF.2022.STS.45.

REFERENCES

[1] V. A. Tran, S.-W. Lee, pH-triggered degradation and release of doxorubicin from zeolitic imidazolate framework-8 (ZIF8) decorated with polyacrylic acid, RSC Advances 11 (2021) 9222–9234

[2] V. T. Le, Y. Vasseghian, V. D. Doan, T. T. T. Nguyen, T.-T. Thi Vo, H. H. Do, K. B. Vu, Q. H. Vu, T. Dai Lam, V. A. Tran, Flexible and high-sensitivity sensor based on Ti$_3$C$_2$–MoS$_2$ MXene composite for the detection of toxic gases, Chemosphere 291 (2021) 133025

[3]   P. Ambhorkar, Z. Wang, H. Ko, S. Lee, K.-I. Koo, K. Kim, D.-I. D. Cho, Nanowire-based biosensors: From growth to applications, Micromachines (Basel) 9 (2018) 679

[4]   V. Anh Tran, T. Khoa Phung, V. Thuan Le, T. Ky Vo, T. Tai Nguyen, T. Anh Nga Nguyen, D. Quoc Viet, V. Quang Hieu, T.-T. Thi Vo, Solar-light-driven photocatalytic degradation of methyl orange dye over Co_3O_4-ZnO nanoparticles, Materials Letters 284 (2021) 128902

[5]   X. Li, J. Mo, J. Fang, D. Xu, C. Yang, M. Zhang, H. Li, X. Xie, N. Hu, F. Liu, Vertical nanowire array-based biosensors: device design strategies and biomedical applications, Journal of Materials Chemistry B 8 (2020) 7609–7632

[6]   V. Anh Tran, L. T. Nhu Quynh, T.-T. Thi Vo, P. A. Nguyen, T. N. Don, Y. Vasseghian, H. Phan, S.-W. Lee, Experimental and computational investigation of a green Knoevenagel condensation catalyzed by zeolitic imidazolate framework-8, Environmental Research 204 (2021) 112364

[7]   C. Jianrong, M. Yuqing, H. Nongyue, W. Xiaohua, L. Sijiao, Nanotechnology and biosensors, Biotechnology Advances 22 (2004) 505–518

[8]   V. A. Tran, G. V. Vo, M. A. Tan, J.-S. Park, S. S. A. An, S.-W. Lee, Dual stimuli-responsive multifunctional silicon nanocarriers for specifically targeting mitochondria in human cancer cells, Pharmaceutics 14 (2022) 858

[9]   V. A. Tran, T.-T. Thi Vo, M.-N. T. Nguyen, N. Duy Dat, V.-D. Doan, T.-Q. Nguyen, Q. H. Vu, V. T. Le, T. D. Tong, Novel α-mangostin derivatives from mangosteen (Garcinia mangostana L.) peel extract with antioxidant and anticancer potential, Journal of Chemistry 2021 (2021) 9985604

[10] V. A. Tran, N. H. T. Tran, L. G. Bach, T. D. Nguyen, T. T. Nguyen, T. T. Nguyen, T. A. N. Nguyen, T. K. Vo, T.-T. T. Vo, V. T. Le, Facile synthesis of propranolol and novel derivatives, Journal of Chemistry 2020 (2020) 9597426

[11] V. A. Tran, S. W. Lee, A prominent anchoring effect on the kinetic control of drug release from mesoporous silica nanoparticles (MSNs), J Colloid Interface Sci 510 (2018) 345–356

[12] V. T. Le, V. D. Doan, V. A. Tran, H. S. Le, D. L. Tran, T. M. Pham, T. H. Tran, H. T. Nguyen, Cu/Fe_3O_4@carboxylate-rich carbon composite: One-pot synthesis, characterization, adsorption and photo-Fenton catalytic activities, Materials Research Bulletin 129 (2020) 110913

[13] V. A. Tran, H. H. Do, V. T. Le, Y. Vasseghian, V. Vo, S. H. Ahn, S. Y. Kim, S.-W. Lee, Metal-organic-framework-derived metals and metal compounds as electrocatalysts for oxygen evolution reaction: A review, International Journal of Hydrogen Energy 47 (2021) 19590–19608

[14] S. S. Meenambiga, P. Sakthiselvan, S. Hari, D. Umai, Nanotechnology for blood test to predict the blood diseases/blood disorders, Nanotechnology for Hematology, Blood Transfusion, and Artificial Blood, Micro and Nano Technologies, Elsevier (2022) 285–311

[15] R. Choubey, N. Sonker, J. Bajpai, P. Jain, A. Singh, Synthesis of polymer nanomaterials, mechanisms, and their structural control, Advances in Polymeric Nanomaterials for Biomedical Applications, Elsevier (2021) 41–63

[16] S. B. Betzler, A. Wisnet, B. Breitbach, C. Mitterbauer, J. Weickert, L. Schmidt-Mende, C. Scheu, Template-free synthesis of novel, highly-ordered 3D hierarchical $Nb_3O_7(OH)$ superstructures with semiconductive and photoactive properties, Journal of Materials Chemistry A 2 (2014) 12005–12013

[17] X. Wu, J. S. Kulkarni, G. Collins, N. Petkov, D. Almecija, J. J. Boland, D. Erts, J. D. Holmes, Synthesis and electrical and mechanical properties of silicon and germanium nanowires, Chemistry of Materials 20 (2008) 5954–5967

[18] A. I. Boukai, Y. Bunimovich, J. Tahir-Kheli, J.-K. Yu, W. A. Goddard Iii, J. R. Heath, Silicon nanowires as efficient thermoelectric materials, nature 451 (2008) 168–171

[19] F. Voelklein, M. C. Schmitt, H. Reith, D. Huzel, Characterization and application of thermoelectric nanowires, Nanowires-Implementations and Applications, IntechOpen (2011) 141–163

[20] J. Ma, M. Zhan, K. Wang, Ultralightweight silver nanowires hybrid polyimide composite foams for high-performance electromagnetic interference shielding, ACS Applied Materials & Interfaces 7 (2015) 563–576

[21] G. Shen, P.-C. Chen, K. Ryu, C. Zhou, Devices and chemical sensing applications of metal oxide nanowires, Journal of Materials Chemistry 19 (2009) 828–839

[22] M. Zhang, J. Li, H. Li, Y. Li, W. Shen, Morphology-dependent redox and catalytic properties of CeO2 nanostructures: nanowires, nanorods and nanoparticles, Catalysis Today 148 (2009) 179–183

[23] O. El-Zubir, E. L. Kynaston, J. Gwyther, A. Nazemi, O. E. C. Gould, G. R. Whittell, B. R. Horrocks, I. Manners, A. Houlton, Bottom-up device fabrication via the seeded growth of polymer-based nanowires, Chemical Science 11 (2020) 6222–6228

[24] M. F. Fatahilah, F. Yu, K. Strempel, F. Römer, D. Maradan, M. Meneghini, A. Bakin, F. Hohls, H. W. Schumacher, B. Witzigmann, A. Waag, H. S. Wasisto, Top-down GaN nanowire transistors with nearly zero gate hysteresis for parallel vertical electronics, Scientific Reports 9 (2019) 10301

[25] A. Solanki, H. Um, Top-down etching of Si nanowires, semiconductors and semimetals, PLoS ONE, Elsevier (2018) 71–149

[26] X. Yu, Y. Wang, H. Zhou, Y. Liu, Y. Wang, T. Li, Y. Wang, Top-down fabricated silicon-nanowire-based field-effect transistor device on a (111) silicon wafer, Small 9 (2013) 525–530

[27] L. Falzone, G. Gattuso, A. Tsatsakis, D. A. Spandidos, M. Libra, Current and innovative methods for the diagnosis of COVID-19 infection, International Journal of Molecular Medicine 47 (2021) 1–23

[28] S. Alpdagtas, E. Ilhan, E. Uysal, M. Sengor, C. B. Ustundag, O. Gunduz, Evaluation of current diagnostic methods for COVID-19, APL Bioengineering 4 (2020) 041506

[29] J. M. Vindeirinho, E. Pinho, N. F. Azevedo, C. Almeida, SARS-CoV-2 diagnostics based on nucleic acids amplification: from fundamental concepts to applications and beyond, Frontiers in Cellular and Infection Microbiology (2022) 263

[30] M. De Felice, M. De Falco, D. Zappi, A. Antonacci, V. Scognamiglio, Isothermal amplification-assisted diagnostics for COVID-19, Biosensors and Bioelectronics 204 (2022) 114101

[31] E. Mohit, Z. Rostami, H. Vahidi, A comparative review of immunoassays for COVID-19 detection, Expert Review of Clinical Immunology 17 (2021) 573–599

[32] B. Udugama, P. Kadhiresan, H. N. Kozlowski, A. Malekjahani, M. Osborne, V. Y. Li, H. Chen, S. Mubareka, J. B. Gubbay, W. C. Chan, Diagnosing COVID-19: The disease and tools for detection, ACS Nano 14 (2020) 3822–3835

[33] C. Ihling, D. Tanzler, S. Hagemann, A. Kehlen, S. Huttelmaier, C. Arlt, A. Sinz, Mass spectrometric identification of SARS-CoV-2 proteins from gargle solution samples of COVID-19 patients, Journal of Proteome Research 19 (2020) 4389–4392

[34] M. Jamiruddin, B. A. Meghla, D. Z. Islam, T. A. Tisha, S. S. Khandker, M. U. Khondoker, M. Haq, N. Adnan, M. Haque, Microfluidics technology in SARS-CoV-2 diagnosis and beyond: A systematic review, Life 12 (2022) 649

[35] R. Samson, G. R. Navale, M. S. Dharne, Biosensors: Frontiers in rapid detection of COVID-19, 3 Biotech 10 (2020) 1–9

[36] K.-H. Liang, T.-J. Chang, M.-L. Wang, P.-H. Tsai, T.-H. Lin, C.-T. Wang, D.-M. Yang, Novel biosensor platforms for the detection of coronavirus infection and severe acute respiratory syndrome coronavirus 2, Journal of the Chinese Medical Association 83 (2020) 701

[37] B. A. Taha, Y. Al Mashhadany, M. H. Hafiz Mokhtar, M. S. Dzulkefly Bin Zan, N. Arsad, An analysis review of detection coronavirus disease 2019 (COVID-19) based on biosensor application, Sensors 20 (2020) 6764

[38] C. Weiss, M. Carriere, L. Fusco, I. Capua, J. A. Regla-Nava, M. Pasquali, J. A. Scott, F. Vitale, M. A. Unal, C. Mattevi, Toward nanotechnology-enabled approaches against the COVID-19 pandemic, ACS Nano 14 (2020) 6383–6406

[39] E. Alphandéry, The potential of various nanotechnologies for coronavirus diagnosis/treatment highlighted through a literature analysis, Bioconjugate Chemistry 31 (2020) 1873–1882

[40] A. Poghossian, M. Jablonski, D. Molinnus, C. Wege, M. J. Schöning, Field-effect sensors for virus detection: From Ebola to SARS-CoV-2 and plant viral enhancers, Frontiers in Plant Science 11 (2020) 598103

[41] V. Generalov, O. Naumova, D. Shcherbakov, A. Safatov, B. Zaitsev, E. Zaitseva, G. Buryak, D. Shcheglov, A. Cheremiskina, I. Merkuleva, Indication of the coronavirus model using a nanowire biosensor, Multidisciplinary Digital Publishing Institute Proceedings 60 (2020) 50

[42] R. Antiochia, Nanobiosensors as new diagnostic tools for SARS, MERS and COVID-19: From past to perspectives, Microchimica Acta 187 (2020) 1–13

[43] G. Seo, G. Lee, M. J. Kim, S.-H. Baek, M. Choi, K. B. Ku, C.-S. Lee, S. Jun, D. Park, H. G. Kim, Rapid detection of COVID-19 causative virus (SARS-CoV-2) in human nasopharyngeal swab specimens using field-effect transistor-based biosensor, ACS Nano 14 (2020) 5135–5142

[44] X. Li, M. Geng, Y. Peng, L. Meng, S. Lu, Molecular immune pathogenesis and diagnosis of COVID-19, Journal of Pharmaceutical Analysis 10 (2020) 102–108

[45] F. Patolsky, G. Zheng, O. Hayden, M. Lakadamyali, X. Zhuang, C. M. Lieber, Electrical detection of single viruses, Proceedings of the National Academy of Sciences 101 (2004) 14017–14022

[46] A. Lakhiar, G. Jianmin, T. N. Syed, F. A. Chandio, N. A. Buttar, W. A. Qureshi, Monitoring and control systems in agriculture using intelligent sensor techniques: A review of the aeroponic system, Journal of Sensors 2018 (2018) 1–18

[47] M. Kundu, P. Krishnan, R. K. Kotnala, G. Sumana, Recent developments in biosensors to combat agricultural challenges and their future prospects, Trends in Food Science & Technology 88 (2019) 157–178

[48] N. H. Kumar, M. Murali, H. Girish, S. Chandrashekar, K. Amruthesh, M. Sreenivasa, S. Jagannath, Impact of climate change on biodiversity and shift in major biomes, Global Climate Change, Elsevier (2021) 33–44

[49] Z. Li, F. Qiu, X. Yue, Q. Tian, D. Yang, T. Zhang, Eco-friendly self-crosslinking cellulose membrane with high mechanical properties from renewable resources for oil/water emulsion separation, Journal of Environmental Chemical Engineering 9 (2021) 105857

[50] Y. Chen, S. Yang, Z. Qiu, Y. Li, F. Qiu, T. Zhang, Fabrication of flexible AgNW/cellulose hybrid film with heat preservation and antibacterial properties for agriculture application, Cellulose 28 (2021) 8693–8704

[51] B. M. Sharma, G. K. Bharat, S. Tayal, L. Nizzetto, P. Čupr, T. Larssen, Environment and human exposure to persistent organic pollutants (POPs) in India: A systematic review of recent and historical data, Environment International 66 (2014) 48–64

[52] S. Mostafalou, M. Abdollahi, Pesticides and human chronic diseases: evidences, mechanisms, and perspectives, Toxicology and Applied Pharmacology 268 (2013) 157–177

[53] M. Jokanović, M. Kosanović, Neurotoxic effects in patients poisoned with organophosphorus pesticides, Environmental Toxicology and Pharmacology 29 (2010) 195–201

[54] H.-F. Cui, W.-W. Wu, M.-M. Li, X. Song, Y. Lv, T.-T. Zhang, A highly stable acetylcholinesterase biosensor based on chitosan-TiO_2-graphene nanocomposites for detection of organophosphate pesticides, Biosensors and Bioelectronics 99 (2018) 223–229

[55] C. Zhang, Y. Fan, H. Zhang, S. Chen, R. Yuan, An ultrasensitive signal-on electrochemiluminescence biosensor based on Au nanoclusters for detecting acetylthiocholine, Analytical and Bioanalytical Chemistry 411 (2019) 905–913

[56] Gong, L. Wang, L. Zhang, Electrochemical biosensing of methyl parathion pesticide based on acetyl-cholinesterase immobilized onto Au–polypyrrole interlaced network-like nanocomposite, Biosensors and Bioelectronics 24 (2009) 2285–2288

[57] Zhang, B. Wang, Y. Li, W. Shu, H. Hu, L. Yang, An acetylcholinesterase biosensor with high stability and sensitivity based on silver nanowire–graphene–TiO_2 for the detection of organophosphate pesticides, RSC Advances 9 (2019) 25248–25256

[58] A. D. Chowdhury, K. Takemura, T.-C. Li, T. Suzuki, E. Y. Park, Electrical pulse-induced electrochemical biosensor for hepatitis E virus detection, Nature Communications 10 (2019) 3737

[59] J. H. Lee, E. J. Chae, S. J. Park, J. W. Choi, Label-free detection of γ-aminobutyric acid based on silicon nanowire biosensor, Nano Converg 6 (2019) 13

[60] M.-Y. Shen, B.-R. Li, Y.-K. Li, Silicon nanowire field-effect-transistor based biosensors: From sensitive to ultra-sensitive, Biosensors and Bioelectronics 60 (2014) 101–111

[61] Q. He, J. Liu, X. Liu, G. Li, D. Chen, P. Deng, J. Liang, A promising sensing platform toward dopamine using MnO_2 nanowires/electro-reduced graphene oxide composites, Electrochimica Acta 296 (2019) 683–692

[62] H. Asadi Samie, M. Arvand, RuO_2 nanowires on electrospun CeO_2-Au nanofibers/functionalized carbon nanotubes/graphite oxide nanocomposite modified screen-printed carbon electrode for simultaneous determination of serotonin, dopamine and ascorbic acid, Journal of Alloys and Compounds 782 (2019) 824–836

[63] Jamal, M. Hasan, A. Mathewson, K. M. Razeeb, Disposable sensor based on enzyme-free Ni nanowire array electrode to detect glutamate, Biosensors and Bioelectronics 40 (2013) 213–218

[64] P. Arora, A. Sindhu, N. Dilbaghi, A. Chaudhury, Engineered multifunctional nanowires as novel biosensing tools for highly sensitive detection, Applied Nanoscience 3 (2013) 363–372

24 Semiconducting WO$_3$ Nanowires for Biomedical Applications

Benachir Bouchikhi
Biosensors and Nanotechnology Group, Faculty of Sciences,
Moulay Ismaïl University of Meknes, Zitoune, Meknes, Morocco

Omar Zaim
Biosensors and Nanotechnology Group, Faculty of Sciences,
Moulay Ismaïl University of Meknes, Zitoune, Meknes, Morocco
Biosensors and Nanotechnology Group, Department of Biology,
Faculty of Sciences, Moulay Ismaïl University, Zitoune, Meknes, Morocco

Nezha El Bari
Biosensors and Nanotechnology Group, Department of Biology,
Faculty of Sciences, Moulay Ismaïl University, Zitoune, Meknes, Morocco

Soukaina Motia
Biosensors and Nanotechnology Group, Faculty of Sciences, Moulay
Ismaïl University of Meknes, Zitoune, Meknes, Morocco
Biosensors and Nanotechnology Group, Department of Biology, Faculty
of Sciences, Moulay Ismaïl University, Zitoune, Meknes, Morocco

CONTENTS

24.1   INTRODUCTION

In recent years, there has been considerable interest in the synthesis of nanostructured materials due to their superior and improved functional properties. Among the different nanostructures, nanowires (NWs) are suitable for gas sensing applications because they offer several advantages. These include a high surface-to-volume ratio, improved electron transfer pathway, dimensions comparable to the surface charge region extension, fabrication and handling ease, small size, and low power consumption. For these reasons, many types of sensors based on metal oxide NWs such as tin dioxide (SnO_2) [1], molybdenum trioxide (MoO_3) [2], tungsten trioxide (WO_3) [3], or zinc oxide (ZnO) [4], have been employed for exhaled breath analysis. Besides, exhaled breath analysis using NW sensors for medical applications is a relatively unexplored research area with very high potential.

The idea of non-invasive, painless, and point-of-care health monitoring has always attracted scientists. Nevertheless, the research activity in breath analysis has increased over the past two decades due to recent advances in analytical instruments and also to the development of new sensors. The use of non-invasive methods to monitor disease progression would have more benefits than traditional methods [5]. Numerous works have been done on the identification of the biomarkers of disease related to breathing VOCs of the affected person [6]. In other terms, the onset of different illnesses is linked to a significant variation in the concentration of certain VOCs in exhaled breath [7].

This chapter summarizes some of the key technological developments and characterization of an electronic nose (e-nose) device based on two types of semiconducting, such tin dioxide (SnO_2), and tungsten trioxide (WO_3) nanowires for healthcare and biomedical applications. The first purpose of this chapter is to focus on the sensing measurement's ability of each family of the sensor arrays to discriminate patients with liver cirrhosis (LCi) from healthy subjects (HC) based on exhaled breath analysis. Moreover, this chapter highlights also the impact of some parameters such as the decoration of metal oxide surface with nanomaterials, the illumination of gas sensor surface with UV light, and the optimization of the sensor's operating temperatures to enhance the gas sensing properties of the WO$_3$ nanowire.

24.1.1   GAS SENSORS BASED ON METAL OXIDE NWS

Since the late 1990s, the rapid development of innovative nanomaterials and technologies has advanced gas sensors. Among them, one-dimensional nanowires (NWs) offer numerous advantages for gas sensors in comparison to bulk semiconductor metal oxide (SMO) and their sputtered thin films, such as high surface-to-volume ratios, excellent thermal stability, Debye lengths comparable to target gas molecules for superior sensitivity, relatively low power consumption, and low tendency to form clusters [8,9]. Up to now, physical deposition and chemical vapor/liquid phase processes can be used for the synthesis of SMO-NWs of various sizes, shapes, and compositions [10].

Several types of semiconductor nanowires, such as In_2O_3, SnO_2, TiO_2, ZnO, and so on, have been extensively useful in gas sensors. For example, Wang et al. [11,12] provided porous SnO_2 nanowires made from glycolate precursors under mild conditions that demonstrated good sensitivity to some gases such as H$_2$, CO, and C$_2$H$_5$OH. Comini et al. proved the use of SnO$_2$ nanowires as sensing materials by displaying remarkable current variations for CO and ethanol, respectively, in a synthetic air environment [13], whereas Law et al. showed the first photochemical NO$_2$ nanosensors, based on SnO$_2$ nanowires, working at room temperature [14].

24.1.2 BREATH ANALYSIS THROUGH NW SENSORS FOR DISEASE DIAGNOSIS

NW sensors are very promising for early chronic disorders' diagnosis and the surveillance of physical conditions in a non-invasive, simple, and inexpensive way. Owing to abnormal metabolism, human breath contains CO$_2$, O$_2$, N$_2$, water, inert gases, and hundreds of other VOCs usually present at very low concentrations [15]. These exhaled breath gases can serve as biomarker species, which are typically identified by a traditional chromatography-mass spectrometry (GC-MS) with bench-top sensitivity and selectivity [16]. To date, various biomarker substances have been investigated in clinical trials, like the ethanol tests used by law enforcement for safe driving [17], the CO$_2$ surveillance in critical care and anesthesia [18], and NO to detect asthma [19]. Nevertheless, GC-MS methods are highly expensive, complex to use, time-consuming, and not portable. The utilization of SMO-NWs for breath detection is efficient for its higher sensitivity, chip-scale miniaturization, and low power operation.

An early investigation for breath sensors was done to test blood alcohol levels in 1983 when Watson carried out the detection action using a Figaro TGS 812 sensor [20]. Nanoscale engineering of SMO sensing materials allows for high sensitivity and lower detection limits. Recently, many articles have previously been published, which provide in-depth information on SMO breath sensors. Alizadeh et al. provide a detailed review of breath acetone sensors to monitor diabetes [21]. They focus on different SMO materials such as SnO$_2$, WO$_3$, TiO$_2$, NiO, ZnO, and various sensor technologies. Tai et al. recapitulate the latest studies on wearable breathing behaviors using a portable humidity monitoring device [22]. Nasiri et al. are concerned with the nanodimensional design of the most modern breath sensors [23]. Liu et al. are reviewing flexible breath sensors, comprising sensing materials, mechanisms of sense, and their fabrication methods [24]. Tricoli et al. review the latest advances in point-of-care surveillance of chronic kidney disorders through the detection of nitrogen biomarkers (urea, NH3, and creatinine) [25]. Furthermore, Guntner et al. highlight the major challenges currently impeding the development of breath sensors, including the enhancement of stability and selectivity of gas sensing, the comprehension of the underlying biochemical and physiological mechanisms of breath biomarkers, and the extraction of sufficient data for clinical trials [26].

Effective approaches have been established to improve the sensing capability of sensors, such as functionalization with additives, design of sensor arrays, and the use of filters as size-selective membranes [27]. For example, Saidi et al. have used a sensor matrix composed of interdigitated chemical gas sensors containing pristine or functionalized WO$_3$ nanowires to analyze exhaled breath samples of patients with lung cancer and healthy controls [28].

24.1.3 EFFECT OF UV-LIGHT IRRADIATION AND HEATING OPERATING TEMPERATURE ON NW-BASED GAS SENSORS

The characteristics of NW-based gas sensors are highly dependent on their operating temperature. Different types of gases have been demonstrated to have different reaction rates. For example, hydrogen sulfide and C$_2$H$_5$OH oxidize on the metal oxide surface at low temperatures. Ketones and alcohols oxidize at intermediate temperatures (above 200°C) while alkanes (methane and propane in specific) oxidize at high temperatures (above 400°C) [29]. Besides, the illumination of metal oxide NW sensors with LEDs is an attractive option for activating chemical reactions on the metal oxide NW surface. It can replace the need to operate the sensors at high working temperatures to some extent [30]. This approach is used to ensure sufficient energy for the acceleration of adsorption-desorption processes. Moreover, illumination can dramatically change the detection properties due to the penetration depths related to the wavelength used **[30]**. For example, experimental results obtained by Li et al. **[31]** proved the effect of illumination by improving the selectivity of the rGO/ SnO$_2$ gas sensor designed to detect NO$_2$ and SO$_2$ gases. Research work conducted by Trawka et al. **[32]** showed that the selectivity and sensitivity of WO$_3$ NW gas sensors have been enhanced by illuminating the active surface with LEDs at selected wavelengths of 362 nm and 394 nm. In addition, new approaches for improving the selectivity and sensitivity of resistive gas sensors have also been demonstrated by Smulko et al. by illumination of the active surface of the sensor arrays by combining several LEDs operating at different wavelengths **[33]**.

24.2 EXPERIMENTAL PROCEDURE

24.2.1 BREATH SAMPLING

In this study, 39 volunteers were involved including 22 liver cirrhosis (LCi) patients and 17 healthy controls (HC). Only adult subjects were included. Volunteers who had used medication, tobacco, alcohol, drugs, food, or drinks before noon were excluded automatically from the study.

In the literature, there are several protocols for the analysis of breath samples. Tedlar® bags have been demonstrated to provide satisfactory sample storage properties in comparison to other bags fabricated from different materials **[34]**. The participants were instructed to wash out their mouths before sampling exhaled breath into Tedlar® bags with the mouthpiece. Each volunteer was asked to give three breath samples. The participant blows into Tedlar® bags (Supelco Inc., Bellefonte, PA. USA) by using a one-way valve. The role of this valve is to block the mixing of the outside air with the collected breath.

An agreement was provided by the Biomedical Research Ethics Committee of the Avicenne University Hospital (Mohammed V University, Rabat, Morocco). All participants signed a consent agreement before sample collection.

24.2.2 ELECTRONIC NOSE EXPERIMENTAL SET-UP

Figure 24.1 shows the developed e-nose device for breath analysis. Generally, it is composed of three basic elements, which are the sampling system, sensor array, and artificial intelligence unit. The sensor array is the most important element in an e-nose device. In the beginning, it comprised

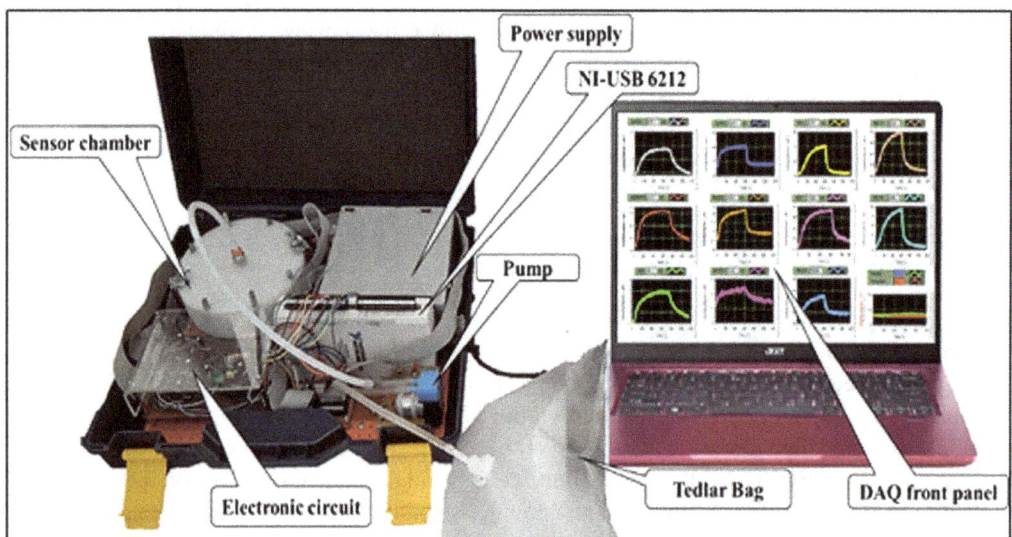

FIGURE 24.1 E-nose device employed for breath analysis.

five commercial SnO$_2$-based sensors. After that, six new gas sensors based on tungsten trioxide (WO$_3$) were elaborated and introduced in the chamber to diversify the sensitivity of the developed device. Hereinafter (in the next sub-sections), we describe both sensor arrays and other elements that composed the e-nose system.

In the software part, a graphical interface programmed by LabVIEW software was developed to acquire the data related to breathing analysis. The interface communicates with the main e-nose board via a USB cable, allowing real-time visualization of sensor conductance values, and storage of raw values for further processing and analysis.

24.2.2.1 Commercial SnO$_2$ Gas Sensors

The choice of a suitable sensor type for a given application should be dictated by the conditions under which the measurement will be made. Detection performance is the main factor to consider when selecting a gas sensor, including detection limit, response time, and sensor durability. In our applications, the selection of commercial SnO$_2$ sensors was made primarily based on their sensitivity to the various chemical compounds that may be present in exhaled breath. To analyze breath VOCs, five SnO$_2$-based gas sensors (Henan Hanwei Electronics Co. Ltd, Zheng zhou - China) were used in the sensor array: MQ-2 (hydrogen, propane, methane), MQ-3 (alcohol), MQ-135 (carbon dioxide, benzene, nitric oxide), MQ-137 (ammonia) and MQ-138 (acetone, toluene, ethanol, methanol). This family of the commercial sensor has been used in several VOC detection applications, including breath analysis [35].

24.2.2.2 Elaborated WO$_3$ Nanowire Gas Sensors

Owing to the broad range of electronic and physical properties, tungsten trioxide (WO$_3$) becomes one of the main n-type metal-oxide-semiconductor used and investigated in the detection of VOCs. However, they also have some drawbacks, such as the deficiency of selectivity and sensitivity for breath analysis, and instability [36]. In this work, we have functionalized WO$_3$ layers by a set of nanoparticles, including noble metal dopants namely platinum (Pt), gold (Au), and an alloy of gold/platinum (Au/Pt), nickel (Ni), and copper (Cu). This functionalization has been done at MINOS of Universitat Rovira I Virgili, Tarragona. The choice of these materials was based on their sensing target species, which were nearby with certain VOCs that could be found in human breath [37-40].

24.2.2.2.1 Sensors' Substrates

Alumina substrates with interdigitated platinum electrodes were employed as transducer elements for resistive gas sensors. The sensor substrate contains the transducer element with a pair of ten interdigitated platinum electrodes (electrode gap is 0.30 mm). The heating resistor is made-up of platinum. Before starting the deposition process, the substrates were respectively cleaned with ethanol, acetone, distilled water, and dried with a nitrogen flow [41].

24.2.2.2.2 Elaboration of WO$_3$ Nanowire Thin Sensing Films

Tungsten hexaphenoxide (C$_{36}$H$_{30}$O$_6$W) is the key precursor to synthesizing tungsten trioxide (WO$_3$) nanomaterials [42]. But, this precursor is not commercially offered, and it was necessary to substitute it with another precursor that allows producing WO$_3$ nanostructures with similar properties. C$_6$O$_6$W is a commercially available precursor, which is recently considered a good candidate for elaborating WO$_3$ layers [43]. Therefore, it was also used for the synthesis of pristine WO$_3$ and WO$_3$-doped with other nanoparticles (NPs) in a single-step process [39]. Gold (III) chloride trihydrate, chloroplatinic acid hydrate, nickel (II) acetylacetonate, and iron (III) acetylacetonate were used as additive precursors for the functionalization of the sensing layers. Table 24.1 shows detailed information about the tungsten hexacarbonyl precursor and other materials used for the functionalization of WO$_3$.

Pristine and functionalized WO$_3$ NWs have been grown using aerosol-assisted chemical vapor deposition (AACVD), in which the precursor's solution is produced ultrasonically, evaporated as aerosol droplets, and transferred via nitrogen carrier gas to the hot wall reactor that contains the

TABLE 24.1
Precursor's Characteristics [41]

Precursor	Molecular Formula	Assay (%)	Molar Mass (g/mol)	Melting Point (°C)
Tungsten hexacarbonyl	C_6O_6W	97	351,9	150
Nickel (II) acetylacetonate	$C_{10}H_{14}NiO_4$	95	256,91	230
Iron (III) acetylacetonate	$C_{15}H_{21}FeO_6$	99,9	353,17	182
Chloroplatinic acid hydrate	H_2PtCl_6	99,9	409,81	60
Gold (III) chloride trihydrate	$HAuCl_4$	99,9	393,83	254

TABLE 24.2
Experimental Conditions Used for the Growth of Pristine and Functionalized WO_3 NWs [41]

Main Precursor: C_6O_6W (mg)	Solvent	Added Materials	Temp (°C)	Obtained Morphology
50	15 mL of acetone + 5 mL of methanol	–	500	Pristine WO_3
		7 mg of $HAuCl_4$	380	NPs (WO_3 + Au)
		7 mg of H_2PtCl_6		NPs (WO_3 + Pt)
		3,5 mg of $HAuCl_4$ + 3,5 of H_2PtCl_6		NPs (WO_3 + Au + Pt)
		10 mg of $C_{10}H_{14}NiO_4$	400	NPs (WO_3 + Ni)
	15 mL of acetone + 5 mL of chloroform	7 mg of $C_{10}H_{14}CuO_4$	380	NPs (WO_3 + Cu)

substrate. AACVD was chosen due to many advantages, such as its high deposition rate [43], a wide range of precursors including the thermally unstable or nonvolatile ones [44], and low operating temperature (usually between 350 and 550°C) [45]. The lower working temperature of the AACVD process makes it convenient for several types of substrates (e.g., silicon, ceramic, glass, and alumina). Furthermore, it is low cost and offers a useful tool for producing sensing films or coatings in a single-step process that saves time and is environmentally advantageous. In our case, six different sensing layers (i.e., pristine WO_3 NWs and metal-doped WO_3 NWs: Pt/WO_3, Au/Pt/WO_3, Au/WO_3, Ni/WO_3, and Cu/WO_3) were deposited using AACVD following the experimental conditions represented in **Table 24.2**.

After growing the films with AACVD, all substrates were subject to a heat treatment at high temperatures in the presence of synthetic air to clean the surface of deposit residues and to stabilize and crystallize the cultivated layers [46].

24.2.2.3 Data Acquisition System

To acquire the data for breath analysis, a data acquisition system is placed between the sensors and the central processing unit (computer). In our applications, the data acquisition system was realized using a NI USB-6212 board from National Instruments (Austin, Texas, USA) with a graphical interface to visualize and save the data provided by the sensor arrays. The choice of this card was accomplished due to its advantageous characteristics, including the following:

- Resolution
  - Analog digital converter (ADC), 16 bits
  - Digital analog converter (DAC), 16 bits

- Number of channels
  - Analog input: eight differential or 16 single-ended
  - Analog output channels: two
- Input range: ±0,2 V, ±1 V, ±5 V, ±10 V
- Input impedance: >10 GΩ in parallel with 100 pF
- Sampling rate: 400 kS/s
- Timing resolution: 50 ns

The monitoring of sensor signals including acquisition and storage of data is controlled via a home-developed GUI programmed under LabVIEW software. The developed program consists of reading and viewing each sensor output in real-time. The acquired raw data of each sensor response is organized and recorded in a separate text file.

24.2.2.4 Measurement Process

Directly after the breath sampling step, the samples were pumped into the sensor chamber at a constant flow rate of 200 mL/min for each Tedlar® bag. The total volume of the sensor chamber is 284 cm^3 made of polytetrafluoroethylene commonly known as "Teflon". The responses of the WO$_3$ functionalized with nanoparticles and SnO$_2$ sensor arrays were acquired for a few minutes during exposure to the breath samples.

The WO$_3$ NW sensors were worked at their optimal functioning temperatures for analysis of breath samples: 120°C (Pristine WO$_3$, Pt/WO$_3$ and Au/Pt/WO$_3$), and 160°C (Au/WO$_3$, Ni/WO$_3$, and Cu/WO$_3$), and they were illuminated with UV light at 394 nm during the experimentations.

24.2.3 DATA PRE-PROCESSING

To minimize the impact of sensor drift, a process for normalizing the response of each gas sensor ($\frac{G-G_0}{G_0}$) is applied where G is the measured conductance and G_0 is the baseline [47].

To facilitate the treatment of multivariate data, only significant features are considered in the initial data. These features keep the maximum of information from these responses. In this work, the processing of the e-nose responses was performed by using the two features mentioned below:

$\Delta G = (G_S - G_0)$: The difference between the stabilized conductance (G_s) and the initial conductance (G_0);

AUC: Area under the sensor response curve calculated by a trapezoidal method. The area examined is defined from 1 to 9 minutes of the measurement time.

24.2.4 DATA ANALYSIS

After feature extraction, data processing methods (PCA, DFA) are applied to discriminate the analyzed breath samples for best representation and interpretation [48]. Pattern recognition methods are essential means to provide a practical system for the characterization of a wide variety of compounds [49]. The purpose of employing these techniques in this study is to investigate the performance of the e-nose in distinguishing the two groups from breath samples by using both supervised and unsupervised techniques.

PCA is one of the most frequently used methods of multivariate data analysis. It is used to study multidimensional data sets with quantitative variables [49]. It is a projection method because it projects the observations from a p-dimensional space with p variables to a k-dimensional space (where k< p) to retain the maximum amount of information (information is measured here by the total variance of the data set) from the original dimensions. In this work, PCA was applied to analyze the multivariate e-nose data to present a clear visualization in a low space projection.

DFA is a supervised chemometric technique that creates an assignment group prediction model. The model consists of a discriminant function (or, for more than two groups, a set of discriminant functions) based on the linear combinations of predictor variables that yield the best discrimination between groups [50]. For the DFA method, the discriminant functions are calculated to minimize the ratio within a group (intraclass variance). Similarly, it maximizes the ratio between each group (interclass variance).

24.3   RESULTS AND DISCUSSION

24.3.1   Comparison of the Gas Sensing Performance of SnO_2 Thin Film and WO_3 Nanowire Sensors

24.3.1.1 WO_3 Nanowires and Commercial SnO_2 Sensor Responses

Figure 24.2 shows the responses of commercial SnO_2 sensors after exposure to the breath of (a) HC, (b) LCi patients; and interdigitated sensors based on WO_3 nanowires after breath VOCs exposure from (c) HC in dark, (d) HC under UV-light, and (e) LCi patients under UV-light. As can be noticed, both families of gas sensors produced higher responses for the LCi patient breath sample compared to the HC. This behavior can be explained by the difference in breath VOC concentration linked to HC and LCi patients [51]. Moreover, the WO_3 nanowire sensor responses towards breath exposure of HC under UV-light showed higher sensitivities than the WO_3 sensor responses in the dark. This confirms that UV-light illumination improves the sensitivity of WO_3 nanowire sensors by accelerating the adsorption and desorption processes in their active area.

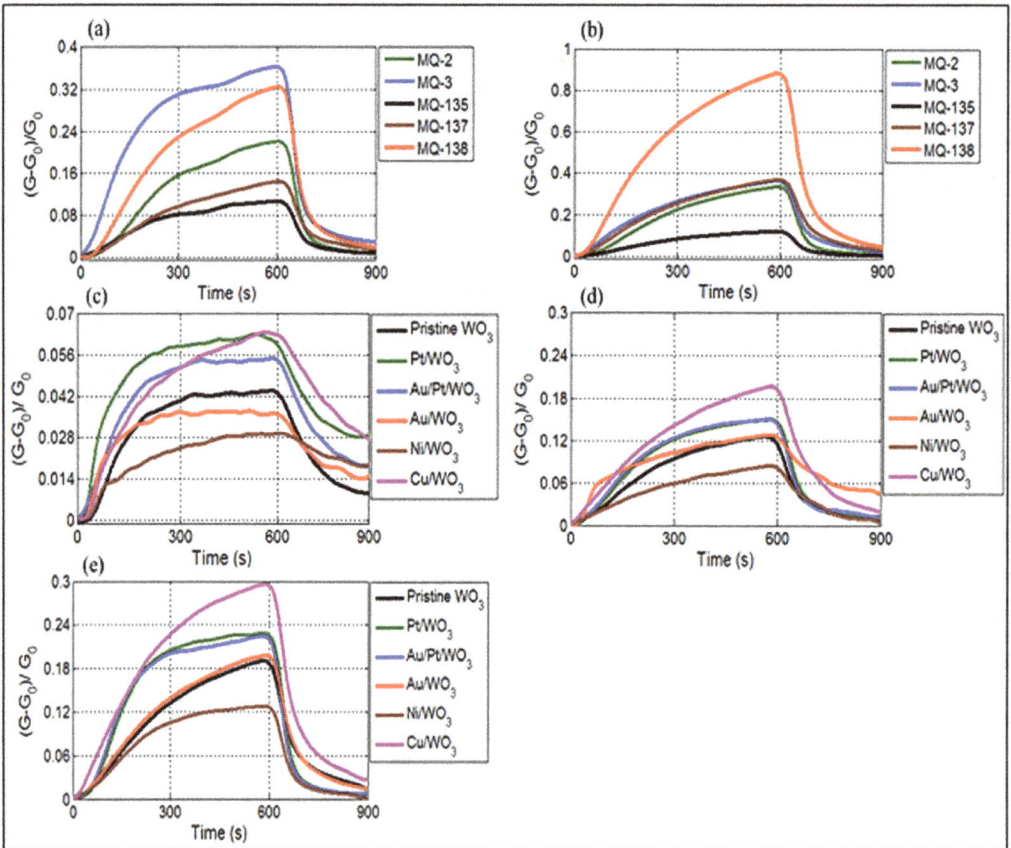

FIGURE 24.2   E-nose responses based on commercial MQ gas sensors after exposure to the breath of (a) HC, (b) LCi patient; e-nose responses based on interdigitated gas sensors when exposed to breath VOCs of (c) HC at dark, (d) HC under UV-light, and (e) LCi patient under UV-light.

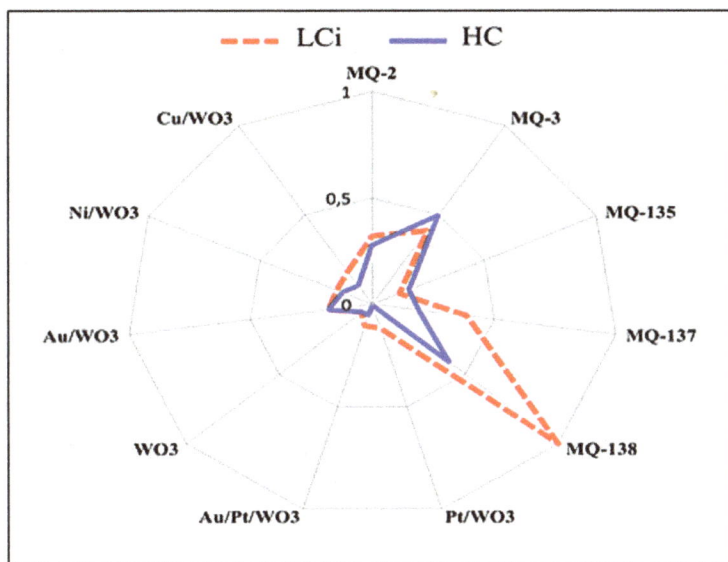

FIGURE 24.3 Radar plots of gas sensor responses to breath samples from patients with LCi and HC.

24.3.1.2 Beverage Responses of the Sensor Arrays Towards Analyzed Exhaled Breath Samples

The sensors are not sensitive to a specific compound. Nevertheless, each sensor is sensitive to various compounds. Combining the responses allows us to trace a typical chemical breath print. After extraction of the variables, the radar plot was employed to check whether there were differences or similitudes in chemical profiles of the several breath samples of LCi patients and HC. **Figure 24.3** displays the radar plot results illustrating the participation of the e-nose sensors. This graph was plotted with area value as a variable. Note that there is a large variation in the pattern among the breath-prints of LCi patients and HC.

24.3.1.3 Discrimination Results by PCA Method

Figures 24.4(a,b) display the projected PCA plot of breath samples from 22 LCi patients versus 17 HCs with data collected by five commercial SnO$_2$ gas sensors and six interdigitated gas sensors based on WO$_3$ NWs, respectively. It can be observed that the breath samples have been partially classified into two groups. Each of these groups corresponds to LCi patients and HC. The results of PCA analysis reveal a data variance of 96.40% and 83.10% that was accounted for by the first three principal components for the SnO$_2$ and WO$_3$ NW sensors, respectively. This shows that SnO$_2$ sensors are more highly correlated than WO$_3$ sensors. Furthermore, the SnO$_2$ sensor matrix provides greater accuracy than WO$_3$ sensors in distinguishing between the two groups related to LCi patients and HC.

24.3.1.4 Discrimination Results by DFA Method

The same dataset was exploited to carry out processing through a supervised technique, called DFA. It was applied to discriminate different breath samples depending on their health status. The DFA results of the commercial SnO$_2$ and WO$_3$ NWs sensors are shown in **Figure 24.5**. The same variables were taken as for the PCA. **Figure 24.5(a)** illustrates perfect separation among breath samples of LCi patients and HC. In contrast, the discriminatory capability of WO$_3$ nanowire sensors, demonstrated in **Figure 24.5(b)**, provides acceptable discrimination among the LCi and HC groups.

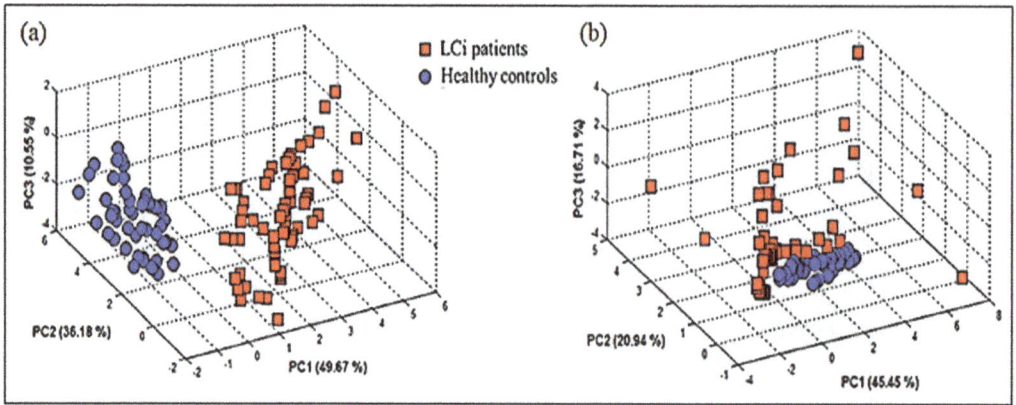

FIGURE 24.4 Unsupervised PCA plot showing data points from breath samples of LCi patients and healthy controls with data expressed by ΔG and area as features of (a) commercial SnO_2 sensor responses and (b) WO_3 nanowire sensor responses.

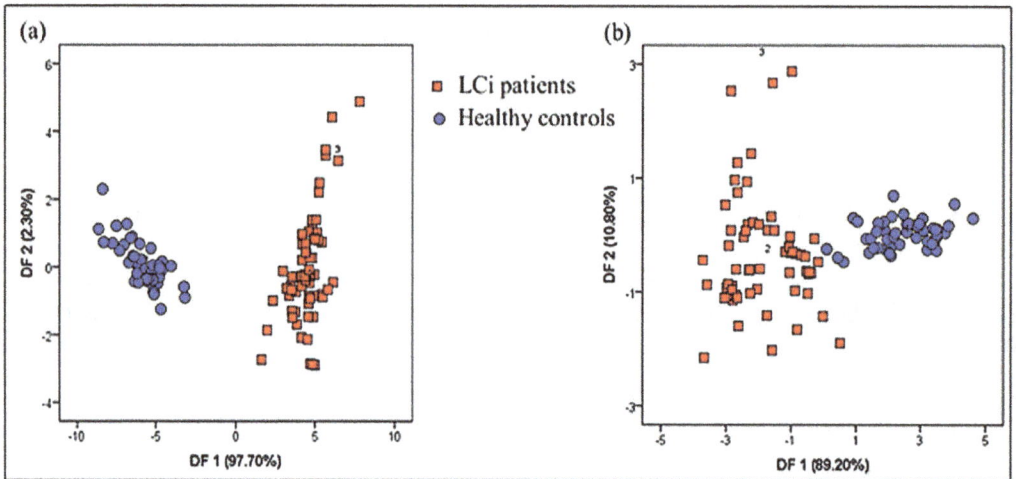

FIGURE 24.5 Supervised DFA plot showing data points from breath samples of LCi patients and healthy controls with data expressed by ΔG and area as features of (a) commercial SnO_2 sensor responses and (b) WO_3 nanowire sensor responses.

24.3.2 OPTIMIZATION OF WO_3 NANOWIRE SENSOR OPERATING TEMPERATURES

Since the sensing surface of the six WO_3 NW gas sensors is based on a different catalytic metal nanoparticle film, their optimal temperature must vary from one sensor to another. For this purpose, we tested all six sensors in the presence of exhaled breath at different temperatures to select the best one.

Figure 24.6 shows the sensor responses for pristine and metal-modified WO_3 nanowires towards breath exposure in dark at room temperature and at different operating temperatures for (a) pristine WO_3, (b) Pt/WO_3, (c) $Au/Pt/WO_3$, (d) Au/WO_3, (e) Ni/WO_3, and (f) Cu/WO_3. We can see that pristine WO_3 (**Figure 24.6(a)**) and Pt/WO_3 (**Figure 24.6(b)**) show their maximum responses in the presence of exhaled breath at a temperature of 120 °C, while $Au/Pt/WO_3$ (**Figure 24.6(c)**) showed a maximum response at 80 °C. For Au/WO_3 (**Figure 24.6(d)**), Ni/WO_3 (**Figure 24.6(e)**), and Cu/WO_3 (**Figure 24.6(f)**), they reached their maximum responses at the temperature of 160°C. From

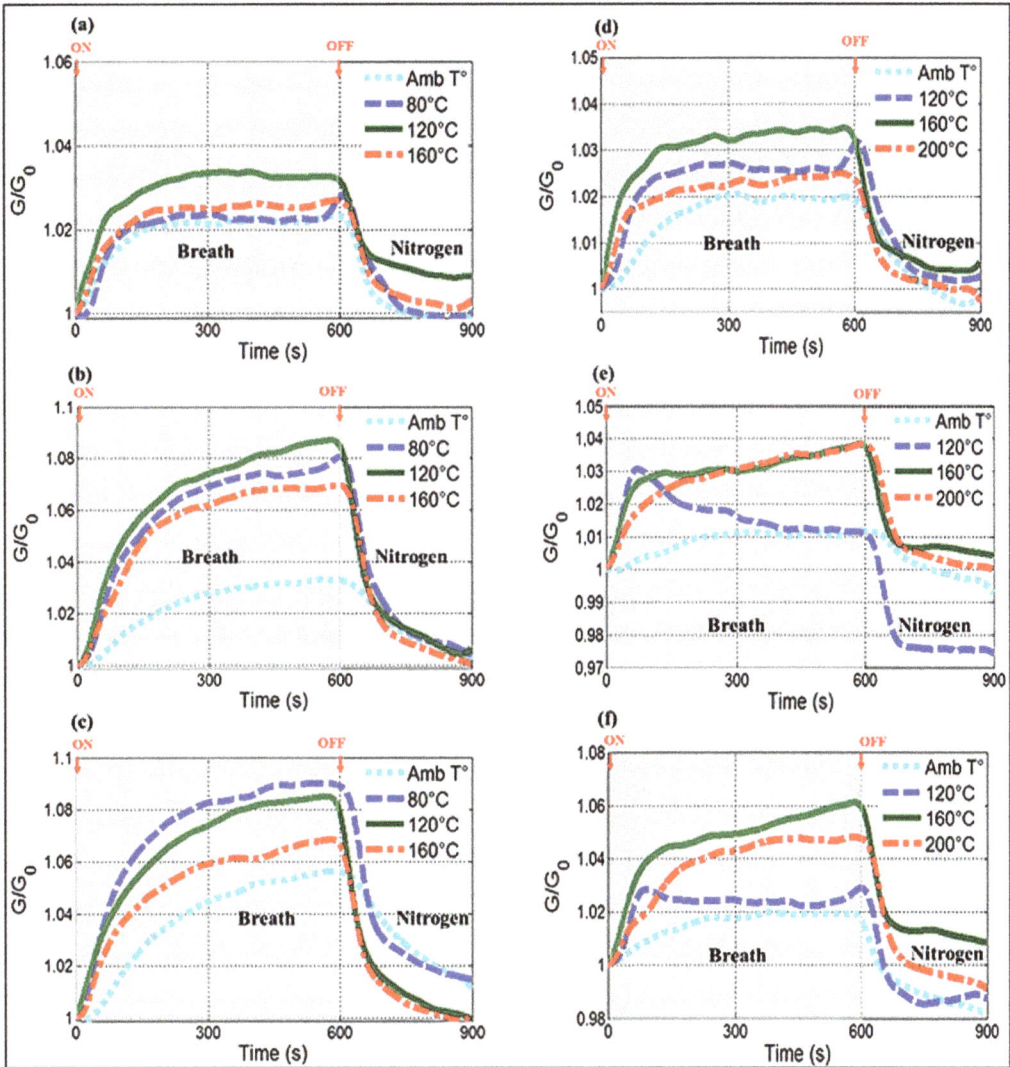

FIGURE 24.6 Sensor responses for a pristine and metal-modified WO$_3$ NWs to breath exposure in dark at different temperatures for (a) pristine WO$_3$, (b) Pt/WO$_3$, (c) Au/Pt/WO$_3$, (d) Au/WO$_3$, (e) Ni/WO$_3$, and (f) Cu/WO$_3$.

the obtained results, we can affirm that the temperature affects the acceleration of the processes of adsorption and desorption in the sensitive zone of the sensor.

24.3.3 Doping Effect of the WO$_3$ Nanowire Sensor on Electrical Conduction Properties

The doping process changes the electrical properties, structural grain alignments, size, shape, surface morphologies as well as catalytic activities of the sensors. Metal doping has several positive effects on sensor characteristics, such as increased sensitivity and selectivity. To demonstrate these advantages, a characterization of exhaled breath from a healthy control by the pure and doped WO$_3$-based sensor array was performed.

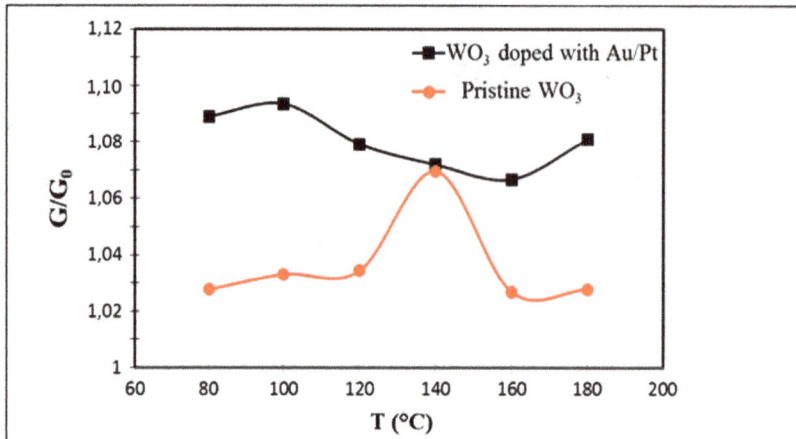

FIGURE 24.7   Responses of pristine and doped WO_3 sensors as a function of operating temperature.

Figure 24.7 shows the response curves of the pristine WO_3 sensor and WO_3 sensor doped with Au/Pt in the presence of exhaled breath from a healthy control at different operating temperatures. As we can see on this graph, the response of the WO_3 sensor doped by Au/Pt is higher than the pristine WO_3 sensor, which translates into an increased concentration of electrons during the phenomenon of adsorption of atoms on the active surface of the sensors. This confirms that doping the semiconductors with foreign impurities can improve the sensitivity of the sensor in detecting VOCs in exhaled breath. However, the maximum sensitivity of the doped sensor operates at a low temperature (100°C), while it was close to 140°C for pristine WO_3. Therefore, the use of the doping process for the semiconductor-based gas sensors may decrease the optimal operating temperature of the sensor, which would lead to a decrease in power consumption.

24.3.4   EFFECT OF ILLUMINATION ON THE RESPONSE OF WO_3 SENSORS OPERATING AT OPTIMAL TEMPERATURES

The density of the free electron-hole pair on the sensing surface of metal-oxide-semiconductors can be increased by using LEDs for illumination. For this reason, the characterization of the breath of a healthy control by WO_3 sensors at optimal temperatures in the dark and under UV-light illumination was done. The objective is to improve the performance of these sensors in terms of sensitivity for VOC detection.

Figure 24.8 shows the sensor responses in the presence of a healthy control's breath in the dark and under illumination. The curves of pristine WO_3, Pt/WO_3, and $Au/Pt/WO_3$, at a functioning temperature of 100°C, are revealed in **Figure 24.8 (a-c)**. **Figure 24.8 (d-f)** shows the sensor responses of Au/WO_3, Ni/WO_3, and Cu/WO_3, at another functioning temperature of 160°C. As displayed in these curves, the conductance of each sensor for the exhaled breath has two different statuses: a low normalized conductance change in the dark and a high normalized conductance change under UV-light illumination with a wavelength of 394 nm. These results show that UV-light illumination improves the sensitivity of the sensor for the detection of VOCs in breath analysis.

24.4   CONCLUSION

Metal oxide NW gas sensors are an important category of sensors, mainly due to their robustness, cost-effectiveness, and convenience. Human breath analysis for illness detection and healthcare screening is a major, state-of-the-art application of NW sensors. To develop efficient sensors, the

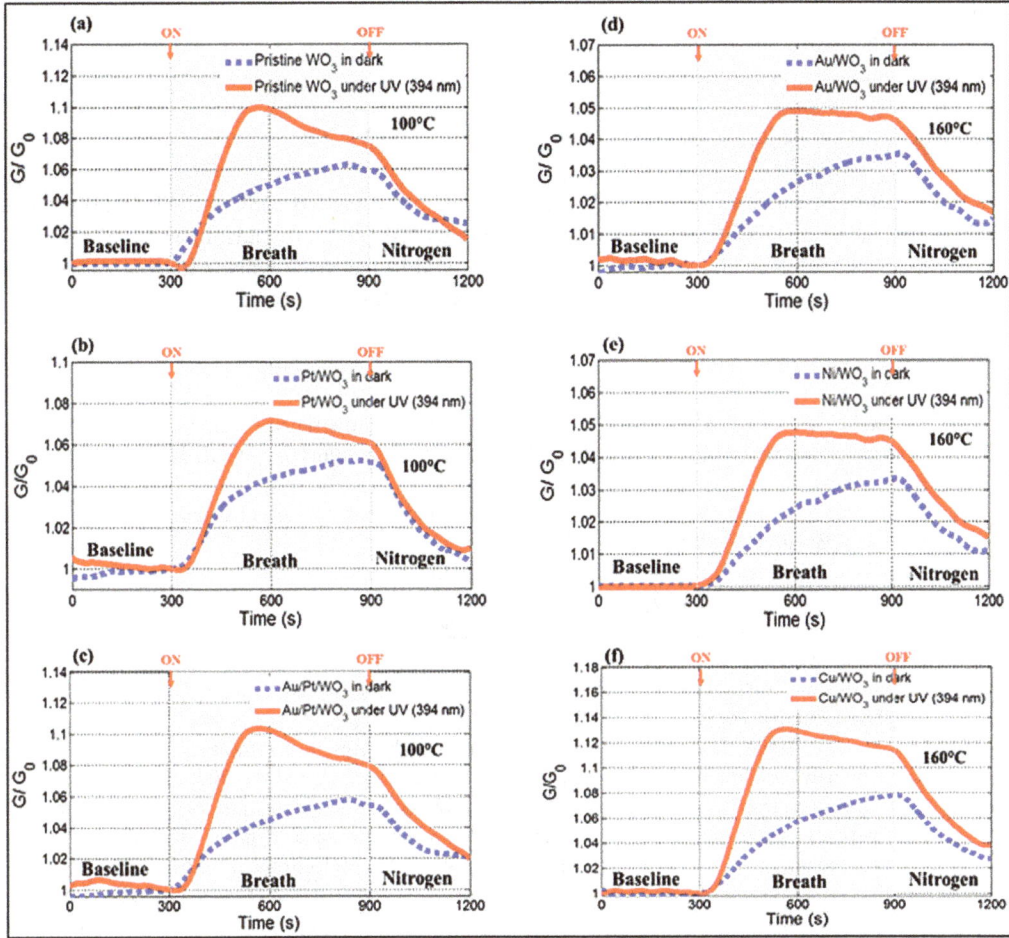

FIGURE 24.8 Sensor responses in the presence of breath in the dark and under UV-light illumination for: (a) pristine WO$_3$, (b) Pt/WO$_3$, (c) Au/Pt/WO$_3$, (d) Au/WO$_3$, (e) Ni/WO$_3$, and (f) Cu/WO$_3$. Sensors (a) to (c) were operated at 100°C, and sensors (d) to (f) at 160°C.

essential attributes of sensitivity, selectivity, and stability. To realize high-performing sensors, the essential parameters such as selectivity, sensitivity, stability, etc. must be optimized. As a first step, a comparative study of WO$_3$ nanowires and commercial SnO$_2$ sensors for the distinction of breath patterns of HC and LCi patients was done. WO$_3$- and SnO$_2$-based sensors display distinct responses between LCi and HC. This means that the VOCs patterns in exhaled breath change with health status. PCA and DFA offer more accuracy for the SnO$_2$ sensor matrix by differentiating the two groups of exhaled breath from LCi and HC patients. Accordingly, the promising results of the proposed gas sensors can be largely extended to other clinical analyses. In a second phase, two methods for enhancing the performance of WO$_3$ NW-based gas sensors under breath exposure were presented. The ability of illumination with UV-light and temperature modulation were examined to enhance the sensitivity of pristine and functionalized WO$_3$ sensors for exhaled breath application. The experimental results achieved by four operating temperature conditions showed that the optimal operating temperature of the first series gas sensors (pristine WO$_3$, Pt/WO$_3$, and Au/Pt/WO$_3$) is 120°C. While the optimal operating temperature of the second series sensors (Au/WO$_3$, Ni/WO$_3$, and Cu/WO$_3$) is 160°C. Besides, the performances of WO$_3$ NW gas sensors can be improved by

UV-light illumination. Based on these results, we can report that the combination of UV-irradiation and sensor functioning temperature can enhance the sensitivity of WO_3 NW gas sensors towards breath exposure.

ACKNOWLEDGMENTS

This work has been funded in part by CANLEISH under the H2020-MSCA-RISE-2020 project, grant agreement ID: 101007653: "Non-invasive volatiles test for canine leishmaniasis diagnosis". The authors wish to express their gratitude to Prof. Eduard Llobet from Université Rovira i Virgili-Tarragone who allowed them to fabricate the WO_3 sensors in his laboratory.

REFERENCES

[1]   S. Salehi, E. Nikan, A.A. Khodadadi, Y. Mortazavi, Highly sensitive carbon nanotubes–SnO_2 nanocomposite sensor for acetone detection in diabetes mellitus breath, Sensors and Actuators B: Chemical, 205, 2014, 261–267.

[2]   P. Gouma, A. Prasad, S. Stanacevic, A selective nanosensor device for exhaled breath analysis, Journal of Breath Research, 5, 2011, 037110.

[3]   D.H. Kim, J.S. Jang, W.T. Koo, S.J. Choi, S.J. Kim, I.D. Kim, Hierarchically interconnected porosity control of catalyst-loaded WO_3 nanofiber scaffold: Superior acetone sensing layers for exhaled breath analysis, Sensors and Actuators B: Chemical, 259, 2018, 616–625.

[4]   G. Katwal, M. Paulose, I.A. Rusakova, J.E. Martinez, O.K. Varghese, Rapid growth of zinc oxide nanotube–nanowire hybrid architectures and their use in breast cancer-related volatile organics detection, Nano Letters, 16, 2016, 3014–3021.

[5]   C. Ghosh, V. Singh, J. Grandy, J. Pawliszyn, Recent advances in breath analysis to track human health by new enrichment technologies, Journal of Separation Science, 43, 2020, 226–240.

[6]   C.M. Durán-Acevedo, J.M. Cáceres-Tarazona, Low-cost desorption unit coupled with a gold nanoparticles gas sensor arrays for the analysis of volatile organic compounds emitted from the exhaled breath (gastric cancer and control samples), Microelectronic Engineering, 237, 2020, 111483.

[7]   G. Lubes, M. Goodarzi, GC–MS based metabolomics used for the identification of cancer volatile organic compounds as biomarkers, Journal of Pharmaceutical and Biomedical Analysis, 147, 2018, 313–322.

[8]   X.P. Chen, C.K.Y. Wong, C.A. Yuan, G.Q. Zhang, Nanowire based gas sensors, Sensors and Actuators B: Chemical, 177, 2013, 178–195.

[9]   E. Comini, Metal oxide nanowire chemical sensors: Innovation and quality of life, Materials Today, 19, 2016, 559–567.

[10] N.P. Dasgupta, J.W. Sun, C. Liu, S. Brittman, S.C. Andrews, J. Lim, H.W. Gao, R.X. Yan, D.P. Yang, 25th anniversary article: Semiconductor nanowires synthesis, characterization, and applications, Advanced Materials, 26, 2014, 2137–2184.

[11] X.C. Jiang, Y.L. Wang, T. Herricks, Y.N. Xia, Ethylene glycol-mediated synthesis of metal oxide nanowires, Journal of Materials Chemistry, 14, 2004, 695–703.

[12] Y.L. Wang, X.H. Jiang, Y.N. Xia, A solution-phase, precursor route to polycrystalline SnO_2 nanowires that can be used for gas sensing under ambient conditions, Journal of the American Chemical Society, 125, 2003, 16176–16177.

[13] E. Comini, G. Faglia, G. Sberveglieri, Stable and highly sensitive gas sensors based on semiconducting oxide nanobelts, Applied Physics Letters, 81, 2002, 1869–1871.

[14] M. Law, H. Kind, B. Messer, F. Kim, P.D. Yang, Photochemical sensing of NO_2 with SnO_2 nanoribbon nanosensors at room temperature, Angewandte Chemie International Edition, 41, 2002, 2405–2408.

[15] Y.Y. Broza, X. Zhou, M.M. Yuan, D.Y. Qu, Y.B. Zheng, R. Vishinkin, M. Khatib, W.W. Wu, H. Haick, Disease detection with molecular biomarkers: From chemistry of body fluids to nature-inspired chemical sensors, Chemical Reviews, 119, 2019, S11761-S11817.

[16] P.E. Leary, B.W. Kammrath, K.J. Lattman, G.L. Beals, Deploying portable gas chromatography-mass spectrometry (GC-MS) to military users for the identification of toxic chemical agents in theater, Applied Spectroscopy, 73, 2019, 841–858.

[17] S.Y. Liu, T.S. Lee, F. Bongard, Accuracy of capnography in nonintubated surgical patients, Chest, 102, 1992, 1512–1515.

[18] M.G. Persson , O. Zetterstrom, V. Agrenius, E. Ihre, L.E. Gustafsson, Single-breath nitric oxide measurements in asthmatic patients and smokers, Lancet, 343, 1994, 146–147.

[19] D. Ludviksdottir, Z. Diamant, K. Alving, L. Bjermer, A. Malinovschi, Clinical aspects of using exhaled NO in asthma diagnosis and management, The Clinical Respiratory Journal, 6, 2012, 193–207.

[20] P. Watson, Simple breath tester checks blood alcohol, Electronics Australian, 45, 1983, 82–83.

[21] N. Alizadeh, H. Jamalabadi, F. Tavoli, Breath acetone sensors as non-invasive health monitoring systems: A Review, IEEE Sensors Journal, 20, 2020, 5–31.

[22] H.L. Tai, S. Wang, Z.H. Duan, Y.D. Jiang, Evolution of breath analysis based on humidity and gas sensors: Potential and challenges, Sensors and Actuators B: Chemical, 318, 2020, 128104.

[23] N. Nasiri, C. Clarke, Nanostructured gas sensors for medical and health applications: Low to high dimensional materials, Biosensors, 9, 2019, 43.

[24] Y. Liu, H. Wang, W. Zhao, M. Zhang, H.B. Qin, Y.Q. Xie, Flexible, stretchable sensors for wearable health monitoring: sensing mechanisms, materials, fabrication strategies and features, Sensors, 18, 2018, 35.

[25] A. Tricoli, G. Neri, Miniaturized bio-and chemical-sensors for point-of-care monitoring of chronic kidney diseases, Sensors, 18, 2018, 18.

[26] A.T. Guntner, S. Abegg, K. Konigstein, P.A. Gerber, A. Schmidt-Trucksass, S.E. Pratsinis, Breath sensors for health monitoring, ACS Sensors, 4, 2019, 268–280.

[27] L. Presmanes, Y. Thimont, I. el Younsi, A. Chapelle, F. Blanc, C. Talhi, C. Bonningue, A. Barnabe, P. Menini, P. Tailhades, Integration of P-CuO thin sputtered layers onto microsensor platforms for gas sensing, Sensors, 17, 2017, 17.

[28] T. Saidi, M. Moufid, K. de Jesus Beleño-Saenz, T.G. Welearegay, N. El Bari, A.L. Jaimes-Mogollon, R. Ionescu, J.E. Bourkadi, J. Benamor, M. El Ftouh, B. Bouchikhi, Non-invasive prediction of lung cancer histological types through exhaled breath analysis by UV-irradiated electronic nose and GC/QTOF/MS, Sensors and Actuators B: Chemical, 311, 2020, 127932.

[29] C. Wang, L. Yin, L. Zhang, D. Xiang, R. Gao, Metal oxide gas sensors: sensitivity and influencing factors, Sensors, 10, 2010, 2088–2106.

[30] E. Espid, F. Taghipour, Development of highly sensitive ZnO/In$_2$O$_3$ composite gas sensor activated by UV-LED, Sensors and Actuators B: Chemical, 241, 2017, 828–839.

[31] W. Li, J. Guo, L. Cai, W. Qi, Y. Sun, J.L. Xu, M. Sun, H. Zhu, L. Xiang, D. Xie, T. Ren, UV light irradiation enhanced gas sensor selectivity of NO$_2$ and SO$_2$ using rGO functionalized with hollow SnO2 nanofibers, Sensors and Actuators B: Chemical, 290, 2019, 443–452.

[32] M. Trawka, J. Smulko, L. Hasse, C.G. Granqvist, F.E. Annanouch, R. Ionescu, Fluctuation enhanced gas sensing with WO$_3$-based nanoparticle gas sensors modulated by UV light at selected wavelengths, Sensors and Actuators B: Chemical, 234, 2015, 453–461.

[33] J. Smulko, M. Trawka, C.G. Granqvist, R. Ionescu, F. Annanouch, E. Llobet, L.B. Kish, New approaches for improving selectivity and sensitivity of resistive gas sensors: A review, Sensor Review, 35, 2015, 340–347.

[34] J. Beauchamp, J. Herbig, R. Gutmann, A. Hansel, On the use of Tedlar® bags for breath-gas sampling and analysis, Journal of Breath Research, 2(4), 2008, 046001.

[35] Y. Adiguzel, H. Kulah, Breath sensors for lung cancer diagnosis, Biosensors and Bioelectronics, 65, 2015, 121–138.

[36] M. Righettoni, A. Tricoli, S.E. Pratsinis, Si:WO$_3$ sensors for highly selective detection of acetone for easy diagnosis of diabetes by breath analysis, Analytical Chemistry, 82, 2010, 3581–3587.

[37] M. Ahsan, M.Z. Ahmad, T. Tesfamichael, J. Bell, P.K. Yarlagadda, Ethanol sensitivity of thermally evaporated nanostructured WO$_3$ thin films doped and implanted with Fe, Applied Mechanics and Materials, 2013, 1938–1945.

[38] S. Vallejos, P. Umek, T. Stoycheva, F. Annanouch, E. Llobet, X. Correig, P. De Marco, C. Bittencourt, C. Blackman, Single-step deposition of Au-and Pt-nanoparticle-functionalized tungsten

oxide nanoneedles synthesized via aerosol-assisted CVD, and used for fabrication of selective gas microsensor arrays, Advanced Functional Materials, 23, 2013, 1313–1322.

[39]   T. Vilic, E. Llobet, Nickel doped WO_3 nanoneedles deposited by a single step AACVD for gas sensing applications, Procedia Engineering, 168, 2016, 206–210.

[40]   C. Piloto, M. Shafiei, H. Khan, B. Gupta, T. Tesfamichael, N. Motta, Sensing performance of reduced graphene oxide Fe doped WO_3 hybrids to NO_2 and humidity at room temperature, Applied Surface Science, 434, 2018, 126–133.

[41]   T. Saidi, Development and applications of electronic sensing systems combined with spectrometric techniques as a robust approach for VOCs analysis of human exhaled breath and urine, Doctoral Thesis, Moulay Ismaïl University of Meknes, Meknes, Morroco, 2018.

[42]   F.E. Annanouch, Z. Haddi, S. Vallejos, P. Umek, P. Guttmann, C. Bittencourt, E. Llobet, Aerosol-assisted CVD grown WO_3 nanoneedles decorated with copper oxide nanoparticles for the selective and humidity-resilient detection of H2S, ACS Applied Materials & Interfaces, 7, 2015, 6842–6851.

[43]   C. Roger, T.S. Corbitt, M.J. Hampden-Smith, T.T. Kodas, Aerosol-assisted chemical vapor deposition of copper: A liquid delivery approach to metal thin films, Applied Physics Letters, 65, 1994, 1021–1023.

[44]   X. Hou, K.-L. Choy, Processing and applications of aerosol-assisted chemical vapor deposition, Chemical Vapor Deposition 12, 2006, 583–596.

[45]   M.U. Qadri, F.E. Annanouch, M. Aguiló, F. Díaz, J.F. Borull, M.C. Pujol, E. Llobet, Metal decorated WO_3 nanoneedles fabricated by aerosol assisted chemical vapor deposition for optical gas sensing, Journal of Nanoscience and Nanotechnology, 16, 2016, 10125–10132.

[46]   E. Navarrete, E. González, T. Vilic, E. Llobet, Cobalt or silver doped WO_3 nanowires deposited by a two-step AACVD for gas sensing applications, Multidisciplinary Digital Publishing Institute Proceedings, 1, 2017, 1–4.

[47]   D. Ahmadou, R. Laref, E. Losson, M. Siadat, Reduction of drift impact in gas sensor response to improve quantitative odor analysis, In IEEE International Conference on Industrial Technology (ICIT), 2017, 928–933.

[48]   O. Zaim, T. Saidi, N. El Bari, B. Bouchikhi, Assessment of "breath print" in patients with chronic kidney disease during dialysis by non-invasive breath screening of exhaled volatile compounds using an electronic nose, IEEE International Symposium on Olfaction and Electronic Nose (ISOEN), 2019, 1–4.

[49]   S. Dragonieri, J.T. Annema, R. Schot, M.P. van der Schee, A. Spanevello, P. Carratú, O. Resta, K.F. Rabe, P.J. Sterk, An electronic nose in the discrimination of patients with non-small cell lung cancer and COPD, Lung Cancer, 64, 2009, 166–170.

[50]   O. Zaim, A. Diouf, N. El Bari, N. Lagdali, I. Benelbarhdadi, F.Z. Ajana, B. Bouchikhi, Comparative analysis of volatile organic compounds of breath and urine for distinguishing patients with liver cirrhosis from healthy controls by using electronic nose and voltammetric electronic tongue, Analytica Chimica Acta, 1184, 2021, 339028.

[51]   R.F. Del Río, M.E. O'Hara, A. Holt, P. Pemberton, T. Shah, T. Whitehouse, C.A. Mayhew, Volatile biomarkers in breath associated with liver cirrhosis—comparisons of pre-and post-liver transplant breath samples, EBioMedicine, 2(9), 2015, 1243–1250.

25 Toxicity and Environmental Impact of Nanowires

Ashokkumar Sibiya and Baskaralingam Vaseeharan
Nanobiosciences and Nanopharmacology Division, Biomaterials and
Biotechnology in Animal Health Lab, Department of Animal Health and
Management, Alagappa University, Karaikudi, Tamil Nadu, India

CONTENTS

25.1 INTRODUCTION

The utilization of nanotechnology-bended materials has been extensively increased in the global developing industries. This widespread use of nanowires (NWs) differs from those of bulk materials and ions due to their inimitable characters, for instance, their minor size (1–100 nm), excessive exterior zone to extent ratio, increased reactivity, virtuous transporter dimensions, and easy variation of outward properties [1]. Nanowires are defined as a wire with a nanoscale width with budding demands in the biomedical and clinical trades [2, 3, 4, 5], as antibacterial agents [6], as sensors of metals [7], in solar cells [8], in the removal of metals [9], and energy storage [10]. Owing to the extensive use, nanowires are released into the environment, when employed via the straight route (feed garnish, nutraceuticals, and medications) or incidental (water purification and disinfection, biofilm and clogging control, wrapping, barcoding, and labeling) methods, and also through nano-based drug delivery in aquatic organisms (fish, shrimp, among others) [11], along the seepages and remainders from nearshore residents and trade [12] (**Figure 25.1**). When nanowires reach the environment, they might cause unembellished vigor hazards to the wildlife. It has been proved that nanowires can tempt cytotoxicity (triggering programmed cell death, autophagy, and mitoptosis), genotoxicity (ensuing in mutagenicity, clastogenicity, and aneugenicity), and epigeneticity [13]. Nanowires exert severe toxic effects on every single creature in an aquatic ecosystem from single cells to multicellular [14, 15]. NWs also amend biological mechanisms such as an organism's ingestion or oxidative status. Although studies regarding the advantages of nanowires are increasing, data concerning the toxic impact of nanowires on the environment is still meager. To the best of our knowledge, no chapters explain the toxic mechanism of nanowires and their adverse effect on the environment so far. Hence this present chapter emphasizes the toxic mechanism of engineered nanowires on the environment.

DOI: 10.1201/9781003296621-25

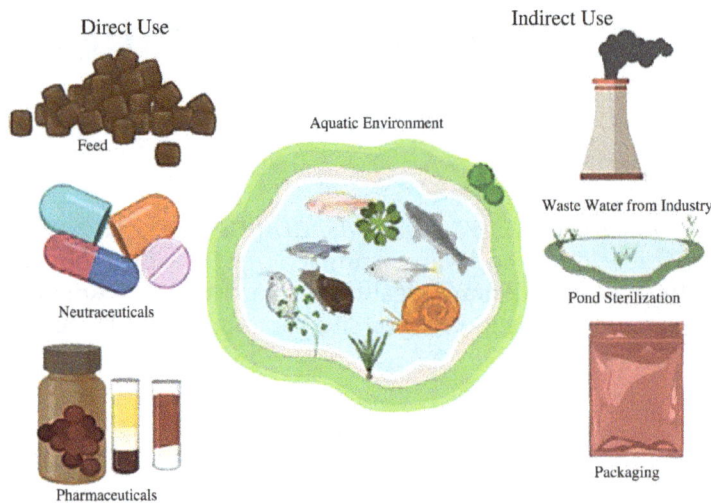

FIGURE 25.1 Various application routes of nanowires in the environment.

25.2 IMPACT OF NANOWIRES ON TERRESTRIAL ECOSYSTEM

Organisms living in the terrestrial ecosystem were affected badly by nanowire pollution, which is then transferred towards the aquatic source. The admission of nanowires within the organs of humans happens using breath, cutaneous communication, and assimilation when humans were bare to nanowires. NWs can puncture the cell wall and interrupt the cellular mechanism, hindering numerous biological progressions [16]. NWs might pass in numerous biological organizations and endure biochemical variations, which may lead to prolonged damaging effects on human well-being, distressing frequently the liver and spleen since they are the foremost poisonous accumulation spots. Exposing the HCT 116 human cell line to nickel nanowires (Ni NWs) of 5.4 μm in length for up to 72 h, cellular uptake of NWs occurs that results in cell viability, cell membrane damage, induced apoptosis, and cell death [17]. Moon et al. [18] experimentally proved that exposing 20 μm silver nanowires for 40 h affects the development and multiplication of the free-living soil nematode, *Caenorhabditis elegans*. After exposing the soil-dwelling earthworm *Eisenia Andrei* to soil containing Ag NWs, results show shorter Ag NWs (10 μm) were more toxic than lengthier Ag NWs (20 μm) and cause cytotoxicity [19]. Adolfsson et al. [20] observed no changes during 88–99 h in genetic factor expression or immune responses, after exposing *Drosophila melanogaster* to bare and hafnium oxide-coated gallium phosphide nanowires (GaP NWs). In addition, the life expectancy, fecundity, frequency of inheritable factor mutation, and insusceptible response of the fruit fly, *Drosophila melanogaster* decrease when treated persistently for 49–58 days. Demir [21] says that when 72 h old third instar larvae of *Drosophila melanogaster* were fed with TiO_2 NWs ranging from 0.01 to 10 mM a noteworthy intracellular oxidative stress and genotoxicity were distinguished in addition to intracellular reactive oxygen species (ROS) production, and genotoxicity. The presence of NWs in terrestrial ecosystems can cause contrary effects to residing creatures together with amendments to the animal's antioxidant defense system, nourishing behavior, reproduction ability, oxidative status, metabolic capacity, development, maturing frequency, and even mortality. Research regarding the ecotoxic impact of nanowires on terrestrial organisms is very scanty hence numerous research in this field is necessary for the future to detect the acute and chronic ecotoxicity of nanowires in terrestrial species.

25.3 IMPACT OF NANOWIRES ON AQUATIC ECOSYSTEM

Nanowires in the terrestrial ecosystem along with other pollutants were washed away into the aquatic environment. Subsequently, nanowires might interrelate using aquatic creatures, transfer to various organs and accumulate in their body, and cause toxicity (**Figure 25.2**). They pass through the body of aquatic organisms, which happens through straight absorption or permits transversely epithelial margins, for instance, branchias, olfactory organs, or by endocytosis and aim principally at the cellular resistant system, gut epithelium, hepatopancreas, and causes a fatal effect on the organisms [22]. Nogueira et al. [23] studied the potential impacts of Ni NWs of different sizes (≤ 1.1, ≤ 11, and ≤ 80 μm) in various aquatic animals including the bacterium *Aliivibrio fischeri*, the algae *Raphidocelis subcapitata*, the macrophyte *Lemna minor*, the crustacean *Daphnia magna*, and the zebrafish *Danio rerio*. The results show low toxicity with no lethal effects but their feeding activity was badly affected, suppressing the nutrition profile and causing retardation in growth. Experimental results of Chae et al. [24] proved that Ag NWs directly hinder the development of algae *Chlamydomonas reinhardtii* and demolished the digestive organs of water fleas *Daphnia magna*. Moreover, lengthier Ag NWs (20 μm) were more toxic than shorter ones (10 μm) to both algae and water fleas, while shorter Ag NWs accumulate in the body of zebrafish *Danio rerio*. When saltwater microcrustacean *Artemia salina* were treated with 0.01, 0.1, 1, 10, 50, and 100 mg/L Ag NWs for 72 h, the rate of immobilization increased significantly in a concentration-dependent manner that induce ROS production with an EC10 and EC50 value of 0.03 ± 0.02 and 0.43 ± 0.04 mg/L [25]. Nelson et al. [26] proved that when *D. rerio* embryos were treated with Si NWs it induces birth defects (teratogenicity) by disturbing neurulation and altering the expression resulting in mortality of the developing embryo at an LD50 of 110 pg/g. Park et al. [27] studied the generative toxicity of silver nanowires (Ag NWs) in *Daphnia magna* over dual successive generations. Here, *Daphnia magna* were exposed to various concentrations of Ag NWs (0, 0.4, 2, 10, and 50 μg/L) and observed delayed egg development, first brood reproduction along with declining the entire progenies production, ecdysis incidence in F0 and F1 generations and disrupt reproductive health. F0 generation increases the number of male newborns and infertile females, however, the number of male newborns increases in the F1 generation.

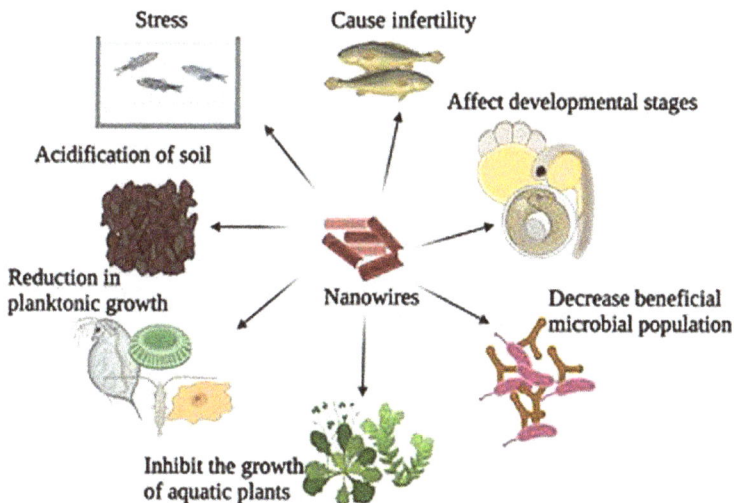

FIGURE 25.2 Impact of nanowires on aquatic organisms.

25.4   TOXIC MECHANISM OF ACTION OF NANOWIRES

The threat assessment associated with the introduction of nanomaterials is essential to be harmless from unanticipated hazards of noxiousness of nanowires. Contact of nanowires to the water bodies and their association with aquatic organisms are of chief concern. The dispersal of nanowires in water can hinder the life of the whole aquatic organisms. Hence, there is an urgent need to estimate the negative toxic mechanism of action of nanowires. The toxic mechanism of nanowires in the aquatic organism was illustrated as a pictorial representation in **Figure 25.3**. The cell or tissue that uptake nanowires shows impacts like necrosis, mitochondrial and DNA damage, reduced growth, deadliness, compact self-defense, altered reproduction, and cell death [28]. Jiang et al. [29] experimentally analyzed whether Ag-doped trimolybdate nanowires generate excess reactive oxygen species (ROS) in bacteria causing damage to cell membrane and cytotoxicity. *Drosophila melanogaster* larvae when exposed to TiO_2 NWs cause mitochondrial impairment, which results in the formation of reactive oxygen species (ROS) following intracellular oxidative stress and genotoxicity due to DNA damage [21]. Park et al. [30] proved that hydrogen titanate ($H_2Ti_3O_7$) NWs activate the formation of cellular fragments in cell lines, which increases the creation of autophagosome-like vacuoles in the cytosol causing cytotoxicity.

Nataraj et al. [31] detected that titanium oxide nanowires (TiO_2 NWs) interrupted the cell membrane of *S. aureus*. Further, Fellahi et al. [32] proved that silicon nanowires (Si NWs) provoked the leakage of sugars and proteins from the cell membrane of *Escherichia coli*. The cell membrane possesses a negative charge and if nanomaterials with a positive charge interact with the cell membrane, then they bind electrostatically to the cell membrane and inhibit cell proliferation leading to cell death [28]. George et al. [33] explain that superoxide ions are generated by the gill cells of *O. mykiss* after exposure to Ag NWs and also affect the survival rates of *D. rerio* embryos. Experimentation of *D. rerio* with Si NWs induced malformations by disturbing neurulation alters auditory expression and causes a lethal effect on the developing embryo [26]. Similar to NW nanoparticles and metals also cause adverse life-threatening effects to various aquatic organisms. Exposure of selenium (Se) to Mozambique tilapia, *Oreochromis mossambicus* alters antioxidant defense mechanisms and inhibit Na+/K+-ATPase activity [34]. Similarly, acute exposure to Se generates free radical that leads to oxidative stress, which varies enzymatic (SOD, CAT, GPx, and GST) and non-enzymatic (GSH and MT) antioxidant enzymes such as superoxide dismutase (SOD), glutathione peroxidase

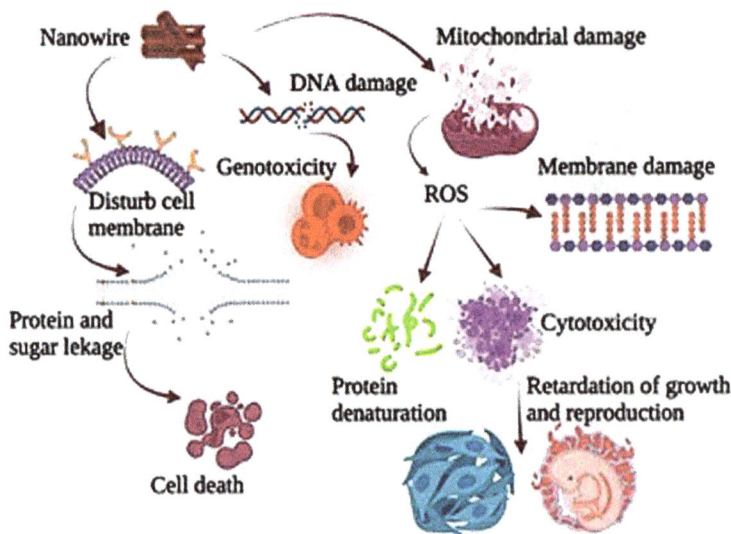

FIGURE 25.3   Mechanism of action of nanowires.

(GPx) and glutathione-s-transferase (GST) and inhibits acetylcholine esterase (AchE) activity [35]. In addition, exposure of *Oreochromis mossambicus* to Ag NPs (silver nanoparticles) results in oxidative stress (LPO and PCA) causing cellular damage, alteration in enzymatic and non-enzymatic antioxidant enzymes, oxidative stress, and neurotoxicity [36].

25.5 CONCLUSION

Based on the information presented in the current chapter, understanding the sources, fate, and effects, nanowires in the environment have made significant progress. The foremost anxiety about the probable toxicity of nanowires to the aquatic environment is an essential concern at this moment. Appropriate strategies are desirable to prevent environmental and health hazards caused by nanowires to aquatic organisms. Current toxicity assessment approaches frequently emphasize in vitro surveys, however in vivo studies will provide appropriate awareness of the nanowires' fate and behavior in aquatic organisms. From the chapter, it is clear that NWs are transported from the terrestrial environment to the aquatic ecosystem resulting in aggregation/agglomeration, which alters the toxicity of NWs in the aquatic environment. NWs can cross cell membrane generating ROS, cytotoxic, and DNA damage. Moreover, they accumulate in mitochondria and can induce inhibition of metabolic activity and changes the permeability of the cell membrane. There are limited studies regarding the ecotoxicological studies of nanowires and still, further ecotoxicological testing of nanowires is needed to safeguard the environment before nanowires are extensively marketed.

ACKNOWLEDGMENTS

The authors would like to thank and acknowledge the Ministry of Human Resource Development, Government of India, and Alagappa University, Karaikudi, for providing financial support in the form of the RUSA Phase 2.0 grant sanctioned No. F.24-51/ 2014-U, Policy (TNMulti-Gen), Dept. of Edn. Govt of India.

REFERENCES

1. Khan I, Saeed K, Khan I. Nanoparticles: Properties, applications and toxicities. Arabian Journal of Chemistry, 2017, 908–931.
2. Brammer KS, Choi C, Oh S, Cobb, CJ, Connelly LS, Loya M, Kong SD, Jin S. Antibiofouling, sustained antibiotic release by Si nanowire templates. Nano letters, 2009, 3570–3574.
3. Johansson F, Jonsson M, Alm K, Kanje M. Cell guidance by magnetic nanowires. Experimental Cell Research, 316(5), 2010, 688–694.
4. Nataraj N, Anjusree GS, Madhavan AA, Priyanka P, Sankar D, Nisha N, Lakshmi SV, Jayakumar R, Balakrishnan A, Biswas R. Synthesis and anti-staphylococcal activity of TiO$_2$ nanoparticles and nanowires in ex vivo porcine skin model. Journal of Biomedical Nanotechnology, 10(5), 2014, 864–870.
5. Singh M, Movia D, Mahfoud OK, Volkov Y, Prina-Mello A. Silver nanowires as prospective carriers for drug delivery in cancer treatment: An in vitro biocompatibility study on lung adenocarcinoma cells and fibroblasts. European Journal of Nanomedicine, 5(4), 2013, 195–204.
6. Fellahi O, Sarma RK, Das MR, Saikia R, Marcon L, Coffinier Y, Hadjersi T, Maamache M, Boukherroub R. The antimicrobial effect of silicon nanowires decorated with silver and copper nanoparticles. Nanotechnology. 24(49) 2013, 495101.
7. Luo L, Jie J, Zhang W, He Z, Wang J, Yuan G, Zhang W, Wu LC, Lee ST. Silicon nanowire sensors for Hg^{2+} and Cd^{2+} ions. Applied Physics Letters. 94(19), 2009, 193101.
8. Garnett E, Yang P. Light trapping in silicon nanowire solar cells. Nano Letters. 10(3), 2010, 1082–7.
9. Youssef AM, Malhat FM. Selective removal of heavy metals from drinking water using titanium dioxide nanowire. Macromolecular Symposia, 337, 2014, 96–101.

10. Tiwari JN, Tiwari RN, Kim KS. Zero-dimensional, one-dimensional, two-dimensional and three-dimensional nanostructured materials for advanced electrochemical energy devices. Progress in Materials Science. 57(4), 2012, 724–803.

11. Shah R, Mraz JB. Advances in nanotechnology for sustainable aquaculture and fisheries. Reviews in Aquaculture, 2019, 1–18.

12. Sinouvassane DJ, Wong LS, Lim YM, Lee P. A review on bio-distribution and toxicity of silver, titanium dioxide and zinc oxide nanoparticles in aquatic environment. Pollution Research. 35, 2016, 701–12.

13. Kanwal Z, Raza MA, Manzoor F, Riaz S, Jabeen G, Fatima S, Naseem S. A comparative assessment of nanotoxicity induced by metal (silver, nickel) and metal oxide (cobalt, chromium) nanoparticles in *Labeo rohita*. Nanomaterials. (2), 2019, 309.

14. Leite C, Coppola F, Monteiro R, Russo T, Polese G, Lourenço MA, Silva MR, Ferreira P, Soares AM, Freitas R, Pereira E. Biochemical and histopathological impacts of rutile and anatase (TiO$_2$ forms) in *Mytilus galloprovincialis*. Science of the Total Environment. 719, 2020, 134886.

15. Zhang W, Xiao B, Fang T. Chemical transformation of silver nanoparticles in aquatic environments: Mechanism, morphology and toxicity. Chemosphere,191, 2018, 324–334.

16. Kwak JI, An YJ. A review of the ecotoxicological effects of nanowires. International Journal of Environmental Science and Technology. 12(3), 2015, 1163–1172.

17. Perez JE, Contreras MF, Vilanova E, Felix LP, Margineanu MB, Luongo G, Porter AE, Dunlop IE, Ravasi T, Kosel J. Cytotoxicity and intracellular dissolution of nickel nanowires. Nanotoxicology. 10(7), 2016, 871–80

18. Moon J, Kwak JI, An YJ. The effects of silver nanomaterial shape and size on toxicity to *Caenorhabditis elegans* in soil media. Chemosphere. 215, 2019, 50–56.

19. Kwak JI, Park JW, An YJ. Effects of silver nanowire length and exposure route on cytotoxicity to earthworms. Environmental Science and Pollution Research. (16) 2017, 14516–14524.

20. Adolfsson K, Schneider M, Hammarin G, Häcker U, Prinz CN. Ingestion of gallium phosphide nanowires has no adverse effect on Drosophila tissue function. Nanotechnology. 24(28) 2013, 285101.

21. Demir E. An in vivo study of nanorod, nanosphere, and nanowire forms of titanium dioxide using *Drosophila melanogaster*: Toxicity, cellular uptake, oxidative stress, and DNA damage. Journal of Toxicology and Environmental Health, Part A. 83(11–12), 2020, 456–469.

22. Sibiya A, Jeyavani J, Santhanam P, Preetham E, Freitas R, Vaseeharan B. Comparative evaluation on the toxic effect of silver (Ag) and zinc oxide (ZnO) nanoparticles on different trophic levels in aquatic ecosystems: A review. Journal of Applied Toxicology. 2022.

23. Nogueira V, Sousa CT, Araujo JP, Pereira R. Evaluation of the toxicity of nickel nanowires to freshwater organisms at concentrations and short-term exposures compatible with their application in water treatment. Aquatic Toxicology. 227, 2020, 105595.

24. Chae Y, An YJ. Toxicity and transfer of polyvinylpyrrolidone-coated silver nanowires in an aquatic food chain consisting of algae, water fleas, and zebrafish. Aquatic Toxicology. 173, 2016, 94–104.

25. An HJ, Sarkheil M, Park HS, Yu IJ, Johari SA. Comparative toxicity of silver nanoparticles (Ag NPs) and silver nanowires (Ag NWs) on saltwater microcrustacean, *Artemia salina*. Comparative Biochemistry and Physiology Part C: Toxicology & Pharmacology. 218, 2019, 62–9.

26. Nelson SM, Mahmoud T, Beaux Ii M, Shapiro P, McIlroy DN, Stenkamp DL. Toxic and teratogenic silica nanowires in developing vertebrate embryos. Nanomedicine: Nanotechnology, Biology, and Medicine, 6, 2010, 93–102.

27. Park HS, Tayemeh MB, Yu IJ, Johari SA. Evaluation of silver nanowires (Ag NWs) toxicity on reproductive success of *Daphnia magna* over two generations and their teratogenic effect on embryonic development. Journal of Hazardous Materials. 412, 2021, 125339

28. Sardoiwala MN, Kaundal B, Choudhury SR. Toxic impact of nanomaterials on microbes, plants and animals. Environmental Chemistry Letters. (1), 2018, 147–160.

29. Jiang Y, Gang J, Xu S-Y. Contact mechanism of the Ag-doped trimolybdate nanowire as an antimicrobial agent. Nano-Micro Lett 4, 2012, 228–234.

30. Park EJ, Shim HW, Lee GH, Kim JH, Kim DW. Comparison of toxicity between the different-type TiO$_2$ nanowires in vivo and in vitro. Archives of Toxicology. 87(7), 2013, 1219–1230.

31. Nataraj N, Anjusree GS, Madhavan AA, Priyanka P, Sankar D, Nisha N, Lakshmi SV, Jayakumar R, Balakrishnan A, Biswas R. Synthesis and anti-staphylococcal activity of TiO_2 nanoparticles and nanowires in ex vivo porcine skin model. Journal of Biomedical Nanotechnology, 10, 2014, 864–870

32. Fellahi O, Sarma RK, Das MR, Saikia R, Marcon L, Coffinier Y, Hadjersi T, Maamache M, Boukherroub R. The antimicrobial effect of silicon nanowires decorated with silver and copper nanoparticles. Nanotechnology, 24, 2013, 495101

33. George S, Lin S, Ji Z, Thomas CR, Li L, Mecklenburg M, Meng H, Wang X, Zhang H, Xia T, Hohman JN, Lin S, Zink JI, Weiss PS, Nel AE. Surface defects on plate-shaped silver nanoparticles contribute to its hazard potential in a fish gill cell line and zebrafish embryos. ACS Nano, 6, 2012, 3745–3759.

34. Gopi N, Rekha R, Vijayakumar S, Liu G, Monserrat JM, Faggio C, Nor SA, Vaseeharan B. Interactive effects of freshwater acidification and selenium pollution on biochemical changes and neurotoxicity in *Oreochromis mossambicus*. Comparative Biochemistry and Physiology Part C: Toxicology & Pharmacology. 2021, 250, 109161.

35. Gobi N, Vaseeharan B, Rekha R, Vijayakumar S, Faggio C. Bioaccumulation, cytotoxicity and oxidative stress of the acute exposure selenium in *Oreochromis mossambicus*. Ecotoxicology and Environmental Safety. 2018, 162, 147–159.

36. Sibiya A, Gopi N, Jeyavani J, Mahboob S, Al-Ghanim KA, Sultana S, Mustafa A, Govindarajan M, Vaseeharan B. Comparative toxicity of silver nanoparticles and silver nitrate in freshwater fish *Oreochromis mossambicus*: A multi-biomarker approach. Comparative Biochemistry and Physiology Part C: Toxicology & Pharmacology. 2022,109391.

26 Future Perspectives of Nanowires

G. Anand
Department of Mechanical Engineering, Achariya College of Engineering Technology, Pondicherry, India

N. Santhosh
Department of Mechanical Engineering, MVJ College of Engineering, Near ITPB, Whitefield, Bangalore, India

S. Vishvanathperumal
Department of Mechanical Engineering, S.A. Engineering College, Thiruverkadu, Tamilnadu, India

CONTENTS

DOI: 10.1201/9781003296621-26

26.1   INTRODUCTION

Wires of different shapes and sizes have been an essential part of human society for centuries. Nanowires are used for a variety of applications, viz., the transmission of electricity, transfer of mechanical forces and motion, transfer of sound, light, and actuation signals. In the past century, researchers developed a brand-new method for creating nanoscopic wires, which are a thousand times thinner than human hairs and can be used in advanced applications in computers, photonics, and related biomechanical devices. With conventional cross-sectional dimensions that may be varied from 1 to 250 nm and lengths ranging from several hundred nanometers to millimeters, semiconductor nanowires are a brand-new elegance of semiconductors [1]. Currently, many nanowire investigations are based on the successful methodology developed from Wagner's research based on the "vapor-liquid-solid transition" process. In 1964, Wagner referred to the growth of silicon microwires or whiskers and said that gold was one of the catalysts that may be utilized in this vapor-liquid-stable process. In addition to aluminum, nickel, copper, titanium, and silicon, gold is now frequently utilized to create nanowires. Wagner's works have set the stage for several nanowire researches being carried out today [2]. The nanowire research began in the late nineties when many researchers were working closely on carbon nanotubes, and some of the research groups started their research on the synthesis and subsequent characterization of nanowires, which include oxides and elemental semiconductors [3–5]. Wagner's vapor-liquid-stable crystal evolution strategies served as the foundation for the evolution of the nanowire evolution strategies. Additionally, the discovery of metal nanoparticles has laid a solid framework for the synthesis of nanowires, which have a wide range of uses [6–7].

In 2001 [8], the vapor-liquid-stable nanowire manufacturing process underwent its first direct in situ investigation. The development of stable liquid interfaces is what led to the synthesis of one-dimensional crystals, which supports the crucial nanometer-scale nucleation step for all vapor-liquid-stable processes. The evolution of the nanowire synthesis mechanism has resulted in the development of nanoparticles. The use of new techniques to synthesize semiconductor nanowires predictably has led to the synthesis of nanowires for advanced applications. These early research outcomes have led to the evolution of nanowires for different compositions and combinations. In the subsequent decades, the research on nanowires has led to dynamic, interdisciplinary studies in the frontier areas with scientists and engineers from many exceptional groups accomplishing their work in the domain of chemistry, physics, mechanical, and electric engineering. The research on nanowires has witnessed the rational synthesis of III–V and II–VI binary and ternary nanowires. Numerous nanoscale axial and radial nanowire hetero-structures have been created and studied for improved functionality as a result of the sub-scalar growth of nanowires with adjustable doping and the fabrication of molecular-scale nanowires [9–11].

The usefulness of rational semiconductor nanowire production and arrangement has produced many nanowires with improved mechanical, electrical, and optical properties. The biocompatibility in addition to superior mechanical, electrical, and optical capabilities and the unique physico-optical capabilities have led to the better performance of the nanowires for advanced applications. These unique pursuits have led to comprehensive developmental opportunities for demanding situations. Additionally, it has been conclusively demonstrated that semiconductor nanowires with predictable electrical characteristics can be rationally designed and synthesized, providing electronically tunable nanoscale building blocks for the synthesis of nanowires. This assessment of the nanowire synthesis is primarily based on colloidal approaches. For instance, the Harvard research teams have developed primary diode systems using crossed p- and n-type nanowires, demonstrating how better voltage difference is produced by planar p-n junctions, which allows for the systematic production of nanowires. They put together powerful bipolar transistors with improved emitter characteristics and collected complementary inverters using p- and n-kind nanowires, which is a crucial step in building logic gates out of nanowires [13]. These nanowires served as crucial building blocks in the bottom-up production of 3D metal-oxide semiconductors with vertical connections. The assessment and development of nanowires primarily for electronics devices have driven the growth of the semiconductor industry. Nanometer-resolution nanowires can also be utilized to measure electrical activity at the neuronal level [14]. By utilizing the brain tissues, they have continued in developing this intriguing field of study while exhibiting amazing spatial and temporal resolution [15]. By using nanowires for prospective optical processes and photonics, this emerging area of the nanowire bio-electric interaction may be included in a brief knowledge assessment [16].

In the early 2000s, this area of nanowire photonics also experienced significant growth. Numerous other optical techniques, such as subwavelength and nonlinear optical mixing with waveguides, have been investigated for those semiconductor nanowire building blocks since the development of the first room-temperature UV nanowire laser [17–19]. The foundation for nanowires based on subwavelength photonic integration, innovative nanowire scanning probe imaging, and spectroscopy, in addition to the conversion of photons from radiations to electricity, was laid by these research endeavors. The studies of nanowire photonics that have resulted from it are now fundamental [20–21]. For optical computers to be achieved, it is necessary to have the ability to control wire behavior at the nanoscale. Chemically produced nanowires are a fundamentally elegant building component of photonic systems with subwavelength optical capabilities. A solid foundation for employing semiconductor nanowires in photovoltaic systems has been established by advancements in nanowire photonics. Various nanowire hetero-structure designs have been shown to capture solar radiations [22–23]. This idea of the usage of nanowires for photovoltaic packages represents the capability to optimize the absorption capability of the photocells and it needs to have a top-notch effect within the subject of renewable energy. Recent studies have put a lot of work into comprehensively examining the photoelectrochemical properties of semiconductor nanowire arrays [24, 25]. Along with the searches for materials for artificial photoconversions, semiconductor nanostructures with complex compositions and heterojunctions were rigorously examined.

For direct solar to photovoltaic systems, semiconductor nanowires are a crucial kind of nanostructure building block. To deterministically contain heterojunctions with stepped forward mild absorption, rate separation, and vectorial representations, nanowires can be easily designed and produced. To mimic the compartmentalized processes, it is also possible to selectively boost extraordinary oxidation or catalytic interfaces onto specific nanowire segments. These nanowires used for photosynthesis reactions collect sunlight and facilitate the photosynthetic procedures to photocatalytically convert the CO_2 and sunlight to energy. A complete assessment of this process leads to the evolution of the nanowires for chemical kinetics and their applications [26, 27]. It is abundantly obvious from the analysis of their research that the nanowires exhibit exceptional thermal conductivity, ion diffusion kinetics, and stress/strain response. Nanowires have excellent functionality for understanding a variety of packages beyond the disciplines of sophisticated technical applications as a result. Nanowires have additionally been tested to exhibit exceptional thermal conductivity [28]. Thin nanowires have thermoelectric properties comparable to those of bulk wires. The electrical conductivity of the silicon nanowires, which have hard surfaces and diameters of about 50 nm, is increased one hundred times, making it easier to use them in real-time applications. Finally, due to the asymmetry shape inside the path of the c-axis, the semiconductors ZnO and GaN owe relatively spontaneous polarization along the longitudinal path of the nanowires. These nanowires' couplings of photoexcitation, piezoelectricity, and optical/electric properties have sparked interest in new research on piezotronics and piezo-phototronics [29].

Now, specific compositions, heterojunctions, and architectural configurations for nanowires can be created. This has already facilitated the use of nanowires in digital and photonic, biomedical, addition to power conversion and automotive applications. Each of these packages has evolved primary interests in the synthesis, characterization, and manipulation of nanowires. **Figure 26.1** gives the schematic outline of the potential that the nanowires offer for their use in future applications. The latest advances in the synthesis of nanowires have resulted in several advantages, especially for chemical, physical, and engineering applications [30, 31]. The interdisciplinary nature of the studies has excited the research interests and thereby inspired this review on the future perspectives of nanowires.

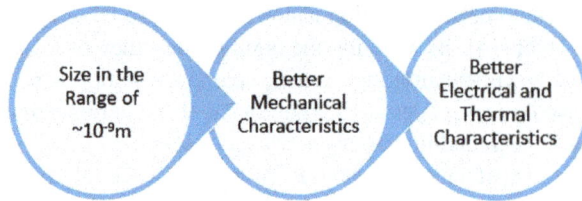

FIGURE 26.1   Potentials of nanowires for future applications.

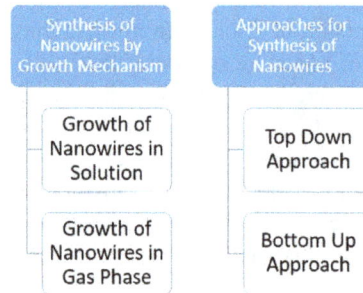

FIGURE 26.2   Schematic representation of the synthesis of nanowires.

26.2   SYNTHESIS OF NANOWIRES

Due to recent advancements in the synthesis processes, the synthesis of nanowires has significantly increased in popularity. The length of nanowires can reach several hundreds of micrometers, while their diameter can reach many tens of nm. Metallic oxides or inorganic non-oxide nanowires may be mono- or polycrystalline. The nanowires must consist of just one element, such as Si or Ge, or they may combine two or more elements, such as Ti, Si, nitrides of boron, and Si carbides, among others. Controlling the composition, crystallinity, etc. is one of the key elements in the synthesis of nanowires [32].

26.2.1   Synthesis Techniques for Nanowires

The synthesis of nanowires depends on the physicochemical characteristics of nanostructures. A few synthesis techniques are reviewed to significantly understand the growth mechanism. However, there is a need to review the growth mechanism and the experimental techniques for the synthesis of the nanowires. In many cases, the synthesis mechanisms have numerous contradictions that may seem to be quite intriguing. This review seems to answer all of them. The strategies of synthesis of nanowires are numerous and come under the two major spectrums of physical and chemical strategies. Physical strategies involve template, patterning, lithography, and electro-filming strategies. The mussel and lithography tactics are bottom-up techniques, whereas the sample and electro-casting techniques are top-down techniques. Precursor decomposition, oxide-assisted growth methods, electrochemical techniques, solvothermal, hydrothermal, and carbothermic procedures are some examples of chemical tactics. Separating the vapor phase from the solution phase is a key component of the second crucial nanowire synthesis approach. Some chemical techniques will work if the catalyst and the substance for the nanowire are eutectic or solubilized in each other. The vapor-liquid-stable process (VLS), which cannot be applied to all material compositions, is an example of this [33]. **Figure 26.2** depicts the general layout of the two main methods used to create nanowires. The different strategies and approaches to the synthesis of nanowires are schematically represented in **Figure 26.2** and reviewed as follows.

26.2.1.1 Growth of Nanowires in Solution

There are numerous techniques for the growth of nanowires in the solution phase and some of them are reported in this section of the chapter.

26.2.1.1.1 Liquid-Solid Mechanism (LS)

Stronger bond anisotropy in a substance makes it possible to generate nanowires utilizing the LS mechanism, which encourages crystallization along the c-axis. This technique allows for the selective growth of nanowires with an extremely high degree of purity and a very low density of structural flaws, but it has the disadvantage of a non-localized growth structure. Sulfide polynitride, (SN) x selenium, tellurium, and molybdenum chalcogenides are all synthesized using this technique.

26.2.1.1.2 Solid-Liquid-Solid Mechanism (SLS)

Catalytic growth mechanisms include the vapor-liquid-solid (VLS) and solid-liquid solid (SLS) processes. The SLS mechanism was developed based on an analogy with the VLS mechanism to acquire quicker growth processes than the VLS mechanism and for lower temperatures [34]. Group III-V semiconductor crystalline nanowires were created using this method [35, 36]. The materials purchased for the formation of nanowires are typically monocrystalline and range in size from 10 to 150 nm [37]. Massive amounts of monocrystalline silicon and germanium nanowires have been produced using the supercritical fluid-solid-liquid (SFLS) approach [38, 39]. This method substitutes a supercritical fluid for the solvent. The SFLS method was used to create nanowires with diameters of several nm and lengths more than several microns. The use of mono-disperse, encapsulated gold nanocrystals in an alkaloid solution is essential to this synthesis. In a traditional method, heated and pressured hexane is used to co-disperse the statically stabilized gold nanocrystals and the Si precursor diphenylsilane. In those circumstances, diphenylsilane can break down and form Si atoms. The Au-Si section demonstrates how silicon and gold can be combined to form an alloy that is in balance with natural silicon. Even if the temperature is higher than 363°C and silicon is more abundant than gold by 18.6 mol%. As a result, until saturation, Si atoms dissolve in gold nanocrystals. As a result of their expulsion from the alloy, the atoms now resemble a typical nanowire. There are several similarities between this mechanism and the SLS system. The supercritical fluid must be used to enable the dissolution and crystallization of silicon. More intriguing is the orientation of Si nanowires, which can be altered by varying the pressure [40].

26.2.1.1.3 Solventothermic Reactions

This approach is one of the crucial steps in the realization of semiconductor nanowires. The fundamental benefit of this approach is that, by applying heat and pressure, most substances may be rendered soluble in an appropriate solvent. The solvent is cooled and drawn after reaching a certain level of evaporation around a heated filament. This creates nanowires. This approach is flexible and has made it easier to create different semiconductor nanowires, nanotubes, and whiskers [41-46]. However, the low yield, low purity, terrible length, and poor morphological homogeneity of the products make them stand out [47, 48].

26.2.1.1.4 Growth from a Mold

There are several sorts of molding strategies (templating), viz., physical and chemical moulding. The chemical molding may be primarily based on a tough mold ("hard templating") or a soft mold, even as the physical moulding leads to the evolution of 1D nanostructures. The nanomaterial may be made round to fit the profile of the mold: there's right away a correlation between the morphology of the mold and that of the nanostructures, leading to the evolution of nanowires.

26.2.1.1.5   Soft Templating Mold

Self-assembled systems from surfactants offer a beneficial pattern for the generation of molds for the 1D nanostructures in huge quantities. Surfactant molecules spontaneously organize into rod-shaped micelles above the critical micellar attentiveness [49]. Soft templating techniques have been used to create nanowires made of CuS, CuSe, CdS, CdSe, ZnS, and ZnSe using this method. In the technique, a physical mold from a porous membrane with nanoscale channels made from mesoporous substances, porous alumina, or polycarbonate membranes is used [50], [51]. Sol-gel or electrochemical methods are used to pack a solution into the nanoscale channels. The resulting nanowires are then separated from the mold. There are two common types of porous membranes: polycarbonate and alumina. To create holes at the layer's bottom, polymer layers (6–20 mm thick) used in polycarbonate membranes are exposed to heavy ions. The polymer layer is then produced with consistent cylindrical pores by chemically advancing these holes. This method creates pores that are randomly distributed at the polymer membrane's bottom [52]. Alumina membranes are created by anodizing aluminum foil in an acidic solution in addition to chemically etched polymer membranes [53]. Further drawing of these membranes into nanowires. Porous alumina membranes are used as a mold to create nanowires from a variety of substances. In addition to polymer membranes, the alumina membrane's pores are the most practical and serve as templates for the creation of nanowires with superior properties. To deposit the catalyst at the base and produce the nanowires, this process must typically be combined with one of the following methods: electroplating, sol-gel deposition, physical vapor deposition (PVD), or chemical vapor deposition (CVD). Furthermore, it is crucial to create monocrystalline materials from polycrystalline solids. The advantage of this templating technique is that the nanostructures with complicated or amorphous levels can be fabricated using a soft mold. The cost of producing the nanowires with the help of this technique is relatively economical for evolving 1D nanostructures and nanowires with superior thermal and electrical characteristics for advanced applications.

26.2.1.2   Nanowire Creation in the Gaseous State

The method that has received the most attention for creating whiskers and nanowires is likely nanowire development from the gas phase. The vapor-liquid-solid (VLS) and vapor-solid (VS) technologies for the formation of nanowires in the gas phase involve evaporation in an environment conducive to the creation of fundamental nanowires or oxides. The most favored method for growing nanowires among the available growth mechanisms is an oxide-supported growth mechanism in the gas phase. No metal catalyst is required for this type of growth. Thanks to the silica powder target containing silicon dioxide, the growth of silicon nanowires are possible [54]. The yield of nanowires is substantially higher than that of the CVD technique, being in the range of a few milligrams. When it comes to oxide-assisted growth, the vapor created by thermal evaporation or laser ablation is crucial. Following the reaction that breaks down silicon oxide into silicon precipitates and oxygen atoms, nucleation takes place on the substrate.

The silicon particles that serve as the core of silicon nanowires are covered in silicon oxide precipitate as a result of the breakdown reaction. In the vicinity of the substrate's cold zone, precipitation, nucleation, and growth take place. This shows that the growth and production of nanowires are driven by the temperature gradient. Additionally, this method makes it possible to create nanowires with 1.3–7 nm diameters. This oxide-assisted growth technique has also been used to create nanowires made of copper and magnesium. These methods with oxide assistance have the primary benefit of not requiring a metal catalyst, which eliminates contamination from the metal atoms in the catalyst. The lower growth temperature seen during MgO production is another advantage.

26.2.1.2.1   Vapor-Solid Mechanisms

Evaporation, chemical reduction, or gaseous processes all contribute to the production of vapors in this process. The substrate is cooled to a lower temperature than the original substance as a result of the vapor being carried and condensed onto it. This method is used in numerous experimental approaches, including laser ablation, thermal vapor deposition, HFCVD (chemical vapor deposition with hot filaments), and MOCVD (chemical vapor deposition from metalorganic precursors). Oxide whiskers and metals with diameters in the micron range have been synthesized using the vapor-solid technique. This technique involved heating commercially available powders of numerous metal oxides, including Si, Zn, and Cu, to high temperatures to produce nanowires of those materials [55]. By adjusting the nucleation and growth conditions, this technique can also create unique structures like nanoribbons. The benefit of this approach is that chemically highly pure nanowires can be produced utilizing a straightforward gas phase process and a basic experimental setup.

26.2.1.2.2 *Vapor-Liquid-Solid Mechanisms*

Among the vapor deposition methods, the VLS method seems to be the most successful in obtaining a relatively large amount of single crystal structure. The Wagner and Ellis team created this procedure in the 1960s to create whiskers that were only a few microns long [56]. The research team headed by Duan and Liever has examined how nanowires can be made from a range of inorganic materials [57]. In a typical VLS process, a gaseous reactant is first dissolved in nanoscale metal catalyst droplets, which is followed by the nucleation and development of nanowires. Droplets are the main culprits in 1D growth. Growth does not result in the catalyst being used up. Each droplet acts as a "guide" for the nanowire as it develops, inhibiting lateral expansion. As was already established, the nanowire component must be soluble in the catalyst or, ideally, form a eutectic for VLS development to occur. The simultaneous existence of the three phases, gas, liquid, and solid, is necessary for this mechanism. The Au-Si mixture and catalyst droplets both become liquid at temperatures above the eutectic temperature when the atomic concentration of gold in silicon surpasses 18.6%. The precursor gas molecules in this scenario are catalytically broken down on the droplet's surface. The catalyst mixture and atoms from the precursor gas build up at the interface at operation temperatures greater than the critical point temperature. Atoms from the gas that make up the nanowires disperse and crystallize inside the droplet. As a result, atoms from the droplets crystallize to form nanowires. The catalytic interaction between these droplets causes nanowire development. Numerous applications that require the ability to manage the location of nanowire can be envisioned thanks to this selectivity. The gaseous feed of molecules needed to grow nanowires has been created using a variety of techniques or combinations of techniques. An approach based on the VLS method, for instance, has produced horizontal growth that corresponds to the path the laser takes on the substrate. By watching two growth occurrences in situ, this mechanism was verified. The first is the growth of nano silicon aggregates, while the second is the growth of germanium nanowires in a TEM chamber with a platinum core. In the first instance, pure germanium served as a potent precursor while gold catalyzed to speed up the reactions. By raising the system's temperature to between 700 °C and 900 °C, the gas is transported in this situation and helps the epitaxial grain growth over the filament core.

26.2.1.3 Top-Down and Bottom-Up Approaches

Further, the synthesis of nanowires is classified as top-down and bottom-up approaches based on the techniques used. In the traditional top-down nanowire synthesis technique, the crystals or bulk wafers are converted into nanowire structures using chemical etching and lithography techniques. The top-down approach is based on the removal of the material from the bulk substrate based on the profile of the nanostructure required.

Unlike the top-down approach, the bottom-up process derives from the chemist's toolkit. In this kit, molecules are assembled one by one and scaled up to produce a material that is thousands of times larger than the average molecule. Bottom-up synthesis is an important technique that uses chemical principles to control the structure and grow the nanostructures from smaller molecules.

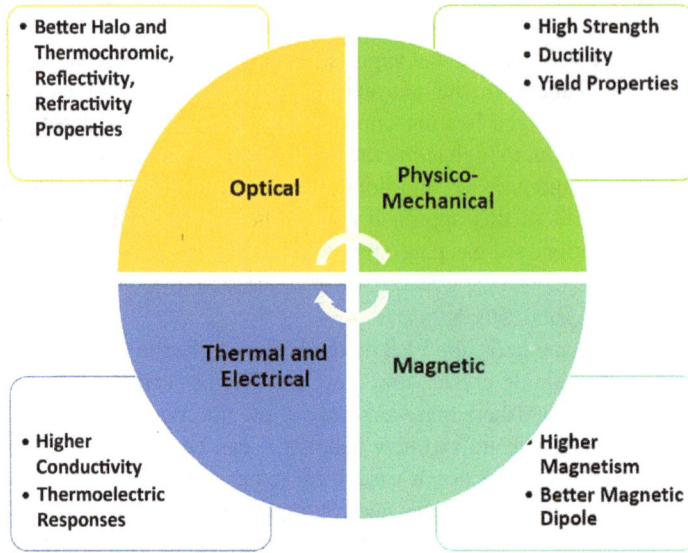

FIGURE 26.3 Schematic of the important properties of the nanowires making them potential materials for use in future applications.

26.3 CHARACTERISTICS OF NANOWIRES

The characteristics of nanowires play an important role in the use of these nanowires in several real applications. Thus, it is important to review the properties of nanowires, viz., mechanical, magnetic, optical, etc. The important characteristics of nanowires that make them potential materials for use in future applications are schematically enumerated in **Figure 26.3**.

26.3.1 PHYSICO-MECHANICAL CHARACTERISTICS

The mechanical properties of nanowires are significant because internal dislocations can modify the conductivity of nanowires as a result of external pressures, induced strains, and temperature changes. During very large scale integration (VLSI) processing, tensile and compressive pressures may cause failures due to electro-migration and delamination. Studies show that nanowires have interesting mechanical properties due to their lower number of defects per unit length and higher aspect ratio compared to bulk materials. Nanowires are one-dimensional systems. Nanowire technology is used in sensors and NEMS (nano-electromechanical systems). They are very sturdy materials because they can store elastic energy and have high modulus and high tensile strength. In addition, silicon nanowires with high vibration frequencies (100 MHz to 1 GHz) can build nanoscale resonators due to their excellent elastic properties.

26.3.2 MAGNETIC PROPERTIES OF NANOWIRES

Understanding the behavior of nanowires in real-time applications depends critically on the magnetic characteristics of nanoparticles. The magnetization vector in nanowires can change thermally due to the low energy of magnetic anisotropy. The term for this is super-paramagnetism. Super-magnetism has been observed in nanowires. A novel class of permanent magnetic materials for usage in nanowires with better magnetic dipole moments can be produced by combining highly anisotropic energy particles with super-magnets.

26.3.3 Thermal Characteristics

A key factor in assessing the performance of nanowires is their thermal properties. The nanowires have a bent shape when used in research or applications. Both the phonon transit and changes in their thermal conductivity may be impacted by the curvature of the material. As phonons may deviate from the critical temperature and pass axially through the nanowire while following the drift direction, nanowire curvature acts as a phonon impedance. Thus, it is crucial to consider the impact of thermal conductivity while examining the impact of nanowires at various temperatures.

26.3.4 Electrical Characteristics

It is critical to comprehend semiconductor nanowires' electrical characteristics since doing so helps determine whether silicon nanowires are appropriate for usage in sensor and electrical device applications. Each unit cell of silicon nanowires has four atoms. There should be three conductance channels if the crystal is flawless. By taking one or two atoms out or adding them back in, you can see how the conductivity changes. Therefore, the conductivity of nanowires is influenced by their crystal structure. Changes in conductivity can be caused by changes in surface finish, such as by scattering charge carriers in nanowires. As the diameter of the nanowires changes, scattering effects are seen. Nanowires' conductivity is extremely susceptible to surface excitation by outside charges because of the high surface-area-to-volume ratio of these materials. They can use silicon nanowires in biosensors and can identify single molecules thanks to this crucial characteristic.

26.3.5 Optical Properties of Nanowires

The polymer's non-aggregated nanoparticles are assigned to optical properties to increase the index of refraction. These techniques can also create materials with non-linear optical or visual characteristics. Glass nanoparticles containing gold and CdSe give semiconductor nanowires a red or orange hue, and as the particle size gets smaller, some oxide polymer nanocomposites display blue-shifting fluorescence. Faraday rotation is one of the highly planned magneto-optical effects on iron fluids that enhances the paramagnetic properties of ferrous nanowires. Also, the nanowires possess better reflectivity, refractivity, and halo-chromic and thermo-chromic characteristics along with thermal absorptivity properties making them suitable for photovoltaic applications.

26.4 APPLICATIONS OF NANOWIRES

Nanowires are based on flat substrates made of semiconductor materials such as silicon and germanium. Nanowires are very thin wires. They are made of metals such as silver, gold, and iron. Nanowires are measured as spatial measurements of approximately 10^{-9} m and are primarily used to create nanomachines and nanostructures in nanotechnology. Small nanowires are made up of nanoparticles with a diameter of only 1 nm, which are used in a vast number of applications.

NWs can be used for electrically conductive applications, especially with semiconductive materials such as Au, Ag, Pt, Al, etc. Bottom-up techniques are used to apply nanowires in real-time, however top-down techniques can be utilized to create nanowires made of semiconductor materials including Si, Ge, and GaN. NWs can be used to make a range of sensors since they have a high elasticity modulus. NWs are also used in the development of NEMS. NWs are recognized as typical nanomaterials because of their exceptional optical, electrical, chemical, and mechanical properties. Conductivity, as a result of their ability to be regulated by a field effect mechanism, NWs are used in sophisticated sensors and next-generation transistors. The International Technology Roadmap for Semiconductors (ITRS) predicts that transistor size can be lowered to 18 nm and gate length can be shortened to around 7 nm. At this dimension, conventional MOSFETs have shown

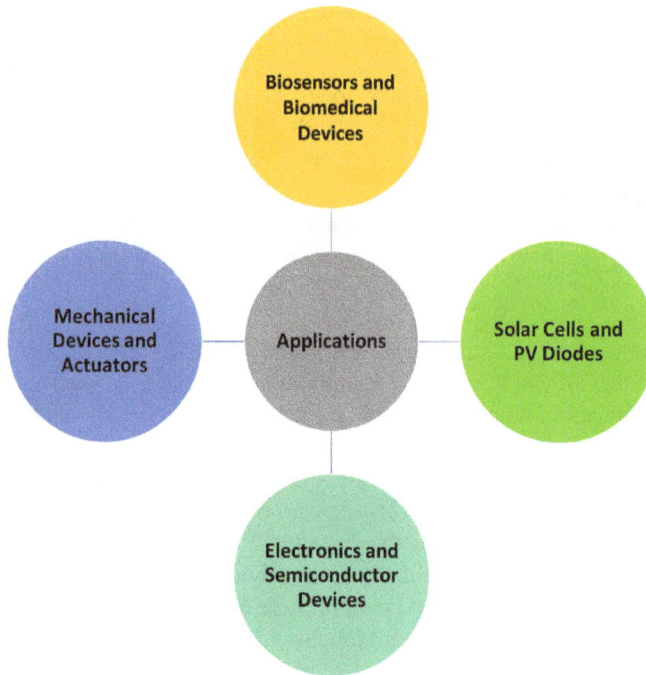

FIGURE 26.4 Schematic outlay of the applications of nanowires in different fields.

to be unsuccessful. MOSFETs based on nanowires are thus possible contenders for such device sizes. Future sensors will also need to react quickly and in real-time to diagnose diseases, therefore nanowires can be employed in a wide range of MEMS and NEMS applications.

Controlling physical and electrical properties requires consideration of the nanowire's size, chemical makeup, and various operating strategies. Applications in electronics, optoelectronics, and photonics are now possible. Due to the demand for higher performance capabilities, the usage of nanowires in field-effect transistors, p-n diodes with crossed nanowires, bipolar junction transistors, and field-effect transistors is expanding in electronic applications. **Figure 26.4** provides a schematic breakdown of the various fields in which nanowires are used.

26.4.1 Use of Nanowires for Thermoelectric Refrigeration

Conventional refrigeration makes use of the principles of compression relaxation cycles comprising mechanical elements that are subject to vibration and noise. The nanowires, on the other hand, work according to the thermoelectricity principle, which is based on the direct conversion of heat energy to electrical energy via the physical phenomena of the nanowires. The moving pieces in the thermoelectric converters are no longer present. It is possible to remember their use for modular refrigeration in a variety of applications due to the lack of vibration. They are incredibly compact and less complicated. Due to manufacturing limitations, thermoelectric refrigerator production is still in its early stages. These brand-new modular thermoelectric converters will help keep electrical equipment cool by being made. An enormous increase in efficiency will be possible with these converters' integration of nanostructures and nanowires.

Thermoelectric refrigeration makes use of thermoelectric nanowires that develop a thermal gradient and result in heat dissipation [58]. This conversion of thermal energy into electrical energy is possible owing to the use of thermoelectric nanowires. These nanowires make use of the principle of the Seebeck and Peltier effect [59].

By the year 1821, Thomas Johann Seebeck had developed the Seebeck effect. An electromotive force is produced when the identical junction of two different substances, A and B, is exposed to a temperature gradient, T (emf). Physically, the Seebeck effect can be identified by the polarization of any material used to make conductive wires when those wires are placed in a temperature region with a gradient that is not zero. From the higher temperature end to the lower temperature end, conduction electrons travel. The temperature has an impact on how quickly conduction electrons flow. Since the electrons at higher temperatures are quicker than the electrons at a colder temperature, the higher temperature end attains a positive charge and the lower temperature end attains a negative charge. The Seebeck voltage relies on the coefficient of thermal expansion, and thus, the nanowires can achieve the required coefficient of thermal expansion within their operational conditions. Once the migration of hot electrons towards the cool zone reaches its saturation phase, a dynamic equilibrium is achieved. The electron diffusion at cold temperatures is offset by the electron diffusion at higher temperatures. As a result, nanowires will be used to activate and monitor the thermoelectric refrigeration process.

The Peltier effect, which is the release or absorption of heat, is brought on when an electric current is passed over the intersections of conductors A and B at the same temperature. This effect was discovered by Jean-Charles Athanase Peltier and involves a very little quantity of heat exchange. The electrical current (I) flowing through the circuit and the conductor's temperature both influence this exchange.

As was already said, the material's nanostructures exhibit traits that are distinct from those of the bulk form. The goal of technological studies in the field of thermoelectricity is twofold: I use a low-dimensional system to improve conversion performance. Because of the remarkable advantages that nanowires provide, the study of low-dimensional systems has proven to be extremely important. (ii) The improvement of phonon dynamics and heat transfer physics in nanostructured systems, has produced encouraging findings [60]. Nanostructures must be structured with one or more dimensions that are bigger than those of electrons and holes and considerably less than the typical phonon-free route. Without affecting electric conductivity, this might likely reduce thermal conductivity [61].

The thermoelectric properties of the nanowires and nanostructures are impacted by two main effects. The first effect is a strong phonon diffusion across the grain limits (limitations between the particular grains making up the nanowires) and grain refining, which results in a decrease in the network's thermal conductivity. The second factor that affects electrical conductivity is the structure and form of the atoms in the nanowires. The behavior of the nanowires is influenced by thin layers of nanometric order. These technologies are now envisioned as being used mostly in low- and medium-temperature applications (between 150 and 200°C). One of the main challenges is to reach a thermoelectric state where the thermoelectric properties do not degrade as the temperature rises [62].

26.4.2   USE OF NANOWIRES IN PHOTOVOLTAICS (PV) APPLICATIONS

The development of clean, renewable electricity is made possible by the development of photovoltaic cells. Photovoltaic cells can be included in nanoelectronic systems as tiny power sources thanks to downsizing. A common method to improve solar efficiency is the introduction of nanostructures or nanowires [63]. Photovoltaic cells must fulfill requirements to be efficient. They must first have the necessary photoconductivity properties. Additionally, they must gather electrons from the electron-hole pairs that are produced as a result of the absorbed photons. These traits of increased radiation absorption and electron conductivity are displayed by nanowires. Additionally, when utilized in solar cells, nanowires offer an exciting alternative to improve the capacity of electron gathering. The electron collector and n-p junction are connected by nanowires, which increase the photocells' conductivity and conductivity properties. As a result, the usage of nanowires for photovoltaic applications has a significant impact on how solar radiation is used to generate power.

26.4.3 Nanowires for Biological Applications

Nanowires are extensively used for biological applications. They have specific electrochemical characteristics that remain unaltered by the changes in their surrounding conditions. Thus the nanowires can be used for Lab on Chip devices to detect vital body conditions easily. This makes them extraordinarily easier to use over the conventional techniques typically used today [64]. Additionally, the nanowires are used for coronary stents and orthodontic braces and as actuators for nanorobots used in medical surgeries. These nanowires are also used for wound healing and post-surgical tissue closure at the subcutaneous level.

26.5 CONCLUSIONS AND FUTURE SCOPE

Nanowires are distinctive nanostructures with diameters between 10 and 9 meters. Alternative terminology for the nanowires includes systems with a thickness or width limited to a few nanometers or less. At those scales the quantum outcomes are important – therefore using the term "quantum wires" is gaining significance. Many sorts of nanowires exist, such as metallic nanowires, viz., Au, Ag, Ni, etc., semiconductor nanowires, viz., Si, Ge, etc., dielectric (e.g., materials, viz., TiO_2, SiO_2, etc.). Molecular nanowires are made from repeating inorganic and organic molecular units. The domains of micro and nanoelectromechanical systems (MEMS and NEMS), biomedical components, photovoltaic devices, microcontrollers, actuators, etc., all heavily rely on these nanometric components. In this chapter, an extensive review of the future perspectives of the synthesis techniques, characteristics, and application of nanowires has been accomplished, and it is herewith reported that the nanowires have huge potential to be used in nanorobots for surgeries, nanoactuators, and MEMS and NEMS devices for advanced engineering and biomedical applications.

REFERENCES

1. S. Zafar, C. D'Emic, A. Jagtiani, E. Kratschmer, X. Miao, Y. Zhu, R. Mo, N. Sosa, H. Hamann, G. Shahidi, H. Riel. Silicon nanowire field effect transistor sensors with minimal sensor-to-sensor variations and enhanced sensing characteristics. ACS Nano, 12(7), 2018, pp.6577–6587.
2. M.S. Gudiksen, L.J. Lauhon, J. Wang, D.C. Smith, C.M. Lieber. Growth of nanowire superlattice structures for nanoscale photonics and electronics. Nature, 415(6872), 2002, pp.617–620.
3. Y. Huang, X. Duan, Y. Cui, L.J. Lauhon, K.H. Kim, C.M. Lieber. Logic gates and computation from assembled nanowire building blocks. Science, 294(5545), 2001, pp.1313–1317.
4. L. Samuelson, C. Thelander, M.T. Björk, M. Borgström, K. Deppert, K.A. Dick, A.E. Hansen, T. Mårtensson, N. Panev, A.I. Persson, W. Seifert. Semiconductor nanowires for 0D and 1D physics and applications. Physica E: Low-dimensional Systems and Nanostructures, 25(2–3), 2004, pp.313–318.
5. Iijima, Helical microtubules of graphitic carbon. Nature, 354(6348), 1991, pp. 56–58.
6. M.I. Khan, X. Wang, K.N. Bozhilov, C.S. Ozkan. Templated fabrication of InSb nanowires for nanoelectronics. Journal of Nanomaterials, 2008, 99, 1–5.
7. R. Bosisio, G. Fleury, C. Gorini, J.L. Pichard. Thermoelectric effects in nanowire-based MOSFETs. Advances in Physics: X, 2(2), 2017, pp.344–358.
8. S. Rahong, T. Yasui, N. Kaji. Y. Baba. Recent developments in nanowires for bio-applications from molecular to cellular levels. Lab on a Chip, 16(7), 2016, pp.1126–1138.
9. D.M. Brodzinsky, J.E. Palacios. Psychological Issues in Adoption: Research and Practice. Praeger Publishers/Greenwood Publishing Group, 2005.
10. X. Kleber, V. Massardier, J. Merlin. The thermoelectric power measurement: an original control technique for metallic alloys. Techniques de l'Ingenieur. Materiaux Metalliques, 2005, pp.RE39-1.
11. R. Venkatasubramanian, E. Siivola, T. Colpitts, B. O'quinn. Thin-film thermoelectric devices with high room-temperature figures of merit. Nature, 413(6856), 2001, pp.597–602.

12. J. Liu, Z. Xu, W. Wu, Y. Wang, T. Shan. Regulation role of $CRTC_3$ in skeletal muscle and adipose tissue. Journal of Cellular Physiology, 233(2), 2018, pp.818–821.

13. V. Vasiraju, D. Norris, S. Vaddiraju. Thermal transport through Zn_3P_2 nanowire-BN microparticle/nanoparticle composites and hybrids. Materials Research Express, 4(7), 2017, p.075041.

14. C.B. Vining. Semiconductors are cool. Nature, 413(6856), 2001, pp.577–578.

15. Y.S Touloukian, R.W. Powell, C.Y. Ho, P.G. Klemens. Thermophysical properties of matter-the TPRC data series. Volume 2. Thermal conductivity-nonmetallic solids (reannouncement). Data book (No. AD-A-951936/4/XAB). Purdue Univ., Lafayette, IN (United States). 1991. Thermophysical and Electronic Properties Information Center.

16. S. Mukherjee, U. Givan, S. Senz, M. De La Mata, J. Arbiol,O. Moutanabbir. Reduction of thermal conductivity in nanowires by combined engineering of crystal phase and isotope disorder. Nano Letters, 18(5), 2018 pp.3066–3075.

17. A.I. Boukai, Y. Bunimovich, J. Tahir-Kheli, J.K. Yu, W.A. Goddard Iii, J.R. Heath. Silicon nanowires as efficient thermoelectric materials. Nature, 451(7175), 2008, pp.168–171.

18. G.S. Nolas, J. Sharp, J. Goldsmid. Thermoelectrics: Basic Principles and New Materials Developments (Vol. 45). 2001, Springer Science & Business Media.

19. Y.S. Ju, K.E. Goodson. Phonon scattering in silicon films with thickness of order 100 nm. Applied Physics Letters, 74(20), 1999, pp.3005–3007.

20. D. Li, Y. Wu, P. Kim, L. Shi, P. Yang, A. Majumdar. Thermal conductivity of individual silicon nanowires. Applied Physics Letters, 83(14), 2003, pp.2934–2936.

21. T.C. Harman, P.J. Taylor, M.P. Walsh, B.E. LaForge. Quantum dot superlattice thermoelectric materials and devices. Science, 297(5590), 2002, pp.2229–2232.

22. D. Li, Y. Wu, R. Fan, P. Yang, A. Majumdar. Thermal conductivity of Si/SiGe superlattice nanowires. Applied Physics Letters, 83(15), 2003, pp.3186–3188.

23. N.A. Melosh, A. Boukai, F. Diana, B. Gerardot, A. Badolato, P.M. Petroff, J.R. Heath. Ultrahigh-density nanowire lattices and circuits. Science, 300(5616), 2003, pp.112–115.

24. B. Tian, X. Zheng, T.J. Kempa, Y. Fang, N. Yu, G. Yu, J. Huang, C.M. Lieber. Coaxial silicon nanowires as solar cells and nanoelectronic power sources. Nature, 449(7164), 2007, pp.885–889.

25. Y. Zhang, H. Liu. Nanowires for high-efficiency, low-cost solar photovoltaics. Crystals, 9(2), 2019, p.87.

26. L. Tsakalakos, J. Balch, J. Fronheiser, B.A. Korevaar, O. Sulima, J. Rand. Silicon nanowire solar cells. Applied Physics Letters, 91(23), 2007, p.233117.

27. G. Andrä, M. Pietsch, T. Stelzner, F. Falk, E. Ose. Nanowires for thin film solar cells. 22nd European Photovoltaic Solar Energy Conference and Exhibition, 2007.

28. K. Bullis. Tiny Solar Cells: Photovoltaics Made of Nanowires Could Lead to Cheaper Solar Panels MIT Technology Review, 2007.

29. M. Bechelany. Nouveau procédé de croissance de nanofils à base de SiC et de nanotubes de BN: étude des propriétés physiques d'un nanofil individuel à base de SiC, Doctoral dissertation, Lyon 1, 2006.

30. B. Gates, B. Mayers, B. Cattle, Y. Xia. Synthesis and characterization of uniform nanowires of trigonal selenium. Advanced Functional Materials, 12(3), 2002, pp.219–227.

31. B. Mayers, Y. Xia. One-dimensional nanostructures of trigonal tellurium with various morphologies can be synthesized using a solution-phase approach. Journal of Materials Chemistry, 12(6), 2002, pp.1875–1881.

32. B. Messer, J.H. Song, M. Huang, Y. Wu, F. Kim, P. Yang. Surfactant-induced mesoscopic assemblies of inorganic molecular chains. Advanced Materials, 12(20), 2000, pp.1526–1528.

33. J.H Song, B. Messer, Y.Y. Wu, H. Kind, P.D. Yang. MMo_3Se_3 (M= Li+, Na+, Rb+, Cs+, NMe4+) nanowire formation via cation exchange in organic solution. Journal of the American Chemical Society 123(39), 2001, pp. 9714–9715.

34. T.J. Trentler, K.M. Hickman, S.C. Goel, A.M. Viano, P.C. Gibbons, W.E. Buhro, W.E. Solution-liquid-solid growth of crystalline III-V semiconductors: an analogy to vapor-liquid-solid growth. Science, 270(5243), 1995, pp.1791–1794.

35. S.D. Dingman, N.P. Rath, P.D. Markowitz, P.C. Gibbons, W.E. Buhro. Low-temperature, catalyzed growth of indium nitride fibers from azido-indium precursors. Angewandte Chemie International Edition, 39(8), 2000, pp.1470–1472.

36. T.J. Trentler, S.C. Goel, K.M. Hickman, A.M. Viano, M.Y. Chiang, A.M. Beatty, P.C. Gibbons, W.E. Buhro. Solution− liquid− solid growth of indium phosphide fibers from organometallic precursors: elucidation of molecular and nonmolecular components of the pathway. Journal of the American Chemical Society, 119(9), 1997, pp.2172–2181.

37. P.D. Markowitz, M.P. Zach, P.C. Gibbons, R.M. Penner, W.E. Buhro. Phase separation in $Al_xGa_{1-x}As$ nanowhiskers grown by the solution−liquid−solid mechanism. Journal of the American Chemical Society, 123(19), 2001, pp.4502–4511.

38. J.D. Holmes, K.P. Johnston, R.C. Doty, B.A. Korgel. Control of thickness and orientation of solution-grown silicon nanowires. Science, 287(5457), 2000, pp.1471–1473.

39. X. Lu, T. Hanrath, K.P. Johnston, B.A. Korgel. Growth of single crystal silicon nanowires in supercritical solution from tethered gold particles on a silicon substrate. Nano Letters, 3(1), 2003, pp.93–99.

40. B. Gelloz, Y. Coffinier, B. Salhi, N. Koshida, G. Patriarche, R. Boukherroub. Synthesis and optical properties of silicon oxide nanowires. MRS Online Proceedings Library (OPL), 2006, p. 958.

41. Y. Xia, P. Yang, Y. Sun, Y. Wu, B. Mayers, B. Gates, Y. Yin, F. Kim, H. Yan. One-dimensional nanostructures: synthesis, characterization, and applications. Advanced Materials, 15(5), 2003, pp.353–389.

42. X. Wang, Y. Li. Selected-control hydrothermal synthesis of α-and β-MnO_2 single crystal nanowires. Journal of the American Chemical Society, 124(12), 2002, pp.2880–2881.

43. J.R. Heath, F.K. LeGoues. A liquid solution synthesis of single crystal germanium quantum wires. Chemical Physics Letters, 208(3–4), 1993, pp.263–268.

44. T. Kasuga, M. Hiramatsu, A. Hoson, T. Sekino, K. Niihara. Formation of titanium oxide nanotube. Langmuir 14(12), 1998, pp. 3160–3163.

45. Q. Zhang, L. Gao, J. Sun, S. Zheng. Preparation of long TiO_2 nanotubes from ultrafine rutile nanocrystals. Chemistry Letters, 31(2), 2002, pp.226–227.

46. Y. Xie, P. Yan, J. Lu, W.Z. Wang, Y.T. Qian. A safe low temperature route to InAs nanofibers. Chemistry of Materials 11(9), 1999, pp. 2619–2622.

47. Y. Li, M. Sui, Y. Ding, G.H. Zhang, J. Zhuang, C. Wang. Preparation of Mg $(OH)_2$ nanorods. Advanced Materials, 12(11), 2000, pp.818–821.

48. H. Ringsdorf, B. Schlarb, J. Venzmer. Molecular architecture and function of polymeric oriented systems: models for the study of organization, surface recognition, and dynamics of biomembranes. Angewandte Chemie International Edition in English, 27(1), 1988, pp.113–158.

49. A. Govindaraj, F. Leonard Deepak, N.A. Gunari, C.N.R. Rao. Semiconductor nanorods: Cu, Zn, and Cd chalcogenides. Israel Journal of Chemistry, 41(1), 2001, pp.23–30.

50. C.N.R. Rao, A. Govindaraj, F.L. Deepak, N.A. Gunari, M. Nath. Surfactant-assisted synthesis of semiconductor nanotubes and nanowires. Applied Physics Letters, 78(13), 2001, pp.1853–1855.

51. M.J. Zheng, L.D. Zhang, G.H. Li, X.Y. Zhang, X.F. Wang. Ordered indium-oxide nanowire arrays and their photoluminescence properties. Applied Physics Letters, 79(6), 2001, pp.839–841.

52. C.R. Martin. Nanomaterials: a membrane-based synthetic approach. Science, 266(5193), 1994, pp.1961–1966.

53. D. Al-Mawlawi, C.Z. Liu, M. Moskovits. Nanowires formed in anodic oxide nanotemplates. Journal of Materials Research, 9(4), 1994, pp.1014–1018.

54. J.C. Hulteen, C.R. Martin. A general template-based method for the preparation of nanomaterials. Journal of Materials Chemistry, 7(7), 1997, pp.1075–1087.

55. G. Espinosa, R.J Silva. CR-39 Nuclear Track Detectors for identification of Pu-239 and Am-241 alpha contamination. Journal of Radioanalytical and Nuclear Chemistry, 194, 1995, pp.185–189.

56. A. Despic, V.P. Parkhulik. Modem Aspects of Electrochemistry, Volume 20, 1989, Plenum Press, New York.

57. X. Duan, C.M. Lieber. General synthesis of compound semiconductor nanowires. Advanced Materials, 12(4), 2000, pp.298–302.

58. A. Huczko. Template-based synthesis of nanomaterials. Applied Physics A, 70(4), 2000, pp.365–376

59. K.B. Shelimov, M. Moskovits. Composite nanostructures based on template-grown boron nitride nanotubules. Chemistry of Materials, 12(1), 2000, pp.250–254.

60. N. Lu, J. Zheng, M. Gleiche, H. Fuchs, L. Chi, O. Vidoni, T. Reuter, G. Schmid. Connecting nanowires consisting of Au55 with model electrodes. Nano Letters, 2(10), 2002, pp.1097–1099.

61. Y. Ma, X. Guo, X. Wu, L. Dai, L. Tong. Semiconductor nanowire lasers. Advances in Optics and Photonics, 5(3), 2013, pp.216–273.

62. S.T. Lee, N. Wang, Y.F. Zhang, Y.H. Tang. Oxide-assisted semiconductor nanowire growth. MRS Bulletin, 24(8), 1999, pp.36–42.

63. N. Wang, Y.H. Tang, Y.F. Zhang, C.S. Lee, S.T. Lee. Nucleation and growth of Si nanowires from silicon oxide. Physical Review B, 58(24), 1998, pp.R16024–16026.

64. N. Wang, Y.H. Tang, Y.F. Zhang, C.S. Lee, I. Bello, S.T. Lee. Si nanowires grown from silicon oxide. Chemical Physics Letters, 299(2), 1999, pp.237–242.

Index

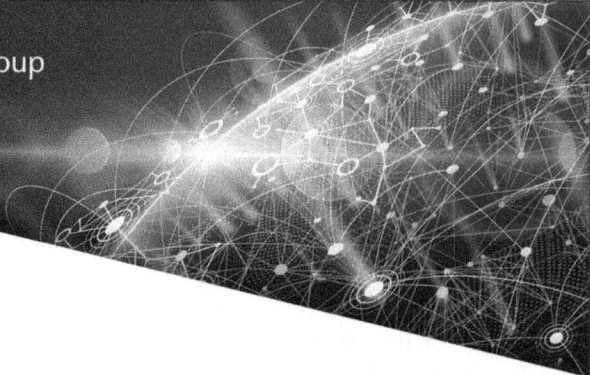

For Product Safety Concerns and Information please contact our EU
representative GPSR@taylorandfrancis.com
Taylor & Francis Verlag GmbH, Kaufingerstraße 24, 80331 München, Germany

www.ingramcontent.com/pod-product-compliance
Lightning Source LLC
Chambersburg PA
CBHW081038220326
41598CB00038B/6915

* 9 7 8 1 0 3 2 2 8 3 9 0 6 *